ATOMIC PHYSICS 3

1969—Atomic Physics 1

Proceedings of the First International Conference on
Atomic Physics, June 3-7, 1968, in New York City
V. W. Hughes, Conference Chairman
B. Bederson, V. W. Cohen, and F. M. J. Pichanik, Editors

1971—Atomic Physics 2

Proceedings of the Second International Conference on
Atomic Physics, July 21-24, 1970, Oxford, England
G. K. Woodgate, Conference Chairman
P. G. H. Sandars, Editor

1973—Atomic Physics 3

Proceedings of the Third International Conference on
Atomic Physics, August 7-11, 1972, Boulder, Colorado
S. J. Smith and G. K. Walters, Conference Chairmen and Editors

ATOMIC PHYSICS 3

Proceedings of the Third International Conference on
Atomic Physics, August 7-11, 1972, Boulder, Colorado

Editors

STEPHEN J. SMITH
Joint Institute for Laboratory Astrophysics
Chairman, Organizing Committee

and

G. K. WALTERS
Rice University
Chairman, Program Committee

with technical assistance by
Lorraine H. Volsky
Joint Institute for Laboratory Astrophysics

PLENUM PRESS • NEW YORK-LONDON • 1973

Library of Congress Catalog Card Number 72-176581

ISBN-13: 978-1-4684-2963-3   e-ISBN-13: 978-1-4684-2961-9
DOI: 10.1007/978-1-4684-2961-9

© 1973 Plenum Press, New York
Softcover reprint of the hardcover 1st edition 1973
A Division of Plenum Publishing Corporation
227 West 17th Street, New York, N. Y. 10011

United Kingdom edition published by Plenum Press, London
A Division of Plenum Publishing Company, Ltd.
Davis House (4th Floor), 8 Scrubs Lane, Harlesden, London NW10 6SE, England

N. V. Fedorenko
(1910-1972)

Nikolai Vasilievich Fedorenko, a member of the Organizing Committee for this Third International Conference on Atomic Physics, passed away in Leningrad on February 2, 1972. A distinguished physicist, well known for his pioneering work in heavy particle collisions, Fedorenko was also a leader on behalf of international cooperation in science. His participation on the Organizing Committee for this conference was but one of many similar roles he has filled in connection with international conferences and organizations over a period of more than a decade. He is remembered, as well, as a kind and gentle man, quietly dedicated, who shared a fundamental optimism about the future of society and the contribution that we in this field of science might make toward that future.

Organizing Committee

P. L. Bender	Joint Institute for Laboratory Astrophysics, Boulder
V. W. Cohen	Brookhaven National Laboratory
R. J. Damburg	Academy of Sciences, Riga
H. Ehrhardt	University of Trier
W. L. Fite	University of Pittsburgh
H. M. Foley	Columbia University
V. W. Hughes	Yale University
R. N. Ilyin	A. F. Ioffe Physico-Technical Institute, Leningrad
P. Jacquinot	Laboratoire Aime Cotton, Orsay
A. Kastler	Ecole Normale Superieure, Paris
H. Kleinpoppen	University of Stirling
D. Kleppner	Massachusetts Institute of Technology
I. Lindgren	Chalmers University of Technology
W. Paul	Deutsches Elektronen-Synchrotron, Hamburg
H. A. Shugart	University of California, Berkeley
K. Takayanagi	University of Tokyo
G. K. Walters	Rice University
G. K. Woodgate	Clarendon Laboratory, Oxford
S. J. Smith, Chairman	Joint Institute for Laboratory Astrophysics, Boulder

Local Committee

James Barnes	Edward U. Condon	Carl Iddings
Peter Bender	John Cooper	Richard Mockler
Harold Boyne	Thomas English	S. J. Smith
Wesley E. Brittin	Roy H. Garstang	

Program Committee

Alec Dalgarno	Harvard University
John L. Hall	Joint Institute for Laboratory Astrophysics, Boulder
William Happer	Columbia Radiation Laboratory
Vernon W. Hughes	Yale University
S. J. Smith	Joint Institute for Laboratory Astrophysics, Boulder
G. King Walters, Chairman	Rice University

CONTENTS

Session C (*W. Paul, Chairman*)

Session D (*H. Ehrhardt, Chairman*)

STATUS OF QED EXPERIMENTS*

Vernon W. Hughes

Gibbs Laboratory, Yale University

New Haven, Connecticut

INTRODUCTION

I shall try to review the status of tests of quantum electrodynamics (QED) through precision measurements on atoms and
elementary particles, with particular emphasis on the experimental
side.[1,2,3] These tests are the ones most appropriate to our
present Conference on Atomic Physics, but it should be remembered
that they are complementary to the tests of QED through less
precise, but at least equally critical or sensitive, measurements
of high energy processes, such as electron or muon pair production,
bremsstrahlung, and electron-electron or electron-positron
collisions in high energy storage rings. I believe that Professor
Kroll will say something about the high energy processes, as well
as about recent theoretical advances in QED.[4]

The precision tests of QED involve electrons, muons, and
photons and their interaction with external electromagnetic
fields.[5,6] In the atom the nucleus provides an external electromagnetic field - a Coulomb field due to a point charge and a
magnetic field due to a point magnetic dipole moment, to first
approximation. In actuality, of course, leptons, photons, and
their electromagnetic interactions do not constitute a closed
world. Photons interact with the strongly interacting particles -
the baryons and mesons - and leptons have weak interactions.
There are small but significant contributions to the measured
energy levels due to strong interaction effects, such as the effect
of proton structure and proton polarizability on the hyperfine
structure interval in hydrogen. In principle, there are contributions due to weak interactions as well. Although these do not
yet appear to be quantitatively significant, they may become

1

observable in somewhat more precise experiments depending on what is the basic theory of weak interactions[7]—what particles are involved, what their properties are, and how the electromagnetic and weak interactions are related. Indeed there are exciting new theories[8,9] which provide a unified, renormalizable, and convergent theory of electromagnetic and weak interactions. These theories introduce as yet unobserved particles — W vector bosons, scalar mesons, or heavier leptons—, which should produce small effects,[10,11] e.g. on the muon g-value and on muonic atom energy levels.

At present, there is excellent agreement between the results of precise measurements and the QED theoretical calculations for a wide variety and number of tests on atoms and elementary particles. In the few cases where there is disagreement, it seems plausible and likely that the cause of disagreement is either unaccounted for systematic errors in the experiment or uncalculated theoretical QED terms. However, we can't be sure, of course, until disagreements have been resolved. Good agreement also exists between experiment and theory for the high energy processes. All of these results have not only confirmed the theory of QED to very high accuracy, but have also proved that the muon is a heavy electron. The sensitivity of the tests can be expressed by stating that the theory is valid to a distance of about 5×10^{-15} cm or a corresponding momentum transfer of 4 GeV/c. These agreements and limits of validity are very valuable in restricting new theories of weak or strong interactions, which would lead to contradictions with QED — e.g. a lower limit can be placed on the mass of the W vector boson from the agreement between experiment and theory on the muon g-value.[10]

In the past two years since our Second International Conference on Atomic Physics, at which Brodsky gave his review[2] of the status of QED (and at about the same time the important review article by Brodsky and Drell[12] appeared), there have been many impressive advances in the experiments and in the theory, including almost all of the principal tests of QED. In this review I will emphasize these advances, some of which have been presented as abstracts to this Conference.[13]

FUNDAMENTAL ATOMIC CONSTANTS

In order to compare the results of a precise measurement of some atomic energy interval with theory, very precise values of certain atomic constants are needed. Figure 1 lists the basic atomic constants and several important derived atomic constants. As you know, direct and precise measurements of these basic atomic constants are usually not possible, and their values are obtained from the measurement of auxiliary constants which are combinations of the basic atomic constants that are convenient to measure. (See Figure 2.) I don't include the fine structure constant α in this

BASIC ATOMIC CONSTANTS

Charge (e)

Mass (m_e, M_p, M_d, M_α)

Magnetic moment (μ_p, μ_d)

Planck's constant (h)

Velocity of light (c)

Derived Atomic Constants

$$R_\infty = \frac{2\pi^2 me^4}{h^3 c}$$

$$\alpha = \frac{e^2}{\hbar c}$$

$$\mu_B = \frac{e\hbar}{2mc}$$

Figure 1: Basic atomic constants.

list because it will be discussed separately later. The accuracy of some of these values as given in the 1969 review of Taylor, Langenberg, and Parker[14] is barely adequate for computing certain measured fine structure and hyperfine structure intervals, in view of the present – and even more so, the projected-experimental accuracies.

There have been recent important advances in determining several of these auxiliary constants.[15] A new measurement of the velocity of light c has been made by measuring the frequency of a 633-nm red He-Ne laser line, whose wavelength has also been measured, to obtain an accuracy in c of 0.06 ppm.[16] An even higher precision of 3 parts in 10^9 in c was announced at this Conference.[17] There is considerable experimental activity to try to improve the accuracy of R_∞ by conventional optical inter-ferometry, and in addition an effort to use the narrow H line obtained by laser saturation spectroscopy which has resolved the Lamb shift in n=3 to n=2 transitions.[18] At present the 1969 value is still the best available published value.

I don't want to raise any doubt in your mind about the reliability of the many numbers I will be giving you. However, I would like to tell you about the following incident. In seeking to get up-to-date information on the value of R_∞, I looked in the Proceedings of the Conference on Precision Measurement and Fundamental Constants held at the National Bureau of Standards in Washington, D.C. in 1970. There was an article

in these <u>Proceedings</u> on the Rydberg by Kessler from
NBS. So I naturally called my old friend Karl Kessler,
who is a well-known expert at NBS in the field of
precision measurement of optical wavelengths. Well,
it turned out he was the wrong Kessler; the article
was by E.G. Kessler, Jr. During our telephone
conversation Karl then gave me a 10 digit number
which I carefully wrote down, planning to get it onto
a slide as the latest NBS value of R_∞. Fortunately
before proceeding much further I found out that this
number was E.G. Kessler Jr.'s telephone number, so it
does not appear in Figure 2.

The value of the magnetic moment of the proton in Bohr magnetons,
μ_p/μ_B, has been improved to a precision of 1 part in 10^8 by measuring
the ratio of Zeeman transition frequencies with the hydrogen maser.[19]

AUXILIARY CONSTANTS

1. Velocity of light

    $c = 2.997\ 925\ 0\ (10) \times 10^{10}$ cm/sec (0.33 ppm)[14]

    $c = 2.997\ 924\ 62\ (18) \times 10^{10}$ cm/sec (0.06 ppm)[16]

2. Rydberg constant for infinite mass

    $R_\infty = 1.097\ 373\ 12\ (11) \times 10^5$ cm^{-1} (0.1 ppm)[14]

3. Magnetic moment of proton in Bohr magnetons

    $\dfrac{\mu_p}{\mu_B} = 1.521\ 032\ 64\ (46) \times 10^{-3}$ (0.3 ppm)[14]

    $\dfrac{\mu_p}{\mu_B} = 1.521\ 032\ 181\ (15) \times 10^{-3}$ (0.01 ppm)[19]

4. Mass ratio of particles

    $\dfrac{M_p}{m_e} = 1\ 836.109\ (11)$ (6.2 ppm)[14]

    $\dfrac{M_d}{m_e} = 3\ 670.40\ (14)$ (37 ppm)[14]

    $\dfrac{M_\alpha}{m_e} = 7\ 294.11\ (53)$ (73 ppm)[14]

Figure 2: Current values of auxiliary constants.

THE FINE STRUCTURE CONSTANT α[20]

An accurate value of the fine structure constant α is
required for all the precision tests of QED. A number of these
measurements which have comparable precision can be regarded as
determinations of α, and a convenient way to compare theory and
experiment for these various measurements is to compare the α
values obtained from them. There are at present six measurements,
each of which determines α to 4 ppm or better: 1) Hydrogen fine
structure, 2) Hydrogen hyperfine structure, 3) Muonium hyperfine
structure, 4) Helium fine structure, 5) Electron g-2, and 6) ac
Josephson effect.

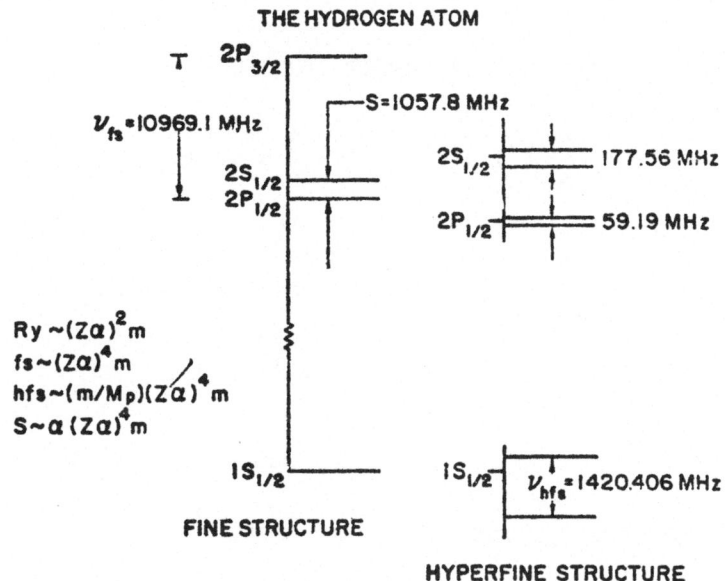

Figure 3: Energy levels of n=1 and n=2 states of hydrogen.[12]

Figure 3 will remind you of the energy levels of the hydrogen
atom, including fine structure and hyperfine structure. Figure 4
shows the value of α obtained from the hydrogen fine structure
interval $2^2P_{3/2}$ to $2^2P_{1/2}$, which has not changed since 1970[14] and
is limited by the experimental error.[21] Figure 5 shows the value
of α from the hydrogen hfs interval, which again is unchanged from
1970.[14] The error in α is due principally to uncertainty in the
theoretical contribution of proton structure and proton polarizability.[3,12] It is perhaps worth noting, but not of relevance to the
determination of α, that the quoted experimental error in Δν has
been increased and the value slightly changed since 1970.[22]

HYDROGEN FINE STRUCTURE

$$\nu_{theor}(2^2P_{3/2} - 2^2P_{1/2}) = \frac{\alpha^2 cR_\infty}{16}\left[(1+\tfrac{5}{8}\alpha^2)(1-\tfrac{m_e}{M_p}) + 2a_e(1-\tfrac{2m_e}{M_p}) + \tfrac{\alpha^3}{\pi}\ln\alpha^2\right]$$

$$\nu_{expt}(2^2P_{3/2} - 2^2P_{1/2}) = 10\ 969.13(10)\text{MHz} \quad (9\text{ ppm})$$

$$\alpha^{-1} = 137.035\ 4(6) \quad (4.4\text{ ppm})$$

Figure 4: Determination of α from a hydrogen fine structure interval. (The most accurate current values of the auxiliary constants given in Figure 2 are used throughout this paper.)

HYDROGEN HYPERFINE STRUCTURE

$$\Delta\nu_{theor} = \left(\frac{16}{3}\alpha^2 c\ R_\infty\ \frac{\mu_p}{\mu_B}\right)\left(1 + \frac{m_e}{M_p}\right)^{-3}\left(1 + \frac{3}{2}\alpha^2 + a_e + \varepsilon_1 + \varepsilon_2 + \varepsilon_3 + \delta_p\right)$$

where

$$a_e = \frac{\alpha}{2\pi} - 0.328\ 48\ \frac{\alpha^2}{\pi^2} + (1.49\pm0.2)\ \frac{\alpha^3}{\pi^3}; \quad \varepsilon_1 = \alpha^2(\ln2 - \tfrac{5}{2})$$

$$\varepsilon_2 = -\frac{8\alpha^3}{3\ \pi}\ln\alpha\ (\ln\alpha - \ln4 + \frac{281}{480}); \quad \varepsilon_3 = \frac{\alpha^3}{\pi}\ (18.4\pm5)$$

$$\delta_p = \delta_p^S + \delta_p^P = \left[(-34.6\pm0.9) + (0\pm4.0)\right]\text{ppm} = (-34.6\pm5.0)\text{ ppm}$$

$$\Delta\nu_{expt} = 1\ 420.405\ 751\ 768\ (9)\text{ MHz} \quad (6 \times 10^{-12})$$

$$\alpha^{-1} = 137.035\ 91\ (34)\ (2.5\text{ ppm})$$

Figure 5: Determination of α from the hyperfine structure interval in the ground state of hydrogen. (The term δ_p^S arises from proton size and recoil; the term δ_p^P arises from proton polarizability.)

There have been substantial advances in experiment and theory for muonium.[23] The existence of a significant dependence of the hfs interval $\Delta\nu$ on the square of the pressure of the stopping gas (a quadratic pressure shift) was discovered by the Yale[24] and Chicago[25] groups, and accounting for this effect yielded a value of $\Delta\nu$ accurate to 2.7 ppm from the Yale group.[24] A major advance in precision was made by the Chicago group[26] by using the method of separated oscillating fields to obtain a resonance line narrower than the natural linewidth[27,28] (see Figure 6) and a value of $\Delta\nu$

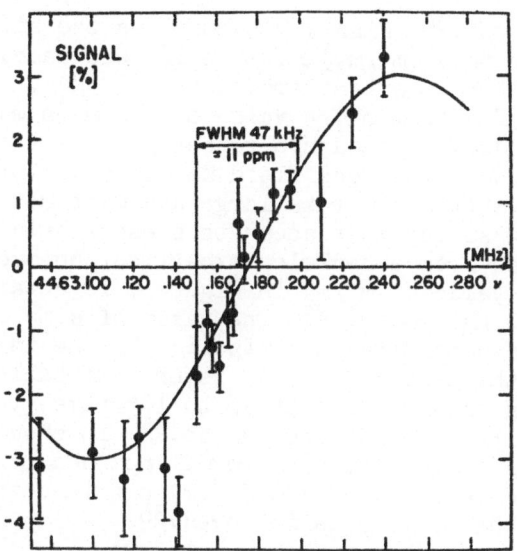

Figure 6: Resonance curve of a transition between hyperfine
structure levels of muonium at very weak magnetic field, obtained
by the method of separated oscillating fields and with a Kr
pressure of 3.6 atm.

MUONIUM HYPERFINE STRUCTURE

$$\Delta \nu_{theor} = \left(\frac{16}{3}\alpha^2 c \ R_\infty \ \frac{\mu_\mu}{\mu_B}\right)\left(1 + \frac{m_e}{m_\mu}\right)^{-3}\left(1 + \frac{3}{2}\alpha^2 + a_e + \epsilon_1 + \epsilon_2 + \epsilon_3 - \delta_\mu'\right)$$

where

$$a_e = \frac{\alpha}{2\pi} - 0.328\ 48\ \frac{\alpha^2}{\pi^2} + (1.49 \pm 0.2)\ \frac{\alpha^3}{\pi^3}$$

$$\epsilon_1 = \alpha^2(\ln2 - \tfrac{5}{2}), \quad \epsilon_2 = -\frac{8\alpha^3}{3\pi}\ln\alpha(\ln\alpha - \ln4 + \frac{281}{480}), \quad \epsilon_3 = \frac{\alpha^3}{\pi}(18.4 \pm 5),$$

$$\delta_\mu' = \frac{m_e}{m_\mu}\left\{ + \frac{3\alpha}{\pi}\left[1 - \left(\frac{m_e}{m_\mu}\right)^2\right]^{-1}\ln\left(\frac{m_\mu}{m_e}\right) - \frac{9}{2}\alpha^2\ln\alpha^{-1}\left[1 + \left(\frac{m_e}{m_\mu}\right)\right]^{-2}\right\}$$

$$\frac{\mu_\mu}{\mu_B} = \left(\frac{\mu_\mu}{\mu_p}\right)\left(\frac{\mu_p}{\mu_B}\right); \quad \frac{\mu_\mu}{\mu_p} = 3.183\ 347\ (9)(2.8\ ppm);$$

$$\frac{m_\mu}{m_e} = 206.768\ 3(6)(2.9\ ppm)$$

$$\Delta\nu_{expt} = 4\ 463.301\ 2\ (23)MHz\ (0.5\ ppm)$$

$$\alpha^{-1} = 137.036\ 40(28)\ (2.0\ ppm)$$

Figure 7: Determination of α from the hyperfine structure
interval in the ground state of muonium.

accurate to 0.5 ppm.[26]   Figure 7 summarizes the situation on $\Delta\nu$.[29,30]
The higher order term $(m_e/m_\mu)\alpha^2\ln\alpha$ in δ'_μ was calculated recently[31]
and contributes the important amount of 0.025MHz. In order to
obtain a value of α from $\Delta\nu$, a value of the muon magnetic moment,
or μ_μ/μ_p, is needed. Although a rather precise value for μ_μ/μ_p
has been obtained from strong field Zeeman transitions in muonium,[32]
the g_J pressure shift effect is large and must be evaluated
theoretically,[33] so the most accurate present value of μ_μ/μ_p is
probably obtained from a Berkeley experiment on muon precession in
liquids.[34] The value of m_μ/m_e in Figure 7 is obtained from the
values of μ_μ/μ_p, g_μ, and g_e.[29] The value of α is accurate to
2.0 ppm which is determined principally by the inaccuracy in
μ_μ/μ_p. Within the next year or so several so-called meson factories producing muon beams with up to 10^4 times the intensity of
present beams will come into operation.[35] It should be possible
with these beams to determine $\Delta\nu$ to 0.1 ppm and μ_μ/μ_p to 1 ppm.
Hence α would be determined to 0.5 ppm, provided the theoretical
calculation of the term ε_3 is improved.[36]

Within the past two years a precision determination of α based
on the fine structure of helium has finally become real after many
years of experimental and theoretical work.[37] Figure 8 shows energy
levels of helium. The fine structure levels involved are the 2^3P_J
levels with J = 0,1, and 2. The principal point about these

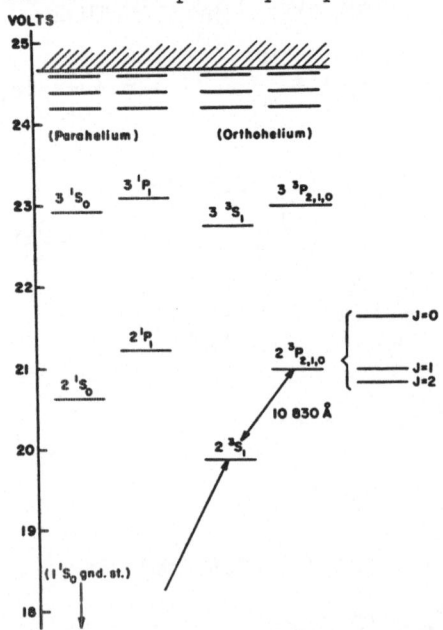

Figure 8: Energy levels of helium atom. The 2^3P fine
structure (not to scale) is shown in detail.

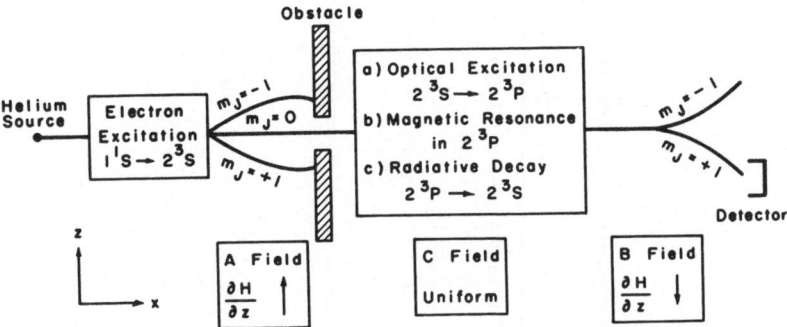

Figure 9: Block diagram of atomic beam optical–microwave magnetic resonance experiment to measure He fine structure.

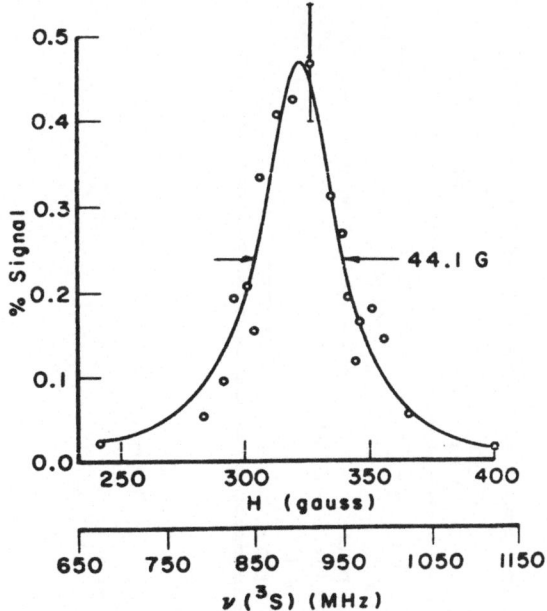

Figure 10: A typical observed resonance line for the transition $(J,M_J) = (1,0) \leftrightarrow (0,0)$ in the $2\,^3P$ state of He. The magnetic field is measured by the Zeeman transition frequency $\nu(^3S)$ in the $2\,^3S$ state of He. The experimental points are indicated by the circles, and a typical statistical error bar is shown. The solid curve is a Lorentzian curve fitted to the data points. The klystron frequency was about 29 600 MHz.

Interval	$\alpha^4 mc^2$	$\alpha^5 mc^2$	$\frac{m}{M_{He}}\alpha^4 mc^2$	Second order	$\alpha^6 mc^2$	ν_{theo}	ν_{expt}	$\nu_{theo} - \nu_{expt}$
ν_{01}	29 564.567 ±0.006 (0.21 ppm)	54.708	−10.707 ±0.000 44 (0.015 ppm)	11.60 ±0.18 (6 ppm)	−3.331 ±0.0039 (0.13 ppm)	29 616.83 ±0.18 (6 ppm)	29 616.864 ±0.036 (1.2 ppm)	−0.03 =1 ppm ±0.18
ν_{12}	2317.103 ±0.0017 (0.76 ppm)	−22.548	1.952 ±0.000 88 (0.39 ppm)	−6.79 ±0.36 (157 ppm)	1.542 ±0.0068 (3.0 ppm)	2291.36 ±0.36 (157 ppm)	2291.196 ±0.005 (2.2 ppm)	+0.16 =70 ppm ±0.36

Figure 11: Theoretical contributions to fine-structure intervals (MHz).[40] The values of α^{-1}, c, and R_∞ are 137.036 02(21) (1.5 ppm), 2.997 925 0(10)x10^{10} cm sec^{-1} (0.3 ppm), 109 737.312 cm^{-1} (0.1 ppm), respectively. Thus $\frac{1}{2}\alpha^2 cR_\infty$ = 87.594 25 x 10^3 MHz (3 ppm).

intervals is that the mean life of the 2^3P state is rather long –
10^{-7} sec – and the larger interval ν_{01} is about 30,000 MHz, so
that the fractional natural width of a resonance line from J = 0 to
J = 1 is 1 part in 10^4, whereas the corresponding value for the
$2^2P_{1/2}$ to $2^2P_{3/2}$ fine structure levels in hydrogen has the much
larger value of 1 part in 10^2. This implies that the fine structure
interval in helium can be measured with higher precision than the
corresponding interval in hydrogen. The measurement for helium has
been made by the atomic beam optical-microwave magnetic resonance
method,[38] illustrated in Figure 9. Figure 10 shows a resonance
line for the transition $(J,M_J) = (1,0) \leftrightarrow (0,0)$. Values for both the
intervals ν_{01} and ν_{12} have been reported from Yale with precisions
of 1 to 2 ppm.[39]

In addition to the experiment, a major theoretical problem
has been how to treat the theory of the fine structure for
the two electrons in helium with sufficient accuracy to determine α.
The theory has required highly accurate and elaborate wave functions,
and the evaluation of relativistic, virtual radiative, and nuclear
recoil contributions. The last important fundamental step was
taken recently by Kroll et al. who put the calculation on a firm
field theoretic basis and evaluated all the remaining $\alpha^4 R_\infty$ as well
as nuclear recoil terms. (See Figure 11.[40]) For the interval ν_{01},
which is the significant one for determining α, the theoretical
value has been evaluated with an estimated error of 6 ppm, and
hence since the experimental error is only 1.2 ppm, α can be
determined to an accuracy of 3 ppm. Further theoretical calculations of the second order term, now fully justified by the field
theoretical calculations, should allow the determination of α to
1 ppm. Further experiments are also planned at Yale to confirm and
improve our knowledge of ν_{01} and ν_{12}.

The famous electron g-2 experiment has been carried through
another stage at Michigan with the recent publication[41,42,43] of a
value for the anomalous magnetic moment factor a to an accuracy of
3 ppm, achieved principally through the use of a larger and more
precise magnetic field and longer trapping times. Figure 12 shows a
schematic diagram of the experiment in which the difference frequency between the spin precession frequency and the orbital
precession frequency in a known magnetic field is measured. Figure
13 shows typical precession data. Other promising experiments to
determine a have been pursued, but as yet their reported accuracies
are not as good as the Michigan experiment.[42] A major theoretical
advance was made in the complete calculation of all the contributions of the sixth order Feynman graphs for the anomalous g-factor.[44]
(See Figure 14.) A summary of experimental and theoretical values
for a is shown in Figure 15. A value of α accurate to 3 ppm can be
obtained. There is considerable effort being devoted to improving
the accuracy in a still further both by the Michigan method and by
resonance methods.

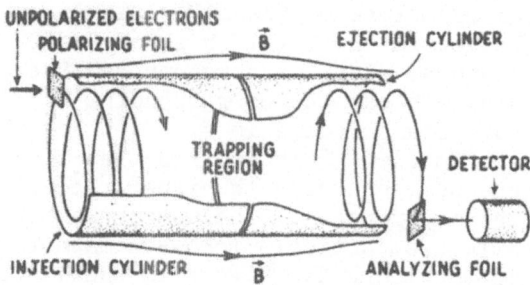

Figure 12: Schematic diagram of the experimental method for
the measurement of the electron g-2 value.

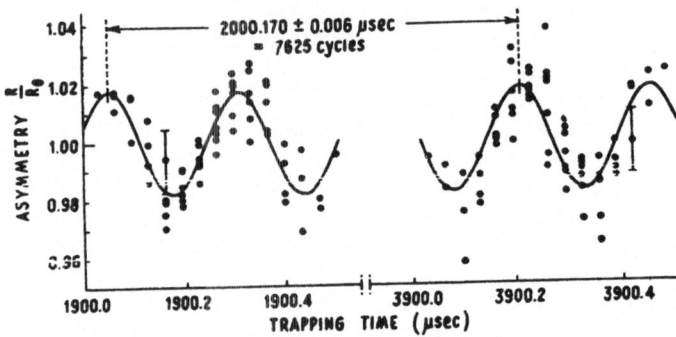

Figure 13: Typical asymmetry data and fitted curves for
the electron g-2 measurement.

Finally with regard to the very important ac Josephson effect,
there have been impressive advances in the accuracy of the determination of e/h, illustrated in Figure 16.[45] An accuracy of 0.12
ppm has been obtained. As you know, this value of e/h is used,
together with certain other constants, to obtain a precise value
for α. The limiting factor is the accuracy in the determination of
the gyromagnetic ratio of the proton γ_p.[15] The best current value
is obtained by the method of free precession in a weak magnetic
field, illustrated in Figure 17. The present precision is about
3 ppm, but there are plans at NBS in Washington to improve the
accuracy of this experiment by an order of magnitude using laser
interferometric techniques to measure the dimensions of the solenoid.
Figure 18 shows the current status on determining α from the ac
Josephson effect.

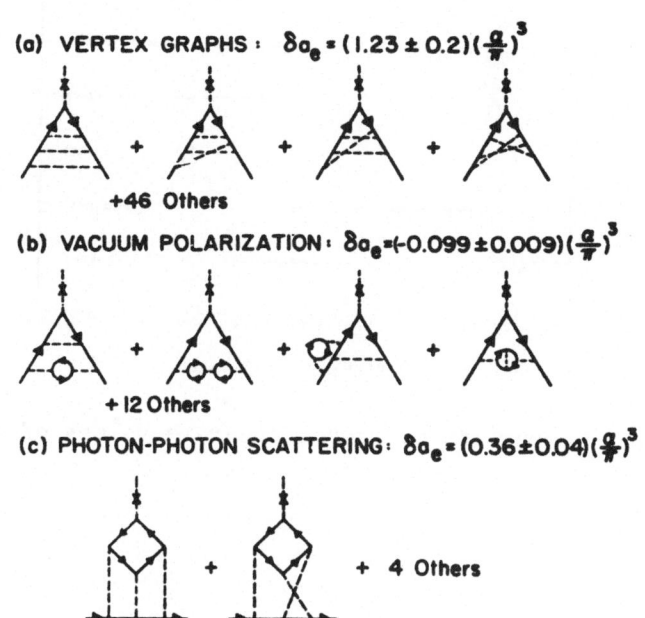

(a) VERTEX GRAPHS : $\delta a_e = (1.23 \pm 0.2)(\frac{\alpha}{\pi})^3$

+46 Others

(b) VACUUM POLARIZATION : $\delta a_e = (-0.099 \pm 0.009)(\frac{\alpha}{\pi})^3$

+ 12 Others

(c) PHOTON-PHOTON SCATTERING : $\delta a_e = (0.36 \pm 0.04)(\frac{\alpha}{\pi})^3$

+ 4 Others

Figure 14: Sixth-order graphs and contributions to the electron magnetic moment anomaly.

ELECTRON g-VALUE

$$g = 2(1 + a)$$

$$a_{theor.} = 0.5 \frac{\alpha}{\pi} - 0.328\ 48 \frac{\alpha^2}{\pi^2} + (1.49 \pm 0.2) \frac{\alpha^3}{\pi^3}$$

$$a_{expt.} (e^-) = (1\ 159\ 656.7 \pm 3.5) \times 10^{-9}\ (3.0\ ppm)$$

$$\alpha^{-1} = 137.035\ 93\ (41)\ (3.0\ ppm)$$

$$a_{expt.} (e^+) = (1\ 160\ 200 \pm 1\ 100) \times 10^{-9}\ (950\ ppm)$$

$$a_{expt.} (e^+) - a_{theor.} (e^+) = (543 \pm 1\ 100) \times 10^{-9}$$

Figure 15: Determination of α from the anomalous g-value of the electron. Theoretical and experimental values of $\underline{a}$ for the positron are compared.

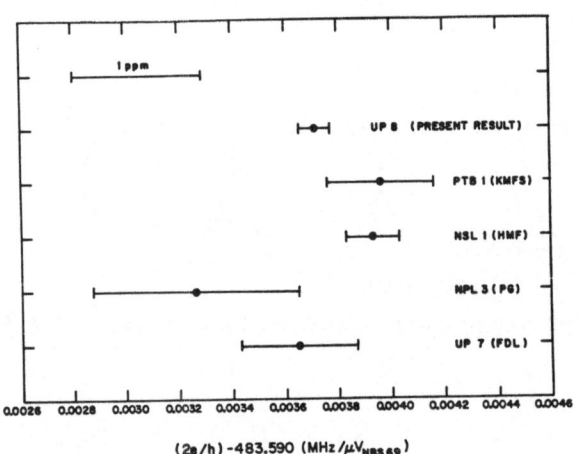

Figure 16:   Comparison of recent accurate values of 2e/h.

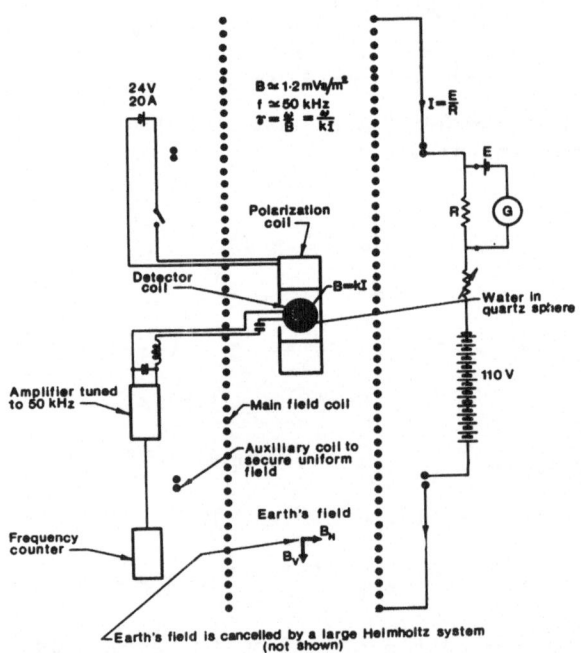

Figure 17:   Measurement of gyromagnetic ratio of the proton by
the method of free precession in a weak magnetic field.
(Schematic diagram).

ac JOSEPHSON EFFECT

$$h\nu = 2eV$$

$$\alpha^{-1} = \left(\frac{1}{4R_\infty \gamma_p} \frac{c\,\Omega_{ABS}}{\Omega_{NBS}} \frac{\mu_p}{\mu_B} \frac{2e}{h} \right)^{\frac{1}{2}}$$

$$\frac{2e}{h} = 483.593\ 718(60)\ \text{MHz}/\mu V_{NBS69}\ (0.12\ \text{ppm})$$

$$\gamma_p = 2.675\ 196\ 5(82)\ \times\ 10^4\ \text{rad} \cdot \text{sec}^{-1} \cdot \text{G}^{-1}\ (3.1\ \text{ppm})$$

$$\frac{c\,\Omega_{ABS}}{\Omega_{NBS}} = 2.997\ 926\ 1(12)\ \times\ 10^{10}\ \text{cm} \cdot \text{sec}^{-1}\ (0.39\ \text{ppm})$$

$$\alpha^{-1} = 137.036\ 10(22)\ \ (1.6\ \text{ppm})$$

Figure 18: Determination of α from the ac Josephson effect.

Figure 19 gives the values of α from the six methods discussed and Figure 20 shows these results graphically. Note that all the values are in very satisfactory agreement. A weighted average yields $\alpha^{-1} = 137.036\ 08\ (14)$ (1 ppm).

FINE STRUCTURE CONSTANT, α

Source	Value of α^{-1} (1 std. dev. error)
Hydrogen fine structure $2^2P_{3/2} \to 2^2P_{1/2}$	137.035 4(6) (4.4 ppm)
Hydrogen hyperfine structure $1^2S_{1/2},\ F = 1 \to F = 0$	137.035 91(34) (2.5 ppm)
Muonium hyperfine structure $1^2S_{1/2},\ F = 1 \to F = 0$	137.036 40(28) (2.0 ppm)
Helium fine structure $2^3P_1 \to 2^3P_0$	137.035 95(41) (3.0 ppm)
Electron g-2	137.035 93(41) (3.0 ppm)
Josephson effect	137.036 10(22) (1.6 ppm)

Weighted average value: $\alpha^{-1} = 137.036\ 08(14)$ (1.0 ppm)

Figure 19: Values of α obtained from various precision measurements.

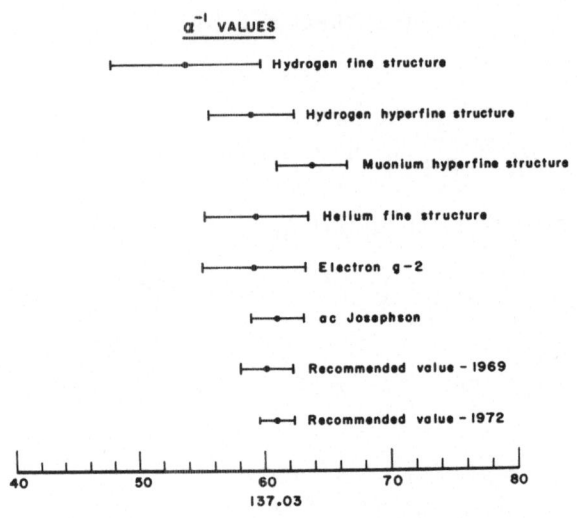

Figure 20: Graph of α^{-1} values.

Description	Order	Magnitude (MHz)
Second order—self-energy	$\alpha(Z\alpha)^4 m\{\log Z\alpha, 1\}$	1079.32 ± 0.02
Second order—vac. pol.	$\alpha(Z\alpha)^4 m$	-27.13
Second order—remainder	$\alpha(Z\alpha)^5 m$	7.14
	$\alpha(Z\alpha)^6 m\{\log^2 Z\alpha, \log Z\alpha, 1\}$	-0.38 ± 0.04
Fourth order—self-energy	$\alpha^2(Z\alpha)^4 m \begin{cases} F_1'(0) \\ F_2(0) \end{cases}$	0.45 ± 0.07 / -0.10
	$\alpha^2(Z\alpha)^5 m$	± 0.02
Fourth order—vac. pol.	$\alpha^2(Z\alpha)^4 m$	-0.24
Reduced mass corrections	$\alpha(Z\alpha)^4 m/Mm\{\log Z\alpha, 1\}$	-1.64
Recoil	$(Z\alpha)^5 m/Mm\{\log Z\alpha, 1\}$	0.36 ± 0.01
Proton size	$(Z\alpha)^4 (mR_N)^2 m$	0.13

$\alpha^{-1} = 137.03608(26)$

$$\mathcal{L} = \Delta E(2S_{1/2} - 2P_{1/2}) = 1057.91 \pm 0.16 \ (\text{L.E.})$$
$$\Delta E(2P_{3/2} - 2S_{1/2}) = 9911.12 \pm 0.22 \ (\text{L.E.})$$
$$\Delta E(2P_{3/2} - 2P_{1/2}) = 10969.03 \pm 0.12 \ (\text{L.E.})$$

Figure 21: Various theoretical contributions to the Lamb shift
in H(n=2). [L.E. designates limit of error.[12]]

FURTHER TESTS OF QED

Other critical precision tests of QED require an accurate
value of α but do not themselves determine accurate values of α.

With regard to the Lamb shift in hydrogen, there have been no
fundamental changes in the experimental or theoretical values,[12]

and the agreement between theory and experiment is satisfactory within the error of about 0.1 MHz or 1 part in 10^4 of the Lamb shift. Figures 21 and 22 give the theoretical contributions to the Lamb shift in hydrogen and the comparison of theory and experiment respectively.

In the past two years there have been two advances with regard to the general topic of the Lamb shift in hydrogenic atoms. The first is the successful application[46,47] of the method of separated oscillating fields to the measurement of hydrogen fine structure,[28] and the achievement of a linewidth narrower than the natural radiative linewidth. Figure 23 shows the experimental arrangement.[46] A fast monoenergetic beam of excited state hydrogen atoms is produced by charge capture by 20 KeV protons in a foil or gas. Rf transitions are produced by two separated rf fields, and transitions are observed by a change in the optical radiation. Figure 24 shows observed lineshapes and indicates the line narrowing or interference pattern obtained with separated rf fields. A new accurate measurement was obtained of the Lamb shift in the n=3 state of H.[47] This method gives promise eventually of yielding an improvement over existing values of the Lamb shift in the n=2 state of H.

The second advance is the determination of Lamb shift intervals of the n=2 state of hydrogenic atoms with high Z. Figure 25 shows the results. The newer data[48,49,13] are for C^{5+} and O^{7+}. These measurements have been done with MeV beams from an electrostatic accelerator, the use of a foil to produce the 2S state, and the determination of the Lamb shift from the observed quenching of the state due to the motional Stark effect in a magnetic field. The experimental accuracies are about 1%. Theoretical calculations have been extended[50] to be valid for large Z. Agreement of the experimental values with the considerably more accurate theoretical values is quite satisfactory.

Reference	$\mathcal{L}_{exp}(\pm 1\sigma)$	Old theory ($\pm$ L.E.)	Old exp-th ($\pm 1\sigma$)	Revised theory ($\pm$ L.E.)	Revised exp-th ($\pm 1\sigma$)
H(n=2)		1057.56±0.09		1057.91±0.16	
Triebwasser, Dayhoff & Lamb	1057.77±0.06		0.21±0.07		−0.14±0.08
Robiscoe	1057.90±0.06		0.34±0.07		−0.01±0.08
Kaufman, Lea, Leventhal & Lamb	(1057.65±0.05)		0.09±0.06		−0.26±0.07
[(ΔE−$\mathcal{L}$)$_{exp}$=9911.38±0.03]					
Shyn, Williams, Robiscoe & Rebane	(1057.78±0.07)		0.22±0.08		−0.13±0.09
[(ΔE−$\mathcal{L}$)$_{exp}$=9911.25±0.06]					
Cosens & Vorburger	(1057.86±0.06)		0.30±0.07		−0.05±0.08
[(ΔE−$\mathcal{L}$)$_{exp}$=9911.17±0.04]					
D(n=2)		1058.82±0.15		1059.17±0.22	
Triebwasser, Dayhoff & Lamb	1059.00±0.06		0.18±0.08		−0.17±0.09
Cosens	1059.28±0.06		0.46±0.08		+0.11±0.09

Figure 22: The Lamb shift in hydrogenic atoms (in MHz).[12]

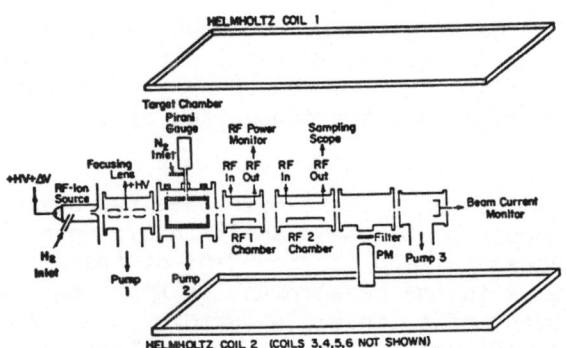

Figure 23: Schematic diagram of apparatus used for observation of the fine structure resonances using a fast hydrogen beam and separated oscillating fields.

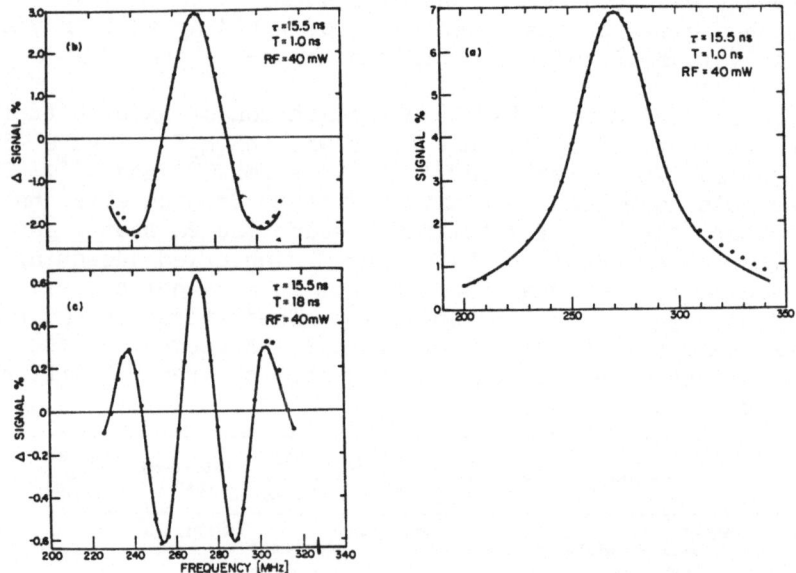

Figure 24: Typical line scans using apparatus shown in Figure 23. For each of these scans the counting rate was approximately 1 MHz and the counting time at each point was 50 seconds. The constant τ is the time spent in each of the rf fields; the constant T is the time required to pass from the first rf field to the second rf field. (a) Scan with rf spectroscopy chambers in the "Single Field" mode. (b) Scan of interference term with T = 1.0 ns. (c) Scan of interference term with T = 18 ns.

LAMB SHIFT INTERVALS FOR HYDROGENIC ATOMS WITH Z>1

Atom	n	Experimental Value (GHz)	Theoretical Value (GHz)	Experiment-Theory (MHz)
$^4He^+$	2	14.045 4(12)	14.044 78(61)	+1.6(13)
$^4He^+$	3	4.183 17(54)	4.184 42(18)	-1.25(57)
$^4He^+$	4	1.768 (5)	1.768 34(51)	0(5)
$^6Li^{2+}$	2	63.031(327)	62.771(17)	+260(327)
$^{12}C^{5+}$	2	780(8)	783.7(2)	$-4(8) \times 10^3$
$^{16}O^{7+}$	2	2 202.7(110)	2 205.17(156)	$-2.5(12) \times 10^3$
$^{16}O^{7+}$	2	2 215.6(75)	2 205.17(156)	$10(8) \times 10^3$

Figure 25: Experimental and theoretical values for Lamb shift intervals of hydrogenic atoms with Z>1 [using α^{-1}=137.036 02(21) (1.5 ppm)].

For positronium[51] there has been a considerable increase in both the experimental and theoretical accuracies for the fine structure interval $\Delta\nu$ in the ground n = 1 state. Figure 26 shows the Zeeman energy levels and the transition observed. The improvement in the experimental accuracy has come about through improved instrumentation, including a larger magnet with high homogeneity and a larger number of more stable detectors as shown in Figure 27.[13,52,53] A resonance curve is shown in Figure 28 where the width is determined by the annihilation rate and by microwave power broadening. Data were taken over a substantial range in pressure (Figure 29), and there is an observed dependence of $\Delta\nu$ on nitrogen pressure. A good fit to the data required the use of both a linear and a quadratic coefficient for the fine structure pressure shift. Figure 30 shows the comparison of the experimental and theoretical values.[31,54] The disagreement amounts to 4 times the experimental one standard deviation error, and indicates the need to calculate the $\alpha^4 R_\infty$ term. With the addition of data acquisition equipment and the accumulation of considerable additional data, the experimental error in $\Delta\nu$ could be reduced to about 10 ppm, which would correspond to choosing the resonance line center to 1 part in 10^3 of its FWHM value. Figure 31 shows the fine structure energy levels of the n = 2 state of positronium. Groups at Yale and at Bell Telephone Laboratories have recently been trying again to produce the n = 2 state by optical excitation from the n = 1 state,[55] but as yet no positive results have been reported.

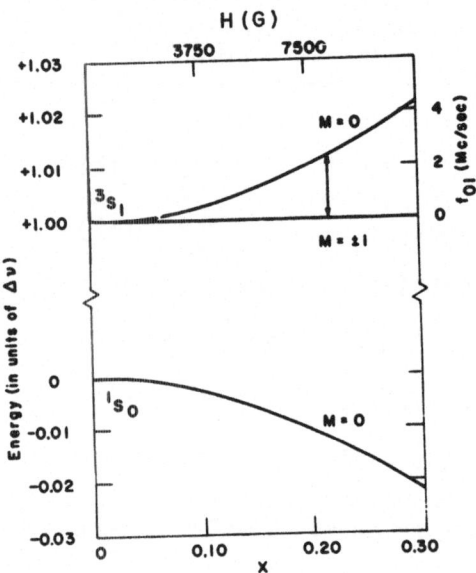

Figure 26:   Zeeman energy levels of positronium in its ground
n=1 state.   The transition observed is indicated by the arrow.

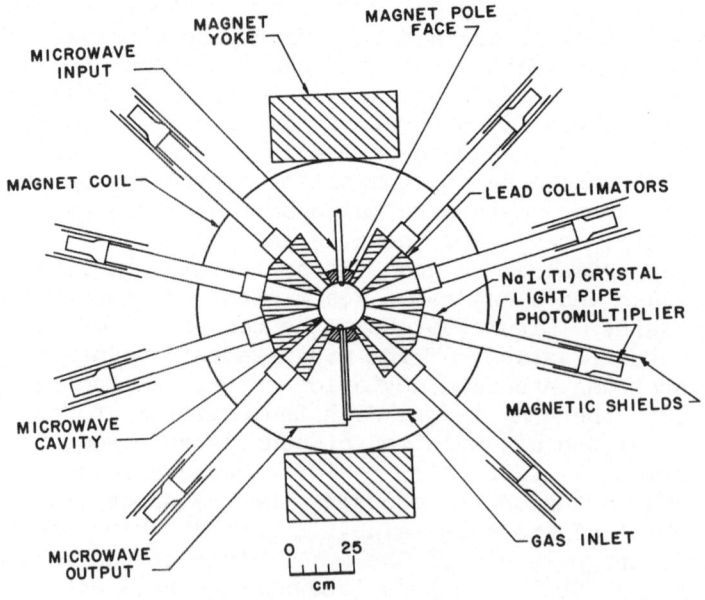

Figure 27:   Positronium fine structure experiment (schematic
diagram).

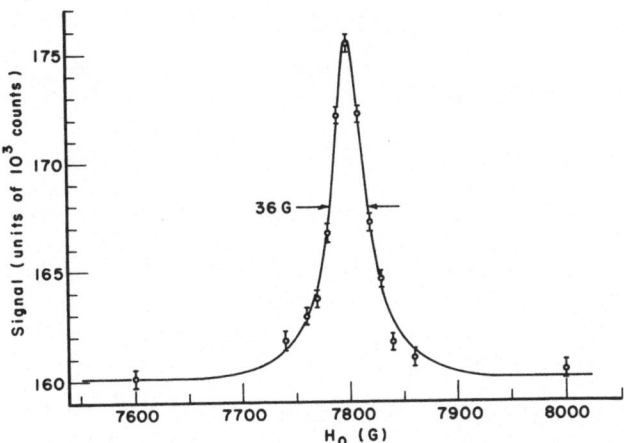

Figure 28: A typical observed resonance line for the Zeeman
transition in positronium. The experimental points are indicated
by the circles, after subtraction of a linear background. The
solid curve is a Lorentzian curve fitted to the data points. The
microwave frequency was about 2300 MHz, and the N_2 gas pressure was
0.3 atm.

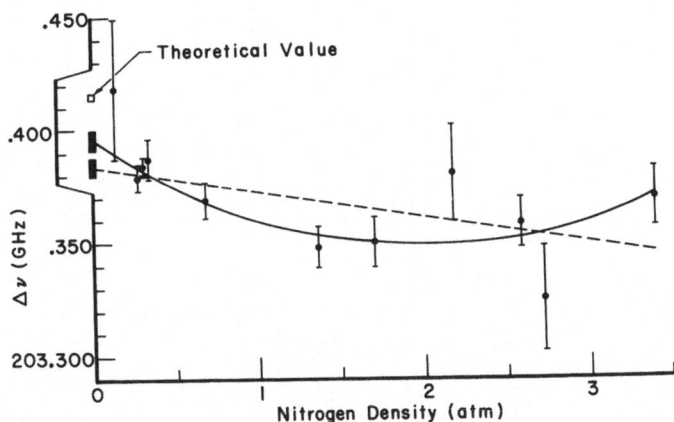

Figure 29: Plot of $\Delta\nu$ for positronium versus N_2 gas density in
units of atm at 23°C. The solid and dashed lines are the
quadratic and linear fits to the data respectively. The solid
bars show one standard deviation errors for the fitted values
$\Delta\nu(0)$.

POSITRONIUM FINE STRUCTURE

$$\Delta \nu_{theor} = \frac{1}{2} \alpha^2 cR_\infty \left[\frac{7}{3} - \frac{\alpha}{\pi} \left(\frac{32}{9} + 2\ln 2 \right) - \frac{3}{2} \alpha^2 \ln \alpha \right]$$

$$= (203.415\ 5 \pm 0.000\ 6)\text{GHz} \ (3.1\ \text{ppm})$$

$$\Delta \nu_{expt} = (203.403 \pm 0.012)\text{GHz} \ (60\ \text{ppm})[53]$$

$$\Delta \nu_{expt} = (203.396 \pm 0.005)\text{GHz} \ (25\ \text{ppm})[13]$$

$$\Delta \nu_{expt} - \Delta \nu_{theor} = (-0.019 \pm 0.005)\text{GHz}$$

Figure 30: Theoretical and experimental values for the fine structure interval in the ground state of positronium [using α^{-1}=137.036 02(21) (1.5 ppm)].

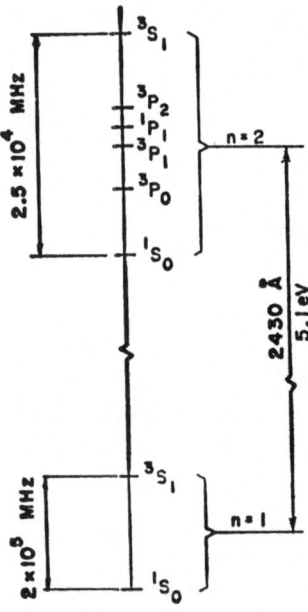

Figure 31: Positronium energy levels, showing the fine structure levels of the n=2 state.

The anomalous magnetic moment factor for the muon constitutes one of the most sensitive tests of QED and probes the validity of QED to very small distances ($\sim$5x10^{-15} cm) because of the small muon Compton wavelength.   Since 1970 the principal advance in the theory has been the completion of the calculation of all α^3 terms for the electron[44] (and muon)[56] anomalous magnetic moments. Figure 32 shows one such numerically important Feynman diagram involving photon-photon scattering where the real particle is the muon and the virtual particles are electrons.   Figure 33 shows the Feynman diagram for hadron vacuum polarization.   Although its contribution is less than the present experimental error, with the expected improvement in accuracy it will be important and represents the effect of strongly interacting particles on a pure lepton property.   The CERN measurement of g_μ-2 reported in 1968 is the most recent value.[57]   Figure 34 shows the current agreement between experiment and theory.[12]   There are ambitious plans to improve the accuracy in the determination of a_μ to about 10 ppm.[42,58]   The experiment will again be a measurement of g-2, similar in principle to the Michigan electron g-2 experiment, but involves both electric and magnetic fields.   Figure 35 shows the magnet planned for the new CERN experiment which will use 3 GeV/c muons.   Figure 36 shows the superconducting muon storage ring for an experiment being planned by a Yale group to utilize 1 to 1.5 GeV/c muons.

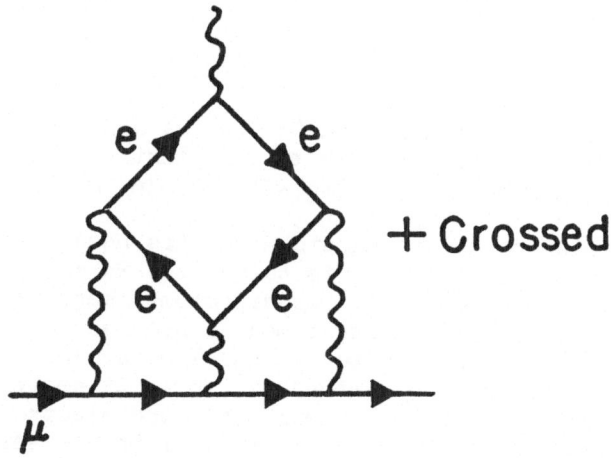

Figure 32: Photon-photon scattering contribution to the sixth-order magnetic moment of the muon.

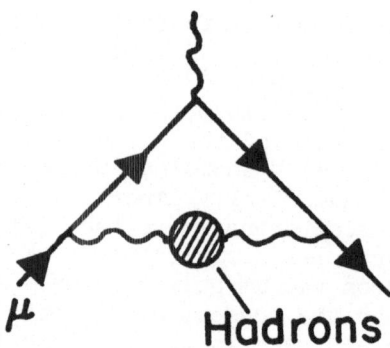

Figure 33: Hadronic vacuum polarization correction to the muon
magnetic moment.

MUON g-VALUE

$$g = 2(1 + a)$$

$$a_{theor.} = 0.5 \frac{\alpha}{\pi} + 0.765\ 78 \frac{\alpha^2}{\pi^2} + (21.8 \pm 1.3) \frac{\alpha^3}{\pi^3} + (65 \pm 5) \times 10^{-9}$$

$$= (1\ 165\ 878 \pm 17) \times 10^{-9} \quad (15\ \text{ppm})$$

$$a_{expt.} = (1\ 166\ 160 \pm 310) \times 10^{-9} \quad (270\ \text{ppm})$$

$$a_{expt.} - a_{theor.} = (282 \pm 310) \times 10^{-9}$$

Figure 34: Theoretical and experimental values for the muon
anomalous magnetic moment factor $\underline{a}$. The last term in $a_{theor.}$ is
the hadronic contribution [using α^{-1}=137.036 08(26) (1.9 ppm)].

     Vacuum polarization in muonic atoms has been a very active
topic recently. The situation is summarized in Figure 37. The
measurements are made on high Z muonic atoms, such as μ^-Pb, with
solid state detectors to an experimental accuracy of about
50 ppm.[59,60,61] In addition to the usual Dirac and reduced mass
terms, the theoretical expression[62,63] includes contributions from
vacuum polarization, finite nuclear size, and electron screening.
The vacuum polarization[64] is a term similar to the self-energy Lamb
shift diagram, and is dominantly due to electron pairs rather than
muon pairs. Agreement between theory and experiment is not good
and further work is clearly needed. It should be noted that the
new unified theories of weak and electromagnetic interactions
predict additional small contributions to the energy levels of

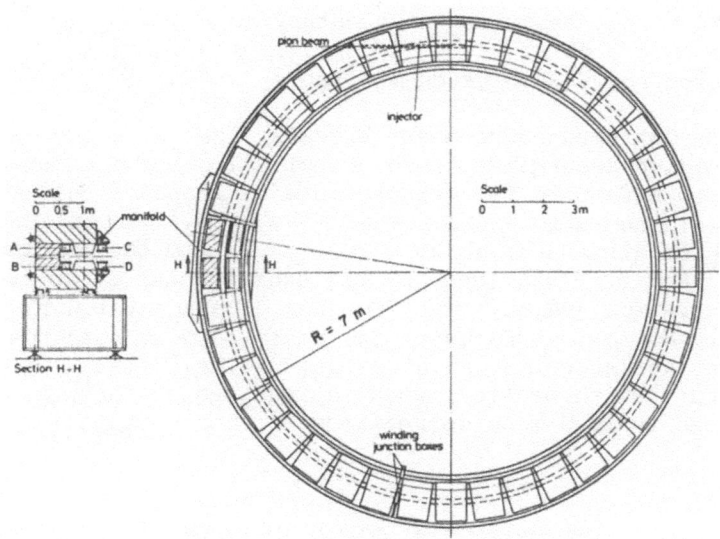

Figure 35: Schematic design of muon storage ring under construction at CERN for muon g-2 measurement, using electromagnets with B ≃ 14kG.

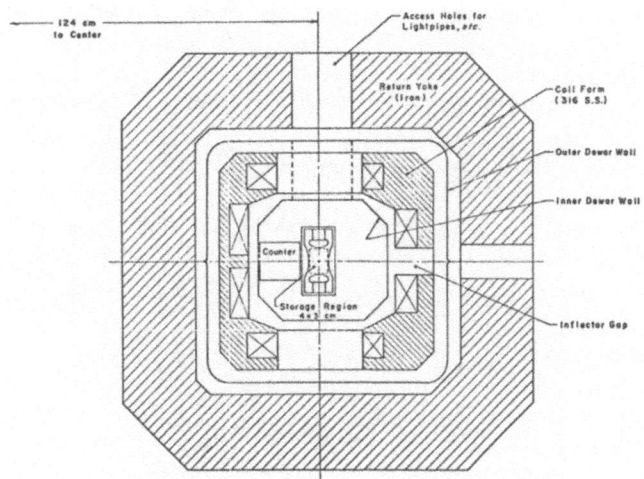

Figure 36: Schematic diagram of a superconducting muon storage ring for muon g-2 measurement. (Yale conceptual design with B ≃ 40kG.)

muonic atoms.[10,11] On the experimental side the use of crystal diffraction spectrometers at the new meson factories will lead to improvements in the experimental accuracy to about 10 ppm. A difficult atomic problem at the 10 ppm level of accuracy is the effect of electron screening on the energy levels of muonic atoms.

At this Conference Professor Brossel[65] has discussed a new[66] higher precision determination in an optical pumping experiment of the Zeeman effect in the metastable $2\,^3S_1$ state of helium, in particular of the ratio $g_J(He,2\,^3S_1)/g_J(H,\ 1S_{1/2})$, to an accuracy of 0.5 ppm, which differs by about 2 std. dev. from the previous experimental value[67] and by 3 std. dev. of the experimental error from the current theoretical value.[13,68] Further experimental work is under way at Ecole Normale, at Berkeley and at Yale to improve the accuracy in the determination of $g_J(He,2\,^3S_1)/g_J(H,1\,^2S_{1/2})$ to about 0.1 ppm. Further theoretical work is required on the higher order radiative and reduced mass corrections to atomic magnetism.[69]

ENERGY LEVELS OF MUONIC ATOMS

$$E_{nj} = -\frac{m_\mu c^2}{2}\left[\frac{Z\alpha}{n}\right]^2\left[1 - \frac{m_\mu}{M_N+m_\mu}\right]\left[1 + \left(\frac{Z\alpha}{n}\right)^2\left(\frac{n}{j+1/2} - \frac{3}{4}\right) + \dots\right]$$

$$+ <\psi|V_{vp}|\psi> + <\psi|V_{fns} + \frac{Ze^2}{r_\mu}|\psi> + \sum_i<\psi|\frac{e^2}{r_{\mu i}}|\psi>$$

$$\Delta E = (\Delta E_{Dirac} + \delta E_{reduced\ mass}) + \delta E_{vp} + \delta E_{fns} + \delta E_{electron\ screening}$$

Theoretical: (μ–Pb, $5g_{7/2} - 4f_{5/2}$, in eV)

$$\Delta E = (\Delta E_{Dirac} + \delta E_{rm}) + \left(\delta E_{vp}^{(1)} + \delta E_{vp}^{(h)}\right) + \delta E_{fns} + \delta E_{es}$$

$$(435,679) + (2190-26)\quad -10\quad -72 = 437,761\pm10\ eV$$

Experimental:[59] (μ–Pb, $5g_{7/2} - 4f_{5/2}$) = $437,687\pm20$ eV

$$\Delta E_{theory} - \Delta E_{exp} = 74 \pm 22\ eV\ (3.4\pm\%\ of\ \delta E_{vp})$$

Figure 37: Theoretical and experimental values for a transition in μ^-Pb. (The subscripts vp and fns refer to vacuum polarization and finite nuclear size, respectively.)

As a final topic, recent progress has been made on muonic hydrogen. There has been theoretical work[70] on calculating the fine and hyperfine structure of μ^-p (and $(\mu^-He)^+$), which is shown in Figure 38. The fine structure is dominated by electron vacuum polarization and we note the inversion of the $2S_{1/2}$ and $2P_{1/2}$ levels as compared to hydrogen. An experiment has been reported at CERN[71] detecting the metastable 2S state of $(\mu^-He)^+$, and there are plans to search for the 2S state of μ^-p at Nevis.[72] Muonic hydrogen would be a very important atom to study from the viewpoint of QED.

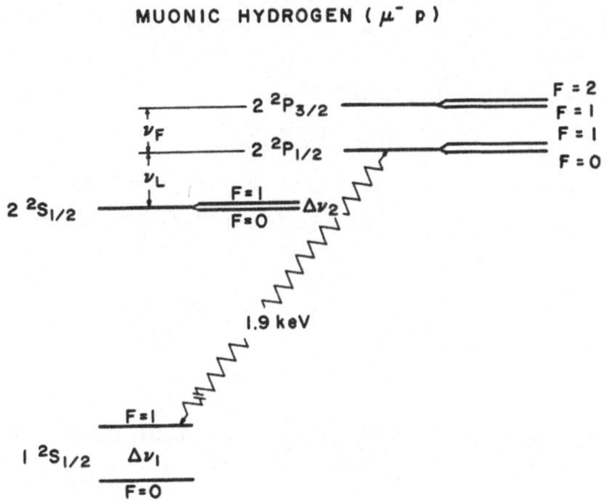

$$\nu_L = O\left[\alpha^3 \frac{m_\mu}{m_e} R_\infty^\mu\right] = 4.9 \times 10^{13} \text{ cps.}$$

$$\nu_F = O\left[\alpha^2 R_\infty^\mu\right] = 0.2 \times 10^{13} \text{ cps.}$$

$$\Delta\nu_1 = O\left[\alpha^2 \frac{\mu_p}{\mu_\mu} R_\infty^\mu\right] = 4.4 \times 10^{13} \text{ cps. } (68{,}000 \text{ Å})$$

$$\Delta\nu_2 \approx \frac{1}{8}\Delta\nu_1 = 0.5 \times 10^{13} \text{ cps.}$$

Figure 38: Energy levels of muonic hydrogen in the n=1 and n=2 states.

In conclusion, we can say that although there are several appreciable discrepancies between theory and experiment, in particular for positronium fine structure, for muonic spectra, and for helium magnetism, agreement is still remarkably good, and there have been important advances in the precision of the tests in the past two years.

As prognosis for our next Conference, on the experimental side
we can certainly expect the impressive recent increases in precision
to continue on most of the experiments I have discussed, and we will
probably learn to study new fundamental systems such as μ⁻p. Presumably the theory as regards calculation of pure QED terms will
keep pace. We will certainly observe then more of the effects of
strong interactions in these simple systems. We might also have
the good furtune to learn something about the weak interactions or
a unified theory of weak and electromagnetic interactions. I think
we should also always keep in mind that there are fundamental
theoretical problems in QED-renormalization with infinite quantities
and the nature (convergence) of the exact solutions-so more sensitive tests of the theory will not necessarily lead to excellent
agreement as a foregone conclusion.

REFERENCES

*The preparation of this report was supported in part by the AFOSR
(F44620-71-C-0042), the AEC (Yale Report COO-3075-26) (AT(11-1)3075),
and the NSF (GP 23722).

1. V.W. Hughes, Atomic Physics I, edited by V.W. Hughes, B.
    Bederson, V.W. Cohen, and F.M.J. Pichanick (Plenum, New York,
    1969) p. 15.

2. S.J. Brodsky, Atomic Physics II, edited by P.G.H. Sandars and
    G.K. Woodgate (Plenum, London, 1971) p. 1.

3. S.J. Brodsky, Proceedings of 1971 International Symposium on
    Electron and Photon Interactions at High Energies, edited by
    N.B. Mistry, (Cornell University, Ithaca, 1972) p. 13.

4. N.M. Kroll, Atomic Physics III, edited by S.J. Smith and G.K.
    Walters, (these proceedings).

5. B.E. Lautrup, A. Peterman, E. deRafael, Phys. Letters 3C,
    193 (1972).

6. N.M. Kroll, Physics of the One- and Two-Electron Atoms, edited
    by F. Bopp and H. Kleinpoppen (North-Holland, Amsterdam, 1969)
    p. 179.

7. T.D. Lee and C.S. Wu, Ann. Rev. Nucl. Sci. 15, 381 (1965).

8. S. Weinberg, Phys. Rev. Letters 27, 1688 (1971).

9. H. Georgi and S.L. Glashow, Phys. Rev. Letters 28, 1494 (1972).

10. J.R. Primack and H.R. Quinn, "Muon g-2 and Other Constraints on
    a Model of Weak and Electromagnetic Interactions without
    Neutral Currents," (Preprint, 1972).

11. R. Jackiw and S. Weinberg, Phys. Rev. D $\underline{5}$, 2396 (1972).

12. S.J. Brodsky and S.D. Drell, Ann. Rev. Nucl. Sci. $\underline{20}$, 147 (1970).

13. Abstracts: Third International Conference on Atomic Physics, (JILA, University of Colorado, 1972).

14. B.N. Taylor, W.H. Parker, and D.N. Langenberg, Rev. Mod. Phys. $\underline{41}$, 375 (1969).

15. D.N. Langenberg and B.N. Taylor, Precision Measurement and Fundamental Constants (NBS Special Pub. 343, U.S. Govt. Printing Office, 1971).

16. Z. Bay, G.G. Luther, and J.A. White, Phys. Rev. Letters $\underline{29}$, 189 (1972).

17. J.L. Hall, Atomic Physics III, edited by S.J. Smith and G.K. Walters, (these proceedings).

18. T.W. Hänsch, I.S. Shahin, and A.L. Schawlow, Nature Phys. Sci. $\underline{235}$, 63 (1972).

19. P.F. Winkler, D. Kleppner, T. Myint, F.G. Walther, Phys. Rev. A $\underline{5}$, 83 (1972).

20. V.W. Hughes, in A Tribute to I.I. Rabi, (Columbia University, New York, 1968), p. 42.

21. J.C. Baird, J. Brandenberger, K.I. Gondaira, and H. Metcalf, Phys. Rev. A $\underline{5}$, 564 (1972).

22. N.F. Ramsey, Precision Measurement and Fundamental Constants, edited by D.N. Langenberg and B.N. Taylor, (NBS Special Pub. 343, U.S. Govt. Printing Office, 1971), p. 317.

23. V.W. Hughes, Ann. Rev. Nucl. Sci. $\underline{16}$, 445 (1966).

24. T. Crane, D. Casperson, P. Crane, P. Egan, V.W. Hughes, R. Stambaugh, P.A. Thompson, and G. zu Putlitz, Phys. Rev. Letters $\underline{27}$, 474 (1971).

25. D. Favart, P.M. McIntyre, D.Y. Stowell, V.L. Telegdi, R. DeVoe, and R.A. Swanson, Phys. Rev. Letters $\underline{27}$, 1340 (1971).

26. D. Favart, P.M. McIntyre, D.Y. Stowell, V.L. Telegdi, R. DeVoe, and R.A. Swanson, Phys. Rev. Letters $\underline{27}$, 1336 (1971).

27. N.F. Ramsey, <u>Molecular Beams</u>, (Clarendon Press, Oxford, 1956).

28. V.W. Hughes, <u>Quantum Electronics</u>, edited by C.H. Townes (Columbia University, New York, 1960), p. 582.

29. W.E. Cleland, J.M. Bailey, M. Eckhause, V.W. Hughes, R. Prepost, J.E. Rothberg, and R.M. Mobley, Phys. Rev. A <u>5</u>, 2338 (1972).

30. R.D. Ehrlich, H. Hofer, A. Magnon, D.Y. Stowell, R.A. Swanson, and V.L. Telegdi, Phys. Rev. A <u>5</u>, 2357 (1972).

31. T. Fulton, D.A. Owen and W.W. Repko, Phys. Rev. A <u>4</u>, 1802 (1971).

32. R. DeVoe, P.M. McIntyre, A. Magnon, D.Y. Stowell, R.A. Swanson, and V.L. Telegdi, Phys. Rev. Letters <u>25</u>, 1779 (1970).

33. J. Jarecki and R.M. Herman, Phys. Rev. Letters <u>28</u>, 199 (1972).

34. K.M. Crowe, J.F. Hague, J.E. Rothberg, A. Schenck, D.L. Williams, R.W. Williams, and K.K. Young, Phys. Rev. D <u>5</u>, 2145 (1972).

35. <u>High Energy Physics and Nuclear Structure,</u> edited by S. Devons, (Plenum, New York, 1970).

36. S.J. Brodsky and G.W. Erickson, Phys. Rev. <u>148</u>, 26 (1966).

37. V.W. Hughes in <u>Facets of Physics,</u> edited by D.A. Bromley and V.W. Hughes (Academic Press, New York, 1970), p. 125.

38. F.M.J. Pichanick, R.D. Swift, C.E. Johnson, and V.W. Hughes, Phys. Rev. <u>169</u>, 55 (1968).

39. A. Kponou, V.W. Hughes, C.E. Johnson, S.A. Lewis, and F.M.J. Pichanick, Phys. Rev. Letters <u>26</u>, 1613 (1971).

40. J. Daley, M. Douglas, L. Hambro, and N.M. Kroll, Phys. Rev. Letters <u>29</u>, 12 (1972).

41. J.C. Wesley and A. Rich, Phys. Rev. A <u>4</u>, 1341 (1971).

42. A. Rich and J.C. Wesley, Rev. Mod. Phys. <u>44</u>, 250 (1972).

43. S. Granger and G.W. Ford, Phys. Rev. Letters <u>28</u>, 1479 (1972).

44. M.J. Levine and J. Wright, Phys. Rev. Letters <u>26</u>, 1351 (1971).

45. T.F. Finnegan, A. Denenstein and D.N. Langenberg, Phys. Rev. B 4, 1487 (1971).

46. C.W. Fabjan and F.M. Pipkin, Phys. Rev. Letters 25, 421 (1970).

47. C.W. Fabjan and F.M. Pipkin, Phys. Letters 36A, 69 (1971).

48. D.E. Murnick, M. Leventhal and H.W. Kugel, Phys. Rev. Letters 27, 1625 (1971).

49. M. Leventhal, D.E. Murnick and H.W. Kugel, Phys. Rev. Letters 28, 1609 (1972); G.P. Lawrence, C.Y. Fan and S. Bashkin, Phys. Rev. Letters 28, 1612 (1972).

50. G.W. Erickson, Phys. Rev. Letters 27, 780 (1971).

51. V.W. Hughes in Physics of the One- and Two-Electron Atoms, edited by F. Bopp and H. Kleinpoppen (North-Holland, Amsterdam, 1969), p. 407.

52. E.R. Carlson, V.W. Hughes, and I. Lindgren, Bull. Am. Phys. Soc. 17, 454 (1972).

53. E.D. Theriot, Jr., R.H. Beers, V.W. Hughes, and K.O.H. Ziock, Phys, Rev. A 2, 707 (1970).

54. E.R. Carlson, V.W. Hughes, M.L. Lewis, and I. Lindgren, "Higher Precision Determination of the Fine Structure Interval in the Ground State of Positronium, and the Fine Structure Density Shift in Nitrogen," (Preprint, 1972).

55. H. Kendall, "The First Excited State of Positronium," MIT Thesis, 1954, unpublished.

56. J. Aldins, T. Kinoshita, S.J. Brodsky, and A. Dufner, Phys. Rev. Letters 23, 441 (1969).

57. J. Bailey, W. Bartl, G. von Bochmann, R.C.A. Brown, F.J.M. Farley, H. Jostlein, E. Picasso, and R.W. Williams, Phys. Letters 28B, 287 (1968).

58. J. Bailey and E. Picasso, Prog. Nucl. Phys. 12, 43 (1970).

59. M.S. Dixit, H.L. Anderson, C.K. Hargrove, R.J. McKee, D. Kessler, H. Mes, and A.C. Thompson, Phys. Rev. Letters 27, 878 (1971).

60. G. Backenstoss, S. Charalambus, H. Daniel, Ch. von der Malsburg, G. Poelz, H.P. Povel, H. Schmitt and L. Tauscher, Phys. Letters 31B, 233 (1970).

61. H.K. Walter, J.H. Vuilleumier, H. Backe, F. Boehm, R. Engfer, A.H.v. Gunten, R. Link, R. Michaelsen, C. Petitjean, L. Schellenberg, H. Schneuwly, W.U. Schroder, and A. Zehnder, Phys. Letters 40B, 197 (1972).

62. M.K. Sundaresan and P.J.S. Watson, Phys. Rev. Letters 29, 15 (1972).

63. B. Fricke, Nuovo Cimento Letters 2, 859 (1969).

64. E.H. Wichmann and N.M. Kroll, Phys. Rev. 101, 843 (1956).

65. J. Brossel, Atomic Physics III, edited by S.J. Smith and G.K. Walters, (these proceedings).

66. M. Leduc, F. Laloe, and J. Brossel, J. Phys. 33, 4 (1972).

67. C.W. Drake, V.W. Hughes, A. Lurio, and J.A. White, Phys. Rev. 112, 1627 (1958).

68. W. Perl and V.W. Hughes, Phys. Rev. 91, 842 (1953).

69. H. Grotch and R.A. Hegstrom, Phys. Rev. A 4, 59 (1971).

70. A. DiGiacomo, Nucl. Phys. 11B, 411 (1969).

71. A. Placci, E. Polacco, E. Zavattini, K. Ziock, G. Carboni, V. Gastaldi, G. Gorini, G. Neri and G. Torelli, Nuovo Cimento 1A, 445 (1971).

72. V.W. Hughes, H. Rosenthal, and C.S. Wu, Bull. Am. Phys. Soc. 16, 617 (1971).

STATUS OF QUANTUM ELECTRODYNAMICS THEORY

Norman M. Kroll

University of California, San Diego

La Jolla, California 92037

INTRODUCTION

The recent history of quantum electrodynamics has been characterized by a steady increase of the precision of experiments which refer to low energy properties of simple quantum electrodynamical systems, and by a continued probing at high energies of the properties of the basic interaction at ever decreasing distances. The high precision experiments have provided a continuing challenge to theorists to deduce consequences of the theory to a sufficient accuracy to maintain a meaningful confrontation between theory and experiments. While puzzling discrepancies have appeared from time to time, it was possible to say[1] at the last conference in this series that no serious discrepancy existed for any of the crucial tests. While this continues to be essentially the case, some recent successes in meeting this challenge have notably increased the precision of agreement.

Perhaps the most striking development in the recent advances in quantum electrodynamics calculations has been the increasing utilization of computers to carry out most, or even all, of the calculation. As experience is gained in the programming of computers to carry out the complex variety of operations required, it is likely that more and more future work will be carried out in this way.

With reference to more general theoretical questions there has been a very interesting, though still speculative, development on the possible linkage between the electromagnetic and weak interactions. These will be discussed in the last part of my report. Other interesting general theoretical questions such as quantum

33

electrodynamics at high energies, and the extent to which quantum electrodynamics may be a self-contained self-consistent theory will be discussed in subsequent papers.

II. COMPUTATIONAL PROBLEMS BEARING ON THE PRECISION TESTS OF QUANTUM ELECTRODYNAMICS

In this section computational advances which have taken place in the past two years will be reviewed. Because of the frequent appearances of reviews of this kind[2] the main emphasis will be placed upon the most recent work.

(a) The 2^3P Fine Structure Splittings in Helium

The fine structure splittings of the 2^3P state of helium have been determined experimentally to 1.2 ppm and 2.2 ppm.[3] Since the determination of these intervals is completely dominated by electromagnetic interactions, these measurements provide the most precise experimental information available for the determination of the fine structure constant. The possibility that such information might become available has been known for many years, and accordingly a rather massive theoretical program directed towards its exploitation was outlined by Schwartz[4] in 1964 and has been in process since that time. Because one is dealing with a two electron atom the theoretical problem is both formidable and of a different character from that involved in the quantum electrodynamics of two body systems such as muonium and hydrogen.

The covariant two particle Bethe-Salpeter equation, with the nucleus treated as a fixed source of Coulomb field,[5] provides the starting point for the general theory.[6] This is conveniently taken in the formally exact form developed by Sucher,[5] which we write as

$$(H_{D1} + H_{D2} + \Lambda_{++}I_c\Lambda_{++} + H_\Delta)\ \phi\ (\underline{p}_1,\underline{p}_2)\ =\ E\ \phi(\underline{p}_1,\underline{p}_2)$$

where H_D is a one electron Dirac Hamiltonian in momentum space (including interaction with the Coulomb potential of the nucleus, V_N) so that

$$H_D = \underline{\alpha} \cdot \underline{p} + \beta m + V_N \quad ,$$

I_c is the Coulomb interaction between electrons and the Λ_{++} are positive and negative energy state projection operators for the energy eigenstates of two non-interacting electrons in the nuclear Coulomb potential. In this form the "relative energy variable," which is characteristic for the Bethe-Salpeter equation, appears only within the definition of the perturbation Hamiltonian H_Δ . H_Δ is defined by

$$H_\Delta = \int \frac{d\epsilon}{-2\pi i} \ (E - H_{D1} - H_{D2}) \ F^{-1} \ G_\Delta \ (F - G_D)^{-1} \ I_c$$

$$+ \ \Lambda_{++} \ I_c \ (1 - \Lambda_{++}) - \Lambda_{--} I_c$$

with

$$F \equiv (\tfrac{1}{2}E + \epsilon - H_{D1}) \ (\tfrac{1}{2}E - \epsilon - H_{D2}) \quad .$$

The quantity, G_Δ, is the usual sum over irreducible two particle graphs describing emission and reabsorption of photons (omitting, of course, the single Coulomb exchange already taken into account by I_c). The aim of the calculation is to include all corrections to the fine structure up to and including order α^2. Apart from effects which can be expressed in terms of the anomalous moments it is sufficient for this purpose to include no more than two photons in the irreducible graphs.

The basic formula for the energy shift is obtained by straightforward application of the Brillouin-Wigner perturbation formula to the fundamental equation, treating H_Δ as the perturbation. Systematic expansion of the various formal operator expressions which make their appearance, and the application of Foldy-Wouthuysen[7] reduction procedures yields an expression which may be interpreted as arising from a reduced four by four Hamiltonian of the Pauli-Schroedinger type. Thus we write

$$H_{reduced} = H_0 + H_2 + H_a + H_4 + \cdots$$

where

> H_0 = non-relativistic Schroedinger Hamiltonian for the interacting electrons;
>
> H_2 = the semi-classical order α^2 Ry corrections (includes contributions from relativistic mass, contact terms, spin orbit, and spin-spin interactions);
>
> H_a = corrections to H_2 due to the anomalous electron moment (order α^3 Ry and α^4 Ry);
>
> H_4 = corrections of order α^4 Ry which are generated by the reduction procedure.

These portions of H_4 which contribute to the fine structure have been reduced to fifteen expressions in coordinate space. Some typical forms are exhibited below.

$$H_{4,1} = \frac{3}{8} \alpha^3 (z\alpha) \nabla_1^2 * \frac{1}{r_1^3} \underline{\sigma}_1 \cdot (\underline{r}_1 \times \underline{p}_1)$$

$$H_{4,5} = -\frac{1}{2} \alpha^4 \frac{1}{r^6} (\underline{\sigma}_1 \cdot \underline{r})(\underline{\sigma}_2 \cdot \underline{r})$$

$$H_{4,12} = -\frac{1}{16} \alpha^4 \frac{1}{r^3} (\underline{\sigma}_1 \cdot \underline{p}_2)(\underline{\sigma}_2 \cdot \underline{p}_1)$$

They are seen to be singular when the electrons coincide or are at the nucleus and for this reason are applicable only when the wave function vanishes at the operator singularities. The fact that one is dealing with a triplet P state effectively confines the quantum electrodynamic effects to low momenta and greatly simplifies the problem. The reduced forms such as those illustrated above are the result of taking advantage of this simplification.

The fine structure is then determined by applying first[4,8] and second[9] order perturbation theory to H_2 and first order perturbation theory to H_4 and H_4.[10] Since H_0 is itself a quite complicated object, the amount of computational atomic physics required is quite extensive; and the various pieces have been evaluated by different individuals as indicated by the various footnotes.

In addition to the above there are corrections of order $\frac{m}{M} \alpha^2$ Ry which must be taken into account.[11] The results[12] of this massive cooperative effort are summarized in Table I. The agreement between theory and experiment based upon the current value of the fine structure constant is seen to be well within the claimed accuracy of the various pieces of the calculation.

The theoretical accuracy is still inadequate to properly exploit the experimental precision. The main obstacle is evidently the second order fine structure. The computational effort which has been devoted to this term is so far comparatively modest and it is hoped that a much more accurate value will become available within a year. Ultimately, terms of order $\frac{m}{M} \alpha^3$ Ry and α^5 Ry should also be estimated. The former are probably trivial while the latter would require an elaborate theoretical effort.

A complete and independent recalculation of the helium fine structure would provide increased confidence. It should be noted, however, that the existence of high precision measurements for both fine structure intervals, provides a valuable check of consistency.

As a concluding comment we note that the helium fine structure calculation is only peripherally dependent upon the validity of the

Interval	ν_{01} (MHz)	ν_{02} (MHz)
$\Delta\langle H_2\rangle$ ($\alpha^4 mc^2$)	29,564.567 ± .006	2317.103 ± .0017
$\Delta\langle H_a\rangle$ ($\alpha^5 mc^2$)	54.708	-22.548
Nuclear Motion ($\frac{m}{M_{He}}\alpha^4 mc^2$)	-10.707 ± .00044	1.952 ± .00088
Second Order ($\alpha^6 mc^2$)	11.60 ± .18	-6.79 ± .36
$\Delta\langle H_4\rangle$ ($\alpha^6 mc^2$)	-3.331 ± .0039	1.542 ± .0068
ν_{theo}	29,616.83 ± .18	2291.36 ± .36
ν_{exp}	29,616.864 ± .036	2291.196 ± .005
$\nu_{th} - \nu_{exp}$	-.03 ± .18	+.16 ± .36
	(-1 ± 6 ppm)	(70 ± 157 ppm)

TABLE I. Summary of theoretical contributions to $3P_1$ fine structure of He.

renormalization program. As such it is a particularly suitable source for a value of α to be used in the determination of quantities such as the electron moment which depend upon the renormalization program in a crucial way.

(b) The Anomalous Magnetic Moment of the Charged Leptons

The magnetic moment anomalies $a = \frac{g-2}{2}$ of the charged leptons have both been experimentally determined with remarkable precision, 3 ppm in the case of the electron[13] and 270 ppm in the case of the muon.[14] These are the only experiments which are sufficiently precise to test renormalization theory to sixth order and thus are of special interest with reference to its validity.

It is traditional to express the electron anomaly a_e as a power series in (α/π), thus

$$a_e = \sum_{n=1}^{\infty} a_e^{(2n)} (\alpha/\pi)^n$$

with the now classical values[15,16,17]

$$a_e^{(2)} = \frac{1}{2}$$

$$a_e^{(4)} = \frac{197}{144} + \frac{\pi^2}{12} - \frac{1}{2} \pi^2 \ln 2 + \frac{3}{4} \zeta (3) \approx 0.32848$$

The principal new development has been the evaluation of the sixth order coefficient. The calculation is one of enormous complexity and has involved a world wide effort.[18] There are 72 diagrams, representatives of which are illustrated in Fig. 1. Various diagrams have been evaluated by different groups (fortunately some by more than one group). The computations have all made use of computer programs to carry out the necessary Dirac algebra and to reduce the Feynman integrals to parametric form. A variety of different programs have been developed for this purpose.[19,20,21,22] The final integrations over the Feynman variables have for the most part been carried out numerically, using two quite different techniques. One is an adaptive Monte Carlo technique,[23] the other[24] a Gaussian method following a variable transformation which smoothes the integrand. The renormalization program as well as intermediate infrared divergences require a substantial amount of manipulation with infinite quantities. The incorporation of the required subtractions into the expressions in such a way as to yield finite integrals for numerical integration is of course essential. Again

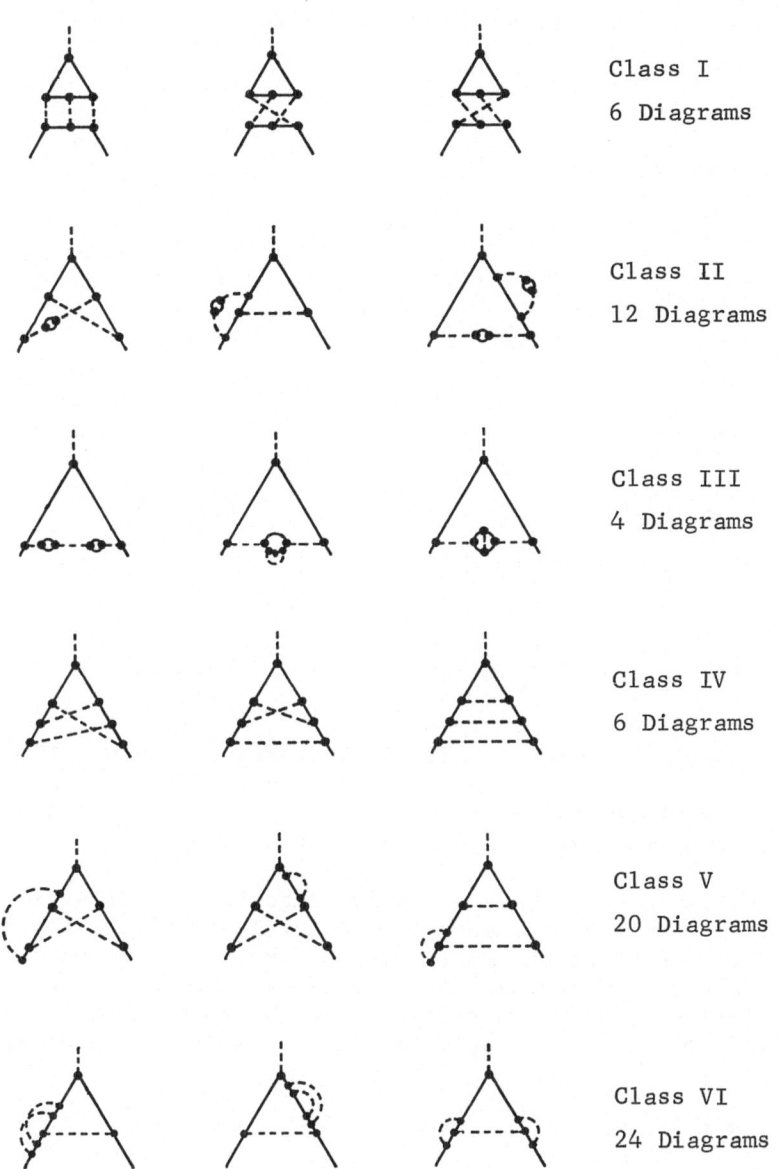

Class I
6 Diagrams

Class II
12 Diagrams

Class III
4 Diagrams

Class IV
6 Diagrams

Class V
20 Diagrams

Class VI
24 Diagrams

Fig. I. Representative Diagrams for the Sixth Order Magnetic Moment Anomaly. The diagram count includes mirror image diagrams.

the different authors have used different techniques. The most automated procedure seems to be that of Levine and Wright.[24] The results of the various authors are summarized below.

CLASS I

$$a_e^{(6)}(I) = 0.36(4) \qquad\qquad \text{Ref. 25}$$

CLASS II

$$a_e^{(6)}(II) = \begin{array}{ll} -0.153(5) & \text{Ref. 26} \\ -0.151(3) & \text{Ref. 27} \end{array}$$

CLASS III

$$a_e^{(6)}(III) = \begin{array}{ll} .055429 & \text{(Analytic) Ref. 28} \\ .05546(6) & \text{Ref. 26} \\ .055(2) & \text{Ref. 27} \end{array}$$

CLASS IV,V,VI

$$a_e^{(6)}(IV) + a_e^6(V) + a_e^6(VI) = .89(10) \qquad \text{Ref. 24}$$

Not shown above are certain individual diagrams in Class IV, V, and VI which have been computed by more than one group. The agreement between the various groups for these diagrams is of the same general quality as that indicated in the compilation above.[18]

It should be noted that the value reported for Class IV, V, VI represents a significant change from the value 1.23(20) reported previously.[24] Part of the difficulty of the numerical work stems from the fact that the final total involves a considerable amount of cancellation between diagrams. The above noted change arose primarily from a five percent shift in the value of one diagram. The final result is

$$a_e^6 = 1.15 \pm .11$$

We thus have

$$\frac{1}{2}(g-2) = (1161\ 409.0\ (2.2) - 1772.3 + 14.4\ (1.4)) \times 10^{-9}$$

$$= 1\ 159\ 651.1\ (2.6) \times 10^{-9} \qquad \text{(Theory)}$$

to be compared with the most recent experimental value[13]

$$\frac{1}{2} (g-2) = 1\ 159\ 656.7\ (3.5) \times 10^{-9}$$

As a consequence of the new theoretical result a discrepancy slightly in excess of one standard deviation has appeared. A completion of the check by independent groups would appear to be mandatory, and a reduction in the numerical error is desirable and probably practical.

The muon anomaly differs from the electron anomaly through the occurrence of diagrams containing closed electron loops. (The contribution of closed muon loops to the electron anomaly is negligible.) Since $a_\mu - a_e$ involves only these electron loop diagrams it is a simpler quantity to evaluate than the anomalies themselves, and its evaluation through sixth order was completed in 1970. Again denoting the coefficients in an $(\frac{\alpha}{\pi})^n$ expansion by a_μ^{2n} we recall

$$a_\mu^{(2)} - a_e^{(2)} = 0$$

$$a_\mu^{(4)} - a_e^{(4)} = \frac{1}{3} \ln \frac{m_\mu}{m_e} - \frac{25}{36} + \frac{\pi^2}{4} \frac{m_e}{m_\mu} + 0 \left(\frac{m_e}{m_\mu}\right)^2 \ln \frac{m_e}{m_\mu}$$

$$= 1.09426 \qquad (\text{Ref. } 29,30,31)$$

$$a_\mu^{(6)} - a_e^{(6)} = 20.3 \pm 1.1 \quad (\text{Ref. } 25,26,32,33,34,35)$$

The large number of references associated with the sixth order results indicate that as in the case of the electron anomaly many groups were involved and extensive use of computers made. Furthermore, with the very recent check[36] of the diagrams of Class I (which contributes 90 percent of the result) all of the diagrams have been done by at least two groups. The total result is

$$a_\mu^{th} = (116\ 581.4 + 413.18 + 27.3 \pm 1.4) \times 10^{-8}$$

$$= 116\ 581.4 \pm 1.4 \times 10^{-8} \quad \text{(leptonic quantum electrodynamics only)}$$

which may be compared with the experimental value[14]

$$a_\mu^{exp} = 116\ 616 \pm 31 \times 10^{-8}$$

The hadronic contribution to a_μ can, in principle, be determined directly from the $e^+ e^-$ annihilation cross section into hadrons. A rigorous lower bound, a_μ (hadron) $\geq 4.36 \times 10^{-8}$ has

recently been given.[37] The Gourdin-De Rafael[38] estimate
a_μ (hadron) = 6.5 ± .5 x 10^{-8} (based upon a combination of VMD and
experiment) or the more recent[39] Bramon, Etim, and Greco value
6.8 ± .9 x 10^{-8} (which uses more data and less theory) brings the
theoretical value within one standard deviation of the experimental
value.

Current experimental proposals envisage an increase in precision by a factor twenty.[40] Such an increase will significantly
sharpen the confrontation with theory and will reveal the hadronic
contribution unambiguously. It would also make reduction of the
uncertainty in a_μ^6 desirable. A speculation on the size of the
eighth order contribution has been given by Lautrup.[41] He has
pointed out that diagrams of the type shown in Fig. II can be
expected to be quite large and estimates that a_μ^8 may be of the
order 100-200. The failure of α/π as a suitable expansion parameter is evident and is due in part at least to the presence of
$\ln^p \dfrac{m_\mu}{m_e}$ (p < n) factors in a_μ^{2n} .

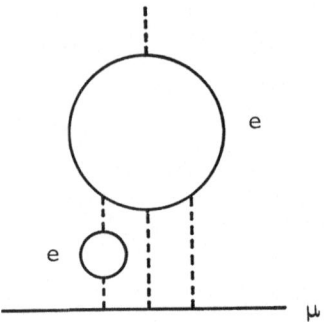

Fig. II. Diagram for an important eighth order contribution to the
muon magnetic moment anomaly.

Weak interaction corrections are also hovering at the limits
of the projected precision. These are quite model dependent. A
new value,[42,43] based upon a renormalizable model of weak interactions is a_μ (weak) = 2 ± .2 x 10^{-9} , depending upon parameters
assigned to the model.

(c) The Lamb Shift

The principal new development is the establishment of an analytic value for the troublesome fourth order term. It is given in
terms of the slope of the Dirac form factor at $q^2 = 0$, thus

$$m^2 \left. \frac{dF_1^{(4)}(q^2)}{dq^2} \right|_{q^2=0} = \frac{\alpha^2}{\pi^2} \left[-\frac{4819}{5184} - \frac{49\pi^2}{432} + \frac{\pi^2}{2} \ln 2 - \frac{3}{4} \zeta(3) \right]$$

$$= \frac{\alpha^2}{\pi^2} \left[0.4699 \right]$$

The contribution to the classical $n = 2$ level shift in hydrogen is .444 MHz. Barbieri, Mignaco, and Remiddi[44,45] have completed an analytic evaluation of all five diagrams, using in their work the dispersion method of Källen and Sabry[46] rather than the more commonly employed Feynman technique of reduction to integrals over auxiliary variables. Agreeing analytic expressions[47,48] for four of the five diagrams have been given by at least one other group. In addition all five diagrams are in agreement with values based upon numerical evaluation[48,49] of the Feynman integrals. Again the role of computers in obtaining this result is noteworthy. It has even proved possible to program the computer to determine the analytic form of the integrals. The extension of this technique to higher order computations may be quite difficult, but would be valuable.

Another significant theoretical advance is an improved treatment of the Z dependence of the second order level shift due to Erickson.[50] The second order level shift, $\Delta E_n^{(2)}$, may be expressed as

$$\Delta E_n^{(2)} = \frac{4\alpha}{3\pi n^3} (\alpha Z)^4 mc^2 \left[C_{41} \ln(\alpha Z)^{-2} + C_{40} + H(\alpha Z) \right]$$

with

$$(\alpha Z)^4 H(\alpha Z) = C_5 (\alpha Z)^5 + C_{62} (\alpha Z)^6 \ln^2(\alpha Z)^{-2} + C_{61} (\alpha Z)^6 \ln(\alpha Z)^{-2}$$

$$+ C_{60}(\alpha Z)^6 + C_7 (\alpha Z)^7 + \cdots$$

Apart from C_{60} and C_7 the indicated coefficients have all been evaluated to sufficient accuracy (mostly exactly), while C_{60} had been estimated to 25%. The new results have somewhat extended the low αZ expansion so as to yield

$$C_{60} = -19.3435 \pm .5$$

$$C_7 \cong 9.56\pi$$

for the n = 1 states (values for other states are available).

The current state of affairs for the $2S_{\frac{1}{2}} - 2P_{\frac{1}{2}}$ separation in hydrogen, including the new results, is summarized in Table II (compilation from Ref. 2).

In addition, Erickson has also evaluated $H(\alpha Z)$ to a good approximation over the entire range of Z from zero to 100. The results are shown in Fig. III.

Order $\times \alpha^2$ Ry	Description	Value (MHz)
α	self energy	1009.920
α	magnetic moment	67.720
α	vacuum polarization	-27.084
$\alpha Z\alpha$	binding	7.140
$\alpha(Z\alpha)^2$	binding	-0.372
$+\alpha(Z\alpha)^3$		
α^2	self energy	0.444
$\alpha^2 + \alpha^3$	magnetic moment	-0.102
α^2	vacuum polarization	-0.239
$\alpha \frac{m}{M}$	recoil (B.S.)	0.359
$(R_e/\lambda_e)^2$	proton size	0.125
TOTAL: Theory	1057.911 ± .012	
Exp.	1057.90 ± .06	(Ref. 51)
	1057.86 ± .06	(Ref. 51,52)

TABLE II.   Compilation of Contributions to Lamb Shift in H(n=2), and Comparison with Experiment. The order α terms include a reduced mass correction.

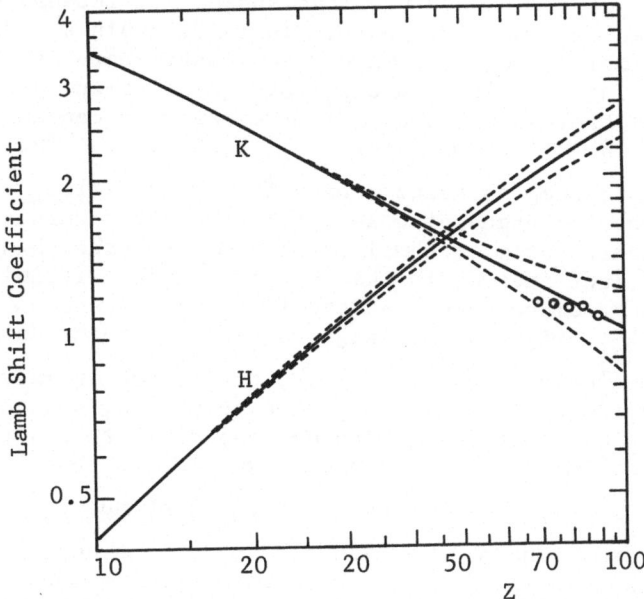

Fig. III. Z-dependence of the Lamb shift coefficient K, defined by $\Delta E_n^{(2)} = \dfrac{4\alpha}{3\pi n^2} (\alpha Z)^4 mc^2 K$, and its higher order in (αZ) part $H(\alpha Z)$, defined by $K = C_{41} \ln (\alpha Z)^{-2} + C_{40} + H(\alpha Z)$. The points shown are those of the earlier numerical work of Desiderio and Johnson[53] for high Z atoms.

So far the reported experiments at higher Z for ionized hydrogenic systems are in good agreement with the theory. Presumably, increasing experimental accuracy will ultimately demand more accurate approximations.

(d) The Hyperfine Structure of Hydrogen

The full exploitation of the extreme accuracy with which this quantity has been known experimentally has been impeded by the uncertainty arising from proton polarization. Omitting this quantity for the moment one has[2]

$$\frac{\Delta \nu_{exp} - \Delta \nu_{th}}{\Delta \nu_{th}} = 2.5 \pm 4.0 \text{ ppm}$$

where the uncertainty is due almost entirely to the fine structure constant. The proton polarization term δ_p, which arises from two

photon exchange diagrams, has been the subject of intense theoretical interest for a number of years. In 1965, Iddings[54] showed that
it could in principle be determined from measurements of the inelastic scattering of polarized electrons on polarized protons.
The analyses of Drell and Sullivan[55] made use of a convenient de-
composition of δ_p into two terms, δ_{p_1} and δ_{p_2}. The term δ_{p_1}
can be related to Compton scattering with physical (i.e. $q^2 = 0$)
photons, and one can argue persuasively[55] (if not rigorously) that
its contribution is no more than 1-2 ppm. On the other hand, it had
been argued,[55] (and indeed, in the face of a then existing substantial discrepancy, emphasized[56] at the first conference on this
series) that δ_{p_2} might be as large as tens of parts per million.
De Rafael[57] has recently shown, however, that existing measurements
on _unpolarized_ inelastic electron-proton scattering set bounds on
the polarized cross sections, which one can exploit to set bounds
on δ_{p_2}. He estimates, using present experiments that
-2 ppm $< \delta_{p_2} <$ 3 ppm. The technique should continue to be helpful
even after information on polarized cross sections begins to appear,
as it can be used to limit the contribution from regions in
scattering parameter space where only unpolarized measurements have
been made.

(e) The Hyperfine Structure of Positronium and Muonium

The most recent theoretical value for the positronium hyperfine splitting is[58]

$$\Delta \nu_{th} = 203,415 \text{ MHz}$$

which exceeds by 19 ± 5 MHz the most recent experimental value[59]
of

$$\Delta \nu_{exp} = 203,396 \pm 5 \text{ MHz} .$$

The theoretical formula may be written

$$\Delta \nu_{th} = \alpha^2 \text{ Ry} \left[\frac{7}{6} - \frac{\alpha}{\pi} \frac{16}{9} + \ell n \ 2) + \frac{3}{4} \alpha^2 \ell n \frac{1}{\alpha} + 0 \ (\alpha^2) \right] .$$

The most recent theoretical correction is a recoil term involving
two photon exchange, evaluated by Fulton, Owen, and Repko.[58] It is
the term

$$\alpha^2 \text{ Ry} \frac{3}{4} \alpha^2 \ell n \frac{1}{\alpha} \quad \sim \quad \alpha^2 \ell n \frac{1}{\alpha} \cdot \Delta \nu$$

which contributes 34 MHz to the difference. Fulton[60] has recently
discussed the unevaluated order $\alpha^2 \Delta\nu$ term and concludes that it
would not be unreasonable for the coefficient to be sufficiently
large to explain the discrepancy. A calculation is clearly urgently
needed.

There is no clear evidence for any discrepancy in the hyperfine interval in muonium but there is some slight indication that
the theoretical value, which already includes the non-recoil part
of the order $\alpha^2 \Delta\nu$ correction, may be somewhat high here as well.
The corresponding recoil part has an extra factor m_e/m_μ and would
be expected to be much smaller. Fulton[60] conjectures, however,
that this term may also contain a $\ell n\ m_\mu/m_e$ factor and hence might
contribute at the current level of precision. This circumstance
urges that the evaluation of order $\alpha^2 \Delta\nu$ recoil corrections not
be specialized to the equal mass case.

(f) The Vacuum Polarization Effect in Muonic Atoms

The higher n circular orbits of muonic atoms are well
placed for exploration of the electronic vacuum polarization potential generated by the nuclear charge distribution as they are well
outside the nuclear charge distribution, well inside the bound
electron distribution and strategically placed with respect to the
vacuum polarization charge distribution. They provide the best
possibility for gaining information about the large Z behavior
of the polarization charge density. For this reason (primarily),
a theoretical investigation[61] of the large Z corrections to the
Uehling potential was carried out in 1956. The corrections were
found to be so small, even at Z = 92, that their detection at that
time seemed to be hopeless.

The remarkable precision which has been achieved in these
measurements, however, now makes it possible to test this prediction. A six standard deviation discrepancy at large Z was recently
reported by Dixit et al.[62] Their work has prompted a reinvestigation
of the vacuum polarization terms as well as other small effects.

The current situation is summarized in Table III. Let us
consider the vacuum polarization terms first. The order $\alpha(\alpha Z)$
term is taken from Ref. 62. The polarization potential is computed from the Uehling-Serber formula taking the finite size
nucleon charge distribution into account (see ref. 63, Eq. (5) for
example for an explicit formula), and then added to the Coulomb
potential for the finite nuclear size charge distribution. The
energy is then determined by solving the Dirac equation for this
total potential. The order $\alpha^2(\alpha Z)$ term is taken from Ref. 64,
Table I, and is based upon the work of Källen and Sabry.[46] As
pointed out by Blomquist,[65] the column marked (b,c,d) in that work

in actuality includes the entire correction, so that only the contribution from that column is included. The order $\alpha(\alpha Z)^3 + \alpha(\alpha Z)^5 + \cdots$ term represents the work of Thomas Bell,[66] based upon the work of Wichmann and Kroll.[61] In carrying out the calculation he has extended the work of Ref. 61 to include a more complete description of the spatial variation of the higher order polarization potential. Similar results have been obtained by Blomquist.[65] The results tabulated here differ both in size and magnitude from those reported earlier.[63] The changes noted in these two terms tend to substantially reduce the discrepancies initially reported.

The screening corrections have recently been reviewed by Vogel[67] and by Fricke, Waber, and Telegdi.[68] Table III contains the values from Ref. 68. There is an inherent uncertainty in the screening correction which arises from lack of knowledge of what electronic energy levels are occupied when the muonic transition takes place. The screening corrections of Ref. 68 are based upon complete self-consistent field calculations for both initial and final levels under the assumption that the lowest ten levels are occupied. Other authors have tended to assume less occupation of these levels, and hence have reported somewhat smaller screening corrections.

The other theoretical entries are taken from Ref. 68 with the following exceptions. The point nucleus values (Column II) have been reduced in the ratio 105.6594/105.6599 to take account of the more recent value of the muon mass,[69] and the error accordingly reduced. The Lamb shift for Pb is taken from Ref. 65, where the not yet published work of Klarsfeld is quoted.

The new theoretical values may be compared with the experimental results of Dixit et al.[62] The worst of the previously reported discrepancies, that in 82^{Pb}, is greatly reduced, but a substantial part of that in 56^{Ba} (4f-3d transitions) remains. Furthermore, all of the differences between theory and experiment have the same sign. Discrepancies similar to those in 82^{Pb} have recently been reported in 80^{Hg} and 81^{Tl} by Walter et al.[70] for the same (5g-4f) transitions as those reported in Pb.

Jackiw and Weinberg[42] have noted that a weakly coupled scalar meson as suggested by recent developments in weak interaction theory could be responsible for the kind of discrepancy that seems to be observed. Sundaresen and Watson[64] suggest a mass of 8 MeV. A mass of 8 MeV together with coupling strength .8 of that suggested by Sundaresen and Watson reduces all of the discrepancies found in Table III to one standard deviation or less.

I element Z Trans.	II Point Nucleus Energy	III Finite Size Effect	IV $\alpha(\alpha Z)$	V $\alpha^2(\alpha Z)$	VI $\alpha(Z\alpha)^3 + \alpha(Z\alpha)^5$	VII Lamb Shift	VIII Elec. Scr.	IX Nucl. Pol.	X Order $\frac{v^2}{c^2}\frac{m_\mu}{M_N}$	XI Energy (Theo)	XII Energy (Exp)
^{47}Ag α_1	306970±2	-29±2	1519	10.5	-10.7	5±2	-12±1	5±2	1	308459±4	308428±19
α_2	303328±2	-11±1	1470	10.2	-10.4	-3±1	-12±1	4±1	1	304777±3	304759±17
^{48}Cd α_1	320422±2	-36±4	1608	11.2	-11.7	6±2	-13±1	5±2	2	321994±5	321973±18
α_2	316457±2	-14±2	1555	10.7	-11.4	-4±1	-13±1	4±1	1	317985±3	317977±17
^{50}Sn α_1	348233±2	-50±5	1795	12.5	-13.9	7±2	-13±1	6±2	2	349979±6	349953±20
α_2	343553±2	-19±2	1731	12.0	-13.5	-4±1	-13±1	5±2	2	345254±4	345226±18
^{56}Ba α_1	439068±2	-140±17	2435	17.0	-22.6	10±3	-17±2	9±3	3	441362±18	441299±21
α_2	431652±2	-53±6	2328	16.2	-21.9	-8±3	-17±2	8±3	3	433907±8	433829±19
β_1	200541±1	0	762	5.2	-9.5	2±1	-30±3	1±0	1	201273±3	201260±16
β_2	199193±1	0	748	5.1	-9.3	-1±0	-30±3	1±0	1	199908±3	199902±15
^{82}Pb β_1	435661±2	-10±1	2190	15.1	-51.2	8±0	-82±7	6±2	2	437739±8	437687±20
β_2	429343±2	-4	2106	14.5	-49.6	-7±0	-83±7	5±2	2	431327±8	431285±17

TABLE III. Muonic Transition Energies in (eV) for the transitions $\alpha_1 = 4f_{5/2} - 3d_{3/2}$, $\alpha_2 = 4f_{7/2} - 3d_{5/2}$, $\beta_1 = 5g_{7/2} - 4f_{5/2}$, $\beta_2 = 5g_{9/2} - 4f_{7/2}$.

III. HIGH ENERGY TESTS

The perfection of verification of quantum electrodynamics in
the precision low energy experiments leads one to ask to what momentum (or to what distances) has the theory been confirmed. Assessments based upon the low energy results are quite model dependent,
as the effects themselves depend upon integrals over momenta of
many particles and therefore do not readily lend themselves to
statements about how well each particle is behaving at some particular energy. (Nevertheless, the magnetic moment of the muon is often
said to confirm the theory to five GeV.)

The less precise high energy experiments are better suited to
dealing with this question as they measure the probability of high
momentum transfers directly. No deviations from the predictions
of quantum electrodynamics have been found in this domain either.
It is conventional to characterize the situation in terms of
characteristic cutoff momenta associated with photon exchange or
with lepton exchange. Current values of these parameters are
exhibited below.

Process	Particle Exchanged	Lower Limit or Cutoff (95% Confidence)
		GeV
$e^+e^- \to \gamma\gamma$	Electron	2.6 (Ref. 71)
$e^+e^- \to e^+e^-$	Photon	6 (Ref. 72)
$e^+e^- \to \mu^+\mu^-$	Photon (Pure Time-like)	10 (Ref. 73)
$\gamma C \to \mu^+\mu^- C$	Muon	2.3 (Ref. 74)

It is noteworthy that all limits except those associated with the
exchange of muons have been provided by colliding beam experiments.
These limits may be expected to improve as colliding beam facilities
of higher energy come into operation. Colliding beams may also be
expected to provide a limit on the muon exchange cutoff via the
process $e^+e^- \to e^+e^-\mu^+\mu^-$, which becomes relatively more prominent
at higher energies.

IV. QUANTUM ELECTRODYNAMICS, GAUGE THEORIES, AND WEAK INTERACTIONS

One of the very striking features of quantum electrodynamics
as it has developed is its apparent independence of the rest of the
world of elementary particles and their interactions. Indeed the
results of the last section show that characteristic hadronic
energies have no particular significance for quantum electrodynamics,
and a paper in these proceedings, by Adler, will suggest that the

fine structure constant itself is determined within the framework
of the theory. There has, however, been a great deal of interest
in the past year in theories which provide a direct link between
weak interactions and quantum electrodynamics, and I would like to
conclude this report with a few brief remarks about this development.

If one describes the weak interactions via the mediation of a
charged vector boson, one finds that analogous to the (e^2/q^2)
interaction between charged particles (in first Born approximations)
one has $(g^2/q^2 + M_W^2)$ between weakly interacting particles, where
M_W is the mass of the charged boson and g^2 is related to the
usual Fermi constant G via $G = \sqrt{2}\, g^2/M_W^2$. It has always been
tempting to link such theories to electromagnetism by assuming g
and e to be related in some simple way dictated by some kind of
family relationship between the photon and the charged vector
boson. The exploitation of this approach has been inhibited, on
the theoretical side, by the fact that both the ensuing weak interaction theory and the electrodynamics of the charged bosons are non-
renormalizable. Experimentally, there has been no sign of the
charged bosons. The assumption $g \sim e$, however, fixes the mass
in the 50 GeV range so that this fact can not be regarded as negative evidence.

The theories to which we refer here start from the premise
that the photon is a member of a Yang-Mills[75] type multiplet of
vector particles, which, in the Lagrangian, all appear with zero
mass. They are introduced in such a way as to provide local
invariance under a specified internal symmetry group. (Isospin
rotations are a typical example.) By <u>local</u> invariance one means
that the Lagrangian must remain invariant even when the symmetry
transformation at different space time points is different. Such
theories have the same ultraviolet behavior as ordinary electrodynamics, but the proliferation of zero mass particles, apart from
its conflict with observation, introduces unsolved infrared
problems. As a second premise, therefore, it is assumed that the
symmetry is spontaneously broken. That is to say, the vacuum or
minimum energy state of the system is not invariant under the
operations of the symmetry group. As a result, all of the vector
particles except one (identified as the photon) acquire mass.[76,77]
Two (or more) of the massive vector particles act as the mediators
of the weak interactions. Others (possibly more massive) may act
as mediators of so far unobserved and even weaker interactions.[78]
While the basic idea described above is not particularly new,
the recent rise in interest has stemmed from the observation that
in the broken symmetry form the "good" ultraviolet properties are
preserved while the "bad" infrared properties are removed.[79,80,81]
Thus one has a theoretical framework within which one may seek a
renormalizable theory of weak interactions.

We illustrate the basic idea by means of the simplest example I know of, a renormalizable quantum electrodynamics of charged vector mesons. The basic fields are A_μ , a three component four-vector field and ϕ , a three-component scalar field, with the Lagrangian density L given by

$$L = -\frac{1}{4} F_{\mu\nu} \cdot F^{\mu\nu} + \frac{1}{2} \left[(\partial_\mu - e A_\mu x) \, \phi \right] \cdot \left[(\partial^\mu - e A^\mu x) \, \phi \right]$$

$$+ \frac{1}{2} m_o^2 \, \phi \cdot \phi - \frac{1}{4} h^2 \, (\phi \cdot \phi)^2$$

with

$$F_{\mu\nu} = \partial_\mu A_\nu - \partial_\nu A_\mu \quad .$$

The "dot" and "cross" in the above expressions are to be interpreted in the usual sense of three dimensional vector algebra, but with reference to the three dimensional internal space. An infinitesimal local rotation of the internal space can be generated by an arbitrary $\varepsilon \, (x_\mu)$ via the transformation relations

$$\phi \rightarrow (1 + \varepsilon \, x) \, \phi$$

$$A_\mu \rightarrow (1 + \varepsilon \, x) \, A_\mu + \frac{1}{e} \partial_\mu \varepsilon \quad .$$

The Lagrangian density is easily seen to be invariant under this transformation. Because the mass term for ϕ has the wrong sign, the minimum energy occurs in a state for which $\langle \phi \rangle \neq 0$. Since $\langle \phi \rangle \neq 0$, the vacuum state is not invariant under the rotations of the internal space. Denoting the components of ϕ by ϕ_i , we choose internal space axes so that $\langle \phi_i \rangle = n \delta_{i3}$, and write $\phi_3 = \chi + n$. One then finds that the fields ϕ_1 and ϕ_2 can be transformed away, and with $W_\mu = (A_{1\mu} + i A_{2\mu})/\sqrt{2}$ the Lagrangian becomes

$$L = L_o (W_\mu , W_\mu^+ , A_{3\mu} , M)$$

$$+ \frac{1}{2} \partial_\mu \chi \, \partial^\mu \chi \; - \; \frac{1}{2} m^2 \chi^2 - \frac{1}{2} m^2 \frac{e}{M} \chi^3 - \frac{1}{8} \frac{m^2 e^2}{M^2} \chi^4$$

$$+ 2 e M \chi W_\mu^+ W^\mu \; + \; e^2 \chi^2 W_\mu^+ W^\mu$$

where L_0 is the ordinary quantum electrodynamic Lagrangian for a charged particle of mass M, spin one, and anomalous magnetic moment corresponding to $g = 2$. (This latter seems to be characteristic of all of the theories.) The additional feature is the presence of a neutral scalar particle of arbitrary mass M coupled to itself and to the charged particles in a manner which is uniquely determined. On the basis of the general considerations mentioned above, it is expected that the various couplings of the scalar particle provide the compensating divergences required to balance those of ordinary spin one (with g - 2) quantum electrodynamics, and that the theory as a whole is renormalizable.

The above example is, of course, not a theory of weak interactions and is intended only to illustrate the basic procedure for generating renormalizable theories. Apart from the fact that one can generate renormalizable weak interaction theories in this way, the approach provides either an a priori interpretation or at least an extended theoretical framework for understanding such things as the shared vector character of electromagnetic and weak interactions, the origin of approximate symmetries, and current algebra relations. In addition, the electron mass, as well as mass differences within an isospin multiplet need not require additional renormalization constants and hence become finite and in principle computable.[82]

Theoretical study in this rapidly developing area includes detailed analyses of the renormalizability problem,[80,81] a search for attractive and realistic models, and applications of simple models to situations which may be accessible to experiment.[83] One example of the latter are the weak interaction corrections to the muon anomaly.[42,43]

Helpful discussion with S. Brodsky is gratefully acknowledged. This work was supported in part by the United States Atomic Energy Commission.

FOOTNOTES

1. S. Brodsky, "The Present and Future State of Quantum Electro-dynamics," Atomic Physics 2 (Proceedings of the Second International Conference on Atomic Physics) Plenum Press, 1971.

2. B. E. Lautrup, A. Peterman, and E. de Rafael, Physics Reports 3, No. 4 (1972). This is the most recent of a number of excellent reviews of the comparison between theory and the high precision experiments. Others are referenced in this work.

3. A. Kponou, V. W. Hughes, C. E. Johnson, S. A. Lewis, and F. M. J. Pichanick, Phys. Rev. Lett. 26, 1613 (1971).

4. C. Schwartz, Phys. Rev. 134, A1181 (1964).

5. J. Sucher, Ph.D. Thesis, Columbia, 1958 (unpublished).

6. M. Douglas, Ph.D. Thesis, University of California, San Diego (1971) (unpublished) and M. Douglas and N. Kroll to be published.

7. L. L. Foldy and S. A. Wouthuysen, Phys. Rev. 79, 29 (1950).

8. B. Schiff, C. L. Pekeris and H. Lifson, Phys. Rev. 137, A1672 (1965).

9. L. Hambro, Ph.D. Thesis, UCRL 19328 (unpublished).
    L. Hambro, Phys. Rev. A5, 2027 (1972).
    L. Hambro, Phys. Rev. A , to be published. (Two papers)

10. James Daley, Ph.D. Thesis, University of California, Berkeley (1971) (unpublished).

11. M. Douglas, Phys. Rev. A , to be published (also Ref. 5).

12. J. Daley, M. Douglas, L. Hambro, and N. M. Kroll, Phys. Rev. Lett. 29, 12 (1972).

13. J. C. Wesley and A. Rich, Phys. Rev. A4, 1341 (1971); Sarah Granger and G. W. Ford, Phys. Rev. Lett. 28, 1479 (1972).

14. J. Baily, W. Bartl, G. von Bochmann, R. C. A. Brown, F. J. M. Farley, H. Jöstlein, E. Picasso, and R. W. Williams, Phys. Letters 28B, 287 (1968).

15. J. Schwinger, Phys. Rev. 73, 416 (1948).

16. C. M. Sommerfield, Phys. Rev. 107, 328 (1957).

17. A. Peterman, Helv. Phys. Acta 30, 407 (1957).

18. For a more detailed summary of this effort see ref. 2.

19. M. Veltman, SCHOONSHIP CERN (1967).

20. A. C. Hearn, Stanford University Reprint No. ITP-247.

21. J. Calmet, These-Marseille 70/p.336 (1970); Proc. Colloquium
     on Computational Methods in Theoretical Physics, Marseilles
     (CNRS) 1970.

22. M. Levine, J. Comput. Phys. 1, 454 (1967).

23. W. Czyz, G. G. Sheppey, and J. D. Walecka, Nuovo Cim. 34,
     420 (1964).
     A. J. Dufner, Proc. Coll. on Comp. Math. in Theoretical
     Physics, Marseilles/70, p. 335 (1971).
     B. E. Lautrup, Proc. of Second Coll. on Comp. Methods in
     Theoretical Physics, Marseilles (1971). (To be published).

24. M. Levine and J. Wright, Phys. Rev. Lett. 26, 1351 (1971);
     M. Levine and J. Wright, Proc. of the Second Colloquium on
     Computational Methods in Theoretical Physics, Centre de
     Physique Theorique de Marseilles (1971). The numerical value
     given in the above references has been modified. Their new
     value, given in the compilation above, is quoted with the per-
     mission of the authors. Its official publication awaits some
     further checks.

25. J. Aldins, S. Brodsky, A. Dufner, and T. Kinoshita, Phys. Rev.
     Lett. 23, 441 (1969); Phys. Rev. D1, 2378 (1970).

26. S. Brodsky and T. Kinoshita, Phys. Rev. D3, 356 (1971).

27. J. Calmet and M. Perrotet, Phys. Rev. D3, 3101 (1971).

28. J. Mignaco and E. Remiddi, Nuovo Cim. 60A, 519 (1969).

29. H. Suura and E. Wichmann, Phys. Rev. 105, 1930 (1957).

30. A. Peterman, Phys. Rev. 105, 1931 (1957).

31. H. H. Elend, Phys. Letters 20, 682 (1966); 21, 720 (1966)
     (correction).

32. B. E. Lautrup, Phys. Lett. 32B, 627 (1970).

33. B. E. Lautrup and E. de Rafael, Nuovo Cim. 64A, 322 (1969).

34. B. E. Lautrup, A. Peterman, and E. de Rafael, Nuovo Cim. 1A,
     238 (1971).

35. B. E. Lautrup and E. de Rafael, Phys. Rev. 174, 1835 (1968).

36. C. T. Chang and M. Levine (Private communication, to be
     published).

37. G. Nenciu and I. Raszillier, Lower Bounds on the Hadronic
     Contribution to the Muon Anomalous Magnetic Moment; Institute
     of Physics, Bucharest, Preprint March 1972.

38. M. Gourdin and E. de Rafael, Nuclear Phys. 10B, 667 (1969).

39. A. Bramon, E. Etim, and M. Greco, Frascati preprint LNF-72/17
     (1972).

40. J. Baily, G. Bassompierre, K. Borer, F. Combley, P. Hatterslee,
     G. Lebee, G. Petrucci, E. Picasso, H. I. Pizer, O. Runolfson,
     and R. Tinguely, The Present Status of the (g-2) Project;
     CERN NP Internal Report 70-13 (April 1970).

41. B. Lautrup, Phys. Lett. 38B, 408 (1972).

42. R. Jackiw and S. Weinberg, Phys. Rev. D5, 2396 (1972).

43. W. Bardeen, R. Gastmans, and B. Lautrup; CERN Th 1485 May 1972
     (to be published); I. Bars and M. Yoshimura, Phys. Rev. D6,
     374 (1972).

44. R. Barbieri, J. A. Mignaco, and E. Remiddi, Nuovo Cimento Lett.
     3, 588 (1970), and Nuovo Cim. 6A, 21 (1971).

45. R. Barbieri, G. Mignaco, and E. Remiddi, Electron Form Factors
     up to Fourth Order, Preprint Dec. 1971. It is worth mention-
     ing that this work includes a study of the fourth order elec-
     tron form factors, both the Dirac and magnetic parts, for
     arbitrary values of the momentum transfer variable q^2. Com-
     plete analytic expressions for the discontinuity across the
     q^2 cut are given, as well as the form of the dispersion
     integral over this discontinuity to be used in determining the
     complete form factor. The form of the dispersion integral is
     modified from the standard form on account of infrared problems.

46. G. Källen and A. Sabry, Dan. Mat. Fys. Medd 29, n. 17 (1955).

47. J. Weneser, R. Bersohn, and N. M. Kroll, Phys. Rev. 91, 1257
     (1953); M. F. Soto Jr., Phys. Rev. A2, 734 (1970). These
     references contain an over all sign error.

48. B. E. Lautrup, A. Peterman, and E. de Rafael, Phys. Letters
     31B, 577 (1970).; A. Peterman, Phys. Lett. 35B, 325 (1971).

49. T. Applequist and S. J. Brodsky, Phys. Rev. Lett. 24, 562
     (1970), and Phys. Rev. A2, 2293 (1970). See reference 2 for a
     detailed discussion and comparison of the results of references
     44, 47, 48, and 49.

50. G. W. Erickson, Phys. Rev. Lett., 27, 780 (1971).

51. R. Robiscoe and T. Shyn, Phys. Rev. Lett. 24, 559 (1970).

52. S. Triebwasser, E. S. Dayhoff, and W. E. Lamb Jr., Phys. Rev.
     89, 98 (1953).

53. A. M. Desiderio and W. R. Johnson, Phys. Rev. A3, 1267 (1971).

54. C. K. Iddings, Phys. Rev. 138, B446 (1965).

55. S. D. Drell and J. D. Sullivan, Phys. Rev. 154, 1477 (1967).

56. S. D. Drell, First Int'l. Conf. on Atomic Phys., Proceedings,
     Ed.: Bederson, Cohen and Pichanick, Plenum Press, 1969.

57. E. de Rafael, Phys. Lett. 37B, 201 (1971).

58. T. Fulton, D. A. Owen, and W. Repko, Phys. Rev. A4, 1802 (1971).

59. V. W. Hughes, Status of Quantum Electrodynamics Experiments,
     this volume.

60. T. Fulton, Johns Hopkins University Preprint, April 1972. At
     the time of Fulton's analysis, the experimental number was
     somewhat lower, so that the discrepancy was 23 ± 4 MHz.

61. E. H. Wichmann and N. M. Kroll, Phys. Rev. 101, 843 (1956).

62. M. S. Dixit, H. L. Anderson, C. K. Hargrove, R. J. McKee, D.
     Kessler, H. Mes, A. C. Thompson, Phys. Rev. Lett. 27, 878
     (1971).

63. B. Fricke, Z. Physik 218, 495 (1969).

64. M. K. Sundaresen and P. J. S. Watson, Phys. Rev. Lett. 29, 15
     (1972).

65. J. Blomquist, Vaccuum Polarization in Exotic Atoms, Research
     Institute for Physics, Stockholm Preprint 1972. The double
     counting of the "double bubble" diagram, pointed out in this
     work, is also present in Refs. 62 and 63. It appears to have
     been due to a misreading of Ref. 46.

66. Thomas L. Bell, Enrico Fermi Institute Preprint EFI 72-38, August 1972, Ref. 64 and 68 also contain re-evaluations of the $\alpha(\alpha Z)^3$ terms with results somewhat different from those of Ref. 65 and 66. The difference is due to a less accurate treatment of the spatial distribution of the polarization charge.

67. P. Vogel, Caltech Report CALT-63-175.

68. B. Fricke, J. T. Waber and V. L. Telegdi, Northwestern University Preprint COO-2127-34, 1972.

69. K. M. Crowe, J. F. Hague, J. E. Rothberg, A. Schenk, D. L. Williams, R. W. Williams and R. K. Young, Phys. Rev. D5, 2145 (1972). See page 2159, subsection B.

70. H. K. Walter, J. H. Vuilleumier, H. Backe, F. Boehm, R. Engfer, A. H. v. Gunten, R. Link, R. Michaelson, C. Petitjean, L. Schellenberg, H. Schneuwly, W. V. Schröder, and A. Zehnder, Phys. Lett. 40B, 197 (1972).

71. G. Bacci, G. Penso, G. Salvini, R. Baldini-Celio, G. Capon, C. Mencuccini, G. P. Murtas, A. Reale and M. Spinetti, Lett. al Nuovo Cim., 2, 73 (1971).

72. R. Borgin, F. Ceradini, M. Conversi, L. Paoluzi, W. Scandale, G. Barbiellini, M. Grilli, P. Spilantini, R. Visentin, and A. Mullachie, Phys. Lett. 35B, 340 (1971).

73. R. Borgia, F. Ceradini, M. Conversi, L. Paoluzi, R. Santonico, G. Barbiellini, M. Grilli, P. Spilantini, R. Visentin, and F. Grianti, Lett. al Nuovo Cim., 3, 115 (1972).

74. S. Hayes, R. Imlay, P. M. Joseph, A. S. Keizer, J. Knowles and P. C. Stein, Phys. Rev. Lett. 24, 1369 (1970).

75. C. N. Yang and R. L. Mills, Phys. Rev. 96, 191 (1954).

76. P. Higgs, Phys. Lett. 12, 132 (1964) and Phys. Rev. 145, 1156 (1956).

77. T. W. B. Kibble, Phys. Rev. 155, 1554 (1967).

78. J. Schwinger, Ann. Phys. (N.Y.) 2, 407 (1957).
    S. L. Glashow, Nucl. Phys. 22, 509 (1961).
    A. Salam and J. Ward, Phys. Lett. 13, 168 (1964).

79.   S. Weinberg, Phys. Rev. Lett. 19, 1264 (1967); ibid. 27, 1688
      (1971).

80.   G. 't Hooft, Nucl. Phys. B35, 167 (1971).

81.   B. W. Lee and J. Zinn-Justin, Phys. Rev. D5, 3121, 3137, 3155
      (1972).

82.   S. Weinberg, Phys. Rev. Lett. 29, 388 (1972).

83.   H. H. Chen and B. W. Lee, Phys. Rev. D5, 1874 (1972).

ATOMIC PHYSICS AND QUANTUM ELECTRODYNAMICS IN THE INFINITE MOMENTUM FRAME*

Stanley J. Brodsky

Stanford Linear Accelerator Center, Stanford University

Stanford, California 94305

I. INTRODUCTION

Over the past few years it has been shown that the use of an "infinite momentum" Lorentz frame[1] has remarkable advantages for calculations in elementary particle physics and field theory, especially in the areas of current algebra sum rules,[2] parton models,[3,4] and eikonal scattering.[5,6] One important advantage is that it allows a straightforward application of the impulse and incoherence approximations familiar in nonrelativistic atomic and nuclear physics to relativistic field theory and bound state problems.

The central idea is this: Suppose we choose a Lorentz frame such that a bound system has momentum $\vec{P}$ in the z-direction. We shall assume that for P chosen large enough, $(P \to \infty)$ all of its constituents will be moving in the positive z-direction; more specifically, we assume the existence of a wave function in the infinite momentum frame:

$$\lim_{P \to \infty} \psi_P(\vec{p_i}) = \psi(\vec{k}_{i\perp}, x_i) \qquad i = 1, \ldots, N$$

where

$$\vec{p_i} \equiv x_i \vec{P} + \vec{k}_{i\perp}$$

$$\sum_{i=1}^{N} \vec{k}_{i\perp} = 0 \; , \qquad \sum_{i=1}^{N} x_i = 1$$

---

*Work supported by the U. S. Atomic Energy Commission.

For example, for a bound state with momentum $\vec{P}$, mass M and N constituents, the characteristic energy denominator of the wave function is

$$\frac{1}{2E} \frac{1}{E-E_i+i\epsilon} \underset{P\to\infty}{=} \frac{1}{2P} \frac{1}{P+\dfrac{M^2}{2P} - \displaystyle\sum_{i=1}^{N}\left[|x_i|\,P + \dfrac{\vec{k}_{i\perp}^2 + m_i^2}{2|x_i|P}\right] + i\epsilon}$$

$$= \frac{1}{M^2 - \displaystyle\sum_{i=1}^{N} \dfrac{\vec{k}_{i\perp}^2 + m_i^2}{x_i} + i\epsilon} \qquad \text{for all } x_i > 0$$

$$= 0 \left[\frac{1}{P^2}\right] \qquad \text{otherwise.}$$

Since $\sum_{i=1}^{N} x_i = 1$, finite contributions are obtained only if $0 < x_i < 1$ for all i. Note that the $P\to\infty$ limit also has the effect of linearizing relativistic square-root phase-space factors. Similarly, one finds that in time-ordered perturbation theory, all diagrams in which intermediate particles are moving backward ($x_i < 0$) can be effectively set to zero, leaving only the relatively few diagrams with forward moving intermediate particles to be considered.[1] [See Section II for examples.] The structure of the $P\to\infty$ wavefunction is formally very similar to nonrelativistic theory; the quantity $k_{i\perp}^2/x_i$ plays the role of the kinetic energy. (More generally the relativistic wavefunction contains arbitrary numbers of constituents, but we may treat each N-particle component state as above.)

Thus the intuition and approximation procedures used in the nonrelativistic problems now becomes applicable to high energy physics and rigorous methods for bound states other than the Bethe-Salpeter formalism now present themselves. Conversely, these techniques indicate a new systematic procedure for handling the relativistic and recoil correction to atomic and nuclear physics problems.

In Section II, we discuss the application of the infinite momentum method to quantum electrodynamics, and the implementation of the renormalization procedure in old-fashioned perturbation theory. In Section III, several applications to problems in atomic physics are outlined. These include inelastic electron-atom scattering, high energy scattering, and rearrangement collisions in atom-atom scattering.

II. QUANTUM ELECTRODYNAMICS AND RENORMALIZATION THEORY IN THE INFINITE MOMENTUM FRAME

(The manuscript for this section was prepared in collaboration with Ralph Roskies)

Recently Ralph Roskies, Roberto Suaya and I have found that time-ordered perturbation theory for quantum electrodynamics evaluated in an infinite momentum reference frame represents a viable, instructive, and frequently advantageous calculational alternative to the usual Feynman diagram approach. The renormalization procedure can be implemented in a straightforward manner. We have calculated the electron anomalous magnetic moment through fourth order in agreement with the Sommerfield-Petermann results, [7] and have calculated representative contributions to the sixth order moment. Our results agree with those of Levine and Wright[8] and represent the first independent confirmation of their result for these contributions.

An outline of our techniques follows; a more complete discussion will be published separately. [9]

The electron vertex in quantum electrodynamics may be computed in perturbation theory using the standard time-ordered momentum space expansion of the S-matrix. Although the final results are independent of the choice of Lorentz frame, it is very convenient to choose a limiting reference frame in which the incident electron momentum P is large. [1] In a general frame, a Feynman amplitude of order e^n requires the evaluation of n! time-ordered contributions, but in a frame chosen such that

$$p = \left(\sqrt{P^2 + m^2}, \ \vec{O}_\perp, P \right) \rightarrow \left(P + \frac{m^2}{2P}, \ \vec{O}_\perp, P \right) \tag{1a}$$

$$q = \left(\frac{q \cdot p}{P}, \ \vec{q}_\perp, O \right) \tag{1b}$$

$(2q \cdot p = -q^2 = \vec{q}_\perp^2)$ only the relatively few time-ordered graphs, in which the momenta of all the internal (on-mass-shell) particles $\vec{p}_i = x_i \vec{P} + \vec{k}_{i\perp}$ have positive components along $P(0 < x_i < 1)$, have a surviving contribution in the limit $P \rightarrow \infty$. In general, the limit $P \rightarrow \infty$ is uniform with respect to the

$$\frac{d^3 p_i}{2E_i} = \frac{d^2 k_{i\perp} dx_i}{2x_i}$$

phase space integrations for all renormalized amplitudes. Thus the order α correction to the anomalous moment $a = F_2(0)$ is obtained from only one forward-moving time-ordered graph[5, 6, 10] (see Fig. 1), up to 3 time-ordered graphs yield the Feynman amplitude for the order α^2 corrections;

between 1 and 15 forward-moving time-ordered graphs contribute to various Feynman amplitudes at order α^3.

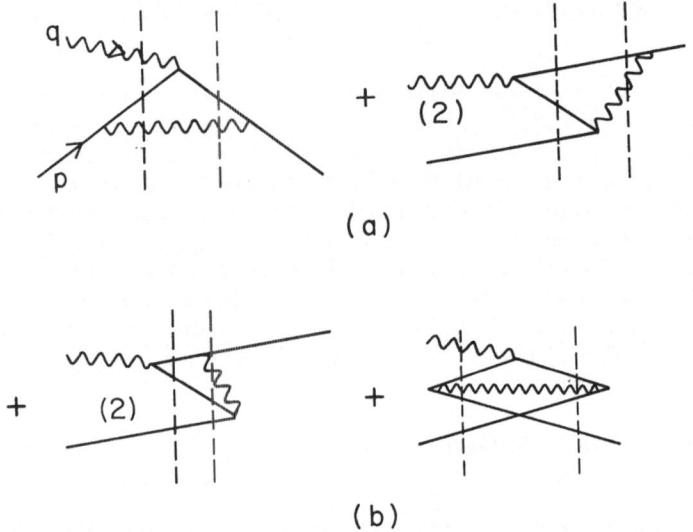

(a)

(b)

Fig. 1--The six time-ordered contributions of the Feynman amplitude for the proper electron vertex Γ_μ in order α. For the components $\mu=0$ or $\mu=3$, only the contribution of the diagram (a) survives in the infinite momentum limit $P \to \infty$ of Eq. (1). In addition, the "Z-graph" contribution for the $\mu=1,2$ components which arises from diagram (b) is automatically included by using the modification of the spinor sum for diagram (a) given in Eq. (2).

___

As emphasized by Drell, Levy, and Yan, [4] time-ordered graphs with backward-moving ($x_i < 0$) internal fermion lines can give surviving P^2/P^2 contributions in the $P \to \infty$ limit if the line extends over only one time interval. These additional contributions (which correspond to contact or "seagull" interactions analogous to the $e^2\emptyset+\emptyset A^2$ interactions in boson electrodynamics) can be automatically included by making a simple modification in the forward-moving contribution: if a forward-moving electron ($x_i > 0$) extends over a single interval I then instead of the usual spin sum

$$\sum_{\text{spin}} u(p_i) \, \bar{u}(p_i) = \not{p}_i + m \, , \qquad p_i^2 = m_i^2 \tag{2a}$$

we write

$$\not{p}_i + \gamma_0 (E_0 - E_I) + m \tag{2b}$$

where E_0 is the total incident energy and E_I is the sum of the energies of all of the particles occurring in the intermediate state I. It is easy to check that this replacement (which corresponds to using energy conservation between the <u>initial</u> and intermediate energies to determine p_i^0 rather than the mass-$\overline{\text{shell}}$ condition) automatically accounts for the contribution of the corresponding negative moving ($x_i < 0$) positron line. A similar modification for the energy of a forward-moving positron (spanning one time interval) accounts for the corresponding negative moving electron line. With these changes all "Z-graph" contributions are accounted for, and one need only consider time-ordered diagrams where all internal lines have $x_i > 0$.

The renormalization procedure for quantum electrodynamics using old-fashioned perturbation theory closely parallels the explicitly covariant Feynman-Dyson procedure. Reducible amplitudes with self-energy and vertex insertions may be renormalized using subtraction terms corresponding to δm, Z_2 and Z_1 counter terms. The integrand for the subtraction term is similar in form to the integrand for the unrenormalized amplitude, except that the external energy used for the denominator for the subgraph insertion is not the external (initial) energy E_0 of the entire diagram but is the energy external to the self-energy or vertex subgraph only. For example, the renormalization of the scattering amplitude shown in Fig. 2a requires δm and Z_2 subtractions (Fig. 2b and Fig. 2c).

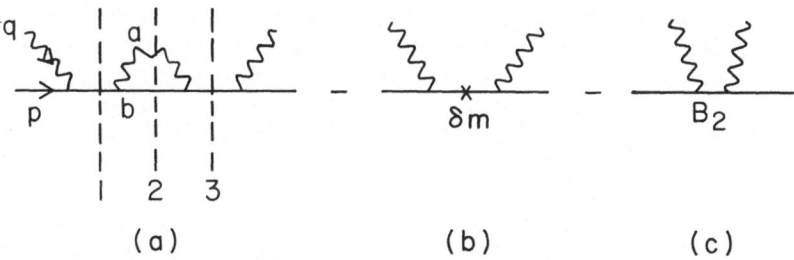

(a) (b) (c)

Fig. 2--Illustration of the renormalization procedure in old-fashioned perturbation theory. (a) A representative time-ordered diagram for the self-energy modification of the Compton amplitude. (b) and (c) The corresponding δm and Z_2 counterterms. The integrand for the δm term is proportional to $(E_1-E_2)^{-1}$.

The integrand of the renormalized amplitude for ϕ^3 theory is constructed from

$$\frac{1}{(E_0-E_1)(E_0-E_2)(E_0-E_3)} - \frac{1}{(E_0-E_1)(E_1-E_2)(E_0-E_3)} + \frac{1}{(E_0-E_1)(E_1-E_2)(E_1-E_2)} \quad (3)$$

where E_i is the total energy of the on-shell particles occurring at interval i. Upon integration over the loop momentum variables (x_i, $\vec{K}_{i\perp}$), the second and third terms yield, by definition, the correct δm and Z_2 counter terms (assuming covariant regularization). On the other hand, if scaled variables

$$\vec{p_a} = x(\vec{p} + \vec{q}) + \vec{k_\perp}$$

$$\vec{p_b} = (1 - x)(\vec{p} + \vec{q}) - \vec{k_\perp}$$

(4)

are chosen to parametrize the momenta of the internal particles, then $\vec{k_\perp} \cdot \vec{q}$ cross terms are eliminated and the integration for the renormalized amplitude from the sum of the three terms is point-wise convergent. In the QED case, the appropriate Dirac numerator must also be constructed such that the (covariantly-regulated) subgraph integration defines the correct counter terms. This procedure leads to finite, renormalized pointwise-convergent (and numerically integrable) amplitudes for the case of all self-energy or vertex insertions.[9,11] The analysis of infrared divergences (via a photon mass regulator) may be carried out in parallel with standard treatments.

In general, we have found that the $P \rightarrow \infty$ limit is uniform (i.e., can be taken before the $d^2k_\perp \, dx$ loop integrations) for the renormalized amplitudes, and there are no subtleties involved at the boundaries of the x_i integration. On the other hand, the evaluation of the (divergent) renormalization constants themselves requires caution. Since covariance is not explicit in this approach, one must be careful to regularize using a covariant procedure, such as the Pauli-Villars method or spectral conditions. The standard covariant expressions for the renormalization constants are obtained if regularization is performed before the $P \rightarrow \infty$ limit is taken.[9]

With the above considerations, it is straightforward to calculate renormalized amplitudes for quantum electrodynamics directly from time-ordered perturbation theory and the interaction density $e : \psi \gamma_\mu \psi A^\mu :$. The covariant Feynman amplitude is obtained from the corresponding (forward-moving) time-ordered graphs with the same topology. The Dirac numerator algebra is the same for each of the time-ordered amplitudes and is identical to the corresponding Feynman calculation. Our techniques also show that quantum electrodynamics may be calculated on the light-cone in the Feynman gauge, rather than the Coulomb gauge.

For the calculation of the lepton vertex, the F_1 and F_2 amplitudes can be obtained simply from standard trace projection operators.[12] The integrand in the variables x_i, $\vec{k_{i\perp}}$ is then obtained from the product of phase space, the numerator trace, and the energy denominators characteristic of old-fashioned perturbation theory.[13] One important feature of this method, besides providing a new and independent calculational technique, lies in the fact that the resulting integrand appears to be much smoother function of the variables x_i, $\vec{k_{i\perp}}$ than the corresponding Feynman parametric integrand obtained by the usual techniques. As a result, the numerical integrations (which are often the most difficult part of higher order calculations in quantum electrodynamics) converge considerably faster.

As an indication, the numerical integration of the contribution of the sixth order ladder graph (Fig. 3a) to the electron's anomalous magnetic moment from old-fashioned perturbation theory required 10^5 evaluations of a smooth well-behaved six-dimensional integrand to obtain a 1% level of accuracy. [14] In contrast, the standard Feynman technique, which involves a five-dimensional integral, required 2×10^6 evaluations of the integrand for comparable accuracy. Our result is

$$\left(\frac{g-2}{2}\right)_{\text{Fig. 2a}} = (1.77 \pm 0.01)\,\frac{\alpha^3}{\pi^3}$$

in precise agreement with the result of Levine and Wright. [8] Our results for the fourth order magnetic moment using $P \to \infty$ techniques agree with the Sommerfield and Petermann calculations;[7] again, the integrands were found to be smooth and rapidly integrable by numerical techniques.

The sixth order ladder graph is a highly reducible graph requiring several vertex renormalization counter terms, but only one time-order survives in the infinite momentum limit. We have also calculated a representative irreducible graph, Fig. 3b, which has eight surviving

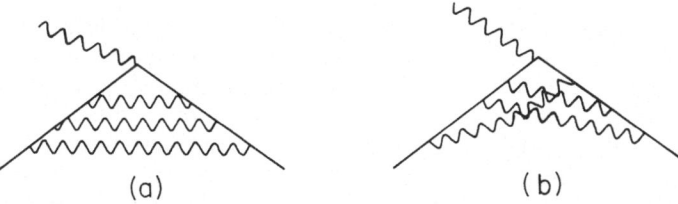

(a)   (b)

Fig. 3--Representative reducible and irreducible contributions to the sixth order magnetic moment of the electron or muon. The ladder graph (a) is obtained from a single time-ordered contribution at infinite momentum (out of a possible 7!), but requires renormalization of the fourth order and second order vertex insertions. The Feynman amplitude for irreducible graph (b) receives contributions from the eight time-ordered graphs with positive moving internal lines.

time orders. In this case there is an eight-dimensional nontrivial integration to be performed and the algebraic work is much more complex. Our result for this graph is $2(1.11 \pm 0.23)\,\alpha^3/\pi^3$ which is consistent with Levine and Wright's result $2(0.90 \pm 0.02)\,\alpha^3/\pi^3$ obtained from a seven-dimensional Feynman parametric integration. Work is continuing to improve the accuracy of our result.

The validity of the infinite momentum reference frame method as a renormalizable calculational procedure in quantum electrodynamics gives field-theoretical parton model calculations a rigorous basis provided that a covariant regularization procedure is used. Our work also demonstrates that the infinite momentum method provides a useful calculational alternative to standard covariant techniques. The $P \to \infty$ method is closely related to field theory quantized on the light cone.[6,15] Our method shows how to renormalize the theory and work in the Feynman gauge.[15]

III. THE ATOM IN THE INFINITE MOMENTUM FRAME

Although the infinite momentum method was developed to treat highly relativistic problems, there are interesting applications to problems of the atom.

An important quantity is the normalized probability distribution

$$f_e(x) = \frac{1}{16\pi^3} \int \frac{d^2k}{x(1-x)} \ |\psi(\vec{k}_\perp, x)|^2$$

$$\int_0^1 f_e(x) \ dx = 1$$

which is the probability for finding an electron moving with momentum xP along the $\vec{P}$-direction in a reference frame in which the atom is moving with momentum $P \to \infty$. The electron wavefunction $\psi(\vec{k}_\perp, x)$ may be found from the solution of the wave-equation of the atom in the infinite momentum frame (see Weinberg[1] and Feldman, Fulton, and Townsend[16]) or by a Lorentz boost of the center-of-mass wavefunction. [For corrections in α, higher particle number (photon, electron-positron pair) states must be included, as in the QED case, see Section II.] Note that $f_e(x)$ is peaked at the value $x = E_e/M_T$ where M_T is the total atomic mass and E_e is the bound electron energy, and that $f_e(x) \to \delta(x - m_e/M_T)$ if the binding energy is taken to zero.

A standard result, derived from parton — constituent field theoretic models[4] — is that the bound electron contribution to deep inelastic wide-angle electron-atom scattering (i.e., $\nu = E_{Lab} - E'_{Lab} \gg$ B.E., $\vec{q}^2R^2 \gg 1$) is given simply by the Mott cross section (for elastic e-e collisions) times $f_e(x)$, with x taken at the value

$$x = Q^2/2M_T\nu \ , \qquad Q^2 = \vec{q}^2 - \nu^2 \quad .$$

Derivations and formulae are given in Ref. 4. This result extends the validity of the impulse and incoherence approximations to the relativistic domain.

A surprisingly simple result can also be obtained for the bound electron contribution to high energy ($\nu \gg$ B. E.) forward photon-atom scattering. One finds[17] that the (spin-averaged) forward Compton amplitude $f(\nu)$ is asymptotically constant and real:

$$f(\nu) \xrightarrow[\nu \gg \text{B. E.}]{} - Z \frac{e^2}{M_T} \int_0^1 \frac{f_e(x)}{x} \, dx = \frac{-Ze^2}{m_{eff}}$$

Note that xM_T plays the role of the effective electron mass; m_{eff} contains corrections from atomic binding and finite nuclear mass corrections. The above result is derived in field theory from the electron-positron z-graph contribution to the electron Compton amplitude, which is effectively a "seagull" diagram in the infinite momentum frame. This result may be compared with the beautiful treatment of high energy photon scattering from an electron bound in a potential that has been given by M. Goldberger and F. Low.[18]

Given the infinite momentum wavefunction we may also determine the electron current contribution to atomic (elastic or inelastic) form factors. Ignoring spin complications, one obtains

$$F(q^2) = \frac{1}{16\pi^3} \int \frac{d^2k_\perp}{x(1-x)} \, \psi^* (\vec{k}_\perp + (1-x) \, \vec{q}_\perp, x) \, \psi(\vec{k}_\perp, x)$$

where $\vec{q}_\perp$ is a vector transverse to $\vec{P}$ with magnitude $\vec{q}_\perp^2 = |q^2|$. Drell and Yan (Ref. 4) have shown that the large q^2 behavior of the elastic form factor $F(q^2)$ is related to the x near one behavior of $f_e(x)$.

A very simple expression may also be given for rearrangement (interchange) collisions in elastic or inelastic atom-atom scattering.[19] The scattering amplitude for elastic H - H rearrangements collisions is proportional to

$$m(t, u) = \int d^2k \int_0^1 \frac{dx}{x^2(1-x)^2} \, \Delta$$

$$\left[\psi(\vec{k}_\perp) \, \psi(\vec{k}_\perp + (1-x) \, \vec{q}) \, \psi(\vec{k}_\perp + (1-x) \, \vec{q}_\perp - x\vec{r}_\perp) \, \psi(\vec{k}_\perp - x\vec{r}_\perp) \right]$$

where

$$\Delta = E_0 - E_{intermediate}$$

$$= 2M_T^2 - \frac{(\vec{k}_\perp - x\vec{r}_\perp)^2 + (\vec{k}_\perp + (1-x)\,\vec{q}_\perp)^2 + 2(xM_p^2 + (1-x)\,m_e^2)}{x(1-x)}$$

The vectors $\vec{q}_\perp$ and $\vec{r}_\perp$ are chosen transverse to $\vec{P}$, with magnitudes

$$\vec{q}_\perp^2 = 2\vec{p}_{c.m.}^2 \ (1 - \cos\theta_{c.m.}) = -t$$

$$\vec{r}_\perp^2 = 2\vec{p}_{c.m.}^2 \ (1 + \cos\theta_{c.m.}) = -u$$

$$\vec{q}_\perp \cdot \vec{r}_\perp = 0 \ .$$

This result ignores the Coulomb interactions between the electrons and between the atoms, but includes the binding forces correctly (including all recoil and relativistic terms). Spin corrections are discussed in Ref. 19.

The corresponding parton-interchange contribution has been shown to agree well with measurements of high energy, large angle, proton-proton scattering (where the proton is regarded as a quark bound state). [19] It would be interesting to measure hard, large angle atom-atom (elastic or inelastic) scattering where the electron exchange contribution is dominant.

Finally, we note that a very hopeful area of application of the infinite momentum method is the spectra of bound states, especially that of positronium and muonium; the infinite momentum old-fashioned perturbation theory approach provides a rigorous alternative to the Bethe-Salpeter formalism, and does have calculational advantages. The work of Feldman, Fulton, and Townsend, [16] who have treated the spin zero bound state problem in the infinite momentum frame, is an important step in this direction.

Acknowledgements

The work in Section II was done in collaboration with R. Roskies and R. Suaya, and the results given in Section III are based on work done with J. Gunion, R. Blankenbecler and F. Close. I would also like to thank S. Drell, J. Bjorken, K. Johnson, J. Kogut, and M. Levine for helpful conversations.

References

1. S. Weinberg, Phys. Rev. 150, 1313 (1966). See also L. Susskind
    and G. Frye, Phys. Rev. 165, 1535 (1968); K. Bardakci and
    M. B. Halpern, Phys. Rev. 176, 1686 (1968).
2. S. Fubini and G. Furlan, Physics 1, 229 (1965); J. D. Bjorken,
    Phys. Rev. 179, 1547 (1969); R. Dashen and M. Gell-Mann, Phys.
    Rev. Letters 17, 340 (1966).
3. J. D. Bjorken and E. A. Paschos, Phys. Rev. 185, 1975 (1969).
4. S. D. Drell, D. J. Levy, and T. M. Yan, Phys. Rev. Letters 22,
    744 (1969); Phys. Rev. 187, 2159 (1969); Phys. Rev. D1, 1035 (1970);
    Phys. Rev. D1, 1617 (1970). S. D. Drell and T. M. Yan, Phys.
    Rev. D1, 2402 (1970); Phys. Rev. Letters 24, 181 (1970).
5. S. J. Chang and S. K. Ma, Phys. Rev. 180, 1506 (1969); 188, 2385
    (1969).
6. J. B. Kogut and D. E. Soper, Phys. Rev. D1, 2901 (1970);
    J. D. Bjorken, J. B. Kogut and D. E. Soper, Phys. Rev. D3, 1382
    (1971).
7. C. M. Sommerfield, Phys. Rev. 107, 328 (1957); Ann. Phys. (N.Y.)
    5, 26 (1958). A. Petermann, Helv. Phys. Acta 30, 407 (1957);
    Nucl. Phys. 3, 689 (1957).
8. M. Levine and J. Wright, Phys. Rev. Letters 26, 1351 (1971);
    Proceedings of the Second Colloquium on Advanced Computing
    Methods in Theoretical Physics, Marseille (1971), and private
    communication.
9. S. Brodsky, R. Roskies, and R. Suaya (in preparation).
10. D. Foerster, University of Sussex preprint (1971).
    Foerster's derivation of the lowest order anomalous moment $\alpha/2\pi$
    is particularly instructive. If the electron interacts with a magnetic
    field (transverse photon polarization), then one finds that the con-
    tribution of diagram 1(a) is negative (but logarithmic divergent) in
    agreement with Welton's classical argument (T. Welton, Phys. Rev.
    74, 1157 (1948)). The surviving Z-graph contribution of diagram 1(b)
    (and its mirror graph) is positive, cancels the divergent term, and
    leaves the finite $\alpha/2\pi$ remainder. Note that diagram 1(b) contains
    the Thomson limit part of the Compton amplitude for the side-wise
    dispersion calculation of S. Drell and H. Pagels, Phys. Rev. 140B,
    397 (1965). The remaining diagrams vanish in the infinite momen-
    tum frame defined in Eq. (1).
11. The renormalization of the vertex insertions is generally algebrai-
    cally more complicated, since, except for ladder graphs, the counter
    term must be rewritten to cancel the contributions of more than one
    time-ordering of the vertex. The procedure for this case is dis-
    cussed in Ref. 9.
12. See, for example, S. J. Brodsky and J. D. Sullivan, Phys. Rev. 156,
    1644 (1967).
13. All of the algebraic steps for our calculations were performed auto-
    matically using the algebraic computation program REDUCE, see
    A. C. Hearn, Stanford University preprint No. ITP-247 (unpublished);

and A. C. Hearn in: Interactive Systems for Experimental Applied Mathematics, eds. M. Klerer and J. Reinfields (Academic Press, New York, 1968).

14. The numerical integrations were performed using the adaptive multi-dimensional integration program developed by C. Sheppey. See J. Aldins, S. Brodsky, A. Dufner, and T. Kinoshita, Phys. Rev. D1, 2378 (1970); A. Dufner, Proceedings of the Colloquium on Computation Methods in Theoretical Physics (Marseille, 1970), and B. Lautrup, op. cit. (1971).

15. T. M. Yan and S. J. Chang, Cornell University preprints (1972).

16. G. Feldman, T. Fulton, and J. Townsend, John Hopkins University preprint (1972).

17. S. Brodsky, F. Close, and J. Gunion, Phys. Rev. D5, 1384 (1972).

18. M. Goldberger and F. Low, Phys. Rev. 176, 1778 (1968).

19. J. Gunion, S. Brodsky, and R. Blankenbecler, Report No. SLAC-PUB-1037 and Phys. Letters (to be published). For a discussion of atom-atom rearrangement collisions in potential theory, see K. M. Watson, in Atomic Physics, Proc. of the First International Conference on Atomic Physics, 1968.

THEORIES OF THE FINE STRUCTURE CONSTANT α

Stephen L. Adler

Institute for Advanced Study, Princeton, N. J. 08540
and
National Accelerator Laboratory[*], Batavia, Ill. 60510

Although the fine structure constant is one of the best determined numbers in physics, the reason why nature selects the
particular value $\alpha = 1/137.03602 \pm 0.00021$ for the electromagnetic
coupling strength is still a mystery, and has provoked much interesting theoretical speculation. For purposes of discussion, the
speculations may be divided roughly into four general types:
(a) Theories in which α is cosmologically determined; (b) Theories in which α is a constant which is determined microscopically through the interplay of the electromagnetic interaction
with interactions of other types, either gravitational, weak or
strong; (c) Theories in which α is microscopically determined
through properties of the electromagnetic interaction alone, considered in isolation from other interactions; and (d) Numerological
speculations.

(a) COSMOLOGICAL THEORIES

A cosmological idea which has received prominent attention
recently is the suggestion that α may vary with the time t
which has elapsed since the beginning of the universe. In one
version[1] of this hypothesis α varies linearly with cosmic time,

$$\alpha \sim t , \tag{1a}$$

---

[*]Operated by Universities Research Associates, Inc. under
contract with the U.S. Atomic Energy Commission.

while in another version[2] (suggested by a possible connection between electromagnetism and gravitation which we discuss below)
the variation is logarithmic,

$$\alpha \sim (\ell n\ t)^{-1}.$$ (1b)

Assuming that the present age of the universe is $2 \cdot 10^{10}$ years, the
rate at which α charges is

$$R = \dot{\alpha}/\alpha = \qquad 5 \cdot 10^{-11}\ year^{-1}\ \text{hypothesis (1a)}$$ (2)
$$-5 \cdot 10^{-13}\ year^{-1}\ \text{hypothesis (1b)}\ .$$

Although these rates of variation are very small, it is remarkable that there is now strong experimental evidence ruling out
both hypothesis (1a) and (1b), as summarized[3] in the following
table:

Source	Limit on R
Fine structure of spectra of distant radio-galaxies	$\lvert R\rvert \leq 2 \cdot 10^{-12}\ year^{-1}$
Nuclear α-decay: geophysical constancy of decay rate of U^{238}	$\lvert R\rvert \leq 2 \cdot 10^{-13}\ year^{-1}$
Spontaneous fission: geophysical constancy of fission decay rate of U^{238}	$\lvert R\rvert \leq 5 \cdot 10^{-13}\ year^{-1}$
Beta decay: agreement of labor-atory and geophysical half-lives for decay Rhenium 187 $\to$ Osmium 187 ($T \sim 5 \cdot 10^{10}$ years)	$\lvert R\rvert \leq 5 \cdot 10^{-15}\ year^{-1}$

The only simple hypothesis which is compatible with these limits
is R=0, which means a fine structure constant which is strictly a
constant over the lifetime of the universe. Such a constancy
could result if α were some sort of cosmological boundary condition, fixed, perhaps, by the detailed structure of the universe
at the beginning of the present expansion phase. Clearly, if α is
determined in such a fashion we could not, with our present
knowledge, hope to calculate it. A more appealing explanation for
the constancy of α is that α is microscopically determined by the
basic particle interaction laws, independent of cosmological considerations.

(b) MICROSCOPIC THEORIES RELATING α TO THE GRAVITATIONAL OR WEAK INTERACTIONS

A basic issue in making a microscopic theory of α is deciding which of the four fundamental interaction types--strong, electromagnetic, weak and gravitational--must be included, and which can be neglected. Because of our very limited theoretical understanding of particle interactions, no systematic discussion of this problem can be given. The best we can do is to review various options which have been seriously studied in the past few years, with the caution that there is at present no proof that any of these theories actually includes the correct combination of interactions.

(b1) Theories Combining Electromagnetism and Gravity

A number of authors[2,4] have speculated that the length

$$a = (G\ \hbar/c^3)^{1/2} = 1.6 \cdot 10^{-33}\ \text{cm} \tag{3}$$

characterizing quantized theories of gravitation may provide a natural cutoff which eliminates the logarithmic divergences of quantum electrodynamics. Apart from details (which in some versions are complicated), the basic consequence of this idea is obtained by requiring that the physical mass of the electron be equal to its electromagnetic self-mass, as calculated with a momentum cutoff Λ of order a^{-1}. Setting $\hbar = c = 1$, this gives

$$m_e = \delta m_e = m_e\, \kappa\, \alpha\, \ln(\frac{\Lambda^2}{m_e^2}) = m_e\, \kappa\, \alpha\, \ln(\frac{\kappa'}{m_e^2 G}), \tag{4}$$

with κ and κ' numerical constants. Thus, Eq. (4) gives us the following relation between the gravitational coupling G and the fine structure constant α,

$$\alpha = \left[\, \kappa\, \ell n\, (\frac{\kappa'}{m_e^2 G})\right]^{-1}. \tag{5}$$

Assuming κ' is of order unity, Eq. (5) is satisfied with $\kappa \sim 137/104 \sim 1$, also near unity, making the relation plausible. The test of a detailed theory of this type would be whether it gives the correct value of κ, not an easy task since lowest order electrodynamic perturbation theory gives $\kappa = 3/4\pi$, which is substantially too small. An experimental test of Eq. (5) may be provided by radar ranging measurements[3] of the rate of change of G. Should Dirac's hypothesis[3] $G \sim t^{-1}$ prove correct, then Eq. (5), which would imply $\alpha \sim (\ell n\ t)^{-1}$, would be ruled out by the stringent limits on the time rate of change of α. A constant G would, of course, be perfectly compatible with Eq. (5).

(b2) Theories Combining Electromagnetism and Weak Interactions

The fact that the electromagnetic and weak interactions both utilize vector type couplings suggests that they may have a common origin. Suppose, for example, that weak interactions are mediated by an intermediate vector boson of mass M_W which couples to leptons and hadrons with the electromagnetic coupling strength e. [5] The effective weak coupling coming from the W-exchange diagram

$$\text{couplings} \propto e \tag{6a}$$

W propagator $\propto (M_W^2 - q^2)^{-1} \sim (M_W^2)^{-1}$
for small momentum transfer q

would be

$$G_F \sim \frac{\alpha}{M_W^2} \; ; \tag{6b}$$

to agree with the experimental value of the Fermi constant $G_F \sim 10^{-5}/M_{proton}^2$ one would need an intermediate boson of mass

$$M_W \sim 30 \, M_{proton}, \tag{7}$$

well beyond the present experimental lower limit on M_W of a few proton masses. A particularly appealing version of this type of theory has been proposed by Weinberg,[6] who constructs a unified, renormalizable theory of weak and electromagnetic interactions. (Unlike the situation in the gravitational cutoff scheme discussed above, the renormalization constants in Weinberg's theory are themselves still infinite.) The basic test of models of this type will of course be the search for heavy intermediate bosons. While the models do not calculate α a priori, if they are proved correct there will be a strong indication that to calculate α one must take the weak interactions, as well as the electromagnetic interactions, into account.

(c) THEORIES IN WHICH α IS DETERMINED BY ELECTROMAGNETISM ALONE

Finally, let us discuss the possibility that α may be determined microscopically by properties of the electromagnetic interaction alone, with the neglect of gravitational, weak and strong interactions. To justify the neglect of gravity we can argue that so far there is no experimental evidence for quantum gravitational effects, and weak interactions may be negligible if they really are weak, rather than being of electromagnetic strength. An argument which may justify the neglect of strong interaction effects will be given later on. The basic requirement which we impose, in an attempt to get an eigenvalue condition for α, is that the renormalization constants of quantum electrodynamics should all be finite. These constants are

m_0 = electron bare mass,

Z_2 = electron wave function renormalization, (8)

Z_3 = photon wave function renormalization;

we require that as the cutoff Λ used to calculate them becomes infinite, m_0, Z_2 and Z_3 should have finite limits. The condition on Z_3 can be stated in the alternative form that the renormalized photon propagator $d_c(-q^2/m^2, \alpha)$ [which is normalized to unity at $q^2=0$] should approach the finite constant $Z_3^{-1} = \alpha_0/\alpha$ as $-q^2/m^2 \to \infty$.

A systematic, non-perturbative attack on the problem of whether Z_3 can be finite was made by Gell-Mann and Low in their classic 1954 paper on the renormalization group.[7] They showed that there is indeed an eigenvalue condition imposed by requiring that Z_3 be finite, but that the condition takes the form

$$\psi(\alpha_0) = 0 \qquad (9)$$

and determines the asymptotic coupling α_0 rather than the physical coupling α. Their analysis leaves α a free parameter of the theory, restricted only by the condition $\alpha < \alpha_0$ coming from spectral-function positivity. This essential conclusion was retained in the subsequent work of Johnson, Baker and Willey (JBW),[8] who made two important advances over the work of Gell-Mann and Low. First, they showed that if Z_3 is finite, then the renormalization constants Z_2 and m_0 can also be finite: The electron wave function renormalization Z_2, which is gauge-dependent, can be made finite by an appropriate choice of gauge (the Landau gauge), while the electron bare mass m_0 takes the simple scaling form

$$m_0 = \text{const} \times m(\frac{\Lambda^2}{m^2})^{-\varepsilon}, \quad \varepsilon = \frac{3}{2}\frac{\alpha_0}{2\pi} + \frac{3}{8}(\frac{\alpha_0}{2\pi})^2 + \dots \tag{10}$$

and therefore vanishes in the limit of infinite Λ provided that $\varepsilon > 0$. (A vanishing bare mass means that the physical mass of the electron arises entirely from its self-interaction.) Second, Baker and Johnson[9] showed that the Gell-Mann Low eigenvalue condition $\psi(\alpha_0) = 0$ implies the much simpler condition $F^{[1]}(\alpha_0) = 0$, where $F^{[1]}(y)$ is a function of coupling y defined as follows. Let us define the photon renormalized proper self-energy $\pi_c(-q^2/m^2, y)$ by

$$d_c(-q^2/m^2, y) = [1 + y\,\pi_c(-q^2/m^2, y)]^{-1}, \tag{11}$$

and let $\pi_c^{[1]}(-q^2/m^2, y)$ denote its <u>single-fermion-loop part</u>,

$$\pi_c^{[1]}(-q^2/m^2, y) = \quad\text{...}\quad + \dots \tag{12}$$

In the limit of asymptotic $-q^2/m^2$ it can be shown that $\pi_c^{[1]}(-q^2/m^2, y)$ grows at worst as a <u>single power</u> of $\ln(-q^2/m^2)$ [higher powers of $\ln(-q^2/m^2)$ can only come from multiple-fermion-loop diagrams where vacuum polarization insertions appear inside fermion loops] ,

$$\pi_c^{[1]}(-q^2/m^2, y) = G^{[1]}(y) + F^{[1]}(y)\ln(-q^2/m^2) + \text{vanishing terms}. \tag{13}$$

The coefficient of the logarithm in Eq. (13) is the function which gives the simplified eigenvalue condition; unlike the Gell-Mann Low function ψ, which involves <u>all</u> vacuum polarization diagrams, the function $F^{[1]}$ involves only a <u>very</u> special subclass of these diagrams.

In addition to showing that $\psi(\alpha_0) = 0$ implies $F^{[1]}(\alpha_0) = 0$, the Baker-Johnson analysis also shows that $\psi(\alpha_0) = 0$ implies $T_{2n}^{[1]}(m=0, y=\alpha_0) = 0$ for $n \geq 2$, where $T_{2n}^{[1]}(m, y)$ is the sum of single-fermion-loop 2n-point functions

$$T_{2n}^{[1]}(m,y) = \left[\; 1 \longrightarrow \bigcirc\;\right] + y\left[\; 1 \longrightarrow \bigcirc\; + \text{permutations}\right]$$

$$+ y^2 \left[\; 1 \longrightarrow \bigotimes\; + \text{permutations}\right] + \ldots \tag{14}$$

Let us now use[10] this powerful result in the following way: We take a single-fermion-loop 2n-point function and contract n-1 pairs of external photon lines with n-1 photon propagators, leaving only two free external photons. The resulting object has the same Lorentz structure as the single-fermion-loop proper self energy $\pi_c^{[1]}$, but some simple combinatorics shows that it is not $\pi_c^{[1]}$ itself, but rather the coupling-constant derivative $(d/dy)^{n-1}\pi_c^{[1]}$. That is, we have

$$(d/dy)^{n-1}\pi_c^{[1]} \propto \boxed{T_{2n}^{[1]}} \tag{15}$$

But since $T_{2n}^{[1]}(m=0, y=\alpha_0)$ vanishes, we learn from Eq. (15) that

$$\frac{d^{n-1}}{dy^{n-1}} F^{[1]}(y)\Big|_{y=\alpha_0} = 0 \qquad n \geq 2, \tag{16}$$

that is, $F^{[1]}$ vanishes with an <u>infinite order zero</u> at $y = \alpha_0$. A similar argument shows that $T_{2n}^{[1]}(m=0, y)$ also vanishes with an infinite order zero at $y = \alpha_0$, and this in turn implies that the Gell-Mann Low function has a zero of infinite order. Hence, <u>if the Gell-Mann Low function ψ has a zero for non-vanishing coupling, it must be a zero of infinite order</u>-- we see that electrodynamics must satisfy an extraordinarily strong condition in order for Z_3 to be finite.

Whether $F^{[1]}(y)$ and $T_{2n}^{[1]}(m=0, y)$ have the required infinite order zero is an open calculational question. There are two possibilities:

(A) $F^{[1]}(y)$ and $T_{2n}^{[1]}(m=0,y)$ do not have the required infinite order zero. Then the renormalization constants of electrodynamics cannot all be finite. [The only way to avoid this conclusion would be if a key technical assumption needed for the renormalization group analysis breaks down. The assumption states that terms which <u>vanish</u> asymptotically in each order of perturbation theory do not <u>sum to give</u> an asymptotically dominant result.]

(B) $F^{[1]}(y)$ and $T_{2n}^{[1]}(m=0,y)$ have an infinite order zero at $y = y_0 > 0$. As we have seen, this allows a class of solutions with finite Z_3, in which α_0 is fixed to be y_0 and $\alpha < y_0$ is undetermined. We will now show[10] that the presence of an infinite order zero allows <u>one additional solution, in which the physical fine structure constant α is fixed to be y_0</u>.

The possibility of an additional solution arises because when an infinite order zero (an essential singularity) is present, different orders of summing perturbation theory lead to inequivalent theories. One natural way of summing perturbation theory is to sum "<u>vacuum-polarization-insertion-wise</u>": One first sums all internal photon self-energy parts, and then inserts the resulting full photon propagators in the vacuum polarization skeleton graphs. This order of summation is the one used by JBW, and leads to their form of the eigenvalue condition $F^{[1]}(\alpha_0) = 0$. To see this we apply "vacuum-polarization-insertion-wise" summation to the single-fermion loop skeleton graphs for the photon proper self-energy, giving

$$\pi_c^{[1]}[-q^2/m^2, \alpha d_c] = $$ $$ (17)$$

where each shaded blob denotes a full renormalized propagator insertion $\alpha d_c/q^2$. Let us now assume that in letting $-q^2/m^2 \to \infty$, we can take the limit <u>inside</u> the infinite sum over insertions represented by Eq. (17), and therefore we replace each blob by its asymptotic limit $\alpha d_c(\infty, \alpha)/q^2 = \alpha Z_3^{-1}/q^2 = \alpha_0/q^2$. This gives

$$\pi_c^{[1]}[-q^2/m^2, \alpha d_c] \sim \pi_c^{[1]}(-q^2/m^2, \alpha_0) \sim G^{[1]}(\alpha_0) + F^{[1]}(\alpha_0)\ln(-q^2/m^2)$$
$$+ \text{ vanishing terms,} \qquad (18)$$

so asymptotic finiteness requires the condition $F^{[1]}(\alpha_0)=0$, i.e., $\alpha_0=y_0$. Similar arguments apply to the multiloop skeleton diagrams, when summed "vacuum-polarization-insertion-wise", and again give the condition $\alpha_0=y_0$.

There is, however, another natural summation order, which is to proceed "loopwise": One first sums all single-fermion-loop vacuum polarization graphs, then one sums all two-fermion-loop vacuum polarization graphs, and so forth. The sum of all single-fermion-loop graphs is just

$$\pi_c^{[1]}(-q^2/m^2,\alpha) \underset{-q^2/m^2\to\infty}{\sim} G^{[1]}(\alpha)+F^{[1]}(\alpha)\ell n(-q^2/m^2) \qquad (19)$$
$$+ \text{ vanishing terms,}$$

so asymptotic finiteness now requires $F^{[1]}(\alpha) = 0$, i.e., $\alpha=y_0$. It is easily seen that the same condition $\alpha=y_0$ guarantees asymptotic finiteness of the multiloop vacuum polarization graphs. So we have found an additional, discrete solution in which α is fixed to have the value y_0. Since $\alpha_0 > \alpha$, for this solution α_0 will be outside the region of analyticity of $F^{[1]}$ and so the interchange of limit with sum used in the "vacuum-polarization-insertion-wise" summation procedure is invalid. Hence the condition $F^{[1]}(\alpha_0)=0$ derived by JBW <u>does not</u> apply to the discrete solution in which α is fixed. (If it did, one would have the contradictory equations $\alpha=\alpha_0=y_0$, $\alpha < \alpha_0$.)

We conclude, then, that requiring the renormalization constants of electrodynamics to be finite, combined with "loopwise" summation, leads to an eigenvalue condition for α. <u>We conjecture that this is the mechanism which fixes the value of the fine structure constant.</u> The eigenvalue condition has the appealing property that it is <u>independent of the number of elementary charged fermion species which are present.</u> To see this, we note that when j species are present, the coefficient of the logarithmic divergence in the single-fermion-loop photon proper self-energy is

$$\sum_{\ell=1}^{j} F^{[1]}(\alpha_\ell), \qquad (20)$$

which vanishes if all $\alpha_\ell = y_0$. The same condition guarantees vanishing of the multiloop vacuum polarization diagrams. So the value of α which is determined is the same as in the one species case, and the j species are all required to have the same basic electromagnetic coupling $\pm\sqrt{y_0}$. Hence charge quantization appears in a natural way.

Let us now give a possible argument for the neglect of the strong interactions. Suppose that elementary charged fermions are present which have strong interactions mediated by neutral boson exchange (the gluon model). Although the bosons do not themselves contribute vacuum polarization loops, they could modify the fermion vacuum polarization loops when they appear as internal radiative corrections, e. g.

gluon   (21)

However, let us now invoke the experimental observation of scaling in deep inelastic electron scattering, one explanation for which[11] is that the exchanges which mediate the strong interactions are actually much more strongly damped at high four-momentum transfer than is the free boson propagator $(q^2+\mu^2)^{-1}$. If this explanation proves correct, then vacuum polarization diagrams with gluon radiative corrections will by themselves be asymptotically finite, and therefore will not contribute to $F^{[1]}$. This means, in turn, that the presence of strong interactions will not alter the eigenvalue condition for α.

What can be said about the prospects of calculating $F^{[1]}(y)$? All that is known at present is the expansion through 6th order in perturbation theory,[12]

$$-y\, F^{[1]}(y) = \frac{2}{3}\left(\frac{y}{2\pi}\right) + \left(\frac{y}{2\pi}\right)^2 - \frac{1}{4}\left(\frac{y}{2\pi}\right)^3 + \ldots \; . \qquad (22)$$

Even though the perturbation theory calculations leading to Eq. (22) are quite horrendous, the resulting coefficients are remarkably simple. A possible clue to the origin of this simplicity may be the fact that $F^{[1]}$ is a property of electrodynamics in the zero fermion mass limit, in which limit the invariance group is the full conformal group, a much larger group than the usual inhomogeneous Lorentz group.[13] Perhaps this fact can be used to develop means for calculating $F^{[1]}$, or at least for approximating it well enough to determine the location of its singularities.

(d) NUMEROLOGICAL SPECULATIONS: WYLER'S FORMULA

So far we have discussed what might be termed[14]"theories in search of number". But no discussion of α would be complete without mentioning a much publicized "number in search of a theory", the formula for α proposed by Wyler,[15]

$$\alpha = \frac{9}{8\pi^4} \left(\frac{\pi^5}{2^4 5!}\right)^{1/4} = 1/137.03608. \tag{23}$$

Whether the agreement of Eq. (23) with experiment has a basis in physics, or is purely fortuitous, remains at present a completely open question.

REFERENCES

1.   G. Gamow, Phys. Rev. Letters 19, 759 (1967).

2.   L. D. Landau, "On the Quantum Theory of Fields," in Niels Bohr and the Development of Physics, W. Pauli, ed. (Pergamon Press, London, 1955); B. S. DeWitt, Phys. Rev. Letters 13, 114 (1964); A. Salam and J. Strathdee, Nuovo Cimento Letters 4, 101 (1970).

3.   For an excellent review see F. J. Dyson, "The Fundamental Constants and their Time Variation", Institute for Advanced Study preprint (1972).

4.   C. J. Isham, A. Salam & J. Strathdee, Phys. Rev. D3, 1805 (1971).

5.   J. Shechter and Y. Ueda, Phys. Rev. D2, 736 (1970); T. D. Lee, Phys. Rev. Letters 26, 801 (1971).

6.   S. Weinberg, Phys. Rev. Letters 19, 1264 (1967) and 27, 1688 (1971).

7.   M. Gell-Mann and F. E. Low, Phys. Rev. 95, 1300 (1954).

8.   K. Johnson, M. Baker and R. Willey, Phys. Rev. 136B, 111 (1964); K. Johnson, R. Willey and M. Baker, ibid 163, 1699 (1967); M. Baker and K. Johnson, ibid 183, 1292 (1969); M. Baker and K. Johnson, ibid D3, 2516, 2541 (1971).

9.   Their result is based on an application of the Jost-Schroer-Federbush-Johnson theorem. See P. G. Federbush and K. Johnson, Phys. Rev. 120, 1296 (1960); and R. Jost, in Lectures on Field Theory and the Many-Body Problem, E. R. Caianiello, ed. (Academic Press, N.Y., 1961).

10.   S. Adler, "Short Distance Behavior of Quantum Electrody-
       namics and an Eigenvalue Condition for α", Phys. Rev. (to
       be published).

11.   See e.g. D. J. Gross and S. B. Treiman, Phys. Rev. D4,
       1059 (1971).

12.   Fourth order calculation: R. Jost and J. M. Luttinger,
       Helv. Phys. Acta 23, 201 (1950). Sixth order calculation:
       J. L. Rosner, Phys. Rev. Letters 17, 1190 (1966) and
       Ann. Phys. (N. Y.) 44, 11 (1967).

13.   For a review of conformal invariance in field theory, see
       e.g. G. Mack and A. Salam, Ann. Phys. (N. Y.)53, 174(1969).

14.   J. A. Wheeler (unpublished).

15.   A. Wyler, C. R. Acad. Sci., Ser. A 269, 743 (1969) and
       271, 186 (1971). For discussions of this formula, see
       B. Robertson, Phys. Rev. Letters 27, 1545 (1971) and
       R. Gilmore, ibid 28, 462 (1972).

$g_J(H)/g_S(e)$ DETERMINATION: PRELIMINARY RESULTS[*]

James S. Tiedeman and H.G. Robinson

Physics Department, Duke University

Durham, North Carolina

Within the last few years, Zeeman spectroscopy of atoms and ions has seen rapid progress in the level of precision attained. Theoretical treatments[1] of the hydrogenic atom in a magnetic field have provided a detailed expression giving the ratio of g-factors of the electron bound in the 1S state to that of the free electron, $g_J(H)/g_S(e)$. Recent experimental activity[2] has centered on determinations of $g_J(H)/g_I(D)$; comparison with theory then concerns only the mass-dependent terms. We have begun an experimental determination of $g_J(H)/g_S(e)$ itself, so that the leading mass-independent correction terms can be checked.

The experimental method is similar to that employed by Balling and Pipkin[3] who determined $g_J(H)/g_S(e)$ to a precision of $\pm$ 1 part in 10^6. Electrons are produced by a discharge in a Ne or He buffer gas in a cell of 100 cm^3 volume. Also present are optically pumped Rb atoms which provide the means to monitor the electron spin resonance through spin-exchange collisions. Zeeman lineshapes in both Rb and electron species are determined in a carefully controlled 50G magnetic field typically using 11 discrete frequencies to examine each resonance lineshape. Data are collected by alternating sequentially between the two resonances to suppress effects of magnetic field drift. After linefitting, the ratio R = $g_J(Rb^{87})/g_S(e)$ is extracted using the Breit-Rabi equation. Finally the desired ratio $g_J(H)/g_S(e)$ is found by use of the previously determined ratio[4] $g_J(H)/g_J(Rb^{87})$.

Among the systematic effects considered are the following.
(a) The ground state hfs of Rb is shifted by the buffer gas: direct
measurement by means of $\Delta F = 1$ transitions of the Rb hfs as
influenced by the environment offers one solution. In addition,
pairs of $\Delta F = 0$ transitions are used for which the shifts in Rb
Zeeman frequencies are essentially of equal magnitude but of
opposite sign, e.g. $(F, m) = (2, 2) \leftrightarrow (2, 1)$ and $(2, -1) \leftrightarrow (2, -2)$.
Upon averaging the calculated R values for both Rb Zeeman transitions,
the error due to hfs error is <1 part in 10^9. (b) The shift[5] in g_J(Rb)
due to buffer gas pressure is ~1 part in 10^8 for He at 100 Torr.
Variation of buffer pressure provides a means of monitoring and
extrapolating such an effect. No shifts of this type have thus far
been detected in this experiment. (c) Rb Zeeman transitions exhibit
frequency shifts[6] as a function of the intensity of Rb light used in
optical pumping. Extrapolation to zero light intensity removes this
shift. The effect is typically ≤ 3 parts in 10^8 at operating light
intensities. (d) Inhomogeneities in the applied magnetic fields
directly affect the resonance lineshape. The 50G field used has a
deviation of ~ 3 parts in 10^7 along the z-axis over a distance of 8 cm.
The homogeneity is taken to be ≤ 7 parts in 10^7 over a 100 cm^3
sample. Runs with different field shimming histories provide the
primary systematic check. (e) The electron spin resonance frequency
is shifted by spin exchange collisions with polarized Rb[7]. A typical
maximum shift is 3 parts in 10^7. The effect changes sign when the
Rb polarization is reversed so that the average electron spin frequency
for both polarizations is unshifted if the magnitude of the Rb polarization is constant. (f) Relativistic corrections[8] to the electron spin
resonance frequency are substantial; e.g. -7.3 parts in 10^8 for 15°C
electrons. Additional heating of the electrons occurs due to applied
rf power and effects associated with e^- production. Shifts of
several parts in 10^7 can be produced. (g) Because macroscopic
volume averaging is prevented by the buffer gas, differences in
either the effective temperature of the electrons or in the applied
magnetic field can occur throughout the sample. Changes in sample
geometry are used to monitor this effect.

The characteristics of the electron spin resonance have been
studied in various cells. Figure 1 shows the electron spin resonance
linewidth as a function of temperature in a 100 Torr He filled cell.
Each point represents the linewidth at zero rf power. As was done
by Balling, Hanson, and Pipkin[7], two additional curves are drawn
representing the expected shapes for either a linewidth dominated by
collisions with the buffer gas ($\propto T^{5/2}$) or a linewidth dominated by

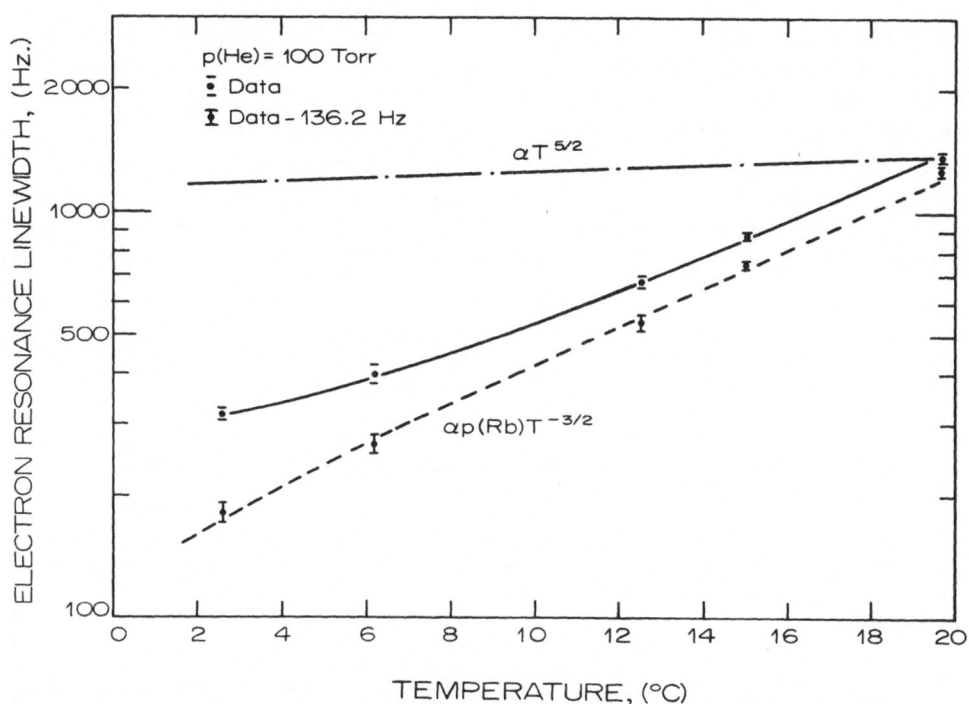

Fig. 1 Electron spin resonance linewidth at zero rf power vs. temperature in a 100 Torr He buffer cell. Linewidth dominated by buffer gas collisions $\propto T^{5/2}$ or by spin exchange with Rb $\propto p(Rb)T^{-3/2}$ is shown with broken lines. The dashed line represents a fit to the data minus a constant (136.2Hz) using the spin exchange form. The vapor pressure of Rb, $p(Rb)$, was calculated using the data of Ditchburn and Gilmore[9].

TABLE I. Sources of electron resonance linewidth in 100 Torr cell

Source	Estimated Contribution (Hz)	
	$T = 15°C$	$T = 6.2°C$
Spin Exchange	730	274
∇H_0	~ 100	~ 100
Buffer Gas	~ 30	~ 30
RF Saturation	60	120
Ion Recombination	1	1
Diffusion	2	2
Total	923	527

spin exchange with Rb ($\propto p(Rb)T^{-3/2}$). A fit has been made to the latter curve by subtracting a constant width from the data points. Table I shows the estimated contributions to the linewidth due to various sources. Figure 2 gives a typical extrapolation of R vs. light intensity. The dominant effect here is the spin exchange shift in the electron resonant frequency. The extrapolated value of R is significantly affected by rf heating of the electron through the relativistic effect mentioned previously. Thus a residual shift remains

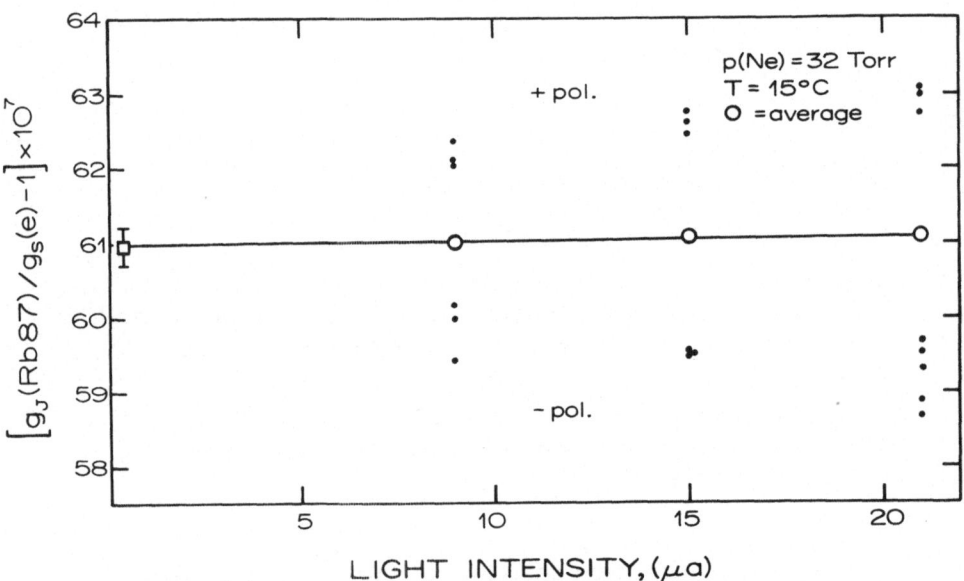

Fig. 2 An extrapolation of $g_J(Rb^{87})/g_S(e)$ vs. the light intensity used in optically pumping Rb. Results at the two light polarizations as well as their averages are shown. A residual shift remains at zero intensity due to rf heating of the electrons.

at zero light intensity. This effect is shown in Fig. 3 as a slope in
the extrapolated average of data taken at both polarizations. When
additional heating of the electrons is present, extrapolation vs.
applied rf power alone is not sufficient. See Fig. 4 and compare
with Fig. 3.

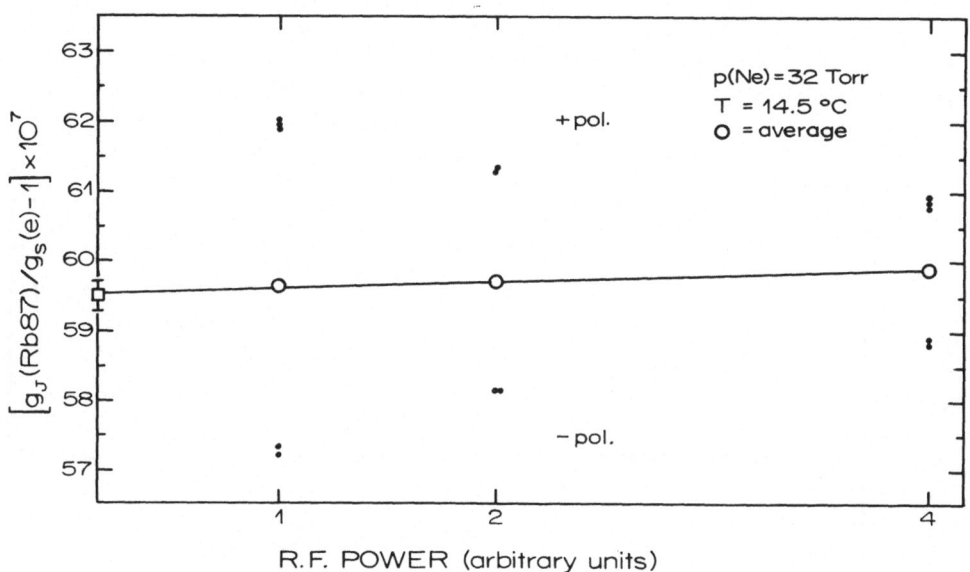

Fig. 3 An extrapolation of $g_J(Rb^{87})/g_S(e)$ vs. rf power in a Ne filled
cell. The slope of the line connecting data point averages at both
polarizations shows the effect of heating the electrons.

Theory[1] yields

$$g_J(H)/g_S(e) = 1 - 17.7051 \times 10^{-6}$$

with mass-dependent terms (through order $\alpha^3 (m/M)^2$) contributing 1.44×10^{-8} to this value. Comparison with theory can be made for R itself by computing

$$R_{th} = \left. g_J(H)/g_S(e)\right|_{th} \times \left. g_J(Rb^{87})/g_J(H)\right|_{exp} = 1 + 58.80 \times 10^{-7}.$$

Using preliminary results for Ne, after making the relativistic correction for $15^\circ C$ electrons the experimental value for R is

$$R_{exp} = 1 + 58.95 \ (1.0) \times 10^{-7}.$$

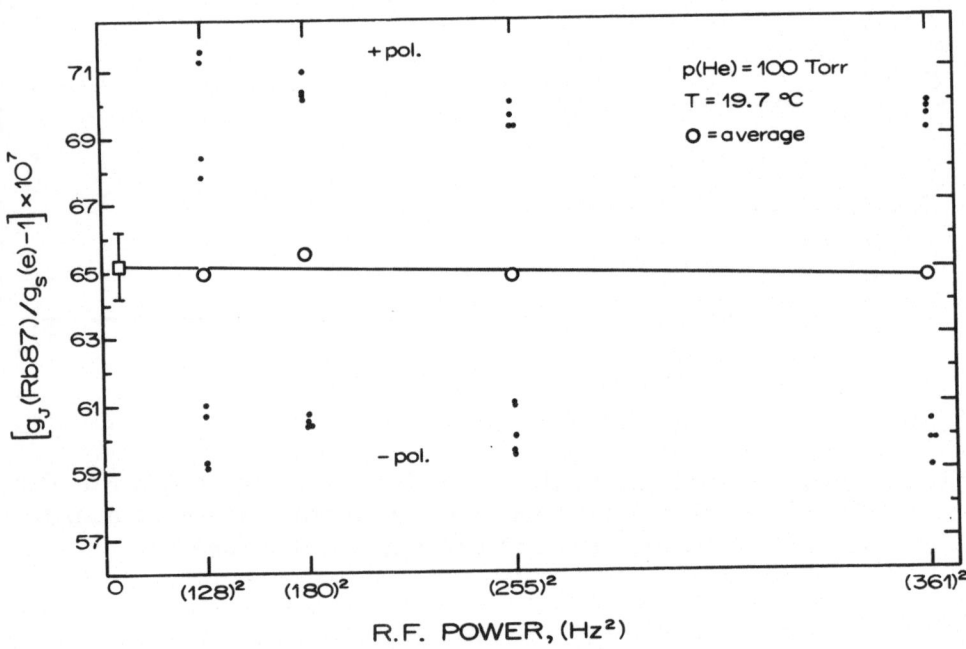

Fig. 4 An extrapolation of $g_J(Rb^{87})/g_S(e)$ vs. rf power in a He filled cell. Through heating, the intense discharge used to create the electrons is believed to cause the offset in the extrapolated value.

Although the present rms error is ~3 parts in 10^8, the quoted error has been increased to 1 part on 10^7 as representative of possible remaining systematic error. The value of R_{exp} is useful as a means of reducing previous g-factor ratios known with respect to $g_J(Rb)$ to g-factors relative to the free electron. Finally, the desired experimental value is obtained

$$g_J(H)/g_S(e)\Big|_{exp} = 1 -17.69 \ (0.10) \times 10^{-6} \ .$$

Agreement with theory is attained. Although it is doubtful that this experiment will eventually be able to criticize the mass-dependent terms in the theory, we are hopeful that a substantial improvement in precision can be made.

REFERENCES

* Work supported in part by the National Bureau of Standards, Grant 1-35856 and by the National Science Foundation, Grant GP-27207.

1. See for example, H. Grotch and R.A. Hegstrom, Phys. Rev. A4, 59 (1971).

2. See for example, F.G. Walther, W.D. Phillips, and D. Kleppner, Phys. Rev. Letters 28, 1159 (1972).

3. L.C. Balling and F.M. Pipkin, Phys. Rev. 139, A19 (1965).

4. W.M. Hughes and H.G. Robinson, Phys. Rev. Letters 23, 1209 (1969).

5. H.G. Robinson and G.S. Hayne, International Conference on Atomic Physics Abstracts (1968) p. 149.

6. B.R. Bulos, A. Marshal, and W. Happer, Phys. Rev. A4, 51 (1971).

7. L.C. Balling, R.J. Hanson, and F.M. Pipkin, Phys. Rev. 133, A607 (1964).

8. G.R. Henry and J.E. Silver, Phys. Rev. 180, 1262 (1969) and G.R. Henry, Private communication.

9. R.W. Ditchburn and J.C. Gilmour, Rev. Mod. Phys. 13, 310 (1941).

EXOTIC ATOMS

C. S. Wu

Columbia University

New York, New York   10027

I.   INTRODUCTION

The name "exotic atoms" is used to designate atomic systems in which any negatively charged leptons or hadrons other than the conventional electrons are bound in the Coulomb field of an atomic nucleus. The exotic atoms are essentially hydrogen-like atoms except for their unusually large energy scales and drastically reduced orbital radii, hence the name. Their life spans from birth to death last only for very brief moments. The transient existence is caused by either the decay of the atomic particle or the capture of the particle by the atomic nucleus.

At present, five different types of exotic atoms have been observed (see Table 1). They are formed by the capture of the slowed negative muons (μ^-), pions (π^-), kaons (K^-), sigma hyperons (Σ^-) and anti-protons ($\bar{p}$) into atomic orbits. The μ^- is a lepton which interacts rather weakly through electromagnetic and weak interactions with the nucleus. The probability of muon capture by the nucleus is very small. The muon is able to reach the lowest orbit (the 1s orbit) and spend a considerable fraction of its life inside the nucleus. For this reason, the muonic atom has been used as a sensitive probe for the study of nuclear structure. On the other hand, the π^-, K^-, Σ^-, and $\bar{p}$ are hadrons which interact strongly with the nucleus. They are generally captured by the nucleus before ever reaching the lower orbits. To single out the exotic atoms formed by the hadrons, we call them the hadronic atoms.

The muonic and pionic atoms have been studied since the early fifties. More precise measurements will be made when the high

Table 1. Long lived negative particles suitable for exotic atoms.
The Ξ^-, Ω^-, and $\bar{D}$ atoms are yet to be observed.

Particle	Spin	Mass (MeV)	Lifetime (sec)	Radius (1s) fm
e	1/2	0.511	∞	52,917/Z
μ^-	1/2	105.6	2.20×10^{-6}	256/Z
π^-	0	139.6	2.60×10^{-8}	194/Z
K^-	0	493.8	1.24×10^{-8}	54.7/Z
$\bar{p}$	1/2	938.3	∞	28.8/Z
Σ^-	1/2	1197.4	1.5×10^{-10}	22.6/Z
Ξ^-	1/2	1321.3	1.7×10^{-10}	20.5/Z
Ω^-	3/2	1672.5	1.3×10^{-10}	16.2/Z
$\bar{D}$	1	1875.6	∞	14.4/Z

intensity proton accelerators (500-900 MeV) will be put into
operation in six months to a year. The K^-'s and $\bar{p}$'s are produced
by protons in the energy region of several tens of GeV. The Σ^-'s
are produced by K^-'s. The observation of their atomic X-rays has
lately increased the family of exotic atoms from two members to
five. The K^- atoms were first observed by C. E. Wiegand and his
collaborators in 1967.[1] The Σ^- and $\bar{p}$ atoms were observed only
one month apart by Backenstoss and his colleagues[2,3] at CERN in
1970. The spectrum of K^- and Σ^- in ^{30}Zn is shown in Fig. 1.[2]

The other hadronic atoms still to be observed are from Ξ^-,
Ω^-, and D. The only qualification for any negative particles to
be eligible for the formation of exotic atoms is to have a life
long enough to survive the life sequences of the atom.

II. FORMATION OF THE EXOTIC ATOMS (SEE TABLE 2)

A. The Prelude

For particles to be captured into the atomic bound states,
they must be slowed down from several hundred MeV to < 2 keV by
ionization collisions in condensed materials. The time required
is about $(10^{-10} - 10^{-9})$ sec.

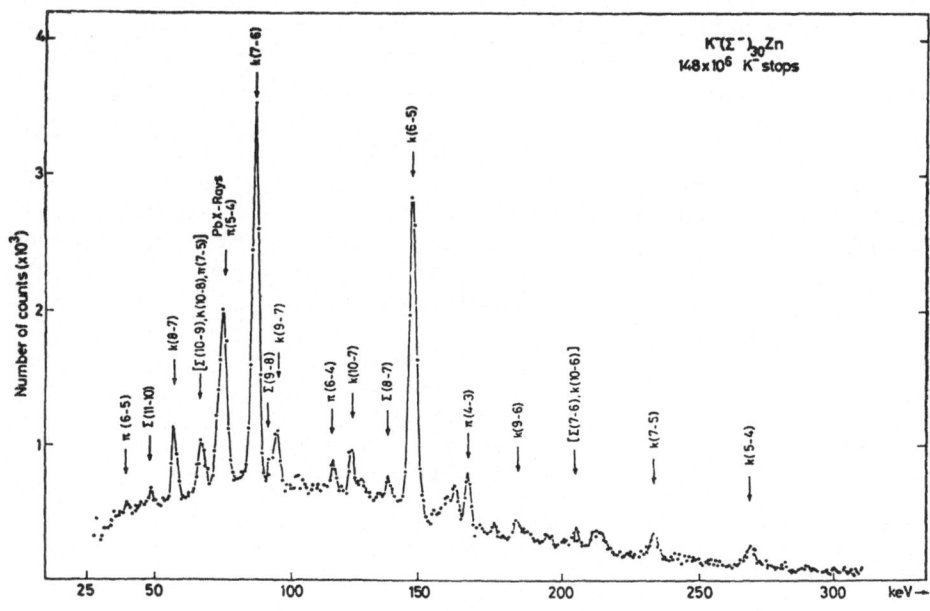

Fig. 1. The X-ray spectrum of K^- (Σ^-) in $_{30}Zn$ (Refs. 2, 9, 13).

B. Birth

The slowed down particles are probably first captured into the highly excited "molecular type" states. The detailed mechanism of capture processes is very involved and very little of it is known. Following the capture, the de-excitation by Auger and radiative transitions brings the exotic atoms from excited states to the orbit close to that of K-electrons. The principal quantum number n of such an orbit, that is. $<r_n>^2_{exotic} \sim <r_1>^2_{K-electron}$, is given by $n \simeq (m_{exotic}/m_e)^{1/2}$. For example, in a muonic atom n is $\simeq (206/1)^{1/2} \cong 14$; in $\bar{p}$ atom $n \simeq (1800)^{1/2} \cong 43$.

C. Life

The most interesting region is inside of the K-electron orbit, where the negative particle finds itself in a nearly unscreened Coulomb field as in a hydrogen-like atom. The de-excitation process is very simple. For high values of n, the initial low energy transitions will interact with outer electrons and yield a strong "Auger effect." However, as n reaches low values, the interaction with the electron structure is no longer important and radiative transitions dominate. The characteristic electric dipole transitions between the circular orbits are

Table 2. From birth to death of an exotic atom.

Prelude ——→	Birth ———	→ Life ———	→ Death
Particles slowing down to 2 keV	Captured into a bound state	Inside of the K-electron orbit: For higher n Auger process dominates. For lower n, via radiative E1 transitions mainly (n, n=n-1 → n-1, n-2)	Particle either decays or is captured by the nucleus
←10^{-10}-10^{-9}→ sec	←——— 10^{-13} ———→ sec		
	Mesochemistry	Fundamental properties of particles, QED, hfs, Nuclear physics, Charge distribution,	
		←———————→ Particle Nucleon Int. Particle-Nuclear Int. Matter Distribution	

$$[(n, \ell = n-1) \to (n-1, \ell-1)].$$

The time for the de-excitation of the atom is given by the transition probability for E1 transitions:

$$W \approx 0.5 \times 10^{10} \; Z^4 \; (m/m_e) \; \sec^{-1} \quad .$$

For a given nucleus Z, the transition rate is proportional to the mass of the atomic particle (m). For example, in muonic lead, the time required for the entire cascade from $n = 14 \to n = 1$ is about 10^{-13} sec.

D. Death

As the negative hadronic particle approaches the nucleus during its cascading from outer to inner orbits, the short-ranged strong interaction between the hadron and the nucleus will manifest itself rather abruptly in the sudden disappearance of radiative transitions. This is the end of a hadronic atom. In fact, hadronic atoms hardly ever reach the lowest states.

The whole process from prelude through birth to the death, in a hadronic atom takes about 10^{-9}-10^{-10} sec. This clearly rules out the formation of the hadronic atoms by the short-lived negative resonance particles which have lifetimes of $<10^{-21}$ sec. It also restricts the possibilities of using a high energy Σ^- beam which may decay before it is slowed down. In Table 1 all eligible negative particles for exotic atoms are summarized; their spins, masses, lifetimes, and Bohr radii are also listed for general reference.

To give some idea of the great reduction in size of these exotic atoms, the calculated first Bohr orbits

$$r_n = \frac{\hbar^2}{\mu e^2} \frac{n^2}{Z}, \qquad n = 1$$

in the Coulomb field of a point nucleus Z = 80 are plotted in Fig. 2. In hadronic-Pb atoms, the K^- and $\bar{p}$ particles are nearly completely captured in orbits of n = 5 and 8 respectively.

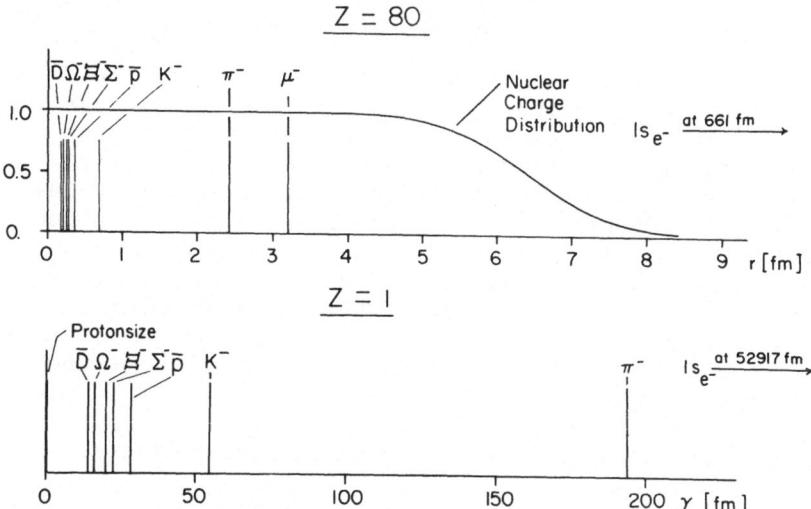

Fig. 2. The calculated positions of the first Bohr orbits of the various long lived negative particles in the field of the point nucleus of Z = 80 and the proton.

III. PHYSICAL INTEREST OF EXOTIC ATOMS

The field of exotic atoms encompasses a tremendously broad scope from chemical effects, atomic physics, nuclear structure to elementary particle physics.

A. Mesochemistry

In the early phase of its atomic formation, the transition energies involved are on the order of chemical bond energies. It has been shown that <u>the initial cascade</u> is influenced by the chemical and physical properties of the material in which the exotic atom is formed. The field of this type of studies is called mesochemistry and is being actively pursued in the U.S.S.R.

B. Atomic Physics and Fundamental Properties of Particles

The most interesting region for the atomic physicists is in the intermediate range of the cascade where both the shielding effect due to the electrons and the strong interaction due to the nucleus are negligible. The transition energies can be calculated to very high accuracy and precision. For this reason it has been applied to determine the fundamental properties of the particles and to test certain aspects of quantum electrodynamics. We will come back to this later.

C. Nuclear Structure Studies

The muonic atom is of particular interest to nuclear physicists. It has been regarded as a sensitive static probe, mapping the charge and magnetic moment distribution of the nucleus. Only in recent years have a series of extremely interesting dynamic properties of the muonic atoms been observed, and they have aroused tremendous interest. In electronic atoms, fine and hyperfine interaction energies are small in comparison with nuclear energies. In muonic atoms, the muonic transition energies range from a few keV to several MeV which is typical of nuclear excitations. The approximate independence of atomic and nuclear degrees of freedom is therefore not always valid here. When the approximation of independence is grossly violated, some interesting dynamic excitations of the muonic atom can be observed. I will give one example of this.

D. Strong Interactions and Nuclear Density Distributions

One of the major objectives of investigating hadronic atoms is to understand the strong interactions between the particle and the nucleus and to eventually use the hadrons as probes of the nuclear periphery. The study of K^-, Σ^-, and Ω^- atoms has just begun and the experimental results are already impressive. However, the theoretical interpretation seems more complicated than was originally anticipated.

IV. SOME RECENT HIGHLIGHTS

For such an extensive field, only a few recent highlights will be presented here.

A. Determination of Masses and Magnetic Moments of Fundamental Particles

1. <u>Masses</u>. The energy of a Bohr orbit is proportional to its mass:

$$E_{Bohr} \propto \mu \frac{(Z\alpha)^2}{2n^2} = \frac{Mm}{M+m} \frac{(Z\alpha)^2}{2n^2} .$$

The uncertainty in the mass determination $\Delta m/m$ is dictated by the precision of the X-ray energy measurement $\Delta E/E$. The bent-crystal diffraction spectrometer offers the best energy resolution. It has an anticipated energy resolution of 40-80 eV and an accuracy of 1-2 eV in the energy range of 50-100 keV. The best measurement of the pion mass (139.577 ± 0.014 keV) (see Table 3) was obtained by Shafer[4] from the measurements of the 4f → 3d transitions in pionic $_{20}$Ca and $_{22}$Ti by this method.

Table 3. The masses of μ^-, π^-, K^-, and $\bar{p}$ as determined by the exotic atom method.

m_μ = 206.76 (2) m_e	by "indirect" measurements (Refs. 5,6)
= 206.767 (3) m_e	from (g-2) and f_μ (Refs. 7,8)
m_π = 139.554 ± 0.008 MeV	Ge(Li) (Refs. 2, 9, 13)
= 139.577 ± 0.014 MeV	bent-crystal (Ref. 4)
m_k = 493.87 ± 0.19 MeV	Ge(Li) (Ref. 10)
= 493.8	
$m_p = m_p ± 0.5$ MeV	Ge(Li) (Refs. 2, 9, 13)

The muon mass is determined by an entirely different indirect method. It happens that the energy of the 4f → 3d transition in $_{15}$P (88.015 keV) lies just slightly above the absorption edge of Pb (at 88.008 keV). So the energies of the three fs components can be obtained by fitting the absorption curve with the appropriate absorption coefficients at these energies which are precisely calibrated on X-ray spectrometers. The muonic mass obtained in these measurements [m_μ = 206.76 (2) m_e][5,6] is in good agreement with the muon mass determined from the g-2 and f_μ experiments [m_μ = 206.767 (3) m_e].[7,8]

However, the chance coincidence, as that found in the case of μ in phosphorus, does not exist for other particles. The luminosity of a bent-crystal spectrometer decreases rapidly with increasing energy. Therefore, its application to high energy is not favorable.

Fortunately, the present Ge(Li) detector can achieve a resolution < 1.5 keV (FWHM) at 500 keV and < 1 keV at 100 keV. If the position of a line can be determined to an accuracy of 1/20 to 1/40 of its width under favorable statistics, a $\Delta E/E \sim 10^{-4}$ or better can be obtained. The X-rays chosen for this type of determination must be from high orbits where the finite size and the strong interaction effects are negligible. Yet they must not be too high; otherwise the uncertainties in the calculated screening effects may affect the accuracy of the results. The other important correction is the vacuum polarization term which has been calculated and tested to higher orders as we will show later. Backenstoss and his colleagues[9] have used the Ge(Li) detectors to determine the m_π and obtained a value of m_π = (139.554 ± 0.008) MeV which is only 23 keV less than that obtained by the crystal method and less than twice the standard deviation. An independent determination of the K^- mass from K^- mesic X-ray transitions was made by Kunselman[10] which gives a value of

$$m_{K^-} = (493.87 \pm 0.19) \text{ MeV}$$

to be compared with the m_{K^-} = (493.84 ± .11) MeV determined from $K^- \to \pi^+ + \pi^- + \pi^-$ decays.[11]

The X-rays of $\bar{p}$ in Tℓ have been used by Backenstoss et al.[2] to determine the $m_{\bar{p}}$. The best value so far for the antiproton mass is $m_{\bar{p}} \geq m_p \pm 0.5$ MeV. The masses of particles determined by exotic atoms are summarized in Table 3.

2. <u>Magnetic Moments</u>. The fine structure splitting of an atomic level in a hydrogen-like atom is proportional to the magnetic moment μ_a of the atomic particle

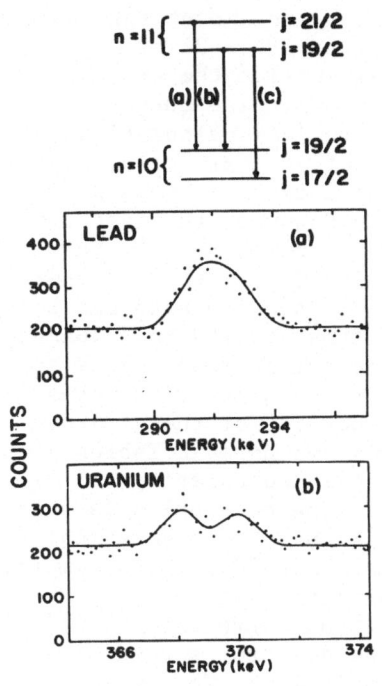

$$\Delta E_{fs} \sim \mu_a \frac{Z^4}{n^5}$$

The splitting is proportional to Z^4 and inversely to n^5. This possibility is of great interest for Σ^-, Ξ^-, and Ω^- as any magnetic moment information on these particles is welcome. Even measurements on 10% accuracy would provide some crucial tests of SU_3 symmetry and various variants of it.

Recently, the fine structure of the $11 \rightarrow 10$ transitions in the $\bar{P}$-Pb and $\bar{P}$-U atoms have been observed by Fox et al.[12] on the AGS at BNL as shown in Fig. 3. From the fine structure splittings of 1.20 ± 0.01 keV in lead and 1.91 ± 0.09 keV in uranium, they obtained an averaged value of the magnetic moment of anti-proton $\mu_{\bar{p}} = -2.83 \pm 0.10$ nm which is in accord with the value $+2.79$ nm for proton.

Fig. 3. The fine structure splittings in $\bar{p}$-Pb and $\bar{p}$-U atoms.

$\mu_{\bar{p}} = -\mu_p$ is expected from CPT theorem and this symmetry is known to be exact to high precision.

B. Tests of Vacuum Polarization

The vacuum polarization is due to the virtual production of $(e^+ e^-)$ pairs in a Coulomb field and results in an increase of the potential produced by the nuclear charge. The range of this correction is of the Compton wavelength $\lambda_e = \frac{\hbar}{mc} \sim 386$ fm.

The Lamb shift (the radiation effects due to the coupling with $\mu^+ \mu^-$ fields etc.) is much smaller due to the large muon electron mass ratio but is not smaller by $(m_e/m_\mu)^2$ as previously assumed.

Recently, several precision measurements have been carried out to test for vacuum polarization corrections in heavy muonic atoms. Backenstoss et al.[13] measured the $4f \rightarrow 3d$ and $5g \rightarrow 4f$ transitions in Pb and Bi and found good agreement (within 2%) with the calculated energies. Dixit et al.[14] showed that the calculated values

are consistently larger than those measured by an amount that varies
with energy or Z from 15 ± 16 eV at 157 keV to 137 ± 22 eV at
438 keV. Such a discrepancy certainly lies outside the expected
validity of quantum electrodynamic calculations. The author
suggested a possible inaccuracy in the theoretical calculations.
Blomqvist[15] recalculated the vacuum polarization correction; besides
the leading term $\alpha Z\alpha$, higher order terms included are $\alpha^2 Z\alpha$ and
$\alpha(Z\alpha)^3$ as well as orders $\alpha(Z\alpha)^5$, $\alpha(Z\alpha)^7$. However, the last two
terms do not influence the final conclusions too much. The term
$\alpha(Z\alpha)^3$ is significant. Blomqvist found the $\alpha(Z\alpha)^3$ term in a previous calculation[16] was in error, giving the opposite sign and too
small a magnitude for the energy shifts. This results in a change
in the higher order vacuum polarization correction from +35 eV to
−35 eV. To give some ideas on the sizes of the correction terms
the various contributions to the binding energies of the $4f_{7/2}$ and
and $5g_{9/2}$ states in muonic ^{208}Pb are reproduced here in Tables 4
and 5.[15] The comparison between theoretical and experimental
results[13,14,17] are summarized in Table 6. The conclusion is that
the experimental energies are slightly less than the theoretical
values by two standard deviations which is about 50 eV. In comparison with the total vacuum polarization correction of 2000 eV, this
implies an accuracy on the vacuum polarization correction of 2-3%
in the heavy nuclei has been achieved. The electron screening
effect is calculated for full electron population. The uncertainty
introduced by the absence of L and K electrons due to Auger processes is around 10 eV. Increasing the theoretical value therefore,
increases the discrepancy.

Table 4. Vacuum polarization energy shifts in eV for the $4f_{7/2}$
and $5g_{9/2}$ states in muonic ^{208}Pb. Constants: α^{-1} = 137.03602;
$\hbar c$ = 197.32891 MeV fm; m_e = 0.5110041 MeV; m_μ = 105.6599 MeV; and
$m_{\mu,red}$ = 105.6023 MeV (Ref. 15).

Contribution		$4f_{7/2}$	$5g_{9/2}$
$\alpha Z\alpha$	Point nucleus	−3652	−1562
	Finite Size	−12	−3
	2nd order	−9	−3
	Total	−3673	−1568
$\alpha^2 Z\alpha$		−25	−11
$\alpha(Z\alpha)^3$		+93	+50
$\alpha(Z\alpha)^5$		+11	+6
$\alpha(Z\alpha)^7$		+2	+1
TOTAL		−3592	−1522

Table 5. Contributions to the energies (eV) of the $4f_{7/2}$ and $5g_{9/2}$ states in muonic ^{208}Pb (Ref. 15).

Contribution	$4f_{7/2}$	$5g_{9/2}$
Dirac energy, point nucleus	−1188316	−758971
Finite size	+4	0
Nuclear motion and relativity	−4	−2
Vacuum polarization	−3592	−1522
Other radiative effects	+10	+3
Electron screening	−77	−155
Nuclear polarization	−4	0
TOTAL	−1191979	−760647

Table 6. A comparison between experimental and theoretical X-ray energies in the vacuum polarization tests.

	Theoretical (eV)	Experimental (eV)	(Theo.−Exp.) (eV)
^{208}Pb			
$5g_{9/2} \rightarrow 4f_{7/2}$	431,332 ± 10	431,295±17 (Ref. 14)	+47
		431,410±40 (Refs.2,9,13)	−78
^{nat}Hg			
$5g_{9/2} \rightarrow 4f_{7/2}$	410,342 ± 10	410,284±24 (Ref. 16)	+58
$5g_{7/2} \rightarrow 4f_{5/2}$	416,146 ± 10	416,087±23 (Ref. 16)	+59
^{203}Tℓ			
$5g_{9/2} \rightarrow 4f_{7/2}$	420,777 ± 10	420,717±23 (Ref. 16)	+60
$5g_{7/2} \rightarrow 4f_{5/2}$	426,881 ± 10	426,828±23 (Ref. 16)	+47

A totally different approach to determine the vacuum polarization correction is to carry out Lamb-Retherford type experiments to determine the energy difference between the 2s → 2p levels in the μ^-p and $\mu^-\alpha$ atoms. If the finite size effect is the only dominating term, then the 2s level lies above the 2p. Due to a much larger vacuum polarization correction in the 2s level than in the 2p, in the μ^-p and $\mu^-\alpha$ atoms the 2s level is shifted below the 2p level as shown in Fig. 4. However, the transitions between the 2s and 2p levels can be induced by tunable lasers at the right frequency and enhancement of the counting rate of the 2p → 1s X-rays (1.9 keV for μ^-p and 8.2 keV for $\mu^-\alpha$) will be detected. Therefore the experiment consists of determining the 2s → 2p resonance frequency by observing a resonant enhancement in the (2p → 1s) X-rays.

Precision measurements of the $2s_{1/2}$ → $2p_{1/2}$ energy differences of μ^-p and $\mu^-\alpha$ muonic atoms will enable us to test the electron

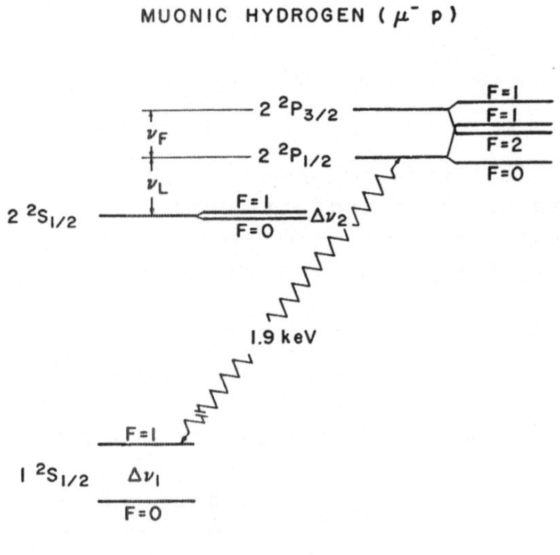

MUONIC HYDROGEN (μ^- p)

$$\nu_L = 0 \left[\alpha^3 \frac{m_\mu}{m_e} R_\infty^\mu \right] = 4.9 \times 10^{13} \text{ cps.}$$

$$\nu_F = 0 \left[\alpha^2 R_\infty^\mu \right] = 0.2 \times 10^{13} \text{ cps.}$$

$$\Delta\nu_1 = 0 \left[\alpha^2 \frac{\mu_p}{\mu_\mu} R_\infty^\mu \right] = 4.4 \times 10^{13} \text{ cps. (68,000 Å)}$$

$$\Delta\nu_2 \simeq \frac{1}{8} \Delta\nu_1 = 0.5 \times 10^{13} \text{ cps.}$$

Fig. 4. The 2s and 2p level diagram of muonic hydrogen (μ^-p) (Ref. 21).

Table 7. The vacuum polarization corrections in $\mu^- p$ and $\mu^- \alpha$ (Refs. 18, 19).

Isotope	Transition	Electron Vacuum Polarization		Lamb Shift	Finite Size Effect
		order α	order α^2		
^{1}H	$2s_{1/2}\ 2p_{1/2}$	-1.5122	-0.0112	0.0049	0.0258 ± 0.0006
^{4}He	$2s_{1/2}\ 2p_{1/2}$	-12.376	-0.086	0.106	2.182 ± 0.078

The energies are in units of $\alpha^2 Ry = 0.13461$ eV.

vacuum polarization to a much better accuracy. This can be seen from Table 7. The accuracy of these tests may reach 0.05% and 0.7% in $\mu^- p$ and $\mu^- \alpha$ atoms respectively. In the case of $\mu^- \alpha$, our present inadequate knowledge of the form factor of α-particles yields an uncertainty in the finite size correction as large as the second order correction in vacuum polarization. However, the situation hopefully will be improved as better electron scattering results become available.

Zavattini et al.[20,21] at CERN have carried out a series of beautiful experiments on the $\mu^- p$ and $\mu^- \alpha$ and now are making the final assault on the Lamb-Retherford type experiment on $\mu^- \alpha$. Figure 5 shows the X-ray spectrum from $\mu^- p$ atoms (4 atm, 493°K) by a gas proportional counter. The observed spectrum is compared with a calculated one assuming a pure radiative de-excitation. The shift of the observed spectrum towards higher energy is probably due to the fact that the Stark mixing cannot be neglected. The experimental distribution requires a relative ratio D of K_α to ΣK

$$D = \frac{K_\alpha}{\Sigma K} = 0.42 \pm 0.10;$$

whereas the calculation gives as much as D = 0.94. A similar X-ray spectrum from $\mu^- p$ was also observed in a liquid hydrogen target.[22] Leon[23] has recently calculated the relative yields for K X-rays from $\mu^- p$ as functions of the hydrogen density. In his calculations, three mechanisms for depopulating the levels were taken into account. They are Stark mixing, radiative de-excitation and Auger de-excitation. The comparison between experiment and calculation is good.

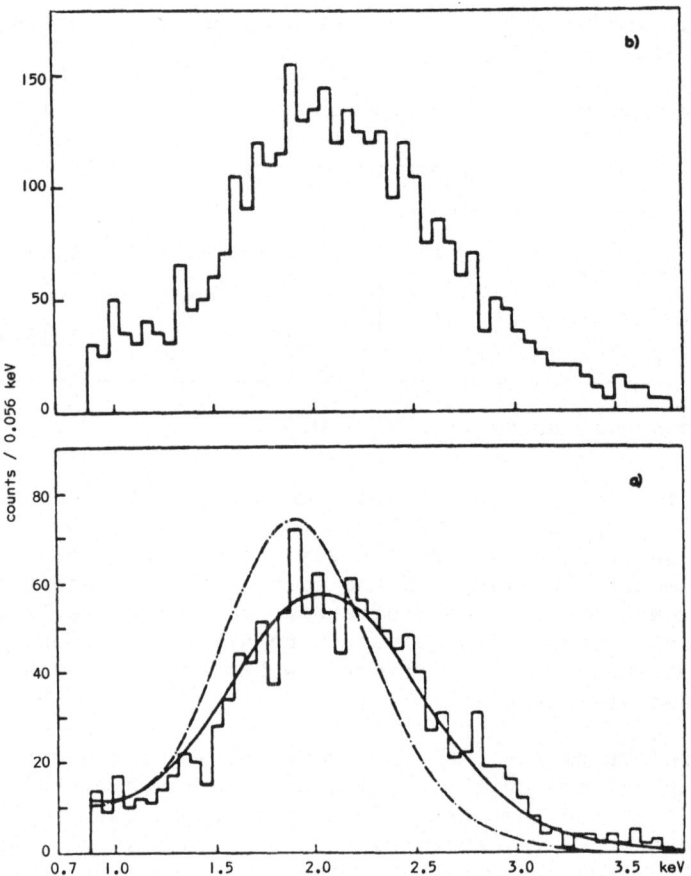

Fig. 5. The spectrum of (μ⁻p) X-rays obtained: (a) requiring the detection of a decay electron, (b) without requiring a decay electron. The solid line of Fig. 5(a) represents the calculated spectrum including Stark mixing; the dashed line without Stark mixing (Ref. 17).

Zavattini et al.[21] have also measured the X-ray spectrum from the μ⁻α system (Fig. 6) (7 atm at 293°K) and sought answers to the following two pertinent questions: (1) What fraction of the μ⁻ stopped in the target He gas form the 2s metastable ionic system $(\mu^-He)^+_{2s}$? (2) What is the lifetime of the metastable 2s state? Figure 6 shows the spectrum of μ⁻α X-rays. From the fit $E_{K_\alpha} = (8.180 \pm 0.070)$ keV, the ratio of K X-rays to the number of stopped muons in gas is

$$R_K = \frac{\text{emitted muonic K X-rays}}{\text{number of muons stopping in gas}} = 0.99 \pm 0.10 \quad .$$

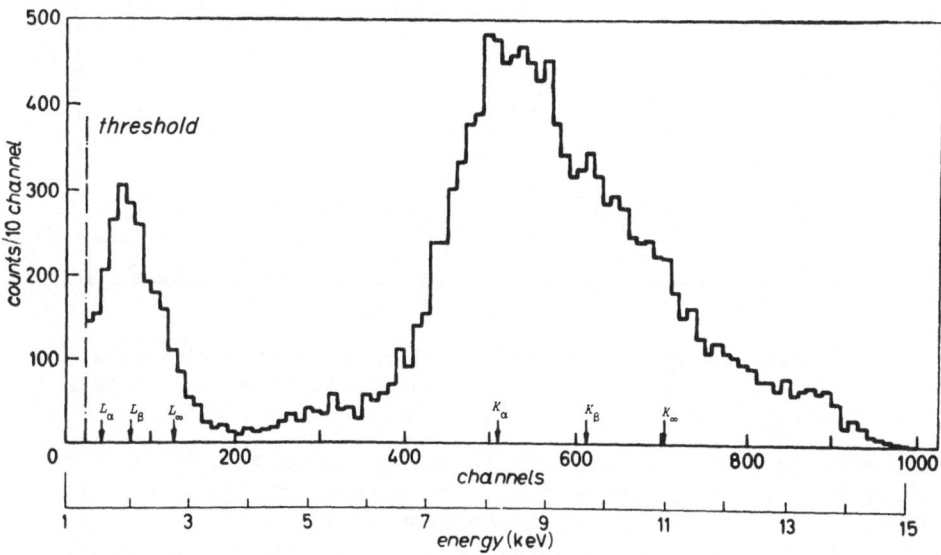

Fig. 6. Muonic X-rays from ($\mu^-\alpha$) system (Ref. 18).

The relative yields of K_α to total K X-rays is

$$D = \frac{K_\alpha}{\Sigma K} = 0.62 \pm 0.08 \qquad .$$

To search for the two-photon emission in the 2s → 1s transition
from the $\mu^-\alpha$ atoms, they looked for the delayed 2s → 1s X-rays
followed by an electron event at a time interval greater than 600
nsec. Figure 7 shows the spectrum of the delayed X-rays which
displays a continuous distribution from the low energy limit to
around 8.2 keV. The dip around 3.2 keV is due to the effect of the
K-edge absorption (3.2 keV) of the minute argon contamination in
the target of He gas (which leaked through the thin window between
the counter and the target chamber). The fraction "ξ" of the
metastable 2s state populated per stopped muon at 7 atm is
$\xi = (3.4 \pm 0.7) \times 10^{-2}$. The lifetime of the two photon emission
from the 2s state

$$(\mu^-\alpha)_{2s} \xrightarrow{2\gamma} (\mu^-\alpha)_{1s}$$

is $\tau = (1.8 \pm 0.4)$ sec. The time distribution of the decay electron ($\mu^- \rightarrow e^- + \nu_e + \nu_\mu$) using the detection of two photon emission
as t = 0 gives a lifetime $\tau = 2.2$ sec which is the well-known
muon decay time. The population of the 2s state in $\mu^-\alpha$ atoms
can be further increased by increasing the He density by between
3 and 5 times the 7 atm before the quenching effect due to Auger

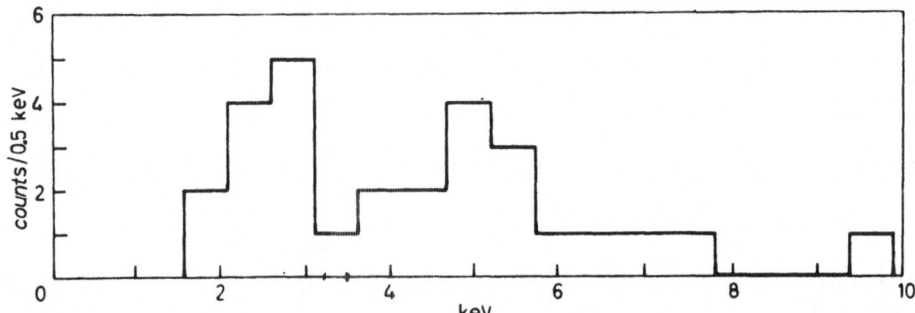

Fig. 7. The energy spectrum of delayed photons from the $(\mu^-\alpha)$ system. These photons are from the two photon emission processes between $(\mu^-\alpha)_{2s} \rightarrow (\mu^-\alpha)_{1s} + 2\gamma$ (Ref. 18).

effects sets in. This experiment is now going on at CERN. They will have results in a few months. The situation in μ^-p is more difficult and also more uncertain because of the softness of the X-rays (1.9 keV) and lack of detailed knowledge of the quenching cross section of the 2s state. An experiment designed to measure the latter has been proposed.[24]

C. Resonance Excitation by Accidental Degeneracy

In the dynamic excitation of the nucleus the muon is playing a role beyond its static probing. It produces nuclear polarization, excites nuclear states, and induces nuclear fission and neutron emission. However, I will show you only one of these excitations, the phenomenon of <u>nuclear resonance excitations by accidental degeneracy</u>. If any of you still remembers the old puzzle of the intensity anomaly observed in muonic ^{209}Bi by Rainwater and his co-workers[25] in the fifties, finally it is solved.

The nuclear resonance excitation by accidental degeneracy happens only in those nuclei where two nearly degenerate states exist. That is

$$E_I + E_{\mu j} \simeq E_{I'} + E_{\mu j'}$$

where $(E_I, E_{I'})$ and $(E_{\mu j}, E_{\mu j'})$ are the nuclear and muonic energies.

Let's take the case of $^{209}_{83}$Bi.[26,27] It is known now that all odd nuclei in the Pb region from Tℓ to Bi have multiplet levels around 2.6 MeV. These multiplets are formed by the coupling between the odd nucleon and the core excitation. For example, the closed core $^{208}_{82}$Pb can be excited to its 3⁻ excited state. The coupling between the $h_{9/2}$⁻ proton with core excitation 3⁻ state splits it into a septuplet as shown in Fig. 8. It turns out there are two sets of almost degenerate systems in ^{209}Bi. One of them is

$$|1> = |3d_{5/2}, \quad I = 9/2^- \quad (G.S.)>$$

$$|2> = |2p_{1/2}, \quad I = 15/2^+ \quad (at\ 2.74\ MeV)>$$

$$\left.\right\} F^{\pi} = 7^- \quad .$$

The energy difference between these two systems is about 2 keV.

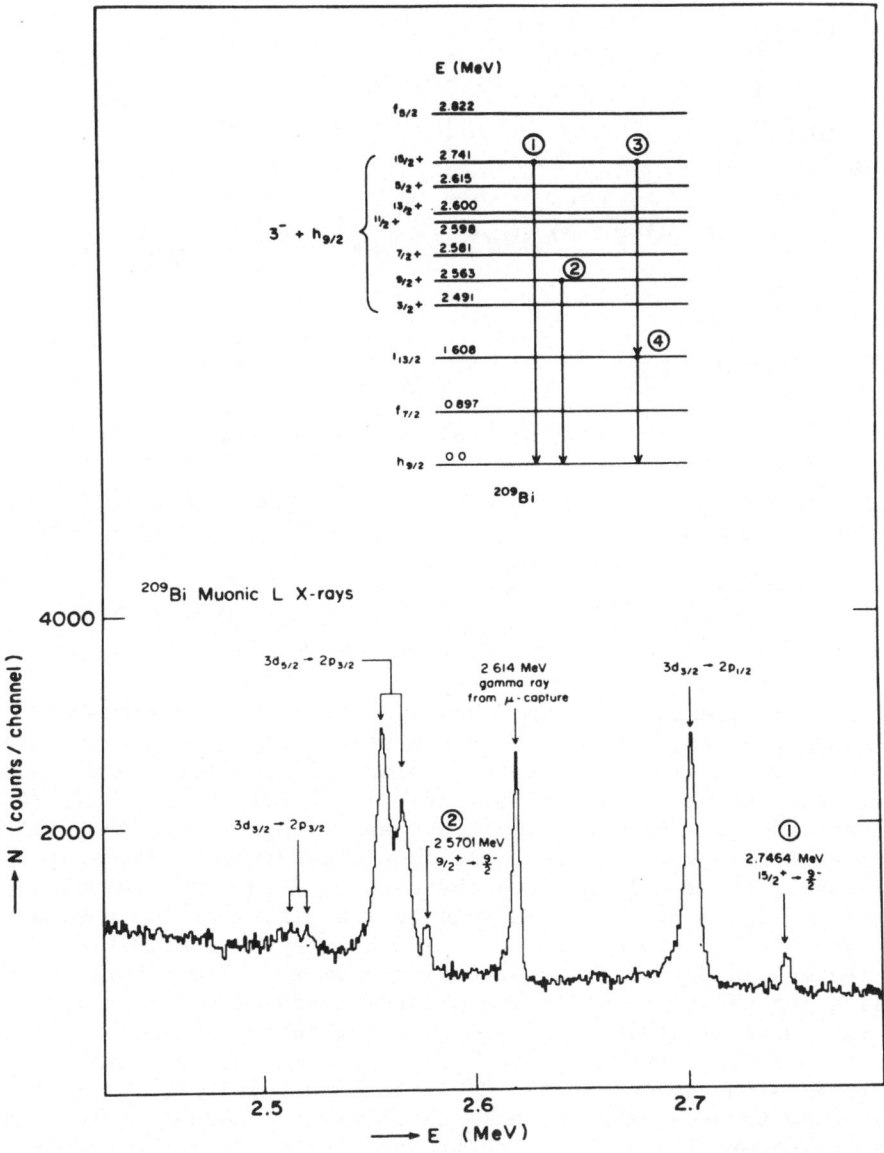

Fig. 8(a). The partial decay scheme of [209]Bi and the prompt γ-rays resulting from the resonance excitations from the 15/2[+] and 9/2[+] states in muonic [209]Bi (Ref. 20).

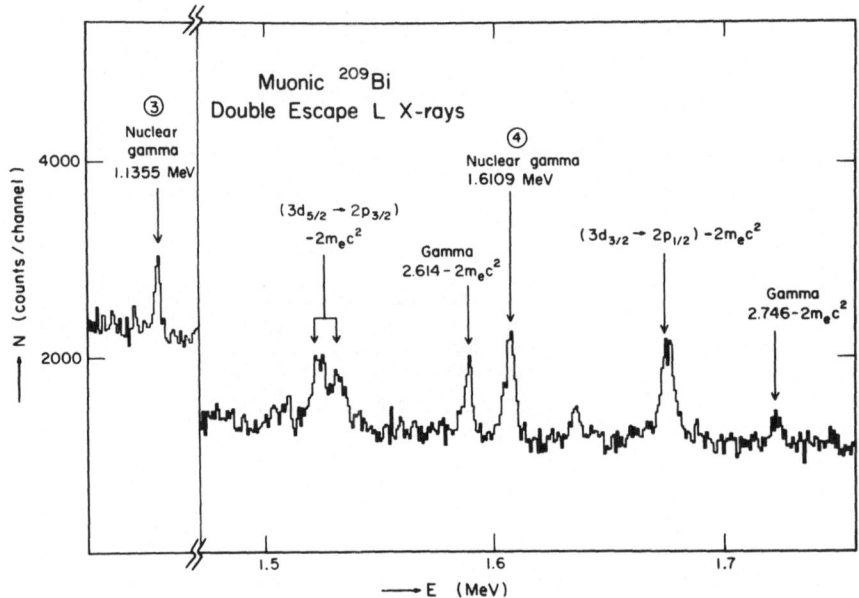

Fig. 8(b). See caption under Fig. 8(a).

The E3 matrix element between them is $\langle 1|H_{\text{int}}|2\rangle = 2.43$ keV. And
the other set is

$$\left.\begin{array}{l} |1'\rangle = |3d_{5/2},\; \text{I} - 9/2^{-} \text{ (G.S.)}\rangle \\[2mm] |2'\rangle = |2p_{3/2},\; \text{I} - 9/2^{+} \text{ (at 2.56 MeV)}\rangle \end{array}\right\} F = 3^{-},4^{-},5^{-},6^{-}.$$

The energy difference is about 9 keV and the E1 matrix element
varies from 0.30 to 1.96 eV. With the presence of these resonances,
one predicts: (1) There are certain probabilities for the nucleus
to be excited to those excited states: ~3.7% for the $15/2^{+}$ (2.74
MeV) state and ~3% for the $9/2^{+}$ (2.560 MeV) state. When the nucleus
de-excites to the ground state, one should observe the de-excitation γ-rays. These γ-lines were observed at CERN[27] and Columbia[26]
as shown in Figs. 8a and b. (2) Due to the mixing of the nuclear
and muonic states, the hfs of the K and L X-rays are both affected,
which results in the observed intensity anomaly. These intensity
anomalies can now be quantitatively interpreted as shown in Table
8, so the puzzle is solved. (3) In a quantitative analysis of the
hfs of muonic M, L, and K X-rays besides taking into account the
contributions due to the static M1 and E2 interactions, one must
also include the resonance effects which mix the same F states and
the large isomer effects which shift the positions of the unperturbed levels due to the presence of the muon. The search for the
best fit to the hfs is carried out on a computer by varying the
static and transition moments and the parameters of the charge

Table 8. The (anomalous) intensity ratios of X-ray doublets of
muonic ^{209}Bi.

Ratio	Experimental		Theoretical	
	Columbia 1971	CERN 1966	without resonance	with resonance
$2p_{3/2} \to 1s$	1.38	1.31	1.92	1.41
$2p_{1/2} \to 1s$	±0.10	±0.2		
$3d_{5/2} \to 2p_{3/2}$	1.52	1.44	1.75	1.48
$3d_{3/2} \to 2p_{1/2}$	±0.10	±0.2		
$4f \to 3d_{5/2}$	1.51	1.18	1.49	1.49
$4f \to 3d_{3/2}$	±0.10	±0.12		
$5g \to 4f_{7/2}$	1.33		1.33	1.33
$5g \to 4f_{5/2}$	±0.10			

distribution. Figure 9 shows the best fit for the ^{209}Bi L X-rays.
Table 9 lists the static moments of the excited states $9/2^+$ and
$15/2^+$ in ^{209}Bi which are involved in the resonances. The M1
moments thus determined are in excellent agreement with those
predicted from a core excitation model.[28] The E3 transition
moments between the nuclear excited state $9/2^+$ and the ground
state $9/2^-$ determined from the hfs best fit are in excellent agreement with those determined from Coulomb excitations[29,30]:

$$B(E3, 9/2^+ \to 9/2^-) = \begin{array}{l} 0.072 \pm 0.014 \; e^2b^3 \; \mu\text{-atom} \\ 0.074 \pm 0.011 \; e^2b^3 \; \text{C.E.} \end{array}$$

and

$$B(E3, 15/2^+ \to 9/2^-) = \begin{array}{l} 0.047 \pm 0.010 \; e^2b^3 \; \mu\text{-atom} \\ 0.048 \pm 0.007 \; e^2b^3 \; \text{C.E.} \end{array}$$

(4) The time which it takes for a muon to cascade from high orbits
to the lowest state (1s) is less than 10^{-15} sec. In the 1s state,
the muon will either decay with a half life of 2.2 μsec or be
captured by the nucleus with a rate never exceeding 10^8/sec. But
the time involved in nuclear γ-transitions is generally much
longer than the muon cascading time but much shorter than the time
muon spends at 1s state. Therefore, the excited nuclear states
will de-excite by emitting a γ-ray while the muon is in the 1s
state. This sequence of events leads to the observation of isomer

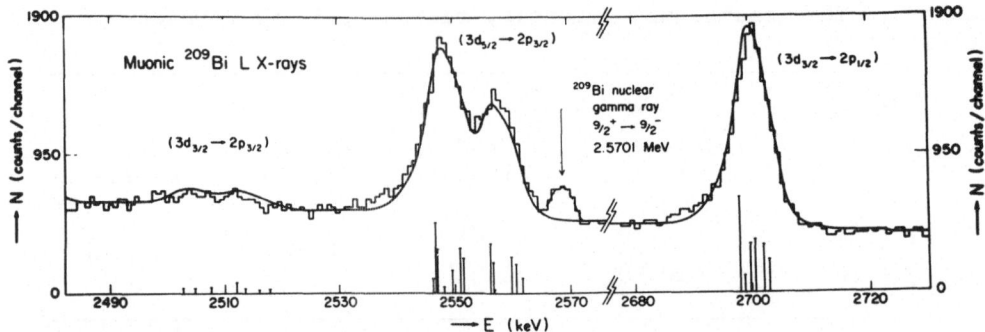

Fig. 9. Best fit to the muonic L X-rays of ^{209}Bi. The histogram represents the experimental data: the smooth curve, the best fit. The vertical bars indicate the location and the relative intensities of the hyperfine components (Ref. 20).

Table 9. The static E2 and M1 moments of the ground and the excited states (15/2$^+$ and 9/2$^+$) of ^{209}Bi.

Level	E2 Moment			M1 Moment		
	This Exp.	Atomic Beam	Core Excitation	This Exp.	Core Excitation	Other Exp.
Ground State 9/2$^-$	-0.37+0.03	-0.31				4.079
Excited 9/2$^-$	+0.11+0.05		+0.30	3.52+0.70	3.75 n.m.	
Excited 15/2$^+$	-0.93+0.40		-1.31	6.2 + 1.2	5.90 n.m.	

shifts and hyperfine effects. The isomer shifts in the case of ^{209}Bi were observed by CERN and Columbia as shown in Table 10. The energy shifts as large as 6-7 keV are observed in this case. This gives a beautiful example of how two otherwise independent systems can be coupled together quantum mechanically.

D. Strong Interactions and Nuclear Matter Distributions

In this section, we will briefly discuss only the K$^-$ atom, as the π$^-$ atom has been extensively reviewed in physics journals.[2,9,13] The absorption of the kaon is quite different from the absorption of the pion. A pion is absorbed predominantly on a pair of nuclei for reasons of energy and momentum conservation; therefore its

Table 10. Isomer shifts in muonic ^{209}Bi (Columbia 1970).

$J_i^\pi - J_f^\pi$	Energy (μ-x Exp.) (keV)	Energy (Coulomb Excitation) (keV)	Muon in the 1s state Isomer shift (keV)		Muon in the 2p state isomer shift (keV)**	Probability of Excitation
			uncorrected	corrected*		
$15/2^+ \to 9/2^-$	2746.41±0.22	2741.4±0.5 (Rochester) 2740.0±0.5 (Minnesota)	+5.7	+6.5	3.4 ± 0.8	3.7%
$9/2^+ \to 9/2^-$	2570.47±0.25	2563.0±1.0 (Rochester) 2563.0±1.0 (Minnesota)	+7.47	+7.47	3.8 ± 0.8	3.1%
$15/2^+ \to 13/2^+$	1135.51±0.13	1132.5±1.0 (Rochester) 1132.0±1.0 (Minnesota)	+3.26	+3.0	---	----
$13/2^+ \to 9/2^-$	1610.97±0.13	1608.9±0.5 (Rochester) 1608.0±1.0 (Minnesota)	+2.52	+3.4	---	----

* corrected for the fast M1 inter-doublet transition

** determined from the best fit to the experimental data

absorption is proportional to ρ^2 and is weak in the nuclear periphery where the density ρ is low. In contrast, the kaon capture
occurs mostly on a single nucleon through the reaction:

$$K + N \rightarrow Y + \pi$$

where Y is a Σ or Λ hyperon and N is a nucleon. As you can see the
Σ^-'s are produced in the absorption process of K^- and will be slowed
down and form Σ^- atoms in the same target of K^- atoms. Thus K^- and
Σ^- X-ray spectra always appear side by side in the same spectrum.

The absorption interaction between the kaon and the nucleus is
expected to be proportional to the nuclear density, rather than the
square of the nuclear density as in pionic atoms. This interaction
is very strong, such that if the kaon penetrates deeply into the
nucleus, there will be so much absorption that sharp and discrete
kaonic levels do not exist. Only if the kaon has a relatively small
overlap with the nucleus, can one expect to have well-defined orbits.
Therefore, the kaon orbits will have appreciable overlap with the
nucleus only on the nuclear surface. Figure 10 shows the overlap
integral

$$\left| \phi_{n_c}(r) \right|^2 r^2 \rho(r)$$

for the critical orbits for the K^-, $\bar{p}$ and Σ^- in $^{184}_{71}W$ using hydrogen wave functions.[31] The critical orbit (n_c, $\ell = n_c-1$) is defined
for a given element as the highest state from which X-ray transitions are unobservable. It can also be seen from Fig. 10 that as
the particle mass increases from K^- to Σ^-, the more distant regions
of the nucleus are explored.

In a crude approximation, the K^- nuclear potential can be
expressed in terms of the free K^-N S-wave scattering lengths for
isospin zero (A_0) and one (A_1) such as in:

$$2_\mu V_N(r) = -4\pi(1 + \frac{m_K}{m_N}) \; [A_1\rho_n(r) + \frac{A_0+A_1}{2} \rho_p(r)]$$

where ρ_n and ρ_p are the neutron and proton densities. Using this
approach several authors[32-34] have shown that by using a K^- bound
nucleon amplitude extrapolated from the low energy K^- free nucleon
scattering amplitude, it is possible to explain Wiegand's experimental results[1,35] on K^- atoms. They emphasized that it is not
necessary to propose the existence of a neutron halo outside the
nucleus as did Wiegand et al.[1,35] Ericson and Scheck[33] calculated
the yields with a neutron distribution which is the same as the
proton distribution, i.e., a two parameter Fermi distribution with
the parameters determined from the muonic X-ray and electron
scattering experiment. Bethe and Siemens[32] used a more complicated

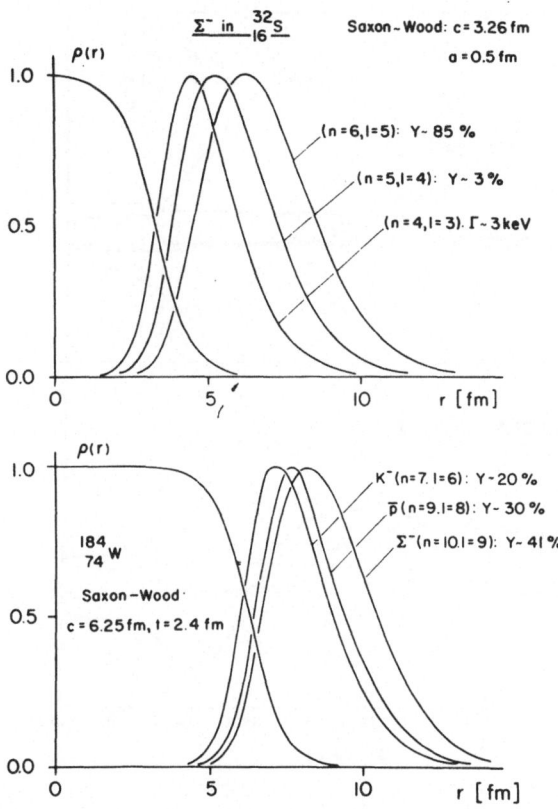

Fig. 10.   Showing the overlap integrals $\left|\phi(r)\right|^2 r^2 \rho_N(r) n_c$
$\ell = n_c - 1$ of Σ^- in $_{16}S$ and also K^-, $\bar{p}$, and Σ^- in $_{74}^{184}W$ (Ref. 28).

neutron distribution which has the same asymtotic exponential tail
as the single particle wave functions. Bardeen and Torigoe[34] used
the independent particle model, and the neutron distribution is
obtained from single particle levels in a potential well. However,
the presence of a Y_o^* (1304 MeV) resonance for K^-p which is located
only 27 MeV below the threshold energy and has a width of 40 MeV
greatly complicates the interpretation (see Fig. 11). The influence
of the Y_o^* resonance will be important if the effective energy of
the K^-N system in the nucleus is way below the threshold. The
problem is: How can one use the _free_ K-N reaction data to determine the K^--N interaction in nuclear media with a Y* strong resonance close by? First of all, one must find kinematically the
distribution of the K^-N system with respect to the total center of
mass energy W.

The strong interaction has both an elastic part and an absorptive part. Therefore, the energy of the level is shifted and the

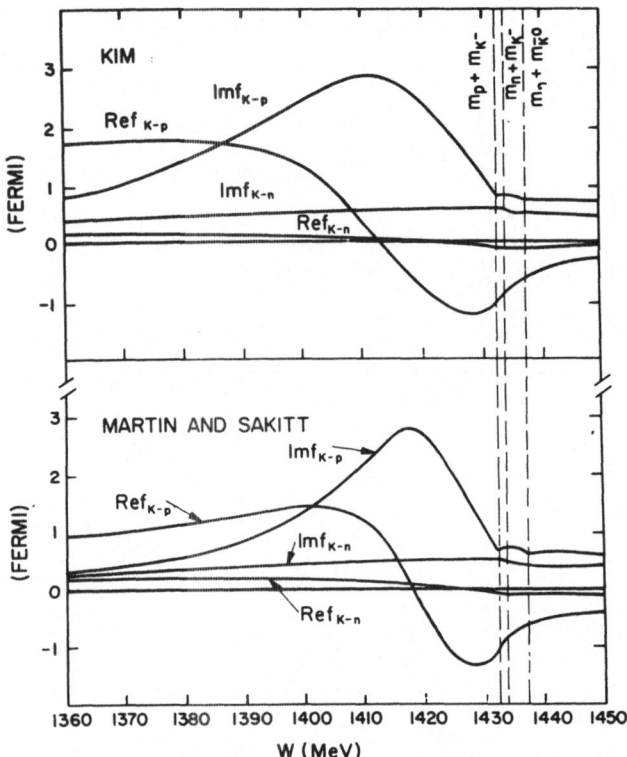

Fig. 11. The energy dependence of the free K⁻ nucleon scattering
amplitudes for the fits of Kim and of Martin and Sakitt (Ref. 36).

width is increased. Furthermore the strongly absorptive potential
has a large distortion effect on the low kaonic orbits and makes
the perturbation treatment unreliable. The precision measurements
of the energy shifts and the line broadening together with the
relative intensities of the last two observable transitions are
extremely important in obtaining the K-nucleus scattering length.
Backenstoss et al.[37] have carried out such measurements with high
precision on a series of elements. The principal facts are:
(1) The energy shifts are all repulsive, and (2) the line broadenings observed are systematically bigger by a factor of two to
three in comparison with the theoretical predictions based on the
approximate potential shown above. No reasonable modifications
of the nuclear density can explain this anomaly.

The unusually large line broadening observed can be due to
several factors. One of them is that it may be due to the influence
of the Y_0^* resonance just below the threshold which makes the

nucleons bound and not free. The other, proposed by Krell[37] is that the real part of the potential V (ReV) is actually attractive not repulsive. The observed repulsive effect could be due to the effect of strong absorption. When the overlap between the particle and the nucleus is strong, the repulsion due to the strong absorption could be dominating and would give a repulsive effect even with the real part of the potential (ReV) positive. Very recently, several theoretical investigations[31],[34] have shown that all these lines of attack are informative, productive and mathematically soluble. It is clear that the problem of the influence of resonances in a bound system on the K^--N scattering length must be understood first. When this knowledge has been attained, then the K^- and $\bar{p}$-atoms should be very useful tools for the study of nuclear structure as we have anticipated for a long time.

This work was supported in part by the National Science Foundation.

REFERENCES

1. C. E. Wiegand and D. A. Mack, Phys. Rev. Letters 18, 685 (1967).
2. G. Backenstoss, T. Bunaciu, S. Charalambus, J. Egger, H. Koch, A. Bamberger, U. Lynen, H. G. Ritter, and H. Schmitt, Phys. Letters 33B, 230 (1970).
3. A. Bamberger, U. Lynen, H. Piekarz, J. Piekarz, B. Povh and H. G. Ritter; G. Backenstoss, T. Bunaciu, J. Egger, W. D. Hamilton, and H. Koch, Phys. Letters 33B, 233 (1970).
4. R. E. Shafer, Phys. Rev. 163, 1451 (1967).
5. J. Lathrop, R. Lundy, V. L Telegdi, R. Winston, D. D. Yovanovich and A. J. Breaden, Nuovo Cimento 17, 109, 114- (1960).
6. S. Devons, G. Gidal, L. M. Lederman, and G. Shapiro, Phys. Rev. Letters 5, 330 (1960).
7. F. J. M. Farley, J. Bailey, R. C. A. Brown, M. Giesch, H. Joestlein, S. van der Meer, E. Picasso, and M. Tannebaum, Nuovo Cimento 15A, 281 (1966).
8. D. P. Hutchinson, J. Menes, G. Shapiro, A. M. Patlach, and S. Penman, Phys. Rev. Letters 7, 129 (1961).
9. G. Backenstoss, Ann. Rev. Nucl. Sci. 20, 467 (1970).
10. R. Kunselman, Phys. Letters 34B, 485 (1971).
11. A. Rittenberg et al., Rev. Mod. Phys. (Suppl.) 43, S1 (1971).
12. J. D. Fox, P. D. Barnes, R. A. Eisenstein, W. C. Lam, J. Miller, R. B. Sutton, D. A. Jenkins, R. J. Powers, M. Eckhause, J. R. Kane, B. L. Roberts, M. E. Vislay, and R. E. Welsh, and A. R. Kunselman, Phys. Rev. Letters 29, 193 (1972).

13. G. Backenstoss, S. Charalambus, H. Daniel, Ch. Von Der Malsburg, G. Poelz, H. P. Povel, H. Schmitt, and L. Tauscher, Phys. Letters 31B, 233 (1970).

14. M. S. Dixit, H. L. Anderson, C. K. Hargrove, R. J. McKee, D. Kessler, H. Mes, and A. C. Thompson, Phys. Rev. Letters 27, 878 (1971).

15. J. Blomqvist, "Vacuum Polarization in Exotic Atoms," 1972, preprint.

16. B. Fricke, Z. Physik 218 495 (1969).

17. H. K. Walter, J. H. Vuilleumier, H. Backe, F. Boehm, R. Engfer, A. H. V. Gpunten, R. Link, R. Michaelsen, C. Petitjean, L. Schellenberg, H. Schneuwly, W. U. Schröder, and A. Zehnder, Phys. Letters 40B, 197 (1972).

18. A. Digiacomo, Nucl. Phys. B 11, 411 (1969).

19. E. Campani, Lett. Nuovo Cimento 4, 512, 982 (1970).

20. A. Placci, E. Polacco, E. Zavattini, K. Ziock, C. Carboni, U. Gastaldi, G. Gorini, and G. Torelli, Phys. Letters 32B, 413 (1970).

21. A. Placci, E. Polacco, E. Zavattini, K. Ziock, G. Carboni, U. Gastaldi, G. Torini, G. Neri, and G. Torelli, Nuovo Cimento 1A, 445 (1971).

22. B. Budick, J. R. Toraskar, and I. Yaghoobia, Phys. Letters 34B, 539 (1971).

23. M. Leon, Phys. Letters 35B, 413 (1971).

24. V. Hughes and C. S. Wu, et al., Proposal to Nevis Laboratory, Columbia University (1971).

25. S. Koslov, V. L. Fitch, and J. Rainwater, Phys. Rev 95, 291 (1954).

26. W. Y. Lee, M. Y. Chen, S. C. Cheng, E. R. Macagno, A. M. Rushton, and C. W. Wu, Nucl. Phys. 3 C, 14 (1972).

27. R. Engfer, in High Energy Physics and Nuclear Structure, ed., S. Devons (Plenum, New York, 1970) p. 115.

28. A de-Shalit, Phys. Rev. 122, 1530 (1961).

29. J. W. Hertel, D. G. Fleming, J. P. Schiffer, and H. E. Gove, Phys. Rev. Lett. 23, 488 (1969).

30. R. A. Broglia, J. S. Lilley, R. Derazzo, and A. W. Phillips, Phys. Rev. C 1, 1508 (1970).

31. T. E. O. Ericson, Lectures delivered at the Joensuu Summer School on Nuclear Physics, July, 1971 (notes prepared by H. A. Schmitt).

32. H. A. Bethe and P. J. Siemens, Nucl. Phys. B 21, 589 (1970).

33. T. E. O. Ericson and F. Scheck, Nucl. Phys. B 19, 450 (1970).

34. W. A. Bardeen and E. W. Torigoe, Phys. Rev. C 3, 1785 (1971).

35. C. E. Wiegand, Phys. Rev. Letters 22, 1235 (1969).

36. W. A. Bardeen and E. W. Torigoe, Phys. Letters 38B, 135 (1972).

37. G. Backenstoss, A. Bamberger, I. Bergström, P. Bounin, T. Bunaciu, J. Egger, S. Hultberg, H. Koch, M. Krell, U. Lynen, H. G. Ritter, A. Schwitter, and R. Stearns, Phys. Letters 38B, 181 (1972).

38. M. Krell, Phys. Rev. Letters 26, 584 (1971).

HIGHLY EXCITED STATES OF HELIUM AND NEON*

W. H. Wing, K. R. Lea† and W. E. Lamb, Jr.

Department of Physics, Yale University

New Haven, Connecticut 06520

I. INTRODUCTION

For the past two years we have been studying the microwave spectra of the high states of singly excited helium[1,2,3] and very recently have observed microwave resonances in highly excited neon as well.[4] The story of how we got into this field, more or less by accident, has been given in Ref. 2, along with a survey of highly excited state phenomena.

To the highly excited electron, the remainder of the atom appears as a core of relatively small dimensions, with charge +e if the atom is neutral, or $(Z-N_e+1)e$ in general. In many respects this atom is similar to a hydrogen atom in a high quantum state. Some relevant properties of such a system are listed in Table I. It is evident that the various quantities depend in very different ways on the principal quantum number n. Thus the behavior of the atom may be expected to change markedly as n increases. On a coarse scale, the energy of the general highly excited atom relative to its ionization limit is usually represented by the Rydberg formula

$$E_n = -R/(n+\delta)^2 \qquad (cm^{-1}) \quad . \tag{1}$$

*Research sponsored by the Air Force Office of Scientific Research under Contract No. F44620-71-C-0042.
†Present address: Dept. of Physics, University College of Swansea, Swansea SA2 8PP, U.K.

TABLE I. Some approximate properties of the hydrogen atom ($\ell \ll n$).

Excitation rates Decay rates Level spacings	$\propto$ $1/n^3$
Number of sublevels	$\propto$ n^2
Number of matrix elements	$\propto$ n^4
Mean atomic radii Electric dipole matrix elements ($\Delta n = 0$)	$\propto$ n^2
Stark shifts	$\propto$ $\mathcal{E}^2 n^7$
Motional field broadening[a]	$\propto$ $(\frac{v}{c})^2 \mathcal{K}^{1-2} n^{4-7}$

Zeeman effect

$$H_L = \mu_o \vec{\mathcal{K}} \cdot (g_L \vec{L} + g_S \vec{S})$$

"Strong-field" electric-dipole rf transitions
        Slopes of 1.4 MHz/Gauss

Quadratic Zeeman effect

$$H_Q = \frac{e^2 \mathcal{K}^2}{8mc^2} r^2 \sin^2 \theta \qquad\qquad \propto \mathcal{K}^2 n^4$$

[a]Exponents of n and $\mathcal{K}$ depend on the relative sizes of $\mu_o \mathcal{K}$
and the fine structure.

Here the Rydberg constant R contains the reduced mass of the elec-
tron-atomic core system. The quantum defect δ is characteristic
of the atom. On this coarse scale, i.e., neglecting relativistic
fine structure, $\delta = 0$ for hydrogen. For other atoms, δ is a slowly
varying function of n or E_n, but decreases rapidly with increasing
orbital angular momentum $\ell = 0,1,2 \ldots n-1$. As $n \to \infty$, δ_ℓ approaches a constant value. According to the one-channel quantum
defect theory,[5] $\pi \delta_\ell = \phi_\ell$, the ℓth partial wave phase shift (relative to hydrogen) in the scattering of a slow free electron by the
charged atomic core.

Figure 1 shows the energy levels of the helium system, and
Fig. 2 an expansion of the region between 20 and 25 eV containing

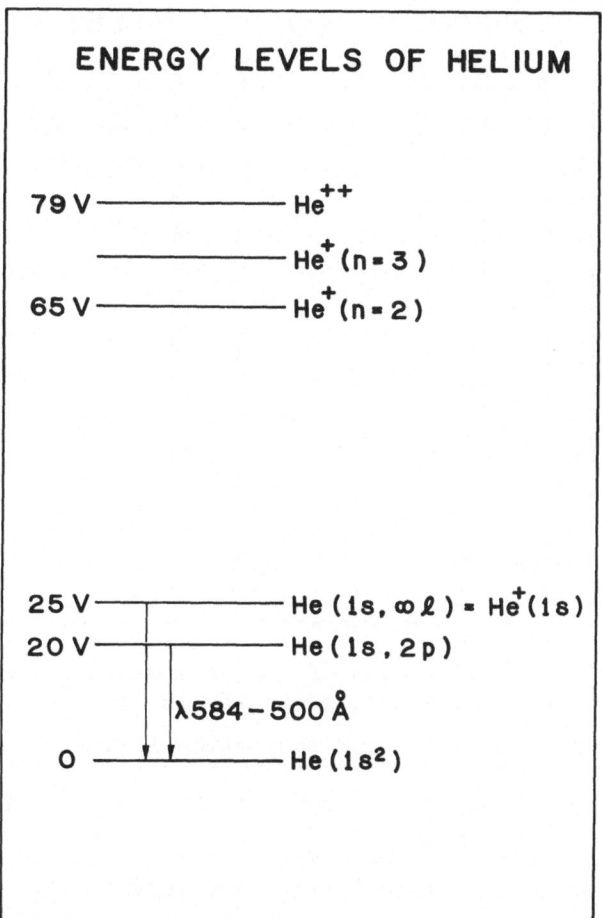

Fig. 1.   Energy levels of the helium atom and ion.   States with two
excited bound electrons are not shown.

the highly excited states 1snℓ of the neutral atom. The dotted
lines in Fig. 2 are the n = 2 and 3 hydrogenic levels. Quantum defects for helium extracted from the NBS energy level tables of
Moore[6] and Martin[7] are shown in Fig. 3 for the known spectral terms.
The erratic behavior of the D, F and G defects is presumably caused
by noise in the data, and the falloff of the S and P defects as
n → ∞ by a slightly incorrect value for the helium first ionization
energy W_∞.

The quantum defects in helium result from electrostatic interactions, the largest of which are: screening of the inner electron
by the outer, electron exchange, and core polarization. Approxi-

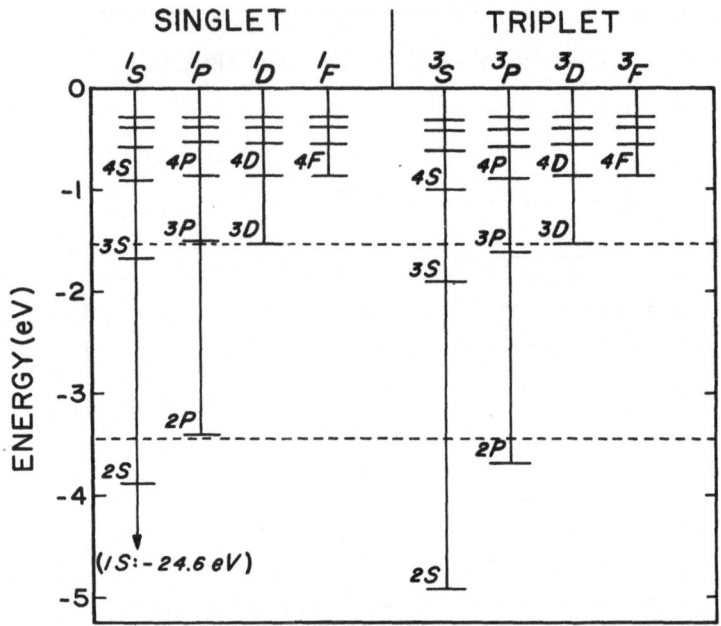

Fig. 2. Singly excited states of the helium atom. The dotted lines indicate two of the hydrogenic levels for Z = 1. Energies in eV are measured from the helium ionization limit (= 1S of He$^+$).

mate asymptotic values for $|\delta|$ and a listing of the contributions to $|\delta|$ for D states appear in Table II. The relative sizes of the effects change greatly with ℓ. For most of the excited states of helium the quantum defect splittings are very small. We have termed these splittings <u>electrostatic fine structure</u>,[2] to contrast them with the usual <u>relativistic fine structure</u>, which includes spin-spin, spin-orbit and quantum electrodynamic contributions. For F states and states of higher ℓ, exchange effects are comparable to or even less than the spin-orbit interactions. Thus, in these states, considerable single-triplet mixing occurs, and the familiar singlet-triplet classification of the levels no longer has much validity. In S, P and D states, however, total spin is a fairly good quantum number.

Relativistic fine structure intervals in helium have been measured directly to at least n = 9 by many workers.[8] By contrast, prior to our work the only electrostatic intervals known from direct observation were the triplet and singlet 2P-2S emission lines at 1 and 2 μm, respectively,[7] and the 95.8-μm $3\,^1P_1$-$3\,^1D_2$

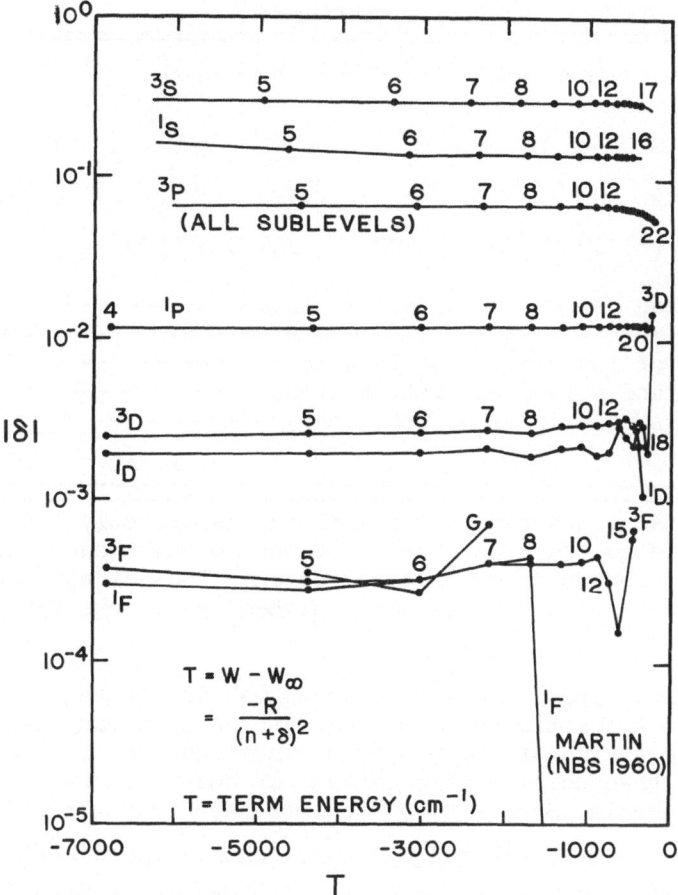

Fig. 3 Summary of quantum defect values for helium obtained from NBS atomic energy tables (Martin, Ref. 7). The data for G states were taken under conditions of strong Stark mixing, and have no fundamental significance.

and 216.3-μm 4^1P_1-4^1D_2 laser transitions.[9] Other intervals must be obtained by taking differences of the wave numbers of optical frequency transitions. These values rapidly lose accuracy as n and ℓ of the excited electron increase and the intervals shrink, as can be seen from the erratic behavior in Fig. 3. The transition energy between two states ℓ and ℓ' of the same n is approximately

$$\nu_{\ell\ell'} = \frac{2cR\,(\delta_\ell - \delta_{\ell'})}{n^3} \qquad (Hz) \qquad (2)$$

TABLE II.   Quantum defects in helium.

(a)   APPROXIMATE VALUES OF $|\delta|$

n^3S	n^1S	n^3P	n^1P	nD	nF	nG ------
.30	.14	.07	.01	.003	.0003	5×10^{-5} --

(b)   CONTRIBUTIONS TO $|\delta|$ FOR D STATES

Polarization of core by outer electron-------------	.0024
Exchange--	.0007
Screening of inner electron by outer----------------	.0002
Spin-spin and spin-orbit interactions---------------	.0001
Other (mass polarization, QED, etc.)-------------- $\leq$	.00002
TOTAL	$\sim$ .003

and thus contains quantum defect differences directly. Above the lowest few states, such as those just mentioned, these intervals lie in the microwave or even radio region. Thus direct microwave experiments can measure these small high-n splittings very accurately.

One serious problem which deserves mention is that of locating the microwave transitions. The linewidths are in many cases so narrow that a search of a few percent in frequency takes a long time. The high-precision calculations of helium energy levels by Hylleraas, Pekeris, Schwartz, and others have concentrated on states of low quantum number (although Accad, Pekeris and Schiff[10] have given theoretical He S and P ionization energies as high as 15^1S, 17^3S, and $5^{1,3}$P to the nearest 300 MHz. The more approximate calculations[11] of quantum defects have an accuracy comparable only to that of conventional spectroscopic experimental methods. Our experimental accuracy, in the best cases, is of higher precision by several orders of magnitude.

For prediction of high-state microwave splittings we have adopted a mixed experimental-theoretical procedure based on the quantum defect model. The quantum defects for S, P and D states are obtained from Seaton's[12] multi-parameter fit to the NBS data of Martin[7], which results in smoothing the spectroscopic data. The quantum defects for the F and higher-ℓ states are obtained from Edlén's[13] polarization model, which, despite its simplifications, is regarded[14] as more reliable than the spectroscopic data for these states. A chart of He singlet transition frequencies pre-

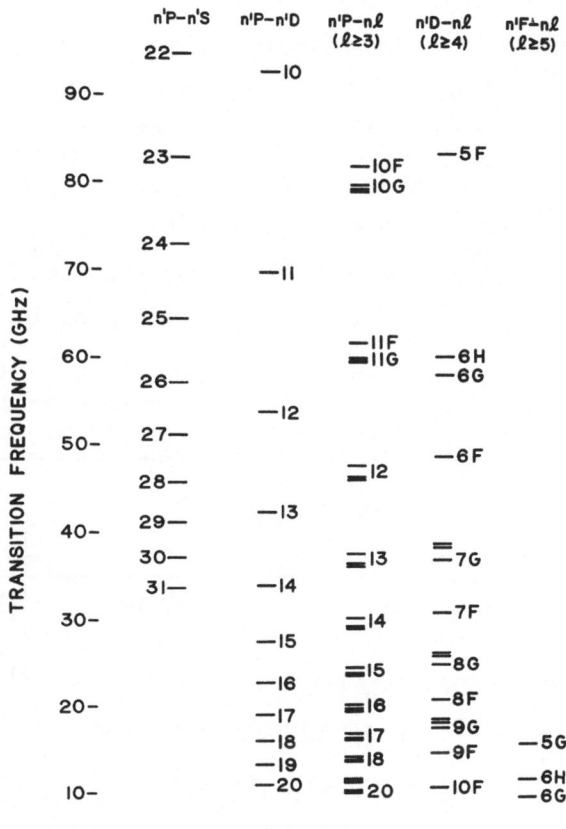

Fig. 4. Expected frequencies for various helium singlet transitions. For S, P and D states, term energies were reconstructed from the spectroscopic data via Seaton's quantum defect fit (Ref. 12). For F and states of higher L, Edlén's polarization model was used (Ref. 13).

dicted by this model is given in Fig. 4. The prediction accuracy can be judged from the results of our work on He, n > 7, shown below in Table IV.

II. THE MICROWAVE-OPTICAL RESONANCE METHOD
AND PRECISION HELIUM RESULTS

The experimental technique we use is the microwave-optical resonance method.[15] The apparatus is similar to that used in recent experiments[16] on the fine structure of the H atom and He^+ ion, and is described in more detail in those publications. We will describe it in the configuration appropriate to our recently completed precision measurement[3] of the intervals $1s7d\,^1D_2 - 1s7f\,^1F_3$ and $1s7d\,^1D_2 - 1s7f\,^3F_3$ in He^4 near 31.5 GHz. The latter transition is spin forbidden, but appears through singlet-triplet mixing in the F states.[17]

The apparatus is shown in Fig. 5, with typical operating conditions listed in the caption. Excited atoms are produced in an X-band waveguide by bombardment from a unipotential-cathode elec-

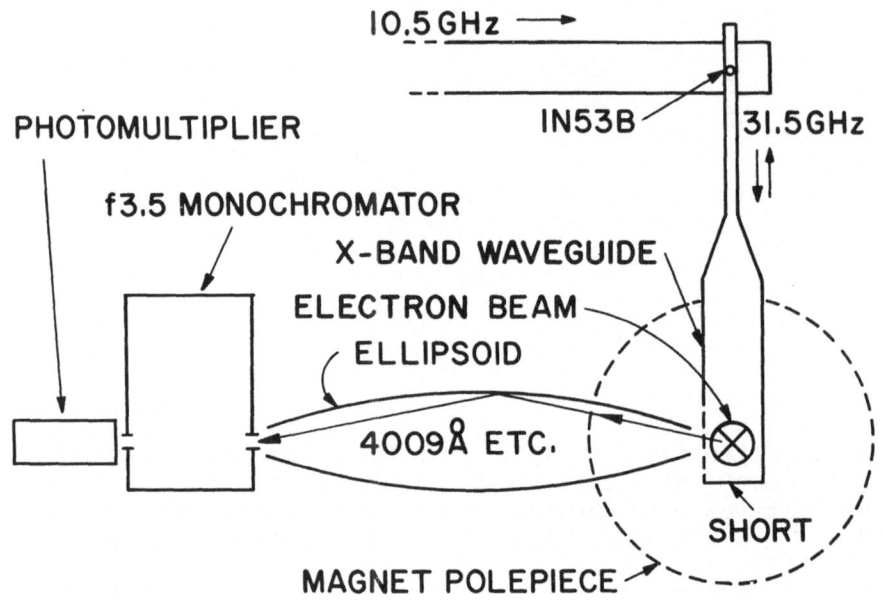

Fig. 5. Apparatus. The magnetic field (when used) and bombarding electron beam are normal to the plane of the figure. For precision work the magnetic field is carefully adjusted to zero. The following operating conditions are typical of precision runs: bombarding current, 200 μA; voltage, 60 V; helium pressure, 1.5 mTorr; incident microwave power, 0.3 μW; residual magnetic field, <0.2 G.

tron gun operated at a voltage near the peak of the 7^1D excitation cross section. Because of the relative excitation and decay rates, 7^1D states are more populated than 7F. Decay light is collected by a hollow ellipsoidal light pipe with polished metal walls. A fast (f3.5) 1/4-m monochromator coupled to a blue-sensitive photomultiplier monitors a single visible emission line. Microwave-induced electric dipole transitions produce a net population transfer out of the 7^1D state, and hence a reduction in the intensity of the 4009-Å line 7^1D-2^1P. The microwave power is switched on and off at 37 Hz, and the resonance signals are synchronously detected.

The microwave source used varies greatly with the application. Since the highly excited states have large electric-dipole matrix elements and long lifetimes, relatively little power is required unless strong broadening mechanisms are present. For the n = 7 work, in the absence of such mechanisms, a harmonic generation source was adequate, which consisted of a 1N53B-diode tripler (we used a commercial R-band harmonic mixer) driven by a stabilized 10.5 GHz X-band oscillator. The 31.5 GHz power (a few micro-watts) was enough to saturate the transitions thoroughly.

A 30.5 cm magnet was used to search for transitions by Zeeman tuning. For singlet states it provides a 24-GHz equivalent tuning range (8500 G at ±1.4 MHz/G). However, at even a few hundred gauss the resonances are broadened greatly by the thermal distribution of motional Stark shifts, as described in Section V; thus much more power is required to exhibit the resonance clearly. For searching we used a 50-mW 32-37 GHz klystron.

For precision work the magnetic field was carefully cancelled. Figure 6 shows a highly power broadened resonance panoramic at zero magnetic field. If power and other broadening effects are reduced, the resonance widths approach an average limiting value of around 1.9 MHz, which is slightly greater than the 1.2 MHz hydrogenic value[11] for the sum of the 7D and 7F natural widths. We do not yet fully understand the excess linewidth.

Figure 7 shows the positions of the He^4, n = 7 sublevels with $\ell \geq 2$, as determined from our mixed experimental-theoretical quantum defect calculation, and the fine structure of the 7F term. The total angular momentum J of the F state can be 2, 3 or 4. The two states with J = L (here 3) are singlet-triplet mixtures, and thus undergo microwave transitions to 7^1D_2; the other two F states are pure triplet, and do not appear in the resonance curve of Fig. 6. It should be noted, however, that in the presence of a magnetic field of a few tens of gauss J is no longer a good quantum number.

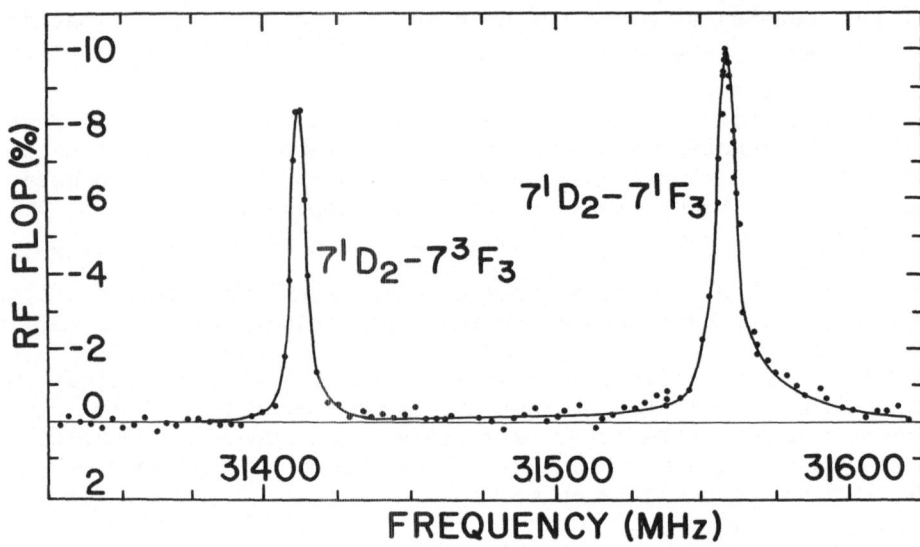

Fig. 6. Microwave resonances observed near 31.5 GHz in the 4009-Å
emission from helium. Conditions for these data were similar to
those listed in the caption of Fig. 5, except for much higher
microwave power (4-6 μW incident) which broadened the resonances to
≈ 8 MHz (full width at half-maximum). A curve has been drawn
through the data points as a guide to the eye.

In this case some of the magnetic sublevels leading to pure triplet
zero-field states will take on a partially singlet character and
can be reached by allowed transitions from 7^1D.

In the Lamb-Sanders-Wilcox model[15] for rf resonances in a decaying two-level system, the change induced by an rf perturbation
$V = <a|e\vec{\mathcal{E}}\cdot\vec{r}|b>$ in the intensity of the observed decay light is a
Lorentzian function. If r_a, γ_a, f_a and r_b, γ_b, f_b are the excitation rates, decay rates, and fractions of detected decays of levels
a and b respectively, and if the level separation and rf frequency
are ω and ν respectively, the observed signal is

$$S_{ab} = \frac{-\frac{1}{4}(f_a - f_b)(\frac{r_a}{\gamma_a} - \frac{r_b}{\gamma_b})(\gamma_a + \gamma_b)\ |V|^2}{(\omega-\nu)^2 + \frac{1}{4}(\gamma_a + \gamma_b)^2(1 + |V|^2/\gamma_a\gamma_b)} \ . \tag{3}$$

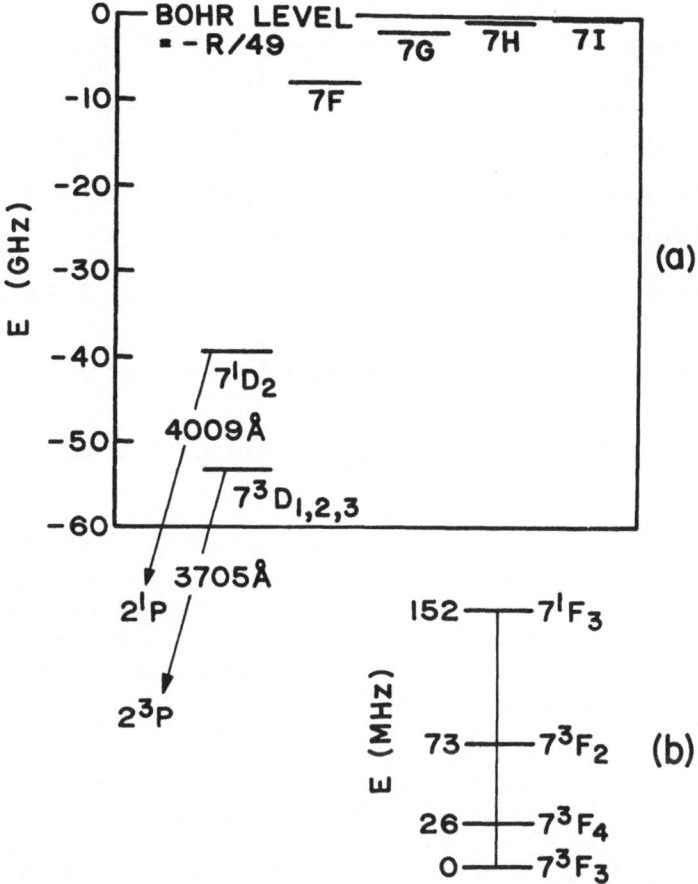

Fig. 7. (a) Structure of singly excited He[4], n = 7, according to a quantum-defect fit for the D states (Ref. 12) and a polarization model for states of higher L (Ref. 13). The S and P states are beyond the boundaries of the figure. (b) Fine structure of the 7F term calculated using the method of Araki (Ref. 17). For L ≤ 2 the exchange splitting is much larger than the spin-orbit interactions, so the 7D term is composed of essentially pure singlet and triplet states. In the 7F term the exchange and spin-orbit interactions are comparable, so the two J = L states become superpositions of 7^1F_3 and 7^3F_3 basis states, while the 7^3F_2 and 7^3F_4 states remain pure triplet. We apply the designation "singlet" to the upper J = L state, which has the larger singlet component.

TABLE III. Various results for 7^1D_2-$7F$ intervals in He^4.
The 7F-term sublevels are not resolved in previous work.
These data have been published (Ref. 3).

Source	Transition	Interval (GHz)
Optical data[a]	7^1D_2-$7F$ mean	33
D-state experimental,[b] F-state theoretical[c] values	7^1D_2-$7F$ mean	31.1
Present work	7^1D_2-7^1F_3	31.55826(10)
	7^1D_2-7^3F_3	31.41207(8)

[a]Ref. 7.

[b]Quantum-defect fit (Ref. 12) to spectroscopic data (Ref. 7).

[c]Polarization model, Ref. 13.

In actual practice the lineshape is complicated by Doppler broadening and, for long-lived states, truncation of the decay exponential as the atom passes out of view of the detector. In addition, the resonance is affected by perturbations. In our experimental conditions these effects were relatively small for n = 7, and in any event the first two complications do not disturb the symmetry of the resonance curve as a function of frequency. To eliminate known perturbations we took 13 runs which were fitted to an equation such as (3), and extrapolated to zero operating conditions (microwave power, bombarding current and pressure). The results appear in Table III together with the best information from previous work. It can be seen that the mixed experimental-theoretical quantum defect model value is somewhat closer than the raw optical data value, presumably because of the smoothing effect of Seaton's least-squares analysis[12] of the D state data.

The 7^1F_3-7^3F_3 splitting of 146.19(13) MHz has not been resolved previously. The best theoretical estimate of this quantity appears to be implicit in the 1937 work of Araki[17], from which T. O. Siu[18] has kindly calculated for us the value 152 MHz, as shown in Fig. 7(b). However, a calculation by Parish and Mires[19] with a similar theoretical basis has given approximately 285 MHz, which is considerably farther off. More theoretical work along this line seems desirable.

TABLE IV. Summary of precision results in He^4. The n=7 results have been reproduced from Table III.

Optical Line Monitored (Å)		Microwave Transition Observed (GHz)		Quantum Defect/ Polarization Theory Value
7^1D–2^1P	4009	7^1D$_2$–7^1F$_3$ 7^1D$_2$–7^3F$_3$	31.55826(10) 31.41207(8)	31.1
10^1D–2^1P	3834	10^1D$_2$–10^1F$_3$ 10^1D$_2$–10^3F$_3$	10.91650(100) 10.86520(100)	10.9
11^1D–2^1P	3806	11^1D$_2$–11^1F$_3$ 11^1D$_2$–11^3F$_3$	8.21300(100) 8.17350(100)	8.3
9^3D–2^3P	3587	9^3D–9F av.(?)	21.76	21.8
20^1P–2^1S	3148	20^1P$_1$–20^1D$_2$(?)	11.01	10.6

In a magnetic field of at least a few hundred gauss, the motional electric field mixes states of different L enough to bring out rf transitions with $|\Delta L|>1$. In this way we have seen the transitions 7^1D$_2$–7G and 7^1D$_2$–7H and have obtained the intervals ν(7F–7G) = 5.60(7) GHz and ν(7G–7H) = 1.06(7) GHz. Here the relativistic fine structure is obscured by the motional Stark broadening of the ensemble.

Recently we have measured several additional transition frequencies in states as high as n = 20. All our precision results to date are summarized in Table IV. The simple fitting formula

$$\nu = \frac{A}{n^3} + \frac{B}{n^5} \tag{4}$$

will reproduce the n^1D–$n^{1,3}$F transition frequencies very closely with A and B as follows (Table V).

TABLE V. Constants for Eq. (4). The values listed were calculated at n = 7 and 11. They give the n = 10 transitions within ±1 MHz.

Transition	A	B
n^1F$_3$ – n^1D$_2$	11004.34	–8812.9
n^1F$_3$ – n^3F$_3$	54.23	–200.2

III. NEON

In neon, a survey of the spectroscopic data compiled by Charlotte E. Moore[6] and more recently by Kaufman and Minnhagen[20] revealed a number of cases where energy levels of opposite parity are separated by only 1 or 2 cm^{-1}. Examples of this are the intervals 7X to $7d_1''$ and 7X to $7d_1'$, using Paschen notation. These levels arise when one of the six electrons in the 2p shell is excited to an f or a d level. The five electrons remaining in the 2p shell couple to form a $^2P_{3/2}$ or $^2P_{1/2}$ state, which describes the "core" about which the excited electron moves. The 7X and 7d levels noted above have in common a $^2P_{3/2}$ core. When the angular momentum of the excited electron is added to that of the core, states of definite J-value are obtained. The state designated 7X in Paschen notation actually comprises two levels, with J values of 1 and 2, which have not been resolved spectroscopically. Table VI summarizes the information pertaining to these levels and their separations.

We initiated a search for the microwave transition $7d_1'$ to 7X in neon, tuning the monochromator (Fig. 5) in turn to some of the known visible wavelengths produced when the $7d_1'$ level decays to the lowest excited group of p-levels. This search was unsuccessful

TABLE VI. 7X and 7d levels in neon.

Energy Level			Displacement of Level from 7X	
Paschen	Configuration	J	Moore[a]	Kaufman and Minnhagen[b]
7X	$(^2P_{3/2})$ 7f	1,2	—	—
$7d_1'$	$(^2P_{3/2})$ 7d	3	1.05 cm^{-1}(31.5 GHz)	1.08 cm^{-1}(32.4 GHz)
$7d_1''$	$(^2P_{3/2})$ 7d	2	1.30 cm^{-1}(39.0 GHz)	1.33 cm^{-1}(39.9 GHz)

[a]Ref. 6
[b]Ref. 20.

Next we looked for the $7d_1$" - 7X transition. A weak microwave resonance was observed as a small decrease in the optical decay radiation $7d_1$" - $2p_7$ at 4636Å. Typical experimental conditions under which this signal was obtained include the following: electron beam current of 0.3 mA at 75 V; neon pressure of 9 millitorr (uncorrected Pirani gauge reading); incident microwave power 40 mW. The signal-to-noise ratio was about 3:1, obtained with a lock-in detector time constant of 10 seconds. Line widths, as far as we could estimate from the weak signals and somewhat erratic line shapes, were 80 to 100 Gauss (FWHM).

A systematic investigation of the position of these microwave signals as a function of magnetic field was undertaken. Figure 8 summarizes the results. The crosses indicate the estimated centers and line widths in gauss. The vertical portion of the cross indicates an uncertainty in the microwave frequency which was in this case measured by a cavity wavemeter. Over the frequency range of klystron oscillators available to us at the present time, the observed centers point to a zero field interval of 38.44(10) GHz. It is likely that the linear extrapolation shown in the figure is not rigorously correct. The ascending dashed lines, originating from the two zero field intercepts, are drawn with slopes of reversed algebraic sign to those of the corresponding solid lines. It is seen that the two resonance centers observed with a klystron at 39.78 GHz do not quite fall on these constructed lines.

The interpretation of these neon data is presently incomplete. It is plausible to associate the extrapolated zero-field interval with the $7d_1$" - 7X interval. In support of this, the signal was lost when the monochromator was detuned from the $7d_1$" decay line.

The raw experimental result of 38.44(10) GHz is to be compared with the values 39.0 and 39.9 GHz (noted in Table VI) derived from earlier spectroscopic data. Whether any significance can be attached to the slightly different extrapolated values in Fig. 8 is doubtful, though they might reflect the different energies of the hitherto unresolved J = 1 and J = 2 7X levels. More work is needed to measure the resonances closer to zero field, and to secure a better signal-to-noise ratio in order to study systematic effects. The use of high microwave powers and quite high neon gas pressures has proven necessary in order to detect the resonances. It would be desirable to try to work at lower pressure, lower microwave power, at weak or zero magnetic fields and with smaller bombarding currents in order to diminish these sources of perturbation on the highly excited atomic states.

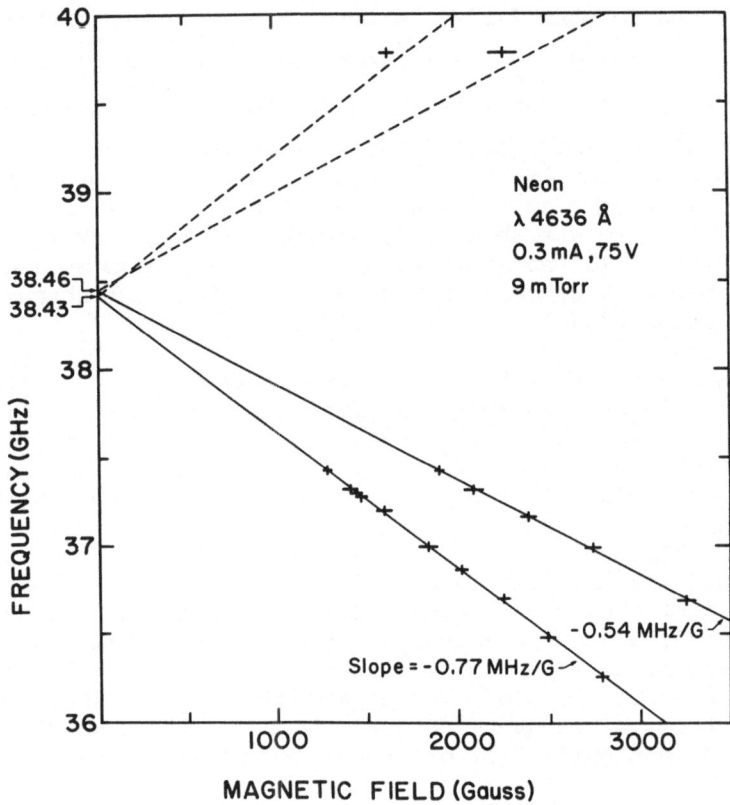

Fig 8. Microwave resonances observed in neon, tentatively identified as the $(^2P_{3/2})7d_2 - (^2P_{3/2})7f_{1,2}$ (Paschen $7d_1'' - 7X$) transition complex. The 0.03 GHz separation of the zero-field intercepts
is probably not significant, although a small f_1-f_2 splitting is
expected. Since the mirror-image connecting lines (shown dashed)
miss slightly the points near 40 GHz, the actual Zeeman transitions
are probably slightly curved.

IV. STRONG PERTURBATIONS

Perturbations of a size which produce only slight shifts in
low-lying energy levels can make gross changes in the appearance of
highly excited state microwave spectra. In our experimental situation, electric fields usually have the largest effect, although
atom-atom collisions can also be important. Electric fields of a
volt per cm or more are produced by macroscopic space charge and
microscopic ion and electron collisions in the excitation region,
and by the thermal motion of helium in magnetic fields

around 1000 G (the motional fields are much smaller for neon). The latter type always, and the former type often, has a range of values of the same order of size as the mean field. Thus both a shift and a broadening of the resonance ensemble result from the second-order Stark effect:[21]

$$\delta(E_a - E_b) = \sum_j \left[\frac{<a|e\vec{\mathcal{E}} \cdot \vec{r}|j>^2}{E_a - E_j} - \frac{<b|e\vec{\mathcal{E}} \cdot \vec{r}|j>^2}{E_b - E_j} \right] = \alpha_{ab} \mathcal{E}^2 \quad . \tag{5}$$

Since the matrix elements $(n,\ell|r|n,\ell-1) = \frac{3}{2} a_0 n\sqrt{n^2-\ell^2}$ behave as n^2 for $n \gg \ell$ and the energy denominators behave as n^{-3}, the Stark shifts increase with n as n^7. Since the natural (radiative) broadening behaves as n^{-3}, the ratio (Stark breadth of ensemble/natural breadth) behaves as n^{10}, with the result that the observed spectrum "blows up", or becomes diffuse, rather suddenly above a certain n.

The motional electric field $\vec{\mathcal{E}} = (\frac{\vec{v}}{c} \times \vec{\mathcal{K}})$ depends on the atom's velocity in the plane normal to the magnetic field $\vec{\mathcal{K}}$. Thus the second order motional Stark shifts depend on the kinetic energy in this plane and have a Boltzmann distribution of values. When the natural breadth may be neglected, the observed microwave resonance intensity (for the case $\alpha_{ab} > 0$) is

$$I(\nu)d\nu = I_0 \left\{ \begin{array}{ll} e^{-\beta(\nu-\nu_0)} d\nu & \nu \geq \nu_0 \\ \\ 0 & \nu < \nu_0 \end{array} \right\} \tag{6}$$

where

$$\beta = \left(\frac{h}{\alpha_{ab}\mathcal{K}^2} \right) \left(\frac{mc^2}{2kT} \right) .$$

The resonance intensity rises abruptly at $\nu = \nu_0$, since no atoms have Stark shifts corresponding to negative kinetic energies. Figure 9 shows such a resonance.

The electric fields also mix sublevels of different ℓ, allowing electric dipole transitions having $|\Delta\ell| > 1$ to occur. Strong rf electric fields can also produce such "multiple quantrum" transi-

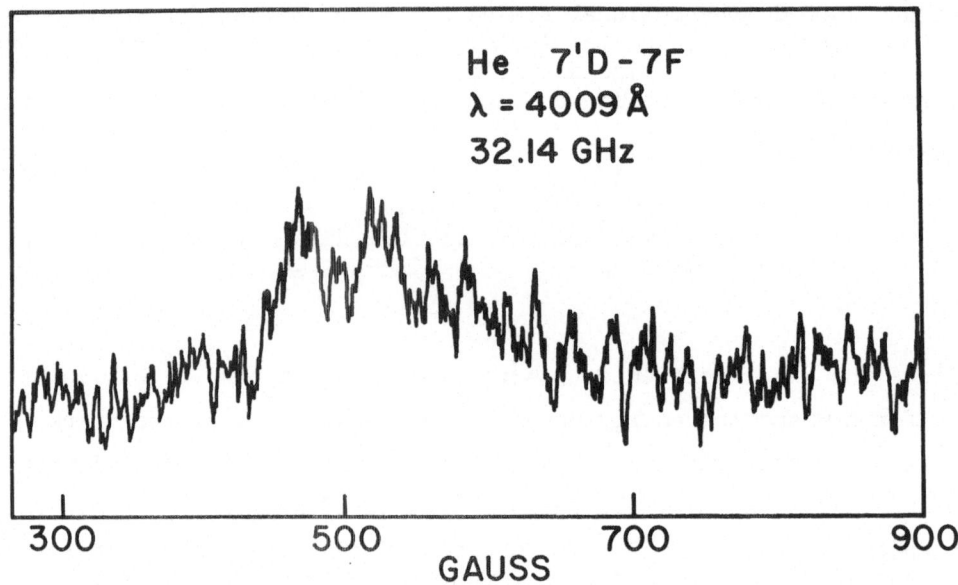

Fig. 9. Section of a magnetic field panoramic of the 7^1D–7F helium resonance having a positive Zeeman slope. The line has an exponential tail on the high–magnetic field side produced by the motional Stark effect (Eq. (6)). This example is also somewhat affected by power broadening and a doublet splitting in the F state.

tions. In a magnetic field a veritable forest of transitions appears, as shown for helium in Fig. 10, which is a composite of a number of recorder traces at various microwave frequencies. These traces were obtained in our early work,[1,2] without a monochromator, in which the photomultiplier was sensitive to all the helium principal series lines n^1P–1^1S at 500–584Å. Figure 11 summarizes all the runs under these conditions. Most of the transitions are presumably of the type n^1P–$n^1\ell$ ($\ell > 1$), with perhaps some n^1P–n^1S also. In addition to Zeeman slopes of ±1.4 MHz/G (g = 1) corresponding to the selection rule $\Delta m_\ell = \pm 1$, g values as high as 8 appear because of the Stark mixing. Another noteworthy feature is an omnipresent background rf–induced signal, or "flop", which appears to be a composite of the infinite number of thoroughly "blown–up" very high state resonances. By comparing Fig. 11 with Fig. 4 we can tentatively identify the group of transitions leading to 29 GHz as 14^1P–14F, G, H,.... and those near 61 and 82 GHz as the corresponding transitions in n = 11 and 10, respectively. The transitions at 72 GHz may be 11^1P–11^1D. The g = 1 transition leading to zero frequency at zero field is again a composite of many high transitions with small splittings. Clearly more work on these levels is needed.

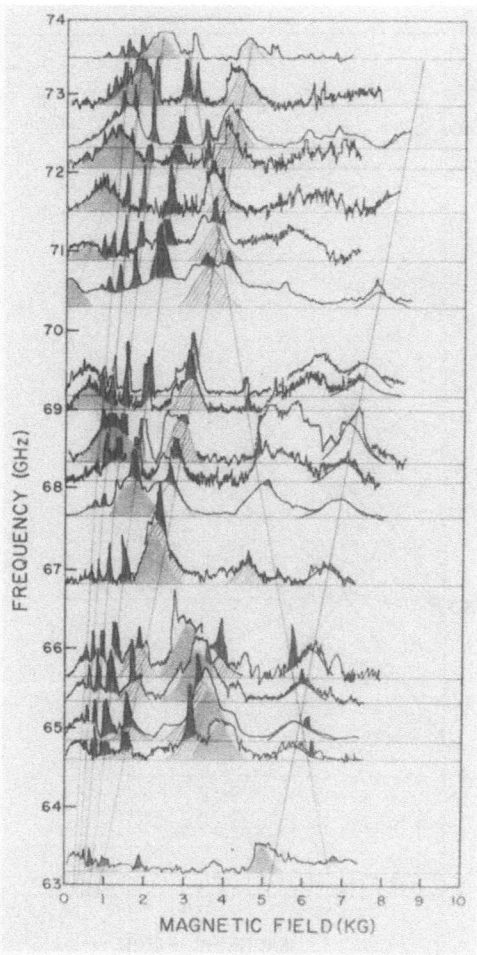

Fig. 10. Composite of a number of magnetic-field sweeps taken in helium with a broadband vacuum-uv detector, at microwave frequencies in the range 63-74 GHz. Some shaded peaks are connected by straight lines which are extrapolated back to zero magnetic field.

The detection scheme for the broadband uv work is interesting (Fig. 12). Although the atom radiates a hard-uv photon whether it is in a n^1P or n^1D state, a rf-induced light intensity change still occurs, presumably because of differential resonant scattering of the uv photons in the 1-meter, 10^{-3}-Torr helium path in the light pipe. In the case of the transitions n^1P-nF, G, H,...., singlet-triplet mixing provides an additional mechanism for diminution of the number of uv photons. The rf transfers singlet atoms into the

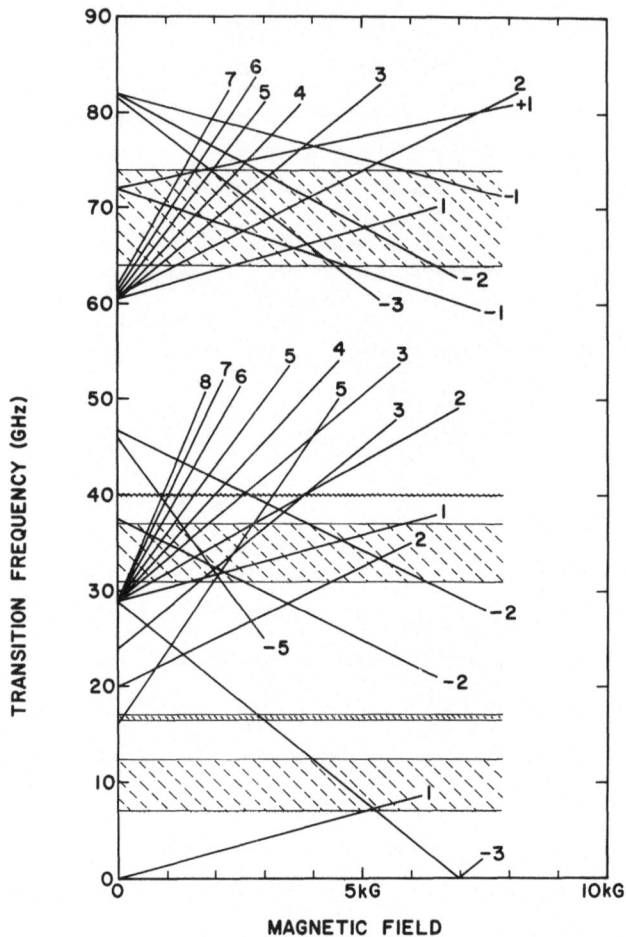

Fig. 11. Summary of results obtained from diagrams like Fig. 10. The frequency ranges explored are shown by shaded bands. The transition lines are numbered by the nearest integral slope in units of μ_o/h.

triplet sequence, where they cascade to 2^3S and ultimately de-excite on the walls of the apparatus without radiating.

V. CONCLUSION

The microwave-optical resonance method is evidently quite versatile, since the requirements of establishment of a population difference and selective detection of the resonated states can be

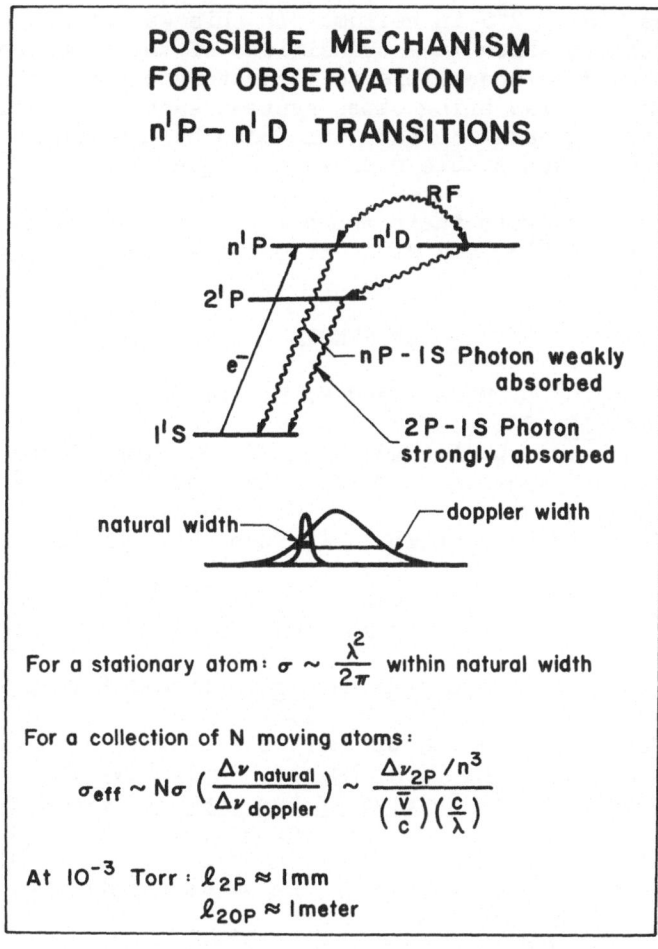

Fig. 12. Possible mechanism for the observation of transitions $n^1P - n^1D$ in highly-excited states of helium. This depends on strong self-reversal of the low members of the principal series by the helium gas in the light pipe shown in Fig. 5.

met in a variety of cases. Its compensating disadvantage, at least in our apparatus, is that it is somewhat "dirty", as compared to, for example, a beam machine.

With some relatively straightforward changes we should be able to make accurate measurements in helium as high as n=15 or 20, and perhaps improve the neon work as well. A great advance should result from the use of a monochromatic laser to transfer upwards to a single high-lying state a portion of the population of a low

state such as 2^1S or 2^3S in helium. In this way we might be able
to reach levels at which $\Delta n=1$ transitions of the type seen by radio
astronomers would be accessible to laboratory experimentation. Such
an approach might also bring other systems such as molecules, which
have much larger quantum defects than does helium, into the range
of experimentally achievable microwave frequencies.

The authors express their appreciation of the collaboration of
D. L. Mader, T. O. Siu, and M. A. Stroscio at various stages of the
work.

REFERENCES

1. W. H. Wing, D. L. Mader, and W. E. Lamb, Jr., Bull. Amer.
   Phys. Soc. 16, 531 (1971).
2. W. E. Lamb, Jr., D. L. Mader, and W. H. Wing, Proceedings of
   the Esfahan Symposium on Fundamental and Applied Laser Physics
   (to be published).
3. W. H. Wing and W. E. Lamb, Jr., Phys. Rev. Letters 28, 265
   (1972).
4. K. R. Lea, unpublished data.
5. M. J. Seaton, Rev. Mod. Phys. 30, 979 (1958); Proc. Phys. Soc.
   London, 88, 815 (1966).
6. C. E. Moore, Atomic Energy Levels (National Bureau of Standards
   Circular No. 467, 1949), Vol. I.
7. W. C. Martin, J. Res. Natl. Bur. Std. (U.S.) A 64, 79 (1960).
8. W. E. Lamb, Jr., and T. H. Maiman, Phys. Rev. 105, 573 (1957);
   J. P. Descoubes, in Physics of the One- and Two-Electron Atoms,
   edited by F. Bopp and H. Kleinpoppen (North-Holland, Amsterdam,
   1969), p. 341; A. Kponou, V. W. Hughes, C. E. Johnson, S. A.
   Lewis, and F. M. J. Pichanick, Phys. Rev. Lett. 26, 1613
   (1971), and many others.
9. J. S. Levine and A. Javan, Appl. Phys. Lett. 14, 348 (1969).
   A precise measurement of the 95.8-μm transition frequency has
   been made by J. S. Levine, A. Sanchez, and A. Javan, to be
   published.
10. Y. Accad, C. L. Pekeris, and B. Schiff, Phys. Rev. A 4, 516
    (1971).
11. H. A. Bethe and E. E. Salpeter, Quantum Mechanics of One- and
    Two-Electron Atoms (Springer-Verlag, Berlin, 1957) and H. A.
    Bethe, "Quantenmechanik der Ein- und Zwei-Elektronen Probleme"
    in Handbuch der Physik, edited by H. Geiger and K. Scheel,
    (Springer, Berlin, 1933), Vol. 24/1.
12. M. J. Seaton, second article in Ref. 5.
13. B. Edlén, in Encyclopedia of Physics, edited by S. Flügge
    (Springer, Berlin, 1964), Vol. 27, Secs. 10 and 33. A more
    detailed treatment of the polarization model for helium has
    been given by C. Deutsch. Phys. Rev. A 2, 43 (1970), and 3,
    1516(E) (1971).

14. W. C. Martin, J. Res. Nat. Bur. Std. (U.S.) A $\underline{74}$, 699 (1970).
15. W. E. Lamb, Jr., and T. M. Sanders, Jr., Phys. Rev. $\underline{119}$, 1901 (1960); L. R. Wilcox and W. E. Lamb, Jr., Phys. Rev. $\underline{119}$, 1915 (1960).
16. S. L. Kaufman, W. E. Lamb, Jr., K. R. Lea, and M. Leventhal, Phys. Rev. A $\underline{4}$, 2128 (1971); R. R. Jacobs, K. R. Lea, and W. E. Lamb, Jr., Phys. Rev. A $\underline{3}$, 884 (1971); D. L. Mader, M. Leventhal, and W. E. Lamb, Jr., Phys. Rev. A $\underline{3}$, 1832 (1971).
17. G. Araki, Proc. Phys. Math. Soc. Jap. $\underline{19}$, 128 (1937).
18. T. O. Siu, unpublished.
19. R. M. Parish and R. W. Mires, Phys. Rev. A $\underline{4}$, 2145 (1971).
20. V. Kaufman and L. Minnhagen, J. Opt. Soc. Am. $\underline{62}$, 92 (1972).
21. At high enough n this approximation is no longer sufficient. A complete solution of the multiple-level time-dependent perturbation problem with damping, rf, and the ensemble distribution of perturbation strengths appears formidable to us. At very high n, the perturbations couple all levels together strongly and the secular equation becomes infinite in dimension.

THEORETICAL STUDY OF ATOMIC RYDBERG STATES

Robert T. Poe

CERN, Switzerland and University of California, Riverside

T. N. Chang

University of California, Riverside

I. INTRODUCTION

Recently, a new microwave-optical method was developed by Wing and Lamb[1] for the experimental study of the atomic Rydberg states. The new technique, with a 10^4-fold improvement in resolution over the conventional spectroscopic methods, makes it possible for the first time to obtain accurate details on the previously unresolved electrostatic fine structures. As emphasized by those authors, the physics of the atomic Rydberg states holds special theoretical interest not only because of its quasi-hydrogenic nature but also because of its close relation to the scattering of slow electrons by the charged ion core. Thus the experimental breakthrough has added great impetus to the search for a parallel progress in theory, the search for a general, first-principle approach, capable of improved quantitative predictions as well as qualitative inter-pretations. As an attempt along this direction, we have carried out a theoretical study on the electrostatic fine structure of the Rydberg series for atomic helium within the framework of the Brueckner-Goldstone[2] many-body perturbation theory approach.

The Brueckner-Goldstone (BG) perturbation theory, developed in the context of nuclear structure theory, was first applied to atomic physics by Kelly[3] in a calculation of the correlation energy of atomic beryllium. Since then, the approach has been fruitfully applied to a wide range of problems in atomic physics. The diverse number of topics applied so far can perhaps be partially exemplified by our own experience with the BG approach at UC Riverside. In the past several years, we have applied the BG approach to : topics for bound state atomic systems which included correlation energies,[4]

polarizability studies[5], and hyperfine structure calculations[6];
topics for electron–atom scattering problems[7]; and topics for
photon–atom reactions such as photoionization[8] and the double
electron ejection by x–ray absorption[9]. Through these calculations,
we have gained extensive working experience with the approach and
familarity with its salient features. This also gives us the
motivating force for applying the BG approach to the present
problem.

II. REVIEW OF THE BG APPROACH

To facilitate subsequent discussions, it is worthwhile to
review briefly some formal as well as practical features of the BG
approach. As we shall see, some of these features shall prove to
be most useful in the present study.

A. Choice of the Single–particle Potential V

The first and foremost task in the BG approach, as in all
perturbational approaches, is the separation of the total
Hamiltonian H of the system into an 'unperturbed' part H_o and the
'perturbing' part H'. For an atomic system, the total Hamiltonian
consists of the single–particle operators T_i, the sum of the
kinetic energy operator K_i and the nuclear Coulomb potential
operator V_i^N for the ith electron, and the Coulomb interactions υ_{ij}
between the ith and the jth electrons. The most general way for
the separation of H_o and H' is through the introduction of an
arbitrary (Hermetian) single–particle potential V_i:

$$H = \sum_i (K_i + V_i^N) + \sum_{i<j} \upsilon_{ij}$$

$$= \sum_i T_i + \sum_{i<j} \upsilon_{ij}$$

$$= \sum_i (T_i + V_i) + (\sum_{i<j} \upsilon_{ij} - \sum_i V_i)$$

$$= H_o + H' . \tag{1}$$

With a given choice of V, one generates a complete set of single–
particle states ϕ_n's

$$(T_i + V_i) \phi_n = \varepsilon_n \phi_n \tag{2}$$

which in turn shall form the basis set for the many–body perturbation theory. The same single–particle potential V, because of its

presence in H', also determines the goodness (i.e. convergence) of the perturbation series.

B. Anti-symmetrization

Since the BG approach is formulated in the second quantization representation, the effects of the Pauli exclusion principle among electrons which is all-important for atomic systems, are automatically taken into account in each other through the built-in anti-commutation relations of the creation and the destruction operators of the second quantization formalism.

C. Linked-cluster Expression

The BG formulation results in the celebrated linked-cluster expansion expression for the energy of the total system

$$E = E_o + \sum_n^{\infty L} <\Phi_o | H' [(E-H_o)^{-1} H']^n | \Phi_o> \quad . \tag{3}$$

Here we sum over only the "linked" terms while the "unlinked" terms containing the spurious divergences simply do not appear.

D. Diagrammatic Representations

Analagous to the Feynmann diagrams in field theory, the terms in the BG perturbation expansion can be represented graphically by topologically distinct diagrams where the "vacuum" is the unperturbed ground state Φ_o while the "particle" and "hole" represent the occupied excited orbital and unoccupied ground state orbital respectively. From a practical point of view, the possibility of this diagrammatic representation of terms offers several most distinct advantages.

Systematic enumeration of contributing terms. By simply drawing all topologically distinct diagrams, systematic and unambiguous enumeration of all contributing terms up to a given order can be readily made in practice.

Interpretation of contributing physical processes. The graphical representation of diagrams enables one to attach a certain physical interpretation to each diagram. Thus considerable understanding of the underlying physical processes can be gained. For instance, this feature was proved to be invaluable in the study of the double electron ejection process.[9] In particular, a "virtual

Auger transition", absent in other theories, was seen to be instrumental in bringing theory and experiment into agreement.

Prior cancellation of opposite terms. Utilizing the diagrammatic approach, one may often make formal cancellation among large,
equal and opposite terms without actually evaluating them. Thus one
needs only to evaluate those diagrams that are directly related to
the very physical quantity of interest. In addition to considerable labor-savings in calculation, this also results in increased
numerical accuracy. For example, in our study of the hyperfine
structure (contact term) of phosphorus,[6] we were able to make
formal cancellations among the spin-up and spin-down terms and
evaluate only those terms which represent the difference between
them. In contrast, all previous attempts in effect evaluated the
large total spin-up contribution and subtracted the large and
nearly equal total spin-down contribution, with the net result
disagreeing with experiment not only in magnitude but also in sign!

III. APPLICATION TO HELIUM (1s,7ℓ) RYDBERG STATES

A. Choice of V

We must first choose the single-particle potential V for the
present calculation. The physics of the helium (1s,7ℓ) atomic
Rydberg states suggests quite naturally the following choice:

$$V = 0, \quad \varepsilon_n = -4/n^2 \quad \text{for } \ell = 0,1 \text{ orbitals}$$
$$V = e^2/r, \quad \varepsilon_n = -1/n^2 \text{ for } \ell = 2,3,\cdots \text{ orbitals.} \tag{4}$$

The above choice will lead to hydrogenic ϕ_n's with $Z_{eff} = 2$ for
s and p basis orbitals and hydrogenic ϕ_n's with $Z_{eff} = 1$ for d,f,g
basis orbitals. Note that two different sets of radial orbitals are
used here. However, orthogonality of the total basis states is still
strictly maintained through the angular momentum eigenfunctions.

The zeroth order energy E_0 is now simply

$$E_0(1s,7\ell) = -(4+1/49) \text{ Ryd} \tag{5}$$

and is the same for all (1s,7ℓ) levels. This is referred to as the
Bohr level, and the level shift from this Bohr level is, in our
case, simply the sum of all higher order energy terms:

$$\text{level shift} = E_1 + E_2 + \cdots \tag{6}$$

We would like to stress the advantage of the present choice of

V. Since the "unperturbed" (1s,7ℓ) levels are "degenerate", we may effectively study the relations of level shifts for different (1s,7ℓ) levels from a common starting point. This important advantage will be lost should one choose instead Hartree-Fock type basis states, as one would normally do in hope of gaining (slightly!) better convergence.

B. Contributing Diagrams and Physical Interpretations

One may now proceed to draw all the topologically distinct diagrams that represent the contributing terms in the perturbation expansion. The present two-electron system of course makes this enumeration particularly simple. The important diagrams up to second order terms are given in Figure 1; other second order diagrams which contribute less numerically are given in Figure 2. The physical interpretations of the diagrams can be given as follows.

<u>Screening effect</u>. The sum of diagrams 1a and 1b (with minus sign) gives the contribution to the level shift due to the incomplete screening of the inner 1s core. That is, since the "unperturbed" 7ℓ hydrogenic orbital assumes complete screening by the 1s orbital, these two diagrams correct for this assumption by taking into account the small penetration of the 7ℓ orbital <u>inside</u> the 1s core orbital electron.

<u>First order exchange term</u>. Figure 1c represents the contribution of the first order exchange between the two electrons. This exchange contribution will have opposite sign for spin-singlet and spin-triplet cases.

<u>Polarization and distortion effects</u>. Diagram 1d corresponds to the dominant second order direct terms of the perturbation expansion. Physically this diagram describes the mutual polarization and distortion between the atomic electrons because of their mutual Coulomb repulsions. As diagram 1d indicates, the 1s electron and 7ℓ electron, through their Coulomb interaction, "excite" into intermediate virtual states p and q respectively before "de-exciting" again back to 1s and 7ℓ orbitals. Thus, this diagram gives the leading contribution of the effects of the interelectronic correlations. By multipole expansion of the Coulomb interaction, we can further separate the polarization effects into contributions from monopole, dipole, quadrupole interactions etc.

<u>Second order exchange effects</u>. Diagram 1e is the exchange counterpart of diagram 1d and represents the additional contribution to the exchange energy when interelectronic correlation and polarization are taken into account. Again the sign of this second order exchange contribution must be opposite for the spin-singlet and spin-triplet cases.

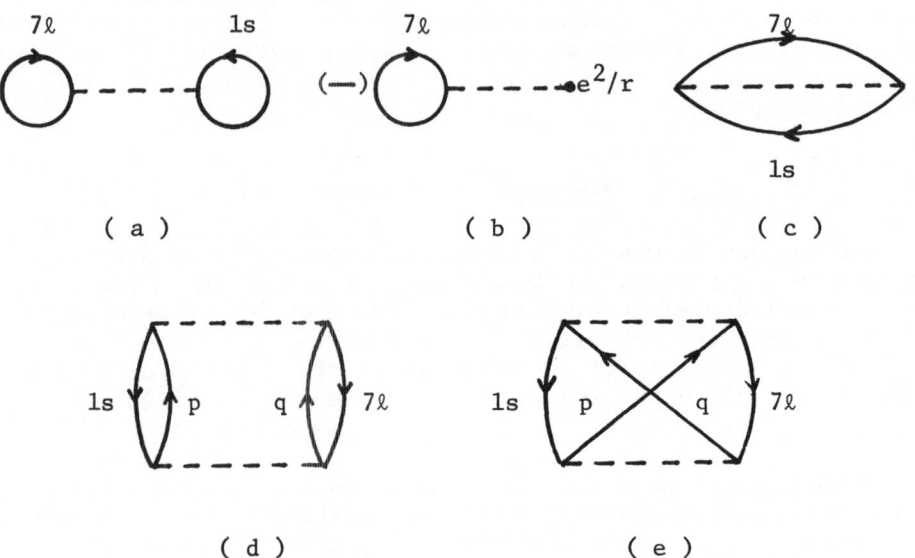

Figure 1.   Leading first and second order diagrams.

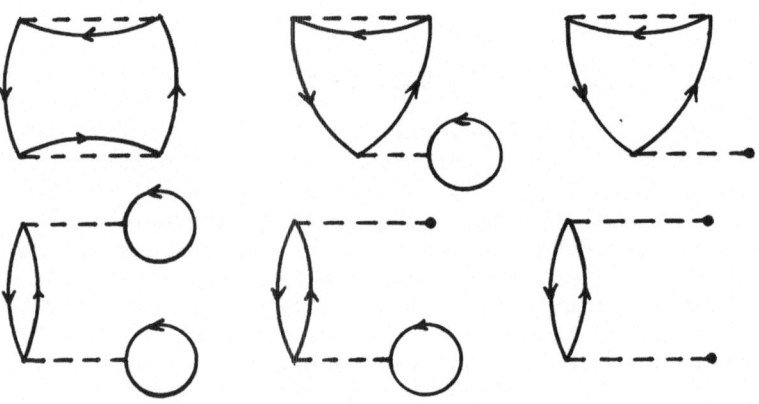

Figure 2.   Other types of second order diagrams.

One can make similar physical interpretations for the diagrams in Figure 2. However, since their contributions are far less than those in Figure 1, we shall not detail them here.

Finally we mention that in order to obtain the <u>average</u> (1s,7ℓ) level shift values, one only needs to evaluate the direct diagrams -- the exchange diagrams cancel formally. On the other hand, to obtain the doublet splitting values between the spin-singlet and spin-triplet levels, one only needs to evaluate the exchange terms -- the direct terms cancel formally. Because of this formal cancellation of terms, improved numerical accuracy (to three significant values) can be achieved, especially in the evaluation of the doublet splittings.

IV.   RESULTS AND DISCUSSIONS

We present the preliminary results of our calculation in the following tables. Table I lists the values of the individual contributions from the leading diagrams. In Table II, the contributions of all diagrams are summed to obtain the total level shifts of the atomic levels. For the (1s,7f) case, the spin-orbit coupling becomes comparable and the effect of this spin-orbit mixing is included in our evaluation.[1] For more direct comparison with experiment, we give the transaction energies and the doublet splitting values in Table III. The results of a polarization model theory by Deutsch[10] are also included wherever applicable.

Examination of Table I allows us to make some general assessments with regard to the relative contributions of various physical effects to the electrostatic fine structure of these Rydberg states. Most prominent is the overall predominance of the dipole polarization effect. For (1s,7d) states, it gives the largest contribution but other effects also contribute appreciably. The screening effect, for example, contributes nearly 10%, indicating that the 7d orbital still makes considerable penetration inside the 1s core. Quadrupole polarization also gives about 5% of the total effect here. For (1s,7f) states, the dipole polarization effect becomes really dominant while screening and other effects diminish rapidly. For the (1s,7g) and (1s,7h) states, the contribution of this long-range dipole term becomes singularly overwhelming.

Next we look at the exchange terms, which are responsible for the doublet splitting between spin-singlet and spin-triplet levels. The most prominent feature here is the importance of the second order exchange. For (1s,7d) levels the ratio of the second order exchange to first order exchange is about 40%, and for (1s,7f)

Table I. Contributions of diagrams to level shifts (in unit of 10^{-6} Ryd). Dipole contribution values in parentheses are evaluated in adiabatic approximation.

Diagrams	Physical effects	(1s,7d)	(1s,7f)	(1s,7g)	(1s,7h)
E_1 1a+1b	Screening effect	-0.9347	-5.767×10^{-3}	-1.6386×10^{-5}	-2.15×10^{-8}
1c	1st order exchange	± 3.4569	$\pm 2.2068 \times 10^{-2}$	$\pm 6.3761 \times 10^{-5}$	$\pm 0.84 \times 10^{-7}$
	2nd order direct (polarization)				
	Monopole (k=0)	-0.04077	-2.47×10^{-4}	insignificant	insignificant
1d	Dipole (k=1)	-5.7261 (-5.4086) -6.4193 (-8.0590)	-1.0360 (-1.0149) -1.2036 (-1.3129)	-0.27061 (-0.26801) -0.28057 (-0.29049)	-0.0865 (-0.0861) -0.0651 (-0.0664)
E_2	Quadrupole (k=2)	-0.18166 -0.3585	-0.01509	insignificant	insignificant
1e	2nd order exchange	∓ 1.30203	$\pm 1.006 \times 10^{-2}$	-----	-----
others		-0.01197 ∓ 0.10007	insignificant	-----	-----

Table II. Level shifts and doublet splittings (in unit of 10^{-6} Ryd).

7^1D_2	-11.618
7^3D_2	-15.728
7F(Ave.)	-2.2612
7G(Ave.)	-0.5512
7H(Ave.)	-0.1516
7^1D_2 - 7^3D_2	4.110
7^1F_3 - 7^3F_3	~ 0.044*

*Spin-orbital interaction included.

Table III. Level transition energies and doublet splittings.

Transition	Experiment (Wing & Lamb)	Pres. Calcu.	Polarization (Deutsch)
7^1D_2 - 7F(Ave.)	31.485 GHz	31.65 GHz*	----
7F(Ave.) - 7G(Ave.)	5.60 GHz	5.625 GHz	6.07 GHz
7G(Ave.) - 7H(Ave.)	1.06 GHz	1.20 GHz	1.38 GHz
7D(Ave.) - 7F(Ave.)	----	38.41 GHz	40.5 GHz
7^1D_2 - 7^3D_2	~ 14 GHz**	13.52 GHz	----
7^1F_3 - 7^3F_3	146.19 MHz	~ 146 MHz	----

*Spin-orbital interaction included.
**Estimated through quantum-defect fit values.

levels the ratio reaches nearly 46%. Moreover, in both cases, the
second order exchange contributes in opposite sign to that of the
first order exchange. It is therefore clear that the effect of the
second order exchange must be properly taken into account if good
value for the doublet splitting is to be obtained. Here it is
interesting to note that the importance of this second order exchange term has previously been observed in electron-atom scattering problems.[7,11] The present calculation suggests that it should
be equally important for bound state system such as the Rydberg
states -- perhaps not too surprising in view of the close relationship between the Rydberg series and the low energy scattering problems.

In Table I we have also presented the values for the dipole
polarization term evaluated with adiabatic approximation, values
in parentheses in Table I. It can be seen that the introduction
of the adiabatic approximation in the dipole term, as often done
explicitly or implicitly in other theoretical models, can appreciably change its value, increasing by about 5% in the (1s,7f) case
and by as much as 10% in the (1s,7d) case. These errors may seem
small but their effects will magnify in proportion in the evaluation of the transition energy between levels.

We now turn to Table III. In all cases where comparison with
experimental results can be made, good agreements are obtained,
both for the transition energy values and for the doublet splitting
values. In comparison, for the average transition energy between
levels, the polarization model theory[10] result is seen consistently
tending to overestimate. We can attribute this at least in part to
the implicit adiabatic assumption of the model. The polarization
model is further limited by its inherent inability to yield doublet splittings.

To summarize, we find the preliminary results of our calculation extremely encouraging. It suggests that the BG approach, in
addition to being a general, first-principle theoretical method,
is capable of markedly improved quantitative predictions. Perhaps
even more gratifying than the good numerical agreement with experiment is the method's unique ability to relate the contributing diagrams to the underlying physical processes involved. To
fully assess the method, more detailed applications must be carried out. Our present calculation, however, does brighten the
hope that the BG approach may complement the recent experimental
breakthrough toward a much better understanding of the physics of
the atomic Rydberg states.

REFERENCES

1. W. H. Wing and W. E. Lamb, Jr., Phys. Rev. Letters 28, 265
    (1972) and references therein.

2. J. Goldstone, Proc. Roy. Soc. (London), A239, 267 (1957).

3. H. P. Kelly, Phys. Rev. 131, 684 (1963).

4. E. S. Chang, R. T. Pu (Poe) and T. P. Das, Phys. Rev. 174, 1
    (1968).

5. E. S. Chang, R. T. Pu (Poe) and T. P. Das, Phys. Rev. 174, 16
    (1968).

6. N. C. Dutta, C. Matsubara, R. T. Pu (Poe) and T. P. Das, Phys.
    Rev. Letters 21, 1139 (1968).

7. R. T. Pu (Poe) and E. S. Chang, Phys. Rev. 151, 31 (1966).

8. T. Ishihara and R. T. Poe, Phys. Rev. A6, 116 (1972).

9. T. N. Chang, T. Ishihara and R. T. Poe, Phys. Rev. Letters 27,
    838 (1971).

10. C. Deutsch, Phys. Rev. A2, 43 (1970).

11. W. M. Duxler, R. T. Poe and R. LaBahn, Phys. Rev. A4, 1935
    (1971).

INNER-SHELL IONIZATION BY HEAVY CHARGED PARTICLES[*]

Werner Brandt

Department of Physics, New York University

4 Washington Place, New York, N.Y.   10003

1.   Introduction

An International Conference on Inner-Shell Ionization Phenomena was held in Atlanta, Georgia, during April of this year. The large number of participants and the sweeping scope of the papers and discussions signify the wide and growing interest in the exploration, understanding and application of inner-shell ionizations. The Proceedings of this conference will appear shortly.[1]

What is it that holds the interest in this class of phenomena? Characteristic target x rays under alpha particle bombardment were first observed by Chadwick[2] just 60 years ago, and one might have expected that the field should have progressed to a mature stage in the intervening years. However, detailed tests of the theory of inner-shell ionization by heavy charged particles have become possible only very recently with the accumulation of wide-ranging accurate ionization cross section measurements.

We summarize the large discrepancies of nearly two orders of magnitude found initially between theoretical and experimental

[*]Work supported by the U.S. Atomic Energy Commission.

cross sections for the Coulomb excitation of K shells by heavy
(compared to the electron) point particles and sketch some of
the very recent developments which have resolved most of the
disparity between the physical phenomenon of K-shell ionization
and its theoretical description. Experiment and theory, as
summarized here, now agree within experimental uncertainties.

We conclude this report by demonstrating that, relative to
the reference frame of Coulomb excitation, a new kind of atomic
excitation emerges at low projectiles velocities when the projectile has an atomic number comparable to that of the target
atom, because then the projectile carries its own electronic
structure into the atomic collision. Inner-shell excitation
and ionization are governed by the Pauli principle in the overlapping electron clouds. As a result excitation and ionization
can be vastly enhanced, as in the case of electron promotion
through quasi-molecular states. We refer to these processes
generically as being caused by Pauli excitation of atoms in
collision. Pauli excitation differs in kind and degree from
the Coulomb excitation underlying most of the discussion in
this paper. The Pauli excitation of inner-shell electrons during
atom-atom collisions is developing into one of the major and most
rapidly advancing new fields in atomic research.

2.   Experimental

Inner-shell excitation cross sections can be determined
experimentally from the yield of either the Auger electrons
or the characteristic x rays emitted from an atom when a hole
is filled in an inner shell that was created by the collision
with a projectile. Projectiles, with properties marked by the
subscript 1, impinge on target atoms fixed in the laboratory
with properties marked by the subscript 2. Prima facie the
distinction between projectile and target is not linked to the
preparation of the projectile beam before entering the target.
Both can be ionized and both will fill their own inner-shell
vacancies with the emission of Auger electrons or characteristic
x rays. The Auger electron or the x ray counted in the laboratory determines the target. In the following this is always
the laboratory target. We concentrate on x-ray emission as the
measure of inner-shell ionization, although some ionization
cross sections derived from Auger yields are also included in
the figures.

Practical limitations lead to two types of measurements of
characteristic x-ray production cross sections. In thick-
target measurements, at projectile energies $E_1 < 1$ MeV/amu,

the target stops all projectiles. The x-ray production cross section, σ_x, follows from the differentiation of the recorded x-ray yield $Y(E_1)$ (x rays/projectile) with respect to E_1,

$$\sigma_x(E_1) = \frac{4\pi}{n_2} \left[\frac{dY(E_1)}{dE_1} S(E_1) + \mu\, Y(E_1) \right], \qquad (1)$$

where n_2 is the atomic density of the target, $S(E_1)$ the stopping power of the target for the projectile and μ the absorption coefficient of the target for the characteristic x ray. Thick-target cross sections can be determined to a relative accuracy of $\pm$ 20%. Absolute values depend on calibration and are less well known.

Thin-target transmission becomes possible at $E_1 > 1$ MeV/amu. The cross section is directly proportional to the thin-target yield,

$$\sigma_x(E_1) = Y(E_1)/tC, \qquad (2)$$

where t is the target thickness and C an instrument constant. Thin-target cross sections have relative accuracies of $\pm$ 2%.

Comparison with theory proceeds via the ionization cross sections of various shells σ_K, σ_L, ..., which are related to the relevant σ_x as, e.g.,

$$\sigma_K = \gamma_K^{-1} \sigma_{Kx}, \qquad (3)$$

where γ_K is the fluorescence yield for K-shell x-ray emission. Current efforts are making available large ranges of accurate fluorescence yields.[1]

3. Theory in Plane-Wave Born Approximation

We refer to projectiles as charged particles when they act as mass points of charge $Z_1 e$ in exciting target K-shell electrons. They can cause only Coulomb excitation of K-shells. Coulomb excitation always dominates at projectile velocities so high that the projectiles move as bare nuclei through the target. But even a slow projectile acts as a bare particle if its K-shell radius, $a_{1K} = a_o Z_{1K}^{-1}$, is large compared to the target K-shell

radius $a_{2K} = a_0 Z_{2K}^{-1}$. At low velocities K-shell ionization
occurs only on deep penetration, to distances $\ll a_{2K}$, and the
projectile electrons which could have caused Pauli excitation
remain outside the interaction region. In short, K-shell Coulomb
excitation dominates at all projectile velocities for projectile-
target pairs such that $Z_1 \ll Z_2$.[3]

The general theory for K-shell ionization by Coulomb
excitation has been developed for some time in the plane-
wave Born approximation (PWBA),[4] and in a semi-classical
formulation.[5] More recently the classical binary-collision
theory was applied to this problem.[6] Its predictions differ
in some respects from those of the quantum mechanical and
semi-classical formulations. The disparities must await
theoretical clarification.

An extensive study links the plane-wave Born approximation
(PWBA) to available experimental data.[7] In using the PWBA one
assumes that (1) the projectile acts as a point charge, (2) the
initial and final particle waves are planar over all space, and
(3) the states of the target electrons are those of the unperturbed target. The exact cross section σ and the approximate
cross section σ^{PWBA} are therefore interrelated as

$$\sigma^{PWBA} = Z_1^2 \Gamma, \tag{4}$$

where

$$\Gamma = \lim_{Z_1 e \to 0} \left[\sigma(Z_1 e)/(Z_1 e)^2 \right] \tag{5}$$

is the exact ionization cross section per particle charge squared
in the limit of vanishingly small projectile charge $Z_1 e$. If
the requirements (1), (2) and (3) are fulfilled, the ability
to calculate realistic σ values is limited only by the quality
of available approximations to the exact eigenfunctions of the
target Hamiltonian. A systematic search for adequate wavefunctions in calculating σ_K^{PWBA} revealed that the conditions in
K shells are sufficiently hydrogenic to make the calculations
invariant, to within a few percent, to the choice of "best"
approximate wavefunctions.

In the laboratory the $Z_1 e \to 0$ limit is usually assumed to
be imitated sufficiently well by restricting the range of target
elements such that $Z_1/Z_2 \ll 1$. While this satisfies requirement

(1), it is not sufficient to satisfy requirements (2) and (3), and discrepancies between experiment and this theory exist.

4. Predictions

The non-relativistic plane-wave Born approximation predicts a universal form for the ionization cross sections, which can be written as

$$\sigma_K^{PWBA} = z_1^2 \, F_2(\xi_K). \qquad (6)$$

The function $F_2(\xi_K)$ depends on the target and the projectile via the minimum momentum transfer $\hbar q_o = \hbar\omega_{2K}/v_1$ through the reduced variable

$$\xi_K \equiv (q_o \, a_{2K})^{-1} = v_1/\tfrac{1}{2}\theta_K \, v_{2K}, \qquad (7)$$

where $\hbar\omega_{2K}$ is the ionization energy of the target K shell of radius $a_{2K} = a_o \, Z_{2K}^{-1}$ with $Z_{2K} = Z_2 - 0.3$, v_1 the velocity of the projectile, and $v_{2K} = v_o Z_{2K}$ that of the target K-shell electrons with $v_o \equiv e^2/\hbar$. The parameter $\theta_K \equiv \hbar\omega_{2K}/Z_{2K}^2 Ry$ measures the non-hydrogenic aspects of the K-shell ionization energy (Ry = 1/2 a.u. = 13.6 eV). It grows slowly with Z_2 from ~ 0.6 for light elements to ~ 0.9 for heavy elements. It is customary to express the projectile-velocity dependence through a variable proportional to the projectile energy,

$$\eta_K \equiv v_1^2/v_{2K}^2 = (\theta_K/2)^2 \, \xi_K^2. \qquad (8)$$

Note that $\eta_K = 40.2 \, E_1$ (in MeV) $/M_1 Z_{2K}^2$, where M_1 is the projectile mass in amu.

Equation (6) states that the cross section divided by z_1^2 is the same for all particles incident with identical velocity on a given target atom. This prediction can be tested, without knowledge of the function F_2, by plotting σ_K^{exp}/z_1^2 for a variety of particles against v_1, or ξ_K, or η_K. Such data should fall along a single curve. The comparison of the locus of this curve with calculations of the function $F_2(\xi_K)$, moreover, tests the quantitative aspects of the theory.

For a study of the dependence of the cross sections on the target, it is natural to scale the cross section with

$$\sigma_{oK} \equiv 8\pi \ a_{2K}^{\ 2} (Z_1/Z_{2K})^2. \tag{9}$$

For example $\sigma_{oK}(Al) = 27 Z_1^{\ 2}$kb and $\sigma_{oK}(Ni) = 1.2 \ Z_1^{\ 2}$kb. Equation (6) can then be written as

$$\sigma_K^{PWBA} = \sigma_{oK} \ f(\eta_K/\theta_K^{\ 2}; \ \theta_K)/\eta_K. \tag{10}$$

One can show that, in the range $\eta_K \lesssim 1$ of interest here, the function f accurately transforms as

$$(\theta_K/\eta_K) \ f(\eta_K/\theta_K^{\ 2}; \ \theta_k) = (\theta_K^{\ 2}/\eta_K) \ f(\eta_K/\theta_K^{\ 2}; \ 1), \tag{11}$$

so that the reduced ionization cross section

$$\sigma_K^{PWBA}/(\sigma_{oK}/\theta_K) = (\theta_K/\eta_K) \ f(\eta_K/\theta_K^{\ 2}; \ \theta_K) \tag{12}$$

depends only on the variable η_K/θ_K^2. The role of θ_K is now merely that of a scale factor.

All K-shell cross sections should follow a universal curve if reduced as indicated on the left-hand side of Eq. (12) and plotted versus ξ_K or η_K/θ_K^2. The assertion of universality can be tested on data measured with different particles impinging with various energies on different targets without explicit knowledge of the function f. Examination of the locus of such reduced data and comparisons with computed values of f assess the validity range of the assumptions leading, from first principles, to the prediction both of universality and of absolute cross sections.

5.   Universality

In Fig. 1 a compilation of the cross section data is plotted as prescribed by Eq. (12) and compared with the computed function f.[3,7] It is apparent that the data do not follow a single curve but scatter widely. Moreover the PWBA

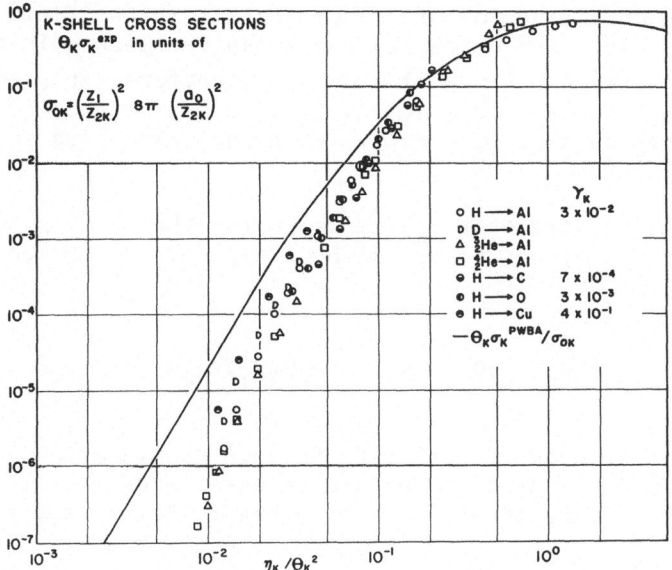

Fig. 1. Cross sections plotted according to Eq. (12).

values far exceed, by orders of magnitude, the experimental
cross sections at low projectile velocities, $\eta_K \ll 1$.

A series of investigations[8-10] has shown that important
physical processes are at the root of the discrepancies. They
derive from the finite charge of the moving projectile and, as
such, do not contribute in the plane-wave Born approximation.[7]
They are the Coulomb deflection of the projectile in the field
of the target nucleus and the perturbations of the target atomic
states by the projectile. A theory of these processes has been
developed in an approximate manner and incorporated in the form
of corrections to the PWBA. For $\eta_K \ll 1$, the result can be
written as [3,7]

$$\sigma_K = 9E_{10}(\pi dq_o \epsilon) \; \frac{\sigma_{oK}}{\epsilon \theta_K} \; \frac{\epsilon \theta_K}{\eta_K} \; f\left(\frac{\eta_K}{(\epsilon \theta_K)^2} \; ; \; \epsilon \theta_K\right). \qquad (13)$$

The exponential integral $E_{10}(y) \simeq (9+y)^{-1}\exp(-y)$ accounts,
through the argument

$$\pi dq_o = (\pi/2) \; Z_1 (m_e/M)\theta_K^{-2}(\eta_K/\theta_K^2)^{-3/2}, \qquad (14)$$

for the Coulomb deflection of the projectile in the field of
the target nucleus and decreases the cross sections. It depends
on the reduced mass $M^{-1} \equiv M_2^{-1} + M_2^{-1}$. Therefore, it predicts an
isotope effect in the cross sections which has been confirmed
by experiments on a given target with projectile pairs such as
(^{1}H, ^{2}D) or (^{3}He, ^{4}He).[10,11]

The scaling parameter θ_K, representing the binding energy
of the target electrons to the K-shell region. is replaced by
$\epsilon\theta_K$, where

$$\epsilon\theta_K \equiv \epsilon_K(\xi_K)\theta_K = \theta_K \left(1 + 2(Z_1/Z_{2K}\theta_K)g(\xi_K) \right) \qquad (15)$$

depends on the projectile velocity through the function $g(\xi_K)$.
This function is given as a certain integral over Bessel
functions and can be computed with errors < 1% from the formula

$$g(\xi_K) = (1+\xi_K)^{-5}(1+5\xi_K+7.14\xi_K^2+4.27\xi_K^3+0.947\xi_K^4). \qquad (16)$$

The binding increases from θ_K to $\epsilon_K\theta_K$ and the cross sections
decrease because the K-shell electrons are bound not only to
their target nucleus of effective charge $Z_{2K}e$ but also, during
the collision, to the deeply penetrating projectile of charge
Z_1e. Both the Coulomb deflection correction, through $9E_{10}$,
and the binding correction, through $\epsilon_K(\xi_K)$, become unity
ξ_K approaches and exceeds unity.

Figure 2 shows the universal graph of the same data as
plotted in Fig. 1, but now prepared according to Eq. (13).[3,7,12]
The influence of the Coulomb deflection has been divided out of
the experimental data and the effective, projectile-velocity
dependent binding energy has been incorporated into the PWBA
treatment through $\epsilon\theta_K$. In effect Fig. 2 represents an extrapolation, based on theory, of the data to the condition $Z_1e \to 0$,
where proper comparison with PWBA can be made. The procedure
may be viewed as providing a test of the PWBA based on data
taken with real projectiles of finite charge, but to which
requirements (2) and (3) for the PWBA now apply. Or it may be
viewed as a test for the approximate Eq. (13) derived to predict cross sections for K-shell ionization by projectiles such
that Z_1/Z_2 is small compared to unity but finite. Data plotted
on the reduced scales prescribed by Eq. (13) form a universal
curve. Its locus is predicted accurately by the PWBA.

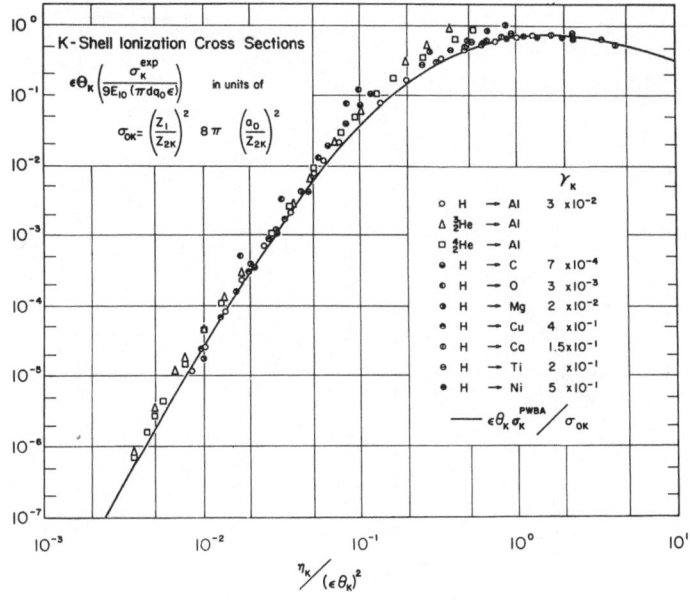

Fig. 2. Cross sections plotted according to Eqs. (13), (15).

When $\eta_K(\epsilon\theta_K)^{-2} > 5 \times 10^{-2}$, the experimental points begin to scatter and to deviate systematically from the curve to higher values. This deviation can be understood in terms recently developed in the study of stopping powers. Precision measurements reveal a correction proportional to $Z_1{}^3$. It has been accounted for theoretically in an approximation which is the classical equivalent of the second Born approximation.[13] One includes the polarization of the bound target states by the particle passing at large impact parameters as a quadrupole contribution. It increases the ionization cross section when $\xi_K \geq 1$.[9] We build a link between low velocities, where binding reduces the cross section, and high velocities, where polarization increases the cross section, by replacing $\epsilon\theta_K$ in Eq. (13) with

$$\zeta\theta_K \equiv \theta_K \left\{ 1 + \frac{2Z_1}{\theta_K Z_{2K}} \left[g(\xi_K) - \frac{1}{2} \frac{1}{\xi_K{}^3} \int_{\omega_{2K}}^{\infty} d\omega \, g_K(\omega) \, I(\frac{\omega a}{v_1}) / (\theta_K/2)^3 G_K(\xi_K) \right] \right\},$$

(17)

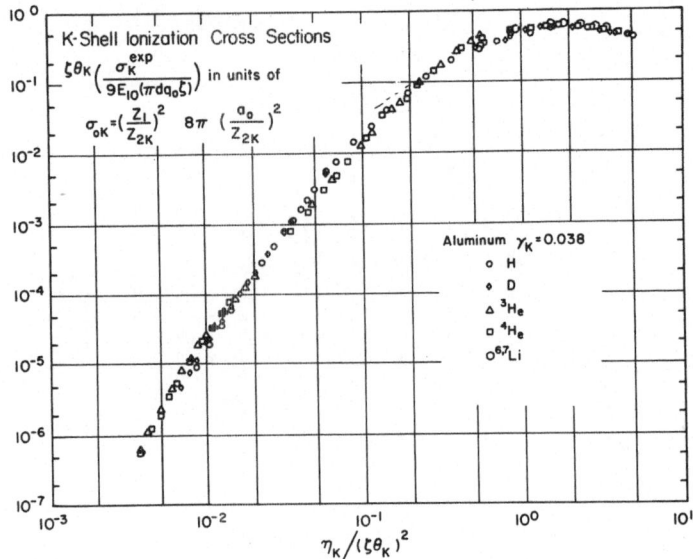

Fig. 3. Cross sections plotted according to Eqs. (13), (17).

where $g(\xi_K)$ is given by Eq. (16), $g_K(\omega)$ is the oscillator strength distribution for the K-shell, $I(\omega a/v_1)$ is a function defined in ref. 13, and $G_K(\xi_K)$ is the K-shell excitation function. Asymptotically $\zeta\theta_K \to \epsilon\theta_K > \theta_K$ for $\xi_K \ll 1$. The function $\zeta \equiv \zeta_K(\xi_K)$ changes from values > 1 to values < 1, and attains the PWBA value of unity at a crossing point near $\xi_K \simeq 1$. This formulation accounts fully for the trends observed in ratios of experimental cross sections of projectiles with different atomic numbers $Z_1 \leq Z_1{}' \ll Z_2$ at identical velocities.[9,10]

Figure 3 is a compilation of all the cross section data measured by our group on Al, plotted as prescribed by Eq. (13) with $\epsilon\theta_K$, Eq. (15), replaced by $\zeta\theta_K$, Eq. (17). It comprises six orders of magnitude in σ_K and three orders of magnitude in E_1/M_1. A similar plot for other target projectile combinations is shown in Fig. 4. No systematic deviations from a common curve remain.[14] The data reach the universal maximum value of the K-shell ionization cross section $\sigma_K \simeq 0.8\sigma_{oK}/\zeta\theta_K$ at $\eta_K \simeq 1.8(\zeta\theta_K)^2$.

The data now define the locus of an experimental curve with a precision comparable to the experimental uncertainties of the data. The PWBA curve agrees with this experimental curve within the same accuracy, as shown in Fig. 5. The order-of-magnitude discrepancies found in Fig. 1 have been resolved. The slightly low values of the experimental curve relative to the theoretical curve in the intermediate range near $\eta_K/(\zeta\theta_K)^2 = 10^{-1}$ may reflect only some minor inadequacies in our still tentative formulation of Eq. (17). Agreement between theory and experiment over the projectile velocity range corresponding to 20 keV/amu to 30 MeV/amu is now close to being limited by the experimental uncertainties.

6.   Atom-Atom Collisions

When the nuclear charge of the projectile is chosen to be comparable to that of the target, $Z_1 \sim Z_2$, low velocity projectiles carry an electron cloud into the encounter: they act as atoms not as particles, and the dominant process leading to inner-shell ionization changes from Coulomb excitation to Pauli excitation. The transient electronic states of the combined projectile and target atoms have been approximated by two-center molecular orbitals. Since they are eigenstates of the system at rest, the nuclear motion during the collision induces transitions between these states. This can create, on separation, inner-shell vacancies in the collision partners.[15,16] Detailed level-crossing schemes are currently being discussed to guide the experiments and their interpretation.[1]

A clear and global demonstration can be given of the shift from Coulomb excitation to Pauli excitation as Z_1 approaches Z_2 by plotting, as in Fig. 6, experimental cross sections for $Z_1 \sim Z_2$ in the reference frame of Coulomb excitation of Figs. 3-5. Figure 6 is an extended version of a graph published earlier[12] with $e\theta_K$, Eq. (15), replaced by $\zeta\theta_K$, Eq. (17). The solid curve represents the experimental and theoretical Coulomb-ionization cross sections of Figs. 3-5 applicable to $Z_1 \ll Z_2$. Clearly, the low-velocity data points for $Z_1 \sim Z_2$ are far from forming a common curve, and can yield cross sections many orders of magnitude larger than the Coulomb cross sections.

When the projectile velocities approach values $v_1 > (v_{1K}, v_{2K})$ the response time of the bound electrons is too long compared to the collision time to allow the Pauli principle to influence the ionization process, and Coulomb excitation dominates once more. Indeed, the experimental points in Fig. 6 asymptotically merge with the Coulomb excitation curve.

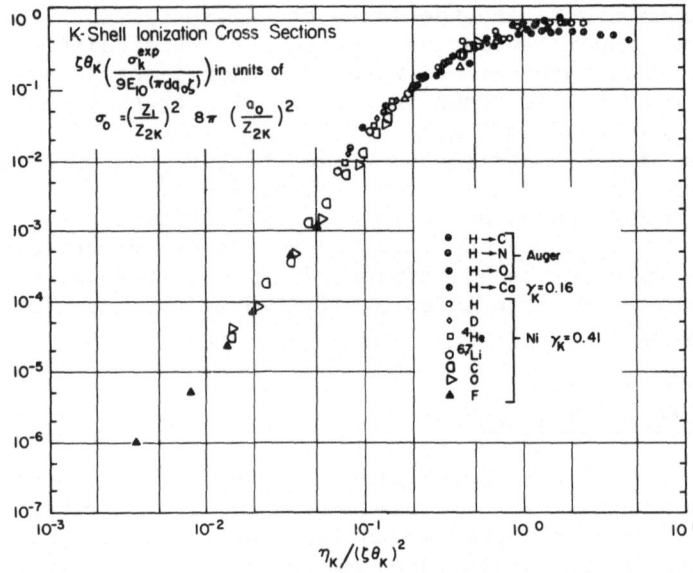

Fig. 4.   Cross sections plotted according to Eqs. (13), (17).

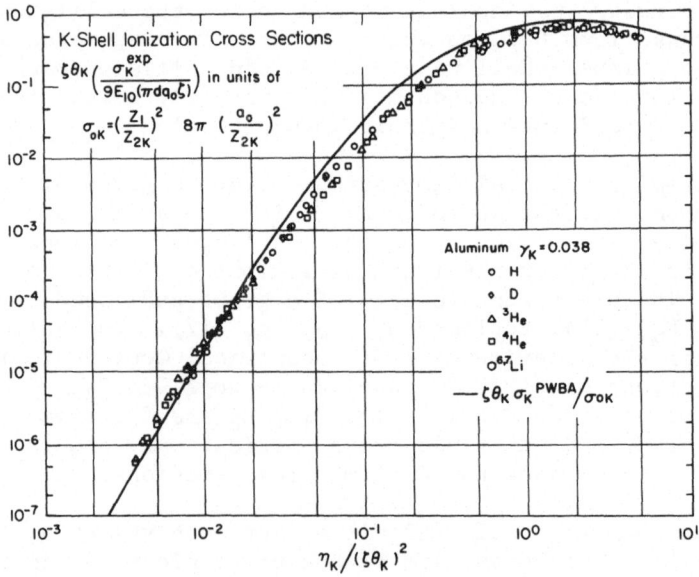

Fig. 5.   Comparison of PWBA with data in Fig. 3.

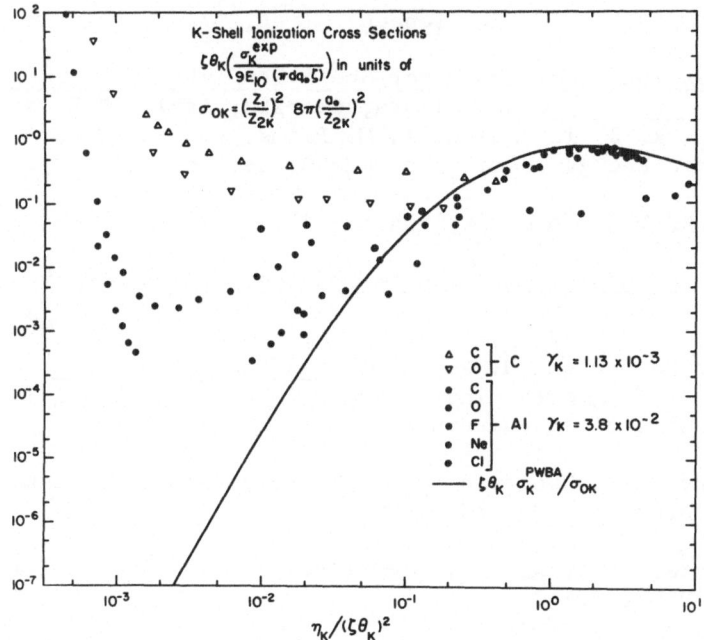

Fig. 6. Cross sections for $Z_1 \simeq Z_2$ compared with
Coulomb excitation curve in Figs. 3-5.

7. Summary

K-shell ionization cross sections for projectile-target
pairs $Z_1 \ll Z_2$ can be predicted comprehensively in terms of
an approximate theory based only on the Coulomb excitation
mechanism which includes effects caused by the finite charge
of the moving point particle. By contrast when the projectile
carries a coterie of electrons into the collision, inner-shell
ionization occurs via complex many-body Pauli excitation processes. Pauli excitation cross sections can be larger by orders
of magnitude than the Coulomb excitation cross sections, and
depend in distinguishable ways on the projectile velocity.

These notes were prepared at the Aspen Center of Physics.
They are based on work performed in collaboration with
George Basbas and Roman Laubert.

References

1. Proceedings Int. Conf. on Inner-Shell Ionization Phenomena and Future Applications, Atlanta, Georgia, April 1972, edited by R. W. Fink, S. T. Manson, J. M. Palms, and P. V. Rao (U.S. Atomic Energy Commission, Oak Ridge, Tenn., 1973).

2. J. Chadwick, Phil. Mag. 24, 594 (1912).

3. W. Brandt, K-Shell Ionization in Ion-Atom Collisions, ref. 1, p. 948.

4. Cf. E. Merzbacher and H. W. Lewis, Handbuch der Physik, edited by S. Flügge (Springer Verlag 1958) Vol. 34, p. 166 ff. and references cited therein.

5. J. Bang and J. M. Hansteen, Kgl. Danske Videnskab. Selskab, Mat.-Fys. Medd. 31, No. 13 (1959).

6. J. D. Garcia, Phys. Rev. A1, 280 (1970); ibid. A1, 1402 (1970); ibid. A4, 955 (1971).

7. G. Basbas, W. Brandt and R. Laubert Phys. Rev. A (in press).

8. W. Brandt, R. Laubert and I. Sellin, Phys. Rev. 151, 56 (1966).

9. G. Basbas, W. Brandt and R. Laubert, Phys. Letters 34A 277 (1971).

10. G. Basbas, W. Brandt, R. Laubert, A. Ratkowski and A. Schwarzschild, Phys. Rev. Letters 27, 171 (1971).

11. W. Brandt and R. Laubert, Phys. Rev. 178, 225 (1969).

12. W. Brandt and R. Laubert, Phys. Rev. Letters 24, 1027 (1970).

13. J. C. Ashley, R. H. Ritchie and W. Brandt, Phys. Rev. B5, 2393 (1972).

14. G. Basbas, W. Brandt and R. Laubert (to be published).

15. U. Fano and W. Lichten, Phys. Rev. Letters 14, 627 (1965).

16. M. Barat and W. Lichten, Phys. Rev. A6, 211 (1972).

FINE STRUCTURE AND HYPERFINE STRUCTURE

OF THE HELIUM NEGATIVE ION*

D. L. Mader and R. Novick

Columbia Astrophysics Laboratory, Columbia University

New York, New York 10027

I. INTRODUCTION

Preliminary fine-structure and hyperfine-structure measurements have been made in the helium negative ion with the isotopes ^{4}He and ^{3}He. Such measurements can be performed to high precision and provide an excellent test of the theory of three-electron atomic systems. Studies of one- and two-electron systems have been very fruitful. The work of Lamb and others on the fine structure of hydrogen has provided a cornerstone of quantum electrodynamics as well as the first precise value for the Sommerfeld fine-structure constant α. Theory and experiment agree to within the experimental accuracy of about 10 ppm for the $^2S_{1/2}-^2P_{3/2}$ interval. Higher precision fine-structure measurements to about 1 ppm have been made in the $(1s2p)\,^3P$ state of helium by Kponou et al.[1] These measurements provide an excellent testing ground for the theory of the two-electron atom and, in particular, of the electron-electron interaction. A recent theoretical work[2] shows promise that in helium the accuracy of the fine-structure calculation can equal that of the measurements. Experiment and the current calculation agree to 1 ppm for the $^3P_0-^3P_1$ interval and to 70 ppm for the smaller $^3P_1-^3P_2$ interval. Such efforts provide a critical test of our understanding of the two-electron system and may lead to an improved value of α.

The three-electron system He$^-$ is amenable to fine-structure measurements accurate to 1 ppm or better. In view of the success of calculations for He, it seems possible that accurate wave functions can be obtained for He$^-$ and that our understanding of three-electron atomic systems can be tested on a fundamental level, comparable to that obtained in hydrogen and neutral helium.

Our preliminary results are of sufficient accuracy to provide a critical test of correlation effects in three-electron wave functions currently available for He⁻. If correlation can be understood in this weakly bound state, it should be well understood in similar states of higher Z. It is also possible that unique three-electron effects[3] can be isolated in He⁻ since it is especially sensitive to correlation.

The helium negative ion is metastable in the $(1s2s2p)^4P$ state, which is the lowest quartet state. This state lies 0.08 eV lower in energy than the 2^3S state of neutral helium.[4] Decay occurs by autoionization to the ground state of He plus a free electron with an energy of 19.7 eV. There are three fine-structure levels in $^4He^-$ with J = 5/2, 3/2, and 1/2. Since the J = 5/2 level is a pure quartet, its decay to the doublet continuum requires a spin flip and ΔL = 2 for parity conservation and proceeds via the tensor part of the spin-spin interaction. The other two J levels decay by this mechanism and also through a slight mixing with the very short-lived doublets of the same configuration. Novick and Weinflash have measured lifetimes for the J = 5/2, 3/2, and 1/2 levels of 500 ± 200, 10 ± 2, and 16 ± 4 μsec, respectively.[5] In the same work a "Zeeman quenching" technique was used to measure the 5/2-3/2 energy interval as $|\Delta_{53}|$ = 1080 ± 270 MHz. The 5/2-1/2 interval, Δ_{51}, was inaccessible to this technique.

The possibility of a high-precision resonance experiment is suggested by the long lifetimes which would yield a natural width of 10 kHz for rf resonance curves.

II.   THEORY

Assuming LS coupling, Manson[6] writes the fine-structure Hamiltonian as

$$\mathcal{H}_{fs} = C_{so}\vec{L}\cdot\vec{S} + C_{ss}[\frac{3}{2}(\vec{L}\cdot\vec{S}) + 3(\vec{L}\cdot\vec{S})^2 - L(L + 1)S(S + 1)] \qquad (1)$$

and predicts values of C_{so} = -1840 MHz and C_{ss} = 342 MHz for the spin-orbit and spin-spin coefficients, respectively. In terms of these constants the fine-structure intervals are:

$$E_{5/2} - E_{3/2} \equiv \Delta_{53} = \frac{5}{2}C_{so} + \frac{15}{2}C_{ss} \quad ; \qquad (2)$$

$$E_{5/2} - E_{1/2} \equiv \Delta_{51} = 4C_{so} - 6C_{ss} \quad . \qquad (3)$$

It can be seen that the difference in sign of C_{so} and C_{ss} will result in a small value for Δ_{53}. Manson's predicted values for the intervals are Δ_{53} = -2030 MHz and Δ_{51} = -9410 MHz.

In ^{3}He$^-$, we have nuclear spin $I = 1/2$ and a hyperfine contribution to the Hamiltonian which Manson writes as

$$\mathcal{K}_{hfs} = a_c \vec{I} \cdot \vec{S} + C_{md} \{ \vec{I} \cdot \vec{L} + \frac{2}{15}[(\vec{I} \cdot \vec{S})L(L + 1)$$

$$- \frac{3}{2}(\vec{I} \cdot \vec{L})(\vec{L} \cdot \vec{S}) - \frac{3}{2}(\vec{L} \cdot \vec{S})(\vec{I} \cdot \vec{L})]\} \quad , \quad (4)$$

again assuming LS coupling. The first term is the Fermi contact term, and the second is the magnetic dipole interaction. Manson estimates $a_c = -3030$ MHz and $C_{md} = 5.6$ MHz. The large value of the contact term is a result of the unpaired s electrons. The fact that most of the hyperfine contribution arises from the (1s2s) core can be seen by comparison with the hyperfine structure in the (1s2s)^{3}S$_1$ state of ^{3}He.[7] If we assume that the (1s) and (2s) electron contact terms are the same in He(^{3}S$_1$) and He$^-$(^{4}P) and if we ignore the 2p electron, then by the method of energy sums we can show that $a_c(^4P) = (4/9)\Delta\nu(^3S_1)$, where $\Delta\nu(^3S_1)$ is the observed hyperfine structure separation in ^{3}He. In this way we estimate $a_c = -2995$ MHz.[8]

It is clear that in ^{3}He$^-$ the hyperfine interaction produces splittings of the energy levels greater than the fine-structure interval Δ_{53}. Thus it is necessary to diagonalize the complete Hamiltonian when calculating energy eigenvalues. In Fig. 1 we show

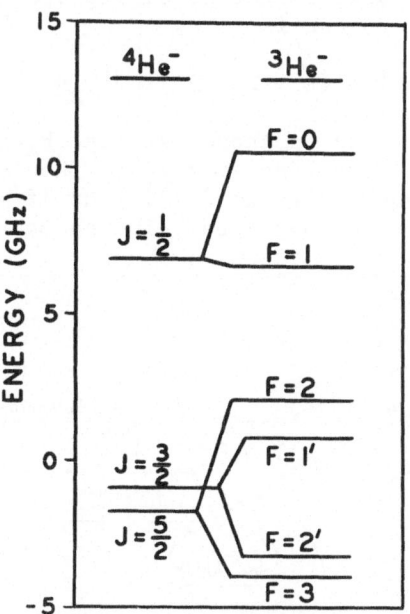

FIG. 1. Energy levels for ^{4}He$^-$ and ^{3}He$^-$.

the zero-field energy levels for $^4\text{He}^-$ and $^3\text{He}^-$. The magnitudes of the constants used in the calculation were measured in the present work, and the signs were predicted by Manson. In Fig. 1 and elsewhere in this paper we denote levels corresponding to the J = 3/2 state in $^3\text{He}^-$ by a prime (') on F, the total angular momentum quantum number.

The energy levels in a magnetic field H are calculated by adding the following term to the Hamiltonian:

$$\mathcal{K}_M = \mu_B(g_L\vec{L}\cdot\vec{H} + g_S\vec{S}\cdot\vec{H}) + \mu_n g_I\vec{I}\cdot\vec{H} \qquad . \qquad (5)$$

The values used for g_L and g_S are single-electron values with the reduced mass correction included in g_L.

The experiment is performed on a beam of He$^-$ ions which passes through an rf interaction region in a time T. Considering two levels a and b with lifetimes τ_a and $\tau_b \gg$ T, the transition rate induced between a and b by an oscillating field of frequency ω is

$$P_{a,b} = \frac{V^2}{V^2 + (\omega_0 - \omega)^2} \sin^2\{\tfrac{T}{2}[V^2 + (\omega_0 - \omega)^2]^{1/2}\} \qquad , \qquad (6)$$

where the transition frequency $\omega_0 = |E_a - E_b|/\hbar$ depends on the static magnetic field, and the perturbation coupling a and b is $\mathcal{K}_{rf} = \hbar \exp(i\omega t)V/2$. Since magnetic dipole transitions are allowed, the rf matrix element is

$$V = |\langle a|(\vec{L} + 2S)\cdot\hat{H}_{rf}|b\rangle|\mu_B H_{rf} \qquad . \qquad (7)$$

The rf field intensity can be chosen so that $P_{a,b}$ attains its maximum value of unity on resonance ($\omega = \omega_0$). Then the resonance curve has a width in frequency units of 0.799/T Hz if T is expressed in seconds. Departures from this ideal line shape arise from a spread

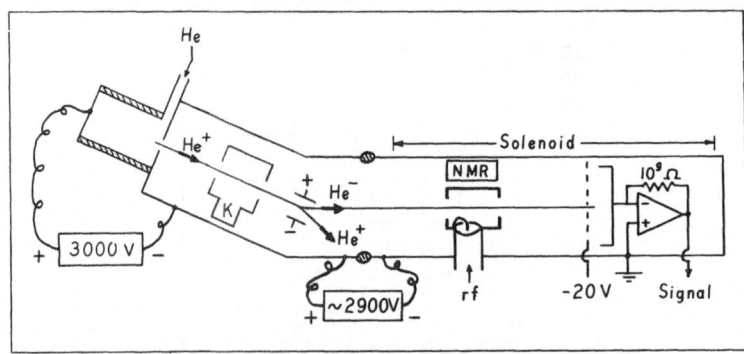

FIG. 2. Schematic diagram of the apparatus.

in beam velocity and a spatial variation in rf field intensity. Additional broadening of the resonance curve is caused by inhomogeneities in the static magnetic field which lead to a spread in ω_0.

In order to detect a transition, a population difference must evolve during the times of flight from the source to the rf region (t_1) and from the rf region to the ion detector (t_2). If the decay rates for the states are written as $\gamma_a = 1/\tau_a$ and $\gamma_b = 1/\tau_b$ and it is assumed that states a and b are both produced at rate r by the source, then the observed change in ion current is

$$\Delta i = er[\exp(-\gamma_a t_1) - \exp(-\gamma_b t_1)]$$
$$\times [\exp(-\gamma_a t_2) - \exp(-\gamma_b t_2)]P_{a,b} \qquad (8)$$

The times of flight depend on the energy (and hence velocity) of the beam. Since the ion detector is movable in vacuum, we have an additional control over t_2.

The decay rates (γ's) depend in general on the static magnetic field. In zero magnetic field and in the absence of hyperfine structure, the levels have a definite value of J. With the application of a magnetic field or hyperfine interaction, states of different J are mixed, and the decay rate is a weighted sum of decay rates of the pure J states. Such mixing produces a significant shortening of the lifetime of the long-lived levels and is the basis of the "Zeeman quenching" technique.[5] Only levels with $M_J = \pm 5/2$ in $^4He^-$ or $M_F = \pm 3$ in $^3He^-$ retain their long lifetime when a static magnetic field is applied.

III. APPARATUS

The apparatus shown in Fig. 2 is essentially that of Ref. 5 with the addition of an rf interaction region. The source consists of a discharge in He gas excited at 80 MHz. A beam of He^+ ions is extracted with an energy of 3 keV. This beam is focused by an Einzel lens to pass through an oven containing potassium vapor. In a nearly resonant charge-exchange process the positive ion is converted into a metastable excited state of the atom. In a second collision the excited He atom absorbs a valence electron from an alkali atom and becomes a negative ion. About 1% conversion of He^+ to He^- can be obtained.[9]

An electrostatic analyzer separates the beam emerging from the oven into its charge components. The negative beam is decelerated to the desired energy and is focused into a 10-m-long drift tube. This tube is wound with a solenoid to provide a static axial magnetic field used to tune the transition frequencies via the Zeeman effect. The rf interaction region is situated 1 m from the end of

the drift tube. Adjacent to this region is an NMR sample which can
be used for precision measurement of the static field above 300 G.
The He⁻ ions are detected with a Faraday cup movable in vacuum.
There is a grid in front of the detector held at -20 V to exclude
electrons released in flight by autoionization.

IV. OBSERVATIONS ON ^{4}He⁻

In Fig. 3, we show the resonant decrease of the ion current as
the magnetic field is swept across the (5/2,-5/2) to (3/2,-3/2)
transition. The rf intensity was adjusted for maximum resonance
strength, seen here as 18% of the total current. With 100% square-
wave amplitude modulation of the oscillator, a lock-in detector can
be used to extract the rf-induced signal from the total ion current.
In Fig. 4, we show the four $\Delta M = -1$ fine-structure transitions from
the 5/2 to 3/2 levels. These resonances have been observed at low
static magnetic field where the various substates of each J level
have nearly the same lifetime. The differing strengths of the
resonances are a consequence of differences among the matrix elements since the rf intensity was too low to cause saturation of the
transitions. These transitions are indicated by the vertical lines
at low field on the energy-level diagram, Fig. 5.

The resonances which were analyzed to provide the fine-structure results of Ref. 10 are shown in Fig. 6. They are identified by
the transitions labeled A, B, C, and D on the energy-level diagram,
Fig. 5. These resonances are observed at relatively high static

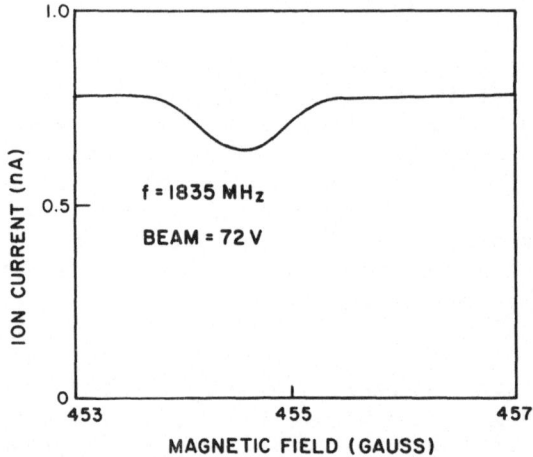

FIG. 3. Resonant decrease in ion current when the (5/2,-5/2)
to (3/2,-3/2) transition is induced by an rf field.

magnetic field, where appreciable mixing of states with the same
magnetic quantum number M occurs.   Transition D is the strongest
because it has the largest rf matrix element and because the M =
-5/2 sublevel retains its zero-field lifetime.   The three smaller
resonances have smaller matrix elements, and the J = 5/2 sublevels
involved have lifetimes shortened from their zero-field values because of mixing with the short-lived J states, leading to smaller
population differences.

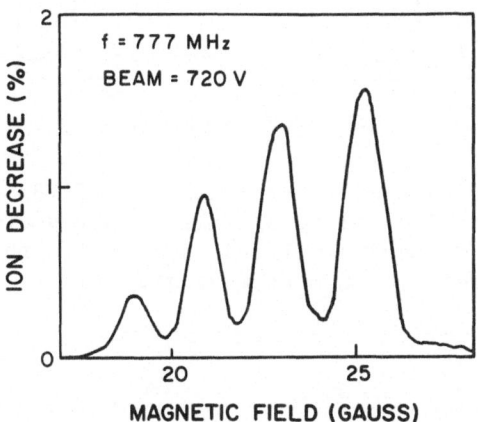

FIG. 4.   Fine-structure transitions with ΔM = -1 from J =
5/2 to 3/2.

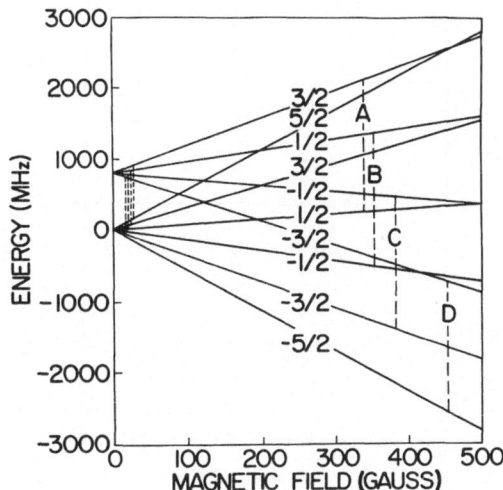

FIG. 5.   Energy sublevels as a function of magnetic field for
the J = 5/2 and 3/2 levels.   Transitions which have been observed
are indicated by vertical dashed lines.

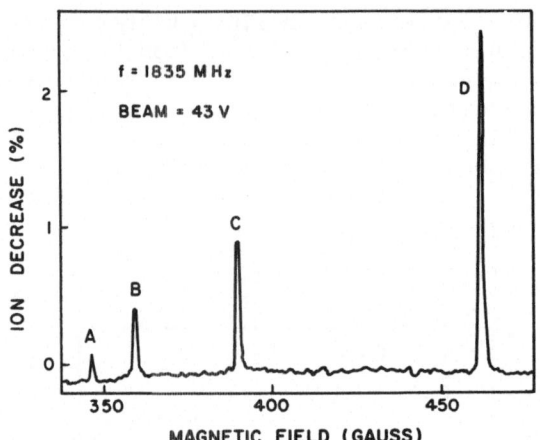

FIG. 6. With an oscillator frequency of 1835 MHz, the $\Delta M =$ +1 transitions from J = 5/2 to 3/2 are observed at high field. Coupling with the distant J = 1/2 levels shifts the three small resonances A, B, and C enough to estimate Δ_{51}.

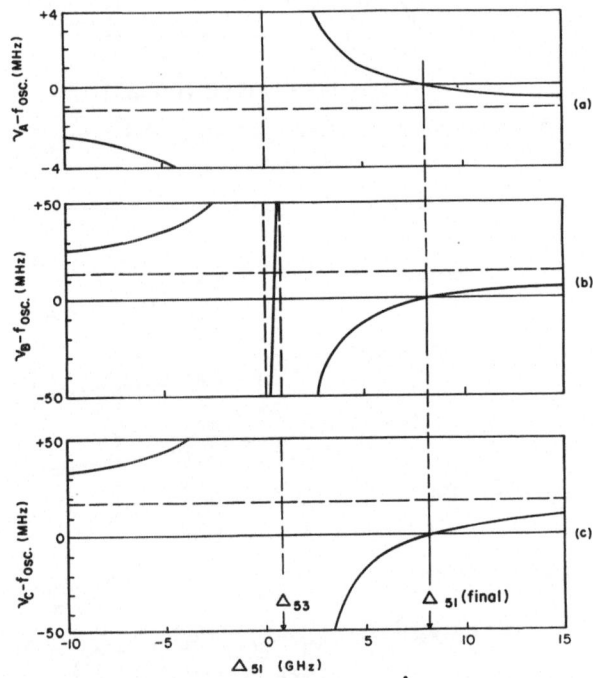

FIG. 7. Calculated transition frequencies for resonances A, B, and C as a function of Δ_{51}. The sign found for Δ_{51} is the same as that assumed for Δ_{53} in this calculation.

In transition D, neither of the participating sublevels has $M = \pm 1/2$, and the transition frequency depends solely on Δ_{53}. However, the remaining three resonances do involve $M = \pm 1/2$ sublevels and have transition frequencies which depend slightly on Δ_{51}. At the magnetic fields where they were observed, the calculated transition frequency for each of the resonances A, B, C, and D changes by -0.2, 1.8, 2.4, and 0 MHz, respectively, if the value of Δ_{51} assumed in the calculation is changed from 8 to 9 GHz, with Δ_{53} fixed at 825 MHz. Thus, A and D serve primarily to establish Δ_{53}. The separation of B and C from these resonances serves primarily to establish Δ_{51}.

Although the rf technique employed here does not establish the sign of the fine-structure intervals, it can be shown that they have the same sign. In Fig. 7, we show transition frequencies for A, B, and C calculated as a function of Δ_{51} at the fields where they were observed. A positive value of 825 MHz was assumed for Δ_{53}. It is seen that there is a unique value of $\Delta_{51} = 8.7$ GHz for which all three calculated transition frequencies agree with the oscillator frequency. Since the intervals have the same sign and $|\Delta_{51}|$ is greater than $|\Delta_{53}|$, we can conclude that the J = 3/2 level lies between the other two. Consequently, the ordering is either the normal structure or fully inverted structure. Manson[6] predicts the fully inverted structure; with increasing energy the order of the states is 5/2, 3/2, and 1/2. The ordering in He⁻ is therefore different from that in the analogous state in excited lithium where it has been shown from studies of the optical quartet spectrum that with increasing energy the order of the states is 3/2, 5/2, and 1/2.[11]

Data were taken on each of the resonances A, B, C, and D under a variety of experimental conditions in a search for systematic effects. The rf power level was varied over a factor of six, beam voltage was changed from 70 to 900 V, and the background pressure of residual gas ranged from 2 to 10×10^{-7} Torr. Forty-one determinations were made of resonance centers using the proton NMR marginal oscillator for field measurements. This body of data was fitted simultaneously with up to 14 parameters: two for fine structure and 12 linear parameters — one for each of the possible shifts with rf power, beam velocity, and background pressure for each resonance. As the number of shift parameters is changed from none to 12, Δ_{53} is bounded by 825.22 and 825.26 MHz, and Δ_{51} has bounds of 8646 and 8672 MHz. These variations are felt to represent the maximum uncertainty due to possible systematic effects with rf power, beam velocity, and background pressure.

A more serious source of possible error is uncertainty in the magnetic field. When resonances are being observed, the NMR probe is displaced from the center of the rf interaction region by 2.5 cm. There are field gradients present which lead to a difference

of ±1/4 G between these two locations. In this preliminary work
we treat this difference as an uncertainty rather than as a correction. The corresponding uncertainties are 0.80 MHz in Δ_{53} and 43
MHz in Δ_{51}. Adding the above uncertainties to these, we obtain

$$|\Delta_{53}| = 825.23 \pm 0.82 \text{ MHz},$$

$$|\Delta_{51}| = 8663 \pm 56 \text{ MHz}.$$

These uncertainties are felt to represent a 68% level of confidence.
We note again that although the sign of these intervals cannot be
established by the technique employed in this work, it has been
shown that they have the same sign. If we adopt the signs found
by Manson, the difference between theory and experiment is about
1 GHz in each case and is within the uncertainty in this simple
calculation.

V. OBSERVATIONS ON ^{3}He$^-$

A helium recirculation system was added to the apparatus for
our work on ^{3}He. In Fig. 8 we show the five $\Delta M = +1$ transitions
from F = 3 to 2' observed at low magnetic field (see Fig. 1). The
zero-field 3-2' interval depends mainly on Δ_{53} and only slightly on
a_c. By estimating the magnetic field from the solenoid current, we
have shown that the value of Δ_{53} is the same to within a few megahertz in ^{4}He$^-$ and ^{3}He$^-$. The three $\Delta M = +1$ transitions between

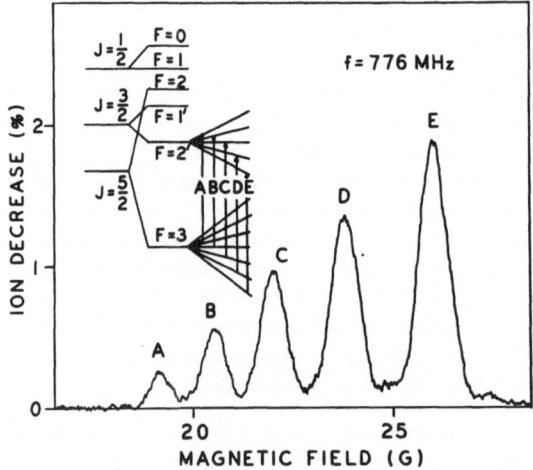

FIG. 8. Transitions in ^{3}He$^-$ between F = 3 and 2' which provide a value for Δ_{53}.

F = 2 and 1' indicated by A, B, and C in Fig. 9 were observed with an oscillator frequency of 779 MHz. Resonance C is shown and was situated at a field high enough to use the NMR gaussmeter. The zero-field 2-1' interval depends strongly on the Fermi contact term. At the present level of data analysis we assume that Δ_{53} equals its value in ^{4}He$^-$ and that $C_{md} = 0$. Using the resonance C, we obtain the Fermi contact term $a_c = -2948.1$ MHz. Due to the present incomplete analysis, we assign an uncertainty of 10 MHz to this quantity, but its accuracy will improve to better than 1 MHz when C_{md} is included in the analysis. A search is under way for the F = 2 to 1 transitions which should lead to a value for Δ_{51} of much improved accuracy.

VI.   CONCLUSION

In the future we plan to make high-precision fine- and hyper-fine-structure measurements. The limitation on accuracy in the work to date has resulted from uncertainties in the magnetic field. This problem can be overcome by observing "field-independent" ΔM = 0 transitions or by making measurements at zero field. If accurate wave functions become available, these measurements will lead to the possibility that our understanding of three-electron atomic systems can be tested on a fundamental level comparable to that obtained in hydrogen and in neutral helium.

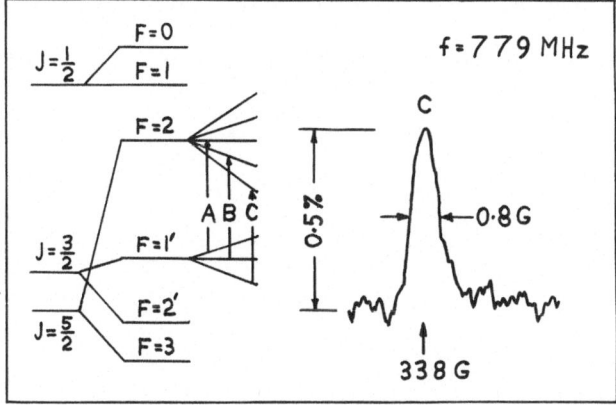

FIG. 9.   Transitions in ^{3}He$^-$ between F = 2 and 1'. These resonances have provided a value for the Fermi contact term.

*Work supported by the National Science Foundation under grant GP 13479. This is Columbia Astrophysics Laboratory Contribution No. 75.

[1] A. Kponou, V. W. Hughes, C. E. Johnson, S. A. Lewis, and F. M. J. Pichanick, Phys. Rev. Lett. 26, 1613 (1971).

[2] M. Douglas, L. Hambro, and N. M. Kroll, Phys. Rev. Lett. 29, 12 (1972).

[3] Ronald J. White and Frank H. Stillinger, Phys. Rev. A 3, 1521 (1971).

[4] B. Brehm, M. A. Gusinow, and J. L. Hall, Phys. Rev. Lett. 19, 737 (1967).

[5] R. Novick and D. Weinflash in Precision Measurement and Fundamental Constants, Proc. Intern. Conf., Natl. Bur. Stds., Gaithersburg, Md., August 3-7, 1970, eds., D. N. Langenberg and B. N. Taylor (Superintendent of Documents, U. S. Govt. Printing Office, Washington, D. C., 1971), NBS Special Publication 343, pp. 403-410.

[6] S. T. Manson, Phys. Rev. A 3, 147 (1971).

[7] S. D. Rosner and F. M. Pipkin, Phys. Rev. A 1, 571 (1970).

[8] W. L. Lichten, private communication, 1972.

[9] B. L. Donnally and G. Thoeming, Phys. Rev. 159, 87 (1967).

[10] D. L. Mader and R. Novick, Phys. Rev. Lett. 29, 199 (1972).

[11] P. Feldman, M. Levitt, and R. Novick, Phys. Rev. Lett. 21, 331 (1968).

STATISTICAL THEORY OF ATOM AND ION POLARIZABILITIES

V. V. Pustovalov, V. P. Shevelko, A. V. Vinogradov

P. N. Lebedev Physical Institute

Academy of Sciences, Moscow, U.S.S.R.

We used the statistical Thomas-Fermi model of atomic structure to describe the existing experimental data on polarizabilities
of atoms and ions with closed electronic shells. Our aim was to
get a theory without any approximations not connected with basic
assumptions of the statistical model.

Atomic electrons in the statistical model are considered as a
degenerate nonuniform Fermi gas. The electron density $\rho(\vec{r})$ and
potential $\phi_0(\vec{r})$ inside the atom satisfy the equations*

$$\nabla^2 \phi(\vec{r}) = 4\pi\rho(\vec{r}) \tag{1}$$

$$\frac{[3\pi^2\rho(\vec{r})]^{2/3}}{2} - \phi(\vec{r}) = E_F \quad , \tag{2}$$

where $\vec{r}$ is the distance from the nucleus, and E_F is the Fermi
energy. These equations are valid both for an isolated atom and
an atom placed into an external field. The potential $\phi_0(r)$ and
electron density $\rho_0(r)$ of an isolated atom are the spherically
symmetrical solutions of equations (1)(2) with $E_F = -(Z-N)/r_0$,
r_0 being atomic radius, Z nuclear charge, and N the number of
electrons (for neutral atom $Z = N$). If the atom is put into the
weak uniform external field $\vec{\varepsilon}$ the electron density and potential
will suffer small deviations

---

*Atomic units are used.

$$\phi(\vec{r}) = \phi_0(r) + \phi_1(\vec{r}), \quad \rho(\vec{r}) = \rho_0(r) + \rho_1(\vec{r}) \quad . \quad (3)$$

The equations for the induced potential $\phi_1(\vec{r})$ and density $\rho_1(\vec{r})$ are obtained by putting (3) into (1)(2) and keeping only the terms of the first order[1]:

$$\nabla^2 \phi_1(\vec{r}) = \frac{4\sqrt{2}}{\pi} \left[\phi_0(r) - \frac{Z-N}{r_0} \right]^{1/2} \phi_1(\vec{r}), \quad r < r_0 \quad (4)$$

$$\rho_1(\vec{r}) = \frac{1}{\pi^2} \left[2\left(\phi_0(r) - \frac{Z-N}{r_0}\right) \right]^{1/2} \phi_1(\vec{r}), \quad r < r_0 \quad . \quad (5)$$

Outside the atom $\phi_1(\vec{r})$ is the sum of potentials of the external field and the induced atomic dipole:

$$\phi_1(\vec{r}) = -(\vec{\varepsilon} \cdot \vec{r}) + \alpha \frac{(\vec{\varepsilon} \cdot \vec{r})}{r^3}, \quad r > r_0 \quad , \quad (6)$$

where α is atomic polarizability. The latter may be found by matching, at the edge of the atom $r = r_0$, the potential $\phi_1(\vec{r})$ inside the atom [which is the solution of (4)] with the potential outside the atom given by equation (6). On doing this one obtains[1]:

$$\alpha = r_0^3 \frac{1 - a/r_0}{1 + 2a/r_0}, \quad a = \left[\frac{\phi_1'(r)}{\phi_1(r)} \Big|_{r=r_0} \right]^{-1}$$

where $\phi_1(\vec{r})$ is defined so that $\phi_1(\vec{r}) = \phi_1(r) \cos(\widehat{\vec{\varepsilon}\vec{r}})$ is the regular solution of equation (4) at $r = 0$.

Formulae (4)(5) show that the density $\rho_1(\vec{r})$ and potential $\phi_1(\vec{r})$ of atomic charge induced by the external field are solutions of a self-consistent problem and cannot be expressed algebraically in terms of the nonperturbed Thomas-Fermi density $\rho_0(r)$. It is at this point that this work differs from the previous statistical calculations[2,3] of atomic polarizabilities.

The curve in the figure presents the dependence of α/r_0^3 on the scaled atomic radius $x_0 = r_0 Z^{1/3}/0.885$ obtained by numerical solution of equation (4). Experimental results are shown in the same figure. The values of atomic radius r_0 were taken from Thomas-Fermi-Dirac model calculations, so the effect of exchange is partly taken into account. For atoms and ions heavier than Ar the agreement between calculated and measured values is reasonable. This fact gives a hope that the curve presented in the figure can

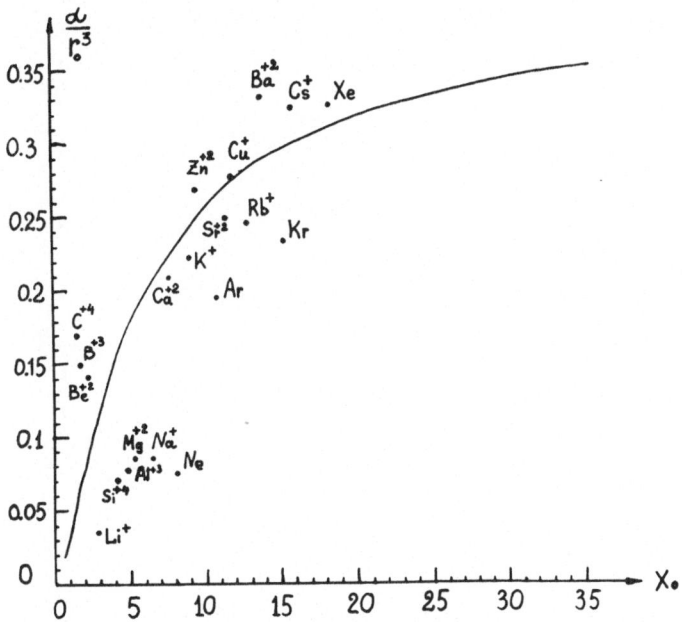

be used for evaluation of polarizabilities of atoms and ions for which the experimental data are not obtained yet.

An analytical expression for polarizability may be obtained for multi-charged ions, when the polarization of atomic electrons just slightly screens the external field. Then it is not necessary to solve (4); one may substitute

$$\phi_o(r) = \frac{Z-N}{r}; \qquad \phi_1(\vec{r}) = -(\vec{\epsilon} \cdot \vec{r})$$

into (5) and find polarizability as it is usually done:

$$\alpha = -\frac{1}{\epsilon} \int \vec{r} \rho_1(\vec{r}) d\vec{r} \simeq \frac{63}{16} \frac{N^3}{Z^4}, \qquad Z \gg N \quad .$$

This expression agrees with the law $\alpha \sim Z^{-4}$ obtained for atoms with all closed electronic shells from exact quantum theory.

References

1. A. V. Vinogradov, V. V. Pustovalov, and V. P. Shevelko, JETP, 1972 (in press).

2. P. Gombas, Die Statistische Theorie des Atoms und ihre Anwendungen (Wien, 1949).

3. R. M. Sternheimer, Phys. Rev. 80, 102 (1950); 96, 951 (1954).

AB INITIO CALCULATIONS OF ATOMIC ENERGY SPECTRA

A. P. Jucys

Institute of Physics and Mathematics of the Academy of

Sciences of Lithuanian SSR, Vilnius, Lithuanian SSR

FOREWORD

In this review paper the main features of ab initio calculations of atomic energy spectra are described. The attention is
concentrated to the methods based on the variation principle. The
theory of using non-orthogonal (Hartree-Fock or analytical) radial
orbitals is indicated. The improved methods of the atomic spectra
theory are characterized. These are: a) method of superposition
of configurations, b) extended method of calculation, and
c) method of incomplete separation of variables. The facts,
which hamper mastering of the theory as well as its further development, are mentioned. The necessity of using as much as possible
unique mathematical apparatus and more close collaboration between
scientists of this field is stressed.

1. INTRODUCTION

The first was Schroedinger(1) who investigated the hydrogen
atom with the help of wave mechanics as early as 1926. This has
been "ab initio calculation" of one-electron atom. It was the
starting point not only for the investigation of atoms on the basis
of new physical theory, but also for the development of the theory
itself. The famous Schroedinger equation has its origin in these
considerations.

.Unfortunately, the Schroedinger equation is exactly solvable
only for one-electron atoms. The attempts to solve it approximately
for two-electron atoms were done by Hylleraas(2,3) and Eckart(4)
as early as 1928 - 1930. These attempts succeeded in devising
elegant methods, which have not been immediately generalized to the

many-electron atom. However, they contained all the main features used later for improving the ordinary many-electron atom theory.

The starting point for the many-electron atom theory was Bohr's(5) concept that approximately every electron of the atom can be regarded as being in a stationary state in the field of the nucleus and of the charge distribution of the other electrons of the same atom. This idea is several years older than quantum mechanics itself. It allowed construction of the approximate wave function of the whole atom with the help of one-electron wave functions, the latter being calculated taking into account the approximate states of all the electrons of the atom in order to bring into self-consistency the field of electrons to each other.

Thus has been born the many-electron atom theory, which we call ordinary. The ways of improving it have been found in the two-electron atom theory mentioned above. The framework of this paper is conditioned by this fact. At first we take notice of methods connected with the proper application of the ordinary theory and afterwards we are going to indicate the ways of refining the theory. We conclude with some considerations of a general character.

2.   HARTREE-FOCK METHOD.   SEPARABLE SPIN-ORBITAL APPROXIMATION

Hartree(6,7) was the first who put into reality Bohr's ideas. He worked out a method for calculating the radial parts of one-electron wave functions, the angle and spin parts having analytical expressions of the same kind as in one-electron atom theory. It was Slater's(8) idea to join the spin part to the orbital part to make the spin-orbital, as it is called nowadays.

Hartree(6) obtained his equations taking into account the screening of the nucleus by the radial distribution of the charge of other electrons. The connection of the Hartree equations with the variation principle has been observed independently by Fock(9) and Slater(10). Fock(9) applying the variation principle to the non-relativistic energy functional obtained the generalized Hartree equations. They are called Hartree-Fock equations.

The Hartree-Fock self-consistent field method has been worked out on the base of the approximation, which, according to Hartree(7), is called the separable spin-orbital (wave function) approximation. It is characterized by:
   2a.   prescription of a definite configuration to the electronic states of the atom,
   2b.   all the electrons of the shell of equivalent electrons being described by the same radial orbital,
   2c.   the wave function of the whole atom not containing the inter-electronic distances.

This approximation is sometimes called the one-particle or independent-particle approximation. Neither the first nor second term is relevant here, because we are almost always dealing with the spin-orbitals in a number exceeding the number of particles under consideration. For example, in order to construct the anti-symmetric wave function of the state ^1S of two p-electrons we must use six spin-orbitals with different sets of $m_l m_s$ (= 1 1/2,-1 -1/2; -1 1/2, 1 -1/2; 0 1/2, 0 -1/2). Thus, the one-electron as well as the independent-electron concept is out of place here. On the other hand, we are dealing here with spin-orbitals, which are separable in some respect, and Hartree's term "separable spin-orbital approximation (SSOA)" is much more appropriate.

Sometimes the term Hartree-Fock approximation is used instead of SSOA. However, this is a little ambiguous, because the main idea of Hartree and Fock was the self-consistent field method (SCF), which is more general than the approximation which was used in those days. For example, the Hartree-Fock SCF method works very well in the approximations far beyond SSOA. When conserving the names of Hartree and Fock for approximation, then it would be relevant to join the names of Bohr and Slater, because there has been Bohr's(5) idea and Slater's(8) antisymmetric combinations of spin-orbitals, on which the method of Hartree(6) and Fock(9) is working so splendidly. In such a case one would say Bohr-Slater-Hartree-Fock approximation on one hand, and HFM of SCF, on the other.

The calculation based on SSOA we shall call the ordinary method of calculation (OMC) in comparison with improved methods of calculation based on the approximations obtained by giving up at least one of the restrictions 2a,b,c quoted above. The Hartree-Fock equations in SSOA we call then ordinary or conventional HF equations when it is necessary to distinguish them from the corresponding equations in approximations other than SSOA.

HF equations mostly are solved numerically by methods worked out by Hartree and Fock themselves. There are a number of programs written for this purpose in different parts of the world. A very widely used program is that of Froese-Fischer(11) in FORTRAN IV language. Almost on the same lines is the program written by Bogdanowich and Karazija(12) for the computer BESM-4 in the machine language. Another method of solving HF equations is due to Roothaan (13), which consists in searching for HF radial orbitals in analytical form.

The procedure for solving HF equations is rather complicated in any case. The main difficulty comes from the exchange terms accounting for the exchange part of the electrostatic interaction between different shells of electrons, the exchange part inside the shells of equivalent electrons being accounted for by the same form of radial integrals as the direct part of interaction; thus they do no harm. Attempts have been made to substitute for the exchange terms of the HF equations simpler expressions accounting for the exchange terms as well as possible. It was Fock(14) who first began to solve this problem by constructing the so-called classical analog of the exchange operator. At the present time Slater's(15) method for averaging statistically the exchange part of the Hartree-Fock equations is used widely. There are a number of programs written for solving such equations, for example, those of Herman and Skillman(16) and Cowan(17).

Besides the solutions of HF equations there are used so-called variational analytical radial orbitals. Those radial orbitals have been used by Hylleraas(2,3), Eckart(4), Zener(18), Morse et al.(19), Fock and Petrashen(20) and so on. The analytical form of these radial orbitals differs more or less from author to author. It has been found by Kupliauskis et al.(21) that the so-called generalized hydrogen-like radial orbitals, having the expressions

$$P(n\ell/r) = N_{n\ell} r^{\ell+1} \left[\exp(-a_{n\ell} r) + C_{n\ell} \exp(-a'_{n\ell} r) \right] , \ell = n-1 \quad (2.1)$$

$$\begin{aligned} P(n\ell/r) = N_{n\ell} r^{\ell+1} & \left[\exp(-b_{n\ell}^{(0)} r) + C_{n\ell}^{(1)} r \exp(-b_{n\ell}^{(1)} r) + \cdots \right. \\ & \left. \cdots + C_{n\ell}^{(n-\ell-1)} r^{n-\ell-1} \exp(-b_{n\ell}^{(n-\ell-1)} r) \right] , 0 \leq \ell < n-1 \end{aligned} \quad (2.2)$$

very well reproduce HF results. If we neglect one exponential function in (2.1) and let all the exponentials equal each other in (2.2), we obtain the hydrogen-like radial orbitals of Fock and Petrashen(20).

All the $a_{n\ell}$'s as well as $C_{n\ell}$'s in (2.1) are to be determined by the variation principle. $N_{n\ell}$ is the normalization factor. $C_{n\ell}^{(i)}$'s in (2.2) are to be found from the orthogonality conditions, when orthogonal radial orbitals are used. On the other hand, these $C_{n\ell}^{(i)}$'s are treated as variational parameters together with $C_{n\ell}^{(i)}$'s when working with non-orthogonal radial orbitals. It has been found by Matulaityte et al.(22) that orthogonal (2.1) and (2.2) give practically the same results as non-orthogonal ones. On the other hand, orthogonal hydrogen-like radial orbitals give much poorer results than non-orthogonal ones.

It stands to reason that the calculations with orthogonal radial orbitals are much easier to carry out than with non-orthogonal ones. However, we cannot always confine ourselves to the orthogonal radial orbitals. This is the case of excited configurations even in the ordinary method of calculation (OMC), saying nothing of improved methods of calculation, where the non-orthogonal radial orbitals are inevitable in many cases.

Fock[9] in his first paper on the SCF problem observed that in the case of excited configurations the radial orbitals (at least some of them) are not orthogonal when there are lower configurations with the same symmetry. Smith[23] in 1932 wrote down HF equations for such a case. Only after 30 years Sharma and Coulson [24] raised this problem anew. Some subsequent papers dealt with some special cases. The more general case has been treated by Prosser and Hagstrom[25]. Quite a general technique to deal with nonorthogonal radial orbitals has been worked out by Jucys et al. [26], who have given methods for constructing the wave functions and expressing the matrix elements of operators without any restriction in the atomic systems under consideration.

When dealing with the excited configurations, the first step is to impose the conditions

$$<K_i LS | K_j LS> = \delta(K_i, K_j) \tag{2.3}$$

on the overlap integral between the configurations K_i and K_j, having the same term LS. This overlap integral is expressed in terms of individual overlap integrals

$$<n_i \ell | n_j \ell> = \int_0^\infty P(n_i \ell/r) P(n_j \ell/r) dr, \tag{2.4}$$

which are not necessarily unity when $n_i = n_j$, nor zero when $n_i \neq n_j$. If, for instance, $K_i = K_1$ is the lowest configuration, having the term LS, then for every configuration K_j with $j>1$ (2.3) must be satisfied. Solving the problem for a definite K_j we must establish the maximal number of conditions

$$<n_j \ell | n_j' \ell> = \delta(n_j, n_j') \,, \tag{2.5}$$

with which the conditions (2.3) do not interfere, in order to work with the maximal number of orthogonal radial orbitals.

When working with the solutions of HF equations the conditions (2.5) are secured with the help of non-diagonal Lagrange multipliers in the same way as in the case of entire orthogonality of radial orbitals. The conditions (2.3) are secured by Lagrange multipliers connecting radial orbitals of different configurations. On the other hand, when work is going on with the help of analytical radial orbitals (2.1) and (2.2), for instance, the conditions (2.3) as well as (2.5) are used to find the expressions for as many of the parameters $c_{n\ell}^{(i)}$ as possible. In any case the conditions (2.3) must be satisfied in the first place.

All the calculations to be carried out in expressing the matrix elements of operators are greatly facilitated by the use of the mathematical apparatus of irreducible tensor operators of Wigner

(27,28) and Racah(29,30) complemented by the graphical methods
worked out by Jucys et al.(31,32) as well as by El-Baz et al.
(33,34). These methods based on the fractional parentage coefficients of Racah(35) tabulated by Nielson and Koster(36),among others,
give a very powerful tool of quantum theory of many-electron atoms.
On these lines are the handbooks written by Judd(37,38) and Wybourne
(39) giving the main principles of mathematical apparatus under
consideration and those by Slater(40), Sobelman(41), Wybourne(42),
Fraga and Malli(43) and Armstrong(44) concerning the practical
application of these methods to atomic spectroscopy. Quite recently
the book by Jucys and Savukynas(45) expounds the methods of the
calculation of matrix elements of operators using the methods just
described.

In practical applications of the theory the first task always
is to perform the calculations with the angle and spin parts of the
matrix elements of operators. Just here the Wigner-Racah formalism
referred to above works very excellently. Then one is left with
radial integrals, which are calculated after the radial orbitals
are estimated in one or another way, except for the semiempirical
approach in which the radial integrals are estimated with the help
of experimental data.

3. IMPROVED METHODS OF CALCULATION. THEORY OF CORRELATION.

The quantum theory of atoms based on the separable spin-orbital
approximation (SSOA) scored a beautiful success. The many-electron
atomic systems have been solved in a rather high degree of accuracy.
However, the precision of experiment in many cases is much higher
than theoretical results. For this reason the semiempirical methods
are used in cases when experimental data are available in a sufficient amount. Nevertheless, semiempirical methods do not contribute
much to the development of the theory itself. On the other hand,
the experimental results are not always available. For this reason
attempts are made to improve the theory in order to obtain good
results by ab initio calculations. This means lifting restrictions
on which SSOA is based.

Disregarding but one of the restrictions in the same order as
they are given by 2a,b,c, we obtain these improved methods (approximations) of calculation:

    3a. method of superposition of configurations (MSC) or multi-
           configuration approximation (MCA);
    3b. extended method of calculation (EMC);
    3c. method of incomplete separation of variables (MISV).

In the first method of improving the theory the one-configura-
tional wave function is substituted for the linear combination of

wave functions differing from each other by some sets of configurational quantum numbers $n\ell$. In the second the number of radial orbitals is increased to be equal to the number of electrons in the shell. In the third method the wave function as constructed in the ordinary method of calculation (OMC) is multiplied by the symmetric function depending on the inter-electronic distances.

The MSC has its origin in the early paper of Hylleraas(2) on the ground state of the helium atom. It is the first method of improving the theory applied to the many-electron atom. It is referred to already in the famous book of Condon and Shortley(46), though on the basis of semiempirical calculations. The first ab initio calculations for many-electron atomic systems have been performed by Hartree and Swirles(47). MSC is often called the configuration interaction approach. Unfortunately, this term is in contradiction with the basic principle of quantum mechanics formulated as the superposition of states. It makes no sense to talk about an interaction of states, because there are physical systems that interact. What is referred to as configuration interaction is responsible only for the non-diagonal matrix elements of the Hamiltonian, having no physical meaning before diagonalization of the matrix.

The EMC takes its origin in the paper of Eckart(4) on the helium atom, as well. He used analytical radial orbitals. By the method of SCF of HF the same problem has been solved by Trefftz et al. (48). In this approach the separability of the angle and spin parts of spin-orbitals is conserved. There are radial orbitals for which the separability is lifted. Thus, the extended method gives the refinement of the radial part of the wave function of the whole physical system.

The MISV takes its origin in the paper of Hylleraas(3), in which the helium atom was treated with the help of the wave function involving the inter-electronic distance. The theory of MISV for many-electron atoms has been extended by Fock et al.(49) and Jucys(50).

Any refinement of the SSOA is generally identified with the calculation of the correlation effects. This term, though not very appropriate, is deeply rooted among the physicists working in the theory of many-particle systems. The refined wave functions are said to be correlated. This last term is especially used in the case of the MISV.

All the improved methods of calculation mentioned above satisfy the standard requirements of the theory. This means that the wave function must be a proper function of the squares of the orbital and spin angular momenta and, besides that, the equivalency condition is to be satisfied (the shell of equivalent electrons is to have vanishing orbital and spin momenta at the number of electrons $4\ell + 2$).

Now we are going to characterize all three improved methods of calculation in the same order as given by 3a,b,c.

<u>Method of Superposition of Configurations (MSC)</u>. In this case the wave function of an N-electron atomic system schematically may be written in the form

$$|K(N)\rangle = \sum_{(u_k)} C_{(u_k)} |K_{(u_k)}(N-k,k)\rangle . \qquad (3.1)$$

Here $K(N-k,k)$ is the configuration built up from two groups of electrons, $N-k$ and k, k taking on values from 0 to N. u_k stands for a set of shells in the configuration. At a given value of k there is left a high degree of freedom in detaching these k electrons from the configuration $K(N,0)$ and selecting a new configuration with a definite u_k. Summation takes place with respect to all the selected configurations at a given value of k, this fact being indicated by taking u_k into brackets.

The expression (3.1) is sometimes referred to as a cluster expansion, two groups of electrons being considered as constituting some kind of cluster. This may be done only very formally, because in atoms the natural clusters are shells of electrons. For this reason it is expedient to couple the spin-orbitals in $K(N-k,k)$ in the same way as it is done in the configuration $K(N,0)$. For instance, let us have $K(N,0) = 1s^2(^1S)2s^2(^1S)2p^{N-4}(LS)$, and let $k = 2$, these two electrons being taken from the first shell. Then one writes $K(N-2,2) = n\ell^2(^1S)2s^2(^1S)2p^{N-4}(LS)$ where $n\ell$ are quantum numbers of two electrons detached from the configuration $K(N,0)$. Thus the other terms of these two electrons are not to be taken into account because they are absent in the configuration $K(N,0)$.

If we say, for instance, that $C_{(u_0)}$ is a maximal one, then the other terms with $k > 0$ in (3.1) give the correction to the one-configuration approximation for the configuration $K(N,0)$ under certain conditions, of course, this correction being called the correlation effect. The selection of configurations $K(N-k,k)$ is dictated by their effects on various physical quantities to be calculated. Generally speaking the sets of configurational quantum numbers $(n\ell)$ nearer to those of $K(N,0)$ are to be taken into consideration at first.

Of particular importance are the configurations with $k = 2$ and 1, the first case being especially important in light atoms. The configurations $K(N-2,2)$ Sinanoglu(51) classifies in the

following manner. If no new set of nℓ appears in K(N-2,2) as
compared with K(N,0) one has internal correlation. On the other
hand, when one new set appears, then one has semi-internal correlation. Finally, when both spin-orbitals have their quantum numbers
nℓ outside the configuration K(N,0), one has the external correlation. In the case k = 1 there may be only internal and external
correlation.

The case k = 2 is said to give pair correlation. This is
the same as the two-electron state approximation of Jucys et al.(52),
this last term not being in use any more. The first ab initio
calculations by MSC by Hartree and Swirles(47) to some extent
belong to this kind of correlation calculation. The problem of
pair correlation calculation is reviewed by Condon(53). It has
been and is being further developed and applied by a great number
of physicists and chemists in different parts of the world.

In pair correlation calculations there appear the so-called
many-fold pairs. They may be linked or un-linked pairs, or
clusters as they sometimes are referred to. We demonstrate how it
looks by the two-fold pair correlation in beryllium-like atomic
systems. From the shell-by-shell improvement in the above-mentioned
two-electron state approximation calculations(52) we taken this
scheme

$$(1s^2 + p_1^2 + \ldots)\ (2s^2 + p_2^2 + \ldots) \approx$$

$$\approx 1s^2 2s^2 + 1s^2 p_2^2 + \ldots + p_1^2 2s^2 + \ldots + p_1^2 p_2^2 + \ldots \qquad (3.2)$$

The last configuration written explicitly is un-linked two
two-electron clusters, as named by Sinanoglu(54). It is evident
that radial orbitals p_1 and p_2 are different because they give
correlations to different shells, and there is no justification to
write p^4 as giving two two-electron clusters linked into one
four-electron cluster.

It has been found that it is expedient to construct the wave
function of the whole system with the help of the two-electron wave
functions as suggested by Fock(55). Such a case is in the approximation (3.2) for the beryllium-like atom. These two-electron wave
functions at present are called geminals. This concept is also used
in MISV discussed below.

If in (3.1) there are used radial orbitals estimated in one-
configuration approximation, the convergence is rather poor.
Hartree et al. (56) proposed to take into account the expression
(3.1) in the very beginning of the application of the Fock variation method. This leads to the HF equations involving the so-called
configurational terms, which sometimes are more important than the

exchange terms. Such an approach is reasonable to call the multi-
configuration approximation (MCA) in order to distinguish it from
the method in which the linear combination (3.1) is used only
after the radial orbitals are determined in separate one-configuration approximations. The MCA speeds up the convergence of the
method considerably, because the configurational terms in the
HF equations displace the radial orbitals of k electrons of the
configuration K(N-k,k) into the direction of the corresponding
radial orbitals in the configuration K(N,0). The result is that
overlapping of these orbitals from two configurations is greatly
increased as is demonstrated by calculations carried out by Jucys
et al.(57) and other recently calculated results. For example,
in the case of the approach schematically written in equation (3.2),
the result of the displacement mentioned is such that it has enough
6 configurations in order to cover 90 per cent of the correlation
energy in the beryllium atom.

The MSC is very widely used for taking into account correlation effects. Many papers are published and are being published
on this problem. To this branch of research belong many papers of
Sinanoglu et al.(51,58 and others), Clementi et al.(59), Harris
et al.(60, and others), Weiss(61, and others) and many, many others.
The approach of Nesbet(62) belongs to the same kind of refinement
of the SSOA. The difference consists only in the way of estimating
the orbitals. For this purpose he is solving the equations similar
to those of Bethe-Goldstone.

Extended Method of Calculation (EMC). For going over from the
OMC to the EMC the radial part of the wave function of a shell
of equivalent electrons is to be subjected to this substitution(63):

$$R(\ell/r_1)R(\ell/r_2) \ldots R(\ell/r_N) \rightarrow$$

$$[N! S_N(\ell)]^{-\frac{1}{2}} \sum_P PR(\ell^{(a)}/r_1)R(\ell^{(b)}/r_2) \ldots R(\ell^{(z)}/r_N), \quad (3.3)$$

$$S_N(\ell) = \sum_P P<\ell^{(a)}|\ell^{(1)}> <\ell^{(b)}|\ell^{(2)}> \ldots <\ell^{(z)}|\ell^{(N)}> . \quad (3.4)$$

The summations take place over all possible permutations

$$P = \begin{pmatrix} 1 & 2 & \cdots & N \\ a & b & \cdots & z \end{pmatrix} , \quad (3.5)$$

the sums themselves being permanents[*] of the radial orbitals

---

[*] Permanents are the symmetric functions which are obtained by
taking a determinant, expanding it in the usual way, and then changing all negative signs to positive signs.

$R(\ell^{(i)}/r_j)$ and overlap integrals $<\ell^{(i)}|\ell^{(j)}>$ of the type (2.4), correspondingly. The permanent in (3.3) is to ensure the symmetry of the radial part of the wave function as it is in the case of the OMC. The antisymmetry is secured by the angle-spin part of the same wave function, which is not subjected to any change in going over from the OMC to the EMC. For this reason the coefficients at the radial integrals are the same as in OMC, the fact which facilitates the application of EMC considerably. The radial integrals themselves undergo this substitution:

$$\binom{N}{q} R(\ell\ell,\ell\ell,\ldots,\ell\ell) \rightarrow S_N^{-1}(\ell) \sum_{i_q>\ldots>i_1} S_N(i_1 i_2 \ldots i_q t,\ell). \quad (3.6)$$

Here $S_N(i_1 i_2 \ldots i_q t,\ell)$ is the permanent of overlap integrals, in which rows i_1, i_2, $\ldots$, i_q are substituted by radial integrals

$$R(\ell^{(i_1)}\ell^{(i_1)},\ell^{(i_2)}\ell^{(i_2)}\ldots\ell^{(i_q)}\ell^{(i_q)}) =$$

$$= \int_0^\infty (q) \int_0^\infty R(\ell^{(i_1)}/r_\ell)R(\ell^{(i_2)}/r_2) \ldots R(\ell^{(i_q)}/r_q)t(r_1,r_2,\ldots,r_q) \times$$

$$\times R(\ell^{(i_1)}/t_1)R(\ell^{(i_2)}/r_2) \ldots R(\ell^{(i_q)}/r_q)r_1^2 dr_1 r_2^2 dr_2 \ldots r_q^2 dr_q, \quad (3.7)$$

$t(r_1,r_2,\ldots,r_q)$ being the radial part of q-electron operator. When q = 2,1, one has two-electron, one-electron radial integrals, correspondingly.

When the EMC is to be applied but to one shell of equivalent electrons, for all other shells conserving the OMC, we can still carry out all the calculations with the help of orthogonal radial orbitals between shells having the same orbital quantum number. To such a case belong the calculations of Jucys et al.(64) as well as the early calculations of Eckart(4) and Trefftz et al.(48), and quite a lot of other calculations referred to in reference 64.

When the EMC is to be applied to several shells of electrons with the same orbital quantum number, the calculations are to be carried out with the help of non-orthogonal radial orbitals between these shells. The general methods of doing such calculations beyond OMC are confined to the operations with the permanents of overlap integrals involving substitutions similar to those indicated in equation (3.6). These operations are very easy to program for computers. Some trouble is in solving the corresponding HF equations because of many second derivatives in the same equation. These equations have been solved only for the helium atom by Trefftz et al. (48). For the time being calculations are carried out mainly with the help of analytical radial orbitals.

The main feature of the EMC is the fact that the substitutes for the radial integrals are less than these integrals calculated in OMC, as expected from the comparison of semiempirical values of these integrals with those calculated in the OMC. A consequence of this is the reduction of the range covered by the energy spectrum of a given configuration as required by experimental data. However, this reduction is not sufficient to obtain a good agreement with experiment.

Further improvement of the theory is attained by a simultaneous application of the EMC and MCA. In such a case there are to be dropped out all the configurations $K(N-k,k)$, in which the fundamental quantum number is changed without changing the orbital quantum number, because this part of the correlation is taken into account by EMC. The EMC reduces the range of the energy spectra, on one hand, and the MCA displaces terms with respect to each other, on the other hand. All these changes are of such a kind that theoretical results are brought closer to experimental data as is seen, for instance, from the calculations of Stasiukaitis et al.(66). In the total, the EMC joined with MCA gives a rather swift convergence of the method. For example, in the case of the helium atom ground state it is enough three-configuration approximation 1s1s——2p2p——3d3d in order to cover 88 per cent of the correlation energy, when the radial orbitals of the type (2.1) are used.

Another method of calculation connected with an increased number of radial orbitals of the shell of equivalent electrons was proposed by Kaldor and Harris(67). That is the so-called Spin-Optimized Self-Consistent Field. They use the wave function of the whole atomic system in the form

$$| K > = AF \sum_{k} t_k Q_k \quad . \tag{3.8}$$

Here A is antisymmetric, F means the product of orbitals and Q_k are linear combinations of all independent spin functions belonging to S and M_S without limitation to the spin compensation in closed shells. The orbitals in F and coefficients t_k and Q_k are optimized.

This method is a generalization of the Spin-Extended Hartree-Fock method as proposed by Löwdin(68) consisting in prescription of "different orbitals for different spins". In this last approach the one-determinantal wave function is spin projected and afterwards orbitals determined by the optimization procedure. The wave functions being proper functions of total spin are not those of the total orbital angular momentum. This method, in its turn, is a generalization of the so-called Unrestricted Hartree-Fock method, in which the wave function is the proper function

neither of orbital nor spin angular momenta. Relations between
methods similar to those just mentioned are briefly discussed by
Kaldor and Harris(67). Almost all of them suffer at least one of
the shortcomings named above, i.e., the equivalence restriction
and the requirement of the wave function to be the proper function
of both orbital and spin angular momenta.

Method of Incomplete Separation of Variables (MISV). In
the MISV the wave function of the atom may be written schematically
in the form

$$|K_{inc}> = V|K> \quad .$$

(3.9)

where $|K>$ is the wave function in OMC and the factor V is a
function involving inter-electronic distances. This function
must be symmetric in electronic coordinates in order not to spoil
the symmetry of the function.

There is a rather high degree of freedom in the choice of the
form of the function V. Widely used is the factor V of Hylleraas
(3) type generalized as well as simplified taking into account
the complexity of many-electron atoms and mathematical difficulties
arising therefrom, because in the case of many-electron atomic
systems there is no possibility to carry out the calculations in
such an extent as it has been done, for instance, by Pekeris(69)
for two-electron atomic systems. Such a factor may be written
in the form

$$V(r_{12}, r_{13}, \ldots) = a_0 + a_1 \sum_i r_i + q \sum_{j>i} r_{ij} + \ldots$$

(3.10)

The terms, not written explicitly, involve squares and higher
powers of r_i and r_{ij} in such a way that the symmetry is ensured.
Parameters a_i are to be found by the variation principle, generally
speaking, together with the radial orbitals involved in $|K>$
However, these last ones very often are taken from the OMC in order
to simplify calculations.

The algebraical calculations in the MISV are greatly simplified
by the use of the mathematical apparatus of Wigner and Racah(27-45)
as proposed by Levinson et al. (70) and used by Uspalis and Ramonas
(71). The main idea of that may be represented by the equality

$$<K_{inc}|H|K_{inc}> = <KV|H|VK> = <K|VHV|K> \quad ,$$

(3.11)

where H is operator of interest. VHV is further regarded as a
new (non-physical) operator. His matrix elements with respect to
$|K>$ as a basis give the matrix elements of H with respect to
$|K_{inc}>$ as a basis.

In order to carry out the calculations involved in (3.11) all the functions of inter-electronic distances in V must be expanded in terms of spherical harmonics. Then the mathematical apparatus of irreducible tensor operators of Wigner and Racah works very well. The result is the expressions for the matrix elements of H in terms of radial integrals. These integrals are far reaching generalizations of Slater-Condon integrals. They are considered, among others, by Jucys(72).

With the MISV is very closely connected the two-electron state approximation or pair correlations mentioned above. Szasz (73) brought the two-electron wave function equation of Fock et al. (49) to the form which corresponds to that of Brenig (74), and, consequently, to that of Brueckner. Attempts to construct the wave function of the whole system with the help of two-electron wave functions or geminals have been made with some success in the theory of atoms and molecules as well as nuclei.

Calculations connected with the practical application of the MISV are rather complicated in any case. The idea of Fock et al. (49) to apply MISV only to one group of atomic electrons, with the rest of the electrons conserving the OMC, helps very much. Some difficulties in securing antisymmetry and orthogonality conditions are still left. Szasz (75) attempts to overcome these difficulties by constructing the corresponding pseudopotential. Simons (76) solves this problem lifting partially the antisymmetry requirement.

Another approach in simplifying the calculations involved in practical applications of the MISV is due to Boys and Handy (77). They factorize V from the functions

$$v_{ij} = \exp\left\{\sum_{\ell} D_{\ell} G_{\ell}(r_i, r_j) + \sum_{p} d_p [g_p(r_i) + g_p(r_j)]\right\} , \quad (3.12)$$

where D_{ℓ} and d_p are adjustable parameters. $G_{\ell}(r_i, r_j)$ and $g_p(r_i)$ are particular sets of expansion functions. Instead of VHV they take $V^{-\ell}HV$ and, consequently, calculations simplify very much, because $V^{-\ell}$ and V cancel out in many cases.

There exists some relation between MISC and MSC. This problem has been considered by Jucys et al. (78) and recently by Roby (79). However, this relation is more algebraic than geometric or physical, because the factor V is invariant under the rotation of the coordinate system, and the configuration does not split up in two parts, not necessarily spherically symmetric, as in the case in (3.1). On the other hand, in the MSC we have the superposition of states in accordance with the main principle of quantum mechanics. It is clear that in the MISV there is no such direct superposition of states.

4.   CONCLUDING REMARKS

Theoretical investigations of atoms and atomic spectra are stimulated by the development of physical science itself as well as by its practical application to spectroscopy, astrophysics, laser technology, plasma physics, and so on. The development of computing techniques favours putting into practice new theories despite very complicated calculations involved. The practical application of the theory contributes very much to its development. However, there are some circumstances hampering the development and, eventually, the practical application of the theory.

In the first place must be mentioned the use of different mathematical apparatus by different working groups. It is confusing that many scientists working on the problem of improving the many-electron atom theory do not take advantage of Wigner-Racah irreducible tensor techniques on the base of coefficients of fractional parentage. Even in the ordinary method of calculation it is not possible to manage without Wigner-Racah techniques in the case of heavy atoms as indicated Wybourne (42). To a much greater extent this is true in the case of improved methods of the theory in view of the mathematical complexity of the problem.

The difference in the mathematical apparatus used hampers the collaboration between the groups of scientists working on closely related problems and does not create favourable conditions to share the programs compiled by different authors. Besides that, the different "mathematical languages" overshadow the resemblances and discrepancies of the methods elaborated by different authors in different parts of the world.

At the present time many people are engaged in the development of the correlation theory. However, there is a lack in the coordination ("Correlation") between the scientists themselves in order to bend efforts to create a highly developed theory on the base of the most powerful and unique mathematical foundation. Removal of these obstacles would greatly facilitate mastering of the theory by young people preparing to work in the corresponding branch of research as well as by experimentalists wishing to become proficient in the theory and, consequently, would further the development of the theory itself and make it easier to apply for practical purposes.

ABBREVIATIONS

EMC -- Extended Method of Calculation
HF -- Hartree-Fock
HFM -- Hartree-Fock Method
MCA -- Multi-Configuration Approximation
MISV - Method of Incomplete Separation of Variables

MSC -- Method of Superposition of Configuration
OMC -- Ordinary Method of Calculation
SCF -- Self-Consistent Field
SSOA - Separable Spin-Orbital Approximation

REFERENCES

1. E. Schrödinger, Ann. d. Phys. 79, 361, 489 (1926), 80, 437 (1926).

2. E. A. Hylleraas, ZS f. Phys. 48, 469 (1928).

3. E. A. Hylleraas, ZS f. Phys. 54, 347 (1929).

4. C. Eckart, Phys. Rev. 36, 878 (1930).

5. N. Bohr, The Theory of Spectra and Atomic Constitution, Cambridge University Press, Cambridge, 1922.

6. D. R. Hartree, Proc. Camb. Phil. Soc. 24, 89, 111 (1928).

7. D. R. Hartree, The Calculation of Atomic Structures, John Wiley and Sons, New York, 1957.

8. J. C. Slater, Phys. Rev. 34, 1293 (1929).

9. V. A. Fock, ZS f. Phys. 61, 126 (1930), 62, 795 (1930).

10. J. C. Slater, Phys. Rev. 35, 210 (1930).

11. C. Froese Fischer, Computer Phys. Commun. 1, 151 (1969).

12. R. Karazija, P. Bogdanovicius, and A. Jucys, Acta Phys. Hung. 27, 467 (1969).

13. C. C. J. Roothaan, Rev. Modern Phys. 23, 69 (1951).

14. V. A. Fock, ZS f. Phys. 81, 195 (1933).

15. J. C. Slater, Phys. Rev. 81, 385 (1931).

16. F. Herman and S. Skillman, Atomic Structure Calculations, Prentice-Hall, Inc., Englewood Cliffs, New Jersey, 1963.

17. R. D. Cowan, Phys. Rev. 163, 54 (1967).

18. C. Zener, Phys. Rev. 36, 51 (1930).

19. M. Morse, L. A. Young, and E. S. Haurwitz, Phys. Rev. 48, 948 (1935).

20. V. A. Fock and M. I. Petrashen, Zh. Exp. i Theor. Phys. $\underline{6}$, 1 (1936).

21. Z. J. Kupliauskis, A. V. Matulaityte, and A. P. Jucys, Liet. fiz. rink., Lit. fiz. sb., $\underline{11}$, 557 (1971).

22. A. V. Matulaityte, Z. J. Kupliauskis, and A. P. Jucys, Liet. fiz. rink., Lit. fiz. sb., $\underline{11}$, 565 (1971).

23. L. P. Smith, Phys. Rev. $\underline{42}$, 176 (1932).

24. C. S. Sharma and C. A. Coulson, Proc. Phys. Soc. $\underline{80}$, 81 (1962).

25. F. Prosser and S. Hagstrom, Intern. J. Quantum Chem. $\underline{2}$, 89 (1968).

26. A. P. Jucys and V. J. Tutlys, Liet. fiz. rink., Lit. fiz. sb., $\underline{11}$, 913, 927 (1971), $\underline{12}$, 5 (1972); A. P. Jucys and J. J. Grudzinskas, Intern. J. Quantum Chem. $\underline{6}$, 455 (1972).

27. E. Wigner, Gruppentheorie und ihre Anwendung auf die Quanten-mechanik der Atomspektren, Friedr. Vieweg u. Sohn, Braunschweig, 1931 (in German), Academic Press, New York and London, 1959 (in English).

28. E. Wigner, On the Matrices which reduce the Kronecker Products of Representations of S. R. Groups, Princeton University, 1940 (unpublished), Reproduced in Quantum Theory of Angular Momentum, Edited by L. C. Biederharn and H. van Dam, Academic Press, New York and London, 1965.

29. G. Racah, Phys. Rev. $\underline{62}$, 438 (1942).

30. U. Fano and G. Racah, Irreducible Tensorial Sets, Academic Press, New York, 1959.

31. A. P. Jucys, J. B. Levinson, and V. V. Vanagas, Mathematical Apparatus of the Theory of Angular Momentum, Vilnius, 1960 (in Russian), Israel Program for Scientific Translations, Jerusalem, 1962 (in English).

32. A. P. Jucys and A. A. Bandzaitis, The Theory of Angular Momentum in Quantum Mechanics, Publishing House "Mintis", Vilnius, 1965 (in Russian).

33. E. El-Baz, Traitment Graphique de l'Algèbre des Moments Angu-laires, Masson, Paris, 1969.

34. E. El-Baz, and B. Castel, Graphical Methods of Spin Algebras in Atomic, Nuclear, and Particles Physics, Marcel Dekker, Inc., New York, 1972.

35. G. Racah, Phys. Rev. 63, 367 (1943).

36. C. W. Nielson and G. F. Koster, Spectroscopic Coefficients for the p^n, d^n, and f^n Configurations, MIT, Cambridge, Massachusetts, 1963.

37. B. R. Judd, Operator Techniques in Atomic Spectroscopy, McGraw-Hill, New York, 1963.

38. B. R. Judd, Second Quantization and Atomic Spectroscopy, Johns Hopkins Press, Baltimore, 1967.

39. B. G. Wybourne, Symmetry Principles and Atomic Spectroscopy, Wiley-Interscience, New York, 1970.

40. J. C. Slater, Quantum Theory of Atomic Structure, McGraw-Hill, New York, 1960.

41. I. I. Sobelman, Introduction to the Theory of Atomic Spectra, State Publishing House of Phys. and Math. Lit. Moscow, 1963 (in Russian).

42. B. G. Wybourne, Spectroscopic Properties of Rare Earths, John Wiley, New York, 1965.

43. S. Fraga and G. Malli, Many-Electron Systems: Properties and Interactions, W. B. Saunders Co. Philadelphia, 1968.

44. Lloyd Armstrong, Jr., Theory of the Hyperfine Structure of Free Atoms, Wiley-Interscience, New York, 1971.

45. A. P. Jucys and A. J. Savukynas, Mathematical Foundations of the Quantum Theory of Atoms, Publishing House "Mintis", Vilnius, 1972 (in Russian).

46. E. U. Condon and G. H. Shortley, The Theory of Atomic Spectra, Cambridge University Press, Cambridge, 1935.

47. D. R. Hartree and B. Swirles, Proc. Camb. Phil. Soc. 33, 240 (1937).

48. E. Trefftz, A. Schlüter, K. -H. Dettmar, und K. Jörgens, ZS f. Astrophys. 44, 1 (1957).

49. V. A. Fock, M. G. Veselov, and M. I. Petrashen, Zh. Exp. i Theor. Phys. 10, 723 (1940).

50.   A. P. Jucys, Zh. Exp. i Theor. Phys. $\underline{23}$, 371 (1952).

51.   O. Sinanoglu, Atomic Physics, Proc. of the First ICAP,
      p. 131, Plenum Press, New York, 1969. Advances in Chemical
      Physics, $\underline{14}$, 237 (1969). Colloque Intern. CNRS, Nr. 194,
      p. C4-89 (1970).

52.   A. P. Jucys, J. J. Vizbaraitė, V. J. Kaveckis, and J. V.
      Batarūnas, Izvestiya Akad. Nauk USSR, Physics Series, $\underline{22}$, 665
      (1958). Liet. TSR MA Darbai, $\underline{B2}$, 3 (1958).

53.   E. U. Condon, Rev· Modern Phys. $\underline{40}$, 872 (1968).

54.   O. Sinanoglu, J. Chem. Phys. $\underline{36}$, 706 (1962).

55.   V. A. Fock, Doklady Akad. Nauk USSR, $\underline{73}$, 735 (1950).

56.   D. R. Hartree, W. Hartree, and B. Swirles, Phil. Trans.
      Roy. Soc. $\underline{A238}$, 229 (1939).

57.   V. V. Kybartas and A. P. Jucys, Zh. Exp. i Theor. Phys. $\underline{25}$,
      264 (1953), A. P. Jucys, V. V. Kybartas, and J. J. Glembockis,
      Zh. Exp. i Theor. Phys. $\underline{27}$, 425 (1954), A. P. Jucys, Advances
      in Chemical Physics, $\underline{14}$, 191 (1969).

58.   I. Öksüz and O. Sinanoglu, Phys. Rev. $\underline{181}$, 42, 54 (1969).

59.   E. Clementi, J. Chem. Phys. $\underline{38}$, 2248 (1963), $\underline{39}$, 175 (1963),
      $\underline{42}$, 2783 (1965), E. Clementi and A. Veillard, J. Chem. Phys.
      $\underline{44}$, 3050 (1966), $\underline{49}$, 2415 (1968), E. Clementi, W. Kraemer,
      and C. Salez, J. Chem. Phys. $\underline{53}$, 125 (1970).

60.   H. F. Schaefer, R. A. Klemm, and F. E. Harris, Phys. Rev. $\underline{181}$,
      137 (1969), J. W. Viers, F. E. Harris, and H. F. Schaefer, Phys.
      Rev. $\underline{A1}$, 24 (1970), F. E. Harris, Colloque Intern. CNRS, Nr.
      194, p. C4-111 (1970).

61.   A. W. Weiss, Phys. Rev. $\underline{122}$, 1826 (1961), $\underline{162}$, 71 (1967), $\underline{188}$,
      119 (1969), $\underline{A3}$, 126 (1971).

62.   R. K. Nesbet, Advances in Chemical Physics, $\underline{14}$, 1 (1969),
      Colloque Intern. CNRS, Nr. 194, p. C4-105 (1970), C. M. Moser
      and R. K. Nesbet, Phys. Rev. $\underline{A4}$, 1336 (1971).

63.   A. P. Jucys, Intern. J. Quantum Chem. $\underline{1}$, 311 (1967).

64.   A. P. Jucys and V. A. Kaminskas, Advances in Chemical Physics,
      $\underline{14}$, 207 (1969), A. P. Jucys and V. J. Stasiukaitis, Intern. J.

Quantum Chem. $\underline{4}$, 333 (1970), Liet. fiz. rink., Lit. fiz. sb. $\underline{11}$, 573 (1971).

65. A. P. Jucys, E. P. Našlenas, and P. S. Zvirblis, Intern. J. Quantum Chem. $\underline{6}$, 455 (1972), Liet. fiz. rink., Lit. fiz. sb. $\underline{12}$, 201 (1972).

66. V. J. Stasiukaitis, V. A. Kaminskas, and A. P. Jucys, Liet. fiz. rink., Lit. fiz. sb. $\underline{12}$, in press (1972).

67. U. Kaldor and F. E. Harris, Phys. Rev. $\underline{183}$, 1 (1969).

68. P.-O. Löwdin, Phys. Rev. $\underline{97}$, 1509 (1955).

69. C. L. Pekeris, Phys. Rev. $\underline{115}$, 1216 (1959).

70. J. B. Levinson, V. V. Vanagas, and A. F. Jucys, Liet. TSR MA Darbai, $\underline{B5}$, 21 (1956).

71. K. K. Ušpalis and A. A. Ramonas, Liet. fiz. rink., Lit. fiz. sb. $\underline{10}$, 341 (1970).

72. A. P. Jucys, Zh. Exp. i Theor. Phys. $\underline{23}$, 357 (1952).

73. L. Szasz, ZS f. Naturforschung, $\underline{14a}$, 1014 (1959).

74. W. Brenig, Nuclear Phys. $\underline{4}$, 363 (1957).

75. L. Szasz, J. Chem. Phys. $\underline{49}$, 679 (1968), L. Szasz and G. McGinn, J. Chem. Phys. $\underline{56}$, 1019 (1972).

76. J. Simons, Intern. J. Quantum Chem. $\underline{6}$, 439 (1972).

77. S. F. Boys and N. C. Handy, Proc. Roy. Soc. $\underline{A309}$, 209 (1969), $\underline{A310}$, 63 (1969).

78. A. P. Jucys, J. B. Levinson, J. V. Batarūnas, and V. V. Vanagas, Liet. TSR MA Darbai, $\underline{B3}$, 35 (1958).

79. K. R. Roby, Intern. J. Quantum Chem. $\underline{5}$, 119 (1971).

BOUND STATES AND CONTINUUM STATES OF ATOMIC SYSTEMS

M. J. Seaton

Department of Physics and Astronomy

University College London

The Schroedinger equation for a system containing N electrons and a nucleus of charge Z is

$$\{H(Z,N) - E\}\, \Psi = 0 \tag{1}$$

where (using Rydberg units),

$$H(Z,N) = -\sum_{i=1}^{N}\left(\nabla_i^2 + \frac{2Z}{r_i}\right) + \sum_{j=i+1}^{N}\sum_{i=1}^{N-1}\frac{2}{r_{ij}}\ . \tag{2}$$

The topic of this paper is concerned with the properties of the solutions of (1), for fixed Z and N and different values of E.

For bound-state solutions of (1) it is required that, for all $i = 1$ to N,

$$\Psi \to 0 \text{ faster than } r_i^{-1} \text{ as } r_i \to \infty\ . \tag{3}$$

Such solutions exist only for discrete values of E, say $E = E_\gamma(Z,N)$. For $Z \geq N$ (neutral atoms and positive ions) we have an infinite number of such solutions, which may be labelled by $\gamma = 1, 2, 3, \ldots, \infty$ and ordered such that $E_1 \leq E_2 \leq E_3 \ldots$. It should be noted that $E_\infty(Z,N) = E_1(Z,N-1)$, since for this level we have one electron with zero binding energy, and the $(N-1)$-electron core in its ground state.

In this talk I am going to introduce two restrictions. Restriction I is to take $Z \geq N$; this excludes discussion of negative

ions and of electron scattering by neutral atoms. Restriction II
is to take $E < E_1(Z,N-2)$; this excludes discussion of states with
two electrons in the continuum. We are then left with the topics:
bound states of neutral atoms and positive ions; and scattering of
electrons by positive ions (excluding ionization). I should, perhaps, confess to a particular interest in electron-positve ion
collisions,* since these processes are of importance for many problems in astrophysics.

Before discussing continuum states it is necessary to clarify
some questions of notation. Let $x_i = (\sigma_i, \hat{r}_i, r_i)$ be the coordinate of electron i (σ_i for spin, $\hat{r}_i$ for angles). We use $\bar{x}_i =$
$x_1, x_2, \ldots, x_{i-1}, x_{i+1}, \ldots, x_N$ for all coordinates except x_i,
and $\bar{r}_i = (\bar{x}_i, \sigma_i, \hat{r}_i)$ for all coordinates except r_i. We introduce functions

$$\psi_\alpha(\bar{r}_i) = \Psi_\gamma(Z,N-1|\bar{x}_i)\chi_{m_s}(\sigma_i)Y_{\ell m_\ell}(\hat{r}_i) \qquad (4)$$

where $\Psi_\gamma(Z,N-1)$ is an eigenfunction for the $(N-1)$-electron problem. In (4), α stands for $(\gamma m_s \ell m_\ell)$; in practice it is convenient to work with vector-coupled states, e.g. $\alpha = (\gamma S_\gamma L_\gamma \ell SLM_SM_L)$.
The total energy for the collision problem is

$$E = E_\alpha(Z,N-1) + k_\alpha^2 \qquad (5)$$

where $E_\alpha(Z,N-1)$ $[\equiv E_\gamma(Z,N-1)]$ is an energy level of the target
ion and k_α^2 is the kinetic energy of the colliding electron
(always using Rydberg units). We have open channels for $k_\alpha^2 > 0$
and closed channels for $k_\alpha^2 < 0$. For closed channels we put

$$k_\alpha^2 = -z^2/\nu_\alpha^2 \qquad (6)$$

where $z = (Z-N+1)$ is the charge on the target ion and ν_α is
the effective quantum number in channel α.

I can now specify S-matrix boundary conditions. For an incoming spherical wave in channel β, outgoing spherical waves in
all open channels, and decaying waves in all closed channels we
have a function Ψ_β with asymptotic form

$$\Psi_\beta \underset{r_i \to \infty}{\sim} (-1)^i N^{-1/2} \sum_\alpha \psi_\alpha(\bar{r}_i) \frac{1}{r_i} \theta_{\alpha\beta}(r_i) \qquad (7)$$

---

*Some aspects of this work were discussed in my paper presented at
the 1st Atomic Physics Conference (Seaton, 1969).

where

$$\theta_{\alpha\beta}(r) = k_\alpha^{-1/2}[\exp(-i\zeta_\alpha) \; \delta_{\alpha\beta} - \exp(+i\zeta_\alpha)S_{\alpha\beta}] \qquad (8)$$

and

$$\zeta_\alpha = k_\alpha r - \frac{z}{k_\alpha} \ell n(r) + c_\alpha \quad , \qquad (9)$$

where c_α is the phase of the regular Coulomb function. For close channels we take

$$i\zeta_\alpha = -|\zeta_\alpha| \quad .$$

The factor $(-1)^i$ in (7) ensures that Ψ_β is anti-symmetric, and the normalizing factor $N^{-1/2}$ is introduced because it simplifies some subsequent expressions.

The probability of a transition $\alpha \to \alpha'$ between two open channels is $P_{\alpha\alpha'} = |S_{\alpha\alpha'}|^2$ and the partial collision strength is $\Omega_{SL} = (1/2)(2S+1)(2S+1)P_{\alpha\alpha'}$. The total collision strength $\Omega(\gamma,\gamma')$ for transitions between two energy levels is obtained on summing over ℓ, ℓ', S and L. The collision cross section is $Q(\gamma \to \gamma') = \Omega(\gamma,\gamma')/[k^2_\gamma(2S_\gamma+1)(2L_\gamma+1)]$ in units of πa_0^2.

One cannot make exact calculations for $N > 1$ and we must therefore base approximate treatments on the use of variational theory. I would like to emphasise a few salient points:

Variational theory for bound states. Consider $(H-E)\Psi = 0$, $(\Psi|\Psi) = 1$, $\Psi' = \Psi + \delta\Psi$, $(\Psi'|\Psi') = 1$. Then, to first order in $\delta\Psi$, $\delta(\Psi|H-E|\Psi) = (\Psi|H-E|\delta\Psi)$ since $(\delta\Psi|H-E|\Psi) = 0$. Integrating by parts and using the boundary condition (3) we obtain $(\Psi|H-E|\delta\Psi) = (\delta\Psi|H-E|\Psi)^*$, which is zero, and hence $\delta(\Psi|H-E|\Psi) = 0$.

Variational theory for continuum states. Consider* $(H-E)\Psi = 0$ where Ψ has S in its asymptotic form, and $\Psi' = \Psi + \delta\Psi$ where Ψ' has $S' = S + \delta S$ in its asymptotic form. If we now integrate by parts and use (7) we obtain $(\Psi|H-E|\delta\Psi) = (\delta\Psi|H-E|\Psi)^* - 2i\delta S$, and hence $\delta\{(\Psi|H-E|\Psi) + 2iS\} = 0$. Given a trial function an improved estimate for S is

$$S_{Kohn} = S_{trial} + \frac{1}{2i}(\Psi|H-E|\Psi)_{trial} \quad . \qquad (10)$$

---

*When subscripts are omitted, matrix expressions are to be understood; thus Ψ is the column vector with components Ψ_β, S the matrix with elements $S_{\alpha\beta}$.

However, an important point should be noted: S_{Kohn} will be correct to first order only if exact target states, $\Psi_\gamma(Z, N-1)$, are
used in the trial function.

A well known approximation in collision theory is to use truncated anti-symmetric expansions of the form

$$\Psi' = N^{-1/2} \sum_i (-1)^i \sum_{\alpha=1}^{M} \psi_\alpha(\bar{r}_i) \frac{1}{r_i} F_\alpha(r_i) \quad . \tag{11}$$

Using the variational principle to determine the "best" function
$F_\alpha(r)$, one obtains a system of coupled integro differential (ID)
equations,

$$\left(\frac{d^2}{dr^2} - \frac{\ell(\ell+1)}{r^2} + \frac{2z}{r} + k^2 \right) F + (V-W) F = 0 \quad . \tag{12}$$

This is the "close-coupling" approximation. In (12), V is a local
potential operator and W is an integral exchange operator.

Quantum defect theory (QDT) is concerned with the analytical
properties of the solutions of (1) as functions of E. The basic
structure of the theory is described in a fairly recent review
(Seaton, 1970) which gives a full bibliography, and need not be
discussed further here. I would, however, like to add some remarks on the present status of the theory.

A complete and rigorous analytical theory has not yet been
developed for the many-particle problem with Coulomb potentials
(consideration of the near-threshold ionization problem shows how
difficult this would be). The procedure used in QDT is to simplify
the problem and then to develop a rigorous theory. The simplifications are:

Simplification 1 is to use, in place of (12), an unsymmetrized
truncated expansion,

$$\Psi' = \sum_{\alpha=1}^{M} \psi_\alpha(\bar{r}_N) \frac{1}{r_N} F_\alpha(r_N) \tag{13}$$

which gives a system of coupled differential equations,

$$\left(\frac{d^2}{dr^2} - \frac{\ell(\ell+1)}{r^2} + \frac{2z}{r} + k^2 \right) F + VF = 0 \quad . \tag{14}$$

The potentials $V_{\alpha\alpha'}$ have asymptotic form for r large $V_{\alpha\alpha'} \sim$

$A_{\alpha\alpha'} \, r^{-m}$ with $m \geq 2$ for $\alpha \neq \alpha'$ and $m \geq 3$ for $\alpha = \alpha'$.

Simplification 2 is to assume that $V_{\alpha\alpha'}(r) = 0$ for $r > r_0$ with $r_0 < \infty$. For $r > r_0$, (14) reduces to the Coulomb equation,

$$\left(\frac{d^2}{dr^2} - \frac{(\ell+1)}{r^2} + \frac{2z}{r} + k^2\right) F = 0 \quad . \tag{15}$$

All required analytic properties of the solutions of this equation are known, and this makes possible the development of a rigorous theory of the analytical properties of the solutions of (14).

Very concisely, the results of the theory may be summarized as follows. When all channels are open we put $S = \chi$, which defines χ. We may now extrapolate χ as a smooth and slowly varying function of the energy (the rigorous theory gives the exact procedure to be used). When some channels are closed the S matrix can vary rapidly as a function of the energy, in the vicinity of resonances and in perturbed Rydberg series. Let us partition S as follows:

$$S = \begin{pmatrix} S_{oo} \begin{array}{c} \text{open-} \\ \text{open} \end{array} & \vdots & S_{oc} \begin{array}{c} \text{open-} \\ \text{closed} \end{array} \\ \text{------------} & \vdots & \text{------------} \\ S_{co} \begin{array}{c} \text{closed-} \\ \text{open} \end{array} & \vdots & S_{cc} \begin{array}{c} \text{closed-} \\ \text{closed} \end{array} \end{pmatrix} \quad , \tag{16}$$

and similarly partition χ. Then

$$S_{oo} = \chi_{oo} - \chi_{oc}\left[\chi_{cc} - \exp(-2\pi i \nu)\right]^{-1}\chi_{co} \quad . \tag{17}$$

Let α and β be two open channels, and γ a closed channel. The energy in channel γ is $k_\gamma^2 = -z^2/\nu^2_\gamma$. As the energy is increased to the point at which γ becomes an open channel, one obtains a Rydberg series of resonances in $S_{\alpha\beta}$, the resonance structure being repeated each time that ν_γ changes by unity. With several closed channels one can obtain many resonances and complicated interference effects.

Bound states occur at poles in the analytic continuation of the S matrix. Expressing S in terms of χ, the condition is

$$\left|\chi - \exp(-2\pi i \nu)\right| = 0 \quad . \tag{18}$$

For the one-channel case we have $\chi = \exp(2i\delta)$, where δ is the phase shift. Solutions of (18) occur at $2\pi\nu = 2\pi n - 2\delta$, where n is an integer, that is to say at $\nu = n-\mu$ where $\mu = \delta/\pi$ is the quantum defect. For the two channel case (18) becomes

$$\begin{vmatrix} \chi_{11} - \exp(-2\pi i \nu_1) & \chi_{12} \\ \chi_{21} & \chi_{22} - \exp(-2\pi i \nu_2) \end{vmatrix} = 0 \quad . \quad (19)$$

The χ matrix is symmetric, $\chi_{12} = \chi_{21}$. If $\chi_{12} = 0$, (19) gives two series of resonances converging to the two levels of the ion core. If $\chi_{12} \neq 0$ one obtains series perturbations.

The theory developed for a simplified model can be used for the real many-electron problem provided that the range of extrapolation is not too large. The most serious limitation of the theory can be appreciated by considering the pure Coulomb problem, equation (15). This has bound-state solutions for $k^2 = -z^2/\nu^2$ where $\nu = n$ and $n \geq (\ell+1)$. That there are no bound-state solutions for $n < (\ell+1)$ is due to the presence of the term $-\ell(\ell+1)/r^2$ in (15). But in the coupled equations (14) we have neglected off-diagonal elements of V which behave like r^{-2}. It follows that the theory becomes unreliable if we extrapolate downwards in energy to a point at which, for any closed channel γ, we have $\nu_\gamma \lesssim (\ell_\gamma+1)$.

I will discuss some of the uses of QDT, and some methods which can be used when QDT fails.

The simplest application is to check the accuracy of computed phase shifts for elastic scattering by positive ions. Figure 1

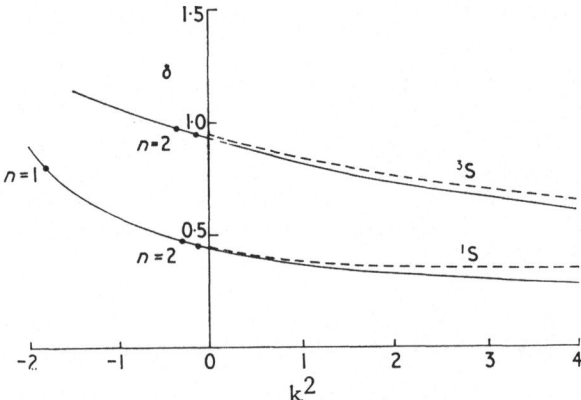

Figure 1. Phase shifts δ for (e+He$^+$) ^{1}S and ^{3}S. For $k^2 < 0$, $\delta = \pi\mu$ where μ is the quantum defect. Full line curves, extrapolated from experimental quantum defects. Dashed curves, polarized orbital calculations of Sloan (1964).

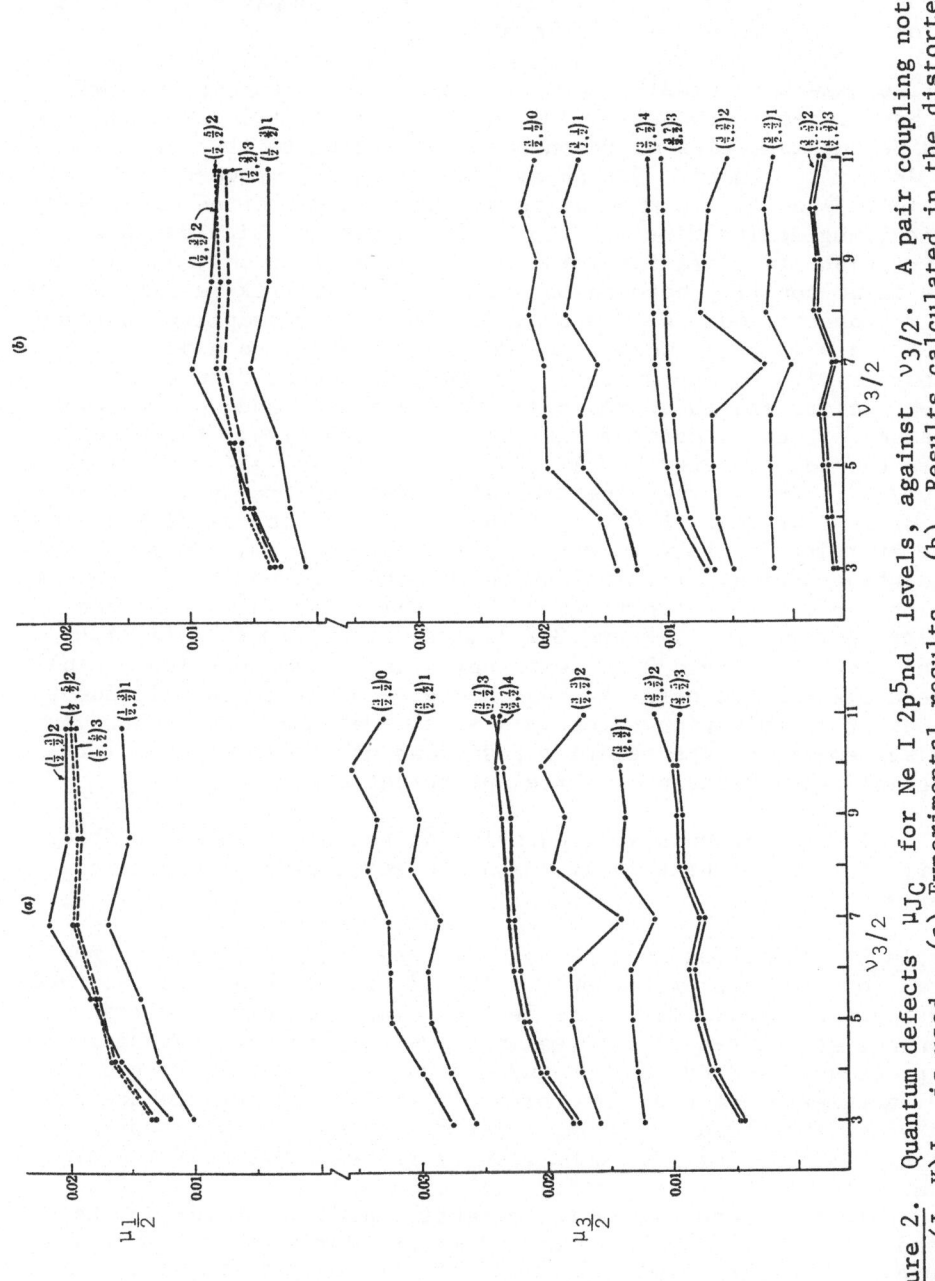

Figure 2. Quantum defects μ_{JC} for Ne I $2p^5nd$ levels, against $\nu_{3/2}$. A pair coupling notation, $(J_C,K)J$, is used. (a) Experimental results. (b) Results calculated in the distorted wave approximation.

shows some results for (e+He$^+$) scattering (Seaton, 1966); the experimental quantum defects μ for He ^{1}S and ^{3}S are extrapolated to give phase shifts $\delta = \pi\mu$, and the results compared with polarized orbital calculations of Sloan (1964).

The case of (e+Ne$^+$) is more complicated, because the Ne$^+$ core has fine-structure, ^{2}P J_C where $J_C = 3/2$ and $1/2$, and this gives perturbed series of resonances converging to the $J_C = 3/2$ and $J_C = 1/2$ limits. Saraph and Seaton (1971) proceed as follows. The (e+Ne$^+$) collision problem is solved in a distorted wave approximation, neglecting fine structure; the representation is then $\alpha = {}^2\text{P}(s\ell)SL$ (s = 1/2 for spin of colliding electron), and since there is no coupling between channels the χ matrix is diagonal in this representation. We now transform to a pair-coupling representation, $\alpha = (J_C, K)J$ where J_C couples with ℓ to give K, and K couples with s to give J. In this representation the two fine-structure states of the core are specified, and we have two different effective quantum numbers, $\nu_{3/2}$ and $\nu_{1/2}$. For bound states the total energy is $E = E({}^2\text{P}_{3/2}) - z^2/(\nu_{3/2})^2 = E({}^2\text{P}_{1/2}) - z^2/(\nu_{1/2})^2$. We use the experimental fine-structure splitting of the core, $E({}^2\text{P}_{1/2}) - E({}^2\text{P}_{3/2})$. The χ matrix is calculated at three energies in the continuum-state region and extrapolated to the bound state region. Figure 2 shows results for the $2p^5nd$ levels. We plot the quantum defects $\mu_{1/2}$ and $\mu_{3/2}$ for the series converging to the $J_C = 1/2$ and $3/2$ limits. Comparison of experimental results [Figure 2(a)] with calculated results [Figure 2(b)] shows a systematic error in the theory -- which is largely due to neglect of polarization potentials in the calculations -- but very good agreement for the relative positions of the levels and for the complicated pattern of series perturbations.

I will now discuss some examples of the use of quantum defect theory for the calculation of resonance structures in collision cross sections.

Example 1. The cross section for O$^+$ $2p^3$ ^{2}D$_{5/2}$ $\rightarrow$ ^{2}D$_{3/2}$ is required for the interpretation of [O II] line intensities observed in gaseous nebulae. The cross section contains very complicated resonance structures in the region below the $2p^3$ ^{2}P threshold. The procedure used by Martins and Seaton (1969) was to obtain χ at 3 energies above the ^{2}P threshold and extrapolate to the region below threshold. A transformation was made to the representation $\alpha = {}^2\text{D}J_C$ $(s\ell j)J$. Some results for the p-waves ($\ell=1$) are shown in Figure 3. The partial collision strengths are plotted against $(\nu-n)$ where ν is the effective quantum number in the closed ^{2}P channels. These calculations took no more than a few minutes of computer time. It is clear that a vast amount of time would have been required if the coupled equations had been solved directly in the resonance region, and no use had been made of analytical theory.

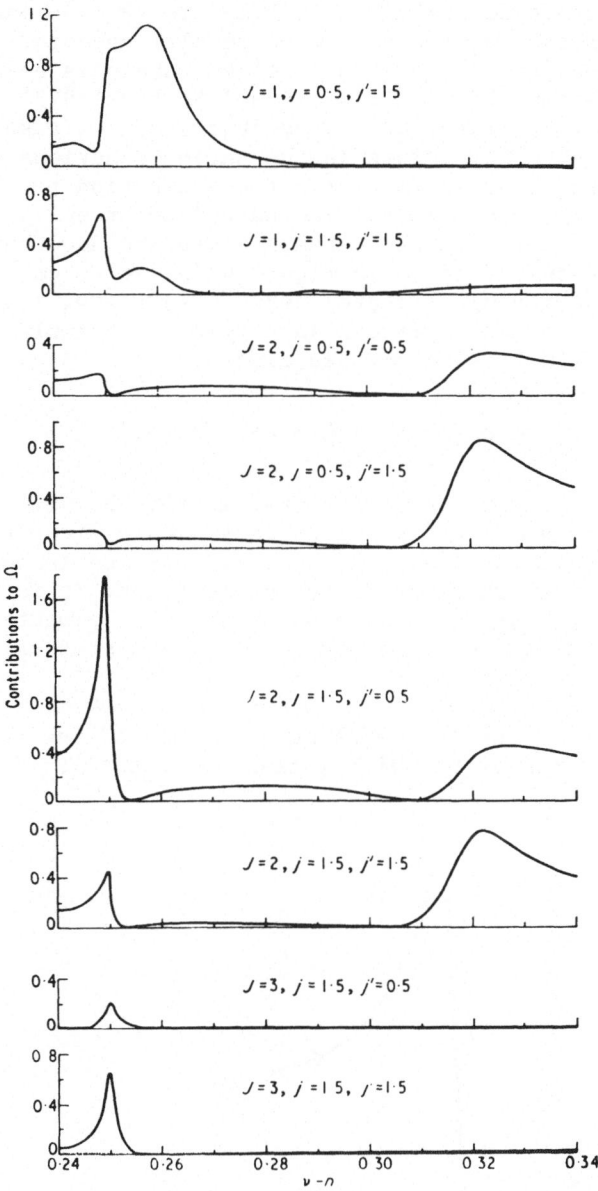

Figure 3. Contributions to $\Omega(^2D_{5/2}, ^2D_{3/2})$ for O^+, at energies below the ^2P threshold. The representation is $\alpha = {^2D_{5/2}}(\ell sj)J$, $\alpha' = {^2D_{3/2}}(\ell sj')J$ with $\ell=1$; ν is the effective quantum in the ^2P channels, and n is an integer. The resonance pattern is repeated when ν changes by unity.

<u>Example 2</u>. The cross sections for O^{2+} $2s^22p^2$ $^3P \rightarrow$ ^1D and ^1S are also required for the interpretation of nebular spectra. They contain near-threshold resonances due to closed channels belonging to the ion configuration $2s2p^3$. Eissner and Seaton (1972) have made calculations using accurate O^{2+} wave-functions. (It should be recalled that first-order errors in the target-functions will give first-order errors in S.) The largest contribution to $\Omega(^3P, ^1S)$ comes from SL = ^2P, odd parity. The calculated results are shown in Figure 4. A number of closed channels are included but the near-threshold resonance shown in Figure 4 is due to only one closed channel giving the resonance state $2s2p^3(^3P^o)3s$ $^2P^o$. Equation (17) has been used as an empirical interpolation formula. In the region of an isolated resonance this gives

$$S_{\alpha\beta} = A_{\alpha\beta} + B_{\alpha\beta}/[C - \exp(-2\pi i\nu)] \qquad (20)$$

where ν is the effective quantum number in the channel giving the resonance, and where $A_{\alpha\beta}$, $B_{\alpha\beta}$ and C vary slowly with E. The procedure is to calculate $A_{\alpha\beta}$, $B_{\alpha\beta}$ and C, assumed to be constants, using values of $S_{\alpha\beta}$ at three different energies; it is then found that (20) gives results agreeing closely with the computed results at the other two energies. The use of (20) is therefore very economical.

<u>Example 3</u>. Dubau and Wells (1972) have calculated the photoionization cross section of Be, with inclusion of the $1s^2$ 2s and 2p

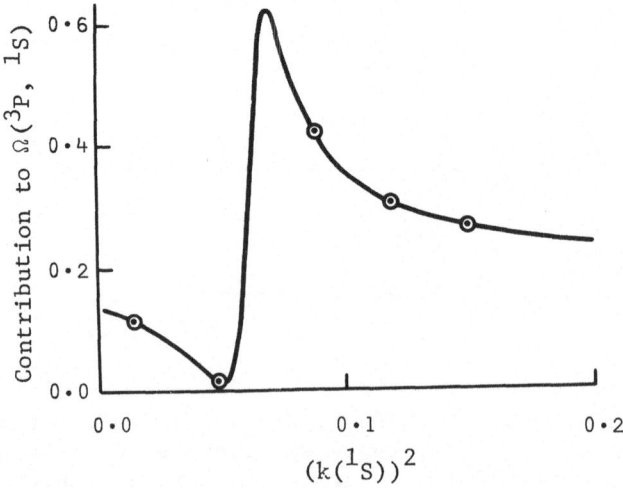

<u>Figure 4</u>. The contribution to the collision strength $\Omega(^3P, ^1S)$ in O^{2+} from $(e+O^{2+})$ states with SL = ^2P, odd parity.

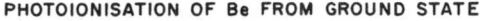

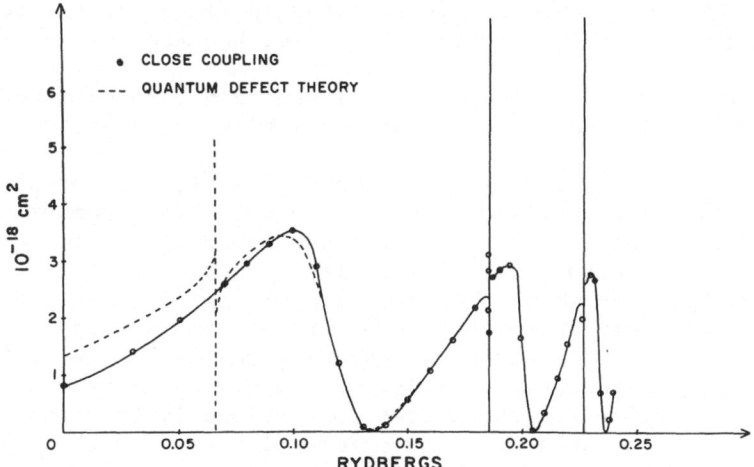

Figure 5. Results obtained by Dubau and Wells (1972) for photo-ionization of Be.

states of Be^+. Their results are shown in Figure 5 for energies below the 2p threshold. There are two series of resonances, 2pns (very broad resonances) and 2pnd (narrower resonances). All calculations are made on solving the coupled ID equations (12). Two curves are plotted. The dashed curve is obtained on solving the problem at three energies above the 2p threshold and extrapolating to the region below threshold using QDT techniques. The full line curve is obtained from solutions of the equations at energies below the 2p threshold (the computed points are marked by small filled circles). It is seen that the extrapolated results are very good at the higher energies, but give a spurious "2d" resonance at lower energies.

We have already seen that QDT may be expected to fail at energies such that $\nu_\alpha \lesssim (\ell_\alpha+1)$ in any closed channel α. Some attempts have been made to generalize the theory, so as to take account of long range coupling potentials, but it then gets much more complicated, and the main practical advantages of the theory are lost.

Another approach is to solve the coupled ID equations at energies for which QDT cannot be relied upon. Some improved numerical techniques have recently been developed (see Burke and Seaton, 1971). It is found that the expansion (11), with functions F satisfying the ID equations (12), which was introduced for studies of the continuum-state problem, can also provide a useful method for bound-state calculations. The simplest example of this approach is

provided by the "frozen core" calculations of Cohen and Kelly
(1966) for He. Only one term is retained in (11), for the configuration 1snℓ. A He^+ 1s function is used, and the nℓ radial
function is a solution of (12). This approximation gives an energy
for the He 1 ^{1}S ground-state which is lower than that obtained in
the restricted (1s^2) Hartree-Fock approximation.

Seaton and Wilson (1972a) have proposed a "frozen cores" approximation which may be used when the ion core has a number of low-
lying levels. As an example they consider bound states in C^{2+}.
This ion is iso-electronic with Be, with low-lying core states
1s^{2}2s and 1s^{2}2p, but whereas states such as 1s^{2}2p3s and 2p3d
are in the continuum for Be they become true bound states for C^{2+}.
Table 1 gives results obtained on including the 2s and 2p C^{3+} core
states and solving the ID equations. A comparison is made with
experimental energies for the states conventionally labelled
2snp(n = 2 to 8) and 2p3s and 2p3d. For all states in the series
there is appreciable configuration mixing, with the conventional
labels indicating the dominant configuration. It is seen that the
difference between calculated and experimental effective quantum
numbers, referred to the appropriate limits, is fairly small.

Some similar calculations have been made by Seaton (1972) for
neutral carbon, including the following states of the C^+ core:
2s^{2}2p allowing for configuration mixing with 2p^3; and 2s2p^2 ^{2}S,
^{2}P, ^{4}P and ^{2}D. Table 2 shows results obtained for the ground-
configuration terms. This includes one-configuration (1s^{2}2s^{2}2p^2)
Hartree-Fock results, the frozen cores results of Seaton, the configuration-interaction Hartree-Fock (CIHF) results of Bagus, Bessis
and Moser (1969) (7 configurations for ^{1}S and ^{3}P, 8 for ^{1}D), and

Table 1. Binding energies ε_1(rydberg units) and effective quantum numbers* ν_i for the $^1P^o$ series in C^{2+}

Identifi-cation	ε_1(exp)	ε_1(calc)	i	ν_i(exp)	ν_i(calc)	ν_i(calc)$-\nu_i$(exp)
2s2p	2.586	2.563	-	---	---	---
2s3p	1.1594	1.1549	1	2.786	2.792	+0.006
2p3s	0.6940	0.6912	2	2.649	2.655	+0.006
2s4p	0.5810	0.5726	1	3.936	3.965	+0.029
2s5p	0.3910	0.3875	1	4.798	4.820	+0.022
2p3d	0.3595	0.3479	2	3.082	3.104	+0.022
2s6p	0.2649	0.2646	1	5.828	5.832	+0.004
2s7p	0.1938	0.1942	1	6.815	6.808	-0.007
2s8p	0.1479	0.1487	1	7.800	7.781	-0.019

*The effective quantum numbers ν_i refer to the 2s limit for i=1,
and the 2p limit for i=2.

Table 2. Energies in Rydberg units for the ground-configuration terms of CI

SL	Hartree-Fock (HF)	Frozen cores	Bagus et al. (CIHF)	Exact non-relativistic
^3P	-75.377	-75.506	-75.518	-75.690
^1D	-75.263	-75.400	-75.418	-75.597
^1S	-75.099	-75.255	-75.308	-75.493

estimates of the exact non-relativistic energy.* The comparatively simple frozen cores approximation is seen to give results not much inferior to the CIHF results (the worst case is ^1S, for which only 3 channels are included in the frozen cores approximation).

Another approach which has given good results in collision theory has been to include "pseudo-states" in the expansion (11) (see Damburg, 1972; and Geltman, 1972). One considers that the ion core is perturbed by the added electron and that, to a first approximation, one can assume this perturbation to be a constant electric field and use first-order perturbation theory. When the resulting "pseudo-states" are included in (11), for low energy scattering one obtains long range potentials with correct α/r^4 asymptotic form. Seaton and Wilson (1972b) consider this method for the calculation of He energy levels. They refer to it as the "polarized core" approximation. Table 3 compares Hartree-Fock, frozen core, and polarized core results for the He ground state; and Table 4 compares frozen core and polarized core results for the excited states. Although the form used for the wave function in the polarized core approximation is most easily justified, in terms of simple physical arguments, for the case of excited or continuum states, it is seen from Table 3 that this approximation

Table 3. Difference between calculated and exact energies for He 1 ^1S

Approximation	{E(calc)-E(exact)} Rydberg units
Hartree-Fock	0.0840
Frozen core	0.0623
Polarized core	0.0275

---

*The differences between the best calculated energies and the exact energies is largely due to neglect of the correlation energy in the $1s^2$ shell.

Table 4. Differences between exact and calculated quantum defects
for excited states of He

SL	n	μ(exact)	{μ(exact) - μ(calc)}	
			Frozen core	Polarized core
^{1}S	2	0.1493	0.0163	0.0088
	3	0.1434	0.0165	0.0093
^{3}S	2	0.3108	0.0048	0.0004
	3	0.3020	0.0041	0.0003
^{1}P	2	−0.0094	0.0113	0.0012
	3	−0.0111	0.0112	0.0010
^{3}P	2	0.0623	0.0136	0.0014
	3	0.0660	0.0128	0.0011
^{1}D	3	0.0018	0.0020	0.0001
^{3}D	3	0.0022	0.0018	0.0001

gives about 2/3 of the correlation energy in the ground state. For
the excited states the polarized core approximation reduces the
error in the calculated energy by a factor of 10, compared with the
frozen core approximation, with the exception of the ^{1}S states
for which the improvement is only by a factor of 2.

In conclusion, I would say that over the years workers on the
continuum state problem have borrowed extensively from techniques
developed for the bound-state problem, but some recent work shows
that a small part of the debt is now being repaid.

REFERENCES

P. S. Bagus, N. Bessis and C. M. Moser, Phys. Rev. 179, 39 (1969).

M. Cohen and P. S. Kelly, Can. J. Phys. 45, 2079 (1965).

R. J. Damburg, Physics of Electronic and Atomic Collisions, eds.
    T. R. Govers and F. J. de Heer (North-Holland, Amsterdam,
    1972), p. 200.

J. Dubau and J. Wells, to be submitted to J. Phys. B: Atom. Molec.
    Phys. (1972).

W. Eissner and M. J. Seaton, Mém. Soc. Roy. Sci. Liège, in press
    (1972).

S. Geltman, Physics of Electronic and Atomic Collisions, eds. T. R.
    Govers and F. J. de Heer (North-Holland, Amsterdam, 1972),
    p. 216.

P. de A. P. Martins and M. J. Seaton, J. Phys. B: Atom. Molec.
    Phys. 2, 333 (1969).

H. E. Saraph and M. J. Seaton, Phil. Trans. Roy. Soc. A271, 1
    (1971).

I. H. Sloan, Proc. Roy. Soc. A281, 151 (1964).

M. J. Seaton, Proc. Phys. Soc. 88, 815 (1966); Atomic Physics, eds.
    B. Bederson, V. W. Cohen and F. M. J. Pichanick (Plenum Press,
    New York, 1972), p. 295; Comments Atom. Molec. Phys. 2, 37
    (1970); J. Phys. B: Atom. Molec. Phys. 5, L91 (1972).

M. J. Seaton and P. M. H. Wilson, J. Phys. B: Atom. Molec. Phys.
    5, L1 (1972a); ibid. 5, L175 (1972b).

RECENT ADVANCES IN THE INTERPRETATION OF COMPLEX ATOMIC SPECTRA

Zipora B. Goldschmidt*

National Bureau of Standards, Washington, D.C. 20234

INTRODUCTION

The conventional energy-level calculations of atomic spectra have usually been carried out only to first order perturbation theory. In this approximation, energy matrices including (only) the electrostatic interaction $G = \sum\limits_{i>j} \frac{e^2}{r_{ij}}$ and the spin-orbit interaction $F = \sum\limits_{i} \zeta_i (\ell_i \cdot s_i)$ within a single configuration, or a very small number of adjacent configurations, were constructed and diagonalized. This led to a rather poor agreement between calculated and observed energy levels, even when all the radial integrals $F^k(n_a \ell_a, n_b \ell_b)$, $G^k(n_a \ell_a, n_b \ell_b)$, $R^k(n_a \ell_a n_b \ell_b, n_c \ell_c n_d \ell_d)$ and $\zeta_{n\ell}$ were considered as adjustable parameters.

The fit between theory and experiment can be improved by adding two types of terms to the conventional Hamiltonian of the investigated configuration. Terms of the first type represent the mutual magnetic interactions. These interactions are provided by the Breit equation,[1] which, in the nonrelativistic limit, breaks up into terms which can be given a simple physical interpretation.[1,2] All of these terms that are spin independent (such as the orbit-orbit interaction, the retardation of the Coulomb interaction, and the relativistic correction due to the mass variation with velocity) are absorbed by the electrostatic parameters (real and effective,[3]

including the additive constant). Into this category also falls
the spin-spin contact term which is diagonal with respect to the
total spin angular momentum of the electrons. The terms which are
not absorbed are the spin-spin (ss) and spin-other-orbit (soo)
interactions, which, respectively, represent the mutual interaction
between the magnetic dipole moments of the electrons and between
the dipole moment of one electron and the orbital motion of another.
Terms of the second type are called "Effective interactions," and
represent, to second order perturbation theory, interactions with
distant configurations. On considering G and F, the two dominant
terms of the Hamiltonian for heavy atoms, we find that the effective
interactions have the form

$$H_{eff} = -(G+F)(G+F)/\Delta E$$

ΔE being the (positive) energy separation between the perturbing
and perturbed configurations. H_{eff} decomposes into three parts:

$$H_A = -GG/\Delta E$$

$$H_B = -(GF+FG)/\Delta E$$

$$H_C = -FF/\Delta E$$

H_A is known as "effective electrostatic interaction," and, in
analogy with the real electrostatic interaction, its tensor-
operator form is spin-independent.[4-6] H_B is called the "effective
electrostatic-spin orbit interaction (effective EL-SO)," and is
spin-dependent, in analogy with the mutual magnetic ss and soo
interactions.[7-11] H_C turns out to have the same angular dependence
as the usual spin-orbit interaction,[7,12] and therefore does not
constitute an independent interaction.

Recently, various systematic investigations have been conducted,
many of them at the Hebrew University of Jerusalem, on the effects
of the mutual magnetic and effective interactions on the energy
level structure of heavy atoms. A review of these will be given
in the present paper.

I. EFFECTIVE EL-SO INTERACTION IN nsnp CONFIGURATIONS.

A systematic investigation has been carried out by the present author on the effects of the spin-dependent interactions on the nsnp configurations in the isoelectronic sequences Zn I - Br VI (n=4), Cd I - Te V (n=5) and Hg I - Bi IV (n=6).

The sp configuration comprises the two terms ^1P and ^3P, which split into the four levels $^1P_1$, $^3P_{0,1,2}$. In the conventional approach, these are described by three parameters: the additive constant F_0, the electrostatic exchange parameter $G_1(sp)$ and the spin-orbit parameter ζ_p. The energy matrices and the energy-level formulas for this configuration are given in table 1.[13] Solving for the parameters so as to obtain the correct over-all triplet width $(=3\zeta_p/2)$, the proper locations of the $^3P_0$ and $^3P_2$ levels (F_0-G_1), and the exact value for the trace of the J=1 matrix $(2F_0-\zeta_p/2)$, one obtains for the parameters the following formulas written at the bottom of this table. However, the use of these parameter values for the diagonalization of the energy matrices results in a calculated $^1P_1-^3P_1$ separation which is greater than the observed one. For example, $(^1P_1-^3P_1)_C-(^1P_1-^3P_1)_O$ ranges from 4.4-40.6 cm^{-1}, from 42.0-163.4 cm^{-1} and from 519.0-1030.0 cm^{-1} for the Zn I 4s4p, Cd I 5s5p and Hg I 6s6p isoelectronic sequences respectively (see table 3). The fact that the calculated $^1P_1-^3P_1$ separation is too big means that the eigenvectors show too big $^1P_1-^3P_1$ mixtures, which in turn result in wrong values for the g-factors and for the lifetime ratios of these levels.

All these discrepancies seem to arise from the fact that the off-diagonal matrix element of the J=1 matrix is too big. A glance at table 1 shows that this matrix element includes only the parameter ζ_p, which was determined so as to give the exact ^3P splitting. The immediate conclusion is that two different spin-orbit parameters are needed, one to determine the ^3P width and the other - the $^1P_1-^3P_1$ separation. The question arises as to what theory will predict such a conclusion.

King and Van Vleck treated this problem in 1939.[14] In their paper they write that "the most natural explanation of the discrepancies is to blame perturbation by other levels, but none seems to

Table 1. CONVENTIONAL ENERGY MATRICES AND EIGENVALUES
OF THE sp CONFIGURATION

$$^3P_0 = F_0 - G_1 - \zeta = E(^3P_0)$$

	$^1P_1$	$^3P_1$
$^1P_1$	$F_0 + G_1$	$\frac{\sqrt{2}}{2}\zeta$
$^3P_1$	$\frac{\sqrt{2}}{2}\zeta$	$F_0 - G_1 - \frac{1}{2}\zeta$

$$\left.\begin{array}{c} E(^1P_1) \\ E(^3P_1) \end{array}\right\} = F_0 - \frac{1}{4}\zeta \pm \sqrt{(G_1 + \frac{1}{4}\zeta)^2 + \frac{1}{2}\zeta^2}$$

$$^3P_2 = F_0 - G_1 + \frac{1}{2}\zeta = E(^3P_2)$$

PARAMETERS:

$$G_1 = \frac{1}{2}\left\{E(^1P_1) + E(^3P_1) - E(^3P_2) - E(^3P_0)\right\}$$

$$\zeta = \frac{2}{3}\left\{E(^3P_2) - E(^3P_0)\right\}$$

$$F_0 = E(^3P_0) + G_1 + \zeta = E(^3P_2) + G_1 - \frac{1}{2}\zeta$$

be located in a suitable position to account for the anomalies."
They suggested solving the difficulty by reasoning that the radial
functions are not quite the same for singlet and triplet states.
To take the difference into account, they replaced ζ by $\lambda\zeta$ in the
off diagonal J=1 matrix element, where λ is a new parameter. Thus,
they solved exactly the problem of four levels with four parameters.
They also showed that their method results in improved intensity
ratios for the $s^2\ ^1S_0$-sp $^1P_1$ and $s^2\ ^1S_0$-sp $^3P_1$ transitions.

In the present paper a different physical approach is proposed,
although, mathematically, it will turn out to be equivalent to the
solution of King and Van Vleck.

Inspection of the deviations between observed and calculated
levels obtained above (which are zero for $^3P_0$, $^3P_2$ and finite and
of opposite sign for $^1P_1$, $^3P_1$) leads to the immediate conclusion that
they are of a spin-dependent character, and therefore spin-dependent
interactions are needed in order to eliminate them. (Indeed, no
additional electrostatic interaction is needed since in the sp
configuration the two terms are already described by two Slater
parameters). The importance of the spin-dependent ss, soo and
effective EL-SO interactions in improving the calculated multiplet
structures has already been established for ℓ^N configurations[8-11]
as will be discussed in Section II of the present paper.

In the present case only one additional parameter is needed in order to obtain an exact description of the four energy levels. On considering the size of the deviations that have to be eliminated, it was decided that from the three spin-dependent interactions mentioned aboved, the interaction mainly responsible in the present case is the effective EL-SO interaction, as will be proved below. This brings us back to the "natural explanation" sought by King and Van Vleck, their "missing" perturbing levels being all the levels belonging to distant odd configurations.

The Hamiltonian representing the effective EL-SO interaction in $\ell^N \ell'$ configurations has therefore been constructed[15,16] and is given below:

$$H_{EL-SO} = \sum_{\substack{k \text{ even} \\ t \text{ odd} \\ i \neq j}} \left\{ S^k(n\ell\,n'\ell' - n\ell\,n''\ell) \left[\frac{1}{3}\ell'(\ell'+1)(2\ell'+1)\right]^{1/2} \left([\underline{u}_i^{(k)} \times \underline{u}_j'^{(t)}]^{(1)} \cdot \underline{s}_j\right) \right.$$

$$\left. + S^k(n\ell\,n'\ell' - n''\ell\,n'\ell') \left[\frac{1}{3}\ell(\ell+1)(2\ell+1)\right]^{1/2} \left(\underline{s}_i \cdot [\underline{u}_i^{(t)} \times \underline{u}_j'^{(k)}]^{(1)}\right) \right.$$

$$+ \sum_{\substack{k,t \\ i \neq j}} \left[T^k(n\ell\,n'\ell' - n''\ell'\,n\ell) \left[\frac{1}{3}\ell'(\ell'+1)(2\ell'+1)\right]^{1/2}(-1)^{\ell+t+k} + T^k(n\ell\,n'\ell' - n'\ell'\,n''\ell) \left[\frac{1}{3}\ell(\ell+1)(2\ell+1)\right]^{1/2}\right]$$

$$\times \left[\left(\underline{s}_i \cdot [\underline{v}_i^{(t)} \times \underline{v}_j^{(k)T}]^{(1)}\right) + \left([\underline{v}_i^{(k)} \times \underline{v}_j^{(t)T}]^{(1)} \cdot \underline{s}_j\right)\right]$$

where $\underline{u}^{(k)}$, $\underline{u}'^{(k)}$, $\underline{v}^{(k)}$ and $\underline{v}^{(k)T}$ are unit tensor operators defined as follows:

$$(\ell\|\underline{u}^{(k)}\|\ell) = 1 \qquad (\ell'\|\underline{u}'^{(k)}\|\ell') = 1$$

$$(n\ell\|\underline{v}^{(k)}\|n'\ell') = (n'\ell'\|\underline{v}^{(k)T}\|n\ell) = 1$$

The parameters $S^k(n\ell n'\ell' - n\ell n''\ell')$, $S^k(n\ell n'\ell' - n''\ell n'\ell')$, $T^k(n\ell n'\ell' - n''\ell'n\ell)$ and $T^k(n\ell n'\ell', n'\ell'n''\ell)$ that represent respectively the direct and exchange parts of these interactions are special cases of

$$Q^k\left(n_a\ell_a n_b\ell_b, n_c\ell_c n_d\ell_d\right) \equiv \left(\ell_a\|C^{(k)}\|\ell_c\right)\left(\ell_b\|C^{(k)}\|\ell_d\right)$$

$$\frac{R^k\left(n_a\ell_a n_b\ell_b, n_c\ell_c n_d\ell_d\right)\zeta\left(n_a\ell_a, n_c\ell_c \text{ or } n_b\ell_b, n_d\ell_d\right)}{\Delta E}$$

where R^k and ζ are respectively Slater and spin-orbit parameters.

For ℓs configurations ($\ell'=0$) the direct and exchange terms including the factor $\sqrt{\ell'(\ell'+1)(2\ell'+1)}$ vanish. In addition, the second direct term turns out to have the same angular dependence as the spin-orbit interaction of the ℓ electron and is therefore dropped. For the remaining exchange term the following results are obtained:

$$k=t=\ell$$

and

$$(s\ell\ SLJ\ ||\ H_{eff\ EL-SO}\ ||\ s\ell\ S'LJ)=0 \begin{cases} S=S'=0 \\ S=0\ \ S'=1 \\ S=1\ \ S'=0 \end{cases}$$

$$(s\ell\ ^3LJ\ ||\ H_{eff\ EL-SO}\ ||\ s\ell\ ^3LJ) = \sqrt{\frac{6\ell(\ell+1)}{(2\ell+1)}}\ T^\ell$$

$=-2T$ for the sp configuration, where $T=\dfrac{T'}{3}$.

Taking into account the J-dependence

$$(-1)^{L+S'+J}\begin{Bmatrix} S & S' & 1 \\ L' & L & J \end{Bmatrix}$$

one obtains the augmented energy matrices and corrected eigenvalues for the sp configuration, as given in table 2. The formulas obtained for the matrix elements in the last table are analogous to those given in table 1, except that the ^3P level formulas now include $\zeta+2T$ instead of ζ. The off-diagonal J=1 matrix element remained unchanged. The purpose of obtaining "two different spin-orbit parameters," one for the ^3P width and one for the $^1P_1-{}^3P_1$ separation, has thus been reached, and the augmented energy matrices are mathematically equivalent to those of King and Van Vleck.

The augmented energy matrices were first diagonalized using the initial values of F_0, G_1, and ζ obtained in table 1 and T=0. Then, using the diagonalization-least squares iterative process, the final values for the parameters given in table 3 were obtained. Those of F_0 and G_1 did not change whereas ζ and T satisfy the equation

$$(\zeta+2T)_{final} = \zeta_{initial}\ .$$

Table 2. AUGMENTED ENERGY MATRICES AND EIGENVALUES OF THE sp CONFIGURATION

$$^3P_0 = F_0 - G_1 - (\zeta + 2T) = E(^3P_0)$$

	$^1P_1$	$^3P_1$
$^1P_1$	$F_1 + G_1$	$\dfrac{\sqrt{2}}{2}\zeta$
$^3P_1$	$\dfrac{\sqrt{2}}{2}\zeta$	$F_0 - G_1 - \dfrac{1}{2}(\zeta + 2T)$

$$\left.\begin{array}{r}E(^1P_1)\\ E(^3P_1)\end{array}\right\} = F_0 - \frac{1}{4}(\zeta + 2T) \pm \sqrt{\left[G_1 + \frac{1}{4}(\zeta + 2T)\right]^2 + \frac{1}{2}\zeta^2}$$

$$^3P_2 = F_0 - G_1 + \frac{1}{2}(\zeta + 2T)$$

PARAMETERS:

$$G_1 = \frac{1}{2}\left\{E(^1P_1) + E(^3P_1) - E(^3P_2) - E(^3P_0)\right\}$$

$$\zeta + 2T = \frac{2}{3}\left\{E(^3P_2) - E(^3P_0)\right\}$$

$$F_0 = E(^3P_0) + G_1 + \zeta + 2T = E(^3P_2) + G_1 - \frac{1}{2}(\zeta + 2T)$$

All parameters vary systematically along each of the sequences. T increases with the degree of ionization and with the atomic number.

The final values obtained for the calculated energy levels are, of course, equal to those of the observed ones; therefore the values of $(^1P_1 - {}^3P_1)_C - (^1P_1 - {}^3P_1)_0$ given in the table for each of the investigated ions reflect the amounts of the corrections for the energy levels obtained through the introduction of T. A comparison of the observed and calculated g-factors and lifetime ratios for Hg I (table 4) shows that the introduction of T so as to fit the energy levels, also results in a very good fit for the g-factors and a great improvement in the fit of lifetime ratios. In order to further improve the fit of lifetime ratios, direct CI with adjacent configurations e.g. d^9s^2p should be included in the calculations.[20,21] Finally, the effects of the mutual ss and soo interactions were accurately evaluated, by calculating both their angular part[15,22,23] and radial part (HF calculation[24]). These were found to be about 1.7% of the effective EL-SO interaction in Br VI and 0.04% in Bi IV, and therefore quite negligible.

In connection with the last conclusion it is worth mentioning that Wolfe[25] obtained formulas for sℓ configurations, which make approximate allowance for the soo interaction. These formulas,

Table 3. The nsnp configuration in the isoelectronic sequences Zn I - Br VI (n=4), Cd I - Te V (n=5) and Hg I - Bi IV (n=6).

n=4	Parameter cm^{-1}	Zn I	Ga II	Ge III	As IV	Se V	Br VI
	F_0	39719.5	59488.0	77585.3	95105.3	112409.5	130022.5
	G_1	7022.5	11198.0	14249.0	16840.0	19183.5	21026.5
	ζ	292.3	800.2	1506.3	2316.1	3356.7	4480.9
	T	46.8	59.9	48.5	68.6	56.6	99.6
	$\zeta + 2T$	386.0	920.0	1603.3	2453.3	3470.0	4680.0
	$(^1P_1-^3P_1)_c-(^1P_1-^3P_1)_0$ (T=0)	4.4	9.0	10.2	18.6	19.2	40.6
n=5		Cd I	In II	Sn III	Sb IV	Te V	
	F_0	37459.5	53783.5	68500.3	82680.2	96679.7	
	G_1	6203.5	9140.5	11165.0	12828.5	14312.0	
	ζ	868.6	2075.2	3467.3	5054.1	6867.8	
	T	136.7	146.4	160.0	181.3	195.4	
	$\zeta + 2T$	1142.0	2368.0	3787.3	5416.7	7258.6	
	$(^1P_1-^3P_1)_c-(^1P_1-^3P_1)_0$ (T=0)	42.0	66.0	93.8	129.6	163.4	
n=6		Hg I	Tl II	Pb III	Bi IV		
	F_0	47806.8	66072.2	82963.5	99507.3		
	G_1	5896.5	8438.5	10174.5	11571.0		
	ζ	3232.9	7130.3	11251.8	15739.4		
	T	516.2	526.2	570.1	616.9		
	$\zeta + 2T$	4265.3	8182.7	12392.0	16973.3		
	$(^1P_1-^3P_1)_c-(^1P_1-^3P_1)_0$ (T=0)	519.0	682.6	859.2	1030.0		

Table 4. g-factors and lifetime ratios for Hg I $^3P_1$ and $^1P_1$ levels.

	Obs	Calc (T=0)	Calc (T≠0)
$g(^3P_1)$	1.48612[17]	1.479	1.487
$g(^1P_1)$	1.013 ± 0.005[18]	1.021	1.013
$\tau(^3P_1)/\tau(^1P_1)$	87 ± 3[19]	60	101

given also by Condon and Shortley,[13] are identical with our augmented matrices (table 2), except for the facts that η appears there instead of T, and a term $(\ell_p \cdot s_p)\eta$ proportional to the spin-orbit interaction has been subtracted. Wolfe therefore believed that the soo interaction was responsible for all the anomalies discussed above.

In conclusion, it has been established for the present problem that the use of effective interactions is physically equivalent to the use of term dependent parameters. In this case the two methods are also mathematically equivalent; however, for more complicated spectra it will be shown below that the method of effective interactions becomes increasingly more efficient than the method of term-dependent parameters.

II. SPIN-DEPENDENT INTERACTIONS IN ℓ^N CONFIGURATIONS

In recent years various successful attempts have been made to obtain a more adequate description of the energy levels of ℓ^N configurations, as compared with the conventional treatment. Substantially better fits between observed and calculated levels have been obtained by the inclusion in their energy matrices of effective electrostatic interactions. These interactions are described by two and three-electron "effective electrostatic parameters," e.g. α, β (two-electron effective parameters), T and T_x (three-electron effective parameters) for d^N configurations, and α, β, γ (two-electron effective parameters) and T_i (i=2,3,4,6,7,8) (three-electron effective parameters) for f^N configurations. The mean error of the energy-level fit was thereby decreased to several tens of cm^{-1}.[26-31] The remaining deviations often were of a pronounced magnetic character, thus suggesting the need for inclusion in the Hamiltonian of the still missing spin-dependent ss, soo and effective EL-SO interactions. A systematic investigation was therefore made of the effects of these interactions on the energy-level schemes of d^N and f^N configurations in heavy atoms. The configurations chosen for this purpose are the $3d^N$ configurations (N=2,3,...,8) in the third and fourth spectra of the iron group and Pr IV $4f^2$ and Pr III $4f^3$.[8,10,11,32] The results obtained are described below.

The operators representing the ss and the soo interactions for

ℓ^N configurations can be described in tensor-operator form by the following formulas:

$$H_{ss} = -\frac{\beta^2}{\sqrt{5}} \sum_k (-1)^k \left(\frac{(2k+5)!}{(2k)!}\right)^{1/2} \times$$

$$\times \sum_{i \neq j} \left[\frac{r_i^k}{r_j^{k+3}} \left(\left[\underline{C}_i^{(k+2)} \times \underline{C}_j^{(k)}\right]^{(2)} \cdot \left[\underline{s}_i \times \underline{s}_j\right]^{(2)}\right) + \frac{r_i^k}{r_j^{k+3}} \left(\left[\underline{C}_i^{(k)} \times \underline{C}_j^{(k+2)}\right]^{(2)} \cdot \left[\underline{s}_i \times \underline{s}_j\right]^{(2)}\right)\right]$$

$$H_{soo} = \frac{2\beta^2}{\sqrt{3}} \sum_k (-1)^k \sum_{i \neq j} \left\{ \frac{r_j^{k-2}}{r_i^{k+1}} (2k+1)(2k-1)^{1/2} \left[\underline{C}_j^{(k)} \times \left[\underline{C}^{(k)} \times \underline{l}\right]_i^{(k-1)}\right]^{(1)} \right.$$

$$\left. - \frac{r_j^k}{r_i^{k+3}} (2k+1)(2k+3)^{1/2} \left[\underline{C}_j^{(k)} \times \left[\underline{C}^{(k)} \times \underline{l}\right]_i^{(k+1)}\right]^{(1)} \right\} \cdot (\underline{s}_i + 2\underline{s}_j) \ .$$

The appropriate radial integrals are those defined by Marvin[33]

$$M^k = \frac{\beta^2}{2} \int_0^\infty \int_0^\infty \frac{r_<^k}{r_<^{k+3}} R_{nl}^2(r_1) R_{nl}^2(r_2) \, dr_1 \, dr_2 \ ,$$

where $r_< = \min(r_1, r_2)$ and $r_> = \max(r_1, r_2)$. k may take the values $0, 2$ and $0, 2, 4$ for d^N and f^N configurations respectively. The expression for the effective EL-SO operator is

$$H_{EL-SO} = -2 \sum_{k \text{ even}} Q^k \left[l(l+1)(2l+1)\right]^{1/2} (2k+1)^{-1/2} \times$$

$$\times \sum_{t \text{ odd}} (2t+1) \begin{Bmatrix} 1 & k & t \\ l & l & l \end{Bmatrix} \left(\underline{U}^{(k)} \cdot \underline{T}^{(1t)k}\right)$$

where the parameters Q^k are defined as follows:

$$Q^k = (l \| \underline{C}^{(k)} \| l)^2 \sum_{n'} \frac{R^k(nlnl, nln'l) \, \zeta_l(nn')}{\Delta E_{nn'}}$$

with $R^k(n\ell n\ell, n\ell n'\ell)$ and $\zeta_\ell(nn')$ being respectively a Slater and a spin-orbit parameter; also,

$$\left(nl \parallel \underline{u}_i^{(k)} \parallel nl\right) = 1$$

$$\underline{U}^{(k)} = \sum_i \underline{u}_i^{(k)}$$

and

$$\underline{T}^{(1t)k} = \sum_i \underline{t}_i^{(1t)k} = \sum_i \left[\underline{s}_i \times \underline{u}_i^{(t)}\right]^{(k)} .$$

k takes on the two sets of values 0,2,4 and 0,2,4,6 for d^N and f^N configurations respectively, but since the k=0 term has the same angular dependence as the spin-orbit interaction, it is dropped from the calculations.

The algebraic matrices of these operators were constructed by using the Racah algebra[34-37] and added to the already existing energy matrices. The appropriate radial integrals were considered as adjustable parameters, and evaluated by means of the diagonalization-least squares procedure.

The introduction of the spin-dependent interactions greatly improved the fit between calculated and observed multiplet splittings in all investigated configurations. Typical examples are shown in figures 1-3. As demonstrated by these figures, the effects of the added interactions is to practically equalize the deviations between calculated and observed energy levels, for all levels belonging to the same multiplet. Thus deviations of magnetic character are eliminated and the remaining deviations are mainly of a purely electrostatic character. These residual deviations prevent the mean error obtained in the least-squares calculations from reflecting the improvement in the fit due to the spin-dependent interactions. A new criterion to measure this improvement has therefore been introduced, which was called "observed minus calculated (O-C) spread." This is defined as the absolute value of the difference between the maximum and minimum deviations for levels belonging to the same multiplet. For the multiplets given in figures 1-3, the reductions in the O-C spread are given below.

1. Cr III $3d^4$ ^3H Mn IV $3d^4$ ^3H
    33 $cm^{-1} \rightarrow 4$ cm^{-1} 56 $cm^{-1} \rightarrow 2$ cm^{-1}

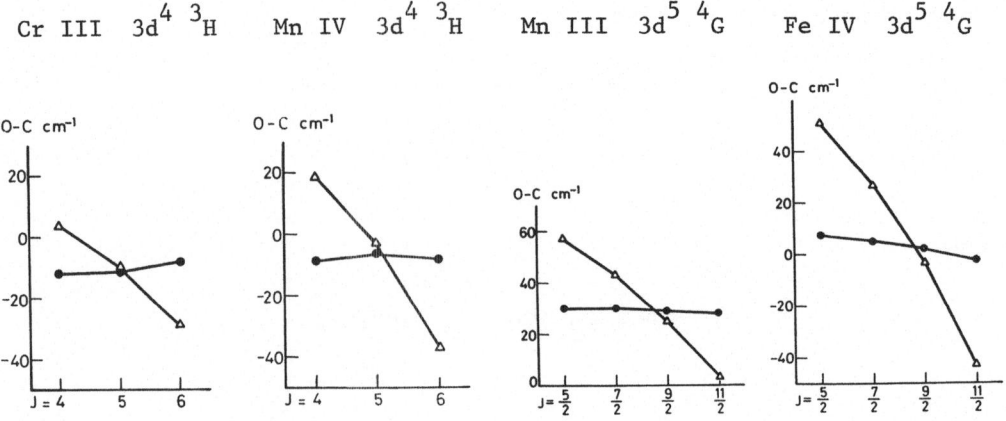

Fig. 1. Improvement in $3d^4\ {}^3H$ multiplet splitting due to the spin-dependent interactions (sdi) ($\triangle$-no sdi; $\bullet$ sdi included).

Fig. 2. Improvement in $3d^5\ {}^4G$ multiplet splitting due to the spin-dependent interactions ($\triangle$-no sdi; $\bullet$ sdi included).

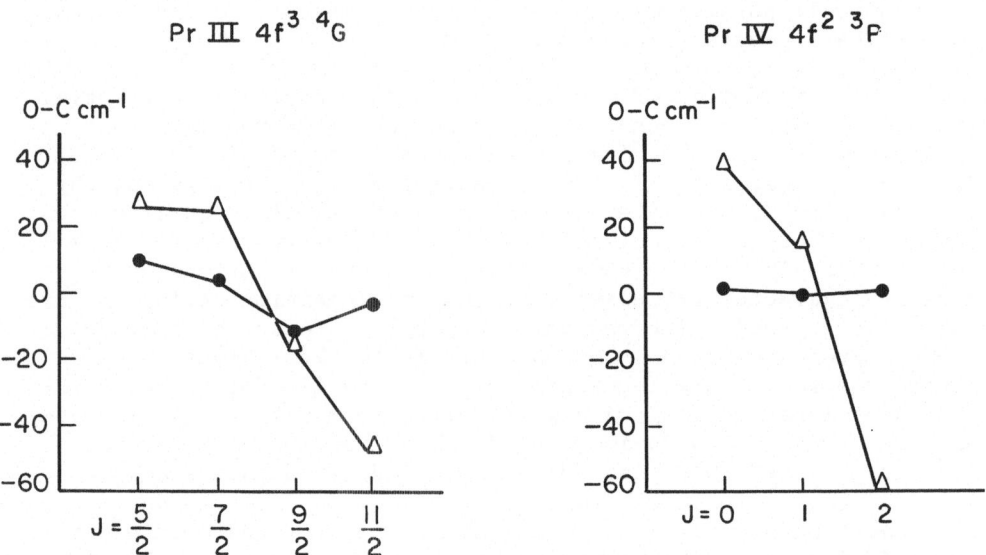

Fig. 3. Improvement in $4f^3\ {}^4F$ and $4f^2\ {}^3P$ multiplet splittings due to the spin-dependent interactions ($\triangle$-no sdi; $\bullet$ sdi included).

2. Mn III $3d^5\ ^4G$

$54\ cm^{-1} \to 2\ cm^{-1}$

Fe IV $3d^5\ ^4G$

$94\ cm^{-1} \to 11\ cm^{-1}$

3. Pr III $4f^3\ ^4G$

$73\ cm^{-1} \to 23\ cm^{-1}$

Pr IV $4f^2\ ^3P$

$97\ cm^{-1} \to 2\ cm^{-1}$

On the whole, the sum of O-C spreads has reduced from 786 cm^{-1} to 274 cm^{-1} for the third spectra of the iron group, from 806 cm^{-1} to 201 cm^{-1} for the fourth iron group spectra, and from 616 cm^{-1} to 324 cm^{-1} in the rare-earths.

The values obtained for the M^ks and Q^ks are given in tables 5 and 6 for the third and fourth spectra of the iron group respectively and in table 7 for the rare-earth spectra. The tables clearly demonstrate the reliability and consistency of the parameter values obtained, which are also in very good agreement with all available theoretical predictions.

The following conclusions concerning these parameters may be drawn:

a) The values of the spin-spin interaction parameters M^k_{ss} turn out to be equal to the values of the corresponding spin-other-orbit interaction parameters M^k_{soo}, as required by their definition.

b) Both the M^ks and the Q^ks are positive. The M^k values form decreasing functions of k, whereas the Q^ks seem to be approximately independent of k.

c) A comparison of the results for the third and fourth spectra, both in transition ions and the rare-earths, shows that the M^ks, like most internal parameters, increase with the degree of ionization, whereas the Q^ks decrease, in common with effective electrostatic parameters.

d) The M^k values are in excellent agreement with the corresponding values obtained by Blume, Freeman, and Watson[38,39] using the Hartree-Fock method. Thus, the results of ab initio calculations are confirmed by semiempirical calculations.

The strengths of the spin-dependent interactions, that is, their contributions to the various multiplet splittings, were found to be: spin-other-orbit and effective EL-SO interactions: several tens of cm^{-1} in the transition ions and up to a few

Table 5. Spin-dependent parameters for $3d^N$ configurations in the third spectra of the iron group (cm^{-1})

Parameter	Ti III $3d^2$	V III $3d^3$	Cr III $3d^4$	Mn III $3d^5$	Fe III $3d^6$	Co III $3d^7$	Ni III $3d^8$
ζ	129	177	239	318	411	520	644
ζ HF	126	184	258	333	426	539	672
$M^0_{ss} = M^0_{soo}$	0.498	0.816	1.134	1.452	1.770	2.088	2.406
M^0 HF	0.704	0.915	1.156	1.433	1.704	2.022	2.375
$M^2_{ss} = M^2_{soo}$	0.229	0.422	0.615	0.808	1.001	1.194	1.384
M^2 HF	0.384	0.499	1.631	0.783	0.930	1.103	1.295
$Q^2 = Q^4$	-1.2	12.7	26.6	40.5	54.4	68.3	82.2

Table 6. Spin-dependent parameters for $3d^N$ configurations in the fourth spectra of the iron group (cm^{-1})

Parameter	V IV $3d^2$	Cr IV $3d^3$	Mn IV $3d^4$	Fe IV $3d^5$	Co IV $3d^6$	Ni IV $3d^7$	Cu IV $3d^8$
ζ	219	287	372	473	593	730	884
ζ HF	219	292	380	486			
$M^0_{ss} = M^0_{soo}$	1.31	1.56	1.82	2.07	2.32	2.58	2.83
M^0 HF	1.14	1.41	1.70	2.04			
$M^2_{ss} = M^2_{soo}$	0.72	0.86	1.00	1.14	1.28	1.42	1.56
M^2 HF	0.63	0.78	0.94	1.12			
$Q^2 = Q^4$	4	19	34	49	64	79	94

Table 7. Spin-dependent parameters for the $4f^N$ configurations in the rare-earths (cm^{-1})

Parameter	Pr III $4f^{3*}$	Pr IV $4f^2$
ζ_{4f}	691	788
ζ_{4f} HF	838	878
$M^0_{ss}=M^0_{soo}$	1.96	2.14
M^0 HF	2.08	2.24
$M^2_{ss}=M^2_{soo}$	1.21	1.37
M^2 HF	1.17	1.26
$M^4_{ss}=M^4_{soo}$	0.23	0.75
M^4 HF	0.79	0.85
$Q^2=Q^4=Q^6$	68.6	50.8

* HF parameters for Pr III $4f^3$ were derived by means of the

equality $\dfrac{P(Pr\ III\ 4f^3)}{P(Nd\ IV\ 4f^3)} = \dfrac{P(Eu\ III\ 4f^7)}{P(Gd\ IV\ 4f^7)}$

hundreds cm^{-1} in the rare-earths. The spin-spin interaction is weaker by one order of magnitude than the other two.

The efficiency of the effective-interaction method can be very well demonstrated in this case by comparing the number of parameters used in the various approximations to the numbers of observed terms and levels. Taking as an example the $3d^N$ configurations of the third spectra of the iron group, fifty-four observed terms, which split into 134 levels, were compared with the results of the theoretical calculations in which all $3d^N$ configurations were treated as a single problem (general least squares calculation).[11] In the conventional treatment 11 electrostatic parameters and 3 spin-orbit parameters were used. Through the addition of the effective and mutual magnetic interactions, the number of parameters increased to a total of 19 electrostatic + 9 spin-dependent parameters. In the "term dependent parameters" method, at least 277 parameters would be needed!

III. TWO AND THREE-ELECTRON EFFECTIVE
ELECTROSTATIC INTERACTIONS IN $\ell^N \ell'$ CONFIGURATIONS

The situation in these configurations up to the beginning of
the present work can be described as follows:

1) Several investigations were conducted on isolated $\ell^N \ell'$ configurations, in which the effective electrostatic interaction of the ℓ^N
core was included in the energy level calculations, in addition to
the electrostatic and spin-orbit interactions.[40,41] These resulted
in some improvement in the fit as compared to the conventional
calculations, but the remaining deviations between calculated and
observed levels were still quite large--up to a few hundreds cm^{-1}.
These results lead to the conclusion that in order to obtain better
fits, the complete effective electrostatic interaction among all
electrons within a configuration should be included in the calculations.

2) In the very complex rare-earth spectra, configuration interaction
(CI) in (large) groups of interacting configurations was treated in
first order perturbation theory. For this purpose a method was
developed by the present author[29,42] according to which the interpretation of several neighboring spectra, both along an isoionic
sequence and of different degrees of ionization of the same element,
is carried out simultaneously, and the values of the parameters
representing the same interactions in the various spectra are
continously compared. This "comparison method" is based on the
fact that the values of the parameters change systematically along
an isoionic row or with increasing ionization of the same element.
By means of this method, the following spectra were investigated
at the Hebrew University:

La II: $5d^2+5d6s+6s^2+4f6p+4f^2+6p^2$ [29,42,43]

Gd II: $4f^7(^8S)(5d^2+5d6s+6s^2),4f^7(^8S)(5d6p+6s6p)$ [44]

Yb II: $4f^{13}(5d^2+5d6s+6s^2),4f^{13}(5d6p+6s6p)$ [45]

Lu II: $5d^2+5d6s+6s^2,5d6p+6s6p$ [29,42]

Ce III: $4f^2+4f6p+5d^2+5d6s,4f5d+4f6s$ [29,42,43]

Pr III: $4f^25d+4f^26s,4f^3+4f^26p+4f5d^2+4f5d6s$ [29,46]

Gd I: $4f^7(^8S)(5d6s^2+5d^26s),4f^7(^8S)(6s^26p+5d6s6p+5d^26p)$ [47]

Yb I: $4f^{14}(5d^2+5d6s+6s^2)+4f^{13}(5d^26p+5d6s6p+6s^26p)$,

         $4f^{14}6s6p+4f^{13}5d6s^2$ [47] .

The following results have been obtained:

a. In all investigated spectra, a very good agreement was obtained
    for both the energy levels and their g-factors.

b. Reliable and consistent values were obtained for the parameters
    representing the various interactions within and between
    configurations.

c. A complete understanding of the effects of the separate inter-
    actions on the energy level structure was achieved.

d. The stage has been reached where it is possible to make
    detailed predictions about the structure of rare-earth spectra,
    even in those cases where the experimental material is very
    poor.

However, the introduction of CI with only adjacent configurations
included (and including effective electrostatic interactions of
ℓ^N cores) was not enough to give satisfactory results in those
spectra where the interacting configurations were largely overlapping and where the density of the energy levels was very high.
In particular, in such spectra it was not possible to obtain, in
the initial stages of the interpretation, an unambiguous correlation of calculated levels and g-factors to the observed levels.
Rare earth spectra having these properties are:

> La II: $4f5d+4f6s+5d6p+6s6p$
>
> Ce II: $4f5d^2+4f5d6s+4f6s^2+4f^26p+4f^3$
>
> $4f^25d+4f^26s+4f5d6p+4f6s6p+5d^3+5d^26s+5d6s^2$
>
> Ce I: $4f5d^3+4f5d^26s+4f5d6s^2$.

These difficulties have been successfully overcome by the inclusion,
in the energy-level calculations, of effective electrostatic interactions between $n\ell-n'\ell'$ non-equivalent electrons.

The results of systematic investigations of the effects of the
effective electrostatic interactions on various groups of configurations of heavy atoms comprising groups of electrons of the types
d^Np and f^Nd^K are now described.

As in the case of ℓ^N configurations, the effective Hamiltonian
for a configuration $\ell^N\ell'$ constitutes two terms: a two-electron
term H^2_{eff} and a three-electron term H^3_{eff}. Each of these comprises,
in analogy to the real electrostatic interaction, both a direct and

an exchange part. They are written below, in tensor-operator
form:[6,7]

$$H_{eff}^2 = -\sum_t F^{(t)} \sum_{i \neq j} (\underline{u}_i^{(t)} \cdot \underline{v}_j^{(t)}) - \sum_t G^{(t)} \sum_{i \neq j} (\underline{z}_i^{(t)} \cdot \tilde{\underline{z}}_j^{(t)})$$

where $\underline{u}^{(t)}$, $\underline{v}^{(t)}$, $\underline{z}^{(t)}$, and $\tilde{\underline{z}}^{(t)}$ are unit operators defined as:

$$(\ell \| \underline{u}^{(k)} \| \ell) = 1 \qquad (\ell' \| \underline{v}^{(k)} \| \ell') = 1$$

$$(\ell \| \underline{z}^{(k)} \| \ell') = 1 \qquad (\ell' \| \tilde{\underline{z}}^{(k)} \| \ell) = 1$$

$F^{(t)}$ and $G^{(t)}$ are respectively the direct and exchange "effective
Slater parameters." The allowed values of t for the direct part are
0 to the smaller of 2ℓ and $2\ell'$, but since the "real" Slater parameters with the even t's contained in this list are already included
in the calculations, only the F^t's having odd t constitute independent parameters. The allowed values of t for the exchange part
are $|\ell-\ell'| \leq t \leq \ell+\ell'$. Here again, only G^t's with t having the
same parity as $\overline{\ell+\ell'+1}$ are to be added as new parameters, those
with t having the opposite parity being already included as "real"
exchange parameters. Table 8 comprises the lists of effective
Slater parameters for the configurations ℓs, dp, fp, fd, and fg.
On the whole, on adding the effective interactions to the energy
matrices, the number of Slater parameters, real and effective, is
equal to the number of LS terms in the $\ell\ell'$ two-electron configuration. In a least-squares calculation including all these interactions the "even" direct and the "$\ell+\ell'$ parity" exchange effective
interactions are absorbed by the original Slater parameters.

The three-electron mixed effective interaction is represented
by H_{eff}^3

$$H_{eff}^3 = -\sum_{k'kt} U(k'kt) \sum_{\substack{ijs \\ \neq}} [\underline{u}_i^{(k')} \times \underline{u}_j^{(k)} \times \underline{v}_s^{(t)}]^{(o)}$$

$$-\sum_{k'kt} V(k'kt) \sum_{\substack{ijs \\ \neq}} (-1)^{k'+k+t} \times$$

$$\times [\underline{u}_i^{(k)} \times \{(\underline{z}_j^{(k')} \times \tilde{\underline{z}}_s^{(t)})^{(k)} + (\underline{z}_j^{(t)} \times \tilde{\underline{z}}_s^{(k')})^{(k)}\}]^{(o)}$$

for $d^N p$ configurations this interaction requires ten independent
parameters. Three of these U(222), U(242), and U(442) represent
the direct interaction and eight V(121), V(122), V(123), V(143),
V(322), V(323), V(342), and V(343) represent the exchange interaction. One parameter with k'+k+t even is dependent on other
electrostatic parameters, real and effective. Their introduction

Table 8. Effective slater parameters (two-electron)

Configuration	Direct Part	Exchange Part
ℓs	---------	---------
dp	F^1	G^2
fp	F^1	G^3
fd	F^1, F^3	G^2, G^4
fg	F^1, F^3, F^5	G^2, G^4, G^6

raises the number of electrostatic parameters in d^2p to 20, against 30 non-vanishing matrix elements. For a d^Np configuration with $N \geq 3$, the number of independent electrostatic parameters is 22.

Complete two-electron effective electrostatic interactions have been included in the energy-level calculations of configurations comprising non-equivalent electrons, e.g. the 3d4p configuration in the isoelectronic sequence Ti III - Mn VI,[6,48] the $3d^9 4p$ configuration in the isoelectronic sequence Zn III - As VI,[6,48] the 4f5d configuration in Pr IV and Ce III and La II,[48,49] the $4f^{13}5d$ configuration in Yb III,[48,49] and the $4f^N 5d^K$ type configurations within the following complexes:

La II: $4f5d+4f6s+5d6p+6s6p$[43]

Ce II: $4f5d^2 4f5d6s+4f6s^2+4f^2 6s+4f^3$[43]

$4f^2 5d+4f^2 6s+4f5d6p+4f6s6p+5d^3+5d^2 6s+5d6s^2$[43]

Pr III: $4f^2 5d+4f^2 6s$[50]

Ce I: $4f5d^3+4f5d^2 6s+4f5d6s^2$[51] .

In the $3d^N 4p$ configurations (N=1,2,...,9) of the third spectra of the iron group, complete two and three-electron effective electrostatic interactions were introduced.[6,52]

The inclusion of these interactions led to a remarkable improvement of the fit between observed and calculated levels in these spectra, as demonstrated by the reduction of the mean errors in tables 9, 10, 11, and 12.

Reliable and consistent values were obtained for all the parameters representing the various interactions in these con-

Table 9. The 3d4p configuration in the isoelectronic sequence TiIII VIV CrV MnVI

Parameter cm^{-1}	TiIII	VIV	CrV	MnVI
F_1	47	50	56	68
G_2	-13.0	-15.9	-23.6	-31.3

Reduction of the mean error: 160 $\longrightarrow$ 20 cm^{-1}

Table 10. The 3d^{9}4p configuration in the isoelectronic sequence ZnIII GaIV GeV AsVI

Parameter cm^{-1}	ZnIII	GaIV	GeV	AsVI
F_1	52.0	55.5	59.0	62.5
G_2	0	0	0	0

Reduction of the mean error: 120 $\longrightarrow$ 35 cm^{-1}

Table 11. Effective Slater parameters for the 4f^{N}5d^K configurations in Pr IV, Ce III, La II, Yb III and Ce II (cm^{-1})

Parameter	Pr IV 4f5d	Ce III 4f5d	La II 4f5d	Yb III 4f^{13}5d	Ce II 5 odd conf 4f5d^2,4f5d6s	Ce II 7 even conf 4f^{2}5d,4f5d6p	La II 4 odd conf 4f5d
F_1	21.2 ±0.6	20.4 ±0.4	84±3	15 ±5	15 ±1	19 ±2	18(fixed)
F_3	3.0 ±1.0	7.3 ±0.7	44±5	9 ±9	6 ±2	7 ±2	7(fixed)
G_2	-38.1 ±0.8	-43.0 ±0.5	-34±3	-32 ±5	-36 ±1	-35 ±1	-36(fixed)
G_4	-1.85±0.06	-1.54±0.04	0	-2.4±1	-1.2±0.1	-1.6±0.1	-1.5(fixed)
r.m.s. decrease	209 $\rightarrow$ 10	227 $\rightarrow$ 7	657 $\rightarrow$ 55	179 $\rightarrow$ 71	Difficult to interpret $\rightarrow$ 38	Difficult to interpret $\rightarrow$ 50	176 $\rightarrow$ 38

Table 12. Two and Three-Electron Effective Parameters for the $3d^N4p$ Configurations of the Third Spectra of the Iron Groups (cm^{-1})

Parameter	Ti III $3d4p$	V III $3d^24p$	Cr III $3d^34p$	Mn III $3d^44p$	Fe III $3d^54p$	Co III $3d^64p$	Ni III $3d^74p$	Cu III $3d^84p$	Zu III $3d^94p$
F_1	58.2	56.7	55.2	53.7	52.2	50.7	49.2	47.7	46.2
U(222)	6858	4523	2783	1640	1092	1141	1786	3028	4865
U(242)	20233	15640	12068	9516	7985	7475	7985	9516	12068
V(123)	5938	3213	1094	-420	-1328	-1630	-1328	-420	1094
V(322)	-890	-890	-890	-890	-890	-890	-890	-890	-890
V(323)	-983	-510	-189	-19	0	-134	-418	-855	-1443
V(342)	8903	5410	2637	584	-749	-1362	-1256	-429	1117
V(343)	11489	6418	2578	-31	-1410	-1559	-477	1835	5378

r.m.s. decrease: 143 cm^{-1} (without 3d–4p effective parameters)

$\xrightarrow{20\%}$ 114 cm^{-1} (through the introduction of F_1)

$\xrightarrow{27\%}$ 83 cm^{-1} (through the introduction of the Us and Vs).

figurations, with the exception of La II 4f5d where direct CI with overlapping configurations has been neglected. The final values of all new effective parameters are given in tables 9-12.

The inclusion of the effective interactions also greatly affected the compositions of the eigenvectors and therefore the g-factors of the energy levels. No experimental g-values are known for the $d^N p$ configurations investigated in our work. A comparison between observed and calculated g-factors could therefore be made only for the rare-earth spectra we calculated. An example of such a comparison is shown in table 13 for the 4f5d configuration of Ce III, which clearly demonstrates the importance of the effective electrostatic interactions in reproducing physical properties of atoms, that is, energy levels and g-factors. Considerable differences can be seen also in the compositions of several eigenvectors; in particular, the lowest level of this configuration has no good "name" in the old calculations, but is 72% ^1G according to the new calculations (and observed and calculated g-factors).

The improvements shown in the last table for the energy levels, eigenvectors, and g-factors of the 4f5d configuration, make it possible to understand the necessity of the inclusion of the effective electrostatic interaction in energy level calculations of the very complex La II, Ce II, and Ce I spectra, which comprise a large number of overlapping configurations of the type $4f^N 5d^K$. The beauty of the "effective interaction method" is reflected in these cases by the fact that the introduction of only four effective 4f-5d Slater parameters allows the theory to give a complete description of each of these spectra.

Table 14 gives an example of the results obtained for the seven (listed above) even configurations of Ce II, the three hundred and five energy levels of which have been calculated in intermediate coupling, with CI and effective electrostatic interactions included. The levels in this table have J=9/2, and range from levels having a pure name in a single configuration, to levels having intermediate coupling names in a single configuration, to levels having no names at all in mixed configurations. The agreements in the fits for the levels and their g-factors speak for themselves. It is left for the reader to calculate the number of parameters needed to describe this spectrum in the "term dependent parameters" method.

Table 13. Energy Levels of Ce III 4f5d (cm^{-1}): I-Without Effective Interactions; II-Effective Interactions Included

Term	J	Obs	Calc I	O-C I	Calc II	O-C II	g-obs	g-calc I	g-calc II	composition percentage I	composition percentage II
^1G	4	3277	3452	-175	3280	-3	0.99	0.904	0.969	43% + ^3H 53%	72% + ^3H 22%
^3F	2	3822	3922	-100	3831	-9	0.76	0.749	0.762	76% + ^1D 23%	72% + ^1D 27%
	3	5502	5570	-68	5506	-4	1.10	1.007	1.073	75% + ^3G 22%	95%
	4	7150	7333	-183	7150	0	1.30	1.148	1.216	51% + ^3G 44%	86% + ^1G 8%
^3H	4	5127	5079	48	5120	7	0.80	0.935	0.859	46% + ^1G 42%	77% + ^1G 18%
	5	6361	6008	353	6368	-7	1.07	1.033	1.033	100%	100%
	6	8350	8137	213	8347	3	1.17	1.167	1.167	100%	100%
^3G	3	6265	5973	292	6262	3	0.76	0.838	0.772	73% + ^3F 25%	93%
	4	7837	7745	92	7839	-2	1.06	1.114	1.056	55% + ^3F 35%	94%
	5	9326	9196	130	9328	-2	1.22	1.197	1.198	99%	99%
^1D	2	6571	7031	-460	6566	5	0.88	0.944	0.924	69% + ^3F 23%	67% + ^3F 28%
^3D	1	8922	8841	81	8913	9	0.52	0.526	0.519	96%	97%
	2	9900	9939	-39	9908	-8	1.18	1.164	1.163	93%	95%
	3	10127	10090	37	10120	7	1.34	1.236	1.243	72% + ^1F 25%	74% + ^1F 23%
^3P	0	11577	11595	-18	11586	-9				100%	100%
	1	11613	11587	26	11615	-2	1.29	1.461	1.469	94%	95%
	2	12642	12689	-47	12644	-2	1.38	1.477	1.484	95%	97%
^1F	3	12501	12621	-120	12504	-3	1.03	1.086	1.080	70% + ^3D 27%	72% + ^3D 26%
^1H	5	16152	16189	-37	16151	1	1.06	1.003	1.002	99%	99%
^1P	1	18444	18410	34	18441	3	0.99	1.013	1.011	93%	94%

Table 14. Ce II: $4f^25d + 4f^26s + 4f5d6p + 4f6s6p + 5d^3 + 5d^26s + 5d6s^2$

J=9/2 Representative Levels

Configuration	Designation	Obs	Calc	O-C	g-obs	g-calc	Composition Percentage
$4f^26s$	$^3H_4$	4166	4150	15	0.949	0.949	93%
$4f^25d$	$^3H\ ^2H$	7012	7023	-11	0.889	0.881	65% + $^3H\ ^4I$ 19% + $^3F\ ^2H$ 12%
$4f5d6p$	$^1G\ ^2H$	24663	24725	-62	0.933	0.922	34% + $^3H\ ^2H$ 25% + $^3H\ ^4I$ 18%
$4f^25d$	$^1I\ ^2H$	27905	28014	-109	0.920	0.919	77% + $^3F\ ^2H$ 7% + $^3H\ ^2H$ 3%
mixed		32198	32133	65	1.195	1.190	$4f^25d\ ^1I\ ^2G$ 28% + $4f5d6p(^3G\ ^4F$ 16% + $^3F\ ^2G$ 15% + $^3H\ ^4G$ 15%)
$4f6s6p$	5/2, $^3P_2$	36202	36191	11	1.186	1.192	43% + 7/2, $^3P_1$ 24% + $4f5d6p\ ^3H\ ^4G$ 14%
$5d^3$	^2G	38542	38433	109	1.097	1.098	35% + $4f5d6p(^1H\ ^2G$ 10% + $^1H\ ^2H$ 9%)

CONCLUSION

All the new interactions described in the present paper can be grouped under the title of "Weak interactions," Yet their great importance in reproducing the physical properties of atoms and ions, such as their energy level structure, g factors, and life-time ratios has been clearly established. The efficiency of the use of effective interactions has been demonstrated.

ACKNOWLEDGMENTS

Part of the present work has been done while spending a sabbatical year at the National Bureau of Standards, Washington, D.C. I am grateful to Dr. W. C. Martin and all staff members of the Spectroscopy Section for their generous hospitality during this year.

REFERENCES

* Permanent address: Racah Institute of Physics, The Hebrew University, Jerusalem, Israel.

1. H. A. Bethe and E. E. Salpeter, Quantum Mechanics of One and Two Electron Atoms (Springer-Verlag, Berlin, 1957).
2. J. C. Slater, Quantum Theory of Atomic Structure (McGraw-Hill, New York, 1960), Vol. II.
3. See definition and explanation below.
4. K. Rajnak and B. G. Wybourne, Phys. Rev. 132, 280 (1963).
5. G. Racah and J. Stein, Phys. Rev. 156, 58 (1967).
6. Z. B. Goldschmidt and J. Starkand, J. Phys. B 3, L141 (1970).
7. J. Stein, Ph.D. Thesis (The Hebrew University, Jerusalem, Israel, 1967) (unpublished).
8. Z. B. Goldschmidt, A. Pasternak and Z. H. Goldschmidt, Phys. Letters 28A, 265 (1968).
9. B. R. Judd, H. M. Crosswhite and Hannah Crosswhite, Phys. Rev. 169, 130 (1968).
10. Z. B. Goldschmidt, J. Phys. (Paris) Suppl. 31, 163 (1970).
11. A. Pasternak and Z. B. Goldschmidt, Phys. Rev. A 6, 55 (1972).
12. K. Rajnak and B. G. Wybourne, Phys. Rev. 134, A596 (1964).
13. E. U. Condon and G. H. Shortley, The Theory of Atomic Spectra (Cambridge University Press, 1951).
14. G. W. King and J. H. Van Vleck, Phys. Rev. 56, 464 (1939).
15. Z. B. Goldschmidt, J. V. Mallow and J. Starkand, European Group for Atomic Spectroscopy, Third Annual Conference 42, 141 (1971).
16. Z. B. Goldschmidt and J. V. Mallow (unpublished).

17. W. W. Smith, Phys. Rev. 137, A330 (1965).
18. M. W. Swagel and A. Lurio, Phys. Rev. 169, 114 (1968).
19. A. Lurio, Phys. Rev. 140, A1505 (1965).
20. E. Caspi and Z. B. Goldschmidt, Bull. Israel Phys. Soc., p. 66 (1971).
21. W. C. Martin, J. Sugar and J. L. Tech, Phys. Rev. A, in press.
22. A. Jucys, R. Dagys, J. Vizbazaite and S. Zvironaite, Trudy Akad. Nauk. Litovsk S.S.R. B 3(26), 53 (1961).
23. Z. B. Goldschmidt and J. Starkand, Bull. Israel Phys. Soc., p. 61 (1971).
24. J. V. Mallow, private communication.
25. H. C. Wolfe, Phys. Rev. 41, 443 (1932).
26. R. E. Trees, Phys. Rev. 83, 756 (1951); 84, 1089 (1951).
27. G. Racah and Y. Shadmi, Bull. Res. Council Israel, 8F, 15 (1959).
28. Z. B. Goldschmidt, unpublished (1961).
29. Z. B. Goldschmidt, Ph.D. Thesis (The Hebrew University, Jerusalem, Israel, 1968) (unpublished).
30. B. R. Judd, Phys. Rev. 141, 4 (1966).
31. Y. Shadmi, E. Caspi and J. Oreg, J. Res. Nat. Bur. Std. (U.S.) 73A, No.2, 173 (1969).
32. Z. H. Goldschmidt, Z. B. Goldschmidt and A. Pasternak, (unpublished).
33. H. H. Marvin, Phys. Rev. 71, 102 (1947).
34. G. Racah, Phys. Rev. 62, 438 (1942).
35. G. Racah, Phys. Rev. 63, 367 (1943).
36. G. Racah, Phys. Rev. 76, 1352 (1949).
37. U. Fano and G. Racah, Irreducible Tensorial Sets (Academic Press, 1959).
38. M. Blume and R. E. Watson, Proc. Roy. Soc. London, A271, 565 (1963).
39. M. Blume, A. J. Freeman, and R. E. Watson, Phys. Rev. 134, A 320 (1964).
40. C. Roth, J. Res. Nat. Bur. Std. (U.S.) 72A, (Phys. and Chem.), No. 5, 505 (1968).
41. R. E. Trees, J. Opt. Soc. Am. 54, 651 (1964).
42. Z. B. Goldschmidt, Spectroscopic and Group Theoretical Methods in Physics (North-Holland Publ. Co. Amsterdam, 411, 1968).
43. Z. B. Goldschmidt, unpublished material.
44. Z. B. Goldschmidt and S. Nir, Physica 51, 222 (1971).
45. G. Racah, Z. B. Goldschmidt and Y. Bordarier, unpublished material.
46. Z. B. Goldschmidt and A. Pasternak, unpublished material.
47. S. Nir and Z. B. Goldschmidt, unpublished material.
48. Z. B. Goldschmidt, D. Salomon and J. Starkand, Bull. Israel, Phys. Soc, p. 63 (1971).
49. H. M. Crosswhite, Phys. Rev. A 4, 485 (1971).
50. S. Feneuille and N. Pelletier-Allard, Physica 40, 347 (1968).
51. Z. B. Goldschmidt and D. Salomon, Bull. Israel Phys. Soc., p. 64 (1971).
52. Z. B. Goldschmidt and A. Edwards, unpublished material.

MODEL POTENTIAL CALCULATIONS FOR TWO-VALENCE ELECTRON
SYSTEMS

C. Laughlin and G. A. Victor
Department of Mathematics, The University of Nottingham,
Nottingham, England
Smithsonian Astrophysical Observatory and Harvard College
Observatory, Cambridge, Massachusetts, U.S.A.

1. INTRODUCTION

Model potential and pseudo-potential methods have been employed by many workers to study the structure and properties of one- and two-valence electron atomic and molecular systems. [1-7] The formalism and many of the earlier results have been reviewed by Weeks, Hazi, and Rice. [1] We are developing consistent semiempirical models for the prediction of the energy spectrum, oscillator strengths, transition probabilities, and other properties for atomic systems with two-valence electrons outside a closed shell core. A feature of our model, which distinguishes it from conventional pseudo-potential methods, is that the solutions to the corresponding one-electron problem, which are used as a basis for solving the two-electron problem, are chosen to have the correct number of radial nodes making orthogonalization to the occupied core orbitals of the same symmetry unnecessary. Long-range polarization terms are included explicitly in the potential, and short-range correction terms are determined so that the observed eigenenergies of the one-valence electron system are reproduced to high accuracy.

In a recent note[7] the method was used to calculate the electron affinity of the lithium atom. In this paper we report some results for beryllium and magnesium.

2. THE MODEL POTENTIAL

Consider an $(N+2)$-electron system having two valence electrons, with position vectors $\underline{r}_1$, and $\underline{r}_2$, outside a closed-shell polarizable core. If we adopt an adiabatic model, assuming that the core electrons follow instantaneously the motion of the valence electrons, and neglect exchange between the valence electrons and the core electrons, the total wave function for the system may be written as[4, 5]

$$\Phi = \Phi_c(\underline{R}|\underline{r}_1, \underline{r}_2) \, \Phi_v(\underline{r}_1, \underline{r}_2) \quad , \tag{1}$$

where the core wave function Φ_c depends parametrically on the positions of the valence electrons $\underline{r}_1$ and $\underline{r}_2$, and $\underline{R}$ labels collectively the position vectors of the N core electrons $\underline{r}_3, \ldots, \underline{r}_{N+2}$. Holding $\underline{r}_1$ and $\underline{r}_2$ fixed, we can obtain $\Phi_c(\underline{R}|\underline{r}_1, \underline{r}_2)$ perturbatively by adopting for H_0 the Hamiltonian for the N-electron core in the absence of the valence electrons and for the perturbation H_1 the valence-core interaction. Allowing $\underline{r}_1$ and $\underline{r}_2$ to vary, assuming $\underline{r}_1$ and $\underline{r}_2$ are both larger than $\underline{r}_i$, $i = 3, \ldots, N+2$, one then obtains the equation for the valence electrons[7]

$$[h(1) + h(2) + V_{12}] \, \Phi_v(\underline{r}_1, \underline{r}_2) = E_v \, \Phi_v(\underline{r}_1, \underline{r}_2) \quad . \tag{2}$$

The resulting one-electron operator h and the interaction term V_{12} in (2) are only correct asymptotically and must be modified for small $\underline{r}_1$ and $\underline{r}_2$ by introducing cut-off functions. Thus we take

$$h(i) = -\frac{1}{2} \nabla_i^2 - \frac{Z}{r_i} + V_c^{HF}(r_i) - \frac{\alpha_d}{2r_i^4} \, W_6\left(\frac{r_i}{r_0}\right) - \frac{\alpha_q'}{2r_i^6} \, W_8\left(\frac{r_i}{r_0}\right) + U(r_i|a), \tag{3}$$

where α_d is the static dipole polarization of the core and the cut-off function W_n is

$$W_n(x) = 1 - \exp(-x^n) \quad . \tag{4}$$

In equation (3) $V_c^{HF}(r)$ is the potential of the core computed from Hartree-Fock ground-state core orbitals and $U(r|a)$ is a predominantly short-range correction term depending on a set of adjustable parameters $a_1, \ldots, a_p$.[8] These parameters, as well as the cut-off radius r_0, are determined so that the one-electron equation

$$h \, \phi_j(\underline{r}) = \epsilon_j \, \phi_j(\underline{r}) \tag{5}$$

reproduces the spectrum of the singly excited states of the positive ion to a high degree of precision.

Further, we take[7]

$$V_{12} = \frac{1}{r_{12}} - \frac{\alpha_d}{r_1^2 r_2^2} \, P_1(\cos\theta_{12}) \, W_3\left(\frac{r_1}{r_0}\right) W_3\left(\frac{r_2}{r_0}\right) - \frac{\alpha_q'}{r_1^3 r_2^3} \, P_2(\cos\theta_{12})$$

$$W_4\left(\frac{r_1}{r_0}\right) W_4\left(\frac{r_2}{r_0}\right) \quad , \tag{6}$$

where $P_\ell(x)$ is the Legendre polynomial of degree ℓ and $\theta_{12} = \mathbf{r}_1 \cdot \mathbf{r}_2 / r_1 r_2$. In our adiabatic model the quantity α_q' in (3) and (6) is the static quadrupole polarizability of the core α_q. However, taking account of the first-order non-adiabatic correction[9] results in $\alpha_q' = \alpha_q - 6\beta_1$, where β_1 is a correction.

3. ENERGY LEVELS

We adopted a value of the cut-off parameter r_0 of the order of twice the core radius and, in the present calculations, we took $\alpha_q' = 0$. Then, using the Hartree-Fock representation of $\Phi_c^{(0)}$ to construct V_c^{HF},[10, 11] the one-electron equation (5) was solved variationally by using expansions of the type

$$\phi_j^{(\ell)}(\mathbf{r}) = f_j^{(\ell)}(r)\, y_{\ell m}(\hat{\mathbf{r}}) \quad , \tag{7}$$

where $f_j^{(\ell)}$ is a linear combination of nodeless Slater-type atomic orbitals

$$f_j^{(\ell)}(r) = \sum_m e^{-\alpha_m r} \sum_{m'} c_{j m m'}^{(\ell)}\, r^{m'} \quad . \tag{8}$$

The potential correction term $U(r|a)$ was chosen to have the form

$$U(r|a) = (a_0 + a_1 r + a_2 r^2)\, e^{-kr} + (a_0' + a_1' r)\, e^{-k' r} \quad . \tag{9}$$

The parameters in $U(r|a)$ were determined by using a least-squares technique due to Peckham[12] to fit several of the lowest eigenvalues of each symmetry of equation (5) to the observed Be^+ and Mg^+ spectra.[13, 14] The resulting parameter values are given in table 1. The energy eigenvalues computed from the resulting one-electron potential differ from the experimental data by less than 0.1% for Be^+ and by less than 0.5% for Mg^+.

A configuration-interaction method is employed to solve equation (2). By use of a LS-coupling scheme the two-electron eigenfunctions are constructed from antisymmetrized products of the one-electron eigenfunctions (7). With up to 40 configurations included in the expansion of the two-electron eigenfunctions the calculation of the eigenvalues and eigenvectors required typically of the order of 5 min on a CDC 6400.

Calculations were carried out for several symmetries for both beryllium and magnesium, and the results are presented in tables 2-5 for singly excited states and in tables 6 and 7 for doubly excited autoionizing states. The accuracy of the model potential eigenvalues is generally of the order of 0.5% for the most penetrating states, and it improves rapidly, as expected, for the Rydberg and autoionizing states. Our calculations

Table 1.   Values of the one-electron potential parameters [in atomic units].

Parameter	Value	
	Be^+	Mg^+
α_d	5.221×10^{-2} [24]	0.48 [25]
α'_q	0.00	0.00
r_0	0.90	1.70
a_0	-3.93	-1.9169
a_1	0.00	-3.4719
a_2	0.00	1.6935
k	4.55	2.00
a'_0	-4.8576×10^{-3}	-4.7906×10^{-3}
a'_1	1.6706×10^{-3}	1.6480×10^{-3}
k'	0.56	0.33

Table 2.   Be I ^1S and ^3S energy eigenvalues [in atomic units].

State	^1S			^3S	
	Model potential	Expt. [17]	Other calculations	Model potential	Expt. [17]
$2s^2$	-1.00828	-1.01191	$-1.0108,$ [26] -1.01179 [27] $-1.0101,$ [28] -1.0007 [3] -1.0139 [1]	–	–
$2s3s$	-0.76181	-0.76276		-0.77389	-0.77460
$2s4s$	-0.71420	-0.71461		-0.71759	-0.71798
$2s5s$	-0.69578	-0.69604		-0.69719	-0.69746
$2s6s$	-0.68672	-0.68698		-0.68744	-0.68766
$2s7s$	-0.68143	-0.68179			
$2p^2$	-0.65510		-0.6608 [15]		

Table 3. Be I $^1P^0$ and $^3P^0$ energy eigenvalues [in atomic units].

State	$^1P^0$			$^3P^0$		
	Model potential	Expt. [17]	Other calculations [15]	Model potential	Expt. [17]	Other calculations [15]
2s2p	−0.81049	−0.81796	−0.81543	−0.90717	−0.91175	−0.91032
2s3p	−0.73499	−0.73766	−0.73644	−0.74200	−0.74349	−0.74294
2s4p	−0.70542	−0.70646	−0.70601	−0.70677	−0.70747	−0.70718
2s5p	−0.69181	−0.69240		−0.69223		
2s6p	−0.68461			−0.68459		

Table 4. Mg I ^1S and ^3S energy eigenvalues [in atomic units].

State	^1S			^3S		
	Model potential	Expt. [18]	Other calculations	Model potential	Expt. [18]	Other calculations
$3s^2$	−0.82809	−0.83355	−0.82905 [29] −0.84075 [3] −0.8331 [30]			
3s4s	−0.63347	−0.63533	−0.63465 [29]	−0.64523	−0.64583	−0.64495 [29]
3s5s	−0.59329	−0.59408		−0.59695	−0.59719	
3s6s	−0.57711	−0.57754		−0.57874	−0.57888	
3s7s	−0.56891	−0.56923		−0.56980	−0.56993	
$3p^2$	−0.51755					

Table 5. Mg I $^1P^0$ and $^3P^0$ energy eigenvalues [in atomic units].

State	$^1P^0$			$^3P^0$		
	Model potential	Expt. [18]	Other calculations [29]	Model potential	Expt. [18]	Other calculations [29]
3s3p	−0.66971	−0.67384	−0.67175	−0.72970	−0.73380	−0.73405
3s4p	−0.60735	−0.60870		−0.61462	−0.61553	
3s5p	−0.58374	−0.58428		−0.58597	−0.58636	
3s6p	−0.57254	−0.57285		−0.57353	−0.57375	
3s7p	−0.56643	−0.56663		−0.56685	−0.56710	

Table 6. Autoionizing levels of Be I: $^1P^0$ [in eV].

State	Model potential	Configuration interaction[31]	Close coupling[32]	Experiment[16]
2p3s	10.70	10.77	10.99	10.71
2p4s	12.05	12.07	12.13	11.97
2p5s	12.56	12.60	12.60	12.53
2p6s	12.81	—	—	12.78
2p3d	11.83	11.86	11.93	11.86
2p4d	12.46	12.47	12.52	12.47

Table 7. Autoionizing levels of Mg I: $^1P^0$ [in eV].

State	Model potential	Experiment[16]
3p4s	9.62	9.86*
3p3d	10.61	10.65
3p5s	10.90	10.93
3p4d	11.25	11.26
3p6s	11.38	11.38
3p7s	11.60	11.62

*The experimental peak is very broad and flat-topped.

Table 8. Oscillator strengths in Be I.

Transition	Model potential $f^{(0)}$	Model potential $f^{(0)} + f^{(1)}$	Other calculations*	Experiment
$2s^2\ ^1S - 2s2p\ ^1P^0$	1.379	1.372	1.424(ℓ),[21] 1.386(v),[21] 1.41,[15] 1.24,[33] 1.254,[34] 1.93[35]	1.38 ± 0.12[22] 0.99[37] 1.02[38]
$2s2p\ ^3P^0 - 2s3s\ ^3S$	0.0910	0.0914	0.0709[36]	0.08,[37] 0.089[38]
$2s3p\ ^1P^0 - 2p3p\ ^1P^e$	0.263	0.261	0.247(ℓ),[15] 0.278(v)[15]	0.31[38]

*ℓ denotes dipole length calculation, v denotes dipole velocity calculation.

Table 9. Oscillator strengths in Mg I.

Transition	Model potential $f^{(0)}$	$f^{(0)} + f^{(1)}$	Other calculations	Experiment
$3s^2\ ^1S - 3s3p\ ^1P^0$	1.750	1.717	2.36,[35] 1.55,[39] 1.74,[29] 1.92[40]	1.850 ± 0.007[41] 1.810 ± 0.005[42] 2.40[43]
$3s3p\ ^3P^0 - 3s4s\ ^3S$	0.147	0.148	0.12,[39] 0.14,[29] 0.10[40]	0.10[43]

eventually become less accurate owing to the limited variational basis used to generate the one-electron eigenfunctions (typically the lowest 6 or 7 one-electron energy eigenvalues are predicted accurately) but this could easily be overcome by using, for example, numerical one-electron solutions. We agree with Weiss' prediction that the $2p^2\ ^1S$ state in beryllium is autoionizing[15] and find that it lies 0.38 eV above the ionization limit as compared to the value of 0.23 eV obtained by Weiss.

The $^1P^0$ autoionizing levels in Be I lying between the 2s ^2S and 2p ^2P limits were obtained by simply excluding the 2s orbital from the product trial functions. The agreement with the experimental data of Mehlman-Balloffet and Esteva[16] is excellent.

4. OSCILLATOR STRENGTHS

Oscillator strength results for Be I and Mg I are presented in tables 8 and 9 together with other accurate theoretical and experimental values. We have used experimental term values[17, 18] to compute the energy differences.

It has been pointed out by several authors that a consistent application of the theory imposes a modification on the transition operator.[19, 5] Thus, in tables 8 and 9, we list oscillator strengths $f_{ik}^{(0)}$ computed from the zero-order transition matrix element

$$I_{ik}^{(0)} = \langle \Phi_v^i(\underline{r}_1, \underline{r}_2) | \underline{r}_1 + \underline{r}_2 | \Phi_v^k(\underline{r}_1, \underline{r}_2) \rangle \tag{10}$$

and oscillator strengths correct through first order in the valence-core interaction where the first-order correction $f_{ik}^{(1)}$ is computed from

$$I_{ik}^{(1)} = \left\langle \Phi_v^i(\underline{r}_1, \underline{r}_2) \left| -\underline{r}_1 \frac{a_d}{r_1^3} W_3\left(\frac{r_1}{r_0}\right) - \underline{r}_2 \frac{a_d}{r_2^3} W_3\left(\frac{r_2}{r_0}\right) \right| \Phi_v^k(\underline{r}_1, \underline{r}_2) \right\rangle \cdot \tag{11}$$

Weisheit and Dalgarno[4] have discussed the sensitivity of dipole transition matrix elements with respect to different cut-off functions and, guided by their investigations, we choose $W_3(r/r_0)$. Test calculations revealed that the results were not sensitive to the choice of cut-off function. The correction to the oscillator strengths, introduced by the modification (11) to the transition operator, was found to be less than 1% for all strong Be I transitions; in Mg I it was, in some cases, as large as 2%.

Our predicted value of 1.372 for the $2s^2\ ^1S - 2s2p\ ^1P^0$ resonance transition oscillator strength in Be I is in close agreement with the value 1.41 obtained by Weiss[20] using configuration interaction wave functions and the dipole length and velocity values, 1.424 and 1.386, respectively, computed by Burke, Hibbert, and Robb[21] using configuration interaction wave functions. It might be noted that all these numbers are in satisfactory agreement with the measured value of 1.38 ± 0.12 recently obtained by Hontzeas et al.[22] using the beam-foil technique.

The calculated value of the dipole polarizability of beryllium is dominated by the contribution of the resonance transition. Our result of 5.61 $Å^3$ is in harmony with the number 5.49 $Å^3$ obtained by Kolker and Michels.[23]

Acknowledgments

We wish to thank Professor A. Dalgarno for helpful conversations and the Nottingham Algorithms Group for the provision of numerical software. The work has been partly supported by the United States Air Force under AFOSR grant 71-2132 A.

References

1. J. D. Weeks, A. Hazi and S. A. Rice, Adv. Chem. Phys. 16, 283 (1969).
2. L. Szasz, J. Chem. Phys. 49, 679 (1968).
3. L. Szasz and G. McGinn, J. Chem. Phys. 56, 1019 (1972).
4. J. C. Weisheit and A. Dalgarno, Phys. Rev. Letters 27, 701 (1971).
5. T. C. Caves and A. Dalgarno, to be published in J. Quant. Spect. and Rad. Transf. (1972).
6. J. C. Weisheit, Phys. Rev. A5, 1621 (1972).
7. G. A. Victor and C. Laughlin, Chem. Phys. Letters 14, 74 (1972).
8. C. Bottcher, J. Phys. B: Atom. Molec. Phys. 4, 1140 (1971).
9. U. Öpik, Proc. Phys. Soc. 92, 573 (1967); C. J. Kleinman, Y. Hahn and L. Spruch, Phys. Rev. 165, 53 (1968).
10. C. C. J. Roothaan, L. M. Sachs and A. W. Weiss, Rev. Mod. Phys. 32, 186 (1960).
11. E. Clementi, IBM J. Res. Develop. Suppl. 9 (1965).
12. G. Peckham, Computer J. 13, 418 (1970).

13. L. Johansson, Arkiv. Fysik 20, 489 (1962).
14. G. Risberg, Arkiv. Fysik 9, 483 (1955).
15. A. W. Weiss, private communication (1972).
16. G. Mehlman-Balloffet and J. M. Esteva, Astrophys. J. 157, 945 (1969).
17. L. Johansson, Arkiv. Fysik 23, 119 (1962).
18. G. Risberg, Arkiv. Fysik 28, 381 (1964).
19. I. B. Bersukev, Opt. Spectrosc. 3, 97 (1957); S. Hameed, A. Herzenberg and M. G. James, J. Phys. B: Atom. Molec. Phys. 1, 882 (1968).
20. A. W. Weiss, see ref. 22.
21. P. G. Burke, A. Hibbert and W. D. Robb, J. Phys. B: Atom. Molec. Phys. 5, 37 (1972).
22. S. Hontzeas, I. Martinson, P. Erman and R. Buchta, to be published.
23. H. J. Kolker and H. H. Michels, J. Chem. Phys. 43, 1027 (1965).
24. G. A. Victor and A. Dalgarno, unpublished.
25. U. Öpik, Proc. Phys. Soc. 92, 566 (1967).
26. A. W. Weiss, Phys. Rev. 166, 70 (1968).
27. D. F. Tuan and O. Sinanoglu, J. Chem. Phys. 41, 2677 (1964).
28. R. K. Nesbet, Phys. Rev. 155, 51 (1967).
29. A. W. Weiss, J. Chem. Phys. 47, 3573 (1967).
30. J. D. Weeks and S. A. Rice, J. Chem. Phys. 49, 2741 (1968).
31. P. L. Altick, Phys. Rev. 169, 21 (1968).
32. D. L. Moores, Proc. Phys. Soc. 91, 830 (1967).
33. C. Nicolaides, see ref. 22.
34. H. P. Kelly, Phys. Rev. 136, B896 (1964).
35. S. Hameed, J. Phys. B: Atom. Molec. Phys. 5, 746 (1972).
36. C. Froese, J. Chem. Phys. 47, 4010 (1967).
37. T. Andersen, K. A. Jessen and G. Sorensen, Phys. Rev. 188, 76 (1969).
38. J. Bromander, Physica Scripta 4, 61 (1971).
39. R. N. Zare, J. Chem. Phys. 47, 3561 (1967).
40. B. Warner, Mon. Not. Roy. Astron. Soc. 139, 115 (1968); 140, 53 (1968).
41. A. Lurio, Phys. Rev. A136, 376 (1964).
42. W. W. Smith and A. Gallagher, Phys. Rev. 145, 26 (1966).
43. H. G. Berry, J. Bromander and R. Buchta, Physica Scripta 1, 181 (1970).

COMPLEX ROTATIONS IN ATOMIC SCATTERING THEORY

G.Doolen, M.Hidalgo, J.Nuttall[†] and R.Stagat

Dept. of Physics, Texas A & M University

College Station, Texas 77843

There has recently been interest shown in the possibility of using the theory of analytic functions of a complex variable to overcome some of the difficulties encountered in calculating atomic scattering amplitudes. In this paper we discuss the complex rotation method, in which all coordinates are transformed by $r_j \to r_j e^{i\alpha}$. Previous authors [1] have pointed out that this transformation makes it possible to describe the (complex) energy of a resonance as a discrete eigenvalue of the rotated Hamiltonian H_α which has the form

$$H_\alpha = e^{-2i\alpha}T + e^{-i\alpha}V \tag{1}$$

for purely Coulomb potentials.

An obvious way of finding an approximation to the eigenvalues of H_α is to construct a finite matrix $H_\alpha^{nm} = (u_n, H_\alpha u_m)$ from a set of normalizable basis functions u_n, and then find the eigenvalues of this matrix. It has not yet been proved that this procedure works in the general case, although this may be done in some simple models [2].

To shed more light on this question, we have applied this method to a simple model with a long range potential.

† Present address: Dept. of Physics, University of Western Ontario, London, Ontario, Canada.

Table 1. Estimates of the value of the complex
          energy E of the resonance in the model
          described in the text obtained with N
          basis functions.

N	Re E	Im E
10	1.01869	-0.745641
20	1.03990	-0.759711
22	1.03971	-0.759856
24	1.03961	-0.759879
26	1.03956	-0.759869
28	1.03955	-0.759858
30	1.03954	-0.759858

The Hamiltonian, defined for $0 \leq r \leq \infty$ was chosen
to be

$$H = -\frac{d^2}{dr^2} + \frac{10}{(r-1)^2 + 10} \qquad (2)$$

and the basis functions were given by $u_n = r^n e^{-ar}$,
$n = 1, \ldots, N$, $a = 2.5$. The positions of the discrete
eigenvalue of H_α^{nm} for different values of N and $\alpha = 30^o$
are shown in Table 1. The rate of convergence appears
to be satisfactory.

In this calculation, the other eigenvalues of H_α^{nm}
lie close to the continuous spectrum of H_α , a line
in the complex plane given by arg $z = -2\alpha$. In the case
of a many body system the continuous spectrum of H_α is
given by $\arg(z - E_k) = -2\alpha$, where E_k, $k = 1, \ldots,$
are the values of the total energy at which new channels
open up. (e.g. for the e - H system $E_k = -1/k^2$).

For atomic systems we see from (1) that the
calculation of H_α^{nm} involves only a knowledge of T^{nm} and
V^{nm} , and these have often been found in connection with
bound state calculations. However, it would not be

surprising if, using u_n as employed in bound state calculations, a slow rate of convergence were found for resonances located near the thresholds E_k. This problem could probably be overcome by adding to the set (u_n) a few basis functions with asymptotic behavior appropriate to the channel in question, but this would mean the evaluation of more integrals.

For a suitable class of short-range potentials the complex rotation method may be used to calculate scattering amplitudes at all energies without the need for an explicit description of every open channel [3]. Unfortunately, this is not the case for Coulomb and other infinite range potentials. However, the method can still be used, if just one channel is open, to remove the spurious singularities in the Kohn method [4]. The approach, which will be published in detail elsewhere, involves the construction of a variational principle for the Jost function, in which wave functions are integrated over the rotated coordinates. The method applies to many-body systems, but we have tested it for the simple model with Hamiltonian

$$H = -\frac{d^2}{dr^2} + \frac{10}{(r + 1)^2 + 2}, \qquad (3)$$

using basis functions $u_n = (r + 2)^{-n}$, $n = 1, \ldots N$ and $\alpha = 30^{\circ}$. The results for the Jost function (whose phase is the negative of the phase shift) are shown in Table 2.

The complex rotation method can also be applied to the Fredholm method as used by Reinhardt et al [5] to remove the need for numerical extrapolation, but without the inclusion of an open channel wave function in the basis set (which was included in the work described above) the rate of convergence is slow.

The conclusions from this work are that it is worth pursuing the complex rotation method for finding resonance positions in atomic systems using computer programs already written for bound state work, but that further matrix elements might be required to obtain a satisfactory convergence rate. Scattering amplitudes with one (or perhaps a few) open channel may also be calculated this way, but the challenge is to use analyticity in a more complicated way to calculate atomic scattering amplitudes at higher energies.

Table 2.   Estimates of the value of the Jost
            function J for the model described in
            the text at energy E = 25 obtained
            with N basis functions.

N	Re J	Im J
2	-0.243328	-0.014441
4	-0.227425	-0.021298
6	-0.228739	-0.020154
8	-0.228580	-0.020310
10	-0.228600	-0.020281
12	-0.228598	-0.020286

References

1. C. Lovelace, in Strong Interactions and High
    Energy Physics, ed. R.G. Moorhouse (Oliver and
    Boyd, London, 1964); E. Balslev and J.M. Combes,
    Commun. Math. Phys. (to appear); B. Simon, pre-
    print; J. Nuttall, Bull. A.P.S., 17, 598 (1972).

2. G.D. McCartor, private communication.

3. F.A. McDonald and J. Nuttall, Phys. Rev. (to be
    published).

4. C. Schwartz, Ann. Phys. 16, 36 (1961).

5. W.P. Reinhardt, D.W. Oxtoby and T.N. Rescigno,
    private communication.

    Work supported in part by the U. S. Air Force
Office of Scientific Research under Grant No. 71-1979A.

PHOTODETACHMENT OF Li⁻ AND Na⁻

D. W. Norcross and D. L. Moores[*]

Joint Institute for Laboratory Astrophysics[†]

The University of Colorado, Boulder, Colorado

INTRODUCTION

Reliable estimates of the photodetachment cross sections and electron affinities of the negative ions of the alkali metals are required for interpretation of the properties of low-temperature plasmas, and in the fields of upper-atmosphere physics and astrophysics. Experimental data are rather scarce and uncertain, the electron affinities being too low ($\sim$0.5 eV) for application of standard photoabsorption techniques near threshold. Even if this were possible, the cross section is expected to display a k^3 dependence at threshold appropriate to an $s \to p$ transition, thereby precluding a sharp onset. The electron affinity of lithium has been measured by Ya'akobi,[1] in experiments with electrically exploded lithium wires, and by Sheer and Fine,[2] who studied the positive and negative surface ionization of a lithium beam. Two measurements of the electron affinity of sodium and the heavier alkali metals using resonant charge exchange are cited in the review of Smirnov.[3] No measurements of the photodetachment cross sections have been made.

Calculations of electron affinities have relied principally on two methods; extrapolation of known spectral data along an isoelectronic sequence (see, for example, Geltman[4] and Edlen[5]), and an <u>ab initio</u> variational method, using configuration-interaction (CI) trial wave functions (see, for example, Weiss[6] and Fung and Matese[7]). The results of a number of calculations using the first method show considerable scatter; but the CI results are in excellent agreement for both Li⁻ and Na⁻. Photodetachment cross sections have been calculated for the lighter alkali negative ions, using a variety

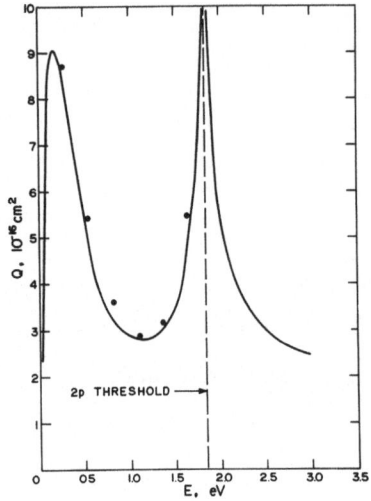

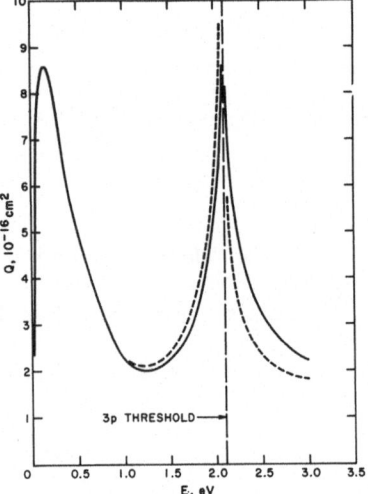

Fig. 1 Partial cross section for
elastic scattering of electrons by
Li in the ^1P partial wave, calculated in the close-coupling
approximation; present results
(——), Burke and Taylor[12] ($\cdot$).

Fig. 2 As fig. 1 for Na;
present two-state (3s,3p)
results (——) and four-state
(3s,3p,3d,4s) results (---).

of simple approximations for the wave functions (see, for example,
Geltman,[4] Ya'akobi,[8] Moskvin,[9] and John[10]).

The results of recent close-coupling calculations for scattering of electrons by neutral lithium[11] and sodium[12] suggest a novel
and perhaps very precise method for measuring their electron affinities. The significant feature of these results is some very pronounced structure in the ^1P partial-wave cross section for elastic
scattering near the first excitation threshold, shown in figures
1 and 2. This structure is partly due to the usual effect observed
when a new channel opens, the so-called Wigner cusp, combined with
what appears to be real resonant behavior in the ^1P phase shift.
With some hesitation this feature might be labelled the nsnp ^1P
autoionizing state of the negative ion, where n=2 for lithium and
n=3 for sodium. (A similar feature obtained in the ^1D partial wave
could be associated with some admixture of np^2 ^1D and nsnd ^1D
states).

The effect of the cusp has been observed in an electron-sodium
scattering angular distribution measurement,[13] and is so prominent
that it has been used to calibrate the electron beam in another
similar experiment.[14] In such experiments, of course, the effect
of all partial waves is observed, but the results are in excellent
agreement with the close-coupling calculations.[12] The suggestion

that the cusp might represent an autoionizing state of the negative ion is supported by recent calculations which yield autoionizing states of both Li⁻ and Na⁻ at even higher energies.[7]

Now consider the photodetachment of Li⁻ and Na⁻. If the negative ion is initially in its ^1S ground state, the final continuum state of the system, according to the dipole selection rules, must be a ^1P state. This final state may also be regarded as the electron-neutral scattering system in the ^1P partial wave. The cusp effect should therefore influence the photodetachment cross section at wavelengths corresponding to ejection of an electron with energy near the first excitation energy, E_1, of the neutral atom. Thus, if the position of this cusp is observed in photodetachment at some wavelength λ_c, the electron affinity I is then related to E_1 and λ_c by

$$I = \frac{1}{R\lambda_c} - E_1 \qquad (1)$$

where R is the Rydberg constant in cm^{-1}, λ_c is in cm, and I and E_1 are in rydbergs. Assuming Weiss' values for the electron affinities, 0.0452 Ry for Li⁻ and 0.0396 Ry for Na⁻, the λ_c values are 5034 Å and 4692 Å, respectively.

THEORY

The photodetachment cross section, expressed in the dipole-length form (neglecting transitions of closed-shell electrons), is given by

$$\kappa_L = 4 \ \pi\alpha a_o^2 \ \frac{(I+k_1^2)}{\omega_i} \sum \left| \int \Psi_f^* (z_1+z_2) \ \Psi_i d\tau \right|^2 \qquad (2)$$

and in the dipole-velocity for by

$$\kappa_V = 16 \ \pi\alpha a_o^2 \ \frac{(I+k_1^2)}{\omega_i} \sum \left| \int \Psi_f^* \left(\frac{\partial}{\partial z_1} + \frac{\partial}{\partial z_2} \right) \Psi_i d\tau \right|^2 \qquad (3)$$

where α is the fine-structure constant, a_o the Bohr radius, ω_i the statistical weight of the initial state, k_1^2 is the ejected-electron energy, in rydbergs, relative to the ground state of the neutral atom and Σ denotes a sum over all initial and final states. Ψ_i is the initial-state wave function, normalized to unity, and Ψ_f is the final-state wave function, normalized to

$$\int \Psi_f^* (k^2) \ \Psi_f(k'^2)d\tau = \pi\delta(k^2-k'^2) \quad . \qquad (4)$$

Both Ψ_f and Ψ_i are antisymmetric in the space and spin coordinates of the electrons.

In the present calculations, the CI wave functions of Weiss were used for Ψ_i. These include fourteen configurations for Li$^-$ and ten configurations for Na$^-$. Solutions of the two-state (three channels, nsk_1p, npk_2d, npk_2s) close-coupling equations for electron scattering by the neutral atom were used for Ψ_f. This approximation allows, therefore, for transitions in which the neutral atom is left in its first excited state. A semi-empirical statistical model potential[15] was used to define the target system in the scattering calculations. The good agreement between the results of these calculations[12] and experiment[13,14] for sodium has already been mentioned. Very little experimental data are available for lithium, but the agreement of the present results with the calculations of Burke and Taylor,[11] who used Hartree-Fock atomic functions, shown in figure 1, is quite good.

As a check on the calculations, in addition to comparison of the length and velocity results, the continuum oscillator strength sum

$$N_o = \frac{1}{4\pi^2 \alpha a_o^2} \int_o^\infty \kappa \, dk_1^2 \qquad (5)$$

may be computed. If there is only one bound state of Li$^-$ or Na$^-$, an assumption verified by the calculations of Fung and Matese, N_o should be equal to 2 in the limit of exact wave functions. Since allowance is not made in the present calculations for transitions to excited states of the neutral other than the first, to higher autoionizing states of the negative ion, or to the continuum of the neutral, we expect to obtain $N_o \lesssim 2$ if the wavefunctions are sufficiently accurate. We may also evaluate

$$\alpha_p = \frac{1}{\pi^2 \alpha a_o^2} \int_o^\infty \frac{\kappa}{(I+k_1^2)^2} \, dk_1^2 \qquad (6)$$

which should be an approximate lower bound on the dipole polarizability of the negative ion.

RESULTS AND DISCUSSION

Photodetachment cross sections computed from (2) and (3) are shown in figures 3 and 4, with the theoretical[6] electron affinities being used for I. The predicted cusp does indeed appear at the excitation threshold. Because of limitations of the computational technique, we were not able to obtain results within less than ~ 15 meV of the np threshold, so it cannot definitely be maintained that there is a cusp, rather than a maximum with continuous derivative, in the cross section. The former does, however, seem the more likely in view of the fact that the np threshold interrupts the phase shift for elastic scattering before it reaches the value $\pi/2$ near which a smooth maximum would occur in the partial scattering cross section.

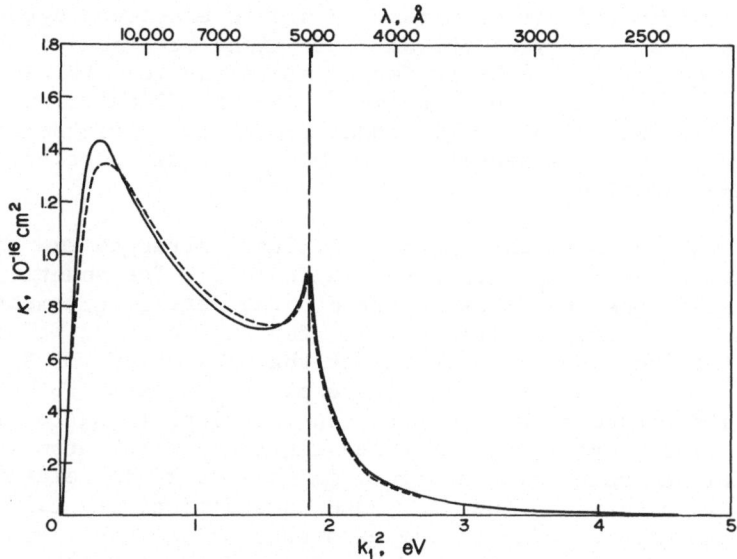

Fig. 3 Photodetachment cross section of Li⁻; in dipole-length
(———) and dipole-velocity (---) .

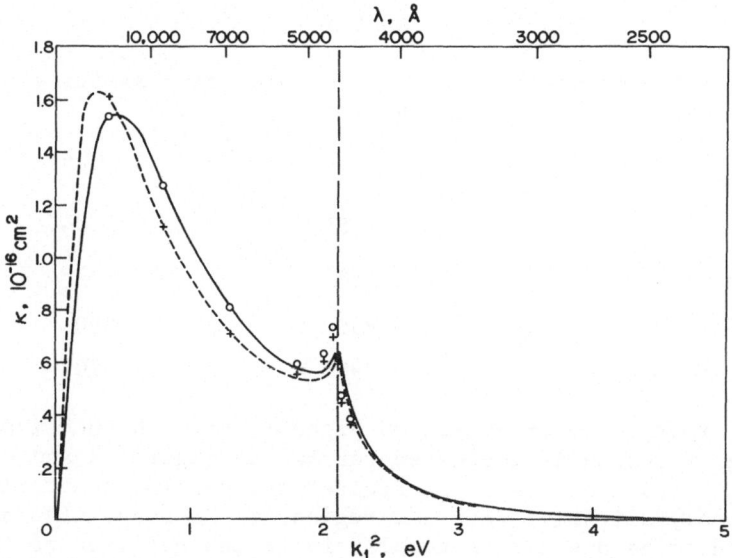

Fig. 4 As fig. 3 for Na⁻. Also shown are the results of three-
state (3s,3p,3d) length (o), and velocity (+) calculations.

The length and velocity results are in excellent agreement for Li⁻ and in good agreement for Na⁻. It might be argued that the velocity results are to be preferred since they emphasize more interior parts of the wave functions, and the CI calculations for the ions were based on a variational minimization of the total energy, which is less sensitive to the long-range part of the trial wave functions.

The values obtained for the oscillator strength sum N_0 and the polarizability α_p are given in Table I. The quantity N_0 is generally less than 2 as expected. The better agreement of the length and velocity results for N_0 for Li⁻ than for Na⁻ is perhaps fortuitous, in view of the fact that the values of N_0 for Li⁻ are considerably less than 2. The explanation seems to be the greater importance of high-energy contributions to N_0 for Li⁻ than for Na⁻. The cross section for leaving the neutral atom in its first excited state, also obtained in this work, is relatively more important for Li⁻ than Na⁻, and in addition, the calculations of Fung and Matese indicate the existence of one other autoionizing ¹P state of Li⁻ (but none for Na⁻), the effect of which we have not attempted to include. Finally, for wavelengths less than 2000 Å, the values of κ used in (5) were obtained by monotonic extrapolation to zero. For Li⁻, however, calculations at a few additional wavelengths below 2000 Å indicate the existence of a small broad maximum (also obtained by John), the contribution of which to N_0 is estimated to be of order 0.1.

TABLE I

The oscillator strength sum N_0 and dipole polarizability of the negative ion α_p.

		Li⁻	Na⁻
N_0	L	1.85	2.04
	V	1.83	1.98
α_p	L	830	990
	V	790	1070

The importance of neglected higher states in the close-coupling expansion is indicated for Na⁻ in figures 2 and 4. Addition of the 3d and 4s states affected the partial scattering cross section only slightly, near the region of the cusp. The difference was found to be due almost completely to the addition of the 3d state.[12] A few points on the Na⁻ photodetachment curve were obtained with the addition of the 3d state, and again only a minor change is observed. The cross section for excitation of the neutral, however, is more strongly affected than is the total. A sensitive test of theory would be provided by using energy analysis of the ejected electrons to study this contribution to the total cross section.

CONCLUSION

The results obtained confirm the existence of a sharp feature in the photodetachment cross section of both Li⁻ and Na⁻. The wavelength at which it occurs and its width are such as to make feasible an experimental determination of their electron affinities, using present dye laser techniques. Preliminary results of a very recent study[16] of Na⁻ photodetachment in the range λ = 4500 Å to 4900 Å using this technique clearly exhibit the predicted cusp-like behavior in the cross section, with the maximum occurring with ± 7 Å of 4687 Å, which if interpreted as λ_c leads to an experimental value for the electron affinity of sodium of 0.0398 ± 0.0003 Ry, in excellent agreement with the CI calculations.[6,7]

REFERENCES

* Visiting Fellow, 1971-72. Permanent address: University College London, Gower Street, London, U.K.

† Operated jointly by the National Bureau of Standards and the University of Colorado, Boulder, Colorado 80302.

1. B. Ya'akobi, Phys. Letters 23, 655 (1966).
2. M. D. Sheer and J. Fine, J. Chem. Phys. 50, 4343 (1969).
3. B. M. Smirnov, Teplofiz. Vysh. Temp. 3, 775 (1965).
4. S. Geltman, Phys. Rev. 104, 346 (1956).
5. B. Edlén, J. Chem. Phys. 33, 98 (1960).
6. A. W. Weiss, Phys. Rev. 166, 70 (1968) and private communication.
7. A. C. Fung and J. J. Matese, Phys. Rev. A 5, 22 (1972).
8. B. Ya'akobi, Phys. Rev. 184, 246 (1969).
9. Yu. V. Moskvin, Teplofiz. Vysh. Temp. 3, 821 (1965).
10. T. John, J. Phys. B (Atom. Molec. Phys.) 5, L121 (1972).
11. P. G. Burke and A. J. Taylor, J. Phys. B (Atom. Molec. Phys.) 2, 869 (1969).
12. D. L. Moores and D. W. Norcross, J. Phys. B (Atom. Molec. Phys.), in press.
13. D. Andrick, M. Eyb and H. Hofmann, J. Phys. B (Atom. Molec. Phys.) 5, L15 (1972).
14. W. Gehenn and E. Reichert, Z. Phys., in press.
15. D. W. Norcross, J. Phys. B (Atom. Molec. Phys.) 4, 1458 (1971).
16. H. Hotop, T. A. Patterson, and W. C. Lineberger, private communication.

RADIATIVE DECAY OF THE METASTABLE STATES OF THE H AND HE SEQUENCES

- THEORY

Gordon W. F. Drake

Department of Physics, University of Windsor

Windsor 11, Ontario, Canada

1. INTRODUCTION

The 2s $^2S_{1/2}$ state of hydrogen and the 1s2s $^3S_1$ and $^1S_0$ states of helium are said to be metastable because angular momentum and parity selection rules prevent the usual electric dipole transitions to lower states. However the metastable atoms, if left to themselves, will eventually decay through some higher order radiative process. Recent developments in both the theory and the experimental techniques used to measure the metastable lifetimes have made possible several new high precision comparisons between the predicted and measured decay rates. In addition to their fundamental interest, the results have important applications under the low density conditions found for example in planetary nebulae. This paper reviews the theory describing the spontaneous radiative decay of the metastable states and the effects induced by a perturbing electric field.

The formal theory of radiative transitions in atomic systems is fully developed in the basic reference work by Heitler[1]. The higher order effects to be considered include two-photon processes and relativistic contributions to single-photon processes. The latter problem can be approached from one of two directions. In the first, the relativistic Hartree-Fock (RHF) equations are solved to obtain approximate relativistic wavefunctions, and matrix elements of the relativistic radiation transition operator $-e\underset{\sim}{\alpha}\cdot\underset{\sim}{A}$ are calculated (see e.g. Rosner and Bhalla[2] for recent work). However the RHF method does not lend itself easily to high precision calculations since it is particularly difficult to include the correlation corrections omitted from the theory. This approach has been reviewed recently[3,4] and will not be further discussed. In the second

269

approach, one attempts to find an equivalent non-relativistic operator (ENO) whose matrix elements between non-relativistic wave-
functions equal those of the relativistic operator $-e\underline{\alpha}\cdot\underline{A}$ up to terms
of some sufficiently high order in the fine-structure constant α.
Although more formal analysis is required to derive the ENO, the
great advantage is that the non-relativistic Schrödinger equation is
much easier to solve and accurate solutions are known for many states
of the simpler atomic systems. The latter approach will be described
in some detail.

2. FUNDAMENTAL EQUATIONS

This section contains the fundamental relations required in
the subsequent discussion. It is convenient to start from the
scattering-matrix formalism of quantum electrodynamics[5]. The
spontaneous emission of a single photon of frequency ω linear
momentum $\underline{k}$ ($|\underline{k}| = \omega/c$) and polarization $\hat{e}$ is described by the first-
order S-matrix element

$$S_{if} = -e \int \overline{\psi}_f(x)\, \hat{A}*(x)\, \psi_i(x)\, d^4x \tag{1}$$

where, in 4-component notation, $\hat{A} = A_\mu \gamma_\mu (\mu = 1,\ldots,4)$ and $\overline{\psi} = \psi^* \gamma_4$.
The ψ's are Dirac spinors and the γ's are Dirac matrices. If $\underline{A}$
and A_0 are the vector and scalar potentials for the electromagnetic
field, then the four components of A are $(\underline{A}, iA_0)$. Similarly x
denotes the 4-vector $(\underline{r}, it)$ and $\psi_n(x) = \psi_n(\underline{r})\, e^{-iE_n t}$ describes a
stationary state of the electron with energy E_n. For spontaneous
photon emission, $A_0 = 0$ in the Coulomb gauge and

$$\underline{A} = \hat{e}(2\pi/\omega)^{1/2}\, e^{ikx} \quad \text{with} \quad kx = \underline{k}\cdot\underline{r} - \omega t. \tag{2}$$

Then (1) becomes

$$S_{if} = -2\pi i U_{if}\, \delta(E_i - E_f - \omega), \tag{3}$$

$$U_{if} = -e(2\pi/\omega)^{1/2} \int \psi_f^*(\underline{r})\, \underline{\alpha}\cdot\hat{e}\, e^{-i\underline{k}\cdot\underline{r}}\, \psi_i(\underline{r})\, d\underline{r} \tag{4}$$

where U_{if} is the matrix element of the effective interaction energy
of the electron with the electromagnetic field. U_{if} is related to
the spontaneous decay rate w_{if} by

$$dw_{if} = 2\pi |U_{if}|^2\, \delta(E_i - E_f - \omega)\, \rho(\omega) \tag{5}$$

where $\rho = d\underline{k}/(2\pi)^3$ is the density of final photon states per unit
volume in the frequency interval $d\omega$ and solid angle $d\Omega_k$.

The plane wave in (2) describes a photon state of definite linear momentum and polarization. Since field-free atomic states are characterized by angular momentum and parity quantum numbers, it is convenient to expand (2) in spherical waves

$$\hat{e} \, e^{-i\mathbf{k}\cdot\mathbf{r}} = \sum_{LM\lambda} \hat{e} \cdot \underset{\sim}{Y}_{LM}^{(\lambda)} \, (\hat{k}) \, g_L(kr) \, Y_{LM}^{(\lambda)*} \, (\hat{r}) \tag{6}$$

where $\underset{\sim}{Y}_{LM}^{(\lambda)}$ is a vector spherical harmonic of angular momentum L, component M, parity $(-1)^{L+\lambda+1}$, and

$$g_L(kr) = (2\pi)^{3/2} \, (-i)^L \, (kr)^{-1/2} \, J_{L+1/2}(kr) \quad . \tag{7}$$

In the long wavelength approximation ($kr \ll 1$), the Bessel function $J_{L+1/2}$ is expanded in a power series to obtain

$$g_L(kr) = \frac{4\pi(ikr)^L}{(2L+1)!!} \left[1 - \frac{(kr)^2}{2(2L+3)} + \dots \right]. \tag{8}$$

Since $kr \equiv \omega r/c$ is $0(\alpha)$, it is usually necessary to retain only the leading term of (8). However the leading two terms will be required in the discussion of relativistic magnetic dipole transitions to follow.

Each term in (6) vanishes on integration over angles in (4) unless angular momentum and parity selection rules are satisfied. If $J_i M_i \pi_i$ and $J_f M_f \pi_f$ denote the angular momentum, z-component and parity of the initial and final states respectively, then only those terms contribute for which

$$M_i - M_f = M$$

$$|J_i - J_f| \leqslant L \leqslant J_i + J_f$$

$$\pi_i = \pi_f(-1)^{L+\lambda+1} \tag{9}$$

but $J_i = 0 \quad J_f = 0$. Since $g_L(kr)$ is $0(\alpha^L)$, the dominant contribution comes from the term with $L = |J_i - J_f|$ and the correct parity. In the low velocity limit, the Dirac spinors are replaced by non-relativistic Pauli eigenfunctions and the equivalent non-relativistic operators evaluated by means of the Foldy-Wouthuysen[6] transformation. The results can be expressed in the form

$$w_{LM}^{(\lambda)} = \frac{2(L+1)}{L(2L+1)\left[(2L-1)!!\right]^2} \left(\frac{\omega}{c}\right)^{2L+1} |\langle f|Q_{LM}^{(\lambda)}|i\rangle|^2 \tag{10}$$

where $Q_{LM}^{(1)}$ and $Q_{LM}^{(0)}$ are the electric and magnetic multipole moment operators respectively and are given by

$$Q_{LM}^{(1)} = e(4\pi/(2L+1))^{1/2} \, r^L \, Y_L^{M*}(r) \tag{11}$$

$$Q_{LM}^{(0)} = \left(\frac{4\pi}{2L+1}\right)^{1/2} \left[\nabla \, R^L \, Y_L^{M*}(r)\right] \cdot \left[\frac{e}{mc(L+1)} \, \underset{\sim}{L} + \underset{\sim}{\mu}\right] ; \tag{12}$$

$\underset{\sim}{L} = \underset{\sim}{r} \times \underset{\sim}{p}$ is the orbital angular momentum operator and $\mu = (e/2mc)\underset{\sim}{\sigma}$ is the electron spin magnetic moment operator. Factors of c are included to display the relative magnitudes of the different terms. The calculation of atomic transition probabilities from the above results has recently been reviewed by Dalgarno[7].

The selection rules (9) strictly prohibit single photon transitions from, for example, the 1s2s $^1S_0$ state of helium (in the absence of nuclear spin) and it is necessary to include spontaneous two-photon emission described by the second-order S-matrix element

$$S_{if}^{(2)} = e^2 \iint \overline{\Psi}_f(x_1) \, \hat{A}_1^*(x_1) \, S_c^{(e)}(x_1,x_2) \, \hat{A}_2^*(x_2) \, \psi_i(x_2) d^4x_1 d^4x_2$$

$$+ e^2 \iint \overline{\Psi}_f(x_1) \, \hat{A}_2^*(x_1) \, S_c^{(e)}(x_1,x_2) \, \hat{A}_1^*(x_2) \, \psi_i(x_2) d^4x_1 d^4x_2. \tag{13}$$

The two terms correspond to the diagrams shown in Fig. 1. The vector potentials for the two photons of frequencies ω_1, ω_2 and polarization vectors $\hat{e}_1$, $\hat{e}_2$ respectively are

$$\hat{A}_1 = \hat{e}_1 \, (2\pi/\omega_1)^{1/2} \, e^{ik_1 x} \tag{14}$$

$$\hat{A}_2 = \hat{e}_2 \, (2\pi/\omega_2)^{1/2} \, e^{ik_2 x} \tag{15}$$

and $S_c^{(e)}(x_1,x_2)$ is the electron propagator in the external field of the nucleus with the spectral representation

$$S_c^{(e)}(x_1,x_2) = \frac{1}{2\pi i} \int_{-\infty}^{\infty} d\omega' \, e^{i\omega'(t_1-t_2)} \sum_n \frac{\overline{\psi}_n(\underset{\sim}{r}_1)\psi_n(\underset{\sim}{r}_2)}{E_n(1-i0)+\omega'} . \tag{16}$$

The summation over n includes both positive frequency (electron) and negative frequency (positron) states with the poles lying in the upper half-plane in the former case and the lower half-plane in the latter case. This ensures the correct time ordering for both positive and negative frequency states. When (14) - (16) are substituted into (13), the time integrals collapse to δ-functions.

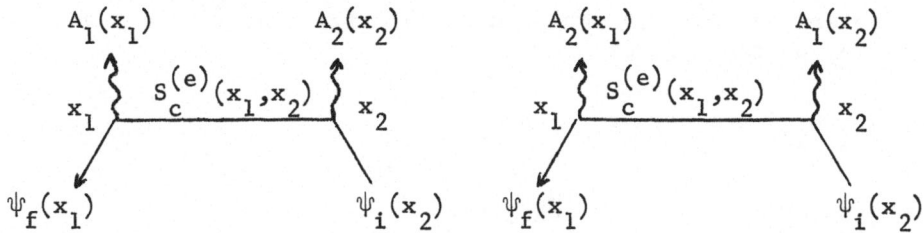

Figure 1.   Diagrams contributing to two-photon emission.
The solid lines represent electrons and the wavy lines
represent photons.

After making the low-velocity approximation, converting the matrix
elements from the dipole velocity to the dipole length form and
performing some standard manipulations, the expression for S_{if} reduces to[5]

$$S_{if} = -2\pi i \, U_{if} \, \delta(\omega_1 + \omega_2 - E_i + E_f) \tag{17}$$

$$U_{if} = -2\pi e^2 \, (\omega_1\omega_2)^{1/2} \sum_n \left[\frac{(\underset{\sim}{r}\cdot\hat{e}_1)_{fn} \, (\underset{\sim}{r}\cdot\hat{e}_2)_{ni}}{E_n - E_i + \omega_2} \right.$$

$$\left. + \frac{(\underset{\sim}{r}\cdot\hat{e}_2)_{fn} \, (\underset{\sim}{r}\cdot\hat{e}_1)_{ni}}{E_n - E_i + \omega_1} \right] \tag{18}$$

and the differential decay rate is

$$dw(\omega_1,\omega_2) = 2\pi |U_{if}|^2 \, \delta(\omega_1 + \omega_2 - E_i + E_f) \, \rho(\omega_1)\rho(\omega_2). \tag{19}$$

The angular correlation between the polarization vectors of
the two photons depends on the angular momenta of the initial and
final states, but for ^1S - ^1S transitions, U_{if} is simply

$$U_{if} = -2\pi(\omega_1\omega_2)^{1/2} \, (\hat{e}_1\cdot\hat{e}_2) \, Q_{if} \tag{20}$$

$$Q_{if} = e^2 \sum_n \left[\frac{(z)_{fn}(z)_{ni}}{E_n - E_i + \omega_2} + \frac{(z)_{fn}(z)_{ni}}{E_n - E_i + \omega_1} \right]. \tag{21}$$

If $a_{||}$ and $a_{\perp}$ denote the polarization vector amplitudes parallel
and perpendicular to the plane defined by the propagation vectors
of the two photons (with $a_{||}^2 + a_{\perp}^2 = 1$), then

$$\hat{e}_1\cdot\hat{e}_2 = a_{1\perp}a_{2\perp} + a_{1||}a_{2||}\cos\theta_k \tag{22}$$

where θ_k is the angle between the propagation vectors $\underset{\sim}{k}_1$ and $\underset{\sim}{k}_2$.

Summing over the linearly independent polarizations $\hat{e}_1 = \hat{e}_{\parallel}$ and $\hat{e}_1 = \hat{e}_{\perp}$ for photon 1 gives

$$\sum_{\hat{e}_1} |\hat{e}_1 \cdot \hat{e}_2|^2 = a_{2\perp}^2 + a_{2\parallel}^2 \cos^2\theta_k \tag{23}$$

and summing over both $\hat{e}_1$ and $\hat{e}_2$ gives

$$\sum_{\hat{e}_1, \hat{e}_2} |\hat{e}_1 \cdot \hat{e}_2|^2 = 1 + \cos^2\theta_k \tag{24}$$

which is the factor correlating the directions of emission if the polarizations are not observed.[8] Zernik[8] obtains an equation similar in form but not equivalent to (23). Finally, the decay rate integrated over angles and summed over polarizations is

$$w(\omega_1,\omega_2)\,d\omega_1 = (8/3\pi)\,(\omega_1^3\omega_2^3/c^6)\,|Q_{if}|^2\,d\omega_1 \tag{25}$$

in atomic units. The frequencies ω_1 and ω_2 are related by the energy conserving requirement $\omega_1 + \omega_2 = E_i - E_f$. The total transition rate is given by

$$w_{if} = \frac{1}{2} \int_0^{E_i - E_f} w(\omega_1,\omega_2)\,d\omega_1 . \tag{26}$$

The factor of $\frac{1}{2}$ is included so that pairs of photons are not counted twice.

The cross sections for stimulated two-photon emission and Raman scattering can be obtained by minor modifications of the results for spontaneous two-photon emission. In stimulated two-photon emission, a photon of frequency ω_1 is absorbed and re-emitted coherently in the original direction, while a second photon of frequency $\omega_2 = E_i - E_f - \omega_1$ is emitted randomly into all available photon modes. If there are N_1 incident photons per unit volume with polarization e_1, then the transition rate for the emission of photon 2 into solid angle $d\Omega_{k2}$ with polarization vector $\hat{e}_2$ is

$$w(\omega_1,\omega_2)\,d\Omega_{k2} = (\omega_1\omega_2^3/c^4)\,|\hat{e}_1 \cdot \hat{e}_2|^2\,N_1\,\rho(\omega_2)\,\delta(\omega_2 - E_i + E_f + \omega_1)$$

$$= (\omega_1\omega_2^3/c^3)\,|\hat{e}_1 \cdot \hat{e}_2|^2\,|Q_{if}(\omega_1,\omega_2)|^2\,N_1\,d\Omega_{k2} . \tag{27}$$

Since $N_1 c$ is the number of incident photons crossing unit area per unit time, the differential scattering cross section is simply

$$d\sigma/d\Omega_{k2} = (\omega_1\omega_2^3/c^4)\,|\hat{e}_1 \cdot \hat{e}_2|^2\,|Q_{if}(\omega_1,\omega_2)|^2 . \tag{28}$$

The cross section can be averaged over $\hat{e}_1$ (initial photon states) and summed over $\hat{e}_2$ (final photon states) as in the previous example, with $a_{\parallel}$ and $a_{\perp}$ being the polarization amplitudes parallel and perpendicular to the scattering plane and θ_k the scattering angle.

Raman scattering differs from the above only in that a photon of frequency ω_1 is absorbed and a single photon of frequency $\omega_2 = E_i - E_f + \omega_1$ is emitted randomly into all available photon modes. The scattering cross section formula is the same as for stimulated two-photon emission except that ω_1 is replaced by $-\omega_1$; i.e.

$$d\sigma/d\Omega_{k2} = (\omega_1\omega_2^3/c^4)\ |\hat{e}_1 \cdot \hat{e}_2|^2\ |Q_{if}(-\omega_1,\omega_2)|^2 \ . \tag{29}$$

The quenching of metastables in a static electric field may be regarded as the zero-frequency limit of the sum of stimulated two-photon emission and Raman scattering by a beam of low frequency photons with $\hat{e}_1$ pointing in the direction of the electric field[8]. Using (27) and the relation $E^2/8\pi = N_1\omega_1$, where E is the electric field strength, the decay rate is

$$w(0,\omega_2)d\Omega_{k2} = (E^2/8\pi)\ (\omega_2^3/c^3)\ |\hat{e}_1 \cdot \hat{e}_2|^2\ |2Q_{if}(0,\omega_2)|^2\ d\Omega_{k2} \tag{30}$$

with $\omega_2 = E_i - E_f$. The two amplitudes are added before squaring because the final states are identical and the two processes contribute coherently[8]. After averaging over $\hat{e}_1$, summing over $\hat{e}_2$ and integrating over angles, the decay rate simplifies to

$$w(0,\omega_2) = (4\omega^3/3c^3)\ |Q_{if}|^2 \ . \tag{31}$$

The above result must be modified if there is a resonance near zero frequency, as is the case for metastable hydrogen. The modifications are discussed in the following section.

3.   THE 2s $^2S_{1/2}$ STATE OF HYDROGEN

The metastability of the $2s_{1/2}$ state of hydrogen was first discussed by Breit and Teller[9] who showed by order of magnitude arguments that, although single photon magnetic dipole (M1) transitions to the ground state are possible, the dominant decay mode is two-photon electric dipole (2E1) emission. However M1 transitions become increasingly important with increasing nuclear charge and will be discussed first. Using (12), the M1 transition operator is, in lowest order

$$Q_{1M}^{(0)} = (e/2mc)L_M^* + \mu_M^* \tag{32}$$

where $L_M = \mp(1/2^{1/2})(L_x \pm L_y)$ for $M = \pm 1$, $L_0 = L_z$, and similarly
for μ_M. The matrix element $\langle 1s|Q_{1M}^{(0)}|2s\rangle$ vanishes if non-relativistic
Schrödinger wavefunctions are used, but a non-vanishing contribution
of relative order α^2 is obtained from relativistic corrections to
the wavefunctions and finite wavelength corrections represented by
the second term in the retardation expansion (8). Returning to (4),
the effective interaction energy of an electron with the transverse
electromagnetic field is given by matrix elements of the operator
$H_{int} = - e\underset{\sim}{\alpha}\cdot\underset{\sim}{A}(\underset{\sim}{r},t)$. In lowest order, the equivalent non-relativistic
operator is $H_{int}(NR) = - (e/2mc)(\underset{\sim}{p}\cdot\underset{\sim}{A} + \underset{\sim}{A}\cdot\underset{\sim}{p}) - \underset{\sim}{\mu}\cdot\underset{\sim}{\mathcal{H}}$, which reduces
to (32) (except for a multiplicative constant) when the vector po-
tential for M1 transitions is used. Here $\underset{\sim}{\mathcal{H}} = \nabla \times \underset{\sim}{A}$ is the magnetic
field. The next higher order corrections obtained by application
of the FW transformation for a vector potential containing the time
dependent factor $e^{i\omega t}$ yield[6,10]

$$
\begin{aligned}
H_{int}(NR) = &- (e/2mc)(\underset{\sim}{p}\cdot\underset{\sim}{A} + \underset{\sim}{A}\cdot\underset{\sim}{p}) - \underset{\sim}{\mu}\cdot\underset{\sim}{\mathcal{H}} \\
&+ (e/8m^3c^3)[\underset{\sim}{p}^2, (\underset{\sim}{p}\cdot\underset{\sim}{A} + \underset{\sim}{A}\cdot\underset{\sim}{p})]_+ + (1/4m^2c^2)[\underset{\sim}{p}^2, \underset{\sim}{\mu}\cdot\underset{\sim}{\mathcal{H}}]_+ \\
&- (e/2mc^2)\underset{\sim}{\mu}\cdot\nabla V \times \underset{\sim}{A} + (e\omega/8m^2c^3)i\nabla\cdot\underset{\sim}{A} \\
&- (\omega/4mc^2)(\underset{\sim}{\mu}\cdot\underset{\sim}{\mathcal{H}} - 2i\nabla\cdot\underset{\sim}{A}\times\underset{\sim}{p})
\end{aligned}
\tag{33}
$$

where $[,]_+$ denotes the anticommutator and $eV = - Ze^2/r$ is the cen-
tral nuclear potential. For M1 transitions, $A = (1/2^{1/2})g_1(kr)[\underset{\sim}{L}\ Y_1^M(r)]^*$ and the $M = 0$ component of $Q_{1M}^{(0)} = - (\tilde{c}/\omega)(3\pi/8)^{1/2}H_{int}(NR)$
reduces to [10,11]

$$
\begin{aligned}
Q_{10}^{(0)} = &(e/2mc)[1 - p^2/(2m^2c^2) - (1/10)(\omega r/c)^2]L_z \\
&+ [1 - 2p^2/(3m^2c^2) - (1/6)(\omega r/c)^2 + Ze^2/(3mc^2)]\mu_z
\end{aligned}
\tag{34}
$$

provided that the leading two terms in the expansion (8) of $g_1(kr)$
are retained. The necessity of including both terms was first
pointed out by Zhukovskii, Kolesnikova, Sokolov and Kherrman[12], but
they did not derive the equivalent non-relativistic transition operator given above. Using non-relativistic hydrogenic wavefunctions
for nuclear charge Z, the matrix element of $Q_{10}^{(0)}$ is

$$
\langle 1s_{1/2}|Q_{10}^{(0)}|2s_{1/2}\rangle = - \alpha^3 ea_0 Z^2 2^{1/2}(2/27 - 2/81)
\tag{35}
$$

where a_0 is the Bohr radius. The first terms of (35), which is the
result obtained by Breit and Teller[9], is associated with relativistic
corrections to the wavefunctions, while the second term comes from
the retardation expansion of $g_1(kr)$. Using (10), the hydrogenic
decay rate summed over final states and averaged over initial states

is[10,11,12]

$$w_{if} = Z^{10}\alpha^9/972 \times (2.4189 \times 10^{-17} \text{ sec})^{-1}$$

$$= 2.4958 \times 10^{-6} Z^{10} \text{ sec}^{-1}. \tag{36}$$

Numerical values are given in Table 1 for several members of the H isoelectronic sequence. The M1 process is important for the ions with $Z > 18$ and becomes dominant for $Z > 45$. The process has never been clearly separated from two-photon emission in a hydrogenic ion, but Boehm[13] has observed the $2s_{1/2} - 1s_{1/2}$ inner-shell X-ray transition in Tm($Z = 69$) with an intensity in reasonable agreement with the relativistic Hartree-Fock calculations of Rosner and Bhalla[2].

The two-photon decay rate of metastable hydrogen was first estimated by Breit and Teller[9] and more accurate calculations were performed later by several other workers[14-17]. Klarsfeld[17] has obtained an analytical expression for Q_{if} (eq. 21) in terms of hypergeometric functions and calculated the non-relativistic decay rate

$$w_{if} = (8.2282 \pm 0.001) Z^6 \text{ sec}^{-1}. \tag{37}$$

The total decay rate, including M1 transitions, is thus $(8.2282 + 2.4958 \times 10^{-6} Z^4) Z^6 \text{ sec}^{-1}$.

The earlier experimental work demonstrated that the decay of metastable H (ref. 18) and He$^+$ (ref. 19) takes place by two-photon emission and established lower limits on the lifetimes. Novick and co-workers also observed spectral[19] and angular[20] distributions in rough agreement with theory (eq. 24 and 25). More recent experiments have yielded definite decay rates which are compared with the theoretical predictions in Table 1. The excellent agreement between theory and experiment implies a limit on the amount of parity violating p state character that may be mixed with the $2s_{1/2}$ state by a hypothetical neutral weak current and/or electromagnetic interaction [24-27]. If the wavefunction is assumed to have the form $\psi(2s) = \phi(2s) + \varepsilon\phi(2p)$, then even a very small value of ε would lead to rapid depopulation of the 2s state by electric dipole transitions to the ground state. The measurement of Prior[21] (see Table 1) implies $|\varepsilon| < 4.7 \times 10^{-5}$.

Both stimulated two-photon emission[28] and Raman scattering[29] from metastable deuterium atoms have recently been detected by Braunlich and Lambropoulos. Although the experimental uncertainties are large, the measured cross sections appear to be in rough agreement with the calculations by Zernick[8] and more recently by Klarsfeld[30].

The depopulation of H metastables by a static electric field

TABLE 1. Comparison of Observed and Theoretical Decay Rates (sec^{-1})
for Metastable Hydrogenic Ions.

Ion	M1[a]	2E1[b]	Total	Experiment
H I	2.496(-6)	8.228(0)	8.228(0)	-
He II	2.556(-3)	5.266(2)	5.266(2)	$(4.91 \, {}^{+\,0.95}_{-\,1.40})(2)^c$
				$(5.20 \pm 0.21)(2)^d$
Li III	1.474(-1)	5.998(3)	5.998(3)	-
C VI	1.509(2)	3.839(5)	3.840(5)	-
O VIII	2.680(3)	2.157(6)	2.160(6)	-
Ne X	2.496(4)	8.228(6)	8.253(6)	-
S XVI	2.744(6)	1.380(8)	1.408(8)	$(1.37 \pm 0.13)(8)^e$
Ar XVIII	8.911(6)	2.799(8)	2.888(8)	$(2.82 \pm 0.20)(8)^e$
Fe XXVI	3.523(8)	2.542(9)	2.894(9)	-

The numbers in brackets are the powers of 10 by which the entries
are to be multiplied.

[a] Magnetic dipole decay rates (refs. 10, 11 and 12).
[b] Two-photon electric dipole decay rates (ref. 17).
[c] Clendenin, Kocher and Novick (ref. 21).
[d] Prior (ref. 22).
[e] Marrus and Schmieder (ref. 23).

(Stark quenching) was first measured and discussed theoretically by
Lamb and Retherford[31] in connection with their Lamb shift measurements. The Stark quench rate has been used in several recent experiments to deduce the Lamb shift in hydrogenic ions[32-34]. Physically, one can think of the decay as an electric dipole transition
occurring through the $2s_{1/2} - 2p_{1/2}$ mixing induced by the electric
field, or as the zero-frequency limit of a two-photon process. In
the traditional Bethe-Lamb (BL) approach, one solves the phenomenological rate equations

$$i\dot{a} = V_{ab}\exp(-i\omega_{ab}t)b - \frac{1}{2}i\gamma_a a$$
$$i\dot{b} = V_{ba}\exp(i\omega_{ab}t)a - \frac{1}{2}i\gamma_b b$$

(38)

where a and b are the $2s_{1/2}$ and $2p_{1/2}$ state amplitudes, γ_a and γ_b
the decay rates in the absence of the external field, V_{ab} the matrix
element of the external field and $\omega_{ab} = E(2s_{1/2}) - E(2p_{1/2})$ is the
Lamb shift frequency. With the initial conditions

$$a = 1, b = 0 \text{ at } t = 0$$

(39)

the solution to (38) leads to the Stark quench rate[31]

$$\gamma_{Stark} = \gamma_a + \gamma_b|V_{ab}|^2/(\omega_{ab}^2 + \gamma_b^2/4) \quad .$$

(40)

If there is a collection of close-lying states (e.g. $2p_{1/2}$, $2p_{3/2}$) then b becomes an index labelling these states and the right-hand side of (40) is to be summed over b.[32,37]

The BL approach has been extended and refined by several authors to include the time dependence and line shape of the radiation,[35-38] but the applicability of the results depends on the experimental conditions. The boundary conditions (39) imply that the field is switched on rapidly relative to the response time of the atom, which is characterized by $h/E(2s_{1/2}) - E(2p_{1/2}) \simeq 10^{-10}$ sec for neutral hydrogen. Since neutral beam velocities are often $\sim 10^6$ cm/sec or less, the above condition may not be satisfied in experiments where a beam of metastables passes into an electric field region, and an adiabatic or intermediate description may be more appropriate. For the hydrogenic ions of nuclear charge Z, the response time is $\sim 10^{-10}/Z^4$ sec and it is even more difficult to switch the field on suddenly.

In the two-photon Raman scattering formalism, the Stark quenching of hydrogen corresponds roughly to the low frequency limit with a resonance near zero frequency. If the field is switched on suddenly, then it has the time dependence

$$E(t) = \begin{cases} 0 & t < 0 \\ E & t > 0 \end{cases} \tag{41}$$

or $\quad E(t) = (1/2\pi) \int_{-\infty}^{\infty} f(\omega'') \exp(-i\omega''t) \, d\omega'' \tag{42}$

with $f(\omega'') = E \lim_{\epsilon \to +0} (i/(\omega' + i\epsilon))$. \tag{43}

In effect, the transition is induced by an incident wave packet of photons with frequency distribution $f(\omega)$ centered about $\omega = 0$. Near the $2s_{1/2} - 2p_{1/2}$ resonance frequency, it is necessary to adopt a modified form of the electron propagator (16) in which the energy shifts and widths are included in the denominator; thus (ref. 5, p. 763)

$$S_c^{(e)}(x_1, x_2) = \frac{1}{2\pi i} \int_{-\infty}^{\infty} d\omega' \, e^{i\omega'(t_1-t_2)}$$

$$\times \sum_n \frac{\bar{\psi}_n(\underline{x}_1)\psi_n(\underline{x}_2)}{E_n - i\gamma_n/2 + \sum_m \frac{|V_{nm}|^2}{-E_m - \omega'} + \omega'} \tag{44}$$

provided that the frequency dependence of V_{nm} is neglected. The symbols in (44) have the same meaning as in (38). If we take

$$\hat{A}_1(x_1) = i\beta(2\pi/\omega_1)^{1/2} \; \underset{\sim}{\alpha} \cdot \hat{e}_1 \; \exp(i\underset{\sim}{k}_1 \cdot \underset{\sim}{r}_1 - i\omega_1 t_1) \tag{45}$$

$$\hat{A}_2(x_2) = i\beta \hat{e}_2 \cdot \underset{\sim}{r}_2 /(2\pi) \int_{-\infty}^{\infty} f(\omega'') \; \exp(-i\omega'' t_2) \; d\omega'' \tag{46}$$

omit the small second term of (13) and retain only the dominant $2s_{1/2} - 2p_{1/2}$ contribution in the first term of (13) and in (44), then the expression for $S_{if}^{(2)}$ reduces immediately to

$$S_{if}^{(2)}(\omega_1) = \frac{e^2 \; (2\pi\omega_1)^{1/2} \; E \; (\underset{\sim}{r} \cdot \hat{e}_1)_{fb} \; (\underset{\sim}{r} \cdot \hat{e}_2)_{ba}}{(E_a - E_f - \omega_1)(E_b - E_f - i\gamma_b/2 - \omega_1) - |V_{ab}|^2} \tag{47}$$

in the non-relativistic dipole length form. The transition probability per unit frequency interval is $P(\omega_1) \; d\omega_1 = |S_{if}^{(2)}|^2 \rho(\omega_1)$, which becomes after summing over polarizations and integrating over angles,

$$P(\omega_1) d\omega_1 =$$

$$\frac{(1/2\pi)\gamma_b |V_{ab}|^2 d\omega_1}{[(E_a - E_f - \omega_1)(E_b - E_f - \omega_1) - |V_{ab}|^2]^2 + (E_a - E_f - \omega_1)^2 \gamma_b^2/4} \; . \tag{48}$$

Eq. (48) is identical with the spectral distribution calculated by Fontana and Lynch[38] (their eq. 27) starting from the BL time dependent formalism. The above derivation demonstrates that the same results can be obtained with ease from the S-matrix formalism for two-photon processes and displays explicitly the dependence of the line shape on the Fourier coefficients describing the time dependence of the external field. When the field is turned on suddenly, the emitted radiation displays two peaks - peak 1 near the $2s_{1/2}$ - $1s_{1/2}$ transition frequency and peak 2 near the $2p_{1/2}$ - $1s_{1/2}$ transition frequency as shown in Fig. 2. Peak 2 results from photon frequencies in the wings of the distribution function $f(\omega)$ at the ω_{ab} resonant frequency. As the field is turned on more slowly, $f(\omega)$ becomes more sharply peaked about $\omega = 0$ and peak 2 becomes progressively weaker. Finally, the time distribution of the radiation and asymptotic decay rate can be obtained from the Fourier transform of $S_{if}^{(2)}(\omega_1)$.

Ott, Kauppila and Fite[39] have shown that mixing with both the $2p_{1/2}$ and $2p_{3/2}$ states must be included to obtain the correct polarization for the quench radiation. The polarization can be obtained by summing the right-hand side of (47) over both members of the $2p_{1/2}$, $2p_{3/2}$ doublet before squaring. As before, e_1 points in the direction of the electric field. It is a simple consequence of the orthogonality of the Clebsch-Gordan coefficients that the resulting

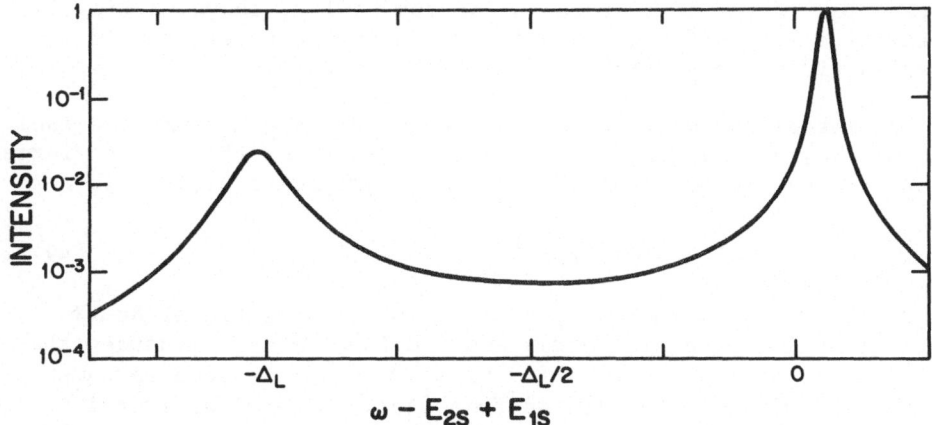

Figure 2. Photon energy distribution for the quench radiation of hydrogen in the 'sudden' approximation with $V_{ab} = 2.2\gamma_b$ (Fontana and Lynch, ref. 38). The integrated distribution is normalized to unity and the energy scale is in units of the Lamb shift $\triangle_L$.

cross terms vanish on averaging over the polarization vector e_2[40,41] for the emitted photon and summing over all magnetic substates. The averaged transition probability is then just a sum of squares. The measured fractional polarization $(I_\sigma - I_\pi)/(I_\sigma + I_\pi)$ is[39] - 0.30 $\pm$ .02 in marginal disagreement with the theoretical value[42] - 0.323. Time dependent calculations by Wooten and Macek[43] show that the instantaneous polarization oscillates in time when the field is turned on suddenly, with the amplitude of the oscillations decreasing to zero as the adiabatic limit is approached.

4.   THE 1s2s $^1S_0$ STATE OF HELIUM

All single-photon transitions from the 2^1S_0 state are strictly forbidden in the absence of nuclear spin, leaving two-photon emission as the dominant decay mode. Even if the nucleus has spin, two-photon emission remains dominant up to very large values of Z. The spectral distribution (eq. 25) and the integrated decay rate (eq. 26) have been calculated to high accuracy by using a discrete variational approximation to the summation over bound and continuum intermediate states in (21)[44], and by summing explicitly over the intermediate states[45]. The two theoretical decay rates for neutral helium are 51.3 sec^{-1} and 50.87 sec^{-1} in reasonable agreement with each other and with the experimental value 50 $\pm$ 2.5 sec^{-1} measured by Van Dyck, Johnson and Shugart[46]. An earlier measurement by Pearl[47] of 26 $\pm$ 5 sec^{-1} is almost certainly too low. The theoretical spectral

distribution of the emitted radiation for He I is shown in Fig. 3.
The relative shapes for H I and for the H and He isoelectronic
sequences differ only slightly from that shown.

The integrated decay rates for helium and the helium-like ions
up to Ne IX are summarized in Table 2. Values for the ions of larger
nuclear charge can be obtained from the approximate asymptotic formula

$$w_{if} = 2(8.228)(Z - \sigma)^6 \text{ sec}^{-1} \tag{49}$$

with $\sigma \simeq 0.797$ as estimated by fitting to the variational Ne IX
result given in Table 2. Marrus and Schmieder[23] have measured the
Ar XVII decay rate to be $(4.3 \pm 0.6) \times 10^8 \text{ sec}^{-1}$ in good agreement
with $4.26 \times 10^8 \text{ sec}^{-1}$ calculated from (49). An earlier identification of the two-photon continuum of Ne IX in a plasma source by
Elton, Palumbo and Griem[48] has been re-interpreted as a blend arising
from a number of double-electron inner shell radiative transitions
of neon ions in various stages of ionization[49].

The electric field quenching of helium 2^1S_0 is described directly by (31) since there is not a resonance near zero frequency in
this case. Specifically, the extra resonance terms included in the
denominator of (44) are very much less than the large electrostatic
splitting between the 1s2s $^1S_0$ and 1s2p $^1P_1$ states and can be neglected. The quench rates have been calculated to high accuracy by
Drake[50] using a discrete variational approximation to the summation
over bound and continuum states in (21), and by Jacobs[51] using an
explicit summation over the intermediate states. The two calcula-
tions yield $0.9320E^2 \text{ sec}^{-1}$ and $0.9315E^2 \text{ sec}^{-1}$, where E is the electric field strength in kV/cm. Both calculations are well within the
experimental uncertainty of the values $(0.926 \pm 0.030)E^2 \text{ sec}^{-1}$ and
$(0.930 \pm 0.005)E^2 \text{ sec}^{-1}$ recently measured by Petrasso and Ramsey[52],
and by Johnson[53] respectively. The quench rates for helium and the
helium-like ions up to Ne IX are summarized in Table 2. The good
agreement between theory and experiment provides some additional support for the theory used to deduce Lamb shifts from the Stark quenching of hydrogenic metastables; however the situation there is
really rather different owing to the presence of a resonance near
zero frequency.

Stimulated two-photon emission and Raman scattering have not
yet been detected from metastable helium-like systems, although some
theoretical results are now available for neutral helium[54]. There
have been several suggestions that stimulated two-photon emission
may be useful in constructing an ultraviolet laser[55-58].

TABLE 2. Spontaneous and Electric Field Induced Decay Rates (sec^{-1}) of the 1s2s $^1S_0$ and $^3S_1$ Sequences.

Z	2^1S_0 2E1 rate[a]	2^1S_0 $\dfrac{quench\ rate}{E^2}$ [b]	2^3S_1 M1 rate[c]
2	5.13(1)	0.9320	1.272(-4)
3	1.95(3)	0.8184	2.039(-2)
4	1.81(4)	0.7492	5.618(-1)
5	9.26(4)	0.7066	6.695(0)
6	3.31(5)	0.6781	4.856(1)
7	9.43(5)	0.6580	2.532(2)
8	2.31(6)	0.6428	1.044(3)
9	5.05(6)	0.6313	3.608(3)
10	1.00(7)	0.6222	1.087(4)
11			2.935(4)
12			7.243(4)
13			1.658(5)
14			3.563(5)
15			7.251(5)
16			1.408(6)
17			2.622(6)
18			4.709(6)
19			8.187(6)
20			1.383(7)
21			2.275(7)
22			3.656(7)
23			5.751(7)
24			8.870(7)
25			1.344(8)
26			2.002(8)

The numbers in brackets are the powers of ten by which the entries are to be multiplied.

[a] Drake, Victor and Dalgarno (ref.44)
[b] Drake (ref. 50). E is the electric field strength in kV/cm.
[c] Drake (ref. 10).

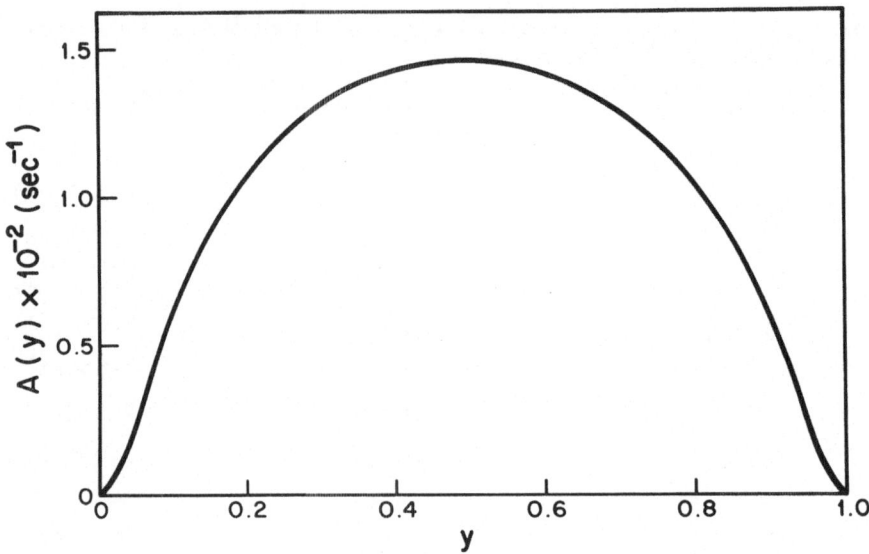

Figure 3. Photon energy distribution for the $2\,^1S_0 - 1\,^1S_0$ two-photon decay of He I (Drake, Victor and Dalgarno, ref. 44). y is the fraction of the transition energy transported by one of the two photons.

5. THE 1s2s $^3S_1$ STATE OF HELIUM

The development of the theory describing the radiative decay of the $2\,^3S_1$ metastable state is a good example of the productive interaction between atomic physics and astrophysics. Breit and Teller[9] in 1940 suggested that the state decays primarily by two-photon emission, incorrectly estimating the $2\,^3S_1 - 1\,^1S_0$ M1 process allowed by the selection rules (9) to be much slower. A calculation by Mathis[59] yielded 2.2×10^{-5} sec^{-1} for the 2E1 decay rate, a value used in the astrophysical literature for many years. However Drake and Dalgarno[60] recently showed Mathis's calculation to be based on an incorrect formulation of the problem, and detailed calculations[44,61] reduced the 2E1 decay rate to 4×10^{-9} sec^{-1}. Gabriel and Jordan[62,63] then stimulated further theoretical work when they identified sharp lines in the spectrum of the solar corona as the $2\,^3S_1 - 1\,^1S_0$ transition of the helium-like ions C V to Si XIII and correctly proposed that the state decays primarily by M1 emission.

The discussion of M1 transitions from the $2\,^3S_1$ state of helium closely parallels that given in Section 3 for the $2s_{1/2} - 1s_{1/2}$ transition of hydrogen. Matrix elements of $Q_{1M}^{(0)}$ (eq. 32), summed over both electrons, vanish in the non-relativistic approximation and it is necessary to include the relativistic correction terms of relative order α^2. Griem[64] obtained the first theoretical estimate of the decay rate with approximate Dirac eigenfunctions, but he did not include the retardation term in (8).

High precision calculations are most easily performed in the Pauli representation since accurate variational solutions to the two-electron Schrödinger equation (including correlation) are readily obtained. The equivalent non-relativistic transition operator, including retardation, was derived almost simultaneously by Drake[10], Schwartz[65], and Feinberg and Sucher[11]; the result for the $2\ {}^3S_1$ - $1\ {}^1S_0$ transition is just the spin-dependent part of the one-electron operator (34) summed over both electrons. In general, the electron-electron interaction terms e^2/r_{12} + B in the Dirac-Breit two-electron Hamiltonian (B is the Breit interaction) lead to terms of relative order α^2 in the equivalent non-relativistic transition operator which depend explicitly on the r_{12} co-ordinate. However these terms are all either independent of the Pauli spin operators $\underset{\sim}{\sigma}_1$ and $\underset{\sim}{\sigma}_2$, or are symmetric in $\underset{\sim}{\sigma}_1$ and $\underset{\sim}{\sigma}_2$, and therefore do not contribute to spin-changing transitions of either the electric or magnetic type[66]. This is in agreement with the general notion derived from the correspondence principle that in lowest non-vanishing order, the N-electron transition operator is just the one-electron operator summed over all N electrons.

The calculated M1 decay rates for the helium-like ions up to Fe XXV are given in Table 2. The value 1.27×10^{-4} sec^{-1} for neutral helium is faster than the values predicted by earlier theories and has important astrophysical consequences to be summarized in the following section. There is very little experimental data for comparison, but Freeman et al.[67] have deduced a C V M1 decay rate of 37 sec^{-1} from observed coronal line intensities in satisfactory agreement with 48.6 sec^{-1} given in Table 2. Also Marrus and Schmieder[23] have measured the $2\ {}^3S_1$ lifetime of Ar XVII to be 172 ± 30 nsec, in apparent disagreement with the predicted value 212.7 nsec. Further relativistic corrections can be expected to decrease the theoretical lifetime by about 2%[65], which is not enough to close the gap. The 2E1 decay rate[44] remains smaller than the M1 decay rate by a factor of 10^3 - 10^4 for all the ions since both processes have the same Z^{10} dependence on the nuclear charge. The discrepancy, if real, is one of the few remaining examples of a disagreement between theory and experiment in one- and two-electron systems and further measurements on other two-electron ions of large nuclear charge would be of considerable interest.

6. ASTROPHYSICAL APPLICATIONS

The radiative decay of the hydrogen and helium metastable states is important under the low density conditions found, for example, in planetary nebulae[68]. Since the time between de-exciting collisions is $\sim 10^4$ sec under typical nebular temperature and density conditions ($T \sim 10^4$ oK, $N \sim 10^4$ cm^{-3}), the $2s_{1/2}$ state of hydrogen and the $1s2s\ ^1S_0$ state of helium depopulate primarily by two-photon emission, but the $1s2s\ ^3S_1$ state depopulates at roughly equal rates by collisions and by radiation. As a result, the ratio of line intensities in the helium triplet spectrum can be used as a density indicator for the nebula[69]. Drake and Robbins[70] have shown that the new M1 decay rate for the $2\ ^3S_1$ state, plus other corrections, substantially resolves a long-standing discrepancy between the measured intensity ratios and those expected by balancing the rates of formation and destruction of the excited triplet states.

Analogous statements hold true for the helium-like ions in the solar corona and solar flares[67,71]. There, the intensity ratio

$$R = \frac{I(2\ ^3S_1 - 1\ ^1S_0)}{I(2\ ^3P_1 - 1\ ^1S_0) + I(2\ ^3P_2 - 1\ ^1S_0)} \tag{50}$$

is particularly useful as a density indicator. The lines in the vacuum ultraviolet and soft X-ray region are now accessible to observation through a number of orbiting observatory programs[72-74]. The technique may also prove useful in the determination of electron densities in thermal cosmic X-ray sources[71].

REFERENCES

1. W. Heitler, The Quantum Theory of Radiation (Oxford U.P., London, 1954).

2. H. R. Rosner and C. P. Bhalla, Z. Physik 231, 347 (1970).

3. R. H. Garstang in Topics in Modern Physics, edited by W. E. Brittin and H. Odabasi, (Colorado Assoc. U. Press, 1971) p. 153.

4. I. P. Grant, Advan. Phys. 19, 747 (1970).

5. A. I. Akhiezer and V. B. Berestetskii, Quantum Electrodynamics (Interscience, New York, 1965).

6. L. L. Foldy and S. A. Wouthuysen, Phys. Rev. 78, 29 (1950).

7. A. Dalgarno in Atomic Physics edited by V. W. Hughes, B. Bederson, V. W. Cohen and F. M. J. Pichanick, (Plenum Press, New York, 1969) p.161; and The Menzel Symposium on Solar Physics, Atomic Spectra and Gaseous Nebulae edited by K. B. Gebbie (N. B.S. Special Publication 353, Washington, 1971) p. 47.

8. W. Zernik, Phy. Rev. 132, 320 (1963) and 133, A117 (1964).

9. G. Breit and E. Teller, Astrophys. J. 91, 215 (1940).

10. G. W. F. Drake, Phys. Rev. A 3, 908 (1971).

11. G. Feinberg and J. Sucher, Phys. Rev. Letters, 26, 681 (1971).

12. V. Ch. Zhukovskii, M. M. Kolesnikova, A. A. Sokolov and L. Kherrman, Opt. Spectrosc. 28, 337 (1970).

13. F. Boehm, Phys. Letts., 33A, 417 (1970).

14. L. Spitzer and J. L. Greenstein, Astrophys. J. 114, 407 (1951).

15. J. Shapiro and G. Breit, Phys. Rev. 113, 179 (1959).

16. B. A. Zon and L. P. Rapaport, Sov. Phys. JETP Letters 7, 52 (1968).

17. S. Klarsfeld, Phys. Letts. 30A, 382 (1969).

18. W. L. Fite, R. T. Brackman, D. G. Hummer and R. F. Stebbings, Phys. Rev. 116, 363 (1959).

19. E. Commins, L. Gampel, M. Lipeles, R. Novick and S. Schultz, Bull. Am. Phys. Soc. 7, 258 (1962); M. Lipeles, R. Novick and N. Tolk, Phys. Rev. Letters, 15, 690 (1965); R. Novick in Physics of the One- and Two-Electron Atoms, edited by F. Bopp and H. Kleinpoppen (North-Holland, Amsterdam, 1969).

20. J. Artura, N. Tolk and R. Novick, Astrophys. J. Letters 157, L181 (1969).

21. J. E. Clendenin, C. A. Kocher and R. Novick, Abstract submitted to the Third International Conference on Atomic Physics (Boulder, 1972).

22. M. H. Prior, Abstract submitted to the Third International Conference on Atomic Physics (Boulder, 1972); Phys. Rev. Letters, to be published.

23. R. Marrus and R. W. Schmieder, Phys. Rev. A 5, 1160 (1972) and earlier papers referenced therein.

24. Ya. B. Zel'dovish and A. M. Perelomov, Zh. Eksperim. i Teor.
     Fiz. 39, 1115 (1960) Sov. Phys. JETP 12, 777 (1961).

25. R. A. Carhart, Phys. Rev. 132, 2337 (1963).

26. J. Bernstein, M. Ruderman and G. Feinberg, Phys. Rev. 132,
     1227 (1963).

27. B. Sakitt and G. Feinberg, Phys. Rev. 151, 1341 (1966).

28. P. Braunlich and P. Lambropoulos, Phys. Rev. Letters 25, 135
     (1970).

29. P. Braunlich and P. Lambropoulos, Phys. Rev. Letters 25, 986
     (1970); P. Braunlich, R. Hall and P. Lambropoulos, Phys. Rev.
     A 5, 1013 (1972).

30. S. Klarsfeld, Phys. Rev. A 6, 508 (1972).

31. W. E. Lamb, Jr. and R. C. Retherford, Phys. Rev. 79, 549
     (1950); W. E. Lamb, Jr., Phys. Rev. 85, 259 (1952).

32. C. Y. Fan, M. Garcia-Munoz and I. A. Sellin, Phys. Rev. 161,
     6 (1967).

33. M. Leventhal and D. E. Murnick, Phys. Rev. Letters 25, 1237
     (1970).

34. D. E. Murnick, M. Leventhal and H. W. Kugel, Phys. Rev. Letters
     27, 1625 (1971).

35. G. W. Series, Phys. Rev. 136, A684 (1964).

36. R. K. Wangsness, Phys. Rev. 149, 60 (1966).

37. O. A. Keller and R. T. Robiscoe, Phys. Rev. 188, 82 (1969).

38. P. R. Fontana and D. J. Lynch, Phys. Rev. A 2, 347 (1970).

39. W. R. Ott, W. E. Kauppila and W. L. Fite, Phys. Rev. A 1, 1089
     (1970).

40. C. E. Johnson, Bull. Am. Phys. Soc. 17, 454 (1972).

41. H. K. Holt and I. A. Sellin, Phys. Rev. A 6, 508 (1972).

42. J. S. Casalese and E. Gerjuoy, Phys. Rev. 180, 327 (1969).

43. J. W. Wooten and J. H. Macek, Phys. Rev. A <u>5</u>, 137 (1972).

44. G. W. F. Drake, G. A. Victor and A. Dalgarno, Phys. Rev. <u>180</u>, 25 (1969).

45. V. Jacobs, Phys. Rev. A <u>4</u>, 939 (1971).

46. R. S. Van Dyck, Jr., C. E. Johnson and H. A. Shugart, Phys. Rev. Letters <u>25</u>, 1403 (1970).

47. A. S. Pearl, Phys. Rev. Letters, <u>24</u>, 703 (1970).

48. R. C. Elton, L. J. Palumbo and H. R. Griem, Phys. Rev. Letters <u>20</u>, 783 (1968).

49. R. C. Elton and L. J. Palumbo, Abstract submitted to the Third International Conference on Atomic Physics (Boulder, 1972).

50. G. W. F. Drake, Can. J. Phys. in press.

51. V. Jacobs, private communication.

52. R. Petrasso and A. T. Ramsey, Phys. Rev. A. <u>5</u>, 79 (1972).

53. C. E. Johnson, Abstract submitted to the Third International Conference on Atomic Physics (Boulder, 1972).

54. V. Jacobs, J. Phys. B, in press.

55. P. P. Sorokin and N. Breslau, I.B.M. J. Res. Develop. <u>8</u>, 177 (1964).

56. R. L. Garwin, I.B.M. J. Res. Develop. <u>8</u>, 338 (1964).

57. A. S. Selivanenko, Opt. Spectrosk. <u>21</u>, 100 (1966), [Opt. Spectrosc (U.S.S.R.) <u>21</u>, 54 (1966)].

58. B. P. Kirsanov and A. S. Selivanenko, Opt. Spectrosk. <u>23</u>, 455 (1967), [Opt. Spectrosc. (U.S.S.R.) <u>23</u>, 452 (1967)].

59. J. S. Mathis, Astrophys. J. <u>125</u>, 318 (1957).

60. G. W. F. Drake and A. Dalgarno, Astrophys. J. Letters <u>152</u>, L121 (1968).

61. O. Bely and P. Faucher, Astron. Astrophys. <u>1</u>, 37 (1969).

62. A. H. Gabriel and C. Jordan, Nature <u>221</u>, 947 (1969).

63. A. H. Gabriel and C. Jordan, Monthly Notices Roy. Astron. Soc. 145, 241 (1969).

64. H. R. Griem, Astrophys. J. Letters 156, L103 (1969) [erratum in 161, L155 (1970)].

65. C. Schwartz, private communication.

66. G. W. F. Drake, Phys. Rev. A 5, 1979 (1972).

67. F. F. Freeman, A. H. Gabriel, B. B. Jones and C. Jordan, Phil. Trans. Roy. Soc. (London), A 270, 127 (1971).

68. D. E. Osterbrock, Ann. Rev. Astron. and Astrophys. 2, 95 (1964).

69. R. R. Robbins, Astrophys. J., 151, 497 (1968); 151, 511 (1968); 160, 519 (1970).

70. G. W. F. Drake and R. R. Robbins, Astrophys. J. 171, 55 (1972).

71. G. R. Blumenthal, G. W. F. Drake and W. H. Tucker, Astrophys. J. 172, 205 (1972).

72. A. B. C. Walker and H. R. Rudge, Astronom. and Astrophys. 5, 4 (1970).

73. W. M. Neupert and M. Swartz, Astrophys. J. Letters, 160, L189 (1970).

74. J. F. Meekins, G. A. Doschek, H. Friedman, T. A. Chubb and R. W. Kreplin, Solar Phys. 13, 198 (1970).

RADIATIVE DECAY OF THE METASTABLE STATES OF THE H AND He

SEQUENCES – EXPERIMENT[*]

Richard Marrus

Lawrence Berkeley Laboratory and Department of Physics

University of California, Berkeley, California 94720

There are four known levels in the spectra of one- and two-electron atoms which can decay predominantly by other than an allowed E1 transition. These are indicated in Fig. 1. The theory of these decays has been described in the preceding paper[1] and we here summarize those properties that are relevant for experiment.

A. Two-photon decay

Both the $2s_{1/2}$ state and the 2^1S_0 state decay to the ground state predominantly by the simultaneous emission of two electric-dipole photons. In this way, both the parity and angular-momentum

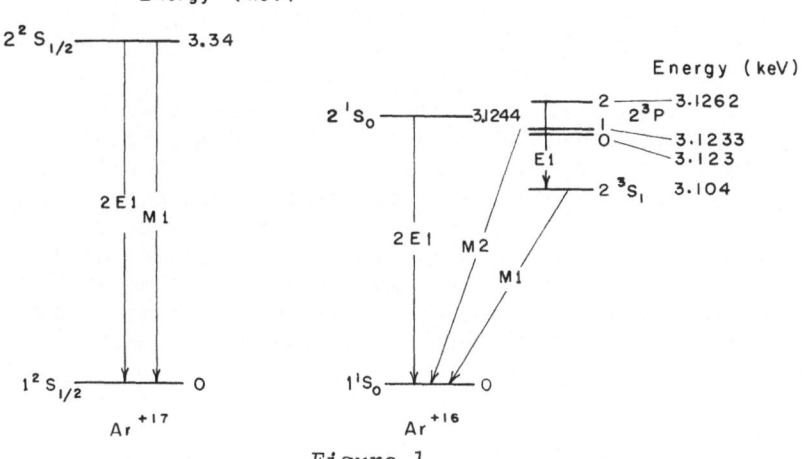

Figure 1

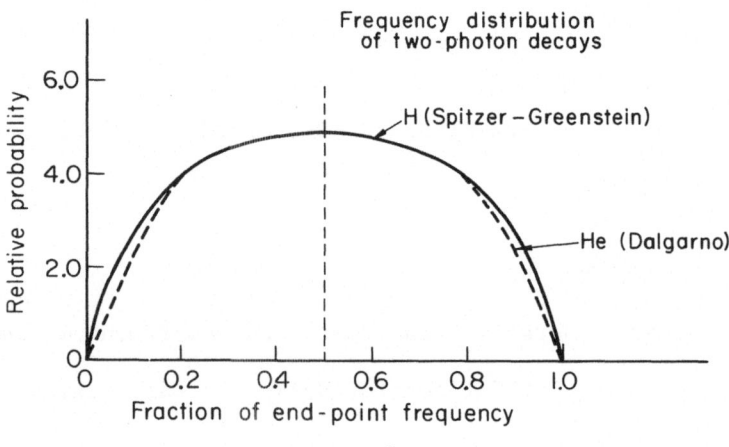

Figure 2

selection rules are satisfied, and an angular correlation function $\omega(\theta) \propto 1 + \cos^2\theta$ exists, where θ is the angle between the two emitted photons. The frequency distribution of the emitted photons has been worked out for both decays[2] and is shown in Fig. 2. They are both broad flat-topped distributions that drop off to half their peak value at about 10% of their endpoint frequency. The lifetimes of both levels for two-photon emission have been worked out.[3] In ordinary hydrogen and helium they are 120 msec and 19.3 msec, respectively, and the asymptotic lifetimes, i.e., valid in the limit of high Z, can be written as:

$$\tau_{2E1}(2s_{\frac{1}{2}}) = \frac{1}{8} Z^{-6} \text{ sec} \qquad \tau_{2E1}(2^1S_0) = \frac{1}{2} \frac{1}{8} Z^{-6} \text{ sec.} \qquad (1)$$

While the 2^1S_0 state can only decay to the ground state (1^1S_0) by the two-photon mechanism because of the 0–0 selection rule, the selection rules permit the $2s_{\frac{1}{2}}$ state to decay by M1 emission. Although the lifetime in ordinary hydrogen for this mode is calculated to be $\tau_{M1}(2s_{\frac{1}{2}}) \approx 4 \times 10^5$ sec, it has a Z^{-10} dependence for one-electron high-Z ions. Hence there will be some value of Z at which the M1 and 2E1 rates become comparable and, if one equates the two lifetimes, the crossover point occurs at about $Z \approx 45$. There is now some experimental evidence that this crossover actually exists.[4]

B. Relativistic M1 decay

The decay of the 2^3S_1 state is primarily by the relativistic M1 mechanism. It is relativistic in the sense that the magnetic dipole decay rate A_{M1} is proportional to

$$A_{M1} \propto \left| <2^3S_1 | \bar{\mu} | 1^1S_0> \right|^2 \quad , \tag{2}$$

where $\bar{\mu}$ is the magnetic dipole operator. In nonrelativistic approximation, $\bar{\mu} = -\mu_B(\bar{\ell} + 2\bar{s})$ and does not involve radial functions. Hence, by the orthogonality of the radial wave functions for the 2s and 1s state, the rate A_{M1} vanishes. However, relativity introduces corrections to the expression for $\bar{\mu}$, and it is as a result of these corrections that A_{M1} is finite. This was first noted by Breit and Teller[5] and the complete expression for the relativistic corrections has been evaluated by many authors.[6] This lifetime has a Z^{-10} dependence in the high-Z limit and in ordinary helium is $\sim 10^3$ sec.

C. Magnetic quadrupole decay

The 2^3P_2 state of ordinary helium decays almost entirely by a fully allowed E1 transition to the 2^3S_1 state with a lifetime of 10^{-7} sec.[7] However, it can also decay via M2 radiation to the 1^1S_0 ground state. That the two rates become competitive in high-Z two-electron ions can be understood as follows.

The E1 and M2 rates are given by

$$A_{E1} \propto \omega_{2p-2s}^3 \left| <2^3P_2 | \bar{r} | 2^3S_1> \right|^2$$

$$A_{M2} \propto \omega_{2p-1s}^5 \left| <2^3P_2 | \bar{r} | 1^1S_0> \right|^2 \quad .$$

Now ω_{2p-2s} is determined primarily by Coulomb repulsion of the electrons and hence is proportional to Z, whereas ω_{2p-1s} is determined by Coulomb interaction with the nucleus and hence is proportional to Z^2. Therefore $A_{E1} \propto Z$, whereas $A_{M2} \propto Z^8$, so that the two rates will become equal at some value of Z. Calculation[8] shows that this occurs at about $Z \approx 18$.

EXPERIMENTS

The first laboratory experiment to study any of the forbidden decay modes was reported in 1960 by the Columbia group of Lipeles, Novick and Tolk.[9] They succeeded in observing and unambiguously identifying the two-photon decay of the $2s_{1/2}$ state of He^+ with the apparatus shown in Fig. 3. The vacuum system consists of three chambers: an ion source chamber, an intermediate differential-pumped chamber and a detection chamber. The metastable He^+ beam is excited by electron bombardment and is then extracted and passed through a microwave cavity. This cavity is tuned to the

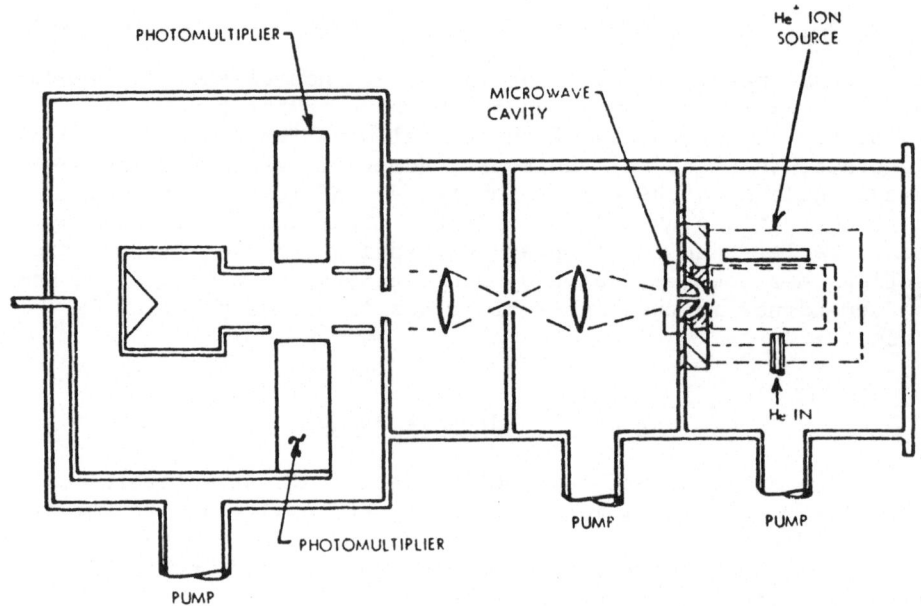

Figure 3

Lamb shift frequency for He$^+$ and is capable of quenching the
metastable state. Since this process is unique for the metastable
state, it can be used to identify any radiation observed in the
detector chamber as arising from this state. The beam is then
focused by a system of electrostatic einzel lenses into the detection chamber. Here, a pair of photomultipliers view the photons
emitted by decays-in-flight of the metastable ions.

With the above apparatus several pieces of evidence were
obtained that convincingly demonstrate the two-photon decay of the
He$^+$ metastable state.

(A) Figure 4 shows the time distribution of the emitted photons. The observed peak is very clear evidence that real coincidences are being observed.

(B) Figure 5 shows the angular correlation of the emitted
photons obtained by varying the angle between the detectors. It is
in good agreement with the theoretically predicted 1+cos$^2\theta$ distribution.

(C) Figure 6 shows the intensity of the single and coincidence
rates as a function of electron bombardment energy. The observed
threshold at 64 eV is in agreement with the theoretical excitation
energy.

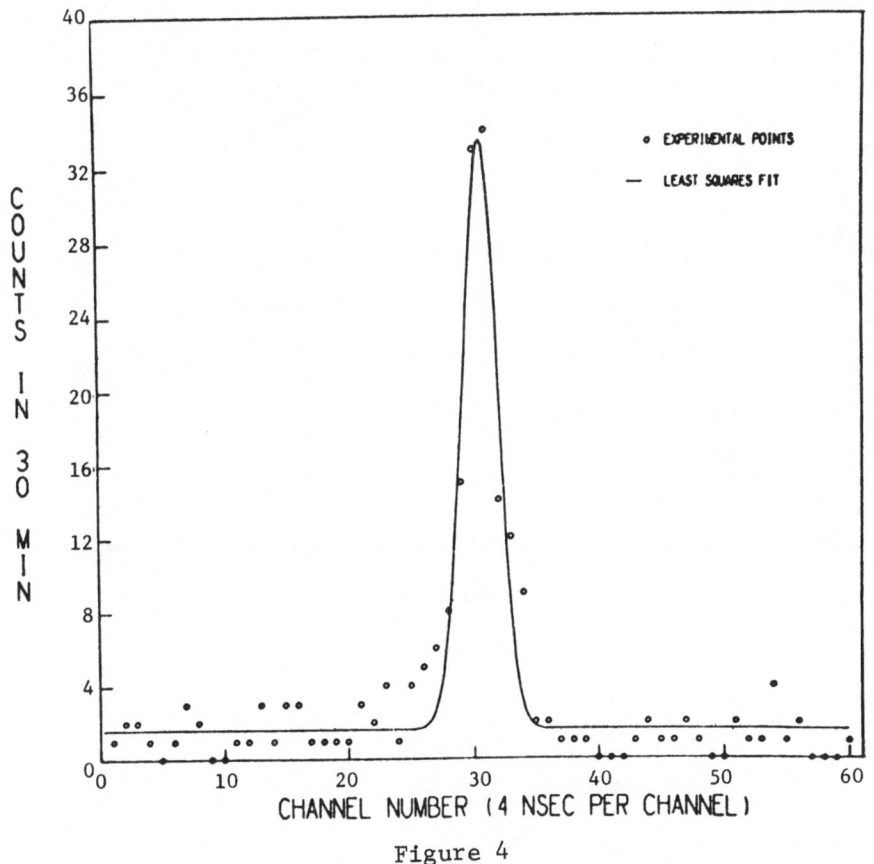

Figure 4

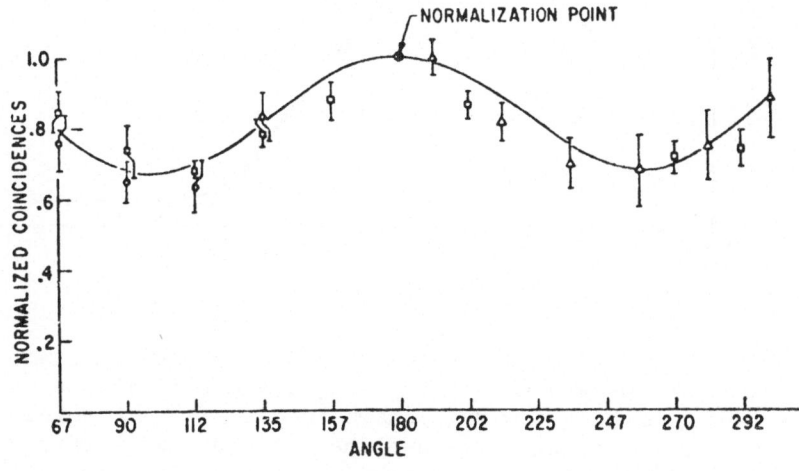

Figure 5

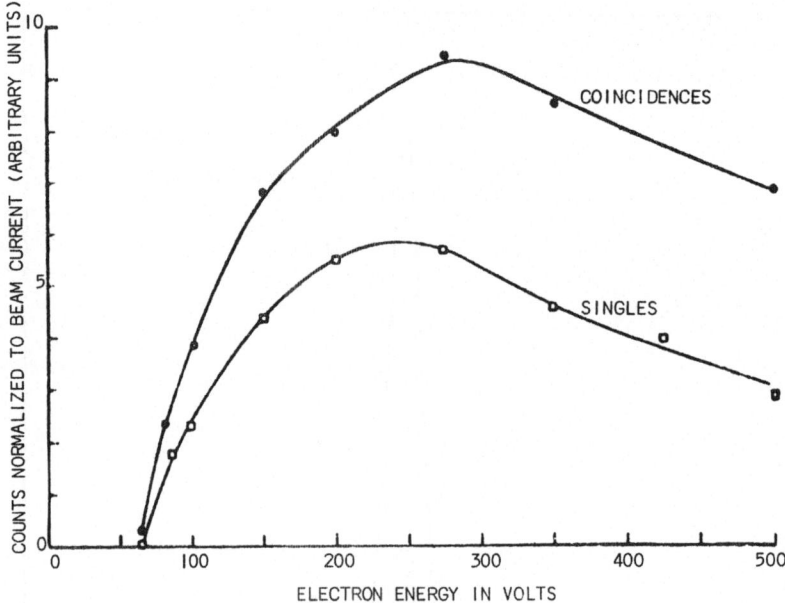

Figure 6

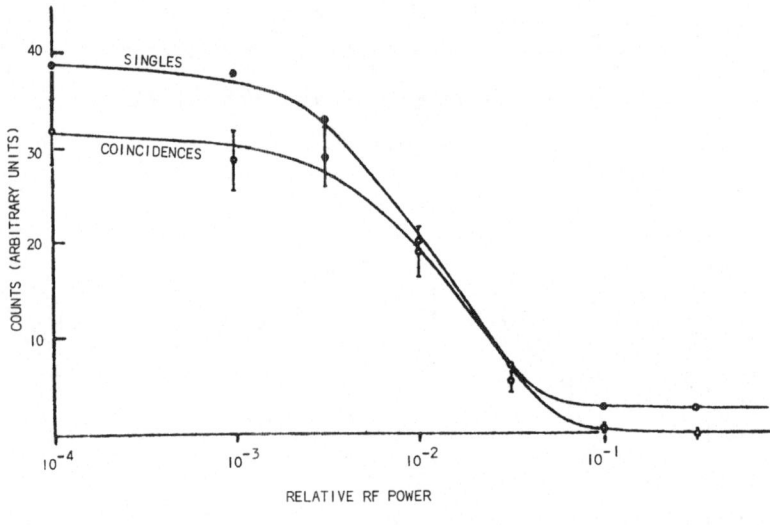

Figure 7

(D) The dependence of the single and coincidence rates on microwave quenching power is shown in Fig. 7. The points are fitted to a quenching curve and the agreement with theory is good. This is perhaps the most convincing evidence that it is the $2s_{\frac{1}{2}}$ state under observation.

In later work a crude energy distribution of the emitted photons was obtained using broadband filters, as was an upper limit to
the lifetime. All of the evidence taken together constitutes a
thoroughly convincing case that the two-photon decay mode has been
observed and its basic properties verified. Work by the Columbia
group has since been directed toward obtaining an accurate lifetime
measurement, and a result has recently been obtained.[10]

Prior,[11] and Prior and Shugart[12] have used a trapping technique
to measure the lifetimes of $2s_{1/2}$ of He^+ and 2^1S_0 of Li^+, respectively.
The schematic of the apparatus for Li^+ is shown in Fig. 8. A beam
of neutral Li atoms directed out of the plane of the paper is excited
to the metastable state by electron bombardment. The electrodes
1, 2, 3, and 4 are so biased as to create a potential well which
traps the ions between electrodes 2 and 3. Trapping in the plane
perpendicular to the paper is accomplished by a magnetic field.
The ions have very low energies and hence they will move in circles
of small radius in the presence of the magnetic field. Typical
numbers reported for the 2^1S_0 experiment are about 10^2 metastable
Li^+ ions stored in a region about 3 cm^3. The trapping region is
viewed by a pair of photomultipliers sensitive to the predominantly
uv radiation expected from the decay.

An important source of photon background was found to be due
to metastable neutral molecules in the background gas excited during

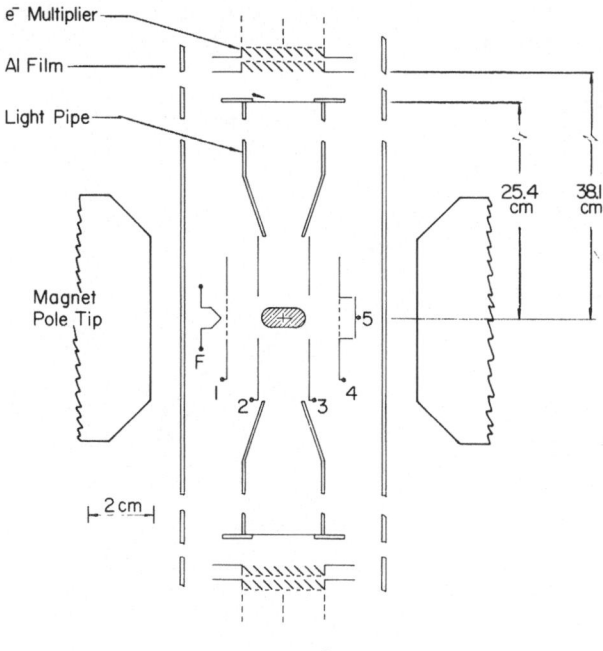

Figure 8

the electron bombardment process. These molecules are free to
drift to the photomultiplier and will give rise to counts as a
result of Auger de-excitation on the dynodes. To eliminate this
background, 800-Å thick aluminum films were used to cover the
multipliers. These films effectively eliminated the metastable
molecule background, but transmitted a reasonable fraction of the
uv photons. It was estimated that the Aℓ film-CuBe photomultiplier
combination responds to ~2% of the radiation over the region between
200 and 500 Å.

Data are accumulated in a cyclical fashion. A typical duty
cycle consists of the following sequence: a fill period lasting
22 μsec during which the electron beam is gated on and the detector
signal is gated off, a storage period of 2 msec during which the
electron beam is gated off and counts from the detectors are stored,
and a dump period of 60 μsec during which residual ions are swept
from the storage region.

Figure 9 shows a representative Li^+ 2^1S_0 decay curve. The
actual result is derived by averaging the results of 34 such curves.
Systematic effects were searched for by varying the following parameters: trapping voltage, magnetic field, electron impact, and Li
beam intensity.

The success of this relatively simple and straightforward
technique gives good hope that it can be extended to other atomic
properties associated with the metastable states, and perhaps to
other metastable states.

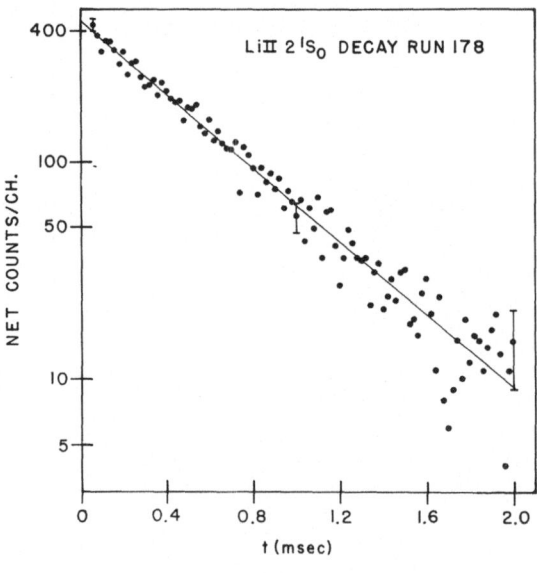

Figure 9

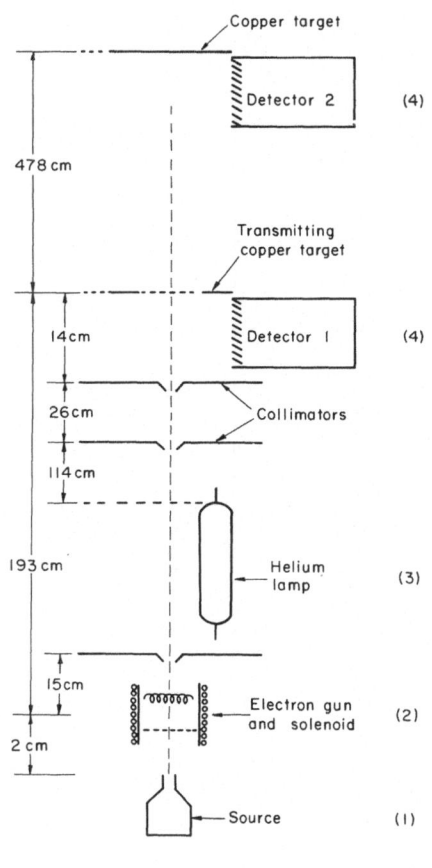

Copper target

Detector 2   (4)

478 cm

Transmitting copper target

14 cm   Detector 1   (4)

26 cm   Collimators

114 cm

193 cm   Helium lamp   (3)

15 cm

Electron gun and solenoid   (2)

2 cm

Source   (1)

Figure 10

The lifetime of the 2^1S_0 state of neutral helium has been measured in a time-of-flight experiment by Van Dyck, Johnson and Shugart.[13] This experiment differs from the ones previously described in that the metastable state is directly detected, rather than the decay photons. A schematic of the apparatus used is shown in Fig. 10. Helium atoms effuse from the source and are excited to the metastable states by electron bombardment. The atoms drift down a long tube to detector 1 where approximately 40% of the beam is stopped by a copper target. The remainder of the beam continues to drift to detector 2 where it is stopped by a solid copper target. A metastable helium atom will cause an electron to be emitted from the copper surface with high efficiency as the result of an Auger process. The Auger electron is then counted by an electron multiplier.

Cyclical data accumulation is also used in this experiment. The electron beam is pulsed on for a 100-μsec period. As a result of the velocity distribution of the metastable atoms thus created, they spread out in space and give rise to a time distribution of the Auger electrons when they strike the fixed detectors. The pulse arising from a particular electron is recorded in a PDP-8 computer memory location corresponding to its arrival time after the initial electron gun pulse. A sample distribution obtained in this way is shown in Fig. 11.

Metastable pulses arise from atoms in both the 2^1S_0 and 2^3S_1 states. The pulses arising from 2^3S_1 atoms can be subtracted by the following technique. A helium lamp placed between the electron excitation region and detector 1 is pulsed on and off. With the lamp on, atoms in the 2^1S_0 state are pumped to the 2^1P_1 state where they decay immediately to the ground state 1^1S_0. However, atoms in the 2^3S_1 state are excited to the 2^3P states from which they return to 2^3S_1. Hence the effect of turning on the lamp is to effectively return all atoms from 2^1S_0 to the ground state, but to leave the population of 2^3S_1 unchanged. Thus, subtracting the two

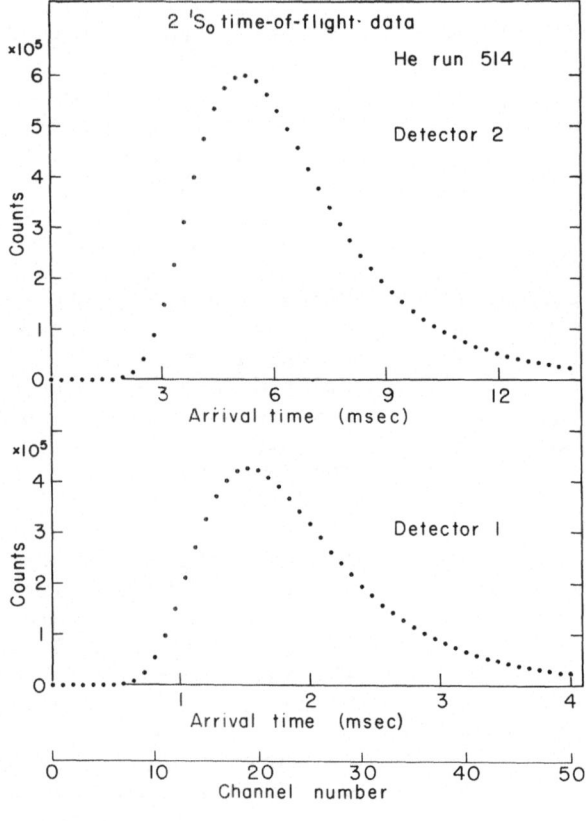

Figure 11

distributions with lamp off and lamp on gives a distribution which depends only on singlet atoms. Let

$$N_i(v) = \int_{surface} \varepsilon_i(v) \, \eta_o(v) \, e^{-t_i/\tau} \, dS \quad ,$$

where $N_i(v)$ is the number of counts registered at detector i due to 2^1S_0 atoms in the velocity interval between v and v + dv; the lifetime of atoms in 2^1S_0 and $\eta_o(v)$ is the initial velocity distribution. Assume that the efficiency ε_i is independent of v and $\eta_o(v)$ is uniform over the detector surface; then write:

$$R = \frac{N_2(v)}{N_1(v)} = \frac{\varepsilon_2 \eta_0(v) e^{-(t_2-t_1)/\tau}}{\varepsilon_1 \eta_0(v)} \quad .$$

Hence the slope of a semi-log plot of R will yield directly the lifetime.

Marrus and Schmieder[14] have taken advantage of the strong
Z dependence of the forbidden-decay transition rates to study all
four forbidden modes in hydrogenlike and heliumlike argon (Z = 18).
At Z = 18 the theoretical lifetimes of the 2^1S_0, 2^3P_2, and $2s_{\frac{1}{2}}$
are in the few-nanosecond range, while the lifetime of 2^3S_1 is
212 nsec. Prior to its recent modification, the old Berkeley
Hilac was capable of delivering argon beams with velocity of
4.4×10^9 cm/sec. This corresponds to very convenient decay lengths
for experimental study.

The experimental apparatus is shown in Fig. 12. Argon ions
from the Hilac are magnetically bent into the experimental apparatus. The ion beam is passed through a foil mounted on a movable
track, the role of the foil being to excite the one- and two-electron ions to the metastable states. The foil-excited beam then
passes in front of a pair of fixed Si(Li) detectors which are
sensitive to x-rays in the energy range >500 eV, with a resolution
of about 175 eV. The endpoint energy associated with the $2s_{\frac{1}{2}}-1s_{\frac{1}{2}}$
and $2^1S_0-1^1S_0$ two-photon transitions are 3.34 keV and 3.12 keV,
respectively. The energies associated with the M1 decay $2^3S_1-1^1S_0$
and the M2 decay $2^3P_2-1^1S_0$ are 3.104 keV and 3.126 keV, and they
are therefore not resolved by the detectors.

A two-field technique was developed for obtaining a beam that
was either predominantly one-electron or two-electron. The one-
electron beam was obtained by passing the Hilac beam through a foil
of about 100 μg/cm^2 thickness placed in front of the bending magnet.
About 25% of the beam emerging from this foil was fully stripped
argon, which was selectively bent by the steering magnet into our
apparatus pipe. This fully stripped beam was passed through a
thin foil (~5μg/cm^2) mounted on the movable track. It was found
that the ratio Ar^{+17}/Ar^{+16} emerging from this foil was $\gtrsim 18$.
Similarly, the two-electron beam was made by passing the Ar^{+14}
beam which emerged from the Hilac directly through the thin foil.
The ratio Ar^{+16}/Ar^{+17} was found to be ≈ 7.

A coincidence spectrum taken with the one-electron beam is
shown in Fig. 13. The time distribution of the arriving photons
shows a very pronounced peak at zero time delay with a width of
about 1 μsec. This width is in good agreement with the instrumental width and the peak is very strong evidence that real coincidences are being observed. The energy sum of those coincident
photons arriving within 1 μsec of each other is also in evidence.
These are seen to peak very strongly at 3.34 keV, which is the
theoretical $2s_{\frac{1}{2}}$ state energy. Also in evidence is the energy distribution of these photons. This distribution is consistent with
the theoretical distribution, although it is highly distorted by
the counter efficiency. This evidence indicates very strongly that
what is being observed is the two-photon decay of the $2s_{\frac{1}{2}}$ state of
Ar^{+17}.

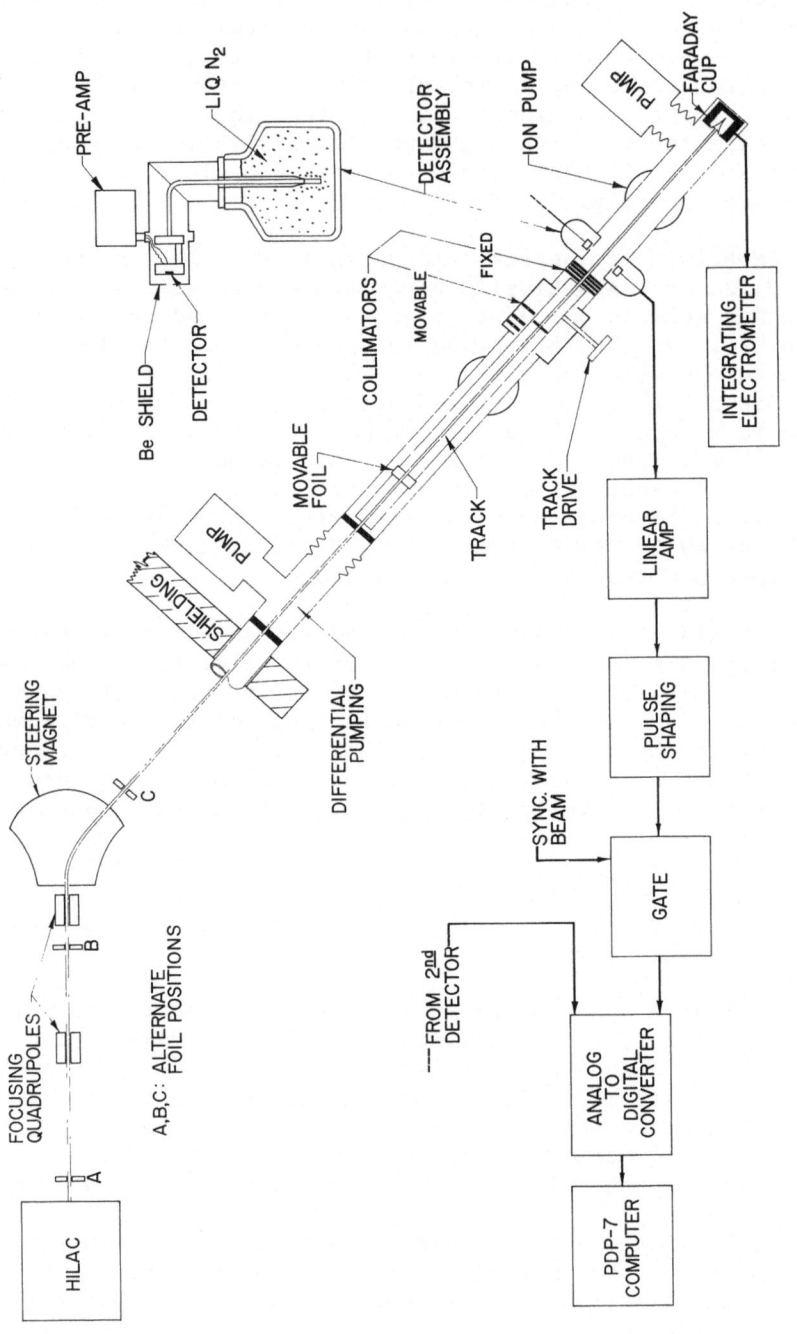

Figure 12

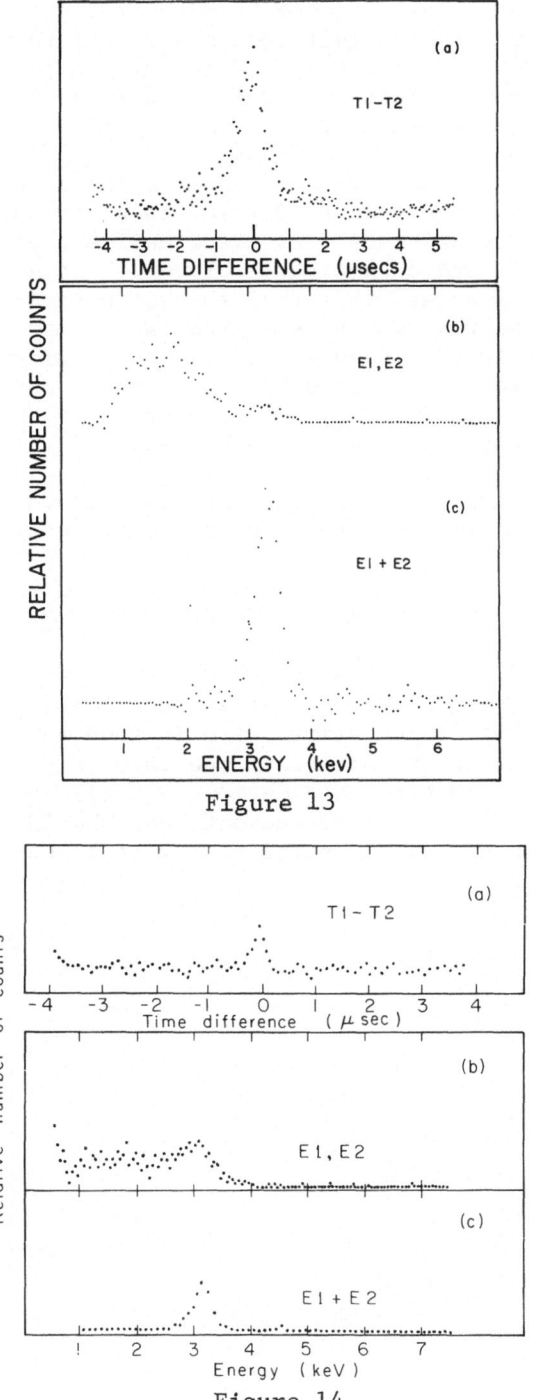

Figure 13

Figure 14

Similar data are shown for the two-electron beam in Fig. 14.
Here the results are similar (although the signal-to-noise is not
as good), with the difference that the energy sum peaks at 3.10 keV.
Hence these data confirm that the two-photon decay of the 2^1S_0 state
of Ar^{+16} is under observation.

A typical energy spectrum obtained from the detectors is shown
in Fig. 15. The line observed at 3.1 keV is a composite of counts
from the M2 and M1 transitions. As noted earlier, these are separated by only 22 eV, and cannot be resolved by the detectors. The
low-energy continuum arises from the two-photon decays, and the
distortions from the theoretical spectrum can be ascribed to detector efficiency. Lifetime measurements can be obtained by taking
such spectra at several foil-detector separations normalized to
the integrated beam current. By integrating under the curves, the
total number of normalized counts is obtained.

Lifetime data are shown in Figs. 16, 17, and 18. It is seen
that this experiment yields results on all four decay modes.

COMPARISON WITH THEORY

Tables I, II, and III compare the experimentally measured
rates with the theoretical rates. It is seen that, for the two-
photon decays, theory and experiment are in good agreement.
However, it is important to note that the theoretical rates are
all based on nonrelativistic calculations. Until the relativistic
and radiative corrections are worked out, one should not be too
sanguine about this agreement, particularly at higher Z.

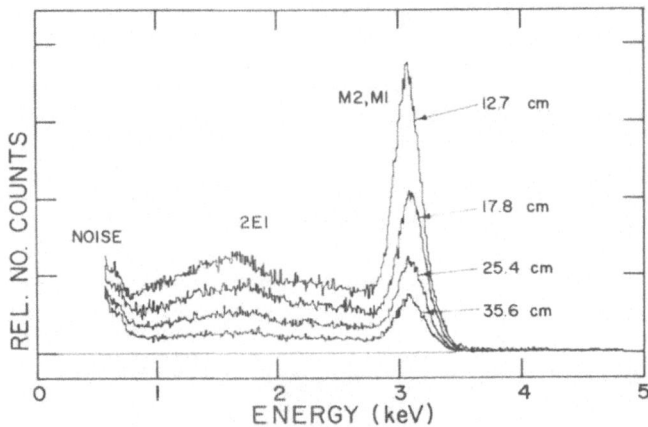

Figure 15

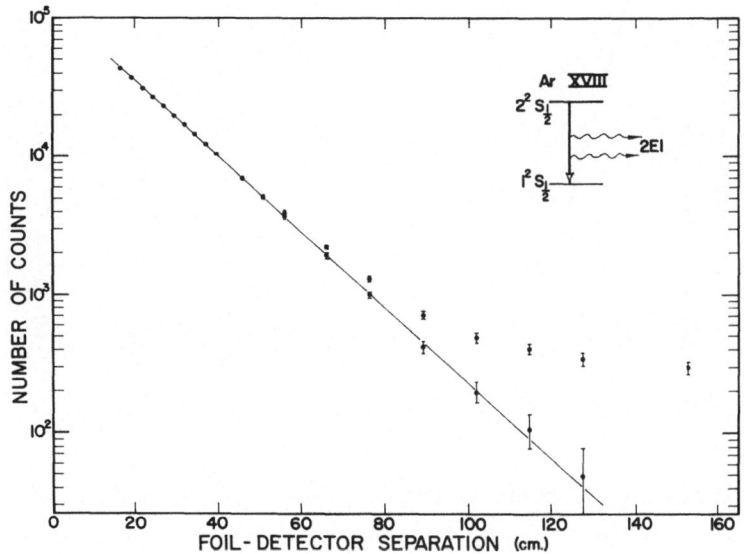

Figure 16

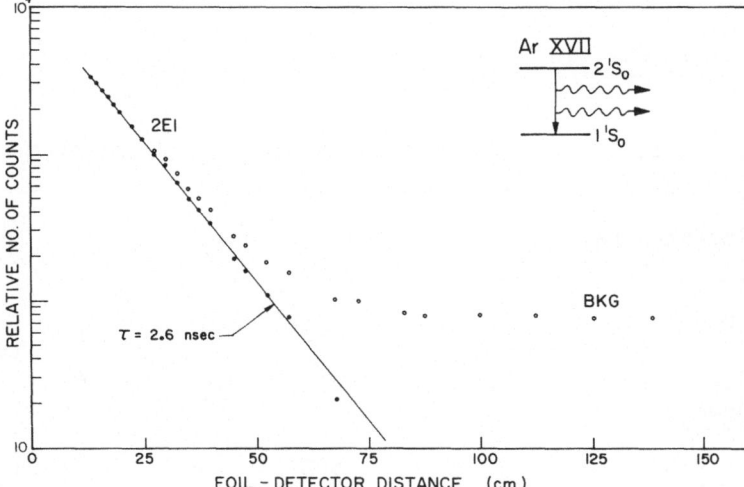

Figure 17

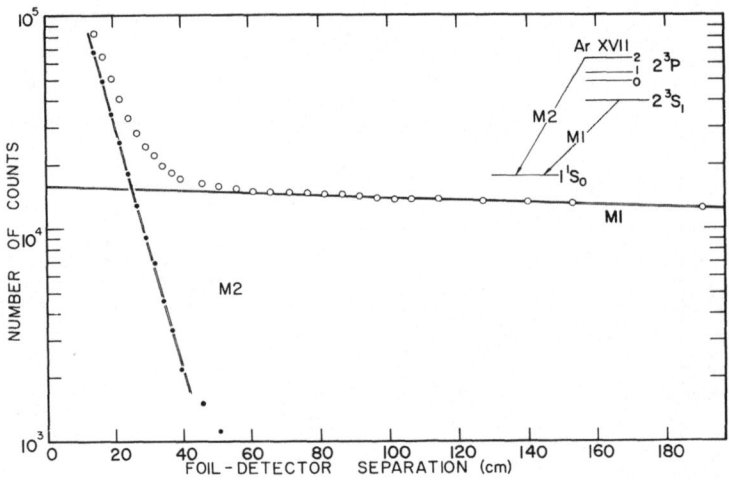

Figure 18

Table I. Experimental Results on $2^2S_{\frac{1}{2}}$ Lifetime

Researchers	Atom	Result	Theory
M. Prior	He II	1.95(9) msec	1.90 msec
Marrus & Schmieder	S XVI	7.3(7) nsec	7.11 nsec
Marrus & Schmieder	Ar XVIII	3.54(25) nsec	3.46 nsec

Table II. Experimental Results on 2^1S_0 Lifetime

Researchers	Atom	Result	Theory
A. S. Pearl	He I	38(8) msec	19.5 msec
Van Dyck, Johnson & Shugart	He I	19.7(1.0) msec	
Prior & Shugart	Li II	503(26) μsec	513 μsec
Marrus & Schmieder	Ar XVII	2.3(0.3) nsec	2.35 nsec

Table III.   Measured and Theoretical Lifetimes

Ion	State	τ_{expt} (nsec)	τ_{theor} (nsec)
Ar^{+17}	$2^2S_{\frac{1}{2}}$	3.54 ± 0.25	3.46
Ar^{+16}	2^1S_0	2.3 ± 0.3	2.35
	2^3S_1	172 ± 30	210
	2^3P_2	1.7 ± 0.3	1.49

A discrepancy appears to exist between the experimental and theoretical value of $\tau(2^3S_1)$ for Ar^{+16}. There is clearly need for further work here. Corrections to the existing theory might be of order 20%. Moreover, it should soon be possible to remeasure this lifetime in two-electron atoms of even higher Z. If there is indeed a discrepancy, then such experiments could establish the Z dependence.

References

*   Work supported by U. S. Atomic Energy Commission.

1.   G. W. F. Drake, preceding paper.

2.   L. Spitzer and J. L. Greenstein, Astrophys. J. 114, 407 (1951); G. A. Victor and A. Dalgarno, Phys. Rev. Letters 25, 1105 (1967).

3.   J. Shapiro and G. Breit, Phys. Rev. 113, 179 (1959); G. W. F. Drake, G. A. Victor and A. Dalgarno, Phys. Rev. 180, 25 (1969).

4.   F. Boehm, Phys. Rev. Letters 33A, 417 (1970) and references therein.

5.   G. Breit and E. Teller, Astrophys. J. 91, 215 (1940).

6.   G. W. F. Drake, Phys. Rev. A 3, 908 (1971); G. Feinberg and J. Sucher, Phys. Rev. Letters 26, 681 (1971).

7.   D. A. Landman, Bull. Am. Phys. Soc. 12, 94 (1967).

8.   R. H. Garstang, Publ. Astron. Soc. Pac. 81, 488 (1969); G. W. F. Drake, Astrophys. J. 158, 119 (1969).

9.   M. Lipeles, R. Novick and N. Tolk, Phys. Rev. Letters 15, 690 (1965).

10.   C. Kocher, Columbia University (private communication).

11.   M. H. Prior, to be published in Phys. Rev. Letters.

12.   M. H. Prior and H. A. Shugart, Phys. Rev. Letters 27, 902 (1971).

13.   R. S. Van Dyck, Jr., C. E. Johnson and H. A. Shugart, Phys. Rev. A 4, 1327 (1971).

14.   R. Marrus and R. W. Schmieder, Phys. Rev. A 5, 1160 (1972).

IONIZATION OF EXCITED ATOMIC PARTICLES BY ELECTRIC FIELDS

R. N. Il'in

A. F. Ioffe Physico-Technical Institute

Leningrad, U.S.S.R.

Electric field ionization of atoms is applied to investigation of surface phenomena (field-ion microscopy[1,2]), as well as to physical chemical analysis (field ionization mass-spectrometry[3,4]). Recently this phenomenon has been used in atomic physics itself, namely, to detect excited long-lived states of atoms and ions,[6] to determine the population of these states[5-7] and to measure the bound energy of electrons in these states.[6,8-10] We can speak of this method of investigating atomic particles as field ionization spectroscopy (FIS).

The mechanism of the electric field ionization of an atomic particle (an atom or an ion) can be illustrated by Figure 1. The external electric field F causes lowering of the potential barrier. As a result an electron from a certain appropriate state can

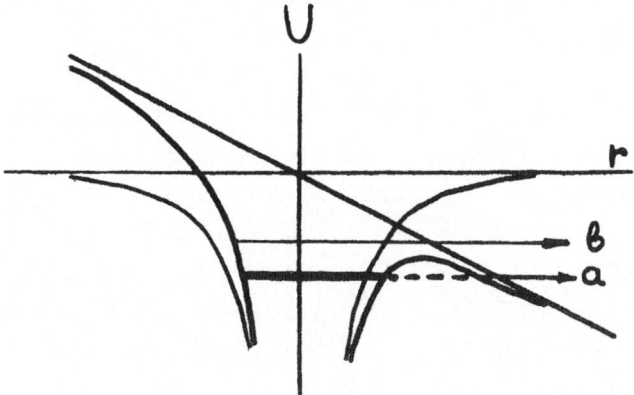

Figure 1. Potential barrier lowering in an electric field.

be transferred to the continuum either through (a) or over the barrier (b). The probability W of tunneling depends on the electric field intensity and the state characteristics of the atomic particle. Transitions caused by fast electric field changes will not be discussed here.

The field ionization of excited hydrogen atoms was first observed by Traubenberg[11] from H_γ - H_ϵ lines quenching. The quenching of the lines took place when the field ionization probability exceeded the de-excitation probability.

In the beginning of the sixties Riviere and Sweetman suggested a new method for the investigation of excited atoms and ions.[5] In this method atomic particles ionized by an electric field were detected directly. Recently a very interesting review concerned with field ionization technique has been written by Riviere[12] -- that is why we touch upon technique problems but briefly.

A schematic representation of the experimental apparatus used by Riviere and Sweetman[5] is shown in Figure 2. The beam of fast atoms or ions ordinarily formed by charge exchange passed through an electric field region with field intensity F. Then the beam was charge-analyzed and detected. The total attenuation of the beam in the electric field was measured as a function of the field intensity. The fraction of the beam transmitted is

$$I(F) = \frac{i(F)}{i_o} \tag{1}$$

where i_0 and i are the primary and attenuated fluxes of particles. The attenuation of the beam is related to the field ionization probability as

$$I(F) = \sum_k f_k \exp\{-tW_k(F)\} \tag{2}$$

where f_k is the fraction of particles in k-state, $W_k(F)$ is the

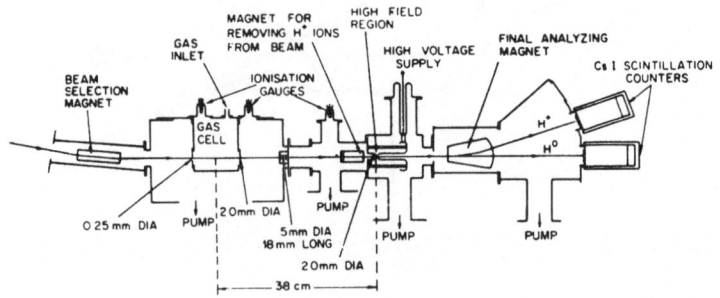

Figure 2. The experimental system used to study electric field ionization of excited hydrogen atoms.

field ionization probability for this state and t is the time of
flight of the particles through the field. Adding to the ionizing
field F a weak alternate field ΔF it is possible to detect by
means of synchronous detection the corresponding value of ΔI. The
differential dependence

$$D(F) = \frac{\Delta I(F)}{\Delta F} \simeq \frac{dI(F)}{dF} \tag{3}$$

is the field ionization spectrum.

The width and the position of the spectral line depends on the
time of flight of particles through the electric field, as is seen
from Figure 3. This figure shows also the connection between
W(F), I(F) and D(F).

The spectrum of excited hydrogen atoms with the principal
quantum number n ≥ 9 obtained by Riviere and Sweetman[13] is given
in Figure 4. The lines of the spectrum correspond to the states
with different n.

At present for FIS electric fields $F \leq 5 \times 10^5$V/cm are
used. This field can ionize particles with electron binding energy

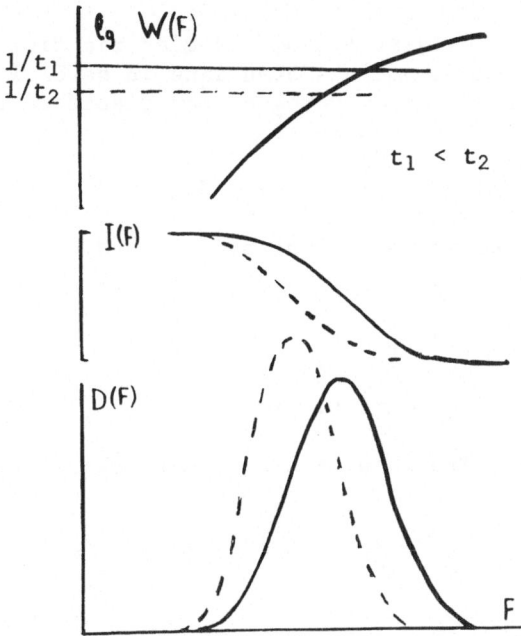

Figure 3. The probability of ionization W(F) and field spectral
line D(f) for flight times t_1 (solid curves) and t_2 (dashed
curves), $t_1 < t_2$.

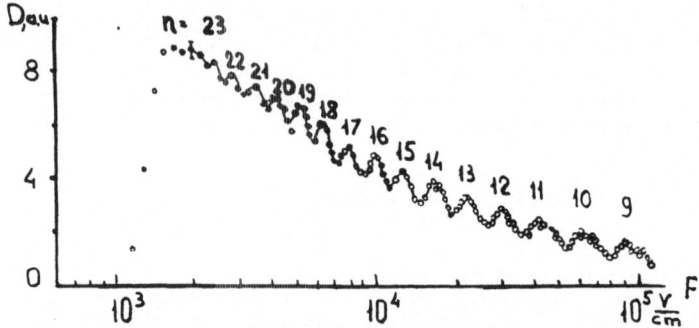

Figure 4. Electrical spectrum of excited hydrogen atoms.[13]

$\epsilon \leq 0.1$ eV. The particles having such value of ϵ may be excited atoms with $n \geq 10$ or excited negative ions. The difference between these two species is as follows. The atoms have many levels with energies well known from optical spectroscopy but the FIS lines are weakly resolved. The negative ions have very few long-lived excited states and it is difficult to measure the binding energy of electrons in these ions by known methods of electron affinity measurements. Because of these differences and the differences in application of the FIS method we shall consider these two types of particles separately.

The theoretical consideration of the electric field ionization of excited hydrogen atoms has been done in several works.[14-18] The Schrödinger equation for a hydrogen atom placed in the electric field F applied in z direction is

$$\left(\tfrac{1}{2}\nabla^2 + \epsilon + \frac{1}{r} - Fz\right)\Psi = 0 \quad . \tag{4}$$

The variables can be separated in parabolic coordinates

$$\xi = r + z \tag{5}$$

$$\eta = r - z \tag{6}$$

$$\phi = \text{arc tg } y/x \quad . \tag{7}$$

After appropriate substitutions we have the system of equations[18]

$$\frac{d^2\chi_1}{d\xi^2} + \Phi_1(\xi)\,\chi_1 = 0 \qquad \Phi_1(\xi) = -\frac{1}{4}\gamma^2 + \frac{\beta_1}{\xi} - \frac{m^2-1}{4\xi^2} - \frac{1}{4}F\xi \tag{8}$$

$$\frac{d^2\chi_2}{d\eta^2} + \Phi_2(\eta)\,\chi_2 = 0 \qquad \Phi_2(\eta) = -\frac{1}{4}\gamma^2 + \frac{\beta_2}{\eta} - \frac{m^2-1}{4\eta^2} + \frac{1}{4}F\eta \tag{9}$$

where

$$\gamma = \sqrt{-2\varepsilon} \tag{10}$$

$$\Psi = (\chi_1(\xi)/\sqrt{\xi}) \cdot (\chi_2(\eta)/\sqrt{\eta}) \cdot e^{im\phi} \tag{11}$$

m is the magnetic quantum number and β_1 and β_2 are the separation parameters.

The electric field causes splitting of the state of a given n into n^2 Stark states with energies[18]

$$\varepsilon_{n,n_1,n_2} = -\frac{1}{2}\frac{1}{n^2} + \frac{3}{2} Fn(n_1-n_2) - \frac{1}{16} F^2 n^4 [17n^2 - 3(n_1-n_2)^2 - 9m^2 + 19] \tag{12}$$

where n_1 and n_2 are the parabolic quantum numbers so that

$$n_1 + n_2 + m + 1 = n \tag{13}$$

$$n_1 = \beta_1/\gamma + (m+1)/2 \quad . \tag{14}$$

The η coordinate increases in the anode direction. The state which corresponds to the electron orbit stretched in this direction i.e., with the highest n_2 and the lowest n_1 has the highest probability of ionization. One can consider $\Phi_2 - \varepsilon/2$ as the potential energy $V_\eta(\eta)$ at the point η.[17,18] The $V_\eta(\eta)$ dependence is given in Figure 5. For a state with m = 1 the barrier disappears at

$$F_o = \frac{\varepsilon^2}{4\beta_2} \quad . \tag{15}$$

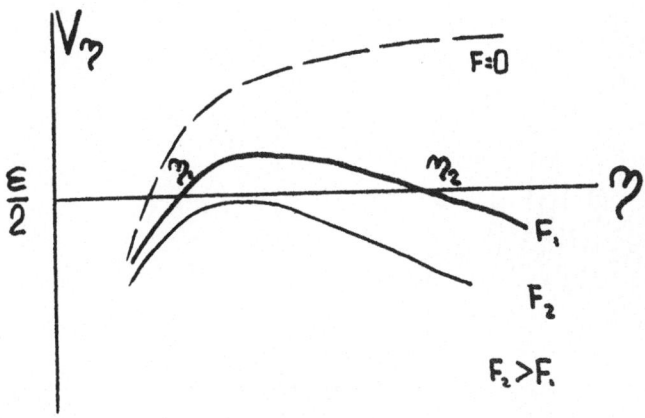

Figure 5. Potential energy $V_\eta(\eta)$ for different electric field values.

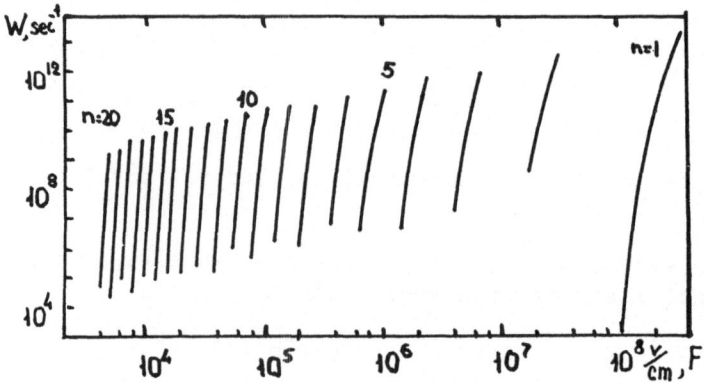

Figure 6. Electric field ionization probability for the hydrogen atoms with $1 \leq n \leq 20$.

In most works mentioned above the WKB approximation was used to obtain the probability $W(F)$. The computations useful for experimental purposes were done by Bailey, Hiskes and Riviere.[16] They computed $W(F)$ for hydrogen atoms with $1 \leq n \leq 20$ for central Stark substates (Figure 6) and components with $n_1 = n - 1$ and $n_2 = n - 1$ and in some cases for all components. Using these data Riviere[20] computed FIS line shapes for each of 64 Stark substates with $n = 8$, the line envelope for the state with $n = 8$ assuming a statistical distribution among the substates (Figure 7) and the shape of FIS spectrum for $8 \leq n \leq 14$ assuming the distribution

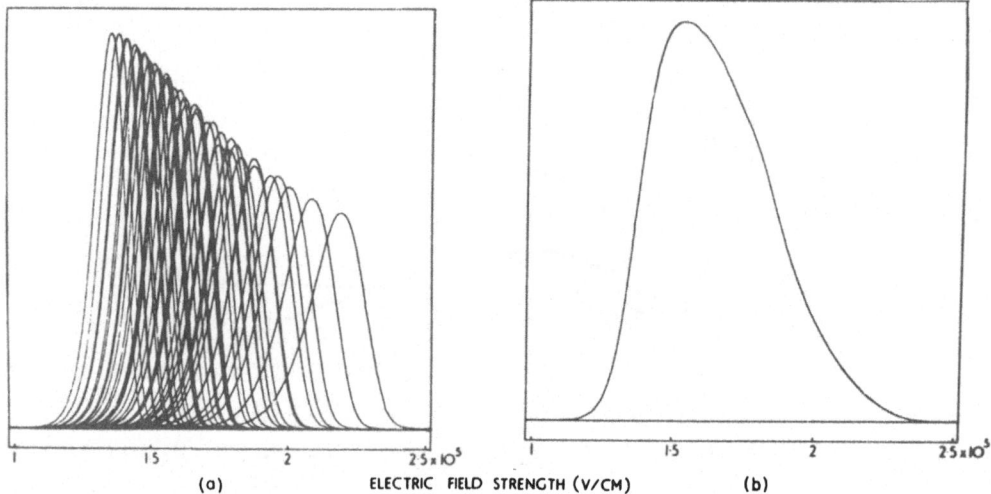

Figure 7. Electrical field spectrum lines for substates (a) and whole level (b) with $n = 8$.[20]

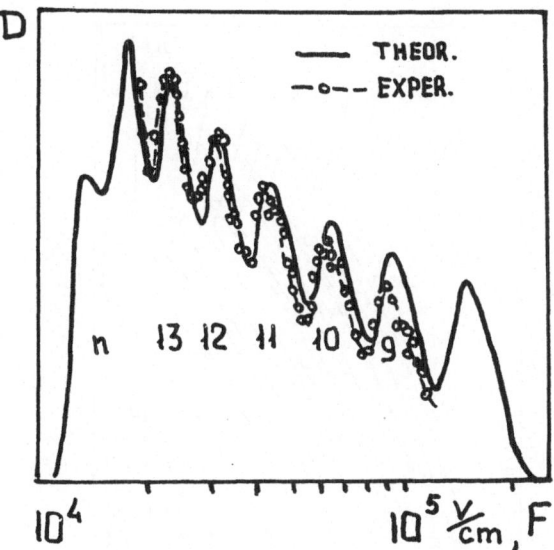

Figure 8. The calculated and experimental field ionization spectra.[20]

of states proportional to n^{-3} (Figure 8). In the latter figure an experimental FIS spectrum is also shown. There exists a satisfactory agreement between the theoretical and experimental results. Some discrepancy may be associated with the following circumstance. The WKB approximation is valid if the barrier is sufficiently wide and high. When the barrier is small and $W(F)$ is large and hence the level has some width it is necessary to consider the width of the level in computations. This was done by Hirschfelder and Curtiss.[17] The comparison of these computations with computations in the WKB-approximation[16] for a hydrogen atom with $n = 5n_1 = 3$ and $n_2 = 0$ is given in Figure 9. One can see that the width of the level is important for $W \geq 10^{11}$ sec^{-1}.

FIS spectra of excited hydrogen atoms were obtained by some other authors, too.[7,21-23] We have also obtained the FIS spectrum of excited helium atoms.[24] The comparison between H and He FIS spectra is given in Figure 10. The lines in the He spectrum are narrower than in H. This difference can be explained by the fact that the hydrogen levels are degenerate and a linear Stark-effect takes place at very small F so that the sublevels are mixed. The helium states, on the other hand, are not degenerate and in weak fields there occurs a square Stark effect which in case of very strong fields becomes a linear one,[25] with no mixing of states taking place. The helium excited atoms were formed by charge-exchange and in this case according to Hiskes[26] the s- and p-sub-states are mostly populated. These substates according to Foster[25] are equivalent to Stark levels with $n_1 = 0$ and $n_2 = n - 1$.

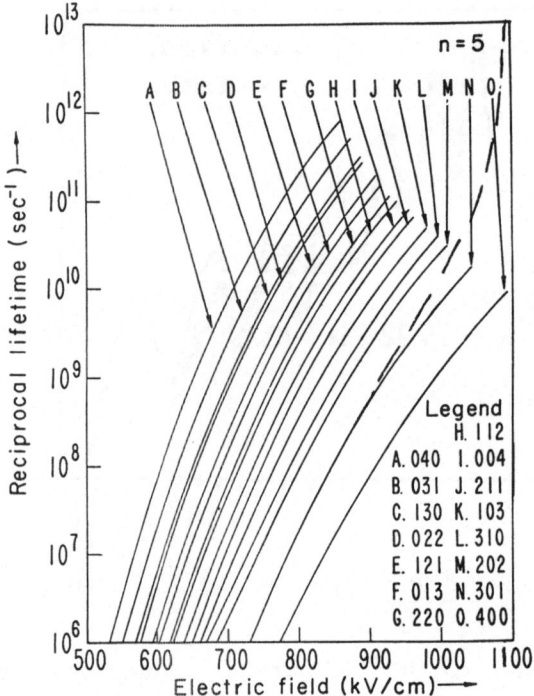

Figure 9. Probabilities of the electric field ionization of substates for n = 5. Solid curves from Ref. 16, dashed curve from
Ref. 17.

In the case of hydrogen atoms the FIS method was applied for
detection of highly excited states and measurement of their population.[5,7] It is convenient to obtain the state population from the
beam attenuation I(F), defined by Eqs. (1) and (2). The distribution of the state populations in the case of charge exchange was
measured in the following way. According to theory[27] the cross section $\sigma_c{}^n$ for electron capture into a state with a sufficiently
large n is proportional to n^{-3}, i.e.,

$$\sigma_c{}^n \Big/ \sum_{n=1}^{\infty} \sigma_c{}^n = a/n^3 \qquad (16)$$

where the sum $\Sigma\sigma_c{}^n$ is the total cross section for an electron
capture and a is characteristic of the state population. It is
easy to show experimentally that for each n the value of the electric field ionizing this state F_n is proportional to n^{-4}

$$F_n = c/n^4 \qquad (17)$$

where c depends on the flight time and the field distribution.

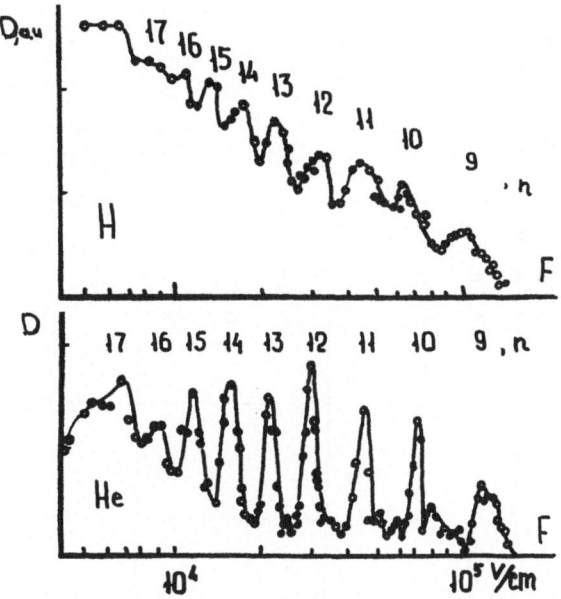

Figure 10. Field ionization spectra for H and He.[24]

Since the lines in the hydrogen FIS spectrum are overlapped we can consider n a continuous variable[7] for the further estimation. Then

$$\frac{dI}{dn} = -\frac{a}{n^3} .$$ (18)

Substituting in (18) F for n in accordance with (17) we can obtain

$$\frac{dI}{dF} = \frac{a}{4\sqrt{c}} \cdot \frac{1}{\sqrt{F}}$$ (19)

$$I(F) = \frac{a}{2\sqrt{c}} \sqrt{F} .$$ (20

The experimental I(F) dependences agree with (20) very well. Measuring separately a and the total charge exchange cross section we can find σ_c^n. The dependence of the σ_c^n on the velocity of H^+ and He^+ ions[28] is given in Figure 11. The works concerned with these measurements were reviewed by Il'in.[28]

The method of FIS may be applied to investigation of long-lived negative ions. Riviere and Sweetman showed that the helium negative ions in the $1s2s2p^4P$ autoionizing state could be ionized by an electric field $F \sim 10^5$ V/cm. Theoretical considerations[31] showed that negative ions of some elements of the III, IV and V groups may have excited states with the same electronic configuration as their ground states. These states are given in Table I.[30]

Table I

Configuration	State	Ion	Excitation energy	Electron affinity	Ion	Excitation energy	Electron affinity
np^2	^1S	B^-	0.99			0.94	
	^1D		0.52			0.39	
	^3P		0	0.33	Al^-	0	0.52
np^3	^2P	C^-	1.46			1.44	
	^2D		1.29			0.88	
	^4S		0	1.24	Si^-	0	1.46
np^4	^1S	N^-	2.6			1.94	
	^1D		1.28			0.82	
	^3P		0	0.05	P^-	0	0.77

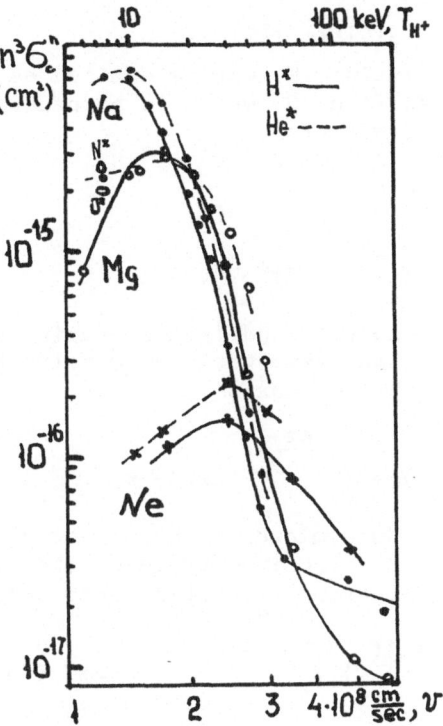

Figure 11. Cross sections for the capture of electrons into excited states of H^+ (solid curves) and He^+ (dashed curves) versus their velocity v. Targets are shown in the figure.

Some experimental results[31,32] suggest the existence of excited C^- ions, but the applied methods of electron affinity determination were not convenient for studying such ions.

Drukarev and Demkov[8] have shown that field ionization can be used for determination of electron binding energy in such ions, but it is essential to know the dependence $W(\varepsilon)$ and to treat the results in terms of a certain model of the ionization process.

Theoretical considerations of field ionization of negative ions were performed for H^- by Darewych et al.,[33] Cahill et al.,[34] and Kaplan et al.[35] and for ions with s-electrons by Demkov et al.[8] Smirnov and Chibisov[9] carried out calculations the results of which could be applied to the ions with p-electrons. They used the one-electron approximation and assumed that the field F was not strong so that there were distances from the nucleus where

$$|\varepsilon| \gg |U(r)| \qquad\qquad (21)$$

$$|\varepsilon| \gg |Fz| \quad . \tag{22}$$

The probability of ionization W was determined as an electron current j through a surface S which is perpendicular to the direction of the electric field F

$$W = \int_S j dS \tag{23}$$

$$j = \frac{1}{2} (\Psi \nabla \Psi^* - \Psi^* \nabla \Psi) \quad . \tag{24}$$

The wave function was obtained from comparing the solutions of Eqs. (8) and (9) with the asymptotic expression for the unperturbed wave function near the z-axis

$$\psi(r) = A r^{\frac{Z_{ef}}{\gamma} - 1} \exp(-\gamma r) \tag{25}$$

where A is a constant determined by the electron behavior inside the atom and Z_{ef} is the effective charge of the atom frame. Then the probability of the electron transition into the continuum is

$$W = \frac{A(2\ell+1)! m! (\ell+m)!}{2\gamma^m (\ell-m)!} \left(\frac{2\gamma^2}{F}\right)^{\frac{2Z_{ef}}{\gamma} - m - 1} \exp\left(-\frac{2}{3} \frac{\gamma^3}{F}\right) \quad . \tag{26}$$

The probabilities of ionization of p-electrons with quantum numbers $m = 0$ or 1 are

$$W_0 = \frac{3}{4} A^2 \frac{F}{\gamma^2} \exp\left(-\frac{2}{3} \frac{\gamma^3}{F}\right) \tag{27}$$

$$W_1 = \frac{3}{4} A^2 \frac{F^2}{\gamma^5} \exp\left(-\frac{2}{3} \frac{\gamma^3}{F}\right) \quad . \tag{28}$$

We undertook an investigation with the aim of searching for metastable negative ions and determining the binding energies of electrons in these ions.[6,10,36] H^-, He^-, C^-, O^-, Al^-, Si^- and P^- ions formed by charge exchange of positive ions with energy 100 keV in nitrogen were investigated. He^-, C^-, Al^-, and Si^- ions were ionized by electric field $F \leq 5 \times 10$ V/cm.

I(F) dependences obtained in this work are given in Figure 12. In accordance with Table I we believe the ^2D state of C^-, ^2P state of Si^- and ^1D state of Al^- to be ionized. It is interesting to note that the value of the ionized fraction of the negative ions is close to the relative statistical weight of the ionized states (Table II).

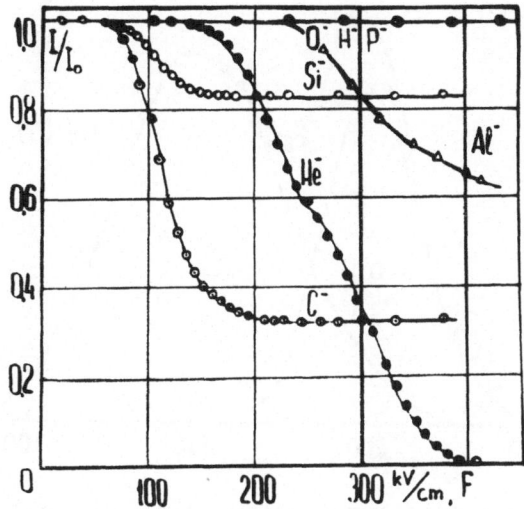

Figure 12. Beam attenuation in the electric field for different negative ions.[36]

Table II

Ion	Ionized state	Nonionzed state	Relative statistical weight of ionized state	Ionized fraction of the negative ion beam
C^-	^2D	^4S	0.71	0.75
Si^-	^2P	$^2D, {}^4S$	0.30	0.2
$A\ell^-$	^1D	^3P	0.36	0.35

The field ionization spectra $D(F)$ for these ions also were measured. The spectra were essential for detection of the sub-states. It was also convenient to use $D(F)$ for obtaining the electron binding energy value.

The spectrum of the ^4P state of the He^- ion is given in Figure 13. It contains two lines. This is easily comprehended if it is considered that the removed p-electron may be in a sub-state with $m = 0$ or 1. Then the attenuation of the ion beam in the field is

$$I(F) = f \exp(-W_0 t) + (1-f)\exp(-W_1 t) \qquad (29)$$

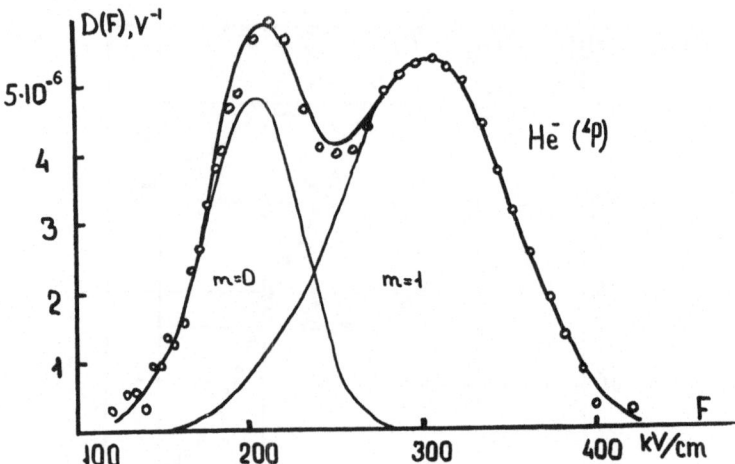

Figure 13.   Field ionization spectrum for He⁻ ion in ^4P state.[10]

where f is the fraction of the ions in the state with m = 0 and W_0 and W_1 are determined by (27) and (28), or else

$$W_0 = 8.2 \cdot 10^7 \alpha_0 F \exp(-6.83 \cdot 10^7 \frac{|\varepsilon|^{3/2}}{F}) \qquad (30)$$

$$W_1 = 0.80 \; \alpha_1 F^2 |\varepsilon|^{-3/2} \exp(-6.83 \cdot 10^7 \frac{|\varepsilon|^{3/2}}{F}) \quad . \qquad (31)$$

Equations (30) and (31) are obtained from (27) and (28) if ε is expressed in eV and F in V/cm.

The computer was used for fitting the experimental values of (F) to the ones calculated by (3),(29),(30),(31). The best fit was obtained at f = 0.335, α_0 = 6.4 × 10^7, α = 0.28, $|\varepsilon_0| = |\varepsilon_1| =$ 0.075 ± 0.004 eV. The value of f is the same as the statistical weight of the substate with m = 0.

It is interesting to note that for autoionizing states there is another mechanism of the field ionization. Khomskii[37] showed that the metastable $2p^2 \; ^3P$ state of the helium atom can be mixed with the unstable $2s2p \; ^3P^0$ state in an electric field F. In such cases the probability of transitions is proportional to F^2. It is not consistent with **our** experimental data for He⁻ ^4P.

The spectrum of the C⁻ ion in the ^2D state is given in Figure 14. The spectral line has an unsymmetrical shape and consists of two or more lines. A very good fit is obtained if the $|\varepsilon|$ values for both states are equal ($|\varepsilon|$ = 0.035 ± 0.004 eV) but α values are different.

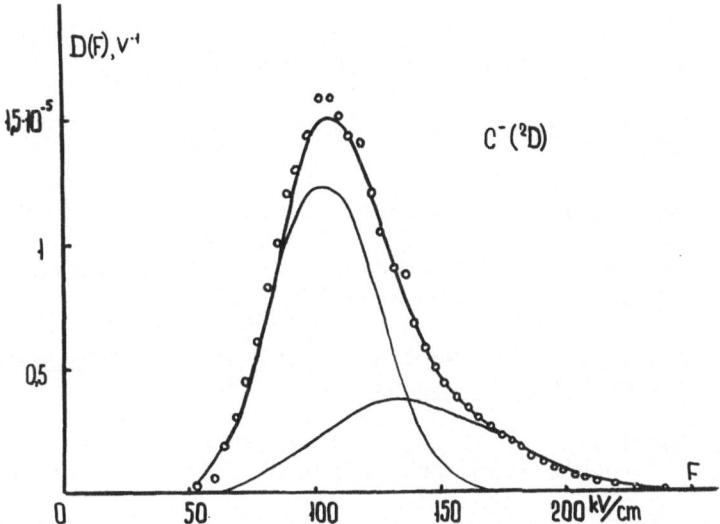

Figure 14. Field ionization spectrum for C^- ion in ^2D state.[6]

The estimations of $|\varepsilon|$ values for Si^- and $A\ell^-$ ions gave $|\varepsilon| \simeq 0.037$ eV for Si^- and $|\varepsilon| \simeq 0.1$ eV for $A\ell^-$.

In conclusion some aspects of the further development of the FIS method should be noted. It is necessary to obtain very strong electric fields with well-known geometry and value in an appropriate volume. It is very desirable to perform precise theoretical calculations of the field ionization taking into account the Stark effect. The FIS method can be applied to electron affinity measurements. Field values for ionization of negative ions with different $|\varepsilon|$ are given in Figure 15.[8]

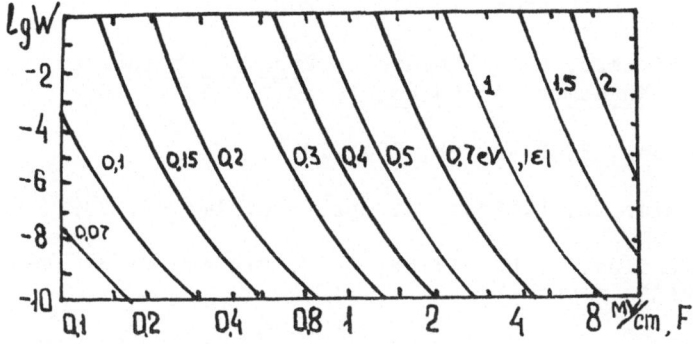

Figure 15. Field ionization probability W vs field strength F for different electron affinities ε.[8]

References

1. E. W. Müller, Z. Phys. 131, 136 (1951).

2. Field-Ion Microscopy. Ed. D. J. Hren, S. Ranganathan (New York 1968).

3. H. D. Beckey, J. Mass. Spectr. Ion Phys. 2, 500 (1969).

4. I. V. Goldenfeld, R. N. Bpndarenko, V. G. Golovaty, Prib. Techn. Eksper., N. 3, 203 (1970).

5. A. C. Riviere, D. R. Sweetman, Proc. 3-rd Intern. Conf. Phys. Electron. Atomic Collisions. London 1963, p. 734.

6. V. A. Oparin, R. N. Il'in, I. T. Serenkov, E. S. Solov'ev, N. V. Fedorenko, Pisma Zh. Eksperim. Teor. Fiz. 13, 351 (1971).

7. R. N. Il'in, B. I. Kikiani, E. S. Solov'ev, V. A. Oparin, N. V. Fedorenko, Zh. Eksperim. i Teor. Fiz. 47, 1235 (1964).

8. Ju. N. Demkov, G. F. Drukarev, Zh. Eksper. i Teor. Fiz. 47, 918 (1964).

9. B. M. Smirnov, M. I. Chibisov, Zh. Eksper. i Teor. Fiz. 49, 841 (1965).

10. V. A. Oparin, R. N. Il'in, I. T. Serenkov. E. S. Solov'ev, Pisma Zh. Eksper. i Teor. Fiz. 12, 237 (1970).

11. R. V. Traubenberg, R. Gebauer, G. Lewin, Naturwiss. 18, 417 (1930).

12. A. C. Riviere, in Methods of Nuclear Physics (Academic Press, New York and London, 1968), p. 208.

13. A. C. Riviere, D. R. Sweetman, Proc. 6-th Intern. Conf. Ionization Phenomena in Gases, Paris, 1963, p. 105.

14. C. Lanczos, Z. Phys. 68, 204 (1931).

15. M. H. Rice, R. H. Good, J. Opt. Soc. Am. 52, 239 (1962).

16. D. S. Bailey, J. R. Hiskes, A. C. Riviere, Nucl. Fusion 5, 41 (1965).

17. J. O. Hirschfelder, L. A. Curtiss, J. Chem. Phys. 53, 1395 (1971).

18. H. A. Bethe, E. E. Salpeter, Quantum Mechanics of One- and Two-Electron Atoms Berlin, 1957.

19. D. Bohm, Quantum Theory (Prentice Hall, New York, 1952) Ch. 12.

20. A. C. Riviere, At. Energy Res. Establ. (Gr. Brit.) Report 4818 (1964).

21. A. C. Futch, C. C. Damm, Nucl. Fusion, 3, 124 (1963).

22. R. Le Doucen, J. Guidini, Abstr. of Papers of 6-th Intern. Conf. Phys. Electron Atomic Collisions. Boston 1969, p. 456.

23. G. A. Khayrallah, R. Karn, P. M. Koch, J. E. Bayfield, Abstr. of Papers of 7-th Intern. Conf. Phys. Electron Atomic Collisions. North-Holland, Amsterdam, 1971, p. 813.

24. R. N. Il'in, V. A. Oparin, I. T. Serenkov, E. S. Solov'ev, N. V. Fedorenko, Zh. Eksperim. i Teor. Fiz. 59, 103 (1970).

25. J. S. Foster, Proc. Roy. Soc. (London), 117, 137 (1928).

26. J. R. Hiskes, UCRL-50602 (1969).

27. J. R. Oppenheimer, Phys. Rev. 31, 66 (1928).

28. R. N. Il'in, in Proc. of Intern. Summer School on Physics of Ioniz. Gas. Ljubljana, 1970, p. 113.

29. A. C. Riviere, D. R. Sweetman, Phys. Rev. Letters 5, 560 (1961).

30. D. R. Bates, B. L. Moiseiwitsch, Proc. Phys. Soc. A68, 540 (1955).

31. M. L. Seman, L. M. Branscomb, Phys. Rev. 125, 1602 (1962).

32. J. F. Paulson, J. Chem. Phys. 52, 5491 (1970).

33. G. Darewych, S. M. Neamwon, Nucl. Instr. Meth. 21, 247 (1963).

34. T. A. Cahill, J. Richardson, J. W. Verba, Nucl. Instr. Meth. 39, 278 (1966).

35. S. N. Kaplan, G. A. Paulikas, R. V. Pyle, Phys. Rev. 131, 2547 (1963).

36. V. A. Oparin, R. N. Il'in. I. T Serenkov, E. S. Solov'ev,
     N. V. Fedorenko, Abstr. of Papers of 7-th Intern. Conf. Phys.
     Electron. Atomic Collisions, North Holland, Amsterdam, 1971,
     p. 796.

METASTABLE STATES OF HIGHLY EXCITED HEAVY IONS[*]

D.J. Pegg, P.M. Griffin, and I.A. Sellin
University of Tennessee, Knoxville, Tennessee 37916 and
Oak Ridge National Laboratory, Oak Ridge, Tennessee 37830

Winthrop W. Smith
University of Connecticut, Storrs, Connecticut 06268

Bailey Donnally
Lake Forest College, Lake Forest, Illinois 60045

INTRODUCTION

Highly stripped heavy ions (i.e. systems with high nuclear charge but a small number of electrons) are of interest from several viewpoints. One reason is that the relativistic magnetic interactions such as the spin-orbit, spin-other-orbit and spin-spin interactions are considerably stronger in these ions than in nearly neutral isoelectronic ions. This situation sometimes allows "forbidden" processes to be experimentally observable, even though the rates for these processes are still very small compared to those for "allowed" processes. A number of recent experiments[1] involving the radiative decay of metastable states of highly stripped ions have advanced our knowledge of atomic structure through a comparison of radiative decay rate measurements with theory. It is also possible to study metastable states in simple heavy ions which do not decay radiatively, but by the autoionization processes instead. Our recent work[2] concerns the study of such states.

Metastable autoionizing states are those which do not auto-ionize via the inter-electron Coulomb repulsion but rather by the

*Research Sponsored in part by the Office of Naval Research, by NASA, and by Union Carbide Corporation and the Oak Ridge Associated Universities under contract with the U.S. Atomic Energy Commission

weaker magnetic interactions. Feldman and Novick[3] showed that
such states were formed in the neutral alkali atoms by a core-
excitation process in which one (or possibly more than one)
electron is excited from previously closed inner shells into
outer orbitals. For example, in lithium and lithiumlike ions
the configuration 1s2s2p will produce two doublets and a quartet
term, the latter being metastable against Coulomb autoionization.
The autoionization process involved in this case consists of
the ejection of one electron into the adjacent continuum while
another electron fills the k-hole. The continuum electron will
have a kinetic energy that is equal to the difference between the
binding energy of the initial state and that of the residual
two-electron ion. The metastability of the initial state implies
that the energy of the emitted electron will be sharp to a very
small fraction of an eV. The selection rules for autoionization
(based upon L-S coupling) are shown in Figure 1. It can be seen

AUTOIONIZATION SELECTION RULES
(LS COUPLING)

	ΔL	ΔS	ΔJ	PARITY
COULOMB	0	0	0	CONSERVED
SPIN-ORBIT	$0, \pm 1$	$0, \pm 1$	0	CONSERVED
SPIN-SPIN	$0, \pm 1, \pm 2$	$0, \pm 1, \pm 2$	0	CONSERVED

EXAMPLE: $(1s\ 2s\ 2p)\ ^4P^o_{5/2} \longrightarrow (1s^2)\ ^1S_0 + k\ ^2F^o_{5/2}$

TO CONSERVE PARITY, L MUST BE ODD

TO CONSERVE J, L = 2, 3

HENCE, $k\ ^2F^o_{5/2}$ IS THE FINAL STATE ;
ONLY THE SPIN-SPIN INTERACTION IS EFFECTIVE

Fig. 1. Autoionization selection rules based upon L-S coupling.
The mode of the spin-spin induced decay of the $(1s2s2p)^4P^o_{5/2}$
is shown.

that a quartet level such as $(1s2s2p)^4P^o_{5/2}$ cannot autoionize via
the Coulomb interaction because of the selection rules on spin and
total angular momentum. This quartet level can however relax via
a spin-spin induced autoionization process, as indicated in the
figure. States other than the lowest lying state of a given spin
system may also radiatively decay to other states of the same multiplicity, since spin is conserved in the process. The lifetimes
of such states will then be determined by the rates for both autoionization and radiation. We turn now to the method we have used
to study these states in highly stripped alkali-like ions.

METHOD

It is known that highly stripped ions are commonly produced by foil-excitation of high energy ion beams. Beams of oxygen, fluorine, chlorine and argon (in the energy range of 2-90 MeV) were accelerated at either the Oak Ridge tandem accelerator or the Oak Ridge Isochronous cyclotron and passed through thin carbon foils (∼15 µg/cm^2). The foil targets serve to both strip and excite the ions of the beam. The beam energies were chosen to maximize the production of the charge state of interest in a particular experiment. The apparatus has been previously described[2] in detail and is shown schematically in Figure 2. Electrons emitted from the foil-excited beam during autoionizing events are collected and energy analyzed using a cylindrical mirror analyzer. Decay studies were made by the usual time-of-flight process in which the relative position of the foil and the spectrometer viewing region were varied and the count rate per unit beam current was monitored as a function of position.

RESULTS

Figures 3 and 4 show autoionization electron spectra obtained using 6 MeV oxygen and fluorine ion beams incident upon

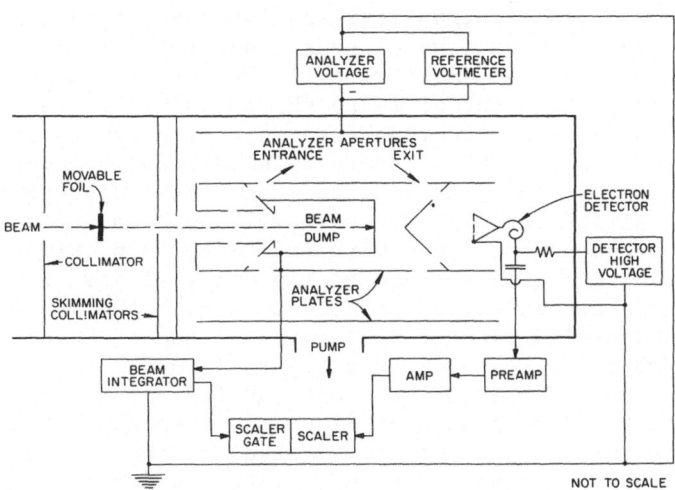

Fig. 2. Diagram of the apparatus. The cylindrical mirror analyzer is shown schematically in a plane containing the beam axis.

a carbon foil target. The vertical scale of Figure 3 is ten times larger than that of Figure 4. Most of the observed features can be identified with quartet states of lithiumlike oxygen or fluorine ions. A number of these states of configurations $1s2sns(n \geq 3)$; $1s2pnp(n \geq 2)$ and $1snsn'p(n, n' \geq 2)$ have been identified including states with $n \geq 5$ all the way up to the series limit established by the two electron ion terms 2^3S^e or 2^3P^0, at which energy there appears a drop-off in intensity. The lowest lying peak is associated with the $1s2s2p$ configuration and the adjacent peak with the $1s2p^2$ configuration. It is also to be expected that states such as $^4D^0$ and $^4D^e$ associated with configurations $1s2snd(n \geq 3)$; $1s2pnp(n \geq 2)$ and $1s2pnd(n \geq 3)$ will be present but there are no known energy calculations for these states that can be used for identification. It should also be noted that states formed from configurations with $n \geq 3$ are well separated in energy from those formed from configurations with $n=2$. The line elements located above the spectra of Figures 3 and 4 indicate the theoretical energies obtained by Holøien and Geltman[4] using variational wave functions and it can be seen that the agreement with these values is, on the whole, very good. The notation used in the figures is also the same as that used by the latter authors. Peak B remains unidentified and could conceivably originate from adjacent metastable ions such as berylliumlike oxygen or fluorine. One possibility that is consistent with electron energetic arguments is that these states may have a configuration of the type $1s2s2pn\ell$ in which the loosely bound outer electron acts as a weakly interacting spectator. This may be true of peak A as well; although Bardsley and Junker[5] and more recently Berry[6] have suggested an alternative assignment of $1s2p^2\ ^2P$ for this feature. This state will not decay by Coulomb autoionization due to the selection rule on parity but there is some doubt however about this assignment since it is expected that the radiative transition probability out of this state will be high. The spectra of Figures 3 and 4 were taken with the target approximately 2 cm from the viewing region so that a time delay >1 nsec is involved. Figure 5 shows a spectrum of oxygen taken with the target directly in view and contains several features that should be pointed out. First, the vertical scale of this figure is much larger than that of Figures 3 or 4. The line element "a" indicates the approximate intensity of the $^4P^0(1)$ peak of Figure 3. Secondly, the peaks A and B can be seen to have grown relative to other features which may be attributed to the different relative lifetimes of the states. Another feature to observe in Figure 5 is the unresolved peaks that were not previously observed (some of these may have the very short lifetimes $\sim 10^{-14}$ sec associated with Coulomb autoionization). Some of the peaks might also again be attributed to spectator states associated with a weakly bound electron attached to three-electron configurations. Bardsley and Junker[5] have calculated the energies of configurations such as $1s2s2p^2$, $1s2p^3$ and

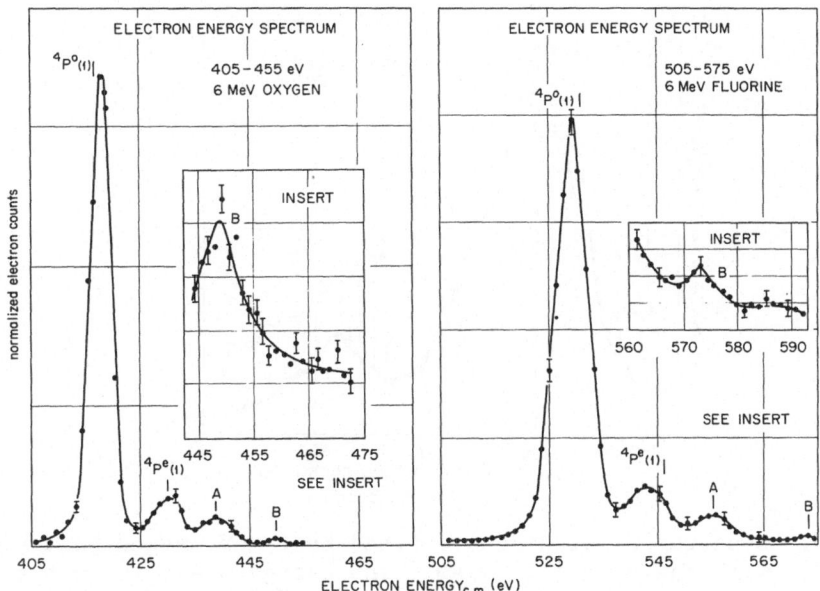

Fig. 3.   Spectra of electrons emitted by 6 MeV oxygen and fluorine
beams.   The features shown are those near the lowest three-
electron quartet state, in the ionic rest frame.

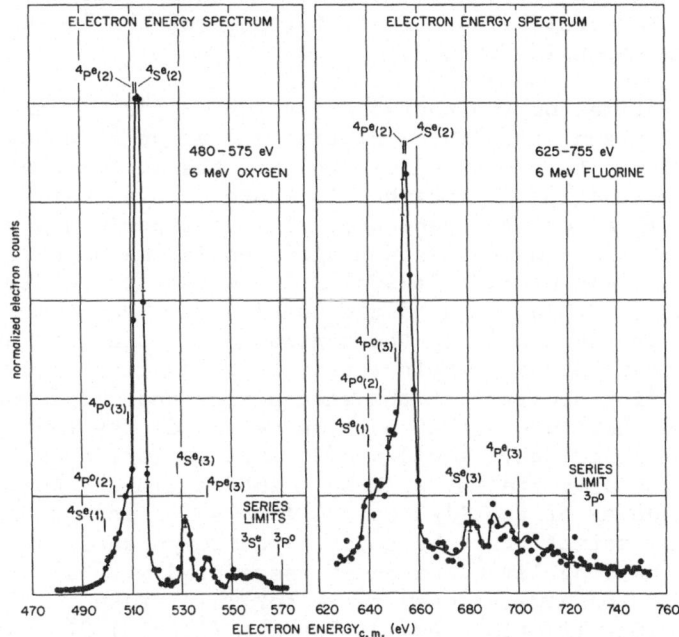

Fig. 4.   As in Figure 3, except the segment shown pertains to the
higher energy three-electron quartet states.

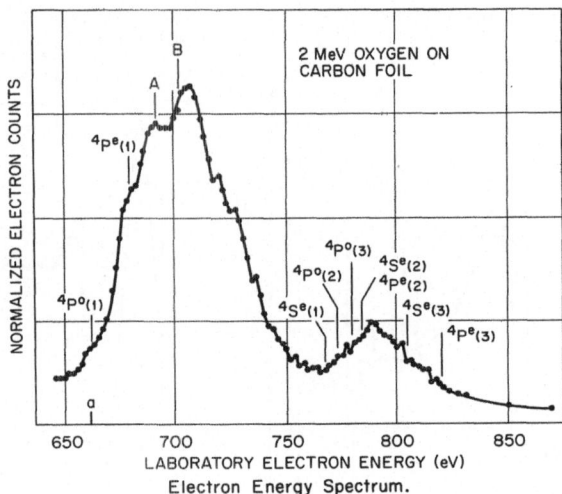

Fig. 5.   Spectrum of electrons emitted by 2 MeV oxygen beam taken
          with the target in view. The energy scale is in the lab-
          oratory frame. The line segment "a" gives the approximate
          intensity of the $^4P^0$ (1) peak of Figure 3.

$1s2p^2 3s$ all leading to quintet states and these energies fall in
the region that could also account for some of the features of
the spectrum shown in Figure 5.

The lifetimes of the nonradiative $(1s2s2p)\,^4P^0_{5/2}$ state in
lithiumlike oxygen, fluorine, chlorine and argon have been
studied in order to try to understand how the strength of the
spin-spin interaction (by which this state decays) varies with
Z along the isoelectronic sequence. In our experiments we have
used the fact that there exists a differential metastability
among the J-levels of the $(1s2s2p)\,^4P^0$ term due to the different
strengths of their coupling to the continuum. Thus the J = 1/2
and 3/2 levels can couple to the continuum via the spin-orbit,
spin-other-orbit and spin-spin interactions and can also mix
with neighboring doublets of the same parity and total angular
momentum and therefore either autoionize rapidly by the Coulomb
interaction or radiate. The J = 5/2 level is more stable since
it autoionizes only via the spin-spin interaction. Thus if one
studies the decay of the $(1s2s2p)\,^4P^0_J$ levels far enough downstream
from the foil target that the short lived components have decayed
away, only the long lived component associated with J = 5/2 remains. Figure 6 shows a decay curve for this state in lithiumlike argon. Our lifetime results for the $(1s2s2p)\,^4P^0_{5/2}$ state in
lithiumlike oxygen, fluorine, chlorine and argon are 25±3 nsec,

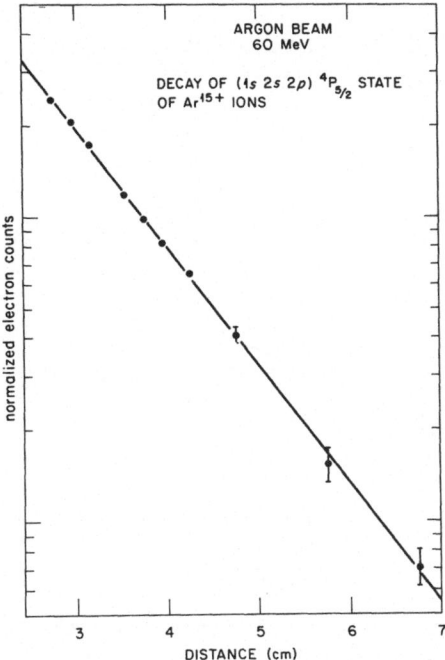

Fig. 6. Decay in flight of the $(1s2s2p)$ $^4P^0_{5/2}$ state of lithiumlike argon obtained at a cyclotron energy of 60 MeV.

15±1 nsec, 0.91±0.04 nsec, and 0.66±0.04 nsec respectively.
Figure 7 provides a summary of our results as well as those of
other investigators [7-9] for the decay rate, γ, of the $(1s2s2p)$ $^4P^0_{5/2}$
state in the lithiumlike isoelectronic sequence. The results
exhibit the Z dependence of the spin-spin interaction (assuming
that no other decay channel competes with this one for the depopulation of the $(1s2s2p)$ $^4P^0_{5/2}$ state). Levitt, Novick and
Feldman[8] have suggested an empirical form, $\gamma = 1.13 \times 10^5 (Z-1.75)^3$,
to fit the data over the range Z = 2-8. In the light of our
recent measurements for Z=17 and 18, it appears that this empirical fit is insufficient to cover the extended range. This can
be seen from Figure 7 where the straight line is the suggested
fit. It seems to us that the new data suggest that the spin-
spin induced decay rate should scale with Z to a higher power
than three.

From the viewpoint of planning future experiments it is of

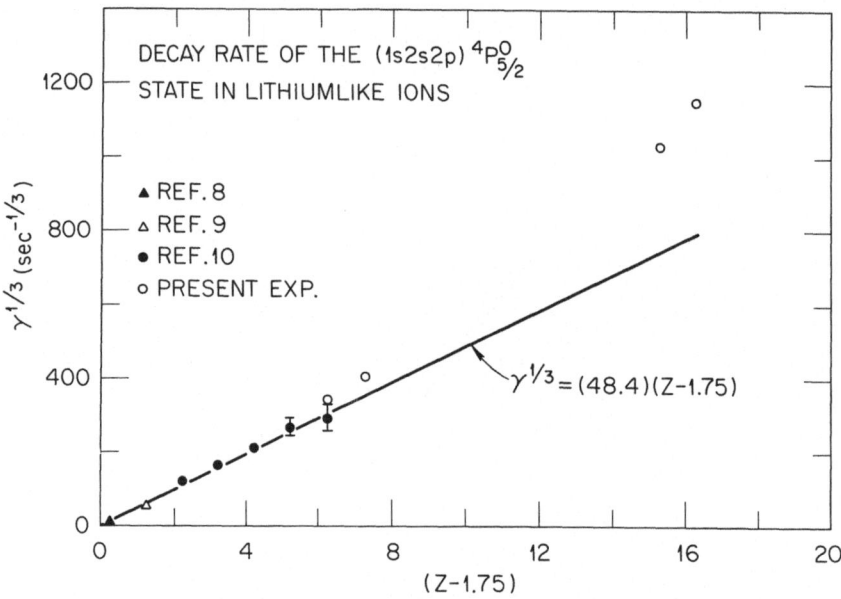

Fig. 7. The decay rate (inverse lifetime) of the $(1s2s2p)$ $^4P^0_{5/2}$
        state in lithiumlike ions. The figure shows the ex-
        perimental decay rates taken to the one-third power
        plotted against an effective nuclear charge, Z-1.75.

interest to ask how commonly these states of high excitation and
angular momentum are formed. Figure 8 shows the measured yield
per incident beam particle of autoionization electrons emitted in
the decay of the $(1s2s2p)$ $^4P^0_{5/2}$ state when oxygen and fluorine ions
at various energies are incident on a carbon target. The results
indicate that for this single state alone, the peak yield is of
the order of 1% per incident ion. It then seems probable that
>1% of the three electron beam fraction is metastable to some de-
gree when other states are considered.

    A more recent study has been made in an attempt to observe
similar metastable autoionizing states in a sodiumlike ion.
Figure 9 shows the autoionization spectrum observed when 5 MeV
chlorine ions are incident on a carbon target. The beam energy
was chosen to maximize the sodiumlike chlorine beam fraction and
we believe that some of the features of the spectrum are due to
the decay of metastable quartet states in this ion. Firm identification of the spectral lines is at present precluded by an
almost complete lack of theoretical calculations of the energies
and lifetimes of such states. It seems plausible however that
many of the states that are observed in decay are quartet states
formed from core-excited configurations such as $2p^5(n\ell)(n'\ell')$

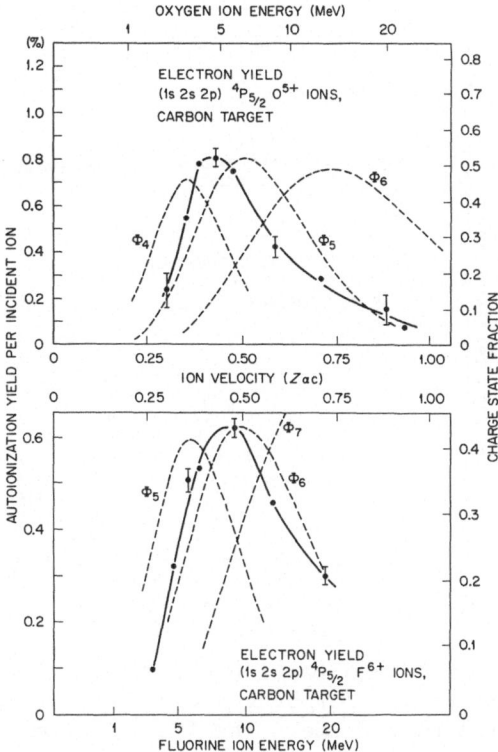

Fig. 8. Energy dependence of the autoionization electron yield from the decay of the $(1s2s2p)$ $^4P^0_{5/2}$ state per incident beam particle for oxygen and fluorine beams incident on a carbon target. Ion velocities are given in units of the k-electron Bohr velocity, $Z\alpha c$. Charge state fractions ϕ_i, where i = ionic charge, are also plotted (dashed curves).

or $2s2p^6(n\ell)(n'\ell')$ with n, n'$\geq$3, in which a single electron has been excited from a previously closed shell or subshell into an outer orbital. It is also conceivable that we are observing the decay of states of even higher multiplicity arising from the simultaneous excitation of two or more inner shell electrons. However, in the spectrum we observe a well-defined drop-off in intensity at an energy coincident with the series limit that would be expected for single core-excited sodiumlike quartet states. The possibility of contamination of our spectrum from single core-excited states of adjacent charge states can be excluded by the fact that the series limits for all such systems restrict maximum electron energies to well below those observed. The possibility of the existence of the previously mentioned spectator states again cannot be ruled out.

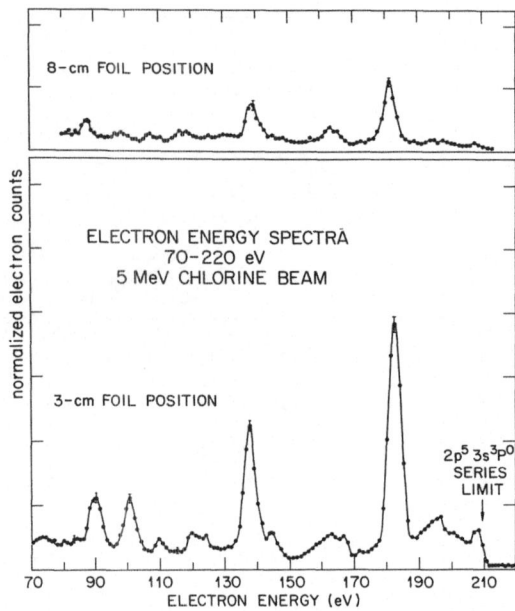

Fig. 9. Segment of autoionization electron spectra from 5 MeV
         chlorine ions undergoing decay in flight, plotted in
         the ionic rest frame. The figure shows data for two
         different target positions, 3 cm and 8 cm from the
         spectrometer viewing region.

    The absolute energies of the four most prominent spectral
features in Figure 9 are 90, 101, 138 and 182 eV respectively with
an estimated uncertainty in each case of ±3 eV. The energies of
these features relative to the assumed series limit are more
accurately measurable. The peak at 101 eV agrees well with a
recent unpublished calculation by Weiss[10] of the energies of the
^{4}S, ^{4}P, ^{4}D states which are formed from the configuration $2p^53s3p$
of sodiumlike chlorine. An estimate of the energies of configurations such as $2p^53sns(n{\geq}4)$, $2p^53snp(n{\geq}3)$ and $2p^53p^2$ can be made by
using simple screening rules. While the accuracy of this method
is not sufficient to positively identify the observed lines, the
results do serve to show that many of the spectral features might
be accounted for by the aforementioned configurations.

    A study of the decay characteristics of the main features of
the presumed sodiumlike chlorine spectrum in Figure 9 indicates
that some of these states are very long lived. We can quote a
lower limit of 43 nsec for the lifetime of the long-lived component associated with the peak at 182 eV, which we note is much
nearer in energy to the so-called series limit than were the

prominent features of the lithiumlike spectra. We thus suspect
that these states are states of high angular momentum which cannot
autoionize via the Coulomb interaction because of the selection
rules mentioned previously. It is particularly interesting to
note that a radiative decay channel which should presumably be
open to the higher energy states (if they are not the lowest lying
states of a given spin system) does not appreciably shorten the
lifetime of the state as much as might be expected. While small
radial transition moments do sometimes occur, it seems more likely
that angular momentum and parity selection rule violations would
be primarily responsible for small radiative transition probabilities. The possibility remains of course that the 182 eV peak,
for example, is the lowest lying, nonradiative state of a system
with multiplicity other than four. Until suitable theoretical
energy and lifetime estimates are available, however, the origin
of these various features must remain something of a mystery.

REFERENCES

1. I.A. Sellin, M. Brown, W.W. Smith, and Bailey Donnally, Phys.
Rev. A2, 1189 (1970); R. W. Schmieder and R. Marrus, Phys.
Rev. Letters 25, 1245 (1970).

2. Bailey Donnally, W.W. Smith, D.J. Pegg, M. Brown and I.A.
Sellin, Phys. Rev. A4, 122 (1971); I.A. Sellin, D.J. Pegg,
M. Brown, W.W. Smith and Bailey Donnally, Phys. Rev. Letters
27, 1108 (1971); I.A. Sellin, D.J. Pegg, P.M. Griffin and
W.W. Smith, Phys. Rev. Letters 28, 1229 (1972); D.J. Pegg,
I.A. Sellin, P.M. Griffin and W.W. Smith,Phys. Rev. Letters,
28, 1615 (1972).

3. P. Feldman and R. Novick, Phys. Rev. 160, 143 (1967).

4. E. Holøien and S. Geltman, Phys. Rev. 153, 81 (1967).

5. N. Bardsley and B.R. Junker, Private Communication.

6. H. G. Berry, Phys. Rev. A6, 514 (1972).

7. L.M. Blau, R. Novick and D. Weinflash, Phys. Rev. Letters,
24, 1268 (1970).

8. M. Levitt, R. Novick, and P. Feldman, Phys. Rev. A3, 130 (1971).

9. I.S. Dmitreav, V.S. Nikoleav, and Ya.A. Teplova, Phys. Letters
26A, 122 (1968).

10. A. Weiss, Private Communication.

DIFFERENTIAL CROSS SECTIONS FOR METASTABLE He AND Ar

H. Haberland*, C. H. Chen and Y. T. Lee[tt]

The James Franck Institute and the Department of Chemistry

The University of Chicago, Chicago, Illinois 60637

The interaction of metastable rare gas atoms with ground state atoms has been the subject of many theoretical and experimental studies. [1,2] Especially the interaction between ground state and metastable He, which seems to be simple enough to be amenable to theoretical calculations, [2] has aroused much interest.

Helium has two metastable states 1s2s, [3]S and 1s2s, [1]S with excitation energies of 19.82 eV and 20.61 eV and lifetimes of $\sim 10^4$s and 19.7 msec respectively. [3] Due to the symmetry of the wavefunction the potentials between metastable and ground state He are split into gerade and ungerade states (see below). All four potentials have a deep inner minimum separated from the van der Waals minimum by a maximum higher than the dissociation limit. The potentials around the deep inner minima are known from spectroscopy, [4] and very recently some data on the inner part of the potentials have also been obtained by metastable He scattering in the 10–20 eV range. [5] Some information on the long range part is available from diffusion and optical pumping experiments. [1] For the He*–Ar scattering only one potential curve is involved. But as the ionization potential of argon (15.76 eV) is much lower than the excitation energy of the metastable He, ionizing collisions can occur even at thermal energies.

A more accurate determination of the potentials, especially near the shallow van der Waals minima, should be possible by differential cross section measurements at thermal energies. Until now only one low resolution angular distribution has been reported for these systems. [6] We report here experimental results and a preliminary analysis of the differential cross

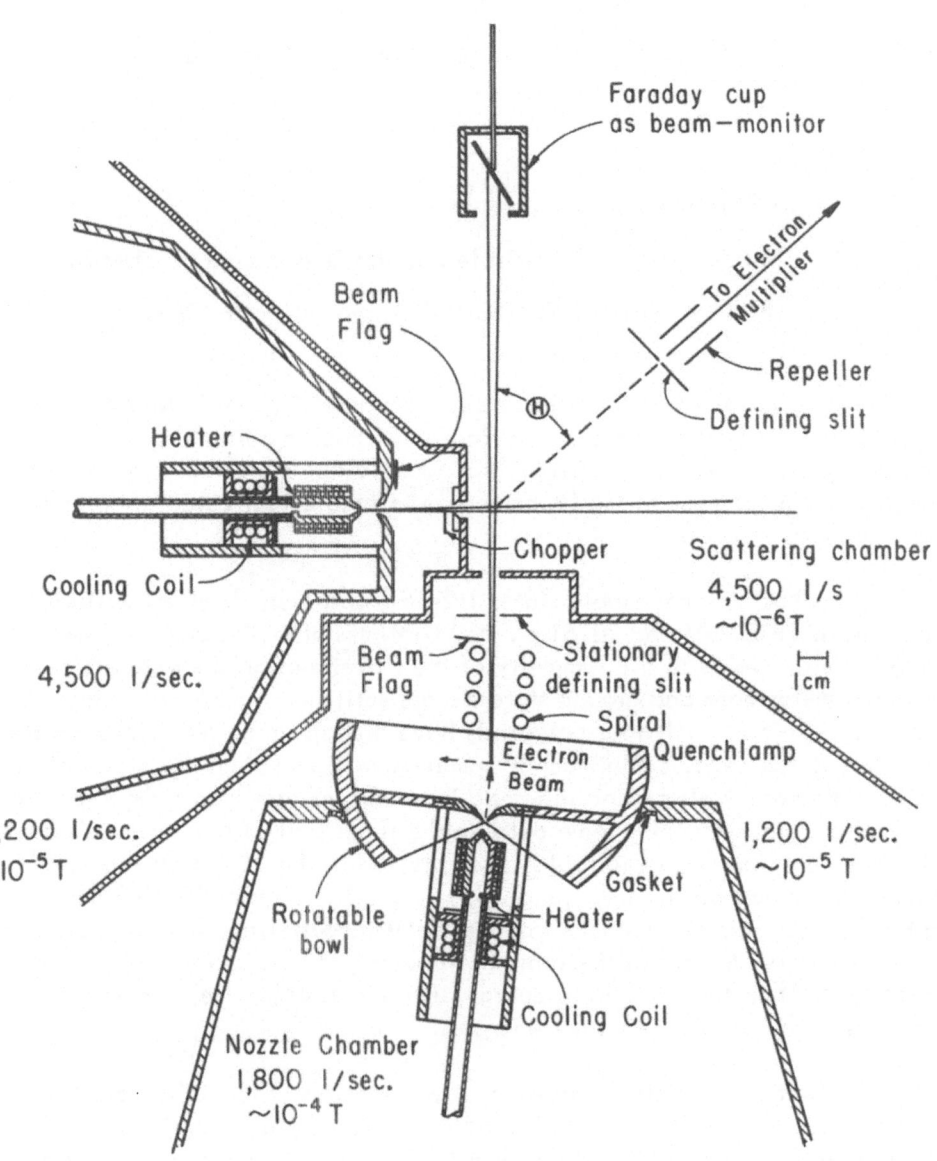

Fig. 1. Schematic of the central part of the apparatus.

section for the systems He*–He, He*–Ar and Ar*–Ar, where * denotes the metastable state.

1. EXPERIMENT

A schematic of the central part of the apparatus is shown in Fig. 1. A supersonic free jet of He or Ar (Mach number ~15) is produced by expanding the gas from ~600 Torr into vacuum through a small nozzle (diameter 0.1 mm). The central part of the beam goes through the skimmer (diameter 1mm) into the second differential pumping chamber. A small fraction of the beam (~10^{10} atoms per sec) is then excited by electrons (250V, 50mA) to two metastable states (^1S, ^3S for He; $^3P_0$, $^3P_2$ for Ar; the $^3P_2$ state of Ar is preferentially populated because of its larger statistical weight). The ^1S state of He can be quenched[7] by a helium discharge lamp so that a pure ^3S beam remains. The momentum transfer to the He during excitation can result in a deflection of the He by several degrees. To compensate for this effect, the nozzle, skimmer, electron beam and quench lamp are rotatable. A time of flight analysis showed that the velocity distribution is not seriously degraded by the excitation. The metastable particles are detected by their Auger emission on the first dynode of an electron multiplier. Both beams are collimated to 2° FWHM.

When the differential cross section experiments were performed the quench lamp had not been installed. Subsequent tests showed that the He* beam consisted mainly of He in the ^1S state. It is estimated that under our operating conditions only $(15 \pm \frac{10}{5})\%$ triplets were in the beam. Thus, the observed features of the differential cross sections are mainly due to the singlet state; the contributions from the triplet state are estimated to be small. Further experiments will be performed in the future with He* in the ^3S state. The center of mass (CM) energies for these experiments are all ~63 meV, which is found to be large enough to overcome the barrier in the $^1\Sigma_u$ state of He*–He.

The experimental results are shown in Fig. 2. The measured differential cross sections are plotted against the CM scattering angle. For He*–He and Ar*–Ar, the differential cross section is seen to fall, oscillate slightly, and then rise towards an exchange peak at 180°. The oscillations, which are antisymmetric about 90°, are for He*–He mainly due to the nuclear symmetry. The differential cross section is not symmetric with respect to 90° as one might expect for identical particles. At low energy the strong interference between direct and exchange scattering destroys this symmetry (see below). For He*–Ar the angular distribution is very similar to that of Ar*–Ar at small angles, but at larger angles a severe loss of intensity is

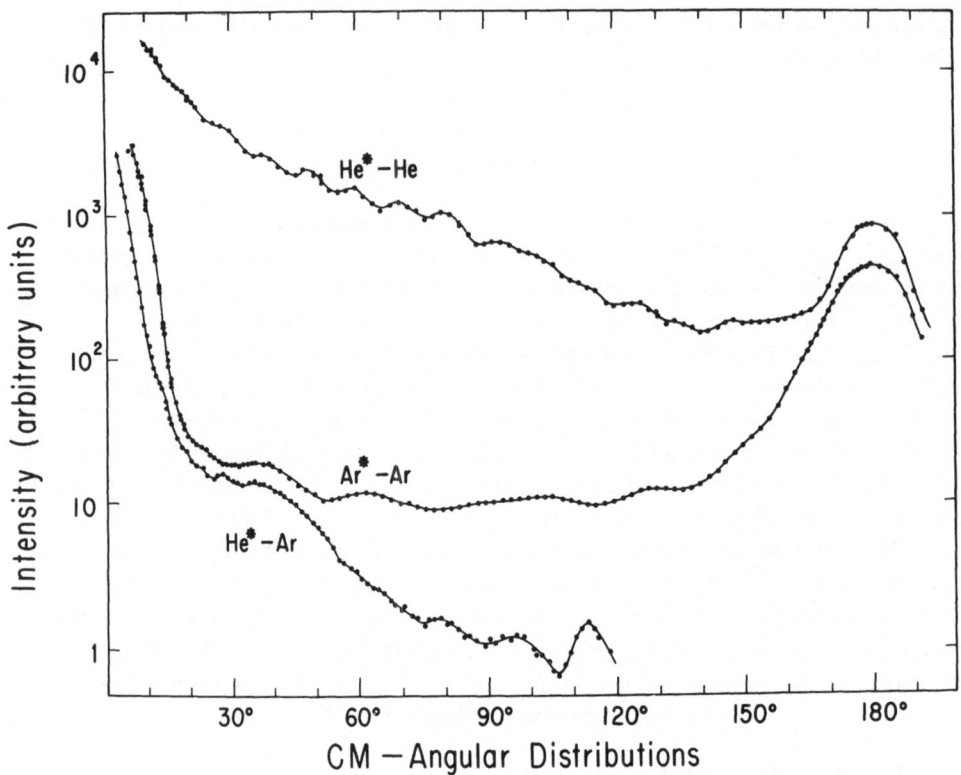

<u>Fig. 2.</u>   Experimental differential cross sections.   The peaks at 180° are
          due to resonant energy exchange.

quite evident.   For low impact parameters (which lead to large scattering
angles) Penning and associative ionization[8] are possible, i. e.,

$$He^* + Ar \rightarrow He + Ar^+ + e \quad \text{(Penning ionization)}$$
$$He^* + Ar \rightarrow HeAr^+ + e. \quad \text{(Associative ionization)}$$

These reactions cause the falloff at large angles, and they also account for
the oscillations around 90°.

2.   THEORETICAL CONSIDERATIONS

In the following theoretical discussion, we will assume that the influence
of the triplets can be neglected.

A.   Optical Model Analysis for He*-Ar

The excitation energy of the metastable He is 4. 9eV above the ionization potential of Ar, so that for all energies ionizing collisions are possible. The system He*-Ar can therefore be described as a bound state (He*+ Ar) embedded into a continuum (He + Ar$^+$ + e).   The configuration interaction between these two states allows Penning and association ionization to occur besides elastic scattering.   A theoretical analysis has been given by Nakamura et al.[9] and Miller.[10] In this experiment we are only observing the elastic channel of this two state problem.   Theoretically, the inelastic channel can be eliminated and its influence accounted for by a complex potential:

$$W(R) = V(R) - \frac{i\Gamma(R)}{2} \quad .$$

This can be done rigorously using the Feshbach operator approach,[11] but the optical potential becomes energy-dependent and non-local. In our case, a local approximation to the optical potential will probably be very good, because the transition is nearly vertical,[9,10] i. e., in the Born-Oppenheimer approximation the nuclei do not move during the transition, and because there is no backcoupling of the ionizing channel.   Once the electron jump has occurred, the electron will not get back into the elastic channel.   The lifetime τ of the initial state is given in terms of the optical potential as

$$\tau(R) = \frac{\hbar}{\Gamma(R)} \quad .$$

Phenomenological optical potentials have been used extensively in nuclear physics.   In atomic and molecular collisions they have been used mainly to describe the elastic scattering of chemically reactive systems.[12,13] For a complex potential the phaseshifts η_ℓ are no longer real

$$\eta_\ell = \eta'_\ell + i\alpha_\ell \quad .$$

The elastic scattering amplitude becomes

$$f(\theta) = \frac{1}{2ik} \sum_{\ell=0}^{\infty} (2\ell + 1)(e^{-2\alpha_\ell} e^{2i\eta'_\ell} - 1) P_\ell(\cos\theta) \quad .$$

The outgoing elastic amplitude is attenuated by a factor $\exp(-2\alpha_\ell)$.   In the limit $\alpha_\ell \to \infty$, the partial wave is absorbed completely and one gets pure shadow scattering of the incoming plane wave.

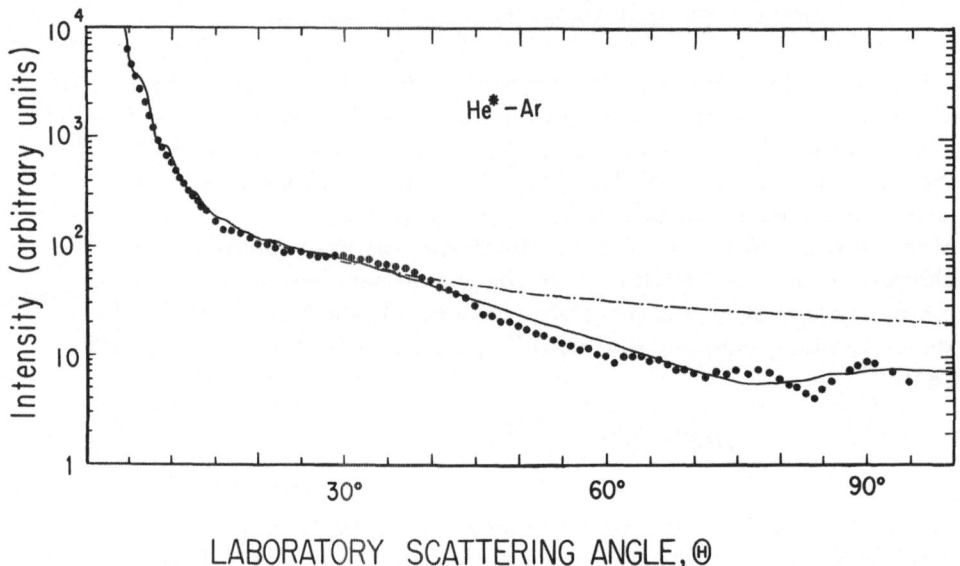

He*–Ar

LABORATORY SCATTERING ANGLE, Θ

Fig. 3.　Comparison of the experimental and calculated results for He*–Ar. The dots give the experimental values. The broken line was calculated assuming only a real potential. The full line is the best fit we have obtained so far from the optical model analysis.

To obtain the complex phaseshifts from a complex potential, one can either integrate the Schrödinger equation numerically, or use the WKB-formula derived by Bottino et al. [12, 14] The usual WKB approximation describes the elastic scattering quite well, even for the scattering of He at thermal energies (error in the phaseshifts ≤5%). An approximation to the complex WKB formula has been found to differ from an exact numerical result by less than ~10%. [15] It is simpler and less ambiguous to analyze experimental data not in terms of $\Gamma(R)$ but instead in terms of the opacity, which is proportional to the flux in the inelastic channel. The opacity 0_ℓ is given in terms of the complex phaseshifts as

$$0_\ell = 1 - e^{-4\alpha_\ell}$$

is approximated assuming simple analytical forms.

We have chosen the following method[12] to analyze our data. The elastic phaseshifts are calculated from a real potential via the WKB approximation. The parameters are chosen so that a good fit to the differential cross section at angles below 30° is achieved.   Semiclassically only large impact parameters $b \geq 3.7$ Å for which the transition probability is vanishingly small lead to the scattering angles below 30°.

In Fig. 3 the best fit we have obtained so far is compared to the experimental results in the LAB system.   The calculated differential cross section was transformed into the LAB-system and averaged over the experimental conditions.   For the elastic potential we used a Morse function smoothly joined by a Spline function to a C/r^6 potential at large internuclear distances.   The parameters used are $\epsilon = 3.9$ meV, $r_m = 5.07$ Å, $\beta = 5$ Å^{-1}, $C = 325$ a. u..   Because of the large reduced energy $E/\epsilon = 16$, the rainbow[16] is at a very small angle.   The observed sharp falloff is the dark side of the rainbow, which is only sensitive to the potential well depth.   Since the ionization potential and polarizability of He* are similar to those of the alkali atoms, r_m and β were chosen to be similar to the alkali-argon potentials. [17]

The analytical form used for the opacity was

$$O_\ell = \frac{a + b \cdot \ell}{1 + \exp[(\ell - \ell_o)/\Delta \ell]}$$

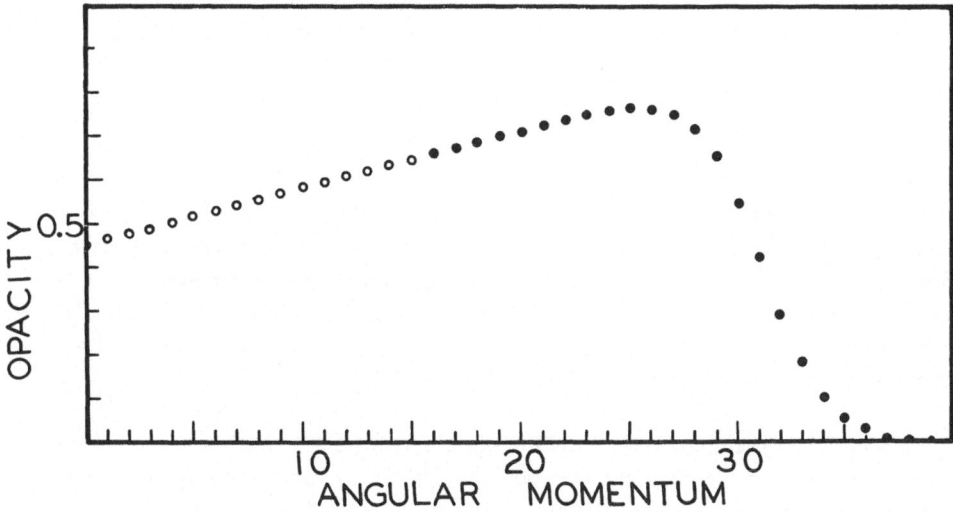

Fig. 4.   The opacity as a function of angular momentum.   The open circles correspond semiclassically to scattering angles larger than 120°, which could not be observed in this experiment.

with a = 0. 45, b = 0. 013, ℓ_o = 31, $\Delta\ell$ = 1. 5. It was observed that a small change in 0_ℓ very often had a pronounced effect on the differential cross section at large angles. For $\Delta\ell \le 1$ oscillations appeared in the angular distribution, but with a much larger wavelength than was experimentally observed.

From the experiments we obtained the opacity as a function of the angular momentum ℓ. The result of theoretical calculations is the auto-ionization width Γ as a function of the internuclear distance R. We have the classical relation[9,10]

$$0_\ell = 1 - \exp\left(-2\int_{r_o}^{\infty} \frac{\Gamma(R)}{\hbar\, v_\ell(R)}\, dR\right)$$

where $v_\ell(R)$ is the relative velocity in the effective potential and r_o is the classical turning point. This relation is only valid if $\Gamma(R)$ is not too large. Several calculations of the $\Gamma(R)$ have appeared.[9,18] They all show a roughly exponential behaviour

$$\Gamma(R) = A \cdot \exp[-R/B] \ .$$

This form was also deduced by Niehaus et al.[19] from their measured energy spectra of Penning electrons.

The total inelastic cross section is given in terms of the opacity as

$$\sigma = \pi \cdot k^{-2} \sum_{\ell=0}^{\infty} (2\ell+1)\, 0_\ell \ .$$

From our fit we obtain $\sigma = 20. 7 \ \text{Å}^2$. This value depends on the value β in the Morse function. For reasonable choices of β we obtain values between 15 and 25 Å^2. There is some discrepancy between different measurements of the total inelastic cross section. Values of 7. 6 Å^2, 9. 0 Å^2 and 16. 4 Å^2 have been obtained[20] at room temperature. Our value is nearest to the value of 16. 4 Å^2 obtained by Schmeltekopf et al.

Olson[21] has assumed A = 1 in the equation for $\Gamma(R)$ and determined B from these measurements. He was able to obtain an opacity with a sharp drop, similar to our results only by assuming $\sigma = 16. 4 \ \text{Å}^2$. For the two lower values of σ the opacity is roughly gaussian-shaped in agreement with the results of Niehaus et al.[19] for metastable He in the triplet state where the total inelastic cross section is smaller than 10 Å^2.

The linear rise of the opacity for small ℓ as shown in Fig. 4 does not seem to be compatible with an exponential dependence of $\Gamma(R)$. We have been unable so far to reproduce the structure in the angular distribution around 90°. The simple analytical form used gives oscillations with a much wider angular spacing even in the limit $\Delta\ell \to 0$. These oscillations can probably be reproduced using a similar opacity with an additional small oscillatory term. This term could be caused by several different effects.

(1) Excitation transfer to nonautoionizing states of argon. In this case the energy could be exchanged between the He* and Ar several times. This would lead to curve crossings and to an oscillatory behaviour of the opacity function. By looking at light emission one could learn something about this excitation transfer.

(2) For associative ionization $He^* + Ar \to HeAr^+ + e$ the width $\Gamma(R)$ might depend on the internal energy state of $HeAr^+$. This would probably also lead to an oscillatory behaviour of the opacity function.

(3) Interference of the transition amplitudes in the incoming and out-going channel (Ref. 10, eq. 33).

Measurements of the energy dependence of the differential cross section and more detailed calculations are necessary to understand these phenomena.

B. . Differential Cross Section for He*-He

The system He*-He consists of two indistinguishable particles. The wavefunction therefore must be either symmetric or antisymmetric with respect to interchange of the atoms.[22,23] Thus, if the atoms were held fixed at a distance R apart, the wavefunction for the electronic motion would be $\chi_{g,u}(R, \vec{r}_a, \vec{r}_b)$ where $\vec{r}_a, \vec{r}_b$ are the coordinates of the electrons in atom A and B relative to their respective nuclei. In the limit $R \to \infty$

$$\chi_{g,u} \to \psi_0(\vec{r}_a)\,\psi_1(\vec{r}_b) \pm \psi_0(\vec{r}_b)\,\psi_1(\vec{r}_a)$$

where ψ_0, ψ_1 are wavefunctions for the ground and excited states concerned. Each of the electronic wavefunctions χ_g and χ_u satisfies its own Schrödinger equation, without any coupling. Therefore we have for all internuclear distances $\langle \chi_u | \chi_g \rangle = 0$. In the Born-Oppenheimer approxima-

tion the total interaction $V(R, \vec{r}_a, \vec{r}_b)$ is averaged over the electronic

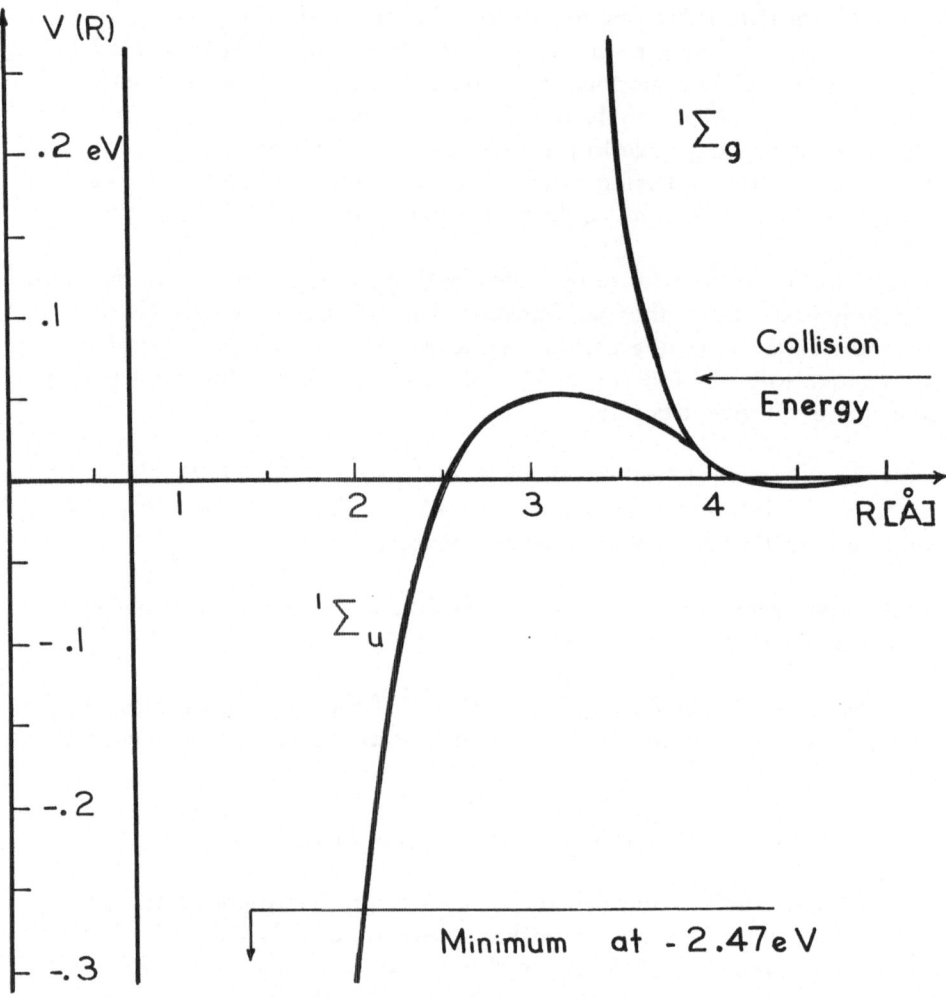

Fig. 5.   Model potentials for He*–He. The assumed splitting between the
            gerade and ungerade curve for large internuclear distances is too
            small to be shown on this scale.

coordinates for fixed R.   So each of the two metastable states is split at
small internuclear distances into a gerade and ungerade state as shown in
Fig. 5. As these states do not interact, the radial wavefunctions and the
scattering amplitudes f_g, f_u can be calculated for each of them independently.
One can define[22, 23] an amplitude for direct f_d and for exchange
scattering f_{ex}.

$$f_d = \frac{1}{2} (f_g + f_u), \qquad\qquad f_{ex} = \frac{1}{2}(f_g - f_u) \ .$$

The total scattering amplitude f is

$$f = f_g + f_u = f_d + f_{ex} \ .$$

If the overlap between f_d and f_{ex} is small it has been argued[23, 24] that one can extract an exchange cross section $\sim |f_{ex}|^2$ from the measurements of the differential cross section. In our calculations the overlap was always non-negligible. At higher energies one would expect a separation of the two contributions. As He is a boson, the overall wavefunction must be symmetric with respect to interchange of the nuclei; therefore the scattering amplitude for the (un) gerade state is (anti) symmetric with respect to interchange of the nuclei. This has the effect that all odd (even) partial waves are missing from the gerade (ungerade) scattering amplitudes. Denoting by capital letters properly symmetrized wavefunctions one obtains

$$f_g \rightarrow f_g(\theta) + f_g(\pi -\theta) = F_g$$

$$f_u \rightarrow f_u(\theta) - f_u(\pi-\theta) = F_u.$$

It follows that

$$F_d(\theta) = \frac{1}{2} (F_g(\theta) + F_u(\theta))$$

$$= \frac{1}{2} (f_g(\theta) + f_g(\pi-\theta) + f_u(\theta) - f_u(\pi-\theta))$$

$$= \frac{1}{2} (F_g(\pi-\theta) - F_u(\pi-\theta))$$

$$= F_{ex}(\pi-\theta).$$

The direct and exchange contributions do not differ for identical particles. One obtains the result that for identical particles and vanishing overlap between direct and exchange scattering the angular distribution is symmetric with respect to 90° CM. This is in contrast with an earlier prediction, where symmetry effects were neglected. [23, 24]

    To obtain the differential cross section the Schrödinger equation was integrated numerically by the Numerov method. No satisfactory fit has been obtained so far, but we can draw a number of conclusions from the calcula-

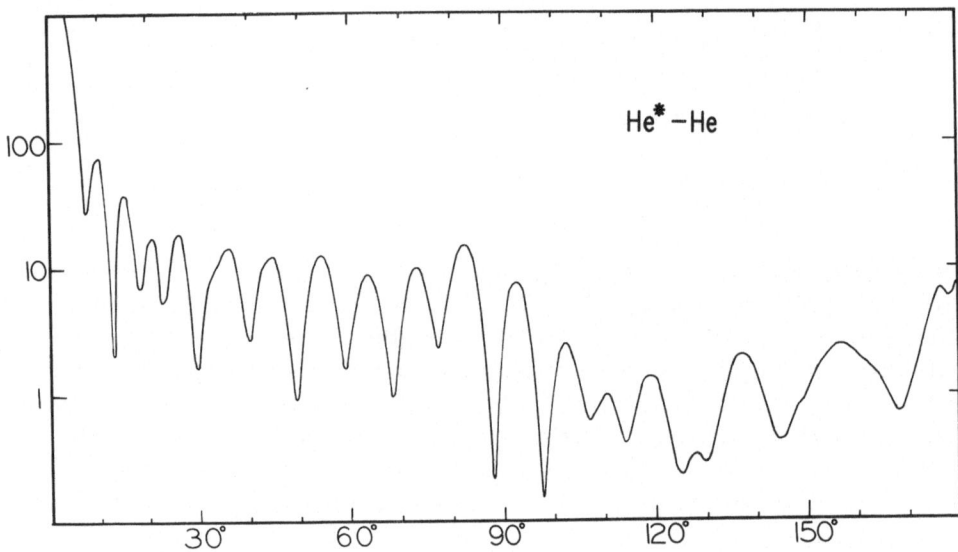

Fig. 6.   Angular distribution calculated from the potentials shown in
           Fig. 5.   The regular oscillations between 30° and 90° are due
           to nuclear symmetry.   At larger angles the orbiting contribution
           from the ungerade potential causes the irregular oscillations.
           The splitting of the two potentials is rather small, giving rise to
           only a small backward peak.

tions.   (1) The oscillations on the experimental curve are heavily damped by
velocity and angular resolution effects, as can be seen from Fig. 6, which
gives a typical computed angular distribution.   (2) The height and form of
the exchange peak at 180° CM is mainly determined by the difference between the gerade and ungerade potential.   This means that this difference
should be small for He and large for Ar.   (3) The general fall-off of the
angular distribution cannot be reproduced if the barrier height in the $^1\Sigma_u$
potential is higher than our collision energy of ~63 meV.   This
probably rules out the value of 80±18 meV for the barrier height, which
has been obtained[2] by combining experimental and theoretical data.   (4)
The regular oscillations on the He*-He differential cross section are mainly
due to nuclear symmetry.   (5) No estimate of the total exchange cross section can be made directly from these experiments.   Assuming distinguishable
particles in the calculations it was observed that non-negligible exchange
scattering occurred even at angles lower than 20° and in some cases even
the direct contribution can dominate at 180°.   But once the potentials have

been determined, one can of course calculate the total exchange cross section assuming distinguishable particles. This cross section has been measured for He^3 in optical pumping experiments.[1] (6) In the deep attractive well of the ungerade potential orbiting occurs. In this case one would expect from the semiclassical scattering theory an exponential dependence of the angular distribution on the scattering angle. From the calculated phaseshifts one sees that indeed orbiting is present for many partial waves. No exponential behaviour was observed when the differential cross section was computed only from the $^1\Sigma_u$ state.

ACKNOWLEDGEMENTS

This research was supported by the U. S. Atomic Energy Commission, and supported in part by the Louis Block Fund, the University of Chicago. We also acknowledge that this research has derived benefit from the general support of Materials Sciences by the ARPA Program at the University of Chicago, and from the Camille and Henry Dreyfus Foundation Teacher-Scholar Grant. The initial design and construction of the metastable beam source is due to Dr. Peter Siska, now at the University of Pittsburgh.

* On leave from the Fakultät für Physik der Universität Freiburg, Freiburg, Germany. Support by the Deutsche Forschungsgemeinschaft is gratefully acknowledged.

†† Alfred P. Sloan Research Fellow.

REFERENCES

1. J. Dupont-Roc, M. Leduc, F. Laloë, Phys. Rev. Letters 27, 467 (1971) and references cited therein.

2. B. Liu, Phys. Rev. Letters 27, 1251 (1971).

3. R. S. Van Dyck, C. E. Johnson, H. A. Shugart, Phys. Rev. A4, 1327 (1971).

4. Spectroscopic Constants Relative to Diatomic Molecules, B. Rosen, ed., Pergamon Press 1970.

5. R. Morgenstern, Stanford Research Institute, private communication.

6.   J. Grosser, H. Haberland, Phys. Letters 27A, 634 (1968).

7.   H. Hotop, A. Niehaus, A. L. Schmeltekopf, Z. f. Phys. 229, 1 (1969).

8.   H. Hotop, et al., VII ICPEAC, p. 1101, and references cited therein.

9.a. H. Nakamura, J. Phys. Soc. JAPAN 26, 1473 (1969).

  b. H. Fujii, H. Nakamura, M. Mori, J. Phys. Soc. JAPAN 29, 1030
      (1970).

10.   W. H. Miller, J. Chem. Phys. 52, 3563 (1970).

11.   S. A. Adelman, W. P. Reinhardt, Phys. Rev. A6, 255 (1972) and
      references cited therein.

12.   J. Ross, E. F. Greene in: Molecular Beams and Reaction Kinetics,
      p. 86, Academic Press 1970, C. Schlier, ed.

13.   G. Wolken, Jr., J. Chem. Phys. 56, 2591 (1972).

14.   A. Bottino, A. T. Longoni, T. Regge, Nuovo Cimento 23, 954 (1962).

15.   J. S. Cohen, N. F. Lane, Chem. Phys. Let. 10, 623 (1971).

16.a. R. M. Eisberg, C. E. Porter, Rev. Mod. Phys. 33, 190 (1961),
      Chapter VII.

  b. R. B. Bernstein, Advances in Chem. Phys. Vol. X, p. 75 (1966).

17.   G. B. Ury, L. Wharton, J. Chem. Phys. 56, 5832 (1972).

18.a. W. H. Miller, C. A. Slocomb, H. F. Schafer, III, J. Chem. Phys. 56,
      1347 (1972).

  b. K. L. Bell, J. Phys. B3, 1308 (1970).

19.   A. Niehaus, et al., a) He (2^3S) + Na, Z. f. Phys. 238, 452 (1970).
                          b) He (2^3S) + Ar, private communication.

20. a. W. P. Sholette, E. E. Muschlitz, Jr., J. Chem. Phys. 36, 3368 (1962).

  b. D. A. MacLennan, Phys. Rev. 148, 218 (1966).

  c. A. L. Schmeltekopf, F. C. Fehsenfeld, J. Chem. Phys. 53, 3173 (1970).

21.   R. Olson, Stanford Research Institute, private communication.

22.   H. S. W. Massey, Electronic and Ionic Impact Phenomena, Vol. 3,
      p. 1870.

23.   R. P. Marchi, F. T. Smith, Phys. Rev. A139, 1025 (1965).

24.   H. L. Richards, E. E. Muschlitz, Jr., J. Chem. Phys. 41, 559 (1964).

Interstellar Molecules

K. B. Jefferts

Bell Laboratories

Murray Hill, New Jersey 07974

The first observation of interstellar molecules occurred in 1937, with the discovery of sharp absorption lines in the ultra-violet spectra of several stars.[1] These features were found to be associated with the presence of interstellar CH^+, CH and CN. These features, which were apparent in the spectra of a large number of bright stars, were essentially the only evidence for the existence of interstellar molecules until 1963, when Weinreb et al.[2] observed absorption at 1665 MHz against several continuum sources. This feature was found to be associated with transitions between the ground state Λ-doublet levels of OH.

In 1968 Cheung et al.[3] observed the inversion spectrum of NH_3 at 1.25 cm wavelength, and the following year saw the discovery[4] of H_2O at 1.3 cm and H_2CO at 6 cm.[5]

All of these observed radio transitions except that due to H_2O involve internal structure transitions of one type or another, i.e. Λ-doublets, K-doublets, or inversion doublets, and that due to H_2O arises from a pair of levels not including the ground state. All of them are also characterized by some form of non-thermal behavior, from the observed maser action of the OH and H_2O, to considerably different temperatures apparent for ortho and para NH_3, to the apparent refrigeration of the H_2CO below the background temperature. This circumstance greatly restricts the utility of these spectral features as astronomical tools, since the mechanisms involved do not seem to be understood.

The desirability of direct observation of lowest rotational transitions in these and other as yet undiscovered molecules was at that time apparent. However, an inspection of rotational

constants for simple molecules disclosed a difficulty: The
frequencies involved are in general much greater than those
quoted above, and in fact, extend well into the never-never land
between the microwave and infrared segments of the spectrum.
Further examination revealed, however, that a technologically
approachable part at the lower end of this spectral range contained $J = 1 \rightarrow 0$ rotational transitions of several highly interesting simple molecules.

An appraisal of available facilities at that time demonstrated that there existed several antennae of suitable characteristics, in particular, the 11 m paraboloid at Kitt Peak, and
that receivers of dismal but useable sensitivity had been constructed in this spectral range. The missing ingredient was frequency control sufficient to examine narrow, $\Delta F/F \sim 10^{-5}$, spectral
features.

At that time, my colleagues A. A. Penzias, R. W. Wilson and
I, in collaboration with S. Weinreb at the National Radio
Astronomy Institute, mounted an effort to generate a stable local
oscillator in this range, and to improve on the noise performance
of the existing state of the receiver art. We were successful in
both generating a system which phase-locked a local oscillator
klystron to a high harmonic of a 100 MHz frequency synthesizer,
and with the aid of C. Burrus of Bell Laboratories, improving the
receiver noise performance by a factor approaching an order of
magnitude. A block diagram of this system is shown in Fig. 1.
These technical successes have been most gratifying especially in
that the frequency control system has to date been used at frequencies ranging from 30 GHz to 173 GHz, and has been instrumental
in the discovery of more than half of the known molecular species.

When we first used this system, in early 1970, we were
instantly successful in finding intense, widespread CO radiation
at 115.271 GHz, and a few days later observed the same $J = 1 \rightarrow 0$
radiation from CN, a transition which, incidently, has never been
observed directly in the laboratory. The discovery of intense
HCN radiation at 88.632 GHz, by Buhl and Snyder[6] quickly followed,
accompanied by the observation of a still unknown transition at
89.190 GHz, termed X-ogen by its discoverers.

The period since then has seen the discovery of CS, SiO,
H_2S and OCS, with several different isotopes of C, O and S being
in evidence, and a host of combinations of H, C, N and O, of
which CH_3OH, methyl alcohol,[7] is probably the most important.

The richest source for molecular transitions has been a
dense cloud in the direction of the galactic center, known as
$SgrB_2$. Most of the known spectral features are observable in
that source which is apparently characterized by a kinetic temperature ~ 50 K, and a neutral hydrogen density 10^6 cm^{-3}. More

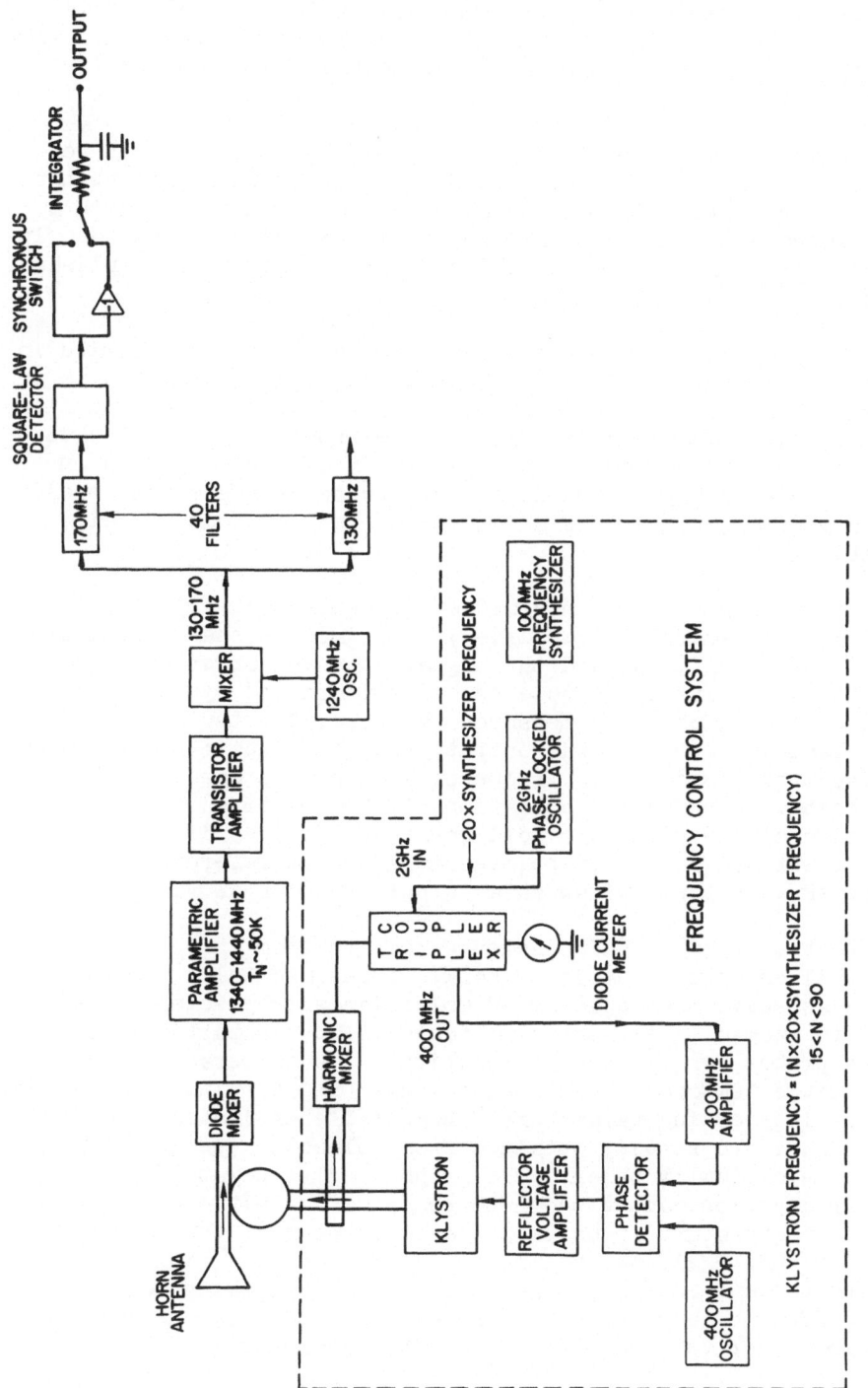

Fig. 1 Block diagram of mm-wave spectral line receiver

typical as a source is the great nebula in Orion, M-42. This
source is characterized by kinetic temperatures of the same order,
but appears to have a somewhat lower neutral hydrogen density of
the order of 10^5 cm^{-3}.

In addition to these regions, CO has been observed in several
dark clouds; cold, dusty regions seen as dark spots against the
normal star field. CO, CS and CN have also been observed in a
shell surrounding the infrared object IRC + 10216. Some description of the fashion in which these observations are accomplished
seems in order.

The microwave portions of our receivers are the ultimate in
simplicity. They consist of a horn antenna, typically 4-5 mm in
diameter coupled to a diode mixer, with a klystron local oscillator coupled through a small hole in the broad wall of the waveguide. The whole assembly is typically 5 cm long. Losses in
even perfect silver waveguide are so severe at these frequencies
that a minimum of guide is an essential for low-noise receiver
operation.

Mixer output at an intermediate frequency of 1390 MHz is
coupled to a parametric amplifier; more gain is then provided by
transistor amplifiers, and the signal is then converted to
150 MHz, where it is fed to a set of filters and synchronous
detectors, each of which measures the power in, typically, a
1 MHz bandwidth. The final output is a set of points depicting
received power plotted vs frequency in, in this case, 1 MHz
increments. Synchronous operation of the system subtracts from
each channel the power provided when the receiver is switched to
a different frequency, or pointed at an empty reference position
in the sky. A typical raw data output is shown in Fig. 2.

What sort of useful knowledge results from this effort? As
the excellent review paper by Rank, Townes and Welch[8] points out,
the relatively close spacing of molecular energy levels as compared to atomic levels makes molecular features particularly useful as probes for regions of relatively low temperature. There
the kinetic temperature and neutral gas (H_2) density are the most
obvious interesting parameters. Reasonable estimates of kinetic
temperature are readily obtained from molecular spectra when the
source is optically thick. The brightness temperature of the
line in question can be shown to be approximately equal to the
kinetic temperature at unit optical thickness of the source,
assuming that exciting collisions occur much more rapidly than
the spontaneous emission time involved. By choosing molecules
with different abundances, e.g. $^{12}C^{16}O$ and $^{13}C^{16}O$, one can, in
principle, derive some idea of thermal gradients in the source.

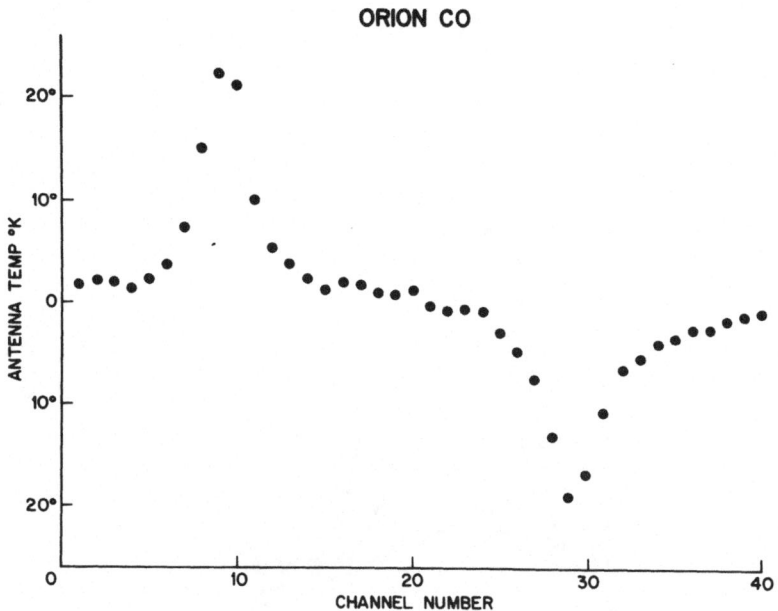

Fig. 2 Raw data output - spectral line receiver, carbon monoxide
in the Orion nebula, rest frequency 115,271 MHz, channel separation 1 MHz. The emission feature is centered in channel 10. The
apparent absorption feature in channel 30 is the result of operating the receiver in a frequency-switching mode, increasing the
frequency 20 MHz for the "reference" half of each synchronous
detector cycle.

Similarly, by choosing molecular lines with much different
dipole moments, and consequently different spontaneous emission
lifetimes, reasonable estimates of collision times, and consequently H_2 densities, can be obtained. As an example, CO,
$\mu \sim 0.1$ Debye, is maintained in thermal equilibrium at H_2 densities greater than about 10^3 cm^{-3}, while CS, or HCN, both of which
have dipole moments of the order of 3 Debye, require H_2 densities
of the order of 10^6 cm^{-3} for equilibration.

Some isotope ratios can be readily and directly derived from
molecular data. See Fig. 3. $^{13}C/^{18}O$ ratios, for example, are
provided by observation of $^{13}C^{16}O$ and $^{12}C^{18}O$ in the same source,
since most sources are not optically thick in either molecular
transition. $^{12}C/^{13}C$ ratios or $^{16}O/^{18}O$ ratios are, unfortunately,
only indirectly obtainable, due to the density of CO in these
clouds. Nevertheless, in the considerable amount of such data
produced to date, there is, I believe, little evidence for ratios
different than terrestrial.

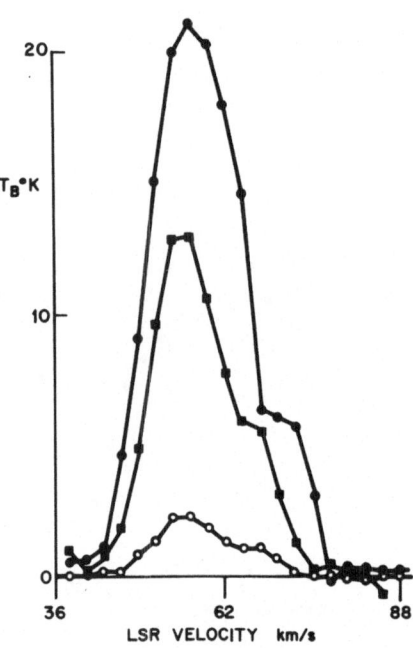

Fig. 3 Carbon monoxide isotopes in W-51 the upper curve corresponds to $^{12}C^{16}O$, lab frequency 115,271.2 MHz, the middle curve to
$^{13}C^{16}O$, 110,201.4 MHz, and the lower curve to $^{12}C^{18}O$, 109,782.2 MHz.

Another unique characteristic of these mm-wave observations
in illustrated by Fig. 4, which is a small map of a part of the
galactic center. The contours are isotherms labelled in brightness temperature of the CO radiation at 115.271 GHz. The fact is
that the 11 m Kitt Peak paraboloid has a factor of 20 higher resolution at this wavelength than, for example, the 150' antenna at
Green Bank, West Virginia has at the hydrogen 1421 MHz line, which
has been most generally used for such mapping. This is the largest
fully steerable antenna in this country which is generally available to astronomers. The resolution of the usual hydrogen 21 cm
map is 50% larger than the whole of Fig. 3. Since CO in particular
seems to be at least as generally distributed as atomic hydrogen
and is comparably bright, it seems that this technique will add
much information related to galactic structure.

To demonstrate that one can not only use molecular spectra
to learn about the interstellar medium, but also the converse,
let us consider briefly Fig. 5. This shows two hyperfine components of the J = 1 → 0 radiation of CN, the source being the
Orion nebula. As I mentioned previously, this transition of this
simple molecule has never been directly observed in the laboratory,

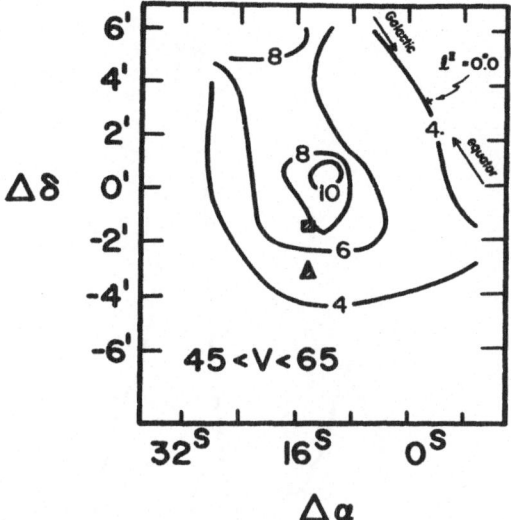

Fig. 4 Small map in the direction of the galactic center.
Contours are isotherms with the labelled brightness temperature.
The scale on the abscissa is in units of seconds of right
ascension; however, its scale is the same as the ordinate, which
is labelled in minutes of arc.

and only rough estimates of the hyperfine structure are available.

We have observed several other transitions of the hyperfine
multiplet, but many of the lines are quite weak and our data are
not yet complete. Nevertheless, we expect that this effort will
yield in the near future values of the rotational constant,
ρ-doubling constant, magnetic and quadrupole hyperfine coupling
constants, accurate to about 1 MHz, for this simple molecule.
These low-density regions, with rare collisions, can obviously be
a very interesting molecular laboratory.

In closing, I would like to illustrate a general shortage of
molecular data with a specific example.

Tables of cosmic abundances show Mg and Fe as reasonably
abundant. More so than, for example, Si. We have observed SiO
in $SgrB_2$, and the desirability of searching for MgO, FeO and
perhaps FeC is apparent. However, measurement of rotational
constants for these molecules is a prerequisite to a search. The

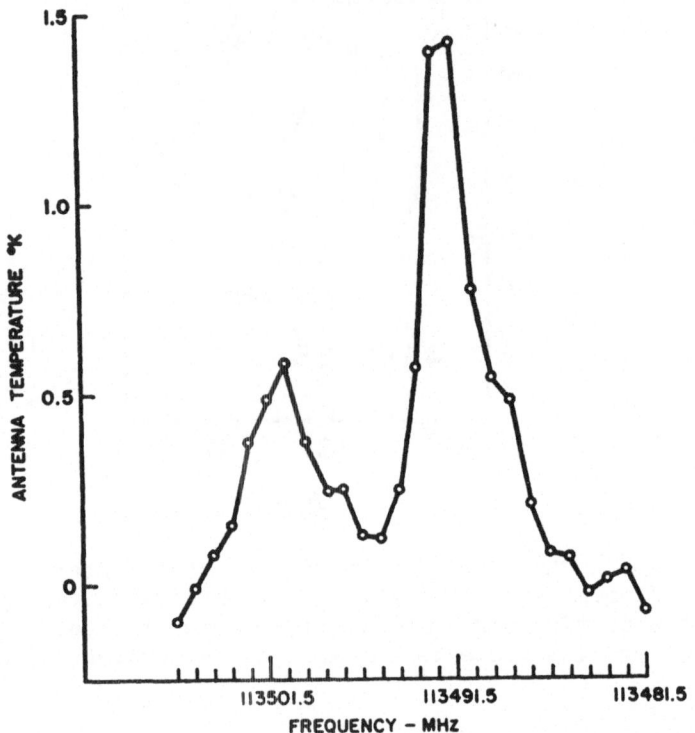

Fig. 5 CN J = 1 → 0 transition in the Orion Nebula, showing two hyperfine components.

refractory nature of these molecules obviously complicates the experimental problem, but I can imagine some ways that these experiments might be done. I would like to interest someone in doing them.

REFERENCES

No effort has been made to provide a complete bibliography. Ref. 8 contains an excellent and timely bibliography.

1. T. Dunham and W. S. Adams, Publ. Astron. Soc. Pac. 49, 26 (1937).

2. S. Weinreb, A. H. Barrett, M. L. Meeks and J. C. Henry, Nature 200, 829 (1963).

3. A. C. Cheung, D. M. Rank, C. H. Townes, D. C. Thornton and W. J. Welch, Phys. Rev. Letters 21, 1701 (1968).

4. A. C. Cheung, D. M. Rank, C. H. Townes, D. C. Thornton and
    W. J. Welch, Nature <u>221</u>, 626 (1969).

5. L. E. Snyder, D. Buhl, B. Zuckerman and P. Palmer, Phys. Rev.
    Lett. <u>22</u>, 679 (1969).

6. L. E. Snyder and D. Buhl, Ap. J. <u>163</u>, L47 (1971).

7. J. A. Ball, C. Z. Gottlieb, A. E. Lilley and H. E. Radford,
    Ap. J. <u>162</u>, L203 (1970).

8. D. M. Rank, C. H. Townes and W. J. Welch, Science <u>174</u>, 1083
    (1971).

INNER-SHELL IONIZATIONS IN THE INTERSTELLAR MEDIUM

Jon C. Weisheit

Lawrence Livermore Laboratory

Livermore, California   94550

I.   INTRODUCTION

An understanding of the dynamics of interstellar matter and of the formation of stars requires in part knowledge of the ionizing processes in interstellar clouds and in the intercloud medium. Observations[1,2] now indicate that the ionization rate per hydrogen atom, ζ_H, is at least 2×10^{-15} sec^{-1} in both interstellar regimes, and that the cloud and intercloud temperatures are typically 100°K and 10,000°K, respectively. Given these temperatures, some non-thermal mechanism must be invoked to explain this large value of ζ_H; however, there exists no direct observational evidence of any such mechanism.

Several theoretical models which predict values of ζ_H in harmony with the observationally determined value have been constructed,[3] and these are summarized in the next section. Future observations of the interstellar medium by satellite-borne ultraviolet spectrographs may resolve which, if any, of the models is appropriate.[4,5] This paper discusses some of the physical processes that must be considered for a correct interpretation of such observations: namely, effects on the ionization equilibria of interstellar carbon, nitrogen and oxygen that result from inner-shell ionization and subsequent transitions. Finally, the relevance of these processes to the understanding of spectra of various types of astronomical objects is noted.

II.   IONIZATION OF INTERSTELLAR MATTER

Table I lists some measured properties that are representative

363

of the interstellar medium. Observations also indicate that most interstellar clouds are not gravitationally bound. This fact has been interpreted as evidence that the clouds and intercloud gas are in pressure equilibrium, a situation not excluded by the observed temperatures and densities given in Table I.

A diffuse background of cosmic X-radiation has been observed at photon energies above 1/4 keV,[6] and cosmic-ray particles have been detected with energies as low as ~100 MeV/nucleon.[7] However, neither the measured fluxes nor fluxes obtained by reasonable extrapolations of them to lower energies are sufficient to ionize hydrogen at the rate inferred from observations.

Field, Goldsmith and Habing[8] constructed a model of the interstellar medium by assuming a uniform gas pressure and by assuming that heating and ionization are in steady-state equilibrium with cooling and recombination. They then showed that a large flux of low energy (~2 MeV) cosmic-ray protons could sustain two thermally stable gas phases in the interstellar medium, and furthermore, that the properties of these two phases were in good agreement with measured values.

Subsequently, Silk and Werner[9] and, independently, Sunyaev[10] demonstrated that a large uniform flux of very soft (<1/4 keV) X-rays also could provide the necessary heating and ionization.

Bottcher et al[11] and McCray and Schwarz[12] recently proposed a time-dependent model in which the ionization of interstellar matter is caused by ultraviolet or soft X-radiation from supernovae outbursts which occur approximately every 10^6 years in a given region of space. Properties of the interstellar medium predicted by the time-dependent model also agree with the general characteristics that are observed.

Silk[4] and Silk and Brown[5] computed the ionization equilibria of several trace elements in an interstellar medium ionized by a steady-state flux of low-energy cosmic rays and soft X-rays,

TABLE I. Observationally Determined Properties of
the Interstellar Medium

	Cloud	Intercloud
neutral density (cm^{-3})	~10	<0.5
fractional ionization	?	0.16
kinetic temperature ($^\circ K$)	70	>500
ionization rate (sec^{-1} H-atom^{-1})	2-4 x 10^{-15}	2-15 x 10^{-15}

respectively. Analogous calculations were also performed by Jura and Dalgarno[13] and Schwarz[14] for the time-dependent model. Significantly different abundances of more highly ionized species are predicted by all three models. These differences in ionization equilibria result in measurable differences in the strengths of interstellar resonance-absorption lines in the spectra of hot stars.

The effect of the diffuse X-ray background on the ionization equilibria of trace elements was neglected in the calculations cited above. However, Weisheit and Dalgarno[15] showed that this radiation strongly affects the ionization equilibrium and, hence, the absorption-line strengths of interstellar carbon. Therefore, ionization due to the diffuse X-ray background is included in the calculations discussed here.

III. INNER-SHELL IONIZATIONS OF CARBON, NITROGEN AND OXYGEN

A. Cosmic-Ray and X-ray Ionization Cross Sections

For the ionization of C, N, and O ions by protons of energy ~1 MeV, L-shell cross sections are larger than K-shell cross-sections by factors ranging up to two orders of magnitude. However, the relative importance of K-shell ionizations increases as the target becomes more highly ionized. In addition, much of the total L-shell cross section is due to the ionization of (inner) L_I-subshell electrons. These points are illustrated by the results presented in Table II. Listed are values of the ratios (σ_V/σ_{TOT}) and (σ_K/σ_{TOT}), where σ_V, σ_K and σ_{TOT} are, respectively, the cross sections for ionization of a valence electron and a K-shell electron, and the total ionization cross section, for collisions of C, N, and O ions with 2 MeV protons.

TABLE II. The Cross Section Ratios (σ_V/σ_{TOT}) and (σ_K/σ_{TOT}) for Ionization by 2MeV Protons

Ion	(σ_V/σ_{TOT})	(σ_K/σ_{TOT})	ION	(σ_V/σ_{TOT})	(σ_K/σ_{TOT})
CI	0.62	0.021	NV	0.75	0.25
CII	0.37	0.052	OI	0.82	0.0084
CIII	0.88	0.12	OII	0.66	0.022
CIV	0.75	0.25	OIII	0.53	0.039
NI	0.72	0.015	OIV	0.34	0.066
NII	0.56	0.033	OV	0.88	0.12
NIII	0.35	0.064	OVI	0.77	0.23
NIV	0.88	0.12			

Proton impact ionization cross sections used herein were computed from the binary-encounter formulae.[16] There are no measurements for ionized targets, but for neutral targets the results are in harmony with experimental data.[17]

The importance of inner-shell photoionization is borne out by the sharp increase of (neutral) atomic X-ray ionization cross-sections observed at subshell edges. This increase is especially pronounced for the K-shell, where Viegele et al[18] have found from mass absorption data that the ratio of the total absorption cross section just to the short-wavelength side of the K-edge to that just to the long-wavelength side behaves as $[3 + 125/\mathcal{Z}]$, where $\mathcal{Z}$ is the nuclear charge. The absorption cross section ratio at the L_I edge is about 1.2 for all atoms.[18]

Viegele et al have tabulated atomic X-ray ionization cross sections for $\mathcal{Z} = 1$ to $\mathcal{Z} = 94$, in the photon energy range of 0.1 keV to 1000 keV. However, so far as the author is aware, there are no mass absorption data for positive ions. The screened-hydrogenic approximation[19] accurately reproduces neutral-atom K-shell photoionization cross sections, and in the present work this approximation was employed to compute K-shell photoionization cross sections of neutral and ionized atoms. Unfortunately, the atomic L-shell cross sections for X-ray ionization that are computed by using the screened-hydrogenic approximation do not agree well with the cross sections determined from mass absorption data. There are, moreover, only a few positive ions for which detailed 2s- and 2p- photoionization cross section calculations have been performed. Hence, the following prescription was devised to obtain the X-ray ionization cross sections of 2s and 2p electrons of C, N, and O ions.

The universal mass absorption curve given by Henke et al[20] was used to generate quickly the total L-shell photoionization cross sections of neutral C, N, and O atoms. Then, it was assumed that the neutral and ionic L-shell photoionization cross sections of a given element are all equal. Finally, at photon wavelengths such that 2s and 2p electrons both may be photoejected, the total L-shell cross section was divided into cross sections for the photoionization of 2s and 2p electrons according to the ratio of screened-hydrogenic 2s and 2p cross sections computed at the same wavelengths.

Figures 1a-1c compare the total photoionization cross sections for neutral C, N, and O determined from mass absorption data[18] with those which Henry[21,22] calculated by using the Hartree-Fock formalism and with those obtained herein. Figures 2a-2c compare the total OII photoionization cross section, the OI(2s) photoioni-

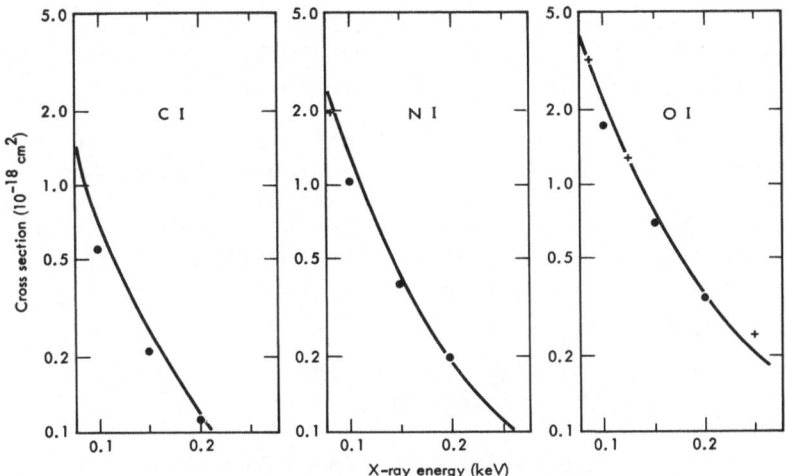

Figures 1a-1c: Total X-ray ionization cross sections of neutral nitrogen (NI), and neutral oxygen (OI) obtained herein are plotted as solid lines. Those computed by Henry (Refs. 21, 22) and those determined from mass absorption data (Ref. 18) are plotted as crosses (+) and filled circles (●), respectively.

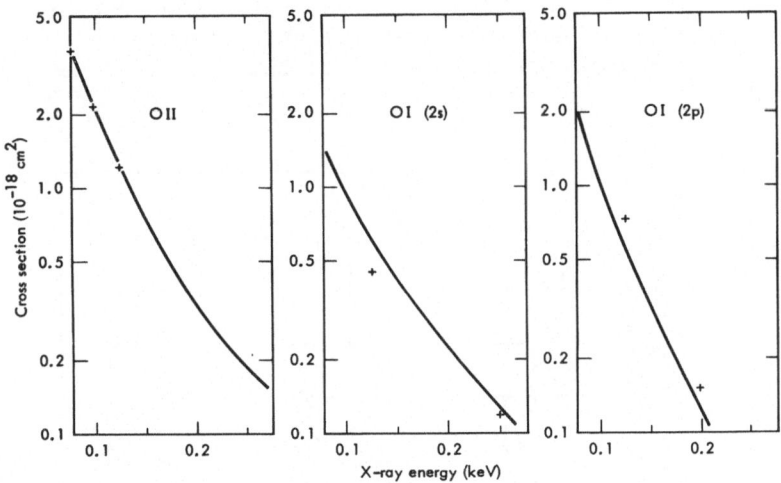

Figures 2a-2c: As in Figures 1a-1c, for the total X-ray ioniza- tion cross section of singly-ionized oxygen (OII) and for the OI (2s) and OI (2p) X-ray ionization cross sections.

zation cross section, and the OI(2p) photoionization cross section, all computed by Henry, with those determined by the present scheme. The agreement is quite good for all these cases.

For the lithium-like ions CIV, NV and OVI, the quantum defect method[23] should give reliable photoionization cross sections near threshold. The prescription used here predicts threshold cross sections for these ions that are less than a factor of 3 greater than the threshold values predicted by the quantum defect method, and at shorter wavelengths the agreement is much better.

B. Decay of Carbon, Nitrogen and Oxygen Ions With an Inner-Shell Vacancy

The three distinct types of transitions that are known to occur in a very light element ($z \leq 10$) with an inner-shell vacancy are depicted schematically in Figure 3, for the case of a K-shell vacancy. For these elements, neutral or ionized, only a radiative transition is energetically possible following the creation of a vacancy in the 2s subshell. However, experiments[24] and calculations[25] have verified that the probability of a radiative transition following the creation of a K-shell vacancy (the K-shell fluorescence yield, ω_K) is less than 0.02 for elements with $z \leq 10$. There are no precise measurements of the probabilities of radiative Auger transitions, but Åberg[26] recently computed the probability of such a transition in each of several light elements and found in all cases that the probability was less than $0.1\,\omega_K$. Hence, the probability of an Auger transition is nearly unity for $z \leq 10$.

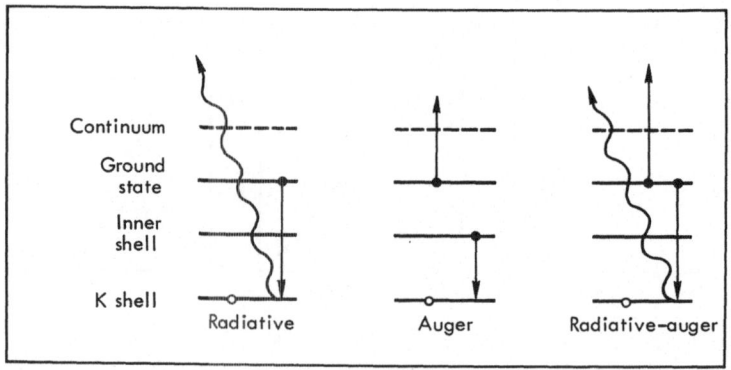

Figure 3: Types of transitions known to occur after the creation of a K-shell vacancy in very light ($z \leq 10$) elements. The straight and wavy arrows represent electron and photon transitions, respectively.

There are no measurements or calculations of the probability of any type of transition following the creation of a K-shell vacancy in a positive ion. A crude (hydrogenic) approximation indicates that Auger yields are not strongly energy dependent but that fluorescence yields vary essentially with the cube of the transition energy. (The two yields are comparable at transition energies of about 10 keV.) In the present investigation it is assumed for each of the elements C, N, and O that the ionic and neutral Auger yields are the same.

Thus, the importance of K-shell ionizations is considerably enhanced by the fact that (if there are at least two L-shell electrons) a K-shell vacancy almost always results in an additional ionization.

IV. IONIZATION EQUILIBRIA OF INTERSTELLAR CARBON, NITROGEN AND OXYGEN

The approximation that $\omega_K = 0$ for the ions CI through CIII, NI through NIV, and OI through OV was made to simplify the equations of ionization equilibrium. (In other words it was assumed that, if an Auger transition can occur, it does.) It then follows that for each element the equations of ionization equilibrium are

$$n_z \zeta_z^{eff} = n_e n_{z+1} \alpha_{z+1} \quad , \tag{1}$$

where n_e is the electron density and where n_z, ζ_z^{eff}, and α_z are, respectively, the number density, the effective ionization rate, and the recombination rate coefficient of the (z-1)-times ionized atom. If the K- and L-shell ionization rates are denoted by ζ_z^K and ζ_z^L, then for an atom of nuclear charge $\mathfrak{z}$ the effective ionization rate is

$$\begin{aligned}
\zeta_z^{eff} &= \zeta_z & , z = 1 \\
&= \zeta_z + n_e \alpha_z (\zeta_{z-1}^K / \zeta_{z-1}^{eff}), & \mathfrak{z} > \mathfrak{z} - z \geq 2 \quad (2) \\
&= \zeta_z & , 2 > \mathfrak{z} - z \geq 0 \quad ,
\end{aligned}$$

where $\zeta_z = \zeta_z^K + \zeta_z^L$. Equation (2) is correct only for $\mathfrak{z} \leq 10$ since, for heavier elements, there may be several possible decay modes following an inner-shell ionization.

Field and Steigman[27] showed that the near-resonant charge transfer reaction OI + HII $\rightleftarrows$ OII + HI strongly affects the ionization equilibrium of interstellar oxygen; hence, this process was included in the oxygen ionization equilibrium equations. The O-H charge transfer rates of Field and Steigman, which were computed by assuming that the reaction cross section is proportional to the

orbiting cross section, were adopted. A similar calculation of
the non-resonant charge transfer reaction rate for NI + HII $\rightleftharpoons$
NII + HI was performed by Steigman et al.[28] However, experimental
data[29,30] at low collision energies do not indicate the enhancement
of non-resonant charge transfer cross sections that is predicted
by the orbiting approximation. Consequently, the charge transfer
reaction was not included in the nitrogen ionization equilibrium
equations.

The set of equations (1) was solved to obtain the ionization
equilibrium of carbon, nitrogen, and oxygen both in an interstellar
cloud and in the intercloud medium. In accordance with measured
values listed in Table I, the following values were adopted for
the cloud and the intercloud medium, respectively: hydrogen densities of 10 and 0.1 cm^{-3}, electron densities of 0.03 and 0.02 cm^{-3},
and kinetic temperatures of 100 and 10000°K. The radiative recombination rate coefficents were computed from the data given by
Tarter.[31]

Three different steady-state ionization models were considered:
a) Ionization only by the measured isotropic X-ray background above
0.15 keV. The X-ray spectrum used in the calculations is $4\pi x 55 E^{-1.4}$
photons cm^{-2} sec^{-1} keV^{-1} for $0.15 < E < 0.30$, and $4\pi x 11 E^{-1.4}$ photons
cm^{-2} sec^{-1} keV^{-1} for $0.30 < E < 18.0$ (Dalgarno and McCray[3]). b) Ionization by the X-ray background and by a δ-function photon flux at
0.1 keV sufficient to produce the measured ionization rate of hy-
drogen, $\zeta_H = 2 \times 10^{-15}$ sec^{-1}. c) Ionization by the X-ray background
and by a δ-function proton flux at 2 MeV sufficient to produce the
measured hydrogen ionization rate.

In models b) and c) ionizations of hydrogen by energetic secondary electrons were considered in determining the ionizing flux.
In addition the ionization of CI by diffuse starlight, at a rate[15]
of 1.7×10^{-10} sec^{-1}, was included in the equilibrium calculations
for all three ionizing models.

The equilibria in the intercloud medium that result from the
hypothetical ionizing fluxes included in models b) and c) are shown
in Figure 4. It is obvious that the X-ray ionization model produces
the higher degree of ionization. This conclusion is also true for
an interstellar cloud, although the cloud's lower temperature se-
verely decreases the abundances of more highly charged ions. Ionization model a) predicts equilibria that are sharply peaked at CII,
NI and OI, and that have percent abundances in the higher stages
of ionization that are approximately 1/10 of those predicted by
ionization model c).

The effect of Auger transitions is most pronounced when the
equilibrium is strongly peaked at one stage of ionization. Auger

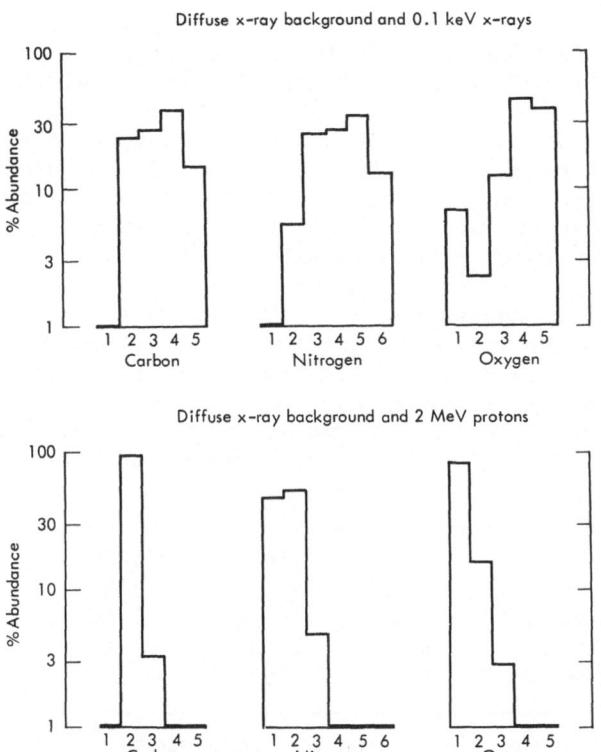

Figure 4: Ionization equilibria of carbon, nitrogen and oxygen (1→neutral atom, etc.) in the intercloud medium that result from different ionizing models.

transitions following K-shell ionizations of that ion then produce much more of the second succeeding stage of ionization than would be produced otherwise. Consider, for example, the ionization of an interstellar cloud by only the diffuse background X-radiation. In this case, the abundances of CIV, NIII and OIII are increased by factors of 2800, 57, and 276, respectively, when Auger transitions are included in the ionization equilibrium equations. However, for the case of ionizations by the diffuse X-ray background and by cosmic rays, the abundances of CIV, NIII and OIII are increased by factors of only 5.9, 1.04, and 1.18, respectively, when Auger transitions are included in the equilibrium equations.

The neglect of all inner-shell ionizations significantly alters the resultant equilibria. For instance, if only valence-electron ionizations of nitrogen are considered, in an intercloud medium ionized by just the diffuse X-ray background the abundances of NII, NIII and NIV are decreased by factors of 4.4, 88, and 2000, respectively.

Table III lists computed strengths of ultraviolet absorption
lines arising from the lowest fine-structure level of each of several interstellar C, N, and O ions. Equivalent widths are given
for all three ionizing models. The results of a previous calculation by Silk[4] for the cosmic-ray ionization model are tabulated for
comparison. A radial velocity dispersion of 8 km sec^{-1} and a path
length of 100 pc were assumed in all of the calculations. The
oscillator strengths of the transitions in Table III were taken
from the recent compilation by Smith.[32]

TABLE III. Equivalent widths* of ultraviolet absorption
lines in the intercloud medium. Ionization models a),
b) and c) are described in the text; ionization model d)
is the model of cosmic-ray ionizations described by Silk.

Transition	Ionization Model			
	a)	b)	c)	d)
CIII (λ977.0)	55.6	169.	130.	144.
CIV (λ1548.2)	51.6	279.	109.	36.
NI (λ1196.6)	193.	9.38	176.	194.
NV (λ1238.8)	$\sim 10^{-3}$	178.	0.34	--
OI (λ1302.2)	212.	139.	211.	--

*All equivalent widths are in units of 10^{-3} Å

It appears that a measurement of the strength of the CIV line
offers the best possibility of determining if either of the proposed steady-state ionizing fluxes exists: Additional ionization
by cosmic rays or by X-rays results in an equivalent width two or
five times larger, respectively, than the width that results from
ionization by only the diffuse X-ray background. The present interpretation of such a measurement, however, would be uncertain
because calculations for the time dependent model, which also predicts a strong CIV line, did not include the ionizations due to
the diffuse X-ray background or Auger transitions following K-shell
ionizations. Hopefully, future calculations for the time-dependent model of the interstellar medium will consider these effects.

V. COMMENTS

Besides the need for revised calculations for the time-dependent model, as mentioned above, other subjects related to ionization equilibria in the interstellar medium now require new attention. These include an amended determination of the rate of
cooling of interstellar matter, which is strongly dependent upon
the abundances of various ions of trace elements.[3]

In addition, the line spectra observed in a variety of astronomical objects, including supernovae, galactic nuclei and quasi-
stellar objects, are poorly understood at present. Highly energetic phenomena are associated with these objects, and so inner-
shell ionizations are probably important in determining the
emergent spectra. Investigations relevant to some of these topics
are now in progress, but clearly much work needs to be done.

With respect to basic atomic processes, there are numerous
problems to be examined. In conclusion, we simply list some of
them: 1) What improvements and extensions can be made for the
prescription used here to obtain X-ray cross sections of positive
ions? 2) What are the decay rates for a multiply-charged ion
with an inner-shell vacancy? 3) What is the net effect on the
ionization equilibria of heavy (ζ >10) systems due to the fact that
several competitive transitions are possible after the creation of
an inner-shell vacancy? 4) Finally, what is the importance of
multiple ionizations, and what transitions follow the simultaneous
creation of more than one inner-shell vacancy?

I wish to thank Prof. A. Dalgarno and Dr. M. Rees for discussions during which many of these ideas developed. Prof. C. F. McKee
is also thanked for a critical reading of the manuscript. This work
was performed under the auspices of the U.S. Atomic Energy Commission.

REFERENCES

1. R. H. Hjellming, C. P. Gordon, and K. J. Gordon, Astron. and
Astrophys. 2, 202 (1969).

2. M. P. Hughes, A. R. Thompson, and R. S. Colvin, Astrophys. J.
Suppl. 23, 323, (1971).

3. A. Dalgarno and R. A. McCray, Ann. Rev. and Astrophys. 10
(Palo Alto, Calif.; Annual Reviews, Inc.) in press; and references cited therein.

4. J. Silk, Astrophys. Letters 5, 283 (1970).

5. J. Silk and R. L. Brown, Astrophys. J. 163, 495 (1971).

6. J. Silk, Space Science Rev. 11, 671 (1970); and references
cited therein.

7. L. Spitzer and M. G. Tomasko, Astrophys. J. 152, 971 (1968).

8. G. B. Field, D. W. Goldsmith, and H. J. Habing, Astrophys. J. 158, 173 (1969).

9. J. Silk and M. W. Werner, Astrophys. J. 158, 185 (1969).

10. R. A. Sunyaev, Astron. Zh. 46, 929 (1969).

11. C. Bottcher, R. A. McCray, M. Jura, and A. Dalgarno, Astrophys. Letters 6, 237 (1970).

12. R. A. McCray and J. Schwarz, The Gum Nebula and Related Problems, ed. S. P. Maran, J. C. Brandt, T. P. Stecher, NASA N-683-71-374, (1971).

13. M. Jura and A. Dalgarno, Astrophys. J. 174, 365 (1972).

14. J. Schwarz, unpublished thesis (Harvard U.), (1972).

15. J. C. Weisheit and A. Dalgarno, submitted to Astrophys. Letters (1972).

16. J. D. Garcia, E. Gerjuoy, and J. E. Welker, Phys. Rev. 165, 66 (1968).

17. J. D. Garcia, Phys. Rev. A1, 1 (1970).

18. W. J. Veigele, E. Briggs, L. Bates, E. M. Henry, and B. Bracewell, Defense Nuclear Agency Report No. DNA 2433F (1971).

19. A. J. Bearden, J. Appl. Phys. 37, 1681 (1966).

20. B. L. Henke, R. White, and B. Lundberg. J. Appl. Phys. 28, 98 (1957).

21. R. J. W. Henry, Planet. and Space Sci. 15, 1747 (1967).

22. R. J. W. Henry, J. Chem. Phys. 48, 3635 (1968).

23. A. Burgess and M. J. Seaton, Mon. Not. Roy. Astron. Soc. 120, 121 (1960).

24. W. Hink and H. Paschke, Phys. Rev. A4, 507 (1971); and references cited therein.

25. D. L. Walters and C. P. Bhalla, Phys. Rev. A3, 1919 (1971); and references cited therein.

26. T. Åberg, Phys. Rev. A4, 1735 (1971).

27. G. B. Field and G. Steigman, Astrophys. J. <u>166</u>, 59 (1971).

28. G. Steigman, M. W. Werner, and F. M. Geldon, Astrophys. J. <u>168</u>, 373 (1971).

29. P. H. Edmonds and J. B. Hasted, Proc. Phys. Soc. A <u>84</u>, 99 (1964).

30. D. Rapp and W. E. Francis, J. Chem. Phys. <u>37</u>, 2631 (1962); and references cited therein.

31. C. B. Tarter, Astrophys. J. <u>168</u>, 313 (1971), and private communication.

32. W. Smith, private communication (1972).

ATOMIC PHYSICS IN THE UPPER ATMOSPHERE

Eldon E. Ferguson

NOAA Environmental Research Laboratories

Boulder, Colorado   80302

INTRODUCTION

The irradiation of a planetary atmosphere by solar ultraviolet light initiates a wide variety of atomic and molecular processes which lead to a multitude of observable atmospheric properties. The branch of geophysics which attempts to arrive at an understanding of this complex situation is called aeronomy. The processes initiated by radiation are fairly obvious; molecules are vibrationally and electronically excited, dissociated, and ionized. A variety of radiative and collisional processes then control the energy degradation of the ions, electrons, and neutrals, and a variety of reaction processes control the ion and neutral composition.

Some of the resulting properties are well measured, e.g. the electron density and electron temperature profiles and certain airglow emission profiles. Some of the properties are essentially unmeasured, e.g. molecular vibrational temperatures and certain minor neutral constituent concentrations. Many of the properties of most interest in connection with atomic physics, e.g. the positive and negative ion concentrations, are known only qualitatively or at least with much less precision than we would like.

In order to obtain the very detailed (yet still incomplete) understanding of the earth's upper atmosphere that now exists, it has been necessary to draw heavily on available knowledge of atomic and molecular processes, both experimental and theoretical. Indeed, a great deal of atomic physics research has been motivated by aeronomy and the tie between aeronomy and atomic physics has historically been a very close one.

Among theoreticians the work of Massey, Bates, and Dalgarno
and their schools has been preeminent over a very wide range of
atmospheric atomic physics. The necessarily more specialized
research of experimentalists has the consequence that experimental
advance in atmospheric atomic physics has been spread over a much
larger number of individuals and laboratories.

It is, obviously, out of the question to give a comprehensive
review of the subject in the space allotted and I will discuss a
few topics of current, and hopefully somewhat general interest.
Similar discussions in similar forums have been given by Donahue[1]
at the Fifth International Conference on the Physics of Electronic
and Atomic Collisions in Leningrad in 1967 and by Dalgarno[2] at the
Seventh ICPEAC in Amsterdam a year ago.

ACCIDENTAL RESONANCE CHARGE-TRANSFER REACTIONS

First consider briefly a problem concerning an accidental
near resonance charge-transfer reaction

$$O^+ + H \underset{k_{-1}}{\overset{k_1}{\rightleftarrows}} H^+ + O \quad . \tag{1}$$

This reaction has been of interest in aeronomy because it is the
major source of H^+ in the ionosphere. It has been widely discussed in atomic physics in connection with the contentious question of whether the cross section of an accidentally resonant
charge-transfer reaction approaches zero or not at very low relative impact energies, specifically thermal. It has been assumed
for many years by most aeronomers that reaction (1) and its inverse
are so rapid that equilibrium exists below ~ 600 km in the ionosphere, leading to

$$[H^+]/[O^+] = R [H]/[O] \quad . \tag{2}$$

The statistical weight ratio $R = 9/8$ is a good approximation at
temperatures greater than $\sim 2000°K$, but R increases to 1.8 at
$300°K$. However, as recently as 1967, Donahue[1] has questioned the
assumption of equilibrium and suggested that atmospheric observations indicated that the theory of Bates and Lynn[3], who proposed
that the cross section would rapidly go to zero at low velocity,
might be correct. Dalgarno[4] has discussed the history of both the
geophysics and the theoretical development on this problem, em-
phasizing the uncertainty in rate constants for asymmetric resonance
charge-transfer reactions at thermal energy. This matter seems to
now be reasonably well resolved experimentally, both in the specific case of reaction (1) and more generally. In the case of (1)
the inverse reaction (-1) has been measured in a thermalized $(300°K)$

flowing afterglow[5] with the result, $k_{-1} = 3.8 \times 10^{-10}$ cm^3/molecule sec, or $\bar{\sigma}_{-1} = 14.5$ Å^2. This value of $\bar{\sigma}_{-1}$ agrees remarkably well with the value 14.1 Å^2 obtained by a long extrapolation of the data obtained by Stebbings, Smith, and Ehrhardt[6] using the formula

$$\sigma^{\frac{1}{2}} = a - b \log E \qquad (3)$$

which they found to fit the data for H$^+$ kinetic energies from 50 - 10,000 electron volts. This seems to establish the disputed validity of the extension of (3) from the symmetric resonance charge-transfer case for which it is derived to an accidental resonance charge-transfer, at least in this case. From use of the calculated equilibrium constant one then obtains $k_1 = 6.8 \times 10^{-10}$ cm^3/molecule sec or $\bar{\sigma}_1 - 27$ Å^2. The value of $\bar{\sigma}_1$ deduced from $\bar{\sigma}_{-1}$ and the equilibrium constant is in remarkably good agreement with a value 20-40 Å^2 obtained by Stebbings and Rutherford[7] by extrapolating beam measurements from 0.6 eV center of mass kinetic energy.

There is of course a broader question involved than this particular reaction, the question of thermal energy charge-transfer rate constants generally. Many measurements of exothermic thermal energy charge-transfer reactions with molecules have been made in the last half-dozen years and they are almost always fast. Presumably accidental resonances are established by vibrational and rotational excitation, although product states have almost never been determined. This clearly implies either that accidental resonance charge-transfers are quite generally fast at 300°K, contrary to at least some theoretical expectation, or that non-resonant charge-transfer is fast which would in effect concede this point as well. Because the ion-induced dipole attraction energy at close approach exceeds kT at 300°K, one does not expect any decrease in rate constant at temperatures below 300°K.

One area in which atomic beam experimentalists could now aid aeronomy is in the determination of rate constants (or cross sections) for charge-transfer of doubly charged ions at as low an energy as possible. Altitude profiles for the ions O^{++} and N^{++} have now been observed with rocket and satellite borne mass spectrometers by Hoffman,[8] but their loss rates by charge-transfer with He, O, O$_2$, and N$_2$ are completely unknown, precluding a detailed analysis of the atmospheric physics.

NEGATIVE ION PROCESSES

Next, I shall discuss in a somewhat general way the atomic and molecular aspects of atmospheric negative ions. This is a topic of greatly renewed current interest because the first successful atmospheric negative ion composition measurements have just recently been reported in the literature.[9,10] Appreciable numbers of negative ions have been observed between 70 and 85 km, the D-region of the ionosphere, and very small but detectable concentrations have been observed at still higher altitudes.

Detailed laboratory studies have preceded the in situ observations so that elaborate reaction schemes are available (Fig. 1) and we now have the prospect of a test of these laboratory derived schemes in the forseeable future. This has not as yet been done due to uncertainties in important atmospheric parameters (particularly minor species concentrations), the complexity of the reaction scheme, and the as yet qualitative nature of the negative ion composition observations.

The major formation process for negative ions, collisional attachment

$$e + 2O_2 \rightarrow O_2^- + O_2 \tag{4}$$

was measured very early, in the laboratory aeronomy time scale, by Chanin, et al.[11] Very recently Spence and Schulz[12] have again studied this three-body attachment, this time in an electron beam experiment in which they were able to show structure (as a function of electron energy) corresponding to the vibrational levels of the O_2^- compound state which serves as an intermediate in the two-stage attachment and stabilization process, first proposed by Bloch and Bradbury[13] in 1935.

The first measured atmospheric loss process for O_2^- was photodetachment measured by Burch, et al.[14] in 1958. Phelps and Pack[15] measured collisional detachment (the inverse of (4)), which is not important at D-region gas temperatures (which are below 300°K), but this did allow a determination of the electron affinity of $O_2(0.46$ eV) which is an extremely important atmospheric quantity. In spite of this early determination by Phelps and Pack there has been a great deal of controversy about the O_2 electron affinity and as recently as 1969 values greater than 1 eV were entering the literature. This matter has now been firmly resolved by the elegant photodetachment electron spectroscopy carried out at JILA by Celotta, et al.[16] which confirmed the Phelps and Pack value.

The most important O_2^- loss in the atmosphere turns out to be associative detachment with atomic oxygen,

$$O_2^- + O \rightarrow O_3 + e \qquad\qquad (5)$$

which was predicted by Dalgarno in 1961[17] and which was measured
in 1966.[18] The measurement requires a capability of reacting
thermal energy ions with unstable neutrals, such as O atoms, and
the flowing afterglow technique developed in Boulder at NOAA[19] is
still unique in this capability.

In addition to O_2^- loss by (5), it has also been found from
laboratory studies that O_2^- charge-transfers to O_3 to produce
O_3^-[20] and from a number of drift tube studies that O_2^- associates
with O_2 to form O_4^-.[21-23] By a series of reactions (that are get-
ting a bit chemical for this forum), O_3^- and O_4^- eventually lead
to NO_3^- for which as yet no reactions have been found that lead to
other negative ions. This reaction scheme is shown in Fig. 1.
It has been found[24,25] that negative ions hydrate, indeed Payzant,
et al.[24] have found the $NO_3^- \cdot H_2O$ bond energy to be 0.54 eV. The
first negative ion observations in the D-region[9] did find a pre-

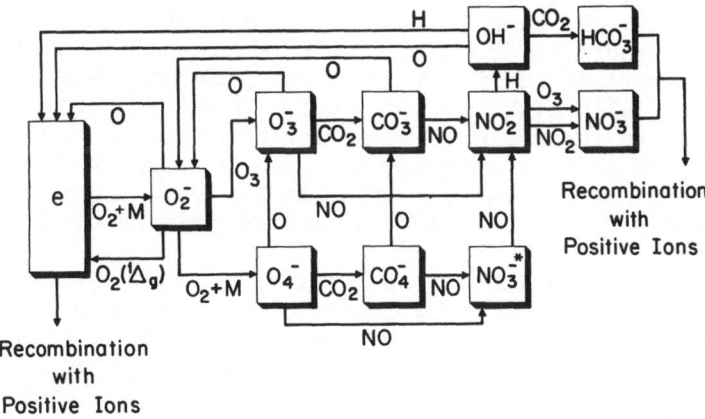

Figure 1.   Schematic Outline of Atmospheric Negative Ion Reactions,
            Excluding Hydration.

dominance of NO_3^- and its hydrates, which was gratifying. However, the second observation[10] found a somewhat different negative ion composition which is confusing.

The question of the NO_3 electron affinity thus is of considerable importance and seems to now be resolved by two quite different current experiments. In our laboratory[26] we have made the chance observation that the reaction

$$NO_3^- + HBr \rightarrow Br^- + HNO_3 \tag{6}$$

is nearly thermoneutral, with the result that $EA(NO_3) = 3.9 \pm 0.2$ eV. This appears to be the largest electron affinity yet established. Payzant, et al.[24] have determined the same value from a measurement of the hydration energy of NO_3^-! They have done this by using a quite remarkable correlation between hydration energy and electron affinity. (This correlation at least seems remarkable to most physicists, although it is more familiar to chemists as an example of the well-known free energy relationships in chemistry.)

In a similar vein, the electron affinity of NO_2 is now reasonably well determined to be 2.38 ± 0.06 eV by several different current experiments, all in good agreement.[24,27-29] The electron affinity of NO_2 has been an intensively studied but very elusive molecular property to determine, with values ranging from 1.6 to 4.0 eV encumbering the literature. The fact that ions like NO_2^- have substantially different geometries than their parent neutral complicates the threshold behavior which is relied on for energy determinations in many experiments. The calculation of electron affinities for molecules as complex as NO_2 still seems to be beyond the state of the art in that rapidly progressing field.

Negative ions which are not observed as such in the atmosphere can also play a powerful role in aeronomy, I refer to the autodetaching resonances which lead to vibrational excitation. It has been known for some time, especially from the work of Schulz,[30] that the cross section for nitrogen vibrational excitation

$$e + N_2 \rightarrow N_2^{-*} \rightarrow N_2(v) + e \tag{7}$$

is very large and this has been well supported by theory.[31] The state of N_2 vibrational excitation is very critical in F-region aeronomy where the reaction

$$O^+ + N_2 \rightarrow NO^+ + N \tag{8}$$

is the rate controlling step for electron loss, reaction (8) being then followed by rapid dissociative recombination of the NO^+.

Schmeltekopf, et al.[32] have shown that the rate constant for (8) increases by a factor of 60 for an increase of N_2 vibrational temperature from 300-6000°K. Vibrational temperatures of N_2 as high as 6000°K have been observed[33] in auroras, and enhanced atmospheric vibrational temperatures undoubtedly occur for other conditions as well. In an abstract submitted to this meeting, R. H. Neynaber and G. D. Magnuson report a study of reaction (8) in a merged beam apparatus over a range of interaction energies from 0.05 to 15 eV, with and without the nitrogen being vibrationally excited. The large vibrational enhancement of the cross sections at low kinetic energy reported in Ref. 32 is confirmed. Neynaber and Magnuson have deduced details of the reaction mechanism from their very interesting measurements.

The process by which negative ions are eventually lost in the atmosphere is of obvious interest. The large electron affinity of NO_3 and its hydrates drastically reduces the solar flux that can be effective for possible photodetachment (the cross sections for which have not as yet been measured!) so that recombination of positive and negative ions is probably the most important eventual loss process.

Remarkable progress has occurred in this field in recent years, largely by the application of merged beam techniques at the Stanford Research Institute.[34-36] The systems $H^+ + H^-$, $O^+ + O^-$, $N^+ + O^-$, $N_2^+ + O_2^-$, $O_2^+ + O_2^-$, $O_2^+ + NO_2^-$, $NO^+ + NO_2^-$, $O_2^+ + O^-$, $NO^+ + O^-$, and $Na^+ + O^-$ have so far been studied in a relative energy range as low as 0.15 eV and up to several hundred volts. The SRI group has developed extrapolation procedures to obtain thermal energy recombination coefficients. The data so far obtained on ion-ion recombination seems to be recapitulating the experience on molecular ion-electron dissociative recombination, namely the recombination coefficient is not drastically sensitive to the ions involved. The range of extrapolated recombination coefficients for the above reactions is only from 10 ± 4 to $51 \pm 15 \times 10^{-8}$ cm^3/sec. The ion-ion recombination coefficients are of about the same magnitude as most of the measured dissociative recombination coefficients of electrons and molecular positive ions. The seeming insensitivity to the molecular nature of the ions is probably more readily understandable in the ion-ion case than in the electron-ion case and indeed at least one example of a very slow dissociative recombination is known (for He_2^+), however it will be surprising if any slow ($k < 10^{-9}$ cm^3/sec) ion-ion recombination is ever found for small ions. Theory and experimental results on ion-ion recombination have been reviewed by Mahan[37] in a review in press.

One important aspect of ion-ion recombination which has received very little experimental study is the question of final products and product states. The only work along this line that I am aware of is the work of Berry and his students[38] in which Na^+ ions neutralize with O^- and H^- to yield Na D-line emission. A similar situation exists for dissociative recombination where only the work of

Zipf[39] on $O_2^+ + e \rightarrow O^* + O$ has measured product states for ionospheric ions.

NO IN THE UPPER ATMOSPHERE

As a final topic I will discuss a major problem of D-region aeronomy, that of the origin of nitric oxide in the lower ionosphere. This problem requires a knowledge of a very wide range of molecular processes for its solution, and thus represents almost a text book case of the interwoven history of aeronomy and molecular physics. This problem is timely since it appears that the key insight into the problem has occurred in the last few years. Some of the supporting atomic physics has already been carried out and some has not.

The minor constituent NO assumes an importance all out of proportion to its relative abundance because, as first suggested by Nicolet[40] in 1945, NO ionization by solar Lyman-alpha radiation constitutes the major D-region ionization source

$$NO + h\nu(1215 \text{ Å}) \rightarrow NO^+ + e \quad . \tag{9}$$

This is a product of several somewhat chance circumstances, the large intensity of Lyman-alpha(the strongest solar line), the low ionization potential (9.25 eV) of NO relative to the major atmospheric constituents and the coincidence of a narrow optical transmission window of O_2 at the Lyman-alpha wavelength.

The concentration of NO in the atmosphere was first measured by Barth[41] who measured the fluorescence of the NO gamma bands using a scanning ultraviolet spectrometer flown on an Aerobee rocket. Barth found very much larger NO concentrations than had been predicted by conventional photochemistry, instigating a search for additional production processes which has carried on to the present. The measurements of Barth have been supported by further experiments by his students, most recently in the work of Meira.[42]

The finding of unexpectedly large NO concentrations gave rise to two perplexing puzzles for aeronomers, the problem of the source of so much NO referred to above and, additionally, the problem of how to dispose of the embarrassingly large supply of NO^+ and electrons provided by its Lyman-alpha photoionization. I am now going to describe the molecular physics of these two problems as it has developed up to the present.

The first obvious NO source was

$$N + O_2 \rightarrow NO + O \tag{10}$$

which is balanced by the fast reaction

$$NO + N \rightarrow N_2 + O \tag{11}$$

leading to $[NO] = k_{10}/k_{11}[O_2]$, an NO concentration which was insufficient by a large factor to account for Barth's observation.

In view of the deficiency of (10), Nicolet[44] in 1965 proposed an ion-molecule reaction source

$$O_2^+ + N_2 \rightarrow NO^+ + NO \qquad (12)$$

which however was found to be very slow in the laboratory[43], $k_{12} < 10^{-15}$ cm^3 sec^{-1} at 300 and 600°K, (corresponding to less than one reaction per million collisions) and which subsequently has been discounted.[45]

Hunten and McElroy[46] in 1968 proposed that metastable $O_2(^1\Delta_g)$, molecules might substantially enhance the rate of reaction (10) and lead to a sufficient NO production. The concentration of $O_2(^1\Delta_g)$ was known directly from measurements of the infrared atmospheric band emission $O_2(^1\Delta_g) \rightarrow O_2(^3\Sigma_g^-) + h\nu$, by rocket borne photometers[47], leading to $[O_2(^1\Delta_g)]/[O_2(^3\Sigma_g)] \sim 3 \times 10^{-5}$ in the 70-80 km altitude range. However, measurements of Clark and Wayne[48] found a rate of only $2.8 \pm 2 \times 10^{-15}$ cm^3 sec^{-1} for

$$O_2(^1\Delta_g) + N \rightarrow NO + O \qquad (13)$$

which is too small by several orders of magnitude to provide the needed NO source.

The final variation on reaction (10) was to let the N atom be excited and Norton[49] first proposed that the reaction

$$N(^2D) + O_2 \rightarrow NO + O \qquad (14)$$

might be the missing NO source. It was subsequently shown in laboratory investigations by Black, et al.[50] that $N(^2D)$ was rapidly destroyed by O_2, $k_{14} = 7 \times 10^{-12}$ cm^3 sec^{-1} and by Lin and Kaufman[51] that the destruction was by reaction to produce NO. The question now is whether the atmospheric concentration of $N(^2D)$ is sufficient to provide the source for (14). The $N(^2D) \rightarrow N(^4S) + h\nu$ (5200°A) emission has been observed in the dayglow[52,53] but too weakly to allow concentration profiles of $N(^2D)$ to be determined. The very long $N(^2D)$ radiative lifetime of 26 hours makes its detection very difficult.

One then asks if sufficient production sources of $N(^2D)$ exist to provide an adequate source. Two principal sources have been proposed, the most important is dissociative recombination of NO^+,

$$NO^+(^1\Sigma) + e \rightarrow N(^2D) + O(^3P) + 0.35 \text{ eV} \quad (15a)$$

$$\rightarrow N(^4S) + O(^1D) + 0.85 \text{ eV} \quad (15b)$$

$$\rightarrow N(^4S) + O(^3P) + 2.75 \text{ eV}. \quad (15c)$$

The rate of NO^+ dissociative recombination is well known[37] but the branching ratio is undetermined. Using plausible values of k_{15a}, several investigations have led to sufficient atmospheric NO production.[45,54-56] These models all require k_{15a} to be 50-80% of k_{15}. The reaction channel (15b) violates spin conservation and can probably be ignored. Bardsley[57] has shown that reaction (15a) is theoretically probable. The measurement of products in reaction (15) clearly has a very high priority in laboratory aeronomy but just as clearly represents a formidable challenge.

The other possible $N(^2D)$ production mechanism is the reaction

$$N_2^+(^2\Sigma_g^+) + O(^3P) \rightarrow NO^+(^1\Sigma) + N(^2D) + 0.51 \text{ eV} \quad (16a)$$

$$\rightarrow NO^+(^1\Sigma) + N(^4S) + 3.08 \text{ eV}. \quad (16b)$$

The total rate constant $k_{16} = 2.5 \times 10^{-10} \text{ cm}^3 \text{ sec}^{-1}$ is known[58] but again the products are unmeasured. Reaction (16) is the major N_2^+ loss process in the upper ionosphere.

Since, according to present models, $N(^2D)$ is the source of ionospheric NO (reaction 14) and $N(^4S)$ is a major loss (reaction (10) it is clear that the branching ratios of processes (15) and (16) are very sensitive atmospheric parameters. Too much $N(^2D)$ production would also upset the scheme! This problem in aeronomy is on dead center pending further laboratory studies of the basic processes. One additional consideration must be made, namely are there other significant loss processes for $N(^2D)$? Lin and Kaufman[51] showed that N_2 deactivation is relatively slow compared to the O_2 reaction and so can be ignored. The other major neutral constituent in the ionosphere, atomic oxygen, has not yet been studied so far as $N(^2D)$ deactivation is concerned and that of course should be done.

The final aspect of this problem, namely disposing of the abundant source of NO^+ and electrons provided by (9) is in an even less satisfactory state. That is, both the concentrations of NO^+ and electrons predicted for the D-region far exceed the measured values. The large production imposed by (9) necessitates our finding larger loss processes than we currently know. It is necessary to have processes which rapidly convert the NO^+ to $H_5O_2^+$ and other water cluster ions below 80 km, because such water cluster ions are observed to be the dominant positive ion species in this region, rather than the primary ion NO^+ and its hydrates.[59] If we can find ways of converting $NO^+ \rightarrow H_5O_2^+$ this also helps alleviate the electron loss problem because Biondi, et al.[60] have found that the electron recombination coefficients for $H_5O_2^+$ and larger water cluster ions ($H_3O^+ \cdot nH_2O$) exceed that for NO^+.

Reaction schemes are known which do convert NO^+ (and also O_2^+) to water cluster ions,[61-67] however they do not seem to be sufficient, i.e. they do not account for the observed ion composition quantitatively or even qualitatively very well. This situation has been reviewed in some detail by Reid[68,69] and Ferguson[70,71] Figure 2 shows the laboratory derived schemes to date. It seems fairly clear that some essential molecular process remains to be recognized before this problem will be solved.

The NO production problem, while possibly solved for the normal undisturbed atmosphere by the $N(^2D) + O_2$ scheme discussed above (depending on the outcome of the necessary laboratory experiments on $N(^2D)$ production), has proven to be a major mystery in certain auroral conditions. Zipf, et al.[72] have found that NO becomes a

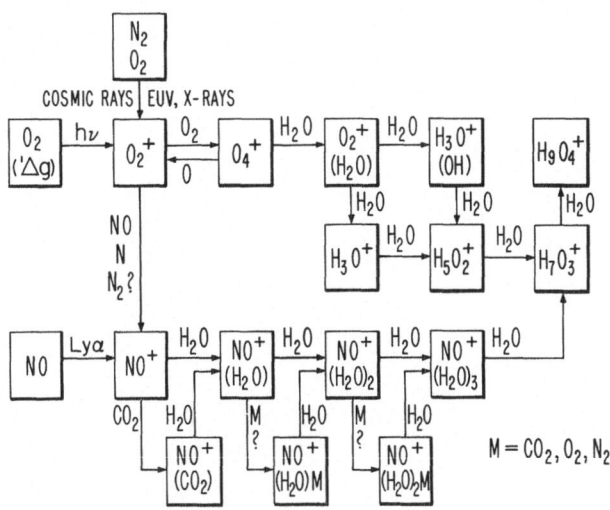

Figure 2.   Schematic Outline of D-Region Positive Ion Chemistry
            [from Ferguson, Rev. of Geophysics and Space Physics,
            9, 997 (1971)].

major atmospheric constituent in certain auroras, actually exceeding the O_2 concentration between 110 and 130 km. The way in which
this happens is quite unknown. Not only is there a problem in
identifying the reactions responsible, there is a serious problem
from an energy conservation point of view, essentially the total
(known) auroral input energy would be required to produce the large
quantities of NO observed! McElroy[73] has recently described this
as the most exciting and perplexing problem in atmospheric physics.

In conclusion, I hope that this discussion has at least indicated the nature of the interaction between atomic physics and one
of its important fields of application, atmospheric physics. I
think it is fair to say that the progress in our understanding has
been great, especially in the last decade. It is clear that much
remains to be done and that the remaining problems are becoming
increasingly difficult ones. The response to this challenge will
almost certainly guarantee another large increase in understanding
in the next decade.

REFERENCES

1. T. M. Donahue, Collision Processes Relevant to Aeronomy, The Physics of Electronic and Atomic Collisions (Joint Institute for Laboratory Astrophysics, Boulder, Colo.) L. M. Branscomb, Editor (1968).

2. A. Dalgarno, Applications in Aeronomy, VII ICPEAC, Amsterdam, July 1971. Invited Talks and Progress Reports, North-Holland Publ. Co., T. R. Govers and F. J. DeHeer, Editors, 1972.

3. D. R. Bates and N. Lynn, Proc. Roy. Soc. (London) A253, 141 (1959).

4. A. Dalgarno, Annales de Geophysique 17, 16 (1961).

5. F. C. Fehsenfeld and E. E. Ferguson, J. Chem. Phys. 56, 3066 (1972).

6. R. F. Stebbings, A. C. H. Smith, and H. Ehrhardt, J. Geophys. Res. 69, 2349 (1964).

7. R. F. Stebbings and J. A. Rutherford, J. Geophys. Res. 73, 1035 (1968).

8. J. H. Hoffman, Science 155, 322 (1967).

9. R. S. Narcisi, A. D. Bailey, L. Della Lucca, C. Sherman, and D. M. Thomas, J. Atm. Terr. Phys. 33, 1147 (1971).

10. R. Arnold, J. Kissel, D. Krankowsky, H. Wieder, and J. Zahringer, J. Atm. Terr. Phys. 33, 1169 (1971).

11. L. M. Chanin, A. V. Phelps, and M. A. Biondi, Phys. Rev. 128, 219 (1962).

12. D. Spence and G. J. Schulz, Phys. Rev. A, 5, 724 (1972).

13. F. Bloch and N. Bradbury, Phys. Rev. 48, 689 (1935).

14. D. S. Burch, S. J. Smith, and L. M. Branscomb, Phys. Rev. 112, 171 (1958); 114, 1652 (1959).

15. A. V. Phelps and J. L. Pack, Phys. Rev. Letters 6, 111 (1961).

16. R. J. Celotta, R. A. Bennett, J. L. Hall, M. W. Siegel, and J. Levine, Phys. Rev. A, July, 1972.

17. A. Dalgarno, Annales de Geophysique 17, 16 (1961).

18. F. C. Fehsenfeld, E. E. Ferguson, and A. L. Schmeltekopf, J. Chem. Phys. 45, 1844 (1966).

19. E. E. Ferguson, F. C. Fehsenfeld, and A. L. Schmeltekopf, Adv. in Atomic and Molecular Phys. 5, 1 (1969).

20. F. C. Fehsenfeld, A. L. Schmeltekopf, H. I. Schiff, and E. E. Ferguson, Planet. Space Sci. 15, 373 (1967).

21. L. G. McKnight and J. M. Sawina, Phys. Rev. A1, 1043 (1971).

22. J. L. Pack and A. V. Phelps, Bull. Am. Phys. 16, 214 (1971).

23. J. P. Payzant and P. Kebarle, J. Chem. Phys. 56, 3482 (1972).

24. J. D. Payzant, R. Yamdagni, and P. Kebarle, Can. J. Chem. 49, 3308 (1971).

25. J. D. Payzant and P. Kebarle, J. Chem. Phys. 56, 3482 (1972).

26. E. E. Ferguson, D. B. Dunkin, and F. C. Fehsenfeld, J. Chem. Phys. 57, 1459 (1972).

27. C. Lifshitz, B. M. Hughes, and T. O. Tiernan, Chem. Phys. Letters 1, 469 (1970).

28. A. P. M. Baede, Physica 59, 541 (1972).

29. D. B. Dunkin, F. C. Fehsenfeld, and E. E. Ferguson, Chem. Phys. Letters, in press.

30. G. J. Schulz, Phys. Rev. 125, 229 (1962).

31. A. Herzenberg and F. Mandl, Proc. Soc. A270, 48 (1962).

32. A. L. Schmeltekopf, E. E. Ferguson, and F. C. Fehsenfeld, J. Chem. Phys. 48, 2966 (1968).

33. K. C. Clark and A. Belon, J. Atm. Terr. Phys. 16, 205 (1959).

34. W. H. Aberth and J. R. Peterson, Phys. Rev. A1, 158 (1970).

35. J. R. Peterson, W. H. Aberth, J. T. Moseley, and J. R. Sheridan, Phys. Rev. A3, 1651 (1971).

36. J. T. Moseley, W. Aberth, and J. R. Peterson, J. Geophys. Res. 77, 255 (1972).

37. B. H. Mahan, Adv. in Chem. Phys., in press.

38. J. Weiner, W. B. Peatman, and R. S. Berry, Phys. Rev. A4, 1824
     (1971); B. L. Blaney, J. C. Shaefer and R. S. Berry, Abstracts
     of this conference.

39. E. C. Zipf, Bull. Am. Phys. Soc. 12, 225 (1967).

40. M. Nicolet, Inst. R. Meteorol. Belg. Mem. 19, 162 (1945).

41. C. A. Barth, J. Geophys. Res. 69, 3301 (1964).

42. L. G. Meira, J. Geophys. Res. 76, 202 (1971).

43. E. E. Ferguson, F. C. Fehsenfeld, P. D. Goldan, and A. L.
     Schmeltekopf, J. Geophys. Res. 70, 4323 (1965).

44. M. Nicolet, J. Geophys. Res. 70, 691 (1965).

45. R. B. Norton and C. A. Barth, J. Geophys. Res. 75, 3903 (1970).

46. D. M. Hunten and M. B. McElroy, J. Geophys. Res. 73, 2421
     (1968).

47. W. F. Evans, D. M. Hunten, E. J. Llewellyn and A. Vallance
     Jones, J. Geophys. Res. 73, 2885 (1968).

48. I. D. Clark and R. P. Wayne, Chem. Phys. Letters 3, 405 (1969).

49. M. B. Norton, Ph. D. thesis, Univ. of Colorado, 1967; ESSA
     Tech. Memorandum IERTM-ITSA 60, 1967.

50. G. Black, T. G. Slanger, G. A. St. John, and R. A. Young, J.
     Chem. Phys. 51, 116 (1969).

51. C. L. Lin and F. Kaufman, J. Chem. Phys. 55, 3760 (1971).

52. L. Wallace and M. B. McElroy, Planet Space Sci. 14,677 (1966).

53. G. Hernandez and J. P. Turtle, Planet. Space Sci. 17, 675
     (1969).

54. D. F. Strobel, D. M. Hunten, and M. B. McElroy, J. Geophys.
     Res. 75, 4307 (1970).

55. D. F. Strobel, J. Geophys. Res. 76, 8384 (1971).

56. D. F. Strobel, J. Geophys. Res. 76, 2441 (1971).

57. J. N. Bardsley, Proc. Phys. Soc.(Atom.Mol.Phys.) 1,3645 (1968).

58. E. E. Ferguson, F. C. Fehsenfeld, P. D. Goldan, A. L. Schmelt-
     ekopf, and H. I. Schiff, Planet. Space Sci. $\underline{13}$, 823 (1965).

59. R. S. Narcisi and A. D. Bailey, J. Geophys. Res. $\underline{70}$, 3687 (1965).

60. M. A. Biondi, M. T. Leu, and R. Johnsen, COSPAR Symposium,
     Urbana, Ill. July 6-8, 1971.

61. F. C. Fehsenfeld and E. E. Ferguson, J. Geophys. Res. $\underline{74}$,
     2217 (1969).

62. E. E. Ferguson and F. C. Fehsenfeld, J. Geophys. Res. $\underline{74}$,
     5743 (1969).

63. W. C. Lineberger and L. J. Puckett, Phys. Rev. $\underline{187}$, 286 (1969).

64. A. Good, D. A. Durden, and P. Kebarle, J. Chem. Phys. $\underline{52}$, 222
     (1970).

65. L. J. Puckett and M. W. Teague, J. Chem. Phys. $\underline{54}$, 2564 (1971).

66. F. C. Fehsenfeld, M. Moseman, and E. E. Ferguson, J. Chem.
     Phys. $\underline{55}$, 2115 2120 (1971).

67. C. J. Howard, H. W. Rundle, and F. Kaufman, J. Chem. Phys. $\underline{55}$
     5772 (1971).

68. G. C. Reid, J. Geophys. Res. $\underline{75}$, 2551 (1970).

69. G. C. Reid, "Mesospheric Models and Related Experiments", ed.
     G. Fiocco, Riedel Press, 1971.

70   E. E. Ferguson, "Mesospheric Models and Related Experiments",
     ed. G. Fiocco, Riedel Press, 1971.

71. E. E. Ferguson, Rev. of Geophys. and Space Phys. $\underline{9}$, 997 (1971).

72. E. C. Zipf, W. L. Borst, and T. M. Donahue, J. Geophys. Res.
     $\underline{75}$, 6371 (1970).

73. M. B. McElroy, Workshop on Dissociative Excitation of Simple
     Molecules, JILA, Boulder, Colo., March 16, 1972.

ROTATIONAL EXCITATION BY RESONANT TRANSFER

OF ELECTRONIC EXCITATION

M. R. Flannery

School of Physics, Georgia Institute of Technology

Atlanta, Georgia   30332

Molecular hydrogen is unique among molecules in that its rotational constant or rotational line separation 2B (=0.0147 eV) is at least an order of magnitude larger than the rotational constants for other molecules (2B = 5 × 10^{-4} eV for N_2, for example). This relatively large rotational line separation for H_2 raises the possibility that rotational excitation of level J with inelastic threshold 2B(2J+3) can occur at thermal energies by transfer of electronic excitation from, say, hydrogen atoms initially prepared in excited states with moderate and large quantum numbers n(~5-20). In the event that such an energy transfer is resonant, i.e., when the electronic energy released via de-excitation exactly balances that required for the rotational excitation, an extremely large cross section would be evident, particularly at very low temperatures. For example, in the following processes,

$$H_2(J=0) + H(9) \rightarrow H_2(J=2) + H(8), \quad \Delta E = 6k = 5 \times 10^{-4} \text{ eV} \quad (1)$$

$$H_2(J=1) + H(12) \rightarrow H_2(J=3) + H(9), \quad \Delta E = -0.6k = 5 \times 10^{-5} \text{ eV} \quad (2)$$

$$H_2(J=2) + H(7) \rightarrow H_2(J=4) + H(6), \quad \Delta E = -31k = -2.6 \times 10^{-3} \text{ eV} \quad (3)$$

almost exact resonance occurs with exothermicity ΔE extremely small in comparison with the energy thresholds[1] of 0.044 eV, 0.074 eV and 0.103 eV required for the J = 2, 3 and 4 excitations respectively. These processes can therefore proceed at low temperatures and their cross sections are expected to attain values considerably greater than those processes involving only electronic de-excitation.

A full quantal treatment of such collisions would be prohibi-
tively difficult and hence, simplified approaches must be sought.
In an effort to examine these resonances, a semi-quantal theory has
been developed for the process,

$$XY(i) + B(n) \rightarrow XY(j) + B(n') \tag{4}$$

in which the electronic energy released (or absorbed) by a target
atom B can be absorbed (or released) via a rotational or a vibrational transition in the molecule XY at thermal energies. The
incoming projectile $X\ddot{Y}$ is assumed to suffer an inelastic collision with the loosely bound electron of atom B such that the decrease (or increase) in the $e-B^+$ system is accompanied by a
simultaneous rotational or vibrational transition in XY. Energy-
change effects that arise from $XY-B^+$ encounters are small by comparison and hence the parent ionic core B^+ is ignored except in-
sofar as it generates a velocity distribution $F(v_1)$ for its orbital electron. The basic approach is similar to that followed by
Bates et al.[2] in their treatment of electron-ion recombination in
the presence of a molecular gas. Such an approach is valid because
of the following considerations:

(a) For high enough n, the valence electron follows a clas-
sical orbit for which the quantal indeterminancies Δp_n and Δr_n in
its linear momentum p_n and in its orbit-radius r_n are $\ll p_n$
and r_n, respectively.

(b) The $e-B^+$ separation $r_n = n^2 a_o \gg a^{\pm}$ the scattering
lengths for $e-XY$ and B^+-XY collisions, i.e. e and B^+ behave as
separate scattering centers because the range of the $e-XY$ interaction is short in comparison with the longer range of the $e-B^+$
Coulombic interaction. For example, in the case of (1), $a^{\pm} \lesssim 0.5a_o$
and $r_n = 81a_o$.

(c) The $e-XY$ collision duration T $\ll$ the orbital period
τ of the electron with speed v_1, about B^+. Therefore, the
electron-ion separation vector does not rotate appreciably during
the encounter. Thus the scattering is instantaneous compared with
the time required for the electron to change direction and therefore, the presence of the binding determines only the electron-
velocity distribution $F(v_1)$ and does not affect the general aspects of the individual scattering. For process (1), $T \approx a^-/v_1 \approx$
$10^{-17}n$ sec and $\tau = 2.4 \ 10^{-17}n^2$ sec and hence $T \ll \tau$.

(d) Also, the $e-XY$ collision time $\ll$ a typical rotational
period of 10^{-12} sec (2×10^{-14} sec for H_2) and is also shorter
than a vibrational period of 10^{-14} sec, i.e. the molecule can be

considered as fixed since it does not rotate or vibrate appreciably during the encounter and hence the non-spherical components of the e-XY instantaneous interaction permit rotational transitions.

The final expression[3] for the cross section Q for the process (4) in which the increase ε_{12} in electronics energy of $B(n)$ is accompanied by an increase Δ_3 in the thermal energy of the incident molecule XY is

$$Q(\varepsilon_{12}, \Delta_3; v_3) = \frac{1}{n'^3 v_3^2} \int_{v_1^-}^{v_1^+} v_1^{-1} F(v_1) dv_1 \int_{g^-}^{g^+} \frac{g^2 dg}{\omega \gamma} (v_1, v_3, g)$$

$$\int_{\psi^-}^{\psi^+} \frac{\sigma_{13}(g, g', \psi) d(\cos\psi)}{[(\cos\psi^+ - \cos\psi)(\cos\psi^+ - \cos\psi^-)]^{1/2}} \quad . \tag{5}$$

Here σ_{13} is the differential cross section for e-XY inelastic (or elastic scattering between the angular limits $\psi^\pm$ which are for a specified energy change

$$\varepsilon = \varepsilon_{12} + \frac{a}{1+a} \Delta_3, \quad a = \frac{M_2 M_3}{M_1 (M_1 + M_2 + M_3)} \tag{6}$$

given by

$$\cos\psi^\pm = \omega^{-1} \gamma^2 \{\alpha(\alpha+\varepsilon) \pm \beta[\omega^2\gamma^2 - (\alpha+\varepsilon)^2]^{1/2}\} , \quad \omega = g'/g . \tag{7}$$

After the collision with the incident molecule of mass M_3 and speed v_3, the electron of mass M_1 is bound to its parent ionic core of mass M_2 with binding energy I/n'^2. The limits $g^\pm$ to the e-XY relative speeds (which are g and g' before and after the collision respectively) are chosen such that $\psi^\pm$ are real and the values are in harmony with the conservation of energy. The lower limit v_1^- to the electronic speed about the $(e-B^+)$ center-of-mass is the minimum of 0 and $-2\varepsilon_{12}/M_1(1+M_1/M_2)$ while v_1^+ is determined from the restriction that the electron cannot approach its parent ion arbitrarily closely and still be regarded as colliding only with the molecule.

The parameters α, β and γ are determined by conditions prior to the collision and are given by

$$\alpha = 1/2 \; M_{13} (v_1^2 - v_3^2 + \frac{1-a}{1+a} \; g^2) \quad . \tag{8}$$

$$\beta = 1/2 \; M_{13} [g^2 (2v_1^2 + 2v_3^2 - g^2) - (v_1^2 - v_3^2)]^{1/2} \tag{9}$$

and

$$\gamma = \frac{M_{13} g}{(1+a)} \; [(1+a)(v_1^2 + av_3^2) - ag^2]^{1/2} \quad . \tag{10}$$

The cross section (5) satisfies the principle of detailed balance.

In order to isolate the resonance effect, we have used the above equations to examine the following processes,

$$H_2 (J=0) + H(10) \rightarrow H_2 (J=2) + H(9), \quad \Delta E = -142k = -0.01 \; eV \tag{11}$$

$$H_2 (J=0) + H(9) \rightarrow H_2 (J=2) + H(8), \quad \Delta E = 6k = 5 \times 10^{-4} \; eV \tag{12}$$

$$H_2 (J=0) + H(8) \rightarrow H_2 (J=2) + H(7), \quad \Delta E = 243k = 0.02 \; eV \tag{13}$$

and

$$H_2 (J=0) + H(9) \rightarrow H_2 (J=0) + H(8), \quad \Delta E = 517k = 0.0445 \; eV \tag{14}$$

which involve transitions between quantum levels n on either side of the resonant 9-8 transition. The elastic and the $J = 1$ and 2 inelastic cross sections for e-H_2 scattering are taken from the close coupling calculations of Henry and Lane[4] who took account of exchange and polarization effects to get agreement with experiment.

The figure displays the computed cross sections for the processes (11) - (13) as a function of the kinetic energy of relative motion. As expected, at low energies, the (9-8) cross sections are indeed much larger (by more than two orders of magnitude) than for the neighboring endothermic and exothermic reactions (11) and (12) respectively. With increasing impact energy however the resonant effect diminishes rapidly and the cross section curve decreases monotonically as $\sim E^{-1}$ to values that eventually lie between the higher (8-7) and the lower (10-9) cross sections. The inset in the figure demonstrates the low-velocity behavior of the cross sections.

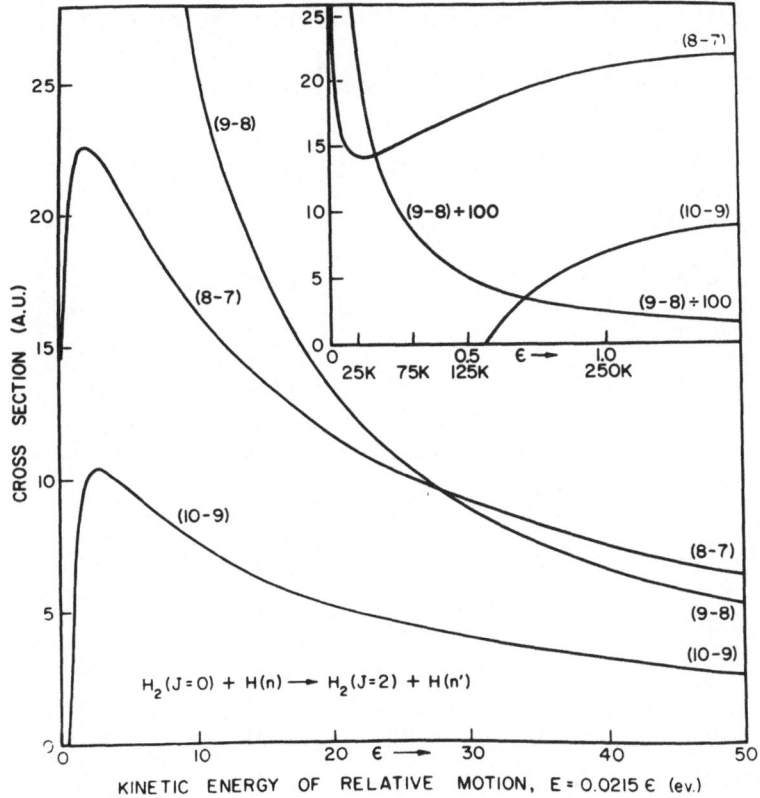

Figure 1. Cross sections for the process $H_2(J=0)+H(n) \rightarrow H_2(J=2)+$ H(n-1) for n = 7, 8 and 9, as a function of impact energy E. Inset shows the threshold and low-energy behavior of the cross sections.

Table I tabulates these cross sections together with those for the comparison case (14) which involves no rotational excitation. Although the e-H_2 elastic scattering cross section is at least an order of magnitude greater than the cross sections for rotational excitations, the resonant process (12) proceeds much faster than (14) except, of course, for the higher impact energies when the resonant effect has considerably diminished.

Table I. Cross sections (a.u.) for the process

$$H_2(J = 0) + H(n) \rightarrow H_2(J') + H(n - 1)$$

ε* \ n	10a	9a	9b	8a
0.001	–	$4.51^{4\dagger}$	5.77	8.15^1
0.01	–	1.26^4	2.06	2.73^1
0.05	–	4.07^3	1.51	1.57^1
0.1	–	2.27^3	1.55	1.43^1
0.6	1.15	4.31^2	2.98	1.93^1
1.0	6.48	2.61^2	4.24	2.10^1
2.5	1.03^1	1.05^2	8.91	2.25^1
5.0	9.64	5.28^1	1.57^1	2.05^1
10.0	7.56	2.64^1	2.59^1	1.64^1
30.0	3.91	8.82	4.01^1	9.06
50.0	2.66	5.30	3.49^1	6.32
80.0	1.81	3.32	2.92^1	4.38
100.0	1.49	2.66	2.67^1	3.65

*The kinetic energy of relative motion is $E = 0.0215 \, \varepsilon$ eV.

†The superscript denotes the power of 10 by which the entry must be multiplied.

$a_{J'} = 2$; $b_{J'} = 0$

- - - - -

Tables II and III display the cross sections for the resonant processes (2) and (3) which involve the J = 3 and 4 rotational excitations respectively. The electron-molecule J = 4 excitation cross sections were taken from the distorted wave calculations of Takayanagi and Geltman.[5] Also shown are the neighboring electronic transitions which dramatize the marked resonance together with the cross sections for electronic de-excitation alone. The overall behavior is similar to that in the figure.

Table II. Cross sections (a.u.) for the process

$$H_2(J = 1) + H(n) \rightarrow H_2(J') + H(n - 3)$$

ε* \ n	13a	12a	12b	11a
0.001	–	–	–	2.51
0.01	–	$8.24^{3\dagger}$	–	8.46^{-1}
0.05	–	1.78^{3}	–	4.81^{-1}
0.5	–	1.79^{2}	1.92^{-2}	6.44^{-1}
1.0	8.72^{-2}	8.95^{1}	4.48^{-2}	8.88^{-1}
2.5	5.34^{-1}	3.58^{1}	1.34^{-1}	1.24
5.0	6.91^{-1}	1.79^{1}	3.08^{-1}	1.36
10.0	1.18	8.96	6.76^{-1}	1.32
30.0	1.16	2.99	1.87	1.68
50.0	8.67^{-1}	1.80	2.68	1.41
80.0	6.25^{-1}	1.13	3.24	1.07
100.0	5.29^{-1}	9.02^{-1}	3.17	9.20^{-1}

*The kinetic energy of relative motion is $E = 0.0215 \, \varepsilon$ eV.

†The superscript denotes the power of 10 by which the entry must be multiplied.

$a_{J'} = 3$; $b_{J'} = 1$.

– – – – –

Similar energy-resonant effects are to be encountered for molecules other than H_2, but, in these instances, the target atom must be in a more highly-excited state since the inelastic thresholds for rotational excitation are then much smaller.

The effect of these rotational transitions at thermal energies may possibly be significant in theoretical investigations of the interstellar medium.

Table III. Cross sections (a.u.) for the process

$$H_2(J = 2) + H(n) \rightarrow H_2(J') + H(n - 1)$$

ε^* \ n	8a	7a	7b	6a
0.15	–	4.50^2	–	2.17^{-1}
0.2	–	5.45^2	–	2.99^{-1}
0.6	–	3.31^2	6.16^{-3}	1.08
0.8	–	2.67^2	6.03^{-2}	1.46
1.0	–	2.24^2	1.39^{-1}	1.81
2.5	4.55^{-1}	1.02^2	9.02^{-1}	3.77
5.0	1.48	5.34^1	2.58	5.51
10.0	1.93	2.74^1	6.52	7.86
30.0	1.70	9.27	2.01^1	5.98
50.0	1.38	5.58	2.84^1	4.94
80.0	1.07	3.49	3.47^1	3.89
100.0	9.30^{-1}	2.79	3.68^1	3.40

*The kinetic energy of relative motion is $E = 0.0215 \, \varepsilon$ eV.

†The superscript denotes the power of 10 by which the entry must be multiplied.

$^a J' = 4$; $^b J' = 2$.

$- - - - -$

REFERENCES

1. K. Takayanagi and Y. Itikawa, Advances in Atomic and Molecular Physics (ed. D. R. Bates and I. Estermann) Vol. 6, p. 128, 1970.

2. D. R. Bates, V. Malaviya and N. A. Young, Proc. Roy. Soc. London A 520, 437 (1970).

3. M. R. Flannery, in preparation.

4. R. J. W. Henry and N. F. Lane, Phys. Rev. 183, 221 (1969).

5. K. Takayanagi and S. Geltman, Phys. Rev. 138, A 1003 (1965).

ELECTRONIC POLARIZATION BEHAVIOR IN COLLISIONS

Benjamin Bederson

New York University

4 Washington Place, New York, N.Y. 10003

1. INTRODUCTION

The literature on electronic polarization behavior in atomic collisions was relatively sparse, up to a few years ago. This situation has now changed, and there is no difficulty in encountering comprehensive review papers on this subject. Some of these are listed in reference 1. These proceedings also contain a review by Professor Kessler on polarized electron sources,[2] which of necessity includes discussion of the physical processes in atomic collisions which underlie most of present polarized beam source technology.

In view of all this recent activity, perhaps the best approach for me to take is to discuss a narrower subject which might, after McDowell and McDaniel, be called a "Case History in Atomic Collisions". This is the continuing, and increasingly intriguing story of low-energy scattering by the alkali elements. It is a subject where polarization effects appear as part of the full story--it is quite impossible to take them out of context of the complete scattering problem, which involves other effects--total, momentum transfer, differential, and inelastic scattering, polarization of resonance radiation and resonances as well.

The study of these electron-alkali collisions appears to be playing an analogous role in collision physics to that played for many years by alkali studies in atomic beam magnetic resonance experiments, many of which were reported upon in those memorable conferences organized and managed by V. W. Cohen for about twenty years, called the "Brookhaven Molecular Beams Conference" (even when held in Heidelberg or Berkeley), and out of which the present

International Conference on Atomic Physics developed. The electron-
alkali collision problem will be the principal topic to be covered
in this paper; other topics will be briefly discussed afterward.

The various experimental aspects of this electron-alkali case
history include:
1. Total electron-alkali cross sections at low energies
2. Momentum-transfer cross sections at very low (i.e., thermal) energies, mostly inferred from measurements of electric
transport properties in alkali metal vapors
3. Spin-exchange cross section measurements in optical pumping and radio-frequency resonance experiments, also at thermal
energies
4. Total spin-exchange cross section measurements at low-
energies from crossed-beam experiments
5. "Full" differential elastic cross sections at low energies,
from crossed-beam experiments
6. Differential spin-exchange cross sections at low energies,
from crossed-beam experiments
7. Measurements of differential cross sections using either
polarized electrons or analysis of electron polarization after
scattering
8. Differential cross sections for the excitation of the
resonance transition of $n^2S_{1/2} \rightarrow n^2P_{1/2,3/2}$, without and with spin
analysis of the target atom
9. Total excitation cross sections for the resonance transition
10. Excitation functions for the resonance transition
11. Polarization of resonance transition.

In addition to the above list, there exist a number of more
ambitious experiments, which are currently in progress but not yet
reported upon, at a number of laboratories. These include differential cross section experiments with polarized electrons and
polarized atoms with and without spin-analysis after scattering.
And finally, there is the class of proposed experiments, mostly not
yet off the drawing boards, which involve either coincidence between scattered electrons and/or atoms and emitted photons with
polarization analysis of all the particles, including the photons.

The relevant theoretical work to be referred to in this
article involves basically two types of calculations. These are:
1) Solution of the Schroedinger equation for the scattering
problem using an adiabatic potential and including effects known
to play an essential role in low-energy electron-alkali scattering,
namely, exchange (i.e. proper symmetrization of the total scattered wave) and polarization effects of the target by the scattered
electron (which is particularly important in the alkalis because
of their extraordinarily high electric dipole polarizabilities).

Such calculations are sometimes called "polarized orbital" calculations, after A. Temkin, who first employed this method to atom-electron problems.[3] The specific calculations which concern us here, for the alkalis, include those of Garrett and Mann,[4] Stone and Reitz,[5] Crown and Russek,[6] Balling[7] and Lan.[8]

2) Close-coupling. Among other reasons for the attractiveness of this approach is the fact that from a single set of numbers, i.e., from the computed elements of the T-matrix, one can calculate all of the observable quantities mentioned in items 1 through 11 above (except for cascade effects). It is in fact because of the partial, though also in some cases quite spectacular, successes of close-coupling theory in predicting experimental results in the light alkalis that one is being led to pursue this approach with ever increasing vigor.[9] After the seminal work of Massey, Seaton, Burke, Smith and their colleagues,[10] the principal numerical calculations in the alkalis have been those of Karule,[11] Karule and Peterkop,[11] Burke and Taylor,[12] Norcross,[13] and Moores and Norcross.[13] The Karule-Peterkop calculations were the first close-coupling calculations done in the alkalis. Despite the relatively less elaborate forms of wave functions then available (e.g. non-inclusion of core-polarization) and despite the relatively smaller computer capability available to these authors, their early successes with lithium, sodium and potassium have been a major stimulant to this entire area.

The most recent, and elaborate, calculations of this sort are those of Moores and Norcross,[13] on lithium and sodium, employing scaled Thomas-Fermi potentials with inclusion of core polarization. Except at very low energies, the elastic phase shifts calculated by all of these workers are in excellent agreement.

The various observables, i.e., cross sections, are constructed from the appropriate scattering amplitudes. For elastic scattering, the singlet and triplet scattering amplitudes, f^+, f^-, obtained from the calculated phase shifts or equivalently from appropriate elements of the T-matrix, can be combined in a variety of ways to yield all the observables described in 1 - 7 above. These are, for example

$$\sigma(\theta) = (1/4) \left| f^+ \right|^2 + (3/4) \left| f^- \right|^2 \tag{1}$$

$$\sigma_{ex}(\theta) = (1/4) \left| f^+ - f^- \right|^2 \quad , \tag{2}$$

where $\sigma(\theta)$ and $\sigma_{ex}(\theta)$ are the "full" elastic differential and differential elastic exchange cross sections respectively. The total elastic and total elastic exchange cross sections are, of course, obtained by integrating over all solid angles. Similar formulae obtain for momentum-transfer and viscosity cross sections.

For inelastic scattering, ns→ np, the various scattering amplitudes, (in the ℓ, m_ℓ· representation) are given by[14]:

$$f^{\pm}(\theta, E_o; n_o 00 \to n\ell m) = (\frac{\pi}{kk_o})^{\frac{1}{2}} \Sigma_L \left[C_{m,-m,0}^{1,L+1,L} Y_{L+1}^{-m} T_{12}^{\pm L} \right.$$

$$\left. - C_{m,-m,o}^{1,L-1,L} Y_{L-1}^{-m} T_{13}^{\pm L} \right] (2L+1)^{\frac{1}{2}} \quad (3)$$

where for a given polar scattering angle θ, incoming electron energy E_o and incoming and outgoing electron momenta k_o, k, the initial and final atomic quantum numbers are $n_o, 0, 0$ and n, ℓ, m. The C's are Clebsch-Gordan coefficients; the Y's are associated spherical harmonics, and $T_{12}^{\pm L}$, $T_{13}^{\pm L}$ are the Lth partial wave singlet and triplet transition-matrix elements, where the total electron+atom angular momenta are L±1 respectively.. Similarly, the elastic scattering amplitudes are

$$f^{\pm}(\theta, E; n_o 00 \to n_o 00) = \frac{1}{k_o} \sum_L T_{11}^{\pm L} P_L(\cos \theta) (2L+1) \quad , \quad (4)$$

with

$P_L(\cos \theta)$ the Lth spherical harmonic.

For brevity, the inelastic scattering amplitudes for s→ p transitions can be written

$$f^{\pm}_{0,1}$$

where the subscript refers to the excited state left with m=0, ±1 respectively. We will see later how these are related to the observables mentioned in experiments 8 - 11 above.

The relation of photoionization to polarized electrons, i.e., the Fano effect,[15] is not directly involved in this discussion, although there is a close indirect relation, since the same potentials may be employed in both types of calculations.[16] The adiabatic-type calculations of course can only be employed in elastic scattering.

Attempts have also been made to improve the few-state close-coupling approximation, particularly as applied to excitation processes, by the inclusion of long-range polarization effects. This results in a sort of close-coupling-polarized-orbital synthesis, which could be expected to be an improvement over the normal close-coupling expansion.[17]

2. ELASTIC AND TOTAL SCATTERING IN THE ALKALIS

To begin this comparison between experiment and theory, I would like to show three figures, none of which refer to polarized systems. The first is a comparison of the measured and calculated (by Karule, and Karule and Peterkop[11]) "total" cross sections for scattering of electrons by potassium in the energy range 0.4 to 9eV. The experiments of Collins et al,[18] Visconti et al,[19] and Kasdan et al[20] were all performed using the atomic beams recoil technique,[21] in which the scattered atoms, rather than the electrons, are observed. Particularly to be noted are the facts that, first, the early Brode[22] measurements, which were for many years the only total cross sections reported in the alkalis, appear to be incorrect; second, the three recoil experiments are absolute, i.e., the ordinate scale is obtained solely from knowledge of experimental parameters (not including the atom-beam density, knowledge of which is not required in a recoil total cross section measurement). The measurements, of course, refer to

$$\sigma_{TOT} = \sum_n \int \sigma_{on}(\theta)d\Omega$$

where the sum is over all final states and the integration is over all scattering angles. Since only the resonant excitation cross section can be calculated from a two-state close-coupling expansion, comparison with theory is obtained by replacing the above sum by

$$\sigma_{TOT} = \int \sigma_{elastic}(\theta)d\Omega + \int \sigma_{s \to p}(\theta)d\Omega \quad .$$

In Fig. 1 the Karule and Peterkop points represent the sum of the total elastic and total 4S - 4P excitation cross sections. The impressive agreement between the earlier recoil measurements (Collins et al and Visconti et al) and the Karule-Peterkop calculations were among the first indications that a very promising situation was developing, in which a relatively modest few-state close-coupling expansion, on one-electron atoms, could find a common meeting ground with a comparably favorable experimental situation.[23]

The other two figures not directly related to polarization are shown below. Figure 2 shows differential elastic cross sections in potassium, including the data of Collins et al,[18] at N.Y.U., the recent angular distribution data of Gehenn and Wilmers at Mainz and Karule's calculation at about 1.1 and 0.9 eV. Figure 3 shows some representative recent differential measurements of Hils et al[25] at JILA and Gehenn and Wilmers compared to Karule and Peterkop, at about 3 eV. The recoil data (Collins et al) and, of course, the theory, are absolute, while the other experiments require normalization. It is seen that agreement between theory and experiment is quite remarkable, all things considered.[26,27]

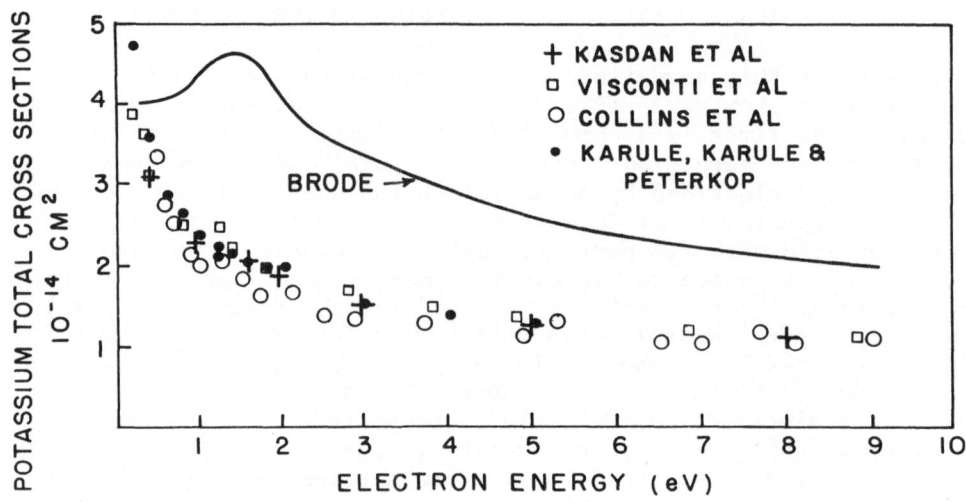

Fig. 1. Comparison of experimental and theoretical total cross
sections for scattering of electrons by potassium (Kasdan, Miller
and Bederson[20]).

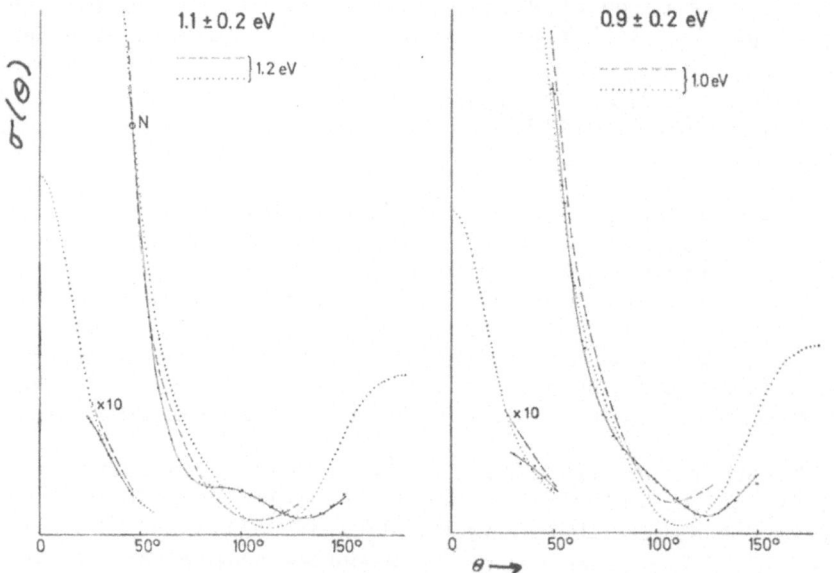

Fig. 2. Plot of elastic differential cross sections for electrons
scattered by potassium. Solid curve: Gehenn and Wilmers; dashed
curve: Collins et al; dotted curve: Karule. The Gehenn and Wilmers
curve is normalized at point N to the other curves. The absolute
value at N is about $28 \times 10^{-16} cm^2/sr$ (from Gehenn and Wilmers[24]).

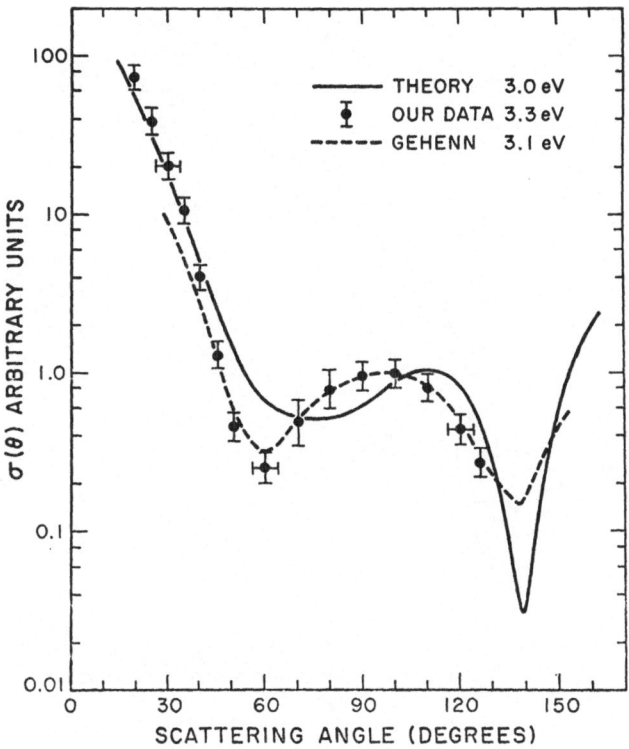

Fig. 3. Normalized elastic differential cross sections for elec-
tron-potassium scattering, from Hils et al[25].

Now, it is no accident that all three experimental groups
represented in Figs. 2 and 3 are engaged in polarized beam work.
The Mainz group is attempting a <u>triple</u> scattering experiment.[28]
Their earlier and continuing work on polarization of electrons
scattered from resonance doublets in the rare gases is also well
known.[29]

The JILA group[25] has recently published the first results of
elastic differential measurement of $|f|^2/\sigma(\theta)$, where f, the "direct
scattering" amplitude, is defined as $f = \frac{1}{2}(f^+ + f^-)$. In this experiment unpolarized electrons are scattered by polarized atoms, with
Mott-analysis performed on the scattered electrons. These are the
first experiments reported in low-energy electron-alkali collisions
with spin analysis of scattered electrons (rather than atoms). The
experimental setup is shown in Fig. 4, and the experimental results
are shown in Fig. 5. It can be seen that agreement with theory is

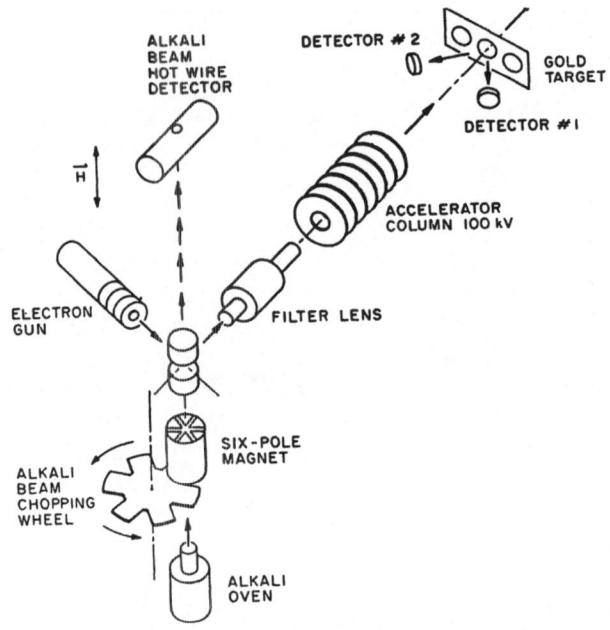

Fig. 4. Experimental setup of the JILA experiment to measure
$|f|^2/\sigma(\theta)$ (from Hils, McCusker, Kleinpoppen and Smith[25]).

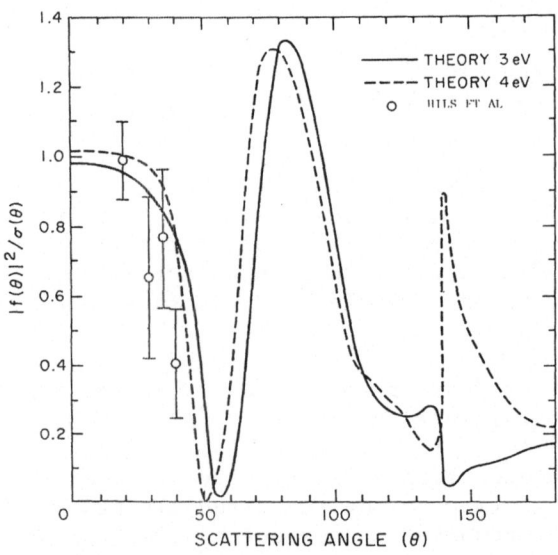

Fig. 5. Plot of experiment and theory (Karule and Peterkop) of
the JILA group's measurements of $|f|^2/\sigma(\theta)$ (Hils et al[25]).

marginal. It is probably too early to say whether there is a significant discrepancy here or not.

A block diagram of the basic recoil-type experimental setup[21]
is shown in Fig. 6. An atom beam is velocity-and spin-state selected by the Stern-Gerlach magnet, before being cross-fired by the
electron beam. After scattering the atom beam is spin-analyzed by
an E-H gradient balance magnet. The entire analyzer-detector
assembly rotates about the scattering center so that selected portions of the differentially scattered atoms can be studied. A
suitable transformation is then made, using simple kinematic relations, to obtain the electron polar scattering angle. Normally
the polarizer and analyzer are set to transmit opposite spin-states,
so that only spin exchanged (or spin-flipped) atoms can reach the
detector when the analyzer is operative. Of course without the
analyzer operative one obtains full differential cross sections.

As in optics experiments with lenses and polarization filters,
one must properly take into account the transmission T and residual
depolarization D of the polarizer and analyzer. This is done
experimentally by setting the polarizer-analyzer to transmit the
same spin-states. In Fig. 7a, α, β are the spin-states and I the
atom current. Setting the source into the up position selects the
$\beta(\downarrow)$ state and measures T; setting the source into the down position selects the $\alpha(\uparrow)$ state and measures D. The composition of
the scattering signal with and without spin-analysis is shown in
Fig. 7b.

Differential elastic spin exchange cross sections for potassium have been obtained by Collins et al[18] using this technique.
These are obtained from measurements of $R(\theta) \equiv \sigma_{ex}(\theta)/\sigma(\theta)$, combined with direct measurements of $\sigma(\theta)$ normalized to the total cross
section at 1 eV. $R(\theta)$ is simply the ratio of atom beam current at
a given detector position with and without the spin analyzer operative, corrected for transmission and residual depolarization.
Partial depolarization caused by hyperfine coupling is avoided in
this work by the use of a high (> 1000 gauss) magnetic field in
the interaction region.[30] Figure 8 shows these results, compared to
Karule's calculation. The bump in the 1.2 eV experimental curve in
the vicinity of 90° is attributable to spin-flip caused by inelastic scattering from the high-energy tail of the electron energy
distribution. A by-product of the large exchange cross sections
at low and high scattering angles is the use of exchange to obtain
polarized electrons.[31]

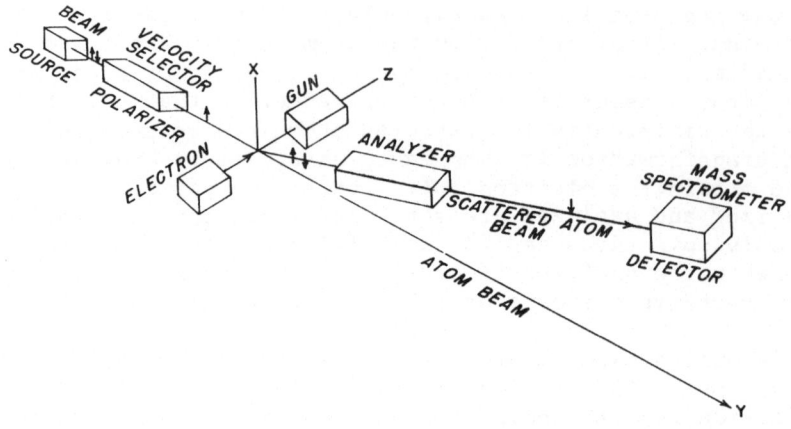

Fig. 6. Schematic diagram of recoil experiment (N.Y.U. group).
↑ indicates direction of spin-polarization. The electron gun is
rotatable in X-Z plane about Y-axis. The analyzer-detector is
rotatable in Y-Z plane about scattering center, and the detector
is also translatable in ±Z-direction.

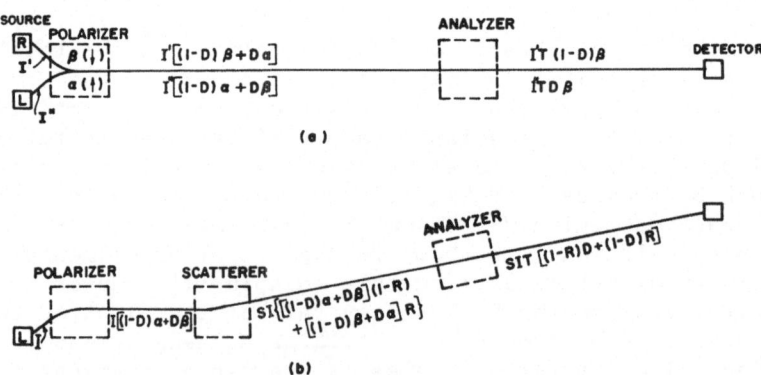

Fig. 7. Means of correcting for imperfect transmission and polarization of polarizer and analyzer in recoil experiment--see text
(from Collins, Bederson and Goldstein[18]).

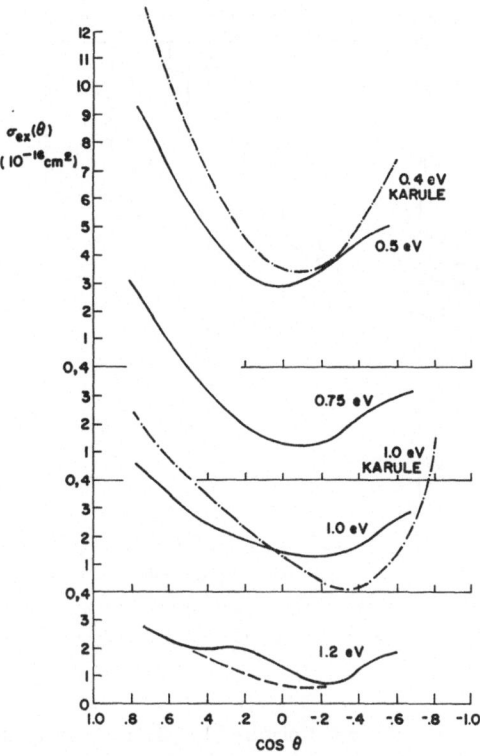

Fig. 8. Differential spin-exchange cross sections (solid curves) at 0.5, 0.75, 1.0 and 1.2 eV. Dot-dashed curves are Karule's values at 0.4 and 1.0 eV. The overall error estimate in $\sigma_{ex}(\theta)$ is ±27% (from Collins, Bederson and Goldstein[18]).

Figure 9 shows measurements of the ratio $\sigma_{ex}(180°)/\sigma(180°)$ over the energy range 0.4 to 1.65 eV, where it can be observed that, first, a maximum occurs at about 1.25 eV, and second, that this ratio actually exceeds unity over a range of energies centering at 1.25 eV. Preliminary measurements of $R(\theta)$ on sodium by Kasdan and Miller at N.Y.U. show similar behavior and similar good agreement with theory.

The situation regarding <u>total</u> elastic exchange cross sections is not quite as favorable as in differential scattering. With regard to beam experiments, such cross sections can be obtained by integrating the results of Fig. 8 over all angles, and extrapolating to small and large angles in some fashion. Such determinations have been made by Collins et al, but the absolute error estimates are quite high (± 35%). These are shown, along with

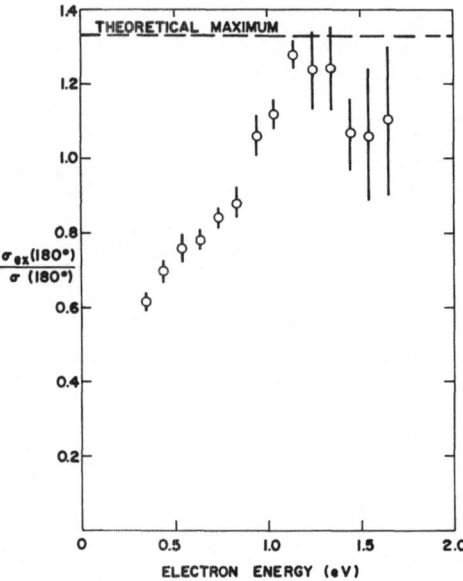

Fig. 9. $\sigma_{ex}(\theta)/\sigma(\theta)$ for $\theta = 180°$ (backward scattering) as a function
of electron energy from 0.5 to 1.65 eV (from Collins, Bederson and
Goldstein[18]).

Karule's values and those of Campbell, Brash and Farago,[32] in
Fig. 10.

     Campbell et al use an electron trap, with low energy electrons
oscillating through a polarized atom beam; the exchange collisions
gradually polarize the trapped electrons, which are extracted after
a suitable time. From the degree of polarization, as measured by
a Mott-analyzer, the total exchange cross section can be inferred.
(This is of course similar to ion and electron trapping schemes in
optical pumping experiments.[33]) One of the difficulties with obtaining quantitative values by this method is the fact that, because
of the relatively wide electron energy spread in the trapping well,
inelastic collisions also occur, and these could contribute significantly to the net polarization. Thus, although the Campbell et
al total cross sections appear roughly a factor or two higher than
the Karule curve, these authors remark "the discrepancy cannot be
considered of much significance in the absence of a rigorous
energy analysis of the extracted electrons".[32]   All the above considered, it appears, at least to this author, that theory and
experiment cannot be said to be in significant disagreement.

     A somewhat more serious discrepancy exists at thermal energies,
in the optical pumping domain (Item 3 in the list of Section 1).
The basic experimental techniques and theoretical analysis necessary
to relate observed optical pumping transmission signals to the

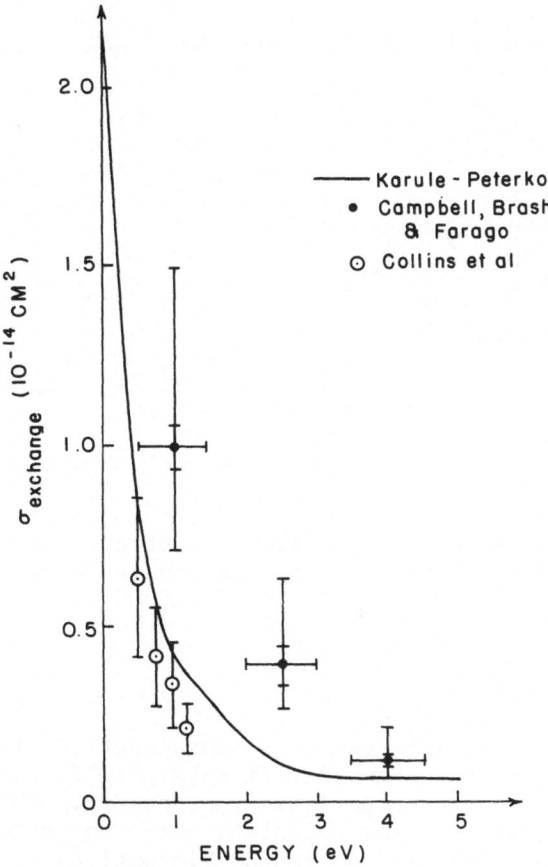

Fig. 10. Total spin-exchange cross section $\sigma_{exchange}$ versus electron energy E (from Campbell, Brash and Farago [32]).

total elastic spin exchange cross section are thoroughly described in a series of papers by Dehmelt, Balling, and others.[34] The values reported at about 400°K by Balling[34] while in reasonable agreement with Karule's values, are over a factor or two higher than the Norcross result (Fig. 11).[35] These values may even have to be raised as a result of nuclear-spin effects.[36] Since Norcross has shown that at these very low energies the high Karule values are attributable to a computational difficulty (lack of uniqueness in the asymptotic solutions of the coupled equations), this appears to be a real discrepancy which clearly requires further investigation. Davis and Balling report on new results including temperature dependence of the spin-exchange cross section in rubidium in this volume.

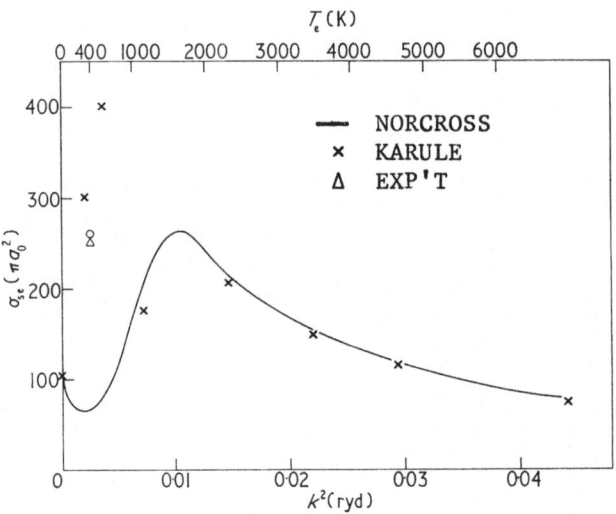

Fig. 11.   Spin exchange cross section for electron-sodium
ing.   The experimental value is from reference 34(from Norcross[13]).

A final remark on spin-exchange cross sections from optical
pumping experiments in the heavier alkalis should also be made.
Such measurements have been performed on rubidium and cesium.[38]
The Balling polarized orbital calculation for rubidium[7] is in not
unreasonable agreement with the recent Visconti et al measurement
(see reference 19), and likewise for Crown and Russek's[6] cesium
calculations.   However, the measured spin-exchange cross section
is much larger than the calculated one for rubidium.

Finally, some recent data by Andrick et al[39] on sodium are
worth special mention.   They have observed a cusp in the elastic
differential cross section near 110°, at the 2.1 eV excitation
threshold, with weaker structure at other angles at the same energy.   The Moores and Norcross calculations reproduce this structure
quite well.   On the other hand, as McCusker[40] has pointed out, the
relative cross sections $|f|^2/\sigma(\theta)$ and $|g|^2/\sigma(\theta)$, where g is the
exchange scattering amplitude, $g=\frac{1}{2}(f^+-f^-)$, exhibit quite different
behavior, primarily because the structure is all in the singlet
phase shifts.   The calculated values for these cross section ratios
show more structure at small, rather than large angles.   Differen-
tial measurements of either $|f|^2$, $|g|^2$, or both would be very
helpful in unravelling this problem, and both the JILA and N.Y.U.
groups are at work on these.

3. EXCITATION: DIFFERENTIAL EXCITATION
CROSS SECTIONS FOR RESONANCE TRANSITION

We now consider excitation events, of the resonance transitions ns→ np. As already discussed, since most of the close-coupling calculations include only s and p target states, it is only these transitions that could be considered in this discussion. Thus we consider experiments of types 8 - 11 listed in Section 1. We first observe that Slevin et al[27] have used the recoil technique to obtain some absolute, differential excitation cross sections for 4s→ 4p in potassium. Figure 12 shows their values, compared to Karule and Peterkop[41] at 3 and 4 eV. In this comparison the <u>relative</u> shapes of theory and experiment agree very well, but there appears to be a serious question concerning absolute values, particularly at 4 eV. Of course it should be noted that absolute experimental cross sections are always very difficult to obtain,[42] although ability to determine absolute values without knowledge of the neutral beam density in the scattering region is supposedly one of the advantages of the recoil method. If the discrepancy is caused by the lack of convergence of the two state calculation, it is difficult to understand why the relative shapes are in such good agreement. Clearly it would be very helpful were conventional electron scattering measurements made of the relative cross sections for these transitions (i.e., using electron spectrometers).

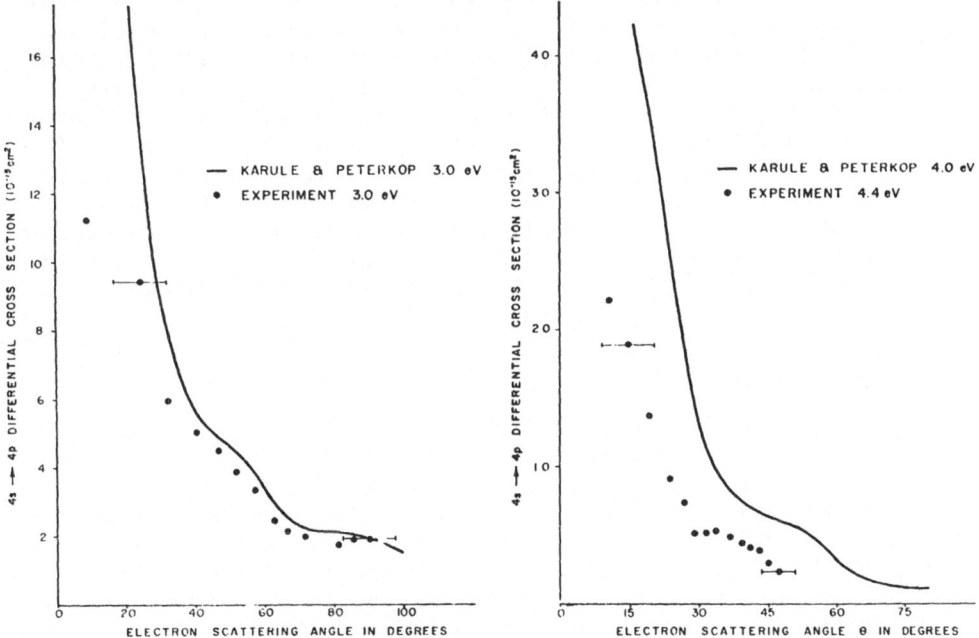

Fig. 12. 4s→ 4p differential excitation cross sections (from Slevin, Visconti and Rubin[27]).

4. EXCITATION AND POLARIZATION OF RESONANCE RADIATION

We now take a rather different tack and consider experiments which measure the optical radiation, associated with excitation to the first p-states (Items 9 - 11 of Section 1). Thus far all reported work has been without coincident detection of photon and scattered electron. As a result, this work necessarily yields total excitation cross sections, since one is observing radiation effects caused by all exciting electrons, regardless of scattering direction, i.e., one must integrate appropriate partial cross sections over all angles. In addition, since no energy-analysis of scattered electrons is involved, one is really observing excitation functions, so that cascading must be corrected for before results referring only to s-p excitation can be presented. Finally, such experiments are normally performed in weak magnetic fields, where nuclear spin couples strongly with the electronic spin. Corrections for this coupling must therefore be applied.[43]

The newest resonance excitation cross sections measured in the alkalis by this means are those of Enemark and Gallagher[44] and of Gould.[45] Their results are in very good agreement with both the Karule and Peterkop and Moores and Norcross calculations.

Regarding the polarization of the resonance radiation, there have been some difficulties although these appear to be now resolved. Following the pioneering experiment of Hafner and Kleinpoppen[46] on measurements of polarization of resonance radiation near threshold for lithium and sodium, it was found that the close-coupling calculations gave rather poor agreement.[47] An attempt was made to add a polarization potential, i.e., to empirically modify the close-coupling equations, in order to obtain better agreement with experiment.[48] However, the Enemark and Gallagher results near threshold are substantially lower than Hafner and Kleinpoppen, and agreement with close-coupling theory is now pretty good (Fig. 13). The discrepancy between the two experiments is unexplained at this time.

5. EXCITATION WITH SPIN-ANALYSIS DIFFERENTIAL
SPIN-FLIP CROSS SECTIONS

Several recent articles have described the relation between scattering amplitudes and a number of possible experiments involving polarized electrons, photons and atoms.[1] Of these experiments perhaps the simplest, apart from those described in the last section, involves observation of the ratios of spin-flip to full differential cross sections, and, in fact, such measurements obtained using the recoil technique are the only ones thus far

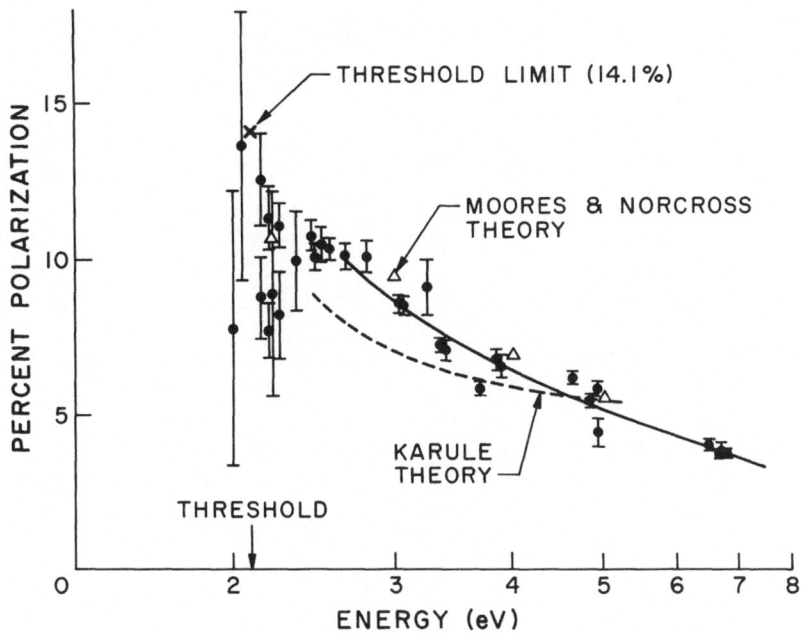

Fig. 13. Polarization of resonance radiation in sodium (from Enemark and Gallagher[44])

performed for differential inelastic scattering involving spin-analysis.[49] Rubin et al,[21] and Bederson[1] show that this ratio, called $R(\theta)$ is given by

$$R(\theta) \equiv \frac{\sigma_{sf}(\theta)}{\sigma_{s \to p}(\theta)} = \frac{4}{9} + \frac{10|g_1|^2 + |g_0|^2 - 8\sigma_1(\theta)}{8\sigma_{s \to p}(\theta)} \tag{5}$$

where $g_{0,1} = \frac{1}{2}(f_{0,1}^+ - f_{0,1}^-)$, $\sigma_{sf}(\theta)$ is the "spin flip" cross section,

$$\sigma_1(\theta) = (1/4)|f_1^+|^2 + (3/4)|f_1^-|^2$$

and the subscripts 0, 1 refer to s-p excitation with magnetic quantum number of the excited state $M_\ell = 0, 1$ respectively. Many of the experimental hazards involved in obtaining differential cross sections by the recoil, or any, technique are avoided in the ratio experiment, where primarily one requires knowledge of the beam velocity distributions and of the transmission and depolarization properties of the polarizer-analyzer combination. Figure 14 shows some experimental results of Goldstein et al,[50] again compared to Karule and Peterkop, whose values for $R(\theta)$ are calculated using their R-matrix elements and Eq.(3). In the figure $R(\theta)$ is plotted against angle, from about 0 to 20°, for energies of 3 and 5 eV.

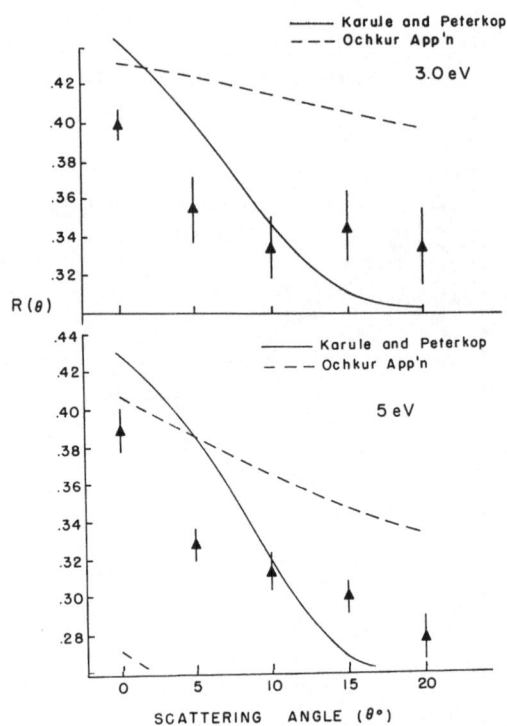

Fig. 14. Observed ratios of differential spin-flip to full differential cross sections for 3 and 5 eV, for 4s - 4p excitation in
potassium, compared to Karule and Peterkop, and the Ochkur approximation (from Goldstein, Kasdan and Bederson[50]).

Similar experiments are now in progress at N.Y.U. for sodium, for
which Moores and Norcross have obligingly printed out, not only
$|f_{0,1}|^2, |g_{0,1}|^2, |f_1 - g_1|^2$ and $\sigma(\theta)$, but $R(\theta)$ as well, over a wide
range of energies. Also shown in the figure is a comparison to the
Ochkur approximation,[51] which is similar to the Born approximation
but also includes an approximate exchange term.

6. SUMMARY OF ELECTRON-ALKALI POLARIZATION STUDIES

I hope that this "case history" has shown how, at least in
one area of electronic collisions, the use of polarized systems in
collision studies--along with other approaches--is helping produce
a unified and generally very encouraging situation. Before ending
this discussion however, it should be noted that the situation in
the heavier alkalis, (rubidium and cesium) where core-correlation
effects are more important, and in general many-body effects play

a bigger role, is not nearly as good at this time. The total and momentum-transfer cross sections, both theory and experiment, are in rather a state of confusion at the moment,[52] particularly in the very-low energy domain. More work on these systems is clearly needed.[53]

7. POLARIZATION EFFECTS IN OTHER TYPES OF ATOMIC COLLISIONS

Although it was not the purpose of this paper to discuss atom-atom collisions, it should be noted that the elastic electron-atom collision work discussed above has an analogy in alkali-alkali collisions at thermal energies, when at least one of the colliding partners is spin-polarized and analyzed. Such experiments yield information concerning the singlet and triplet interatomic potentials. Kleppner, Pritchard and their students have employed a technique similar to the recoil technique, and have exploited it fruitfully for precisely that purpose.[54]

Other collision experiments which involve similar techniques to polarize one of the collision partners have recently been performed. These are designed primarily to study anisotropies in the interaction potentials, which are generally assumed isotropic in alkali collisions. Such studies will become increasingly important as one moves to more complex atomic systems, and especially, into molecular collisional studies.[55]

Still another valuable application of polarization analysis in collisions is the study at Rice University of the polarization of electrons extracted from optically-pumped rare gases, both during the active regime of a glow-discharge, or in the afterglow.[56] Apart from its use as a source of polarized electrons, such experiments add a new dimension to the experimental untangling of complex discharge chemistry. Hill et al,[57] for example, have successfully demonstrated conservation of spin angular momentum in rare-gas metastable - metastable deexcitation - ionization reactions of the type, for example, $He^M + He^M \rightarrow He(1^1S_0) + He^+ + e^-$, where He^M is either the singlet or triplet metastable. Spin-dependence of Penning reactions (where one of the metastable colliding partners is replaced by another atom or molecule) is yet another process which can be studied by this means. Fast polarized metastable hydrogen atom beams have also been used in collision experiments, for example, in the measurements of cross sections for stripping of hydrogen atoms in metastable and ground states.[58]

Finally, I would like to note one important aspect of polarization effects which has not received much attention as yet, and understandably so since laboratory technology has only recently measured up to the challenge of such refined experiments. This concerns the use of polarization-type experiments in the testing of fundamental concepts in collision theory, and in quantum mechanics itself. The spin-conservation study mentioned in the previous paragraph is an example of the former. As another example, it should be noted that the electron-electron collision problem,[59] with spin analysis, has not been studied experimentally.

Still another intriguing and even practical problem concerns the over-determination of the scattering parameters in a number of independent collision experiments. As an example, consider elastic electron-alkali collisions. Time-independent collision theory, non-relativistically, i.e., neglecting spin-orbit interactions, requires two scattering amplitudes for a complete description of the scattering process, i.e. f, g (or alternatively $f^{\pm}$). Assuming one indeterminate phase, this implies that three independent experiments are necessary, at a given energy and polar scattering angle, for a complete determination of the observables. In fact, more than three independent experiments are possible (already there have been three for electron-potassium elastic scattering, namely $|g|^2$ [Collins et al] $|f|^2$ [Hils et al] and $(1/4)|f+g|^2 + (3/4)|f-g|^2$ [Gehenn and Wilmers, Collins et al and Hils et al]). Table I lists a number of possible experiments which include the three just mentioned, but describe others as well--a number of which are in progress at various laboratories. When these parameters are over-determined, one will be interested to see whether any inconsistencies implying a breakdown in the simple scattering theory will develop.

It should also be noted that, particularly in the heavy alkalis, relativistic effects are in fact not necessarily completely negligible, in analogy to a similar situation in low-energy scattering by mercury and other heavy elements.[1] When this is the

Expt. No.	Type of experiment	Quantity observed
I	$e(\uparrow\downarrow) + A(\uparrow\downarrow) \rightarrow e(\uparrow\downarrow) + A(\uparrow\downarrow)$	$\frac{3}{4}|f-g|^2 + \frac{1}{4}|f+g|^2$
IIa	$e(\uparrow\downarrow) + A(\uparrow) \rightarrow e(\uparrow) + A(\uparrow)$	$\frac{1}{2}|f-g|^2$
b	$\rightarrow e(\downarrow) + A(\uparrow)$	$\frac{1}{2}|f|^2$
c	$\rightarrow e(\uparrow) + A(\downarrow)$	$\frac{1}{2}|g|^2$
IIIa	$e(\uparrow) + A(\uparrow\downarrow) \rightarrow e(\uparrow) + A(\uparrow)$	$\frac{1}{2}|f-g|^2$
b	$\rightarrow e(\uparrow) + A(\downarrow)$	$\frac{1}{2}|f|^2$
c	$\rightarrow e(\downarrow) + A(\uparrow)$	$\frac{1}{2}|g|^2$
IV	$e(\uparrow) + A(\uparrow) \rightarrow e(\uparrow) + A(\uparrow)$	$|f-g|^2$
V	$e(\uparrow) + A(\downarrow) \rightarrow e(\uparrow) + A(\downarrow)$	$|f|^2$
VI	$e(\uparrow) + A(\downarrow) \rightarrow e(\downarrow) + A(\uparrow)$	$|g|^2$

Table I. Tabulation of possible elastic collision experiments, electron alkali, involving various combinations of polarized and analyzed beams (from Bederson[62]).

case, one requires additional amplitudes to describe the scattering. Such an effect could complicate the "overdetermination" problem in the heavy alkalis, but probably not in lithium, sodium and potassium.

Similar considerations obtain in excitation collisions, although here there are many more observables (seven, for example, in the s-p excitation, non-relativistically).

An example of the use of electron polarization behavior in collisions to test fundamental concepts in quantum mechanics is the suggestion recently made by Gerjuoy and Faisal[60] to employ photo-dissociation of alkali molecules (e.g., NaK) to search for experimental tests of hidden variable theory, as well as of the Einstein-Podolsky-Rosen "effect", for spin-$\frac{1}{2}$ systems, i.e., for systems obeying Fermi-Dirac statistics.[63]Gerjuoy and Faisal propose that two Stern-Gerlach analyzers be employed to study the spin-correlations of the dissociated alkali atoms. A similar proposal has also been made by Golden.[61] It is too early to decide whether such experiments are in fact actually feasible with existing technology, but at this time it appears reasonably hopeful that they are. Perhaps there will be more to say on this subject at ICAP IV.

ACKNOWLEDGMENT

The work described in this paper which was performed at New York University has been supported by grants from the Air Force Office of Scientific Research, the Army Research Office, Durham and the National Science Foundation. I would like to thank my colleagues who sent me results prior to publication, including Drs. Farago, McCusker, and Reichert, and my colleagues at New York University who participated in much of the work described in this article, particularly Professors H. H. Brown Jr. and T. M. Miller.

REFERENCES

1. Some earlier review papers, and papers of general interest in this subject include: "Production of Polarized Electron Beams", K. Jost and H.D. Zeman, HEPL-590, Stanford University, Palo Alto, Cal., April 1969; "Electron Spin Polarization by Low Energy Scattering", J. Kessler, Rev. Mod. Phys., 41, 3 (1969); "Polarized Electrons", W. Raith, Atomic Physics [ICAP-I], edited by B. Bederson, V. W. Cohen and F. M. Pichanick, Plenum Press, pp. 389-415 (1969); "Polarization effects in elastic electron scattering" (in German), W. Eckstein, Institut Fur Plasmaphysik, Garching, Germany IPP 7/1 Feb. 1970; "Electron Spin Polarization", P. S. Farago, Reports on Progress in Physics, 34, 1055 (1971); "Polarized Electrons and Some of

their Uses," V. W. Hughes, International Conference on Polarized Targets,Lawrence Berkeley Lab., Univ. of California Aug. 1971. Review papers on related subjects were presented at one of the precursors of the present Conference series, the International Symposium on the Physics of One- and Two-Electron Atoms, held in Munich 1968. These include articles by P. S. Farago and H. Chr. Siegmann, W. Raith, E. Reichert and H. Kleinpoppen [North-Holland Publ. Co., Amsterdam 1969, edited by Bopp and Kleinpoppen]. See also reference 2.
For general discussions of the types of information obtainable in collision experiments in the alkalis involving polarized collision partners, see H. Kleinpoppen, Phys. Rev. A3, 2015 (1971); also a series of articles by B. Bederson in Comments on Atomic and Molecular Physics: I, 41 (1969); I, 65 (1969); II, 160 (1970-71).

2.   "Polarized Electron Sources," J. Kessler, these proceedings.

3.   A. Temkin, Phys. Rev. 107, 1004 (1957); 121, 788 (1961).

4.   W. R. Garrett and R. A. Mann, Phys. Rev. 130, 658 (1963); 135, A580 (1964); W. R. Garrett, Phys. Rev. 140, A705 (1965).

5.   P. M. Stone and J. R. Reitz, Phys. Rev. 131, 2101 (1963).

6.   J. C. Crown and A. Russek, Phys. Rev. 138, A669 (1965).

7.   L. C. Balling, Phys. Rev. 179, 78 (1969).

8.   V. K. Lan, J. Phys. B: Atom. Molec. Phys. 4, 658 (1971).

9.   It should also be noted, as has often been pointed out, that the alkalis do represent a special case where convergence in any excited state expansion is particularly favorable. This is due primarily to the relatively weak binding of the valence electron, which results in the atomic polarizability being attributable almost completely to the first excited (resonant) transition ns → np. Other atomic systems do not possess such a fortuitous property and therefore in general will require more elaborate expansions. We thank Dr. Temkin for his comments on this point.

10.   See, for example, The Theory of Atomic Collisions, Third Edition, N. F. Mott and H. S. W. Massey (Oxford U. Press, 1965), p. 524 ff.; M. J. Seaton, in Atomic and Molecular Processes, ed. by D. R. Bates (Academic Press, N.Y., 1962), pp. 374-420; P. G. Burke and H. M. Schey, Phys. Rev. 126, 147 (1962); P. G. Burke and K. Smith, Rev. Mod. Phys. 34, 458 (1962); P. G. Burke, A. Hibbert and W. D. Robb, J. Phys. B: Atom. Molec. Phys. 4, 153 (1971).

11. E. M. Karule and R. K. Peterkop, Atomic Collisions III, ed.
     by V. Ia Veldre, Latvian Academy of Sciences, Riga, 1965 (Trans-
     lation TT-66-12939 available through SLA Translation Center,
     John Crear Library, Chicago); p. 1-27 (inelastic and elastic
     collisions above ns-np threshold); E. M. Karule, ibid., p. 29
     -48 (elastic collisions below threshold).

12. P. G. Burke and A. J. Taylor, J. Phys. B: Atom. Molec. Phys.,
     2, 869 (1969).

13. D. W. Norcross, J. Phys. B: Atom. Molec. Phys. 2, 1300 (1969);
     Corrig. 4, 628 (1971); 4, 1458 (1971); D. L. Moores and D. W.
     Norcross, J. Phys. B, in press [Tables of Li and Na elastic
     scattering results].

14. See M. J. Seaton in Atomic and Molecular Processes, reference
     10.

15. U. Fano, Phys. Rev. 178, 131 (1969); V. W. Hughes, R. L. Long
     Jr., M. S. Lubell, M. Posner and W. Raith, Phys. Rev. 5, A195,
     (1972); G. Baum, M. S. Lubell and W. Raith, Phys. Rev. Lett.
     25, 267 (1970); also "Polarized Electrons", W. Raith (reference
     1).

16. "Photodetachment of Li⁻ and Na⁻", D. W. Norcross and D. L.
     Moores, these proceedings.

17. N. Fuetrier, H. Van Regemorter and V. K. Lan, J. Phys. B:
     Atom. Molec. Phys. 4, 670 (1971).

18. R. E. Collins, B. Bederson and M. Goldstein, Phys. Rev. 3,
     A1976 (1971).

19. P. J. Visconti, J. A. Slevin and K. Rubin, Phys. Rev. 3, A1310
     (1971).

20. "Absolute Total Cross Sections for Electron Scattering by
     Light Alkalis", A. Kasdan, T. M. Miller and B. Bederson, Book
     of Abstracts, ICAP III (Boulder, 1972).

21. See, for example, K. Rubin, B. Bederson, M. Goldstein and
     R. E. Collins, Phys. Rev. 182, 201 (1969); Methods of Experi-
     mental Physics, edited by B. Bederson and W. L. Fite. (Acade-
     mic Press, Inc., New York 1968) Vol. 7A, pp. 89-95; B. Bederson
     and L. J. Kieffer, Rev. Mod. Phys. 43, 601 (1971) pp. 610-613.

22. R. B. Brode, Phys. Rev. 34, 673 (1929).

23. It should also be noted that recent polarized-orbital calcula-
     tions have also been reasonably successful in predicting total
     cross sections. For example, see Visconti et al [ref. 19]
     for a comparison of total e-Rb scattering with Balling's
     polarized orbital calculation [ref. 17]. However it should
     be noted that the Balling calculation represents elastic scat-
     tering only in this comparison.

24. W. Gehenn and M. Wilmers, Z. Physik 244, 395 (1971).

25. D. Hils, M. V. McCusker, H. Kleinpoppen and S. J. Smith, Phys.
     Rev. Lett. 29, 398 (1972).

26. The Gehenn and Wilmers results also agree very well, when nor-
     malized, to Karule and Peterkop, and to the old recoil data
     of Rubin et al [K. Rubin, J. Perel and B. Bederson, Phys. Rev.
     117, 151 (1960)]. The "fine structure" shown in the 3 and 5
     eV Rubin curves in the Gehenn and Wilmers paper are actually
     the forward scattering peaks in the 4s-4p excitation, and
     should not be included in the elastic plot [see Rubin, et al,
     reference 21]. Recently Gehenn and Reichert [W. Gehenn and
     E. Reichert, Z. Physik, in press] have performed similar
     measurements in sodium, with equally impressive agreement
     with close-coupling theory.

27. Actually the complete story isn't completely rosy. Slevin et
     al [J. A. Slevin, P. J. Visconti and K. Rubin, Phys. Rev. A5,
     2065 (1971)], using a recoil technique involving detection of
     scattered atoms in two dimensions, obtain absolute elastic
     and inelastic cross sections over a wide range of energies
     and angles. The 3eV elastic data do not agree at all well
     with the data shown in Fig. 3.

28. E. Reichert, private communication.

29. E. Reichert and H. Deichsel, Phys. Lett. 25A, 560 (1967).

30. K. Rubin, 4th International Conference on the Physics of Elec-
     tronic and Atomic Collisions, Quebec 1965 (Science Bookcraft-
     ers, Hastings-on-Hudson, N.Y.); A. E. Glassgold, Phys. Rev.
     132, 2144 (1963); A. E. Glassgold and J. F. Walker, Phys. Rev.
     160, 11 (1967).

31. The potential use of spin-exchange as a means of producing
     polarized electrons has been long recognized. See, for exam-
     ple, P. S. Farago in reference 1, also J. Byrne and P. S.
     Farago, Proc. Phys. Soc., Lond. 86, 801 (1965); R. J.
     Krisciokaitis and W. Y. Tsai, Nucl. Instrum. Meth., 83, 45

(1970); B. D. Obedkov and E. X. El-Mosallamu, Vestnik Leningradskogo Universiteta, p. 43, (1971 No. 22); G. Drukarev, Seventh International Conference on Physics of Electronic and Atomic Collisions, (North-Holland Press, Amsterdam, 1971).

32. D. M. Campbell, H. M. Brash and P. S. Farago, Proc. Roy. Soc., in press [I am indebted to Professor Farago for sending me his results prior to publication].

33. F. G. Major and H. G. Dehmelt, Phys. Rev. 170, 91 (1968).

34. H. G. Dehmelt, Phys. Rev. 109, 381 (1958); L. C. Balling and F. M. Pipkin, Phys. Rev. 136, A46 (1964); L. C. Balling, Phys. Rev. 151, 1 (1966).

35. The exchange result reported in the pioneering paper of Dehmelt in 1958 (reference 34) was actually only an estimate of the cross section, and should not be considered to be a quantitative measurement.

36. H. Gibbs, Phys. Rev. 139, A1374 (1965).

37. "The Temperature Dependence of the Electron-Rubidium Spin-Exchange Cross Section", S. J. Davis and L. C. Balling, these proceedings.

38. L. C. Balling, R. J. Hanson and F. M. Pipkin, Phys. Rev. 133, A607 (1964).

39. D. Andrick, M. Eyb and H. Hofmann, J. Phys. B: Atom. Molec. Phys. 5, L15 (1972).

40. M. McCusker, private communication.

41. E. M. Karule and R. K. Peterkop, Latvijas PSR Zinatnu Akademijas Vestis, p. 3 (1971 No. 1).

42. See B. Bederson and L. J. Kieffer, Rev. Mod. Phys. 43, 601 (1971); also L. J. Kieffer and G. H. Dunn, Rev. Mod. Phys. 38, 1 (1966).

43. D. R. Flower and M. J. Seaton, Proc. Phys. Soc. 91, 59 (1967).

44. E. A. Enemark and A. Gallagher, Phys. Rev. A 6, 192 (1972).

45. G. N. Gould, Ph.D. Thesis, U. of New South Wales, 1970 (unpublished).

46. H. Hafner and H. Kleinpoppen, Z. Physik 198, 315 (1967).

47. E. M. Karule, Latvijas PSR Zinatnu Akademijas Vestis, p. 9
     (1970 No. 3).

48. N. Feutrier, J. Phys. B: Atom. Molec. Phys. $\underline{3}$, L152 (1970).

49. See Rubin et al, reference 21; H. Kleinpoppen, reference 1,
     B. Bederson, reference 1.

50. M. Goldstein, A. Kasdan and B. Bederson, Phys. Rev. $\underline{5}$, A660
     (1972).

51. V. I. Ochkur, Soviet Phys. - JETP $\underline{18}$, 503 (1964).

52. See B. Bederson and L. J. Kieffer, reference 42, also, "Sur-
     vey of Electron-Cesium Collision Probabilities: Momentum
     Transfer Collisions", J. A. Dayton, Jr., Lewis Research Center
     NASA report TMX-1897, Oct. 1969.

53. It should be noted that Kasdan, Miller and Bederson report on
     new total cross section measurements at this meeting on sodium
     and lithium. While the sodium results are in almost exact
     agreement with Moores and Norcross, the lithium results lie
     some 20 - 30% above the calculations. Norcross (private
     communication) believes this could be attributed to core
     correlation effects which are particularly difficult to take
     into account in lithium.

54. D. E. Pritchard, D. C. Burnham and D. Kleppner, Phys. Rev.
     Lett. $\underline{19}$, 1363 (1967). For discussion of the theoretical
     problem of spin-exchange in alkali atom-atom collisions, see
     for example A. Dalgarno and M. R. H. Rudge, Proc. Roy. Soc.
     $\underline{A286}$, 519 (1965); H. O. Dickinson and M. R. H. Rudge, J. Phys.
     B: Atom. Molec. Phys. $\underline{3}$, 1448 (1970).

55. H. G. Bennewitz and R. Haerten, Z. Phys. $\underline{227}$, 399 (1969); S.
     Stolte, J. Reuss and H. L. Schwartz, Physica $\underline{57}$, 254 (1972).

56. M. V. McCusker, L. L. Hatfield and G. K. Walters, Phys. Rev.
     $\underline{A5}$, 177 (1972).

57. J. C. Hill, L. L. Hatfield, N. D. Stockwell and G. K. Walters,
     Phys. Rev. $\underline{A5}$, 189 (1972); see also the review talk "Polari-
     zation of Ions and Electrons by Optical Pumping Techniques",
     L. D. Schearer, Atomic Physics 2 [ICAP II], edited by P. G. H.
     Sandars, (Plenum Press, N.Y., 1971), p. 87.

58. B. Donnally and W. Sawyer, Phys. Rev. Lett. $\underline{15}$, 439 (1965);
     ICPEAC VI, Mass. Inst. of Tech. Press (1969) p. 488.

59. P. Stehle, Phys. Rev. 110, 1458 (1958).

60. E. Gerjuoy and F. Faisal, to be published.

61. D. Golden, private communication.

62. B. Bederson, Comments on Atomic and Molecular Physics, 1, 41 (1969).

63. A. Einstein, B. Podolsky and N. Rosen, Phys. Rev. 47, 777 (1935).

BEAM RESONANCE MEASUREMENTS OF ATOMIC QUADRUPOLE MOMENTS

P. G. H. Sandars and A. J. Stewart

Clarendon Laboratory

Oxford University

I. INTRODUCTION

The starting point for this work is the well-known multipole expansion of the interaction between a charge distribution and an externally applied electrostatic field. For an atom in a laboratory field the expansion converges rapidly since it is a power series in the ratio of the size of the atom to the dimensions of the electrodes producing the field. A further simplification occurs when the field is sufficiently weak that it can be treated in first order because then the odd order multipoles vanish for reasons of parity and time-reversal symmetry. In addition, the monopole term is zero because an atom has no net charge. It follows that not only is the quadrupole moment the first non-vanishing multipole moment for an atom, it is also the only one which it is possible to measure directly through its interaction with a laboratory applied field.

While there is obviously considerable interest in direct measurements of a fundamental atomic property such as the quadrupole moment, these measurements present serious difficulties. Not only is the interaction with attainable laboratory fields quite small, but one also has to cope with the complications caused by the dipole interaction taken to second order - the well-known quadratic Stark effect. This is much larger than the quadrupole interaction at the highest attainable fields and it is difficult to separate from it because both effects behave like tensors of rank 2 in their dependence on the atomic quantum numbers.

Recently, we showed that it is possible to overcome these difficulties by using relatively low electric fields and taking

advantage of the high sensitivity possible with the atomic beam
resonance method. We made initial **measurements** of the quadrupole
moments of the ground states of Al and In,[1,2] with the results
given in Table I below:

Atomic State	θ_{exp}	θ_{cf}
Al 3p $^2P_{3/2}$	2.53 ± .15	2.8
In 5p $^2P_{3/2}$	2.94 ± .1	3.32

Table I. Quadrapole moments in atomic units

We have also quoted theoretical values which were calculated using
the central field Hartree-Fock **approximation**. Agreement is quite
close, though in both cases the experimental result is lower than
the theoretical one by more than our quite conservatively estimated experimental error. The discrepancies may be significant
but they are too small for us to be certain.

In the present work we have extended these measurements to
the $^3P_2$ metastable states of the rare gases. Here the situation
is very different; disagreement between our experimental results
and the predictions of the central-field theory is almost complete. But this is of course the justification for the experiments. In suitable cases, the closed shell polarization effects
associated with the name of Sternheimer should make the quadrupole
moment of an atom differ markedly from the predictions of the central field approximation. Measurement of such moments affords us
important direct evidence on the existence and magnitude of these
shielding effects which play an important part both in hyperfine
structure and in the interaction between an ion and the ligand
fields in a crystal.

II. EXPERIMENT

Our atomic beam resonance apparatus for metastable rare-gas
atoms is essentially the same as that used previously in the search
for electric dipole moments[3] except that the uniform electric field
plates in the resonance region have been replaced by a four-pole
system which gives a uniform field gradient over the area of the
beam. The apparatus is illustrated schematically below in figure 1.

In a magnetic field which is sufficiently large that it forms
the axis of quantization, the energy of a Zeeman sub-level is
given by:

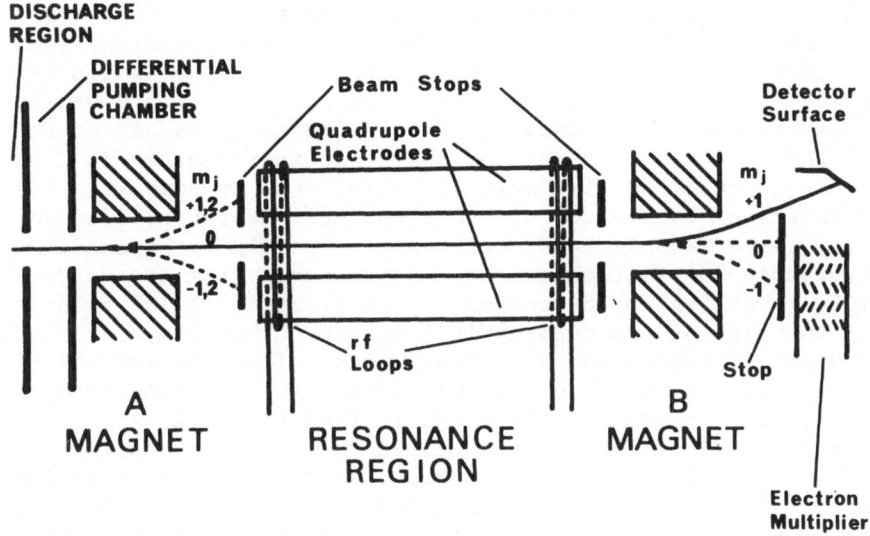

Figure 1. Atomic beam resonance apparatus for metastable $^3P_2$ rare gas atoms. The total length of the apparatus is approximately 3 meters.

$$W = g_J \, \mu_0 \, M_J \, B \, - \, \frac{3M_J^2 - J(J+1)}{2J(2J-1)} \, \theta \, \frac{\partial E_z}{\partial z}$$

$$- \, \frac{3M_J^2 - J(J+1)}{4J(2J-1)} \, (3E_z^2 - E^2) \, - \, g_J \, \mu_0 \, M_J \, (\underline{v} \times \underline{E})_z \quad .$$

The first term is the usual magnetic interaction, the second is the quadrupole interaction which we wish to measure, the third is the quadratic Stark splitting, while the fourth is the interaction between the magnetic moment of the atom and the magnetic field it experiences as it moves with a velocity $\underline{v}$ through the electric field $\underline{E}$.

The beam apparatus was set to observe the $M_J = 0 \rightarrow 1$ transition at a magnetic field of 40 mG and an electric field gradient of $4 \cdot 10^3$ V/cm^2. The resonance line-width was of order 250 Hz and the intensity was 10^5 atoms per second on a background of comparable size. The applied field gradient was limited to this comparatively low value because at higher gradients the range of quadratic Stark shifts experienced by different atoms in the inhomogenious electric field was comparable to the linewidth and the resonance was degraded with consequent loss of sensitivity. At the fields used, the quadratic Stark shift was of order 100 Hz, the $\underline{v} \times \underline{E}$ effect of order 2 Hz and the quadrupole interaction of order 0.5 Hz.

The basic problem in the experiment was to separate out and measure the small quadrupole interaction in the presence of the larger quadratic Stark and $\underline{v} \times \underline{E}$ effects. The method which we used has been developed in our search for electric dipole moments and is described in reference 3. The rf frequency is set on the side of the resonance so that any shift of its frequency is translated into a proportional change in intensity. By comparing resonance shifts on both sides of the resonance one can readily separate resonance dependent effects from changes in background. The quadrupole interaction can be distinguished from the quadratic effect by reversing the direction of the electric field which changes the sign of the quadrupole interaction but leaves the quadratic Stark effect unchanged. The $\underline{v} \times \underline{E}$ effect is more difficult to separate from the quadrupole interaction because both change sign on reversal of the electric fields. To do so, we have to reverse the direction of the magnetic field which changes the sign of the $\underline{v} \times \underline{E}$ effect but leaves the quadrupole interaction unchanged. These reversals of the electric and magnetic fields, together with the comparison between the two sides of the resonance, were controlled by our on-line PDP 8 computer, using the orthogonal square wave method described elsewhere.[4] The computer was also used to lock the beam resonance frequency to a local oscillator, to control a calibration signal and to make an on-line analysis of the resonance signal to yield simultaneous values for the quadrupole interaction, the $\underline{v} \times \underline{E}$ effect, the calibration and their respective statistical errors.

The combined results of a number of experiments on each quadrupole moment are given in Table II below. As a check that we were indeed measuring a true quadrupole interaction, we established the linearity of the quadrupole shift against applied field gradient and we also verified that the predicted dependence on the an angle between the electric field gradient and the magnetic field was obeyed not only by the quadrupole interaction but also by the $\underline{v} \times \underline{E}$ effect. Our main source of error was the difficulty of ensuring that only the $M_J = 0 \rightarrow 1$ transition was observed. The double quantum $M_J = 0 \rightarrow 2$ transition has the same magnetic field frequency but a quadrupole interaction which is four times larger. While our system of deflections discriminates against the double quantum transition, it is difficult to ensure that it is entirely absent and so we have extrapolated our quadrupole measurements to the limit of zero rf power where the single quantum transition should be dominant. The quoted errors are our estimate of the uncertainty in this extrapolation procedure.

III. THEORY

In the central field approximation the quadrupole moments of the metastable $^3P_2$ states of the rare gases are given by

Atomic state	θ_{exp}	θ_{cf}	R_{exp}
Ne $2p^5 3s\ ^3P_2$	$-.048 \pm .005$	$-.234$	0.817
Ar $3p^5 4s\ ^3P_2$	$-.042 \pm .004$	$-.613$	1.016
Kr $4p^5 5s\ ^3P_2$	$+.046 \pm .005$	$-.822$	1.117
Xe $5p^5 6s\ ^3P_2$	$+.30 \pm .03$	-1.15	1.330

Table II. Experimental and theoretical central field quadrupole
moments in atomic units. $R_{exp} = 1 - \theta_{exp}/\theta_{cf}$.

$$\theta_{cf} = \langle np^5(n{+}1)s\ ^3P_2 | \sum_i -r_i^2\ C^2(\theta_i,\phi_i) | \ np^5(n{+}1)s\ ^3P_2 \rangle$$

where $C^2(\theta_i,\phi_i)$ is the standard spherical harmonic and the sum
over i spans the atomic electrons. The angular part of this expression can be readily evaluated and the quadrupole moment expressed in terms of the mean square radius of the np shell,
$\langle r^2 \rangle_{np}$. To evaluate this, we need radial wave-functions and these
should ideally be obtained from a Hartree-Fock calculation. How-
ever, as far as we are aware, no such calculations are available
for these metastable states and we have therefore had to use solutions of the radial Schroedinger equation in the potential given
by Herman and Skillman.[5] We expect them to be of adequate accuracy for our purpose.

Our calculated central field values of the quadrupole moments
are compared with the experimental results in Table II above. As
will be seen, there is an almost complete disagreement; not only
are the experimental values very different in magnitude from the
theoretical ones but in two cases the sign is different. This
change of sign is completely inexplicable on any central field
model. Our results, therefore, constitute an unusually dramatic
disagreement between experiment and the simple central field model
of atomic structure.

The most likely explanation is closed shell polarization of
the type first discussed by Sternheimer in relation to the quadrupole hyperfine structure and later extended by him to the case of
atomic quadrupole moments.[6] Physically, the quadrupole moment of
the np^5 shell interacts with and polarizes the $(n{+}1)s$ shell by
exciting the electron into a d state thus giving rise to a modified quadrupole moment for the atom. More formally, in the next
order beyond the central field approximation, when one takes into
account the first order effects of the residual two particle Coulomb interaction between the electrons, one obtains a contribution
to the atomic quadrupole moment of the form:

$$\theta_{pol} = \sum_{n'} \langle np^5(n+1)s \ | \sum_i -r_i^2 \ C^2(\theta_i, \phi_i) | \ np^5 n'd \rangle \times$$

$$\langle np^5 n'd | \sum_{i>j} \frac{1}{r_{ij}} | \ np^5(n+1)s \rangle \times \frac{1}{w_{(n+1)s} - w_{n'd}} + C.C.$$

The important feature of this expressions is that, to a quite good approximation, we can neglect exchange and assume that the polarized $s \rightarrow d$ electron is outside the np electrons in which case the Coulomb matrix element in the expression above is readily shown to be proportional to the central field quadrupole moment, θ_{cf}. It follows that we can write $\theta_{pol} = -R\theta_{cf} + \theta_{pol} = (1-R)\theta_{cf}$. This form suggests that the main effect of the $s \rightarrow d$ excitation is to shield (or enhance) the original quadrupole moment of the open p shell. If this is so, then it should be interesting to compare θ_{exp} and θ_{cf} and deduce an 'experimental' shielding factor: $R_{exp} = 1 - \theta_{exp}/\theta_{cf}$.

Values of R_{exp} deduced in this way are given in the fourth column of Table II above. We see immediately that, whereas the values of θ_{exp} vary in an apparently erratic manner, the values of R_{exp} form a smooth progression. The change in sign of the observed quadrupole moments between Ar and Kr now has the natural explanation that R goes through unity at that point. This smooth variation of R_{exp} suggests that the explanation suggested above may be along the right lines but confirmation must await detailed calculations of R. Such calculations are now being carried out by Sternheimer.[7] His results, which will be reported elsewhere, are most encouraging; not only are the values of R of the correct sign to give shielding but the computed values of R are in excellent agreement with our experimental values.

REFERENCES

1. J. R. P. Angel, P. G. H. Sandars and G. K. Woodgate, J. Chem. Phys. 47, 1552 (1967).

2. M. A. Player and P. G. H. Sandars, J. Phys. B 3, 1620 (1970).

3. P. G. H. Sandars, M. A. Sheen and G. K. Woodgate, to be published.

4. G. E. Harrison, M. A. Player and P. G. H. Sandars, J. Sci. Inst. 4, 750 (1971).

5. F. Herman and S. Skillman, Atomic Structure and Calculations (Prentice Hall, Englewood Cliffs, New Jersey, 1963).

6. R. M. Sternheimer, Phys. Rev. 146, 140 (1966).

7. R. M. Sternheimer, private communication (1972).

RECENT ADVANCES IN HELIUM OPTICAL PUMPING

J. BROSSEL

Ecole Normale Supérieure - Paris - France

INTRODUCTION

In this paper, I will describe a few experiments dealing with optical pumping in helium. The field was opened more than 10 years ago [1][2] and many people have kept very active in it all this time.

My purpose is not to make a complete review of the many results which have been obtained : the space which is available to me is far too short for that. I will limit myself to the description of a few experiments which have been carried out at the ENS over the past 3 or 4 years (we came to this field rather late !), or which are still under investigation. So, I will not try to be complete in any manner, to give credit and accordingly to be fair to all those who have worked in the field and brought many essential results to light.

The work I am going to discuss deals with conventional optical pumping (ie with a conventional light source) in weak discharges in pure ^{3}He, pure ^{4}He and ^{3}He-^{4}He mixtures. It has already been extended in fact to the other rare gases.

In a first part, I will recall the basic facts about optical pumping in helium, how one can build--via metastability exchange-- a large nuclear orientation $< \vec{I} >$ in the ground state of ^{3}He [2] [3][4] (^{3}He, 1^1S_0) and, from there, an electronic orientation $< \vec{J} >$ in many excited states of ^{3}He [5] and in the ground state $1^2S_{1/2}$ of the helium three ion ^{3}He$^+$ which is produced in the discharge [6][7]. As we shall see all these orientations are strongly coupled to each other, and this allows a large variety of ways to monitor magnetic resonance signals in many levels of ^{4}He, ^{3}He and ^{3}He$^+$. This led us to the study of the relaxation—due to various kinds of collisions in the gas phase—of several of those states. I will then limit myself to a brief description of the following problems :

• relaxation of (^{3}He, 1^{1}S$_0$) (ground state of ^{3}He) [8] due to the radiofrequency field non uniformity.

• relaxation of the 2^{3}S$_1$ metastable state, in pure ^{3}He, and ^{3}He-^{4}He mixtures via metastability exchange [1][2][7][9][10]. I will give a few results about the closely related problem of the

• measurement of the Landé g factor in (^{4}He-2^{3}S$_1$) [11]

• charge exchange collisions in ^{3}He$^+$, ie collisions [6][7][12] between (^{3}He$^+$, 1^{2}S$_{1/2}$) and (^{3}He, 1^{1}S$_0$)

• Finally, I will describe preliminary measurements on the relaxation of a few n^{1}D$_2$ levels in a weak He discharge [13].

OPTICAL PUMPING IN HELIUM

The Optical Pumping Scheme

    Figure 1 shows the energy level diagram of the lowest states of the helium atom.

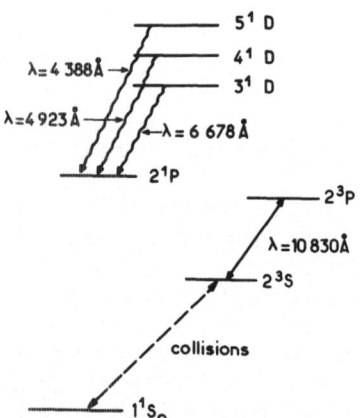

Fig. 1   Energy level scheme of the helium atom . Optical pumping of the 2^{3}S$_1$ metastable level populated by a wcak rf discharge is produced by the λ = 10 830 Å line.

A weak discharge is maintained in He at a low pressure (below a few torr). This raises the atom into all its excited states or even ionises it. This produces also in a steady state manner a small population of 2^3S_1 helium metastables. As is well known [1] [2], these can be oriented in a static field $\vec{B}_0$ by optically pumping this level with the 10 830 Å ($2^3P \rightarrow 2^3S_1$) circularly polarized line.

The most obvious way to monitor this 2^3S_1 orientation is to measure the absorbed light on the pumping beam itself at 10 830 Å : the magnetic resonance between the Zeeman sublevels of (^{4}He, 2^3S_1) -or between those of the hfs F and F' states in (^{3}He, 2^3S_1)- is easily observed. The same goes for transient signals of many different kinds which appear when the pumping beam is switched on and off ... etc. In this way one can study the longitudinal and transverse relaxations of this state, which are produced (in the gas phase) by collisions of different kinds.

Relaxation of the 2^3S_1 State

The basic results (they appear very puzzling at first sight) have been obtained by the people who did the pioneering work in the field [1][2][3]. I will only mention what is of interest to my present purpose. One observes what follows :

In pure ^{4}He, the transition between Zeeman sublevels of 2^3S_1 (g = 2) can be very narrow indeed. In the absence of any trace of ^{3}He, it is essentially determined (at very low rf levels [9] and pumping beam intensities) by the life-time of 2^3S_1 metastables ie by the diffusion time to the walls of the cell [14][15] (where most authors think that they are destroyed). The line gets narrower at higher ^{4}He pressures and allows a very high precision measurement of g which I will describe later on. The result is interesting because this g value can be obtained with comparable precision, from Q.E.D. calculation [16].

In pure ^{3}He, the resonances of the 2 hfs levels of 2^3S_1 F=3/2 and F'=1/2 are found to be much broader than in ^{4}He (under similar conditions). Their width increases at higher ^{3}He pressures (linear dependence). It is determined by binary collisions between (^{3}He, 2^3S_1) and (^{3}He, 1^1S_0) resulting in metastability exchange, ie (^{3}He$_A$, 2^3S_1) + (^{3}He$_B$, 1^1S_0) $\rightarrow$ (^{3}He$_A$, 1^1S_0) + (^{3}He$_B$, 2^3S_1).

Indeed metastability exchange exists also in ^{4}He -ie between (^{4}He, 2^3S_1) and (^{4}He, 1^1S_0)- and takes place at the same rate, but there is no way to detect it in an optical pumping experiment. One has to admit then, that in metastability exchange collisions all the information of electronic character which is carried by the incoming (^{4}He, 2^3S_1) metastable is transferred, during the colli-

sion, to the outgoing one : <u>in fact electron clouds are just ex-</u>
<u>changed ([17]) between the colliding atoms.</u>

As a consequence, in ^{3}He-^{4}He mixtures the electronic information present on the 2^3S_1 metastable atom goes to the other isotope
in a metastability exchange collision. Because the Landé factors
are different in the metastable states of the 2 isotopes, the transverse magnetisation rotates at very different frequency in each.
When fields are high enough the transverse relaxation of 2^3S_1 is
lost for an isotope because of metastability exchange collisions
against another isotope (an isotope relaxes the other). One finds,
for instance, that the (^{4}He, 2^3S_1) linewidth is proportional to
the ^{3}He pressure ([7])([10]).

The Ground State Nuclear Orientation

There is another very remarkable fact which is observed ([2])([3])
in ^{3}He when the 2^3S_1 state is oriented by optical pumping : <u>there</u>
<u>appears a nuclear orientation in the 1^1S_0 ground state</u>. It can reach
very large values (20% at room temperature). It clearly comes from
the oriented 2^3S_1 metastables via metastability exchange collisions.
The ^{3}He nuclear relaxation in the 1^1S_0 ground state has been studied
extensively ([9]). It is very slow and seems to depend on many factors of comparable importance. In the absence of a discharge, it
changes markedly with temperature ([18]) and is very sensitive to
the inhomogeneity of the static field B_0 ([2])([3])([19])([20]) when B_0
is close to zero.

Because the concentration of metastable 2^3S_1 atoms is so low
in weak discharges, the time which is necessary to obtain the full
ground state nuclear orientation < $\vec{I}$ > can be very long. On the
other hand, as we will see, metastability exchange collisions can
transfer back < $\vec{I}$ > to the 2^3S_1 state at a rather fast rate because of the large population of ground state atoms. In other words
the 2^3S_1 and 1^1S_0 <u>orientations are strongly coupled</u> and the 1^1S_0
nmr appears on optical transitions starting from the 2^3S_1 level,
for instance on the 10 830 Å line : this is a very convenient way
to monitor the nmr signal.

Nuclear Spin Conservation in Collisions in the Gas Phase

We look now into the reasons why a nuclear orientation
appears in (^{3}He - 1^1S_0) because of metastability exchange when
there exists an orientation in (^{3}He - 2^3S_1). As is well known
optical pumping of (^{3}He, 2^3S_1) will produce in this state an electronic orientation < $\vec{J}$ > and a nuclear orientation <$\vec{I}$ > as well.
As shown by J.C. Lehmann ([21]) this takes place via the $a\vec{I}.\vec{J}$ hyperfine interactions in the 2 levels involved in the pumping transition (here 2^3P and mainly in 2^3S_1).

A metastability exchange collision is due to an electrostatic interaction which does not act on $\vec{I}$ (or $\vec{S}$) directly. Moreover it is a very short process : the value of the cross section ($\sim 10^{-15} cm^2$) leads to an interaction time θ of the order of 10^{-13} sec (the time it takes for a He atom at thermal velocities to travel 1 Å). If H is the part of the static hamiltonian acting on $\vec{I}$ directly (Zeeman and hfs interactions) $H\theta \ll 1$: the evolution of $\vec{I}$ during the time θ the collision lasts, is $e^{-iH\theta}$. It is completely negligible ([17]) : during the metastability exchange collision itself, the nuclear spins of the 2 colliding atoms remain unaffected. In other words the nuclear orientation $< \vec{I} >$ present in 2^3S_1 goes fully into the ground state 1^1S_0 (the same happens in the reverse process).

The above remarks hold for most collisions in the gas phase, because the electrostatic interaction (which does not act on $\vec{I}$ or $\vec{S}$ directly) is the dominant one in most of those processes : their duration θ is so short that the nuclear spins remain unaffected during the collision itself. On the other hand <u>between</u> successive collisions the evolution of $\vec{I}$ is governed by the static hamiltonian. It is quite clear that the same can be said about the electronic spin $\vec{S}$ but the condition $H'\theta \ll 1$ (H' is the static hamiltonian acting on $\vec{S}$) is much more difficult to meet because of the magnitude of the (electronic) Bohr magneton and of the spin-orbit interaction.

Existence of an Atomic Orientation in Excited Atomic
Levels of the Neutral Helium Atom and in the Helium Ion,
in Optically Pumped Helium Weak Discharges

Quite clearly the 1^1S_0 and 2^3S_1 orientations are not the only ones to exist and to be strongly coupled in an isotropic discharge in optically pumped helium ([5]). The ground state ^{3}He atoms (carrying $< \vec{I} >$) are raised to different excited states by collisions with <u>electrons</u> and <u>ions</u> present in the discharge (note that at the pressures at which one operates the distribution of their velocities is isotropic). Here again, the interaction responsible for the excitation is an electrostatic one of very short duration. As a consequence the ground state orientation $< \vec{I} >$ <u>is fully transferred</u> to <u>all</u> excited states during the excitation process.

During the life-time τ of those states -in weak fields- the dominant hamiltonian acting on $\vec{I}$ is the $a\vec{I}.\vec{J}$ hfs interaction. The corresponding evolution will be appreciable whenever $a\tau \gg 1$. Because $a\vec{I}.\vec{J}$ mixes states with the same m_F value, $< \vec{I} >$ will be partly transformed into <u>electronic</u> orientation $< \vec{J} >$ in these states. Of course, this shows up in the degree of circular polarization of the lines originating in those states. One sees, then, that, in an optically pumped discharge all excited states of the neutral

atom for which $a\tau \gg 1$ <u>are oriented; all these orientations are</u>
<u>strongly coupled to the metastable</u> 2^3S_1 <u>and ground state</u> 1^1S_0 <u>orien-</u>
<u>tations</u> from which they originate. As a consequence one can monitor
the 2^3S_1 and 1^1S_0 resonances <u>by looking at the circular polariza-</u>
<u>tion of any one of the lines emitted by the discharge and coming</u>
<u>from levels where</u> $a\tau \gg 1$ (for instance one can use, among others,
the 6 678 and 5 876 Å lines). This method is very useful and often
leads to a better signal to noise ratio than the one obtained on
the absorption lines starting from 2^3S_1.

In the presence of collisions reducing the lifetime τ to a
smaller value τ' the condition $a\tau' \gg 1$ is more difficult to meet,
and the method we described couldn't be operative at high enough
pressures. It would not work either in high fields, when $\vec{I}$ and $\vec{J}$
are decoupled by $\vec{B}_0$. This last problem has been studied in detail.
One finds at intermediate field values that an electronic alignment
(as well as $< \vec{J} >$) appears, leading to a linear degree of polarization of the emitted lines. This has been used to determine the hfs
constants of several excited levels of ^{3}He ([5])([22]). Another very
interesting example when whatever we have said applies, is "<u>charge</u>
<u>exchange</u>",

$$A + B^+ \rightarrow A^+ + B$$

in particular when atoms A and B are identical (they can be different isotopes though). A process of this type allows one to obtain
<u>an oriented A$^+$ ion</u> when one starts with a neutral atom A on which
a <u>nuclear</u> orientation exists. It is in this way, we believe, that
an orientation appears in the ground state $1^2S_{1/2}$ of ^{3}He$^+$ (and,from
what we said, in its excited states as well) produced in weak discharges in optical pumped ^{3}He. When one takes into account the value of the charge exchange cross sections and when one remembers
the very long relaxation time of the 1^1S_0 ground state orientation,
one finds that within this lifetime a (^{3}He, 1^1S_0) atom can undergo
several charge exchange collisions. As a consequence the orientation
in (^{3}He, 1^1S_0), (^{3}He, 2^3S_1) and (^{3}He$^+$, $1^2S_{1/2}$) are strongly coupled,
and the $\left[^3\text{He}^+, 1^2S_{1/2}\right]$ resonance can be monitored ([6]) on any of
the lines of the <u>neutral atom</u> for which $a\tau \gg 1$ which are emitted
by the discharge, or, on the absorbed light at 10 830 Å ([12]), on
the pumping beam.

The "decoupling" of $\vec{I}$ and $\vec{J}$ during the time θ for which a
collision lasts, has been a great help in the course of study of
<u>many relaxation processes</u> in the gas phase ([23]). P.L. Bender was the
first to suggest it ([24]). If one can work out the theory for an
even isotope (I = 0) the results can be used directly when $I \neq 0$.
One has just to recouple $\vec{I}$ and $\vec{J}$ <u>between</u> collisions. The relaxation
times for $I \neq 0$ bear simple relations with those for I = 0 ([25]).
This has found direct applications in the study of Holtsmark collisions ([26]), in collisions of many excited states against other atomic species ([27])([28])([29]), in (electronic) spin exchange collisions ([30]), etc., etc.

The Experimental Set-Up

We will finally briefly describe the experimental set up (fig. 2). The helium lamp L is followed by a 1,1 µ filter and by a circular polarizer P_1. The cell containing ^{3}He (or ^{4}He, etc.) is excited by a weak rf discharge and placed in a uniform longitudinal magnetic field $\vec{B_0}$. It is placed in a radiofrequency field $\vec{B_1}\cos\omega t$, linearly polarized at right angles to $\vec{B_0}$.

On the other side of the cell one finds a rotating quarter wave plate λ/4 followed by a linear polarizer P_2. This is alternately (depending on the position of the slow axis of the λ/4 plate) a σ^+ or σ^- analyser. The photomultiplier PM detects then an ac current at twice the frequency of rotation of the λ/4 plate, whose intensity is proportional to the difference of the σ^+ and σ^- intensities emitted by the line λ (isolated by filter F_2) ie proportional to the amount of orientation < J > present in the atomic state under study. Most of the measurements were made at 6 678 Å (3^1D - 2^1P) and at 5 876 Å (3^3D - 2^3P).

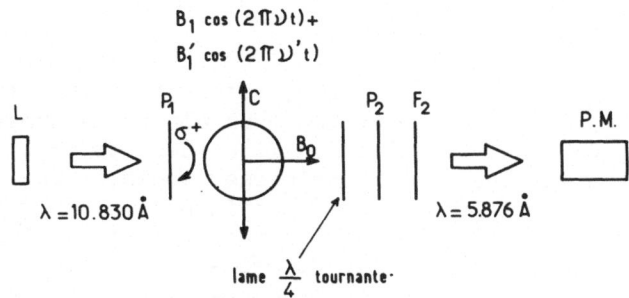

$B_1 \cos(2\pi\nu t) +$
$B_1' \cos(2\pi\nu' t)$

L P_1 C P_2 F_2 P.M.

σ^+ B_0

λ = 10.830 Å   λ = 5.876 Å

lame $\frac{\lambda}{4}$ tournante·

Fig. 2   The experimental set up.
L : helium lamp
P_1 : circular polarizer
C : discharge cell containing helium
P_2 : linear polarizer
F_2 : optical filter passing 5 876 Å
PM : phototube.

A spectrum obtained in this way, in ^{3}He with a very small amount of ^{4}He impurity, is shown on fig. 3. The Landé g factor scale covers values between 4 and 4/3. One can see 2 resonances corresponding to the hfs levels F = 3/2 and F' = 1/2 of (^{3}He, 2^3S_1).

In addition to this one sees a weak resonance at g = 2. It becomes much more intense in pure ^{4}He where its width is much reduced and gets narrower at higher ^{4}He pressures. It belongs to (^{4}He, 2^3S_1) and it is the line on which metastability exchange collisions between ^{4}He and ^{3}He have been studied.

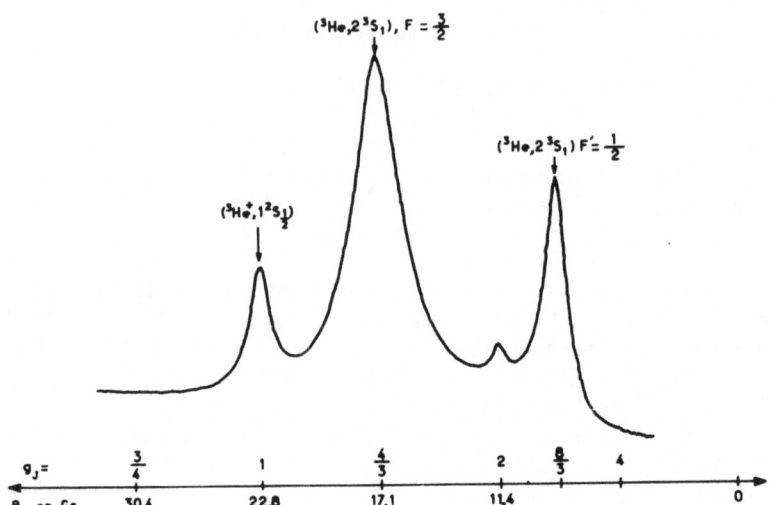

Fig. 3   Resonances observed at 32 Mc/s in ^{3}He containing a small
           impurity of ^{4}He. The resonance at g = 2 is the (^{4}He-2^{3}S$_1$)
           resonance.

Because g = 2, one might wonder whether this line might not
come from the resonance of polarized free electrons present in
the discharge. In fact, the electron line (because of the high
temperature of electrons in the discharge) would be very broad ;
it has been found ([31])([32]), though, but in post discharges only.
Moreover we have measured the Landé g factor of this line with
fairly high precision (as we will describe in a later section).
The value we find identifies this line unambiguously as being due
to (^{4}He - 2^{3}S$_1$).

The line at g = 1 is the resonance due to the hfs level F=1
of the ground state 1^{2}S$_{1/2}$ of ^{3}He$^+$. We identified it ([6])([7]) by measuring the splitting into 2 components it undergoes in fields high
enough to partially decouple $\vec{I}$ and $\vec{J}$ in this state : the 2 Zeeman
transitions ($m_F \rightarrow m'_F$),(1 $\rightarrow$ 0) and (0 $\rightarrow$ -1) have different frequencies then. This allows a determination of the hfs constant : it
coincides to better than 10^{-4} with the very well known ([33]) value
pertaining to (^{3}He$^+$, 1^{2}S$_{1/2}$). The identification of this line is
then unambiguous.

The splitting is shown on fig. 4 and 5, on which one notices
the very narrow and intense lines of the nmr of ^{3}He in its ground
state 1^{1}S$_0$. These numbers are used to calibrate the field.

These markers are produced by an rf field whose frequency is
chosen so that the nmr in (^{3}He,1^{1}S$_0$) will occur for static field

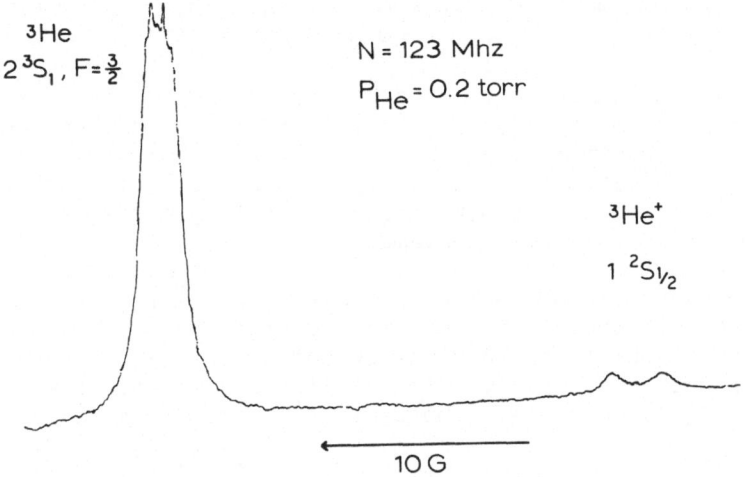

Fig. 4   Splitting of the ($^3\text{He}^+$, $1\,^2S_{1/2}$) resonance in high fields.

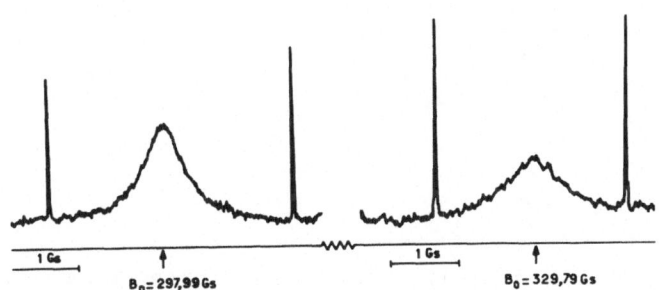

Fig. 5   Splitting of the ($^3\text{He}^+$, $1\,^2S_{1/2}$) resonance with the nmr of (^{3}He, $1\,^1S_0$) markers.

values close to those for which the paramagnetic resonances under study (for instance the (^{4}He, $2\,^3S_1$) resonance) happen to fall. This rf field is produced by harmonic generation from a very stable quartz controlled, continuously running oscillator, whose frequency is multiplied by known amounts. The sweeping of the resonance under study, for instance (^{4}He, $2\,^3S_1$), is slow enough to allow one to switch from one harmonic to the other and get the successive nmr markers. The resonance under study is produced by a second set of rf coils which is fed by a voltage whose frequency is controlled by the same quartz controlled oscillator as above.

NUCLEAR MAGNETIC RELAXATION OF (^{3}He, 1^1S_0). INFLUENCE
OF THE rf FIELD $\vec{B_1}\cos\omega t$ NON UNIFORMITY (^{8})

In the experiments that we have described, the rf field
$\vec{B_1}\cos\omega t$ is produced by a set of 2 small rf coils, 25 centimeters in
diameter in approximately Helmholtz position. Their axis is at
right angles to $\vec{B_0}$ (also produced by a large pair of Helmholtz
coils). The center of the spherical ^{3}He cell is placed at the
common center of the coils systems.

Under those circumstances, the nmr (^{3}He, 1^1S_0) line has its
usual Lorentzian shape (fig. 6) (as expected for a 2 level atomic
system) and its width is very narrow indeed. It depends on many
parameters but is typically a few milligauss. It has been
shown that T_2 is essentially determined by metastability exchange.

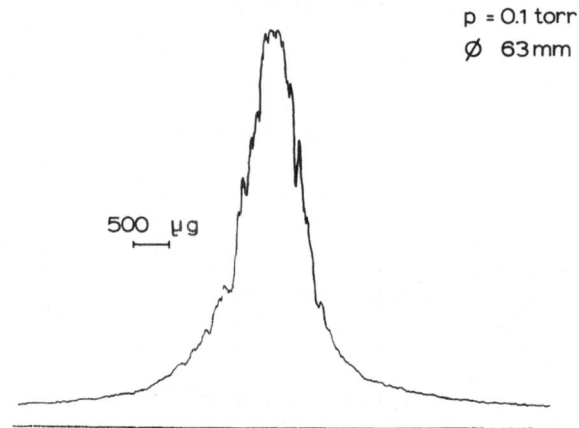

p = 0.1 torr
Ø 63 mm

500 μg

Fig. 6   The nmr (^{3}He, 1^1S_0) resonance. Notice the width of
500 microgauss.

During the course of our experiments, we unexpectedly observed
on some occasions a very odd shape for the nmr line. It appeared,
as shown on fig. 7, as the sum of 2 Lorentzian curves, with the
same axis of symmetry, <u>but very different widths</u>. The narrow one is
just the usual resonance we just described. It would appear first,
at low rf amplitudes. When increasing $\vec{B_1}$, the "broad" resonance
-a few <u>tenths</u> of a gauss large- would appear as shown on fig. 8.
The saturation behaviour of these 2 resonances looked very puzzling.
The narrow one would saturate first, but at higher rf amplitudes,
it would be "eaten up" by the broad curve, but the total intensity
(at resonance) would remain constant. On some occasions the narrow
resonance would not appear at all.

It took us some time before we actually proved that the above
effects are observed when one operates in the presence of large rf

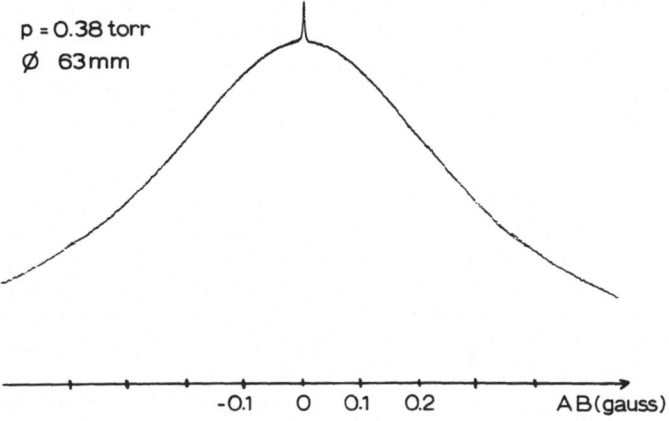

Fig. 7   The nmr (^{3}He, $1\,^1S_0$) resonance in the presence of a large
rf field inhomogeneity.

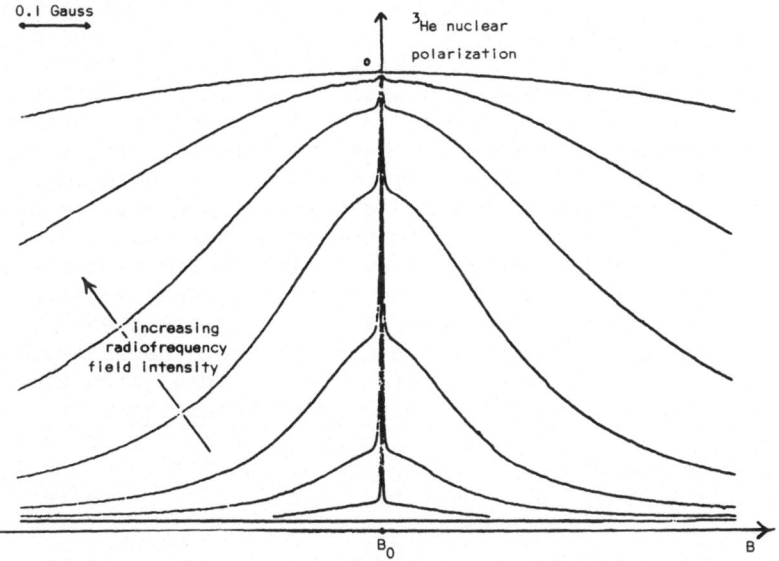

Fig. 8   Behaviour of the large and narrow resonances at increasing rf field amplitudes.

field gradients (in fact, we had connected by mistake the 2 coils
of our rf Helmholtz set in opposition so that the fields they
produced subtracted).

    Before we knew this we had wondered whether we were observing
the nmr in the J=0, $2\,^1S_0$ metastable state of helium. Indeed, the
width would be determined, then, by the diffusion time to the walls

and orders of magnitude were very satisfactory. Taking into account
the necessary coupling with the ground state 1^1S_0 orientation, we
could obtain a picture explaining qualitatively in a satisfactory
manner the observed facts. But the values we had to admit for the
metastability exchange cross section between 2^1S_0 and 1^1S_0 were
much too small.

How does one understand the existence of the "broad resonance"
when a non uniform rf field is used? We believe it is produced by the
slow diffusion of the spins in the presence of the non uniform part
$\overrightarrow{\Delta B_1}$ of the rf field, whereas the "narrow resonance" is due to the
uniform part $\overrightarrow{B_1}$ of it (in other words, with this definition, $\overrightarrow{\Delta B_1}$
varies from point to point in the cell and $< \overrightarrow{\Delta B_1} >$ (over the cell
volume) is zero, whereas $\overrightarrow{B_1}$ is the same in every point). If $\overrightarrow{B_1}$ is
zero, the narrow resonance doesn't appear at all. Going into the
rotating coordinate system at frequency ω in a uniform static field
$\overrightarrow{B}$, the spins "see" a static field in the xOy plane, at every point
in the cell : it is the sum of $\overrightarrow{B_1}$ and of the component $\overrightarrow{\Delta_1 B_1}$ in the
xOy plane, of $\overrightarrow{\Delta B_1}$. To make matters simpler, we will assume that $\overrightarrow{B_1}$
is zero. In the rotating coordinate system, the spins diffuse in a
static field $\left[\overrightarrow{B} - \overrightarrow{B_0} \text{ along } Oz, \text{ (O at resonance)} \right]$ in the presence
of a slowly changing field $\overrightarrow{\Delta_1 B_1}(t)$: clearly, $\overrightarrow{\Delta_1 B_1}$ changes in direction and in modulus over the cell volume. If the direction of
the spins is initially along Oz, this time depending perturbation
$\overrightarrow{\Delta_1 B_1}(t)$ (or rather, the rotating part of it) will flip them if its
Fourier spectrum contains the eigenfrequency of the atoms. In other
words, in the rotating coordinate system, we are then in a situation
which is similar to the one which exists in the laboratory system
(in the absence of a radiofrequency field) in the presence of sta-
tic field inhomogeneities $\overrightarrow{\Delta B_0}$: as we have already mentioned T_1
decreases then markedly in the presence of $\overrightarrow{\Delta B_0}$. A theory of this
effect can be worked out. A few assumptions are necessary. $\overrightarrow{\Delta_1 B_1}(t)$
is supposed to be a random function with a correlation time τ_c.
In our experiments, the "motion narrowing" condition $\overline{\omega_1}^2 \tau_c^2 \ll 1$ is
satisfied ($\overline{\omega_1}$ is in fact a complicated quantity measuring the
"strength" of the interaction and proportional to the amplitude of
$\overrightarrow{\Delta_1 B_1}$, ie, for a given cell geometry, to the voltage applied to the
rf coils).

At the pressures at which we operate, the magnetisation obeys
a diffusion equation. Considering one diffusion mode one finds

$$(1) \quad \begin{cases} \dfrac{1}{T_1} = \dfrac{\overline{\omega_1}^2 \, \tau_c}{1+(\omega_0-\omega)^2\tau_c^2} \\[2em] <M_z> = \dfrac{T_p^{-1}}{T_p^{-1}+T_1^{-1}} M_0 = M_0 \left[1 - \dfrac{T_p \overline{\omega_1}^2 \tau_c}{1+T_p\overline{\omega_1}^2\tau_c+(\omega-\omega_0)^2\tau_c^2} \right] \end{cases}$$

T_p is the M_0 pumping time.

If several modes are present, <u>one has to sum up</u> over them.

Clearly τ_c is of the order of the diffusion time τ_d associated with the corresponding diffusion mode. τ_c can be computed, knowing the diffusion equation, and the $\overrightarrow{\Delta B_1}$ field configuration.

When $\overrightarrow{\Delta B_1}$ is produced by a set of Helmholtz coils as mentioned above, and for spherical cells, one finds that τ_c is just equal to τ_d for the lowest diffusion mode (for more sophisticated $\overrightarrow{\Delta B_1}$ configurations higher modes appear and have been observed).

Formulas (1) have been checked in some detail. They describe very well our observations.

<u>T_1 has been measured by transient techniques</u>. One finds that T_1^{-1} is a linear function of $\overline{\omega_1}^2$. At a given rf amplitude, T_1^{-1} depends on $(\omega_0-\omega)$ (fig. 9); the curve is Lorentzian with a half-width $2/\tau_c$ which is indeed independant of the rf power (fig. 10).

One also finds that τ_c is equal to τ_d : for a given cell radius R, τ_c is proportional to the ^{3}He pressure P and at a given P, it is inversely proportional to R^2 (fig. 11). Finally, one can deduce from these measurements the value of the self diffusion coefficient D_{33} (^{3}He in ^{3}He) at room temperature. One finds

$$D_{33} = (1440 \pm 90)\ \text{cm}^2\ /\text{sec at 1 torr}.$$

This is in excellent agreement with the value one can deduce with the help of the theory from 2 previous measurements ([34])([35]) of D_{34} (^{3}He in ^{4}He) using the difficult technique of the 2 bulbs

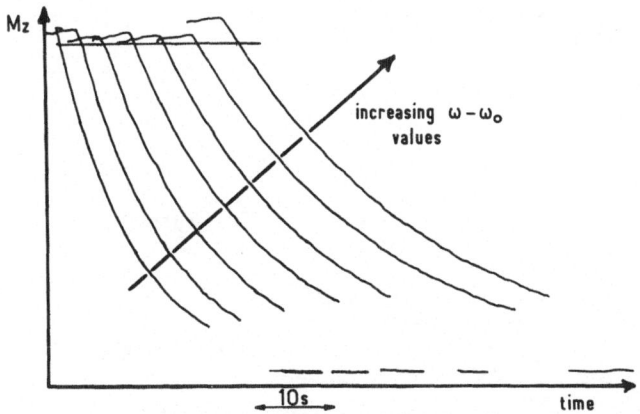

Fig. 9   Relaxation of M_z produced by rf field inhomogeneities, T_1 becomes longer at increasing $\omega - \omega_0$ values.

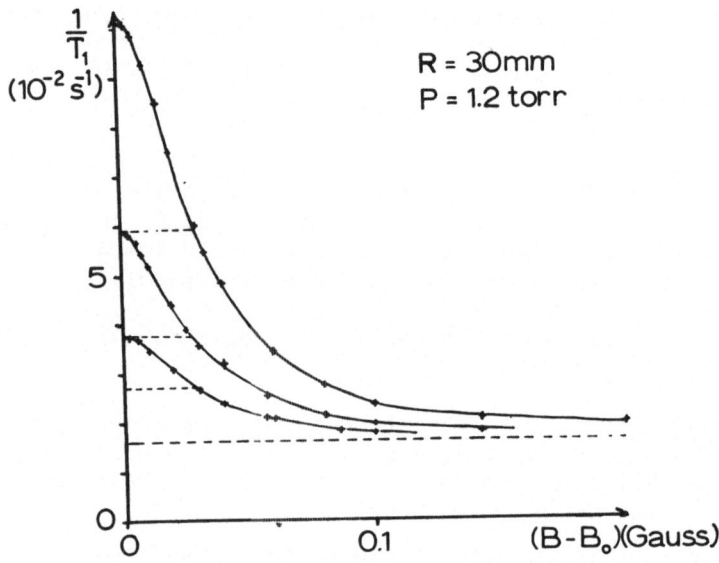

Fig. 10   Curves giving $1/T_1$ (due to rf field inhomogeneities)
           versus $(B-B_0)$ at increasing rf field levels ($B = \omega/\gamma$; B_0
           is the value of the static field at which one operates).
           The curves are Lorentzian. Their width does not depend
           on the rf field amplitude.

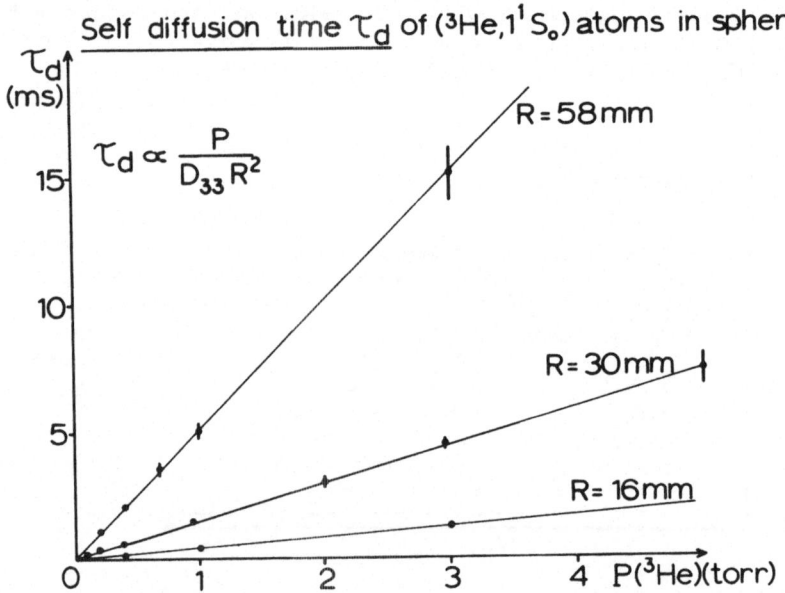

Fig. 11   Variation of the self-diffusion time of ^{3}He with ^{3}He
           pressure in spherical cells of different diameters.

method and mass spectrograph analysis. This confirms the interpretation we just gave. In the determination of the diffusion coefficient, the method above requires only the knowledge of the pressure and of the inside diameter of the cell : the uncertainties we have on those 2 parameters set the limit of the precision we give (τ_d is in fact measured to about 1%). This method is then extremely simple and can be used at low pressures and temperatures (it is the first time, to our knowledge, that D_{33} is directly measured at a few torr). It seems to have many advantages over existing methods (spin echoes) and might prove interesting in other fields.

Coming back to M_z (ie to the nmr curve) we see that the theory predicts the following (formula 1) : the resonance should be Lorentzian (this has been checked satisfactorily), but, this time, its width markedly depends on the rf amplitude $\bar{\omega}_1$ (this appears in formula 1). At low rf levels, the width should extrapolate to $2/\tau_d$ and this is indeed the case.

Finally at high rf levels, at resonance, $< M_z >$ saturates to zero. This means that when the rf field gradient $\vec{\Delta B}_1$ is large enough the slow diffusion of the spins in it will be sufficient to produce their complete disorientation : accordingly in this case, the "narrow resonance" will not show up on top of the "large one".

On the other hand, this is not the case any more at lower rf levels when $T_p\bar{\omega}_1{}^2\tau_d < 1$. The slow diffusion in $\vec{\Delta B}_1$ is not sufficient any more (even at resonance, $B = B_0$, $\omega = \omega_0$) to destroy completely the nuclear polarization $< \vec{I} >$. The intensity of the "broad line" is then $T_p\bar{\omega}_1{}^2\tau_d$ and its width is close to $2/\tau_d$.

If now, one adds a uniform rf field $\vec{B}_1$, the slow diffusion in it will not change anything or affect in any way the broad line, but within a frequency interval of the order of $2/T_2$, it will manage to destroy (completely or partially, depending on its amplitude $\vec{B}_1$), whatever remained of $< \vec{I} >$ after diffusion through $\vec{\Delta B}_1$ ($\tau_d \ll T_2$).

We would like to make a final remark : the real movement of the spins is in fact a rapid random movement, with an amplitude of the order of the mean free path, superimposed on the slow diffusion movement we just considered. To the very fast random time dependent perturbation associated with it corresponds a correlation time much shorter than τ_d . At very large $\vec{B}-\vec{B}_0$ values when one is very far on the wing of the "broad resonance" we have considered here (and when, accordingly, the effect of the slow diffusion becomes negligible) this might become the dominant relaxation mechanism. It seems ([3])([20]) that one meets a situation of this kind when one studies the effect of static field inhomogeneities ($\vec{\Delta B}_0$) in "high" $\vec{B}_0$ fields.

METASTABILITY EXCHANGE RELAXATION

We will give now a brief account of the calculation of the relaxation due to metastability exchange in ^{3}He. The analysis is based on a theoretical model of the collision which can be justified from first principles ([36])([37])([38])([39]) and has been developed by previous workers ([3])([17]).

a) There is a transfer of electronic clouds from one nucleus to the other, so that at the end of the collision itself the outgoing metastable has the same <u>electronic</u> density matrix as the incoming one before the collision (this is due, among other things, to electronic spin conservation).

b) During the collision the nuclear spins of the 2 nuclei remain unaffected.

	^{3}He in the ground state 1^1S_0	^{3}He in the metastable state 2^3S_1
Before a collision	ρ_f	ρ_m
Just after a collision	$\rho'_f = Tr_e\, \rho_m$	$\rho'_m = \rho_f \otimes Tr_n\, \rho_m$
After evolution due to hfs coupling	$\rho''_f = \rho'_f$	$\rho''_m = \sum_F P_F \cdot \rho'_m \cdot P_F$

TABLE I. - Evolution of the density matrices of two ^{3}He atoms due to metastability exchange collisions. Tr_e and Tr_n are trace operations on the electronic and nuclear variables respectively. P_F is the projector on the F hfs level.

Table I indicates the evolution of the density matrices in the ground state (1^1S_0), ρ_f and in 2^3S_1, ρ_m, due to metastability exchange <u>alone</u>. One notices that immediately after the collision there is no correlation between nuclear spin and electronic variables. But the collision induces hfs coherences ; those die out <u>between</u> collisions because of the hfs interaction ($a\tau_e \gg 1$).

From this, the following equations can be derived. They give the evolution of the orientations $< \vec{F} >$ and $< \vec{F'} >$ present in the 2 hfs states $F = 3/2$ and $F' = 1/2$ of 2^3S_1, and of the ground state nuclear magnetisation $< \vec{I} >$

$$(2)\begin{cases} \dfrac{d}{dt} < \vec{F} > = -\dfrac{1}{\tau_e} < \vec{F} > + \dfrac{1}{\tau_e} \left(\dfrac{5}{9} < \vec{F} > + \dfrac{10}{9} < \vec{F'} > + \dfrac{10}{9} < \vec{I} >\right) \\[2mm] \dfrac{d}{dt} < \vec{F'} > = -\dfrac{1}{\tau_e} < \vec{F'} > + \dfrac{1}{\tau_e} \left(\dfrac{2}{9} < \vec{F'} > + \dfrac{1}{9} < \vec{F} > - \dfrac{1}{9} < \vec{I} >\right) \\[2mm] \dfrac{d}{dt} < \vec{I} > = -\dfrac{1}{T} < \vec{I} > + \dfrac{1}{T} \left(\dfrac{1}{3} < \vec{F} > - \dfrac{1}{3} < \vec{F'} >\right). \end{cases}$$

τ_e and T are the average times between 2 metastability exchange collisions in the metastable and ground states respectively (n/τ_e = N/T, n and N being the respective populations). When the different resonances are clearly resolved, one can use the secular approximation. It clearly appears then, in eq. (2) that besides being coupled to $< \vec{I} >$, $< \vec{F} >$ and $< \vec{F'} >$ are <u>directly coupled</u> by metastability exchange : in 2^3S_1 there is some electronic spin orientation and this is not affected by metastability exchange : part of it remains on F,the rest being transferred from F'. As a consequence F and F' Zeeman coherences have different life times, which <u>are longer</u> than τ and longer than if electron transfer didn't exist.

As a consequence, the widths of the F and F' Zeeman resonances are different. When they are measured in gauss the ratio is 1.16 ± 0.02; keeping in mind the values of the g factors, this leads to 0.579 ± 0.012 when they are taken in cycles whereas the theoretical value deduced from II is 0.572. Figure 12 shows that the widths are proportional to the ^{3}He pressure. From the slopes of the F and F' lines one gets the <u>same</u> value of τ_e for both levels and the same value of the cross-section σ = (7.6 ± 0.4) $10^{-16} cm^2$. This is in very good agreement with the value obtained by Greenhow from the

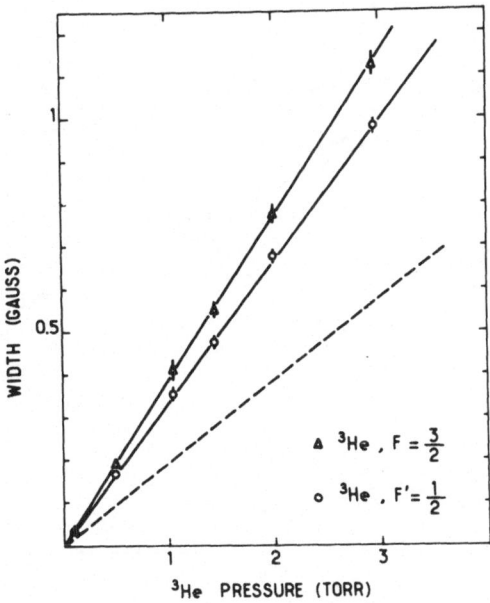

Fig. 12   The widths of the F = 3/2 and F = 1/2 hfs levels of
           (^{3}He, 2^3S_1) due to metastability exchange in pure ^{3}He
           are different and linear functions of ^{3}He pressure.

broadening of the nmr 1^1S_0 line and also with the value of
Colegrove et al.([40]) (deduced from the broadening of the F=3/2
resonance) when corrected by a 9/4 missing factor, and by another
factor 1.3 allowing for the same definition of the cross-section.
It is also consistent with a recent deterimination by Rosner et
al. ([41]). A similar analysis has been done for (^{3}He - ^{4}He) collisions and leads to the interpretation of metastability exchange
collisions in ^{3}He-^{4}He mixtures (eq.(2) are not valid, then). Figure 13 corresponds to a total pressure of 0,1 torr. It gives (full
lines) the result of the theory and shows how the widths of the
different resonances in F and F' (^{3}He, 2^3S_1) and in (^{4}He, 2^3S_1)
vary with the concentration in ^{3}He. In fig. 13-a, the widths are
given in cycles, in fig. 13-b they are in gauss. The plotted points
give the experimental results. The agreement is quite satisfactory.

        We will not make a detailed comment of fig. 13, but one can
see, for instance, that in pure ^{4}He, a trace of ^{3}He will broaden
the (^{4}He - 2^3S_1) line. In pure ^{3}He, on the other hand, if there
is a trace of ^{4}He, every (^{4}He, 2^3S_1) metastable atom will lose its
transverse electronic orientation every τ_e sec whenever a collision

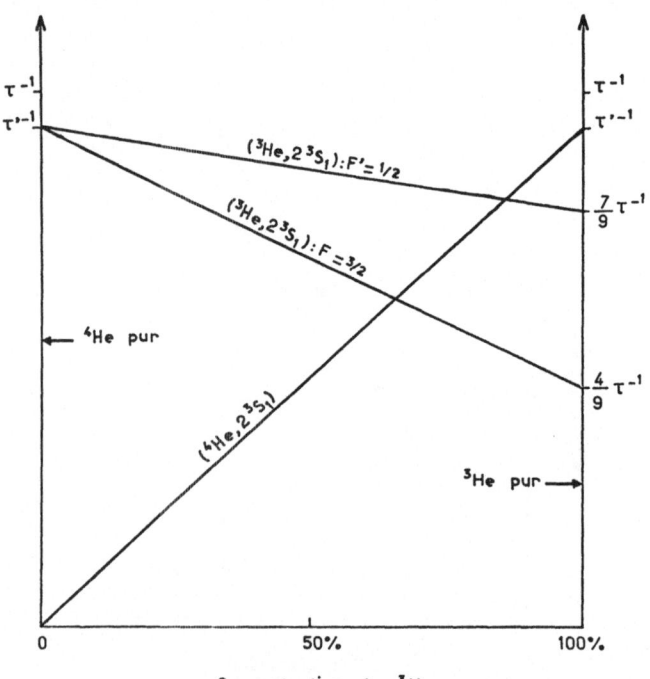

Fig. 13-a Dependence of the widths of (^{3}He, 2^3S_1) F'=1/2 and
          F=3/2 resonances and of the (^{4}He, 2^3S_1) resonance in
          ^{3}He-^{4}He mixtures. The widths are expressed in cycles.

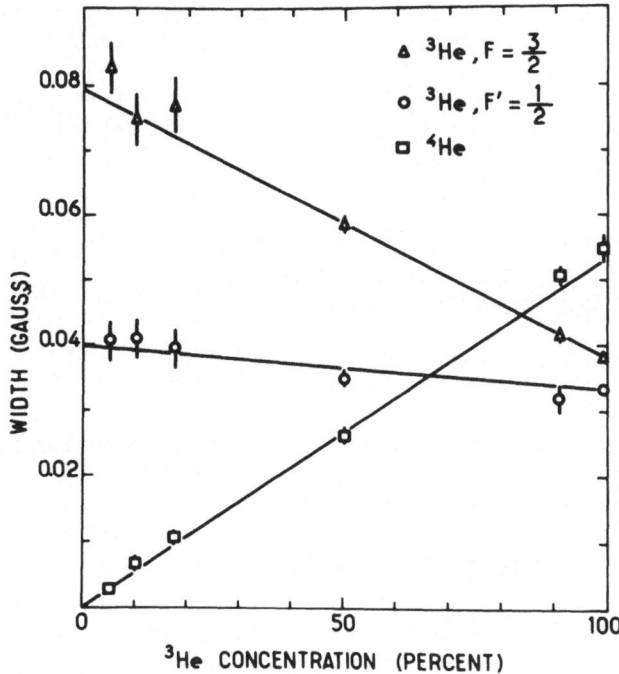

Fig. 13-b The same as 13-a, but the widths are expressed in gauss. Experimental points have been plotted on this figure.

with ^{3}He happens, because once on (^{3}He, 2^3S_1) it rotates at a very different Larmor frequency ($\vec{S}$ and $\vec{I}$ get recoupled). At the next collision, τ_e sec later, the electronic orientation fed back in ^{4}He is completely out of phase (in large enough fields) with the one which stayed all the time on (^{4}He, 2^3S_1).

Equations (2) also give a detailed picture of the behaviour of the nuclear magnetic (^{3}He, 1^1S_0) ground state resonance. They describe also another aspect of metastability exchange collisions: they can produce a shift of the different resonances in (^{3}He, ^{4}He, 2^3S_1) and in the nmr of the ground state [42][43][44]. One can interpret those shifts in a manner very similar to the one given by Cohen-Tannoudji for light-shifts due to real transitions [45]. They are a consequence of the fact that the transverse magnetisation has not   the same Larmor frequency in the 2 states coupled by metastability exchange. As light shifts due to real transitions, they are field dependant and vanish in zero field and in high fields. They have been observed and studied for the ground state nmr. Nonetheless some care must be exercised to get the proper interpretation. One must take into account not only the circulation of coherence between < $\vec{F}$ > (< $\vec{F}'$ >) and the ground state, but also   the

one due to the direct coupling of < F > and < F' >. The correct
formulas describing this effect are those of ref. ([10]).

In the precision measurements we have made of the g factor of
(^{4}He, 2^3S_1) and that we report below, all the shifts -in (^{4}He,
2^3S_1) and in (^{3}He, 1^1S_0)-(used to calibrate the field) were completely negligible as separate checks have shown.

We consider that the above picture gives an essentially correct
interpretation of metastability exchange and of
the behaviour of the magnetic resonance lines in (He, 2^3S_1).

Landé g factor of ^{4}He - 2^3S_1

We measured, by the techniques already described, the
ratio a of the resonating frequencies of (^{4}He - 2^3S_1) and (^{3}He -
1^1S_0) in the same field. We obtain the value

$$a = \frac{1}{2} \frac{\mu_J \; (^4He - 2^3S_1)}{\mu_I \; (^3He - 1^1S_0)} = 864.02392 \pm 0.00006 \; .$$

Figure 14 gives a histogram of about 100 separate measurements.
The quoted error bar is twice the standard deviation Δ' ($\Delta' = \Delta/\sqrt{n}$,
n : number of measurements, Δ, rms deviation from the mean value).
This measurement has been described at length in a separate publication ([11]). We will just make here a few remarks about it.

Many precautions were taken to insure that the above result is
not affected by systematic errors and many checks were made to that
end. Nonetheless systematic errors are the essential difficulty in
a measurement of this kind and the only way to detect them is to

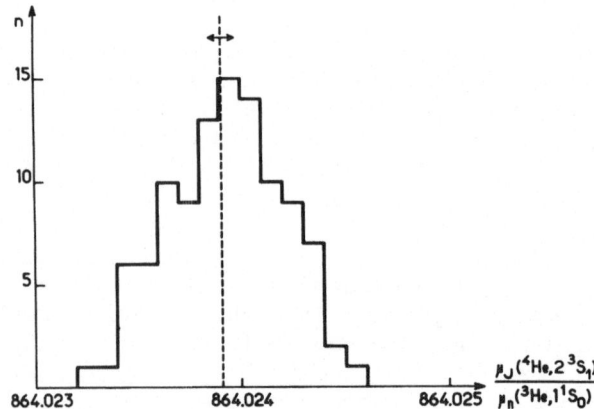

Fig. 14   The histogram of the measurements of the ratio a.

redetermine this ratio in another Laboratory by the same -or by
another- method. We feel it is very much worthwhile to do it.

Indeed, one can combine the above result with others in the
following way.

Drake, Hughes et al.[46] measured

$$x_{exp} = \frac{g_J(^4He-2^3S_1)}{g_J(^1H, 1^2S_{1/2})} = \frac{1}{2} \frac{\mu_J(^4He-2^3S_1)}{\mu_J(^1H, 1^2S_{1/2})} = 1-(23.3\pm0,8) \ 10^{-5} \quad .$$

Using QED, they also made [16] a theoretical estimate of the
same quantity

$$x_{th} = 1 - (23.3 \pm 1) \ 10^{-5} \quad .$$

They assume, among other things, that the anomalous moments of the
2 electrons in He are simply additive. Williams and Hughes [47]
measured the ratio of the nuclear moment of ^3He, to the proton
moment in molecular hydrogen

$$y = \frac{\mu_J(^3He, 1^1S_0)}{\mu_p (H_2)} = 0.76178685 \pm 8 \ 10^{-8} \quad .$$

When one knows

$$z = \frac{\mu_J(^1H, 1^2S_{1/2})}{\mu_p(H_2)}$$

the quantity a we measured is simply

$$a = \frac{xz}{y} \quad .$$

There exist 2 independant and very precise determinations
of z [19][48][49], which are in excellent agreement. The best way
to use our a value is to combine it with the z and y determinations
to get a new and more precise value of x. When this is done, one
gets

$$x = 1 - (21.6 \pm 0.5) \ 10^{-6} \quad .$$

The quoted uncertainty is essentially due to the uncertainty on z.

The difference between this value and x_{exp} of Drake and Hughes
is 7 times the standard deviation of their measurement, ie twice
their error bar. Our value is again somewhat outside the error bar
of Hughes et al. QED estimate of x_{th}. It seems interesting to understand why.

The technique we used can eventually lead to higher precision
on a : the $(^4He - 2^3S_1)$ line gets narrower at lower temperatures.

On the other hand, many factors which were found negligible at the above precision will have to be explored again. In any case, the precision on x_{exp} will not be greatly improved because it is essentially determined by the present precision of the z measurements.

We would like to explore other possibilities which might prove more interesting because they would not necessitate intermediate measurements : it would be very valuable to observe simultaneously in the discharge the $(^4He, 2^3S_1)$ resonance and other resonances of simple systems, directly calculable from QED : $(^4He^+, 1^2S_{1/2})$ is obviously one (we have already observed $(^3He^+, 1^2S_{1/2})$). The free electron is another. It might also prove interesting to investigate He-H2 mixtures, in the hope that by the many kinds of couplings we have described, some species like H might be oriented, leading to the observation of the resonances of $(^1H, 1^2S_{1/2})$, of the proton in H2 or perhaps of the free proton. Indeed we feel that the above experiments definitely show that high precision measurements on simple -and basic- atomic systems can be made in weak discharges.

CHARGE EXCHANGE COLLISIONS BETWEEN $(^3He^+ - 1^1S_{1/2})$ AND $(^3He, 1^1S_0)$

As already mentioned, this process, which is still under investigation, has been studied on the $(^3He^+, 1^2S_{1/2})$ resonance. One finds that its <u>intensity</u> depends markedly on the discharge level (ie on $^3He^+$ concentration), but, on the other hand, at reasonably high ^3He pressures, its width looks rather insensitive to it. Accordingly the discharge level, which is a difficult parameter to control when one changes the ^3He pressure, does not seem to affect appreciably the result of the measurements.

The width of the $^3He^+$ resonance is determined among other things by

a) <u>charge exchange binary collisions</u> between the $^3He^+$ ground state and the large excess of ^3He neutral atoms in their ground state $(^3He, 1^1S_0)$. As for every exchange process, the cross section, as we shall see, is rather large ($\sim 3 \ 10^{-15} cm^2$). This leads to a broadening of the resonance which has a linear dependance on the ^3He pressure at large pressure values.

b) <u>the diffusion time to the walls</u>, where the $^3He^+$ ions recombine with free electrons and disappear; this is inversely proportional to the ^3He pressure and is less and less important the bigger the cell diameter.

In fact this effect is very important because the diffusion time of the ions is determined by ambipolar diffusion, which is very fast and depends on the discharge level.

The measurements we made so far must be considered as preliminary, because a detailed study of the behaviour of the resonances should be made. Moreover their precision is not high. Figure 15 indicates the general behaviour of the curve giving the width of (^{3}He$^+$, 1^2S$_{1/2}$) versus ^{3}He pressure. At high pressures the dependance becomes linear, the cross section can be deduced from the slope.

Charge exchange is also an electrostatic process of very short duration, so that the theory of the relaxation which we outlined for metastability exchanges is valid in the present case. In particular, in charge exchange between identical atoms, the nuclear spins remain unaffected and the electron clouds are just exchanged. Accordingly equations very similar to eq.(2) can be written for this case. The only difference comes from the fact that in the ^{3}He 1^2S$_{1/2}$ ground state J = 1/2 and F = 1, F' = 0. This clearly leads to different numerical factors in the coefficients of < $\vec{F}$ >, < $\vec{F}'$ >, and < $\vec{I}$ >; because < $\vec{F}'$ > = 0, one cannot draw any information from the comparison of the widths of Zeeman resonances in 2 separate hfs levels. One finds

$$\frac{d}{dt} < {}^1\vec{F} > = \frac{1}{\tau_{c.e}} \left[- < {}^1\vec{F} > + \frac{1}{2} < {}^1\vec{F} > + < \vec{I} > \right] .$$

Here again the life-time of < $^1\vec{F}$ > is longer than τ_{ce}(it is $2\tau_{c.e}$) because of electron spin conservation in charge exchange.

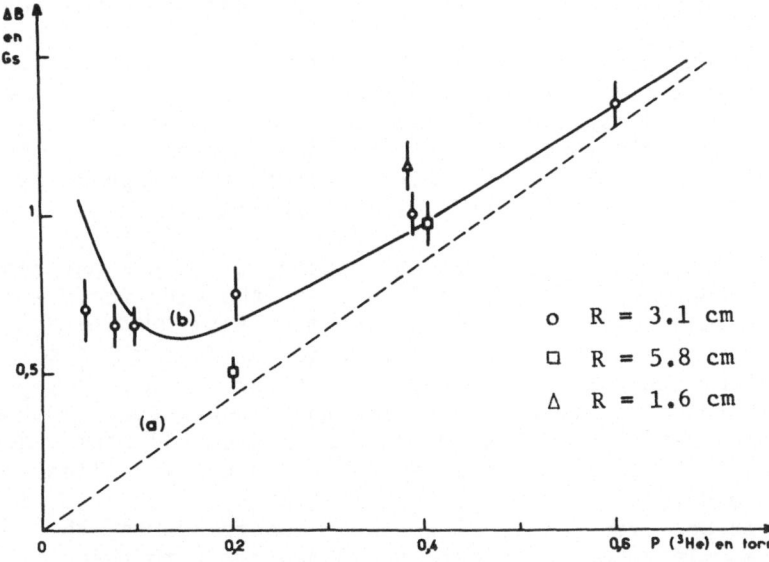

Fig. 15   Variation of the width of the (^{3}He$^+$, 1^2S$_{1/2}$) resonance with ^{3}He pressure (in spherical cells of different diameters).

Keeping in mind this factor 2, one finds for the cross-section the value $(2.8 \pm 0.7) \ 10^{-15} cm^2$, which is in fair agreement with the known extrapolated values at low energies, deduced from measurements [50] on accelerated beams of ions (values between $1.5 \ 10^{-15}$ and $3.5 \ 10^{-15} cm$).

COLLISIONS AFFECTING THE n^1D_2 STATES

Finally, we would like to mention briefly some preliminary experiments which were carried out in oriented ^{3}He weak discharges, using techniques very similar to those we have already described.

I have explained how, through the $a\vec{I}.\vec{J}$ coupling in the excited state, an electronic orientation $< \vec{J} >$ is built in this state, when $a\tau \gg 1$. This appears in the degree of circular polarization $\mathcal{P}(\sigma^+, \sigma^-)$ emitted by the line. For $J = 2 \rightarrow J = 1$ transitions (n^1D_2 levels) one finds

$$\mathcal{P}(\sigma^+, \sigma^-) = 0,24 \ \mathcal{P}_n$$

$\left[\mathcal{P}_n \text{ is the nuclear polarization } \mathcal{P}_n = \frac{1}{2} < I_z > \right]$.

By adding a second probe beam one can measure separately $\mathcal{P}(\sigma^+, \sigma^-)$ and $< \mathcal{P}_n >$ at several pressures and see how collisions affect $< \vec{J} >$, once obtained from $< \vec{I} >$ via the $a\vec{I}.\vec{J}$ interaction.

An effect was mentioned earlier, in which collisions would reduce the life-time τ to some shorter value τ'. In this case the condition $a\tau' > 1$ might not be met, leading to a smaller $< \vec{J} >$ value. On the contrary in the experiments reported below, the condition $a\tau' \gg 1$ is always satisfied, because the range of explored pressures $(50 \ \mu < p < 5 \ torr)$ is not very large.

On the other hand, collisions inducing transfers between Zeeman sublevels will destroy $< J_z >$, and the electronic alignment eventually present in level J; so also will quenching collisions. One can then expect the ratio $\mathcal{P}(\sigma^+, \sigma^-) / \mathcal{P}_n$ to start from its theoretical value $0,24$ for $n^1D_2 \rightarrow {}^1P_1$ transitions and then to decrease with the gas pressure. But one easily sees that it will not go to zero. Indeed all atomic species, with which the atom in the n^1D_2 state can possibly collide, carry with them electronic or nuclear orientation -or both- and will transfer it back (proportional to the pressure) during the collision. As a rule then one expects the above ratio to reach a constant value at high enough pressures (but pressures not large enough for the condition $a\tau' < 1$ to hold). One can show that quenching collisions will produce a similar effect. Indeed, this is actually

observed for levels 4^1D_2 and 5^1D_2. For 3^1D_2 the behaviour is different at low pressures, due to cascades from 4^1F, 5^1F levels which
contribute appreciably to the 3^1D_2 population.

Fitting for 4^1D_2 and 5^1D_2 the theoretical curve with the experimental data one is led to the conclusion that the quenching collision cross-section for those 2 levels is rather small ($\simeq 30\text{Å}^2$).
On the other hand the disalignment and disorienting cross sections
seem rather close to each other, and very large (250 Å^2 and
400 Å^2 respectively). They increase greatly with the principal
quantum number n. These results seem to be in good agreement with
previous measurements wherethe total cross section of these levels
were measured ([51]). As usual, the interpretation of those relaxation data raise difficult problems, because it is so difficult to
identify, in a discharge, among the many processes present, the one
one is really looking at. In particular many difficulties are raised because of the presence, within kT, of several levels in the
immediate vicinity of the n^1D_2 levels. Indeed, it is difficult to
predict what amount of orientation or alignement these (presumably)
oriented levels will feed back to the one under study. Finally,
cascades may have large effects also.

Nonetheless we believe that some information on those processes is indeed welcome, in particular the way those cross-sections behave at lower temperatures should prove interesting.

Conclusion

To conclude, one can say that optical pumping in helium
is a field which is very much alive and in which many interesting
-and, at times, basic problems- can be studied. We made several
suggestions to that end. Finally, most of those problems can be
studied also in other rare gases, where many results have already
been obtained.

References

1.   F.D. Colegrove and P.A. Franken, Phys. Rev. 119, 680 (1960).
2.   F.D. Colegrove, L.D. Schearer and G.K. Walters, Phys. Rev.
      132, 2561 (1963).
3.   L.D. Schearer and G.K. Walters, Phys. Rev. 139, A1398 (1965);
      L.D. Schearer, Thesis, Rice University (1966).
4.   W.A. Fitzsimmons, Master's Thesis, Rice University (1966).
5.   M. Pavlovic and F. Laloë, J. Phys. (Paris), 31, 173 (1970).
      F. Laloë, Thesis, Paris (1970); Annales de Physique, 6, 5
         (1971).

6.   M. Leduc and F. Laloë, Opt. Comm., 3, 56 (1971).

7.. M. Leduc, Thesis, Paris (1972).

8.   R. Barbé, M. Leduc and F. Laloë, Communication to the Confe-
         rence of the European Group for Atomic Spectroscopy,
         Amsterdam (1972); J. Phys. (Paris) to be published.

9.   R. Byerly, Master's Thesis, Rice University (1965); Ph.D.
         Thesis, Rice University (1967).

10. J. Dupont-Roc, M. Leduc and F. Laloë, Phys. Rev. Lett. 27,
         467 (1971); and J. Phys. (Paris), to be published.

11. M. Leduc, F. Laloë and J. Brossel, J. Phys. (Paris), 33, 49
         (1972).

12. M. Pinard and J. Van der Linde, Opt. Comm., to be published.

13. M. Pinard, J. Van der Linde and F. Laloë, Communication to
         the conference of the European Group for Atomic Spectro-
         scopy, Amsterdam (1972).

14. I.Ya. Fugol' and P.L. Pakomov, J.E.T.P. Lett., 3, 254 (1966);
         Soviet Phys., J.E.T.P., 26, 526 (1966).

15. W.A. Fitzsimmons, N.F. Lane and G.K. Walters, Phys. Rev. 174,
         193 (1968).
     W.A. Fitzsimmons, Ph.D. Thesis, Rice University (1968).

16. W. Perl and V.W. Hughes, Phys. Rev., 91, 842 (1953).

17. R.B. Partridge and G.W. Series, Proc. Phys. Soc. 88, 983
         (1966).

18. W.A. Fitzsimmons and G.K. Walters, Phys. Rev. Lett. 19,
         943 (1967).
     W.A. Fitzsimmons, L.L. Tankersley and G.K. Walters, Phys. Rev.
         179, 156 (1969).

19. M. Than Myint, Thesis, Harvard University (1966).

20. R.L. Gamblin and T.R. Carver, Phys. Rev. 138 A, 946 (1965).

21. J.C. Lehmann, J. Phys. (Paris), 25, 809 (1964); Thesis, Paris
         (1967); Ann. de Phys. 2, 345 (1967).

22. F. Laloë, C.R. Acad. Sci. (Fr.), 267 B, 208 (1968).

23. C. Cohen-Tannoudji, Comments on Atomic and Molecular Physics,
         vol. II, n° 1, p. 24 (1970).

24. P.L. Bender, Thesis, Princeton University (1956).

25. J.P. Faroux and J. Brossel, C.R. Acad. Sci. (Fr.), 261, 3092
         (1965); 262 B, 41 and 1385 (1966); 263 B, 612 (1966);
         264 B, 1452 and 1573 (1967); 265 B, 393 (1967).
     J.P. Faroux, Thesis, Paris (1969).

26. A. Omont, J. Phys. (Paris), 26, 26 (1967); Thesis, Paris
         (1967).
     A. Omont and J. Meunier, Phys. Rev., 169, 92 (1968).

27. F.W. Byron and H.H. Foley, Phys. Rev. 134 A, 625 (1964).

28. B.R. Bulos and W. Happer, Phys. Rev. A, 4, 849 (1971).

29. J.F. Papp and F.A. Franz, Phys. Rev. A, 5, 1763 (1972).

30. F. Grossetête, J. Phys. (Paris), 25, 383 (1964); Thesis,
         Paris (1967).

31. H.G. Dehmelt, Phys. Rev. 109, 381 (1958); J. Phys. Rad.
         (Paris), 19, 866 (1958).

32. L.D. Schearer, Phys. Rev. 171, 81 (1968).
33. H.A. Schuesser, E.N. Fortson and H.G. Dehmelt, Phys. Rev.
        187, 5 (1969).
34. P.J. Bendt, Phys. Rev. 110, 85 (1958).
35. G.A. Du Bro and S. Weissman, Phys. of Fluids, 13, 2682 (1970).
36. R.A. Buckingham and A. Dalgarno, Proc. Roy. Soc. A 213, 327
        and 506 (1952).
37. H.J. Kolker and H.H. Michels, J. Chem. Phys. 50, 1762 (1969).
38. M. Kodaira and T. Watanabe, J. Phys. Soc. Jap. 27, 1301 (1969).
39. S.A. Evans and N.F. Lane, Phys. Rev. 188, 269 (1969).
40. F.D. Colegrove, L.D. Schearer and G.K. Walters, Phys. Rev.
        135 A, 353 (1964).
41. S.D. Rosner and F.M. Pipkin, Phys. Rev. A, 5, 1909 (1972).
42. L.D. Schearer, F.D. Colegrove and G.K. Walters, Rev. Sci.
        Instr. 35, 767 (1964).
43. H.G. Dehmelt, Rev. Sci. Instr. 35, 768 (1964).
44. A. Donszelmann , Thesis, Amsterdam (1970); Physica, 56, 138
        (1971).
45. J.P. Barrat and C. Cohen-Tannoudji, J. Phys. Rad. 22, 443
        (1961).
     C. Cohen-Tannoudji, Ann. de Physique, 7, 423 and 469 (1962).
46. C.W. Drake, V.W. Hughes, A. Lurio and J.A. White, Phys. Rev.
        112, 1627 (1958).
47. W.L. Williams and V.W. Hughes, Phys. Rev. 185, 1251 (1969).
48. M. Than Myint, D. Kleppner, N.F. Ramsey and H.G. Robinson,
        Phys. Rev. Lett. 17, 405 (1966)
49. E.B.D. Lambe, Thesis, Princeton University (1959).
50. E.W. McDaniel, "Collision Phenomena in Ionized Gases", Wiley,
        New York.
51. C.W.T. Chien, R.E. Bardsley, and F.W. Dalby, Canad. J. Phys.
        50, 279 (1972).

THE TEMPERATURE DEPENDENCE OF THE ELECTRON-RUBIDIUM

SPIN-EXCHANGE CROSS SECTION

S. J. Davis and L. C. Balling

University of New Hampshire

Durham, New Hampshire   03824

I.   INTRODUCTION

The low-energy scattering of electrons from alkali atoms is of considerable theoretical interest because of the relative simplicity of the scattering problem. Spin-exchange optical pumping experiments provide a convenient means for studying electron-alkali atom collisions at thermal energies. In such experiments, free electrons are polarized in a weak magnetic field by spin-exchange collisions with optically-pumped alkali atoms. If an rf field is applied at the electron spin resonance frequency, the electrons are depolarized and spin-exchange collisions partially depolarize the alkali atoms. The depolarization of the alkali atoms results in a decrease in the intensity of the pumping light transmitted by the optical-pumping cell. In this way, one can observe the free electron spin resonance.

If experimental conditions are adjusted so that the electron-alkali atom spin-exchange collisions are the dominant relaxation mechanism for the free electron spin, then the linewidth $\Delta\nu$ of the spin resonance is given by[1]

$$\Delta\nu = \langle N\ v\ \sigma_{SE}\rangle \quad , \tag{1}$$

where N is the density of alkali atoms in the optical-pumping cell, v is the velocity of the free electrons, and σ_{SE} is the spin exchange cross section for electron-alkali atom collisions. The bracket indicates an average over the thermal velocity distribution of the electrons. The spin-exchange cross section is given by

463

$$\sigma_{SE} = \frac{\pi}{k^2} \sum_{\ell=0}^{\infty} (2\ell + 1) \sin^2 (\delta_\ell^3 - \delta_\ell^1) \quad , \qquad (2)$$

where $\hbar k$ is the momentum of an electron, and δ_ℓ^3 and δ_ℓ^1 are the triplet and singlet scattering phase shifts of the ℓ-th partial wave for the scattering of an electron from an alkali atom.

The spin-exchange collisions also produce a shift $\delta\nu$ in the center frequency of the resonance. The ratio of the shift to the linewidth is given by[1]

$$\frac{\delta\nu}{\Delta\nu} = \frac{<\kappa \; v \; \sigma_{SE}>}{<v \; \sigma_{SE}>} \frac{P}{2} \qquad (3)$$

where

$$\kappa = \frac{\frac{1}{2} \sum_{\ell=0}^{\infty} (2\ell + 1)\sin 2(\delta_\ell^3 - \delta_\ell^1)}{\sum_{\ell=0}^{\infty} (2\ell + 1) \sin^2(\delta_\ell^3 - \delta_\ell^1)} \quad , \qquad (4)$$

and P is the electronic polarization of the optically-pumped alkali atoms.

From Eq. (1), one sees that we can study the temperature dependence of σ_{SE} by measuring the linewidth of the electron resonance at fixed alkali atom density as a function of temperature. Measurements of the frequency shift as a function of temperature also provide information on the energy dependence of the scattering phase shifts.

In the experiments reported here, the linewidth and frequency shift of the spin resonance of electrons polarized by collisions with optically-pumped Rb atoms have been measured as a function of temperature from 200-840°K.

II.   APPARATUS

A block diagram of the apparatus is shown in Fig. 1. In most respects it resembles the standard optical-pumping setup. The optical-pumping cell is situated in a firebrick oven inside two concentric cylindrical magnetic shields. The magnetic field (~ 15 mG) was produced by a Helmholtz coil pair inside the shields. The rf used to drive the electron spin resonance was chopped by a mercury relay at a rate of 10 Hz. At resonance, the intensity of the pumping light transmitted by the optical-pumping cell is

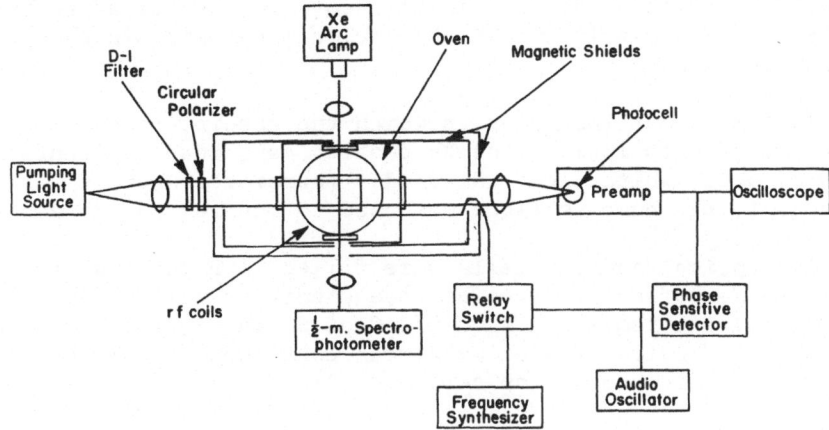

Figure 1. A block diagram of the apparatus.

amplitude modulated at the same rate. The signal was detected
with a photocell, amplified, and displayed on an oscilloscope and
phase sensitive detector.

At room temperature, the density of Rb atoms in the cell was
determined by the vapor pressure of Rb. For measurements at ele-
vated temperatures, the Rb was driven into a small side-arm attached to the bottom of the cell. As the cell was heated by the
oven, the sidearm was cooled by a regulated flow of air. The density of Rb in the cell was determined by the sidearm temperature
which was controlled by the air flow. For measurements at low
temperatures, the oven was cooled by dry nitrogen gas flowing
through liquid nitrogen. The sidearm was heated by hot air, which
drove Rb in controlled amounts up into the cell. Complete des-
criptions of the techniques for optical pumping at high and low
temperatures are given in Refs. 2 and 3.

For measurements at room temperature with cells well coated
with Rb, the density of Rb atoms was taken to be that given by
vapor pressure curves.[4] The density of Rb atoms in the cell for
measurements at high and low temperatures was established by measuring the fractional absorption of a beam of white light travers-
ing the cell at right angles to the magnetic field and comparing
this to the fractional absorption at a known density at room temperature. The white light from a 1000 watt high pressure Xe arc
lamp was collimated in a beam 1.0 cm in diameter. The beam passed
through holes cut in the sides of the magnetic shields. After
passing through the cell, the beam was focused on the slits of a
Jarrell Ash scanning spectrophotometer with the slit opening set
at 40μ. The D-2 absorption line was monitored. The advantage of

white light is that the fractional absorption is not sensitive to temperature changes in the Rb absorption line shape in the presence of a buffer gas.

It turned out that runs in which the absorption of pumping light was used to monitor the Rb density gave results identical to those in which white light was used. Therefore, when it was more convenient, the pumping light was used to monitor the Rb density.

The optical-pumping cells were filled with various pressures of He and Ne buffer gases. The free electrons were produced either by the ionizing radiation from 1/3 C of H^3 gas introduced into the cell with the buffer gas or by a weak continuous rf discharge in a turret above the absorption cell.

III. MEASUREMENTS

The width of the electron resonance, extrapolated to zero rf power, was measured as a function of the fractional absorption of white light or of the pumping light at various temperatures, and the slopes of the linear plots were compared. The ratio of the slopes obtained at temperatures T_1 and T_2 gives the ratio of

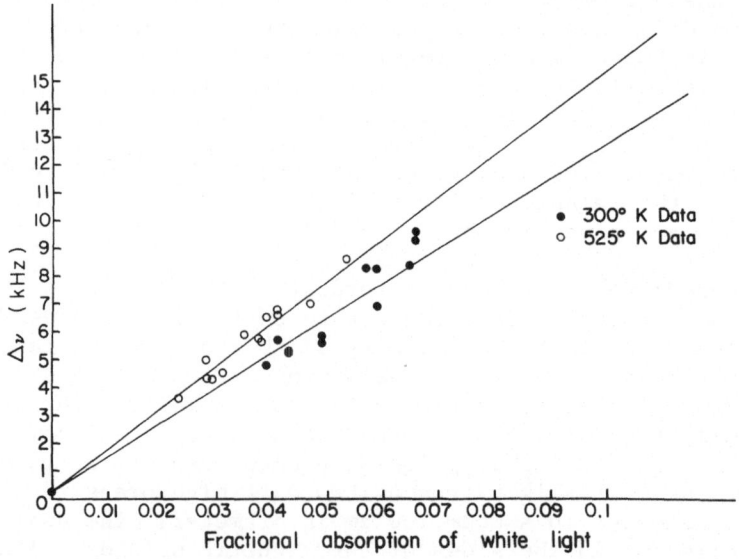

Figure 2. A plot of the electron linewidth versus the fractional absorption of white light at 300°K and 525°K. The cell contained .3C tritium and 52 Torr He as a buffer gas.

$<v \sigma_{SE}>$ at T_1 to $<v \sigma_{SE}>$ at T_2. This is equivalent to the ratio of the electron resonance linewidth $\Delta\nu(T_1)$ at temperature T_1 to the linewidth $\Delta\nu(T_2)$ at T_2 with the Rb density held constant. A plot of linewidth vs fractional absorption at two different temperatures for a typical run is shown in Fig. 2. Results from our most reliable runs are summarized in Table I and Fig. 3.

Table I

A summary of the results of runs in which the electron linewidth at fixed Rb density at various temperatures was compared to the linewidth at 300°K.

Buffer Gas Pressure (Torr)	Source of Electrons	Temperature °K	Density	$\dfrac{\Delta\nu(T)}{\Delta\nu(300°K)}$
Ne 54	Discharge	570 ± 25	White Light	.9 ± .2
Ne 54	Discharge	570 ± 25	White Light	.9 ± .2
Ne 54	Discharge	840 ± 50	White Light	1.1 ± .2
Ne 14	H^3	625 ± 25	White Light	1.0 ± .2
Ne 14	H^3	625 ± 25	Pumping Light	1.2 ± .2
Ne 14	H^3	190 ± 20	Pumping Light	.9 ± .2
He 52	H^3	525 ± 25	White Light	1.2 ± .2

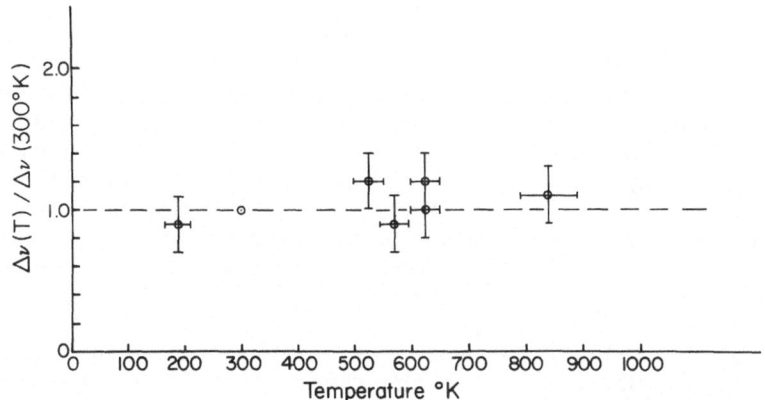

Figure 3. A plot of the ratio $\Delta\nu(T)/\Delta\nu(300°K)$. The values are obtained from Table I.

The error bars on the measured slopes were estimated from the reproducibility of the results from run to run. The error in the temperature measurements was chiefly due to temperature gradients within the oven.

The frequency shifts of the electron resonance due to spin-exchange collisions was measured at fixed Rb density at various temperatures by measuring the center of the resonance first with left and then with right circularly polarized pumping light. The sign of the shift changes with the sign of the Rb polarization. The magnitude of the Rb polarization was estimated by comparing the absorption of pumping light which was circularly polarized (Rb atoms polarized) with the absorption for linearly polarized light (Rb atoms unpolarized). This estimate assumed that the absorption was proportional to [1-P]. The results of these measurements are shown in Table II.

Table II

The ratio of twice the frequency shift $2\delta\nu_0$ to the electron line-width $\Delta\nu$ at different temperatures. The polarization of the Rb atoms was estimated from the absorption of pumping light.

Temperature	$\dfrac{2\delta\nu_0}{\Delta\nu}$	Polarization
190°K	.19 ± .05	.2 ± .1
300°K	.15 ± .01	.2 ± .1
640°K	.055 ± .015	.1 ± .05

IV. SOURCES OF SYSTEMATIC ERROR

Errors due to gradients in the Rb density were eliminated by going to relatively low buffer gas pressures. A stringent check on gradient effect could be made by cooling the cell just below 0°C and driving the Rb out of the sidearm. In this situation the Rb diffuses to the wall and sticks there, setting up a maximum gradient problem. Measurements made under these conditions were required to agree with measurements made under uniform density conditions at 30°C.

Equations (1) and (3) were derived neglecting the effect of nuclear spin. Corrections to the theoretical line shape due to inclusion of nuclear spin effects are completely negligible under the conditions of this experiment. These effects will be discussed in detail elsewhere.

V. CONCLUSION

As one can see from Fig. 1, there is little if any temperature dependence in the electron linewidth at fixed Rb density.

That is, $<v\,\sigma_{SE}>$ is very nearly a constant as a function of temperature. Within experimental error, therefore, the spin-exchange cross section is well approximated by a $1/v$ velocity dependence. Thus, in the temperature range 200 - 840°K we can write

$$\sigma_{SE} = \frac{\sigma_0}{v} \quad .$$

In order to determine the constant σ_0, the electron linewidth was measured at room temperature. The zero rf linewidth of the electron resonance at 20°C in a cell well coated with Rb is 1300 ± 100 Hz. This is in complete agreement with the result obtained in Ref. 2. Using a particular vapor pressure curve[4] to determine the Rb density N at this temperature one finds that

$$\sigma_{SE} = \frac{8.8 \times 10^{-7}}{v} \; cm^2 \quad .$$

From Table II one sees that the ratio of the frequency shift to linewidth decreases with increasing temperature, but that this change may be entirely due to a decrease in the Rb polarization.

REFERENCES

1. L. C. Balling, R. J. Hanson, and F. M. Pipkin, Phys. Rev. 133, A607 (1964).

2. L. C. Balling, R. H. Lambert, J. J. Wright, and R. E. Weiss, Phys. Rev. Letters 22, 161 (1969).

3. J. J. Wright, L. C. Balling, and R. H. Lambert, Phys. Rev. 1, 1018 (1970).

4. R. E. Honig and D. A. Kramer, RCA Review, 285 (1969).

ORIENTATION OF SHORT-LIVED MERCURY ISOTOPES BY MEANS OF
OPTICAL PUMPING DETECTED BY β AND γ RADIATION

J. Bonn, G. Huber, H.-J. Kluge, U. Köpf, L. Kugler,

E.-W. Otten and J. Rodriguez

CERN, Geneva, Switzerland, and

I. Physikalisches Institut, University of Heidelberg,

Germany

The expansion of optical spectroscopy to short-lived isotopes ($T_{1/2} \sim$ sec) far from the valley of nuclear stability encounters several difficulties:

  i) the production mechanism of isotopes far from stability is no longer specific for one isotope only;

 ii) the number of atoms produced in saturation is small;

iii) the preparation time of samples becomes long compared to the lifetime of the isotope in question.

Until now, the application of purely optical techniques seems to be limited to isotopes with half-lives of about 1 hour[1-4]. This limit can be pushed into the region of seconds[5] by (i) using mass-separated isotopes, (ii) increasing the sensitivity by detection of nuclear radiation instead of the photons of the electronic transition, and (iii) applying on-line techniques.

ISOTOPE PRODUCTION

The combination of the 600 MeV Synchro-cyclotron with the on-line mass separator ISOLDE[6] at CERN is a powerful facility for producing long chains of neutron-deficient isotopes. Using a lead target one obtains by spallation Pb(p;3p,xn)Hg up to about 10^8 Hg ions per second in saturation (Fig. 1).

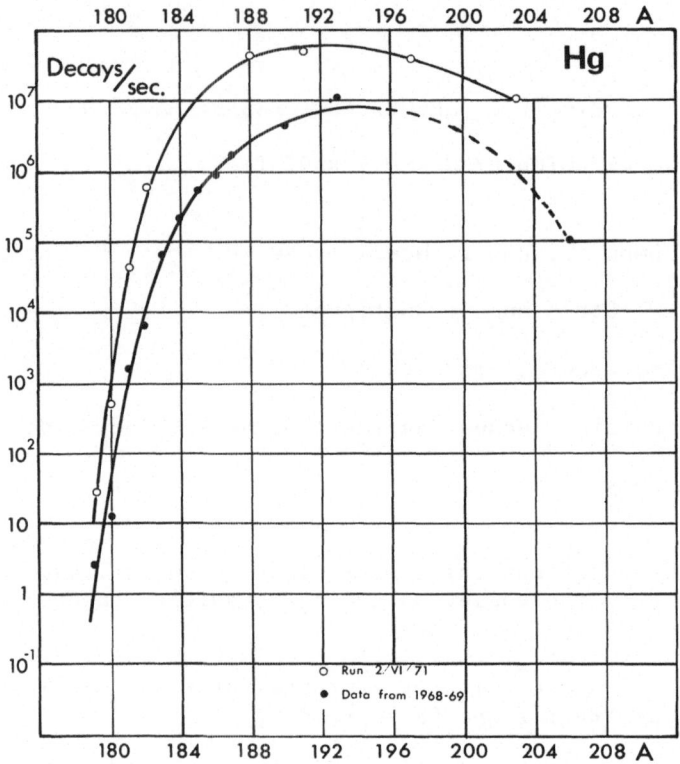

Fig. 1: Production rate of mercury isotopes versus mass number.

EXPERIMENTAL PROCEDURE

An optical pumping (OP) apparatus is connected to the mass separator. By means of an automatic transfer system[7] the isotope under investigation is evaporated into the resonance cell. A ^{204}Hg microwave lamp serves as a light source for OP via the $^3P_1$ state. Since the hfs splitting and the isotopic shift (IS) of the 2537 Å line is known to be much larger than the Doppler width, the light source is subjected to a magnetic field by which the hfs of the radioactive isotope is scanned (Fig. 2). At the matching points OP takes place, leading to a nuclear orientation.

Since the angular distribution of nuclear radiation is sensitive to the spin orientation of the nucleus, the OP signal can be detected either by the asymmetry of the β-decay (Fig. 2) or the anisotropy of γ radiation[8] (Fig. 3).

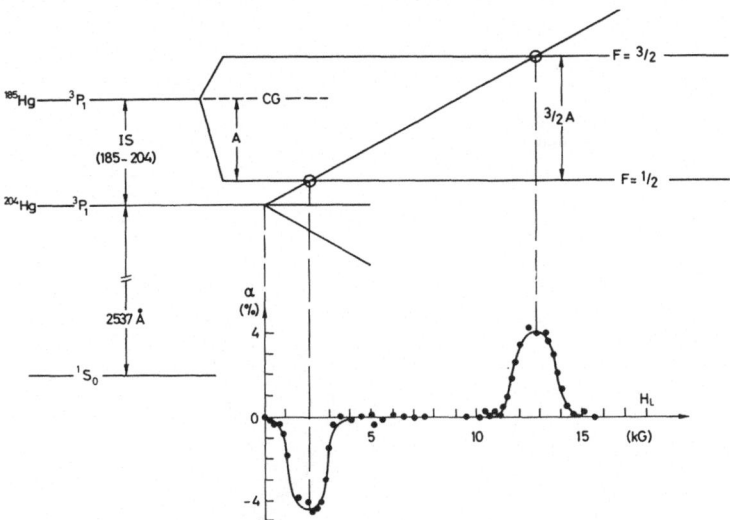

Fig. 2: Bottom: Experimental β–decay asymmetry of ^{185}Hg as a function of scanning magnetic field.

Top: Deduced atomic level scheme of the excited $^{3}P_1$ state of ^{185}Hg. As there are only two scanning signals, the nuclear spin is equal to 1/2. From the distance one obtains the A factor. The centre of gravity (CG) gives the isotopic shift (IS) relative to ^{204}Hg in the light source; g_I can be measured by destruction of the β asymmetry by nuclear resonance in the atomic ground state.

RESULTS

The data of the measured isotopes are compiled in Table 1. The most interesting numbers are those of the IS. In Fig. 4 the IS of all Hg isotopes known today are displayed versus the neutron number. Disregarding the odd-even staggering, the dotted line represents the trend of the IS from ^{205}Hg near the closed neutron shell (N = 126) until ^{187}Hg. Since the quadrupole moments of these nuclei are rather small, the nuclear volume effect of the IS determines the slope of the line in Fig. 4 (2.2 GHz/neutron). From this value one finds the "IS discrepancy"[10] to be ρ = 0.51(3) which increases slightly by allowing for a slowly growing deformation away from the closed n-shell down to ^{187}Hg and is compatible with the number ρ = 0.65(10) extracted out of the whole of known IS.

The IS of ^{185}Hg and ^{183}Hg, however, show a strong deviation from the extrapolated line. The mean square radii of the charge distribution of these nuclei are equal to those of ^{196}Hg, having 11

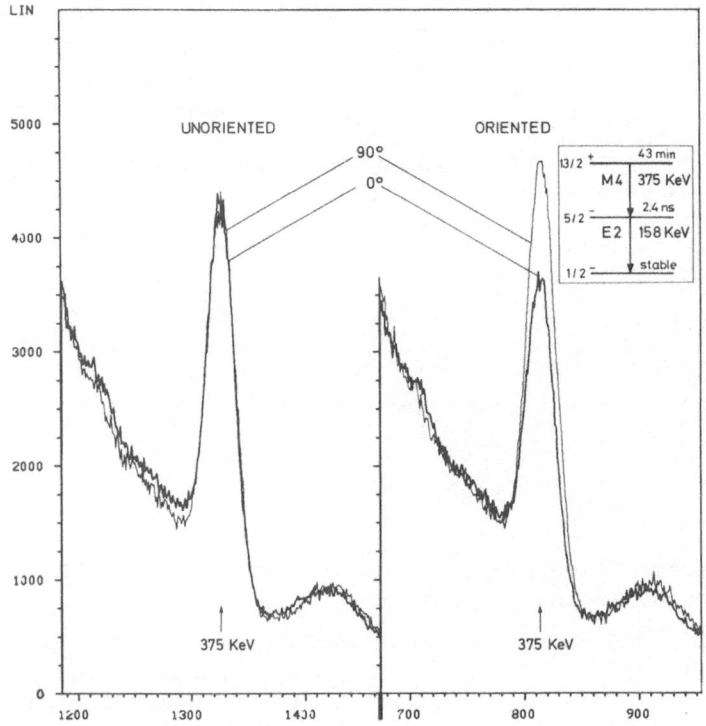

Fig. 3:   Insert: Nuclear decay scheme of the ^{199}Hg isomer. The
          orientation of the 13/2 state was detected by the 0°-90°
          anisotropy of the M4 and E2 transition. The observed effect
          in the M4 transition is shown by a plot of the spectrum of
          the 0° and 90° NaI(Tl) counter in the oriented (right) and
          the unoriented (left) case.

or 13 neutrons more. Since an effect that would change the nuclear
charge density is highly unlikely, the reason for this deviation
should be ascribed to a sudden change in the nuclear shape as big
as $\delta(\langle\beta^2\rangle)^{1/2} = \pm0.23$ for ^{185}Hg and ±0.25 for ^{183}Hg, where β is the
deformation parameter. Such an effect was observed a long time ago
by optical spectroscopy in the middle of a closed proton and neutron
shell by going from N = 88 to N = 90 in the rare earth. It was explained and later confirmed as a sudden increase of static deformation. In the case of Hg there was no prediction of such a change
in nuclear structure near a closed proton shell (Z = 82). Very
recent calculations[11] by the Strutinsky method reproduce the observed effect to some extent, and seem to give a hint that a nuclear
shape transition takes place in the nuclei around ^{185}Hg.

Table 1

Results obtained at the ISOLDE facility. The data of the stable
isotopes are added for comparison; α is the measured asymmetry or
anisotropy corrected for background. The already known IS of
^{192}Hg[1]) could be redetermined off line using a mass-separated sample.

^{x}Hg	$T_{1/2}$	Method	α [%]	I	μ_I [nm]	Q_{Hfs} [b]	IS($^{x}Hg-^{204}Hg$) [GHz]
205	5.5 min	on-line β-asy.	10	$\frac{1}{2}$	0.5911(5)	-	-1.7(5)
199^{m}	43 min	on-line γ-aniso.	32	$\frac{13}{2}$	-0.99862 [a])	+2.0(1.3)	12.3(4)
192	5 h	off-line optical	-	0	-	-·	27.5(3)
187	2.4 min	on-line β-asy.	3	$\frac{3}{2}$	-0.580(6)	-0.3(1.1)	37.1(6)
185	50 sec	on-line β-asy.	-10	$\frac{1}{2}$	0.499(4)	-	19.2(4)
183	8.8 sec	on-line β-asy.	-7	$\frac{1}{2}$	0.513(9)	-	18.9(8)
201	stable	optical	-	$\frac{3}{2}$	-0.5513	+0.50	8.91
199	stable	optical	-	$\frac{1}{2}$	0.4979	-	14.67

a) McDermott, 1971

Fig. 4

Hfs splitting and IS
relative to ^{204}Hg in
the $^{3}P_1-^{1}S_0$ 2537 Å line
of the mercury spectrum

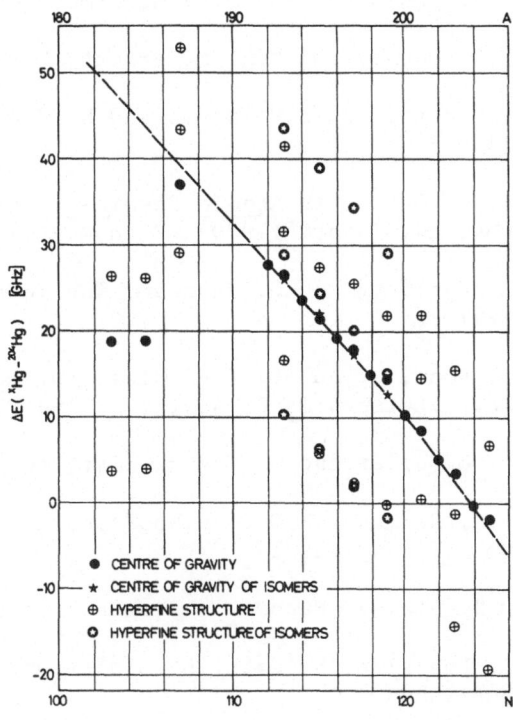

A measurement of the IS cannot distinguish between static or dynamic deformation nor between prolate and oblate shape, hence a quadrupole moment has to be measured. Unfortunately the two exotic isotopes ^{183,185}Hg have I = 1/2; but since an isomer of ^{185}Hg exists, a measurement of its hfs (using the γ anisotropy in the decay of optically pumped ^{185m}Hg) would help to clarify the situation.

REFERENCES

1. P.A. Moskowitz, C.H. Liu, G. Fulop and H.H. Stroke, Phys. Rev. C4, 620 (1971), and references cited therein.

2. N.S. Laulainen and M.N. McDermott, Phys. Rev. 177, 1606 (1969).

3. R.L. Chaney and M.N. McDermott, Phys. Letters 29 A, 103 (1969).

4. D. Goorvitch, S.P. Davis and H. Kleiman, Phys. Rev. 188, 1897 (1969).

5. E.-W. Otten, in Atomic Physics 2, Proc. 2nd Int. Conf. on Atomic Physics, Oxford (1970) (Plenum Press, NY, 1971).

6. A. Kjelberg and G. Rudstam, The ISOLDE collaboration, CERN 70-3 (1970).

7. J. Bonn, G. Huber, H.-J Kluge, U. Köpf, L. Kugler and E.-W. Otter Phys. Letters 36 B, 41 (1971) and Phys. Letters 38 B, 308 (1972).

8. U. Capeller and M. Mazurkewitz, to be published.

9. R.J. Reimann, B.D. Geelhood and M.N. McDermott, Bull. An. Phys. Soc 16, 848 (1971).

10. D.N. Stacey, Report on Progress in Physics 24, 171 (1966).

11. A. Faessler, U. Götz, B. Slalov and T. Ledergerber, to be published.

EXCITED ATOMIC AND MOLECULAR STATES IN LIQUID HELIUM

W. A. Fitzsimmons

Physics Department, University of Wisconsin (Madison)

The microscopic structure of dense gases and liquids and
the nature of excited atomic and molecular states in dense media
has been a subject of long standing scientific interest. For
example the fluorescence of certain compounds when exposed to
sun light has been known for hundreds of years, and very similar
substances are important today as the active chemicals in modern
dye lasers. The investigation into the chemistry of dense
systems is an active area of research where many methods of
initial excitation are used such as u.v. light, γ rays, α and β
particle bombardment, mechanical shock, heat, etc. The first
several spin singlet and spin triplet electronic energy levels
of a complex molecule in a room temperature liquid can be thought
of in terms of the diagram shown in Fig. 1. A recent book by

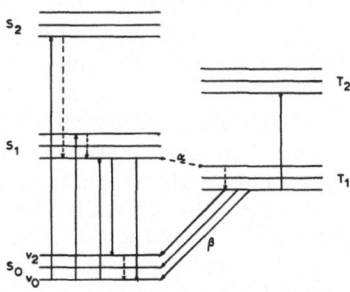

Fig. 1. A schematic energy level diagram for a molecule. S_i and
T_i represent singlet and triplet states, V_i the vibrational levels. Solid arrows represent possible optical
transitions, dashed arrows represent internal conversion
transitions.

I. B. Berlman, in which he discusses the absorption and fluorescence spectra of over 200 aromatic molecules, is testimony to
the quantity of optical data that is available concerning the
ground and first excited singlet states indicated by S_0 and S_1
in the figure.[1]   There is also reference to data which indicates
that the lowest triplet state, T_1, can be very long lived such as
in the case of benzene where the lifetime is about 25 sec at 4.2°K.
However, despite the long historical interest in the optical
properties of liquids, there is really very little known about the
detailed nature of the excited states involved. For example, a
recent paper suggests that the reader might speculate as to which
part of several organic dye molecules is responsible for the intense fluorescence that is pertinent to many dye lasers.[2]   Thus
most physicists would conclude, and perhaps correctly, that this
is a hopelessly complex situation and that the problem is not
particularly amenable to detailed investigation. But perhaps one
should consider a more simple system first. The remainder of this
paper will be a brief review of recent progress that has been
made toward understanding the electronic excitations of a much
simpler fluid -- liquid helium.

     During the past four years, two experimental research groups,
one at Rice University directed by G. K. Walters and my own
group at the University of Wisconsin, have been studying the
electronic excitations that are produced in electron beam bombarded superfluid helium.[2,3,4,5]   Substantial progress has
been made toward understanding the atomic and molecular excitations of liquid helium, and a good deal of this progress can be
attributed to the theoretical work which has been carried out
also at Rice University by Lane et al. [6]   The experimental
arrangement, which includes a 200 KeV electron accelerator, beam
steering apparatus, helium cryostat, and uv spectrometer, is
shown schematically in Fig. 2. An important feature of this
apparatus is the metal foil which separates the liquid sample
from the common vacuum of the accelerator and dewar. The foil
is thick enough to support the mechanical stresses, but thin
enough (0.000 125") to allow the electrons to pass through with
negligible loss in energy. The range of the electrons in the
liquid is about 3 mm, and beam currents are usually between 0.1
and 3.0 µA. Sapphire windows attached to the dewar walls provide
visual access to the excited liquid in directions both parallel
and perpendicular to the incident beam. Until very recently,
all the experiments were carried out on superfluid helium
(T<2.18°K) because the very high thermal conductivity of the
superfluid prevents the liquid from boiling. However, we have
recently operated a sample chamber which incorporates a thin
foil at pressures in excess of 30 atmospheres. Using this cell
we have investigated the spectra of superfluid helium under
pressure as well as the spectra of normal liquid helium. This
type of sample chamber will also allow one to study other more or

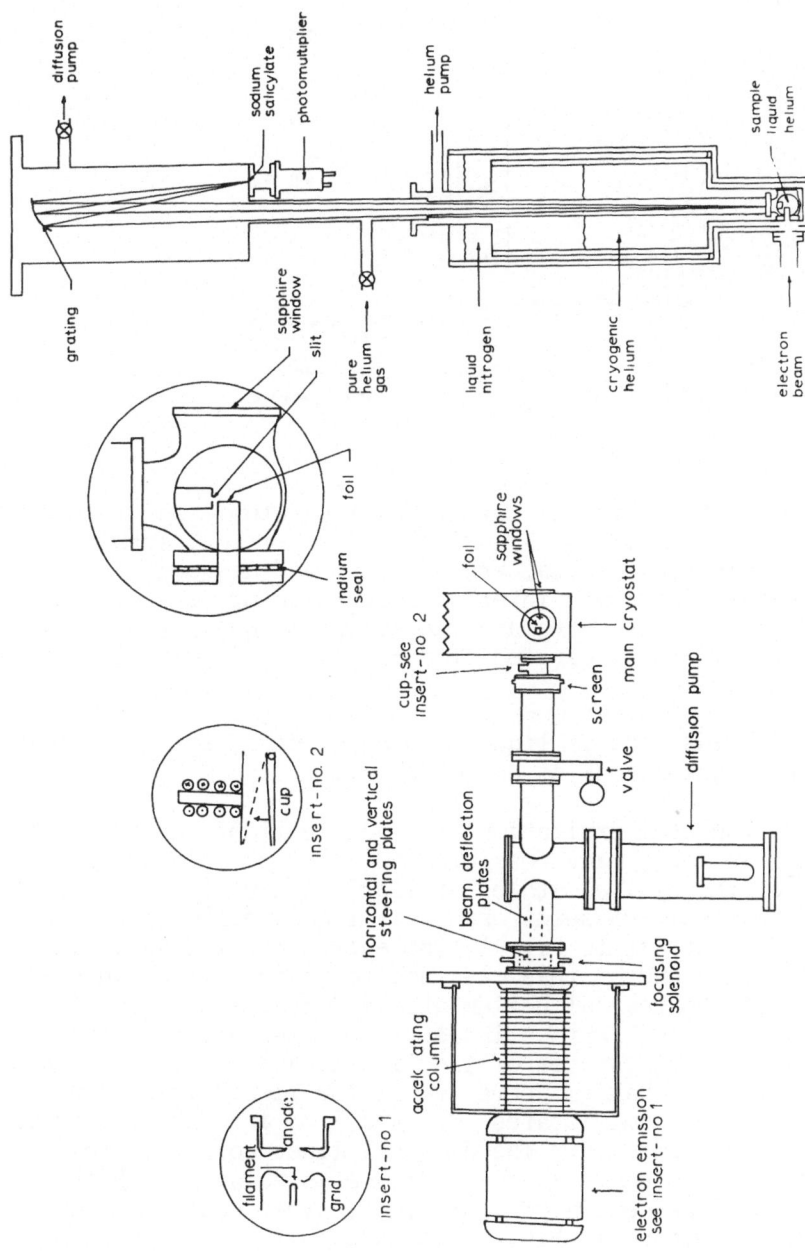

Fig. 2. Experimental apparatus: electron accelerator, beam tube, and lower portion of liquid-helium cryostat; and sample chamber, and vacuum uv spectrograph.

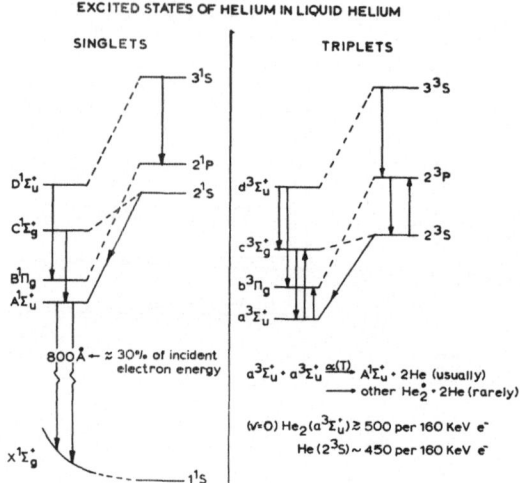

Fig. 3.   Energy level diagram for the excited atomic and molecular
          states of helium in liquid helium. Vertical arrows indi-
          cate observed optical transitions, and diagonal arrows
          indicated observed internal conversion processes. The
          diagonal lines correlate the states of He_2^* with the states
          of He^* in the separated atom limit.

less simple liquids such as liquid hydrogen, neon, etc., and also
the electronic excitations of solid helium.

The electron beam excitation of liquid helium, introduced by
Dennis et al. in 1969, made possible the first spectroscopic investigation of the excited states of liquid helium.[3] A partial
summary of the resulting spectroscopy of liquid helium is shown
in Fig. 3.   In contrast to Fig. 1, the energy level diagram in
Fig. 3 is labelled and in this case according to well known excited states of the neutral helium atom and diatomic helium
molecule. The vertical arrows indicate observed optical transitions between excited states of the liquid, and the two diagonal
arrows indicate a mechanism operative in the liquid which converts
the two metastable atomic states, 2^1S and 2^3S, into the molecules
$A^1\Sigma_u^+$ and $a^3\Sigma_u^+$ respectively. The diagonal lines correlate the
states of the He_2^* molecule which are to be associated with the
states of He^* in the separated atom limit. The identification of
the energy levels shown in Fig. 3 was made on the basis of the
fluorescence spectrum of liquid helium which is shown in Fig. 4.
Under the present experimental conditions, all the significant
fluorescence of liquid helium is encompassed within the wavelength intervals of 600 to 1,100 Å, and 6,000 to 11,000 Å.

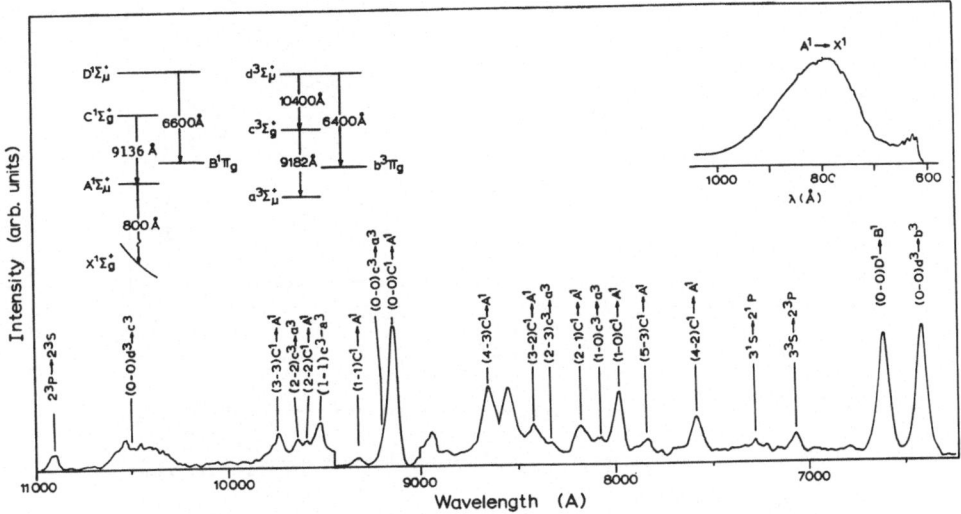

Fig. 4.   The emission spectrum of electron beam excited superfluid
helium.   The states of He$_2^*$ identified on the basis of
this spectrum are indicated by the energy level diagram.

There are several aspects of the excited states of liquid
helium which are apparent in the emission spectrum. First the
threshold for associative recombination in the liquid, He* + He →
He$_2^+$ + e → He$_2^*$, is apparently just above the 3^1S and 3^3S states
since there is no significant emission from more highly excited
atoms.[7]   Secondly, all the observed molecular emissions originate from Σ states of the molecule. This suggests the rapid
non-radiative quenching of non Σ states. More direct evidence of
this quenching is provided by the absence of the $b^3\pi$ → $a^3\Sigma$(2.1μm)
band even though the rate of population due to the d^3 → b^3(6,400 Å)
transition is sufficient to provide a detectable emission at 2.1 μm.
Thirdly, 13 out of the 21 identified infrared bands are due to
significant populations of the first through fifth excited vibrational levels of the $C^1\Sigma$ and $c^3\Sigma$ states. Furthermore, upon
closer examination the d^3 → c^3, D^1 → B^1, and d^3 → b^3 bands reveal
the rotational structure characteristic of a diatomic molecule.
Thus the vibrational and rotational relaxation times for states
of He$_2^*$ in liquid helium are apparently very long when compared to
the typical relaxation time of about 10^{-12} sec for a molecule in
an ordinary liquid.[1]   And finally, the 800 Å continuum due to
the radiative dissociation $A^1\Sigma$ → $X^1\Sigma$ is by far the most intense
emission of electron excited liquid helium. Absolute measurements of the intensity indicate that about 30% of the incident
electron energy is converted into ultraviolet radiation.[4]

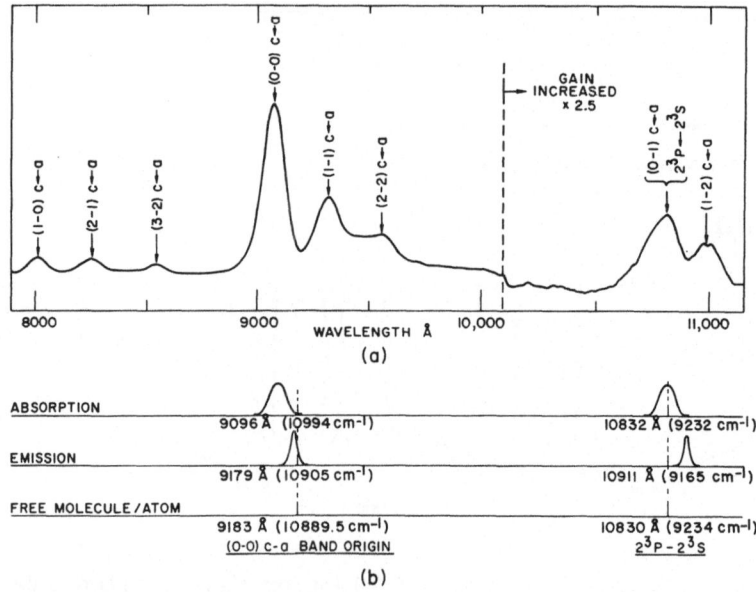

Fig. 5. (a) Absorption spectrum of electron-bombarded liquid
helium from 8000 to 11,000 Å. (b) Comparison of (0-0)
c-a molecular bands and 2^3P-2^3S atomic lines, as observed
in absorption and emission. The free-molecule band origin
and free-atom line positions are also shown.

Another aspect of the emission spectrum is the apparently
rapid radiative population of the two spin triplet metastable
states of helium, He(2^3S) and He$_2$($a^3\Sigma_u^+$). Hill et al. reported
the absorption spectrum of the excited liquid which is due to
these metastable states and they studied the spectral shifts be-
tween emission and absorption for the $c^3 \leftrightarrow a^3$ and $2^3P \leftrightarrow 2^3S$ transitions, as well as the shifts from the resonance values for the
free atom and molecule.[5] Their results are shown in Fig. 5.
Note, the absorption spectrum in Fig. 5 indicates a significant
population of the first and second excited vibrational levels of
the $a^3\Sigma_u^+$ state. (The serious overlap between the atomic and
molecular absorption bands in the neighborhood of 10,800 Å can in
fact be eliminated. This will be discussed shortly.) Hickman
and Lane successfully interpreted the observed blue shift of the
absorption bands with respect to the emission bands, and they
also accounted for the red shift of the $2^3P \to 2^3S$ emission with
respect to the resonance value of 10,830 Å.[6] The essence of
the theory involves using a pseudopotential to calculate the
excited electron-liquid interaction, including the long range
van der Waals interaction, and then adding the energy which is

associated with any changes in the density of the surrounding
fluid. Their results indicate that the excited states reside
within a bubble in the liquid which has a characteristic radius
R_b, with perhaps some non-spherical distortion which depends upon
the type of state inside the bubble. As indicated in Fig. 6, the
reason for the shifts between absorption and emission is due to
the fact that the optical transitions take place in a time short
compared to the relaxation time for the bubble, and thus the
processes of absorption and emission take place in bubbles of
different size as well as shape. Under the conditions of saturated vapor pressure, the energy of the bubble is determined basically by the surface tension energy $\gamma(4\pi R_b^2)$. However at higher
pressures, the pressure term $P(4/3\pi R_b^3)$ dominates and this results
in the additional blue shift for both the $c^3 \leftarrow a^3$ molecular band
and the $2^3P \leftarrow 2^3S$ atomic line shown in Fig. 7. Hickman and Lane
predict a total shift of 18 cm^{-1} for the $2^3P \leftarrow 2^3S$ line between
saturated vapor pressure and 25 atmospheres which agrees well with
the observed shift of 20 ± 5 cm^{-1}.[8] The $c \leftarrow a$ absorption band
exhibits a larger pressure shift which is suggested also by their

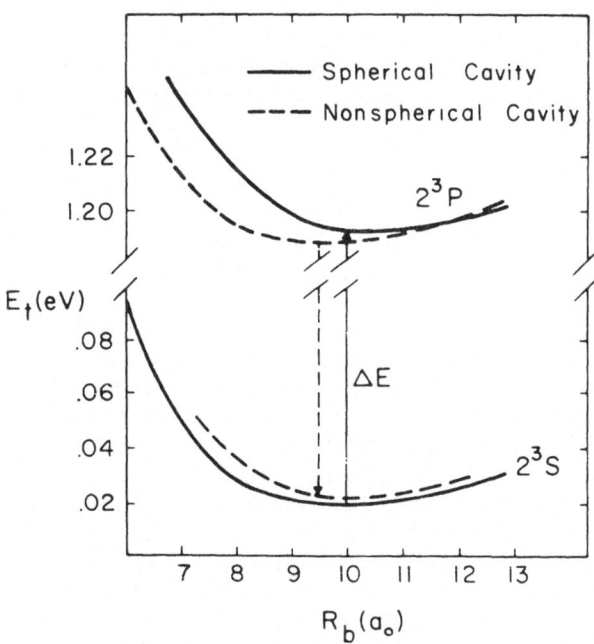

Fig. 6. Total atom plus liquid helium cavity energy versus "bubble
radius" R_b for spherical and non-spherical cavities.

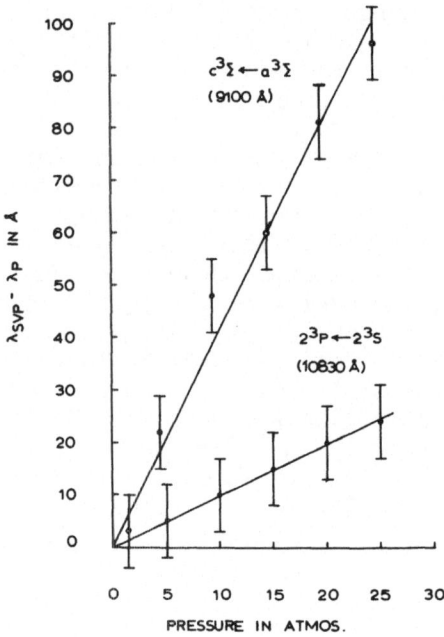

Fig. 7. The observed blue shift as a function of pressure for the
        $(0-0)c^3 \leftarrow a^3$ molecular band and $2^3P \leftarrow 2^3S$ atomic line in
        liquid helium. The measurements are with respect to the
        positions at a saturated vapor pressure of 2 Torr.

theory. Within the near future we will begin to investigate
the emission spectrum as a function of pressure.

        As far as the metastable states are concerned, the Wisconsin
group has taken a somewhat different approach and that is to study
the dynamic or transient behavior of these states. Under the
experimental conditions of 1 μA of 160 KeV electron excitation
there is a steady state concentration of about 2 x 10^{13}/cm^3 and
10^{12}/cm^3 for the $a^3\Sigma$ and 2^3S states respectively and this is more
than enough to study the transient behavior. The metastable $a^3\Sigma$
molecules are more easily studied by monitoring the $b^3\pi \leftarrow a^3\Sigma$
absorption band at 2.1 μm which is shown in Fig. 8.[5] This band
clearly shows that the rotational degrees of freedom of the metastable molecule are not in thermal equilibrium with the surrounding
liquid. The temperature, beam current, and time dependence of the
growth and decay of this signal indicates that the metastable
molecules are destroyed during bilinear collisions between pairs
of the metastable states. In this case, and as shown in Fig. 9,
the steady-state concentration of metastable molecules varies as

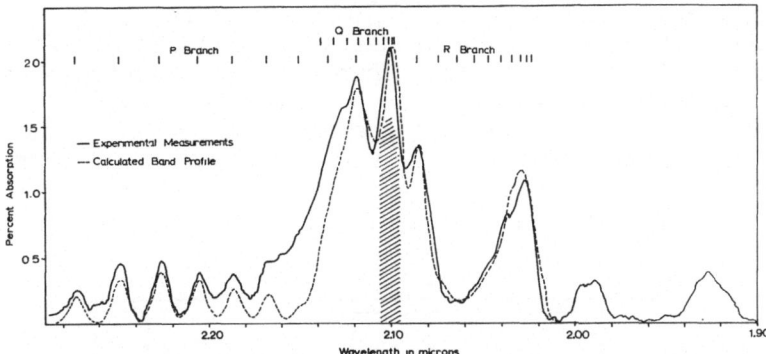

Fig. 8. $b^3\pi_g \leftarrow a^3\Sigma_u^+$ molecular absorption band of electron beam excited superfluid helium. The locations of the first ten members of the R, Q, and P branches of the free molecule are shown. The calculated band profile is based on an assumed Gaussian line shape with a full width at half maximum of 30 cm^{-1} and with the J = 1,3,5,7,9,11,13,15, 17, and 19 rotational levels assumed to be populated in the ratios of 100:13:4.5:5.0:6.4:6.0:9.0:6.4:4.8:2.5 respectively. The interpretation of the 1.92 and 1.99 μm absorption bands is given in the text and also in Fig. 12. The hatched area indicates the spectral region isolated for the transient measurements.

the square root of the beam current and the decay of the concentration following a pulse of electron beam varies as $1/M = 1/M_o + \alpha(T)t$, where M_o is the initial concentration, $\alpha(T)$ is a temperature dependent reaction coefficient, and t is the time.[9] With 1 μA of pulsed electron beam, the characteristic time for the growth and decay of concentration of metastable molecules is about 1 msec.

The temperature dependence of $\alpha(T)$ arises from the fact that the bilinear reaction is diffusion limited and the diffusion coefficient for a $He_2(a^3\Sigma_u^+)$ state in superfluid helium is inversely proportional to the number density of rotons in the superfluid. Thus as shown in Fig. 10, the bilinear reaction coefficient increases as the temperature is reduced from 2.08 to 1.4°K since the number of rotons depends upon the temperature thru $N_R \alpha e^{-\Delta/T}$ where Δ is about 8.6°K.[10] The failure of $\alpha(T)$ to follow $1/N_R$ at the lower temperatures is not understood, although the deviation is not due to minute concentrations of He^3 since the addition of 1% He^3 has only a minor effect. The fluorescence of liquid helium also exhibits a delayed afterglow following a pulse of electron beam. The slow repopulation of the excited states for times as long as several msec into the afterglow has been shown to be due to the bilinear destruction of $a^3\Sigma_u^+$ metastable molecules. In fact practically all the $a^3\Sigma$ bilinear destruction events result in the repopulation of the $A^1\Sigma_u^+$ state. There is also evidence that the optical branching ratio in the case of the $d^3 \to c^3$ and $d^3 \to b^3$ transitions is very sensitive to the surrounding fluid since the ratio of the transition rates v_{dc}/v_{db} is observed to be temperature dependent.

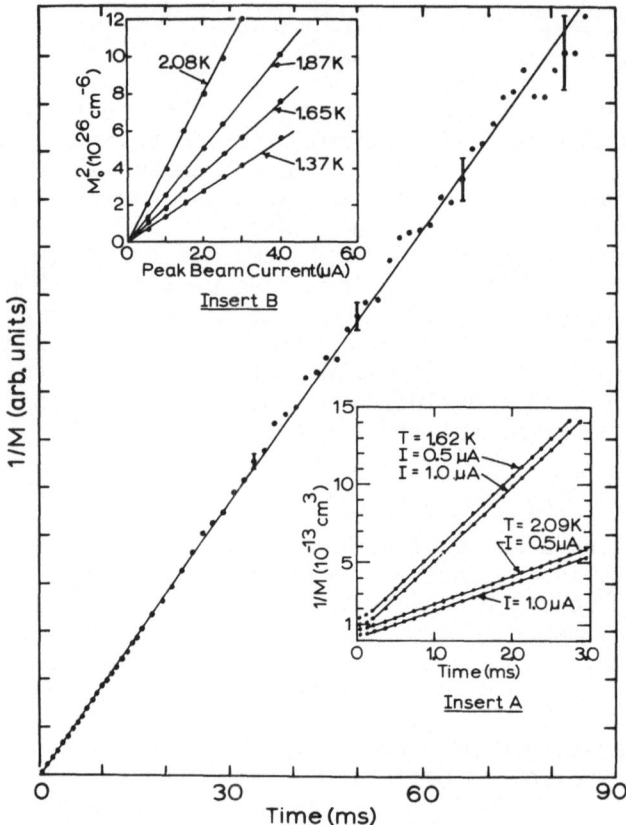

Fig. 9.   Inverse concentration of $(v=0)a^3\Sigma_u^+$ molecules, for T = 2.08°K, as a function of time after the beam is off; insert A shows the temperature and beam current dependence. The initial nonlinear variation is due to the 0.25 msec response time of the PbS detector. Insert B shows the beam current dependence of the steady state concentration at various temperatures.

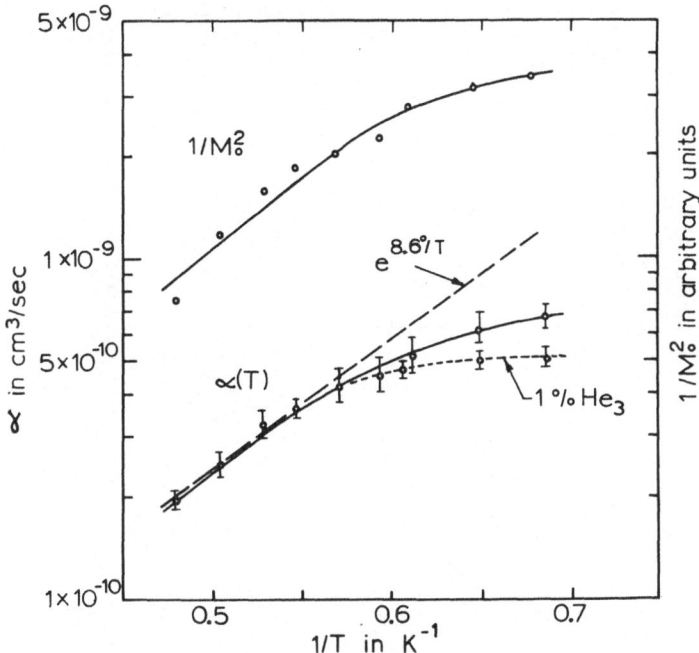

Fig. 10. The bilinear reaction coefficient $\alpha(T)$ as a function
of inverse temperature. The dotted slope corresponds to
the activation energy for rotons. The effect of 1% He3
doping of sample is shown. Also shown in the temperature
variation of $1/M_0^2$ for the case of 1-μA of beam current.

On the other hand, the lifetime and decay mode of the He(2^3S)
metastable atom is distinctly different from that of the corresponding metastable molecule. For example the lifetime of the 2^3S
state in the liquid is about 15 μsec and it is independent of beam
current and temperature. (The rapid response time for the metastable atom can be played off against the much slower 1 msec response
time for the metastable molecules in order to separate the overlapping atomic and molecular absorptions near 10,800 Å. For example
by pulsing the electron beam rapidly, such as 100 μsec on and 100
μsec off, the molecular signal is essentially eliminated leaving
only the atomic absorption line.) As shown in Fig. 11, the destruction of the He(2^3S) atoms can be correlated with the delayed
growth of the previously unidentified 1.92 μm absorption band shown
in Fig. 8. We believe that the 1.92 μm absorption band (and also
the 1.99 μband) is due to the conversion of the 2^3S atoms into a$^3\Sigma$
molecules, resulting in the population of the very highest vibrational levels of this state. The relevant molecular potential
curves and the suggested optical transition is shown in Fig.12.[11]

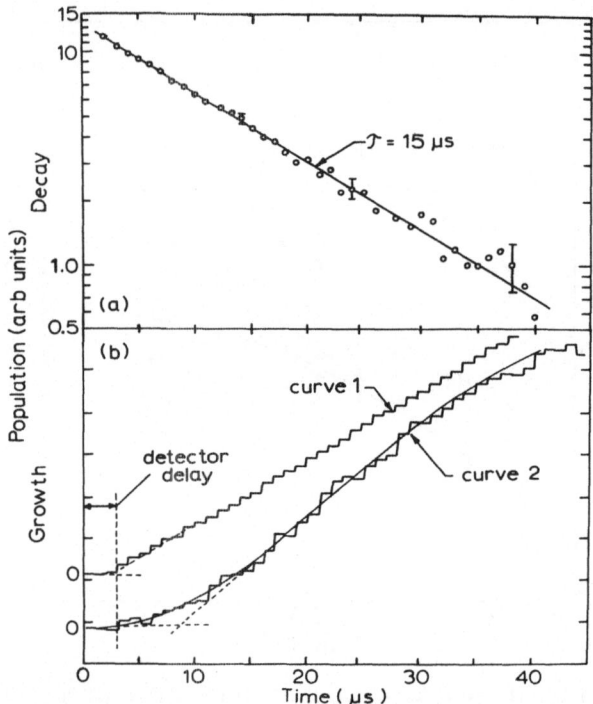

Fig. 11.   (a) Logarithmic plot of the He(2^3S) concentration for
times after the beam is off.   (b) Curve 1 shows the
growth of the (v=0)a$^3\Sigma$ concentration for times after the
beam is turned on.   The apparent delay is due to the PbSe
detector.   Curve 2 shows the delayed growth of the 1.92µm
absorption band.   The solid line is the predicted delay
assuming the absorbing species are formed by the destruction of metastable atoms.

The conversion of metastable atoms into high vibrational
levels of the corresponding molecule is also evident in the case
of the He(2^1S) state.   This mechanism appears to be the source of
the shorter wavelength 600 Å bands in the uv spectrum of the excited liquid that is shown in Fig. 4.   The shortest wavelength
members of the 600 Å bands are shown in more detail in Fig. 13.
The interpretation of these bands, which appear also in the spectrum of a helium discharge, is in terms of the highest two quasibound vibrational levels of the $A^1\Sigma_u^+$ state. [12,13]   A graphical
representation of this interpretation is shown in Fig. 14 and the
locations of the bands as observed in the spectra of the liquid and
gas phases are compared with the theoretical predictions in Table I.

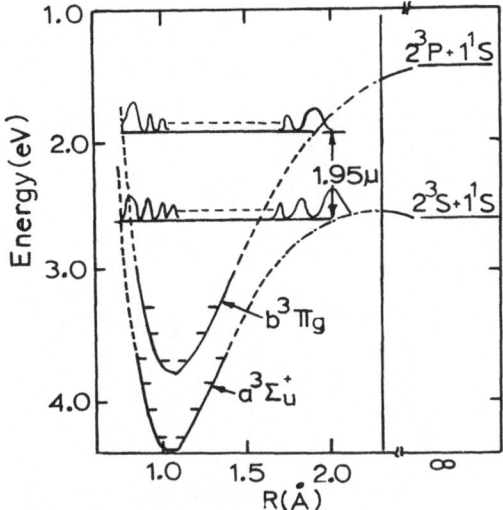

Fig. 12.   Diagram showing the potential curves for the $a^3\Sigma_u^+$ and $b^3\pi_g$ states of He_2. This figure indicates the possibility of an optical absorption near 1.9 μm occurring between the highest vibrational levels of the a^3 state to correspondingly high vibrational levels of the b^3 state.

     The broad 800 Å peak in the uv spectrum of the excited liquid is due to radiative dissociation of the $He_2(A^1\Sigma_u^+)$ molecule which originates from the lowest several vibrational levels of this state. Absolute measurements indicate that about 30% of the incident electron energy is converted into uv radiation and the possibility of developing this into a useful source of light is immediately obvious. We have studied the uv emission spectrum as a function of beam current for input power levels as high as 10 watts. Our results shown in Fig. 15 indicate a slight change in the band profile at the lower wavelengths, but the intensity of major 800 Å remains essentially linearly dependent upon beam current. On the basis of these results we have begun construction of a uv source with the intensity expected to be about 3 watts of uv light, or equivalently an expected photon flux of 10^{10} photons/sec/Å band width at 800 Å exiting the associated monochromator. A diagram of this source is shown in Fig. 16.

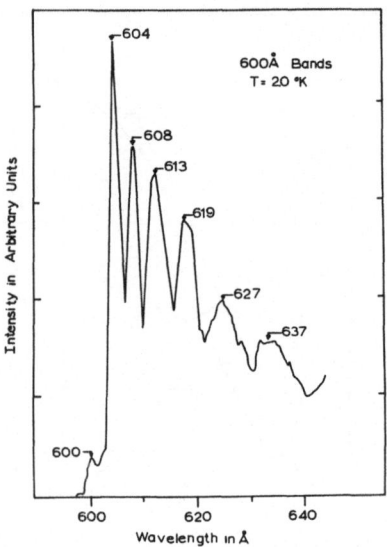

Fig. 13.   600 Å bands emitted by electron bombarded superfluid
helium. The spectrum was observed by locating the free
surface of the liquid sample just above the position of
the incident electron beam; otherwise, the first four
members of this series of bands are absorbed within the
first several mm of liquid.

$\lambda(\text{Å})$-Liquid	$\lambda(\text{Å})$-Gas	$\lambda(\text{Å})$-Theory
	600.07	600.0
(600)	600.06	600.9
	602.3	602.2
604	605.0	604.9
608	608.4	608.4
613	613.0	613.0
619	619.0	618.5
627	627	625
636	636	634
648	648	645
665	662	658
686	685	675
707	708	695

Table I.   The locations of the maxima for the 600 Å bands of He_2
as they appear in the liquid spectrum, in the spectrum
of a gaseous discharge (Ref. 12), and as predicted
theoretically (Ref. 13).

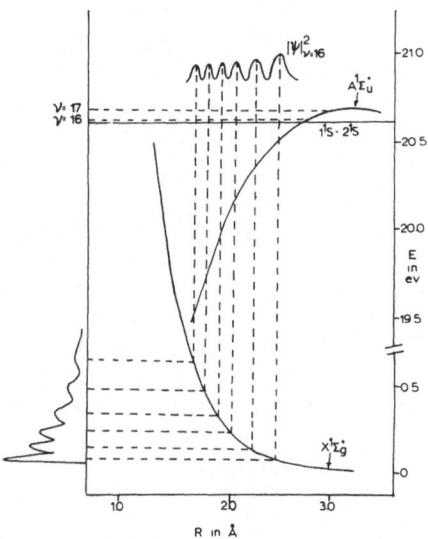

Fig. 14.   Graphical representation of the qualitative features of emission from the second highest ($v=16$) quasibound vibrational level of the $A^1\Sigma^+_u$ state of helium.   This level is estimated to be 1.2×10^{-2} eV above the 1^1S-2^1S separated atom limit (see Ref. 13).

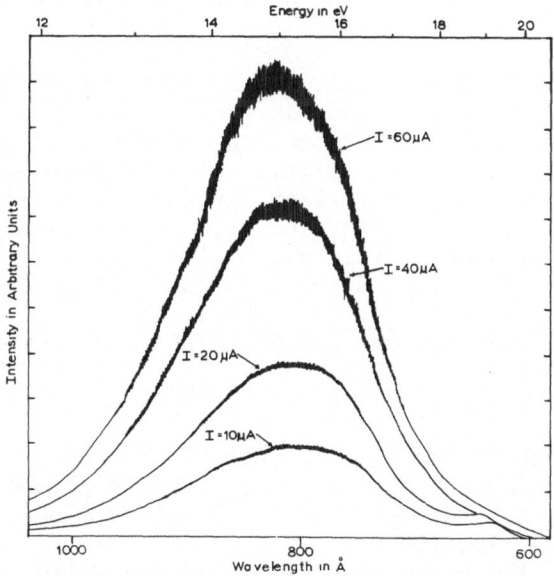

Fig. 15.   U.V. emission spectrum of electron-bombarded superfluid helium versus beam current.   The highest current shown corresponds to an input power of about 10 watts.

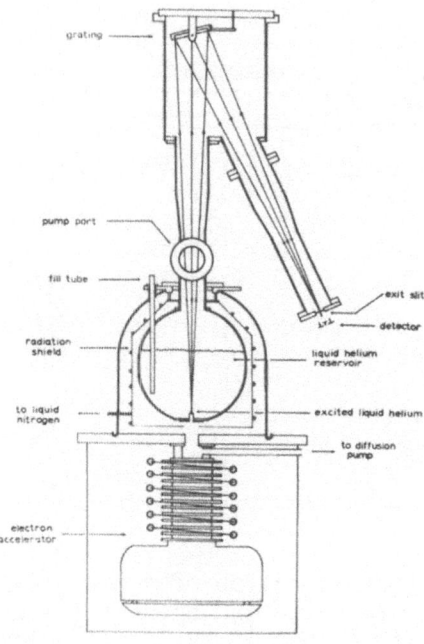

Fig. 16.   Schematic diagram of a high intensity electron pumped-
           liquid helium ultraviolet light source.

1) Isadore B. Berlman, Handbook of Fluorescence Spectra of
    Aromatic Molecules (Academic Press, New York, 1971), 2nd Ed.
2) Benjamin B. Snavely, Proceedings of the IEEE 57, 1374 (1969).
3) W. S. Dennis, E. Durbin, Jr., W. A. Fitzsimmons, O. Heybey,
    and G. K. Walters, Phys. Rev. Letters 23, 1083 (1969).
4) M. Stockton, J. W. Keto, and W. A. Fitzsimmons, Phys. Rev.
    5, 372 (1971) and Phys. Rev. Letters 24, 654 (1970). See
    also C. M. Surko, R. E. Packard, G. J. Dick, and F. Reif,
    Phys. Rev. Letters 24, 657 (1970).
5) J. C. Hill, O. Heybey, and G. K. Walters, Phys. Rev. Letters
    26, 1213 (1971).
6) A. P. Hickman and Neal F. Lane, Phys. Rev. Letters 26, 1216
    (1971).
7) J. A. Hornbeck and J. P. Molnar, Phys. Rev. 84, 621 (1951).
8) A. P. Hickman and Neal F. Lane, private communication.
9) J. W. Keto, M. Stockton and W. A. Fitzsimmons, Phys. Rev.
    Letters 28, 792 (1972).
10) J. Wilks, The Properties of Liquid and Solid Helium (Oxford
    U.P., London, 1967).
11) M. L. Ginter and R. Battino, J. Chem. Phys. 52, 4469 (1970).
12) Y. Tanaka and K. Yoshino, J. Chem. Phys. 39, 3081 (1963).
13) K. M. Sando, Mol. Phys. 21, 439 (1971).

ELECTRON SPECTROSCOPY FOR CHEMICAL ANALYSIS

K. Siegbahn

Institute of Physics, University of Uppsala

Uppsala, Sweden

INTRODUCTION

When atoms are brought close together to form molecules the orbitals of individual atoms are perturbed and replaced by molecular orbitals. Inner orbitals, i.e. with higher binding energies, may still be regarded as atomic and belonging to specified atoms within the molecule, whereas the external orbitals combine to form the valence level system of the entire molecule. These orbitals take a more or less active part in the chemical bonds which are formed between the atoms in the molecule and which specify the chemical properties. The chemical bonds affect the charge distribution so that the original neutral atoms can be regarded as charged to various degrees and with different signs with a net charge of zero for a neutral molecule. We may describe the situation by regarding the individual atoms in the molecule as spheres with different potentials. Inside each charged sphere the atomic potential, set up by the removal of a certain small charge from its surface to the neighbouring atoms taking part in the chemical bond, is constant according to classical electrostatics. The result of this atomic potential is to shift the whole inner level system of any atom by a small amount, each level being shifted an equal amount. Levels belonging to different atoms in the molecule are generally shifted differently, however, and by measuring these "chemical shifts" for individual atoms in the molecule a mapping can in principle be made of the charge or the potential distribution in the molecule. This is then a reflection of the chemical bondings between the atoms which in turn in principle can be described by the bonding orbitals in the external valence level system. Recently, a new spectroscopy has been developed which can handle these problems in a direct and informative way. It is based on a high-resolution study of the energy distribution of electrons expelled from their orbitals by means of some exciting radiation. By means of X-radiation one can reach not only the valence region but also the inner orbitals, the core region, and with this radiation - which was the first to be used - both valence spectra and chemical shifts of atomic core spectra can be studied. Each element in the periodic system has its characteristic binding energies of the core levels, and the elemental composition of a chemical compound is therefore the first result that one obtains from an electron spectroscopical investigation using X-radiation. The next step is to get information about the chemical bonds from the deviations of the atomic core level binding energies by measuring the chemical shifts encountered. The third step is to measure the electron spectrum of the outermost molecular orbitals, the valence spectrum. A lot of chemical information is obtained by a combination of these electron spectroscopic studies. It was therefore appropriate to name the field "Electron Spectroscopy for Chemical Analysis" - ESCA. "Analysis" is then taken in a broader sense than such chemical analysis which is confined to the determination of the chemical composition but includes also, according to the above, molecular properties which are not covered by such other methods. It turns out that ESCA is applicable to the analysis of elements over the whole periodic system and to chemical species both in solid and gaseous form. It is essentially a surface method - the analysis is performed on a layer less than 50 Å, the lower limit being a small fraction of a single atomic layer. It can also be applied to problems concerning the solid state, such as the conduction bands of metals, certain phenomena associated with paramagnetism, plasmons and surface reactions. It is an alternative to other existing spectroscopies in the sense that the study emphasizes the electrons themselves which are taken directly from their molecular bonds whereas the wide scope of photon emission or absorption is covered by other spectroscopies. The information which can be obtained by means of electron spectroscopy is in several respects different from that obtained from these previous spectroscopies and is therefore a complement to them.

EXCITATION OF ELECTRON SPECTRA

Let us consider an atom which is a constituent of a molecule. Schematically, the electron distribution can be separated into two regions, one inner which is the core region and one external which is the valence electron region (Fig. 1). The latter has a depth of some 50 eV and the molecular orbitals within this region are more or less delocalized. The core orbitals, on the other hand, are localized to one particular atom. Within this region the chemical bond and the charge redistribution are associated with a potential change which results in a chemical shift of the binding energies of the core levels. Ultraviolet radiation, in particular the HeI (21.22 eV) and HeII (40.8 eV) resonance lines, can expel monokinetic

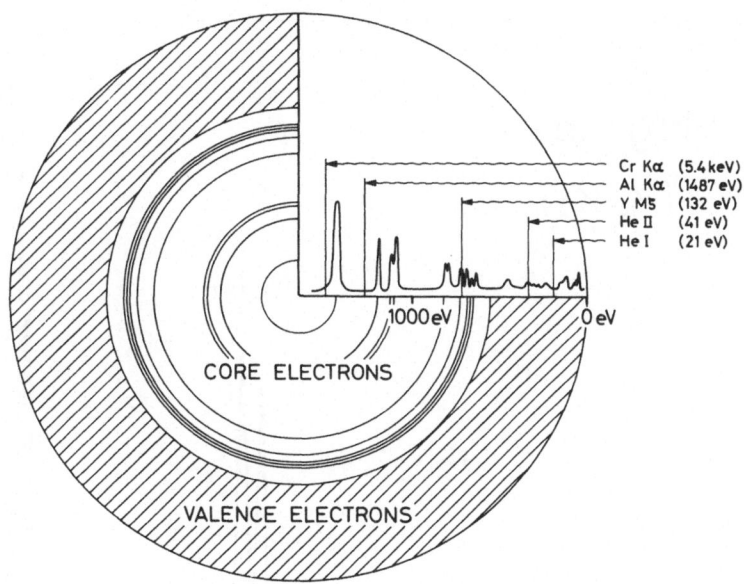

Fig. 1. The level system of an atom in a molecule can be divided into a valence electron region and an atomic core region. Excitation of electron lines from the various regions can be made at different photon energies.

electrons from this region. Because of the narrow UV lines the corresponding electron lines are quite sharp, around 10–25 meV, unless broadened for other reasons (rotational, spin–orbit broadening, etc.). Usually the most external orbitals give rise to the sharpest lines often containing resolved vibrational structures. Further inwards, the electron lines are usually more complex and broader. When the orbitals in the valence region are excited by an X-ray line, for example the AlKα or the MgKα, the electron lines are broader because of the inherent width of these rays. The intensities are also different from the UV-excited lines. Some orbitals, namely those having predominantly s symmetry, which are difficult to excite by UV radiation, are strongly excited by X-ray lines, whereas the orbitals with p symmetry are strongly excited by HeI radiation. The photoelectric cross-section dependence on the level binding energy in the valence region and the energy of the exciting radiation is likely to vary more for exciting radiation which is of the same order of magnitude as the level binding energies because of the stronger dependence on the final states involved in the transitions. Generally, the combined information from both UV- and X-ray excitation gives complementary information concerning the symmetry properties of the levels in the valence region. The valence region can also be covered by some ultra soft X-rays like YMζ radiation (130 eV) which is, like

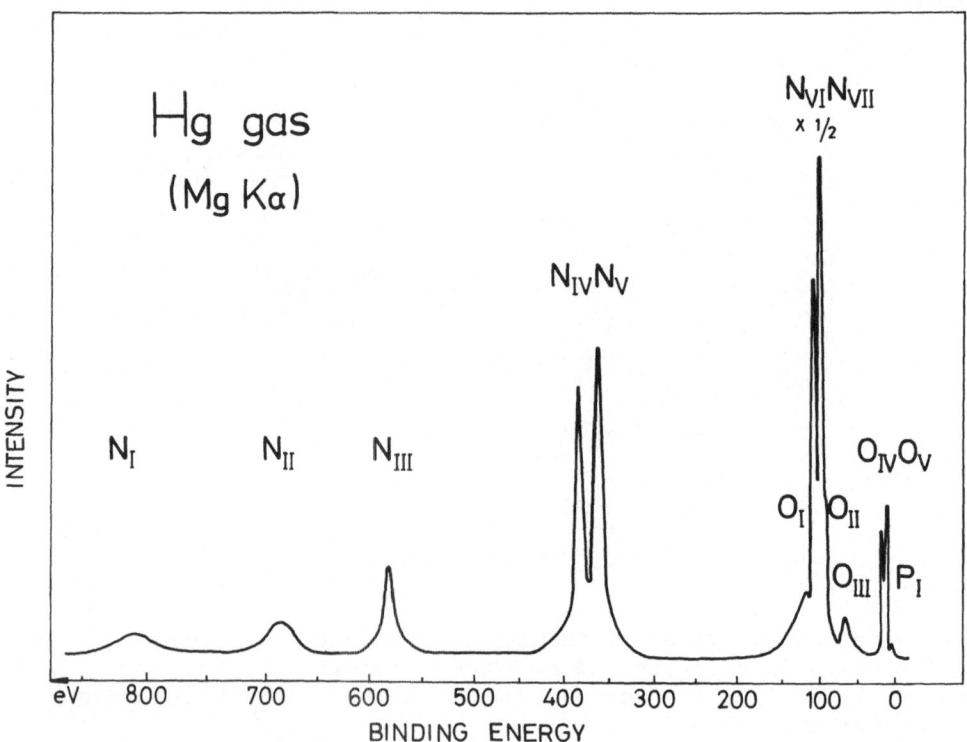

Fig. 2. Electron spectrum of Hg vapour excited by MgKα.

the UV-radiation, too soft to penetrate into the core region. The
1s levels of carbon, nitrogen, and oxygen have binding energies
around 290, 410 and 540 eV, respectively, and are conveniently
excited by Al and Mg Kα radiation, which are capable of exciting
all core levels in elements which have binding energies less than
1485 and 1250 eV, respectively. The chemical shifts are obviously
easiest to study for sharp levels and one therefore prefers the 2p
levels to the 2s levels for the second row elements like sulfur or
silicon, for example.

Fig. 2 shows an electron spectrum of Hg vapour (at ≈ 50 C^o)
excited by MgKα. All sublevels in the N, O and P shells are recorded
in the spectrum with different inherent widths (and naturally also
intensities). This is a scan spectrum and individual lines and line
groups can be recorded in much greater detail.

Fig. 3 is a recording of the outer part of the electron
spectrum of solid gold, excited by MgKα. The strong spin doublet
lines $(N_{VI}N_{VII})$ from the $4f_{5/2}$ and $4f_{7/2}$ orbitals are seen. Of

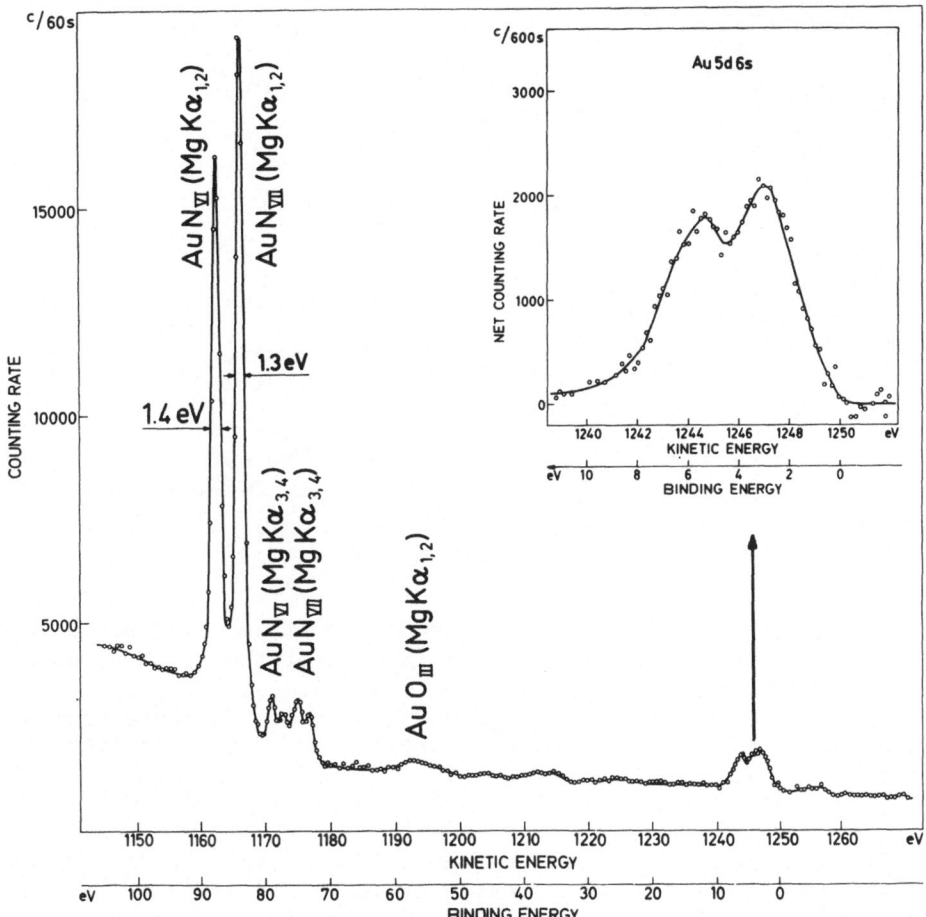

Fig. 3. External part of the electron spectrum of gold excited by MgKα showing the strong $N_{VI}N_{VII}$ doublet and the conduction band.

particular interest is the small intensity electron distribution close to binding energy zero. Inserted in the figure is a study of this distribution, which is the conduction band of gold. It has a characteristic broad double maximum (primarily due to spin orbit coupling) profile (d-band) with a low intensity component (s-band) close to the Fermi level at $E_{bind} = 0$.

Fig. 4 shows a recent, much improved version of the same conduction band. It is taken by a newly designed ESCA instrument provided with an X-ray monochromator system which increases the resolution. The details of the profile coincide quite well with a recently performed APW calculation.

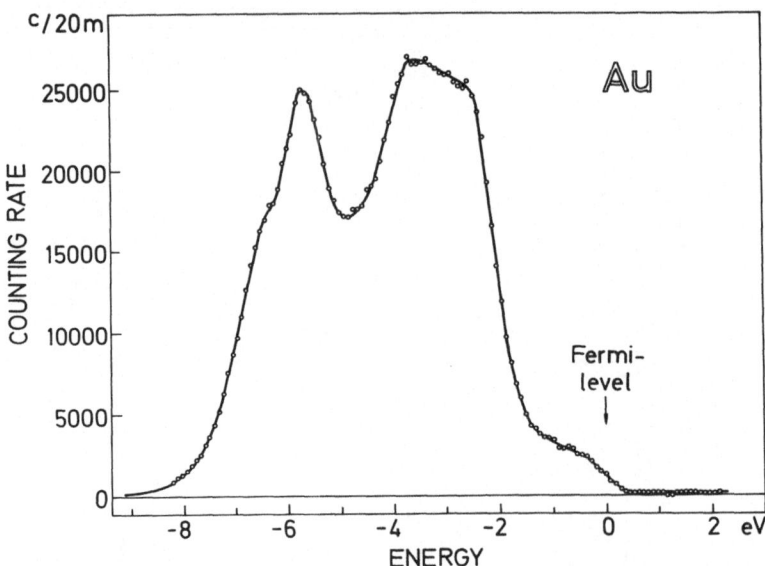

Fig. 4. Conduction band of Au using monochromatized AlKα for excitation of the electron spectrum. The main part of the spectrum comes
from the d electrons. The two-maximum shape is caused by spin-orbit
coupling. The s band which extends to the higher kinetic energies,
is of low intensity in this mode of excitation. The shape fits
recent calculations well.

     It is interesting to follow systematically the electron spectra
of the conduction bands of series of elements and their alloys and
also the dependence on the energy of the exciting radiation. By UV-
excitation the shape is a strong function of hν. The shape in this
region cannot be expected to be representative for the density of
states of the conduction band since the oscillator strength varies
significantly over the band region due to the influence on the
transition rates of the final states, in these cases situated closely above the Fermi level. Gradually, for higher excitation energies,
the shapes of the conduction band electron spectra start to approach
those recorded by means of X-ray excitation. There are other interesting phenomena to be studied in this connection, for example the
shape dependence on the concentration at alloying at high
dilution and the possible shape change at the melting point at
different excitation energies.

     The intensities of electron lines have been shown to depend on
the electron emission angle for single crystals. The diffraction-like
pattern is also dependent on the core level involved.

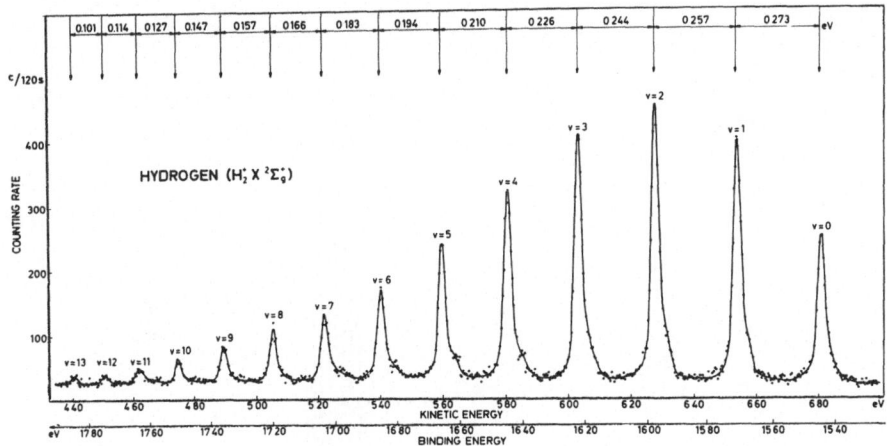

Fig. 5. Electron spectrum of molecular hydrogen, excited by HeI resonance radiation at 21.2 eV. The fourteen peaks correspond to vibrational levels of the molecular ion.

Figures 5, 6 and 7 are examples of electron spectra excited by means of the HeI radiation in a sequence of increasing complexity. Fig. 5 shows a vibrational band in the electron spectrum from H_2 at a binding energy of 15.4 eV. The intensities of the lines are determined by the Franck-Condon factors of the transitions. The widths are here primarily set by the Doppler widths in addition to the rotational fine structure. H_2 is naturally a simple case and already NH_3 shows a more complex spectrum because of the increased number of degrees of freedom for vibrational motion. Fig. 6 shows two main bands, one with a simple vibrational structure and another which is much more complex, and requires high resolution to be analyzed. The former is due to the $(...1e^43a^1)^2A_1 \leftarrow (1e^43a^2)^1A_1$ transition, with an adiabatic ionization energy of ≈ 10 eV and consists essentially of a long progression of vibrational lines with spacing of ≈ 900 cm^{-1}. The vibration has been identified as the ν_2 bending mode which produces the inversion motion of ammonia. Franck-Condon analysis of this band indicates that the ground state of NH_3^+ is planar. The first excited state of NH_3^+ is $(...1e^33a^2)^2E$. The complex band structure at about 15 eV corresponds to the $^2E \leftarrow {}^1A_1$ transition. It appears that the transition is a composite of two overlapping bands whose respective maxima lie at ≈ 15.8 eV and 16.8 eV. Thus, ionization of an electron produces two ionic states that are very close in energy. The vibrational structure is not composed of simple progressions and it appears to be superimposed on a dissociative continuum, most likely due to $NH_3^+ \rightarrow NH_2^+ + H$ and $NH_3^+ \rightarrow NH^+ + 2H$.

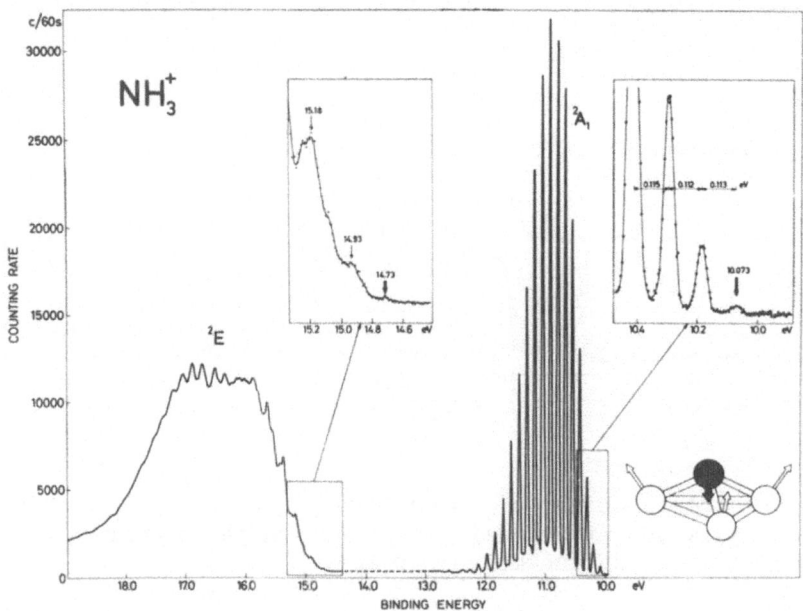

Fig. 6. Valence electron spectrum of ammonia excited by HeI. The left part is the $(...1e^33a^2)^2E \leftarrow (...1e^43a^2)^1A_1$ transition containing several vibrational progressions. The right part of the spectrum is the $(...1e^43a^1)^2A_1 \leftarrow (...1e^43a^2)^1A_1$ transition. This vibration is the ν_2 bending.

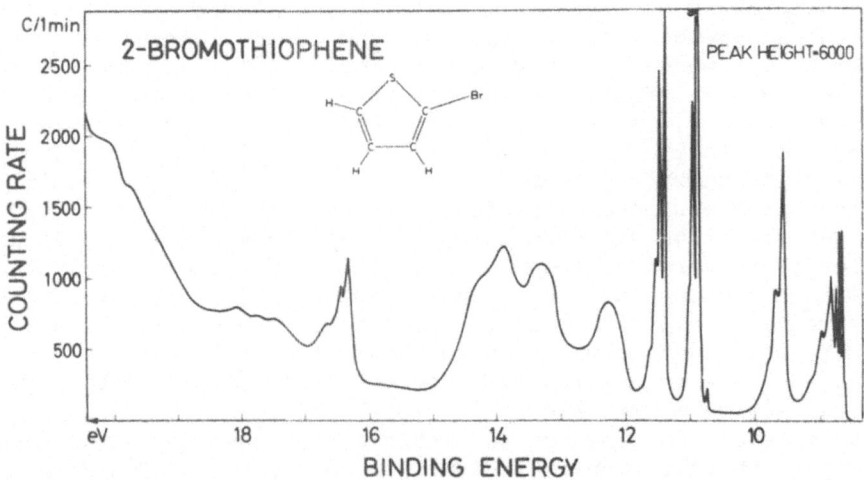

Fig. 7. Valence electron spectrum of 2-bromothiophene excited by HeI. The two high-intensity vibrational bands with binding energies around 11 eV are due to Br lone-pair electrons.

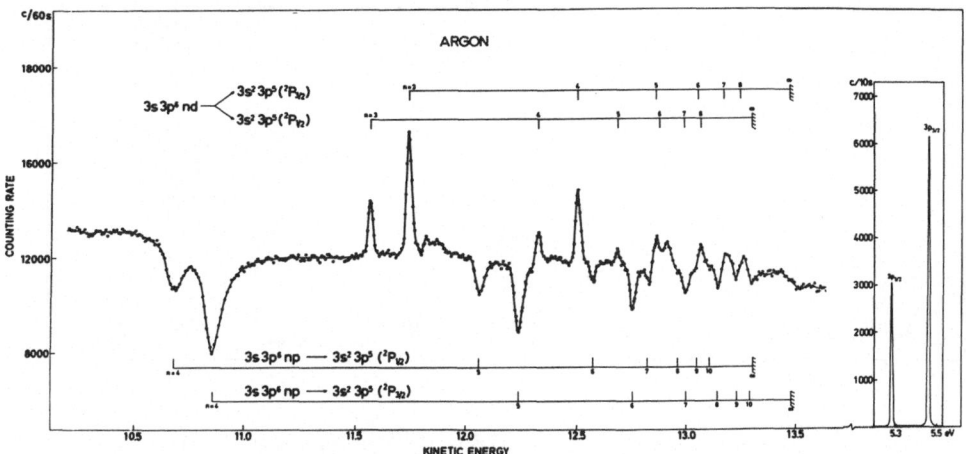

Fig. 8. Autoionization electron spectrum from argon showing four identified Rydberg series of lines. Inserted in the figure is the spin orbit splitting of the $3p^5(^2P)$ term in argon, photoionized by HeI.

The appearance energy of NH_2^+ is 15.7 eV, in agreement with the maximum of the continuum. The closer energy analysis further suggests that Jahn-Teller forces are operative in this ionic state. The shape of this electron band is incidentally remarkably similar to the band resulting from the $(a_1^2t_2^5)$ $^2T_2$ ← $(a_1^2t_2^6)$ $^1A_1$ transition in methane.

Fig. 7 shows the valence electron spectrum of 2-bromothiophene excited by HeI. The two high-intensity vibrational bands with binding energies around 11 eV are due to Br lone-pair electrons. This example is given to indicate that for most moderately sized molecules the valence electron spectra become more complex and contain unresolved structures of ≈ 0.5-1.0 eV even at high instrumental resolution. However, also such structures are informative to analyze in terms of possible molecular states.

The electron excitation processes which have been discussed so far usually are described as due to photoelectron emission. In electron spectroscopy many other lines appear, however, which may even sometimes be mixed with such photoelectron lines and which are due to different dynamics. The most important of these are Auger effect, autoionization effect and shake-up and shake-off processes. The two last mentioned effects result in well defined electron lines as well as continuous electron distributions having intensities which in many cases cannot be neglected, and in particular shake-up lines contribute much useful information about unoccupied molecular orbitals. Electron lines due to Auger effect and autoionization which are also well defined in energy do not require monoenergetic photons for their excitation as in the photoelectron effect but can also

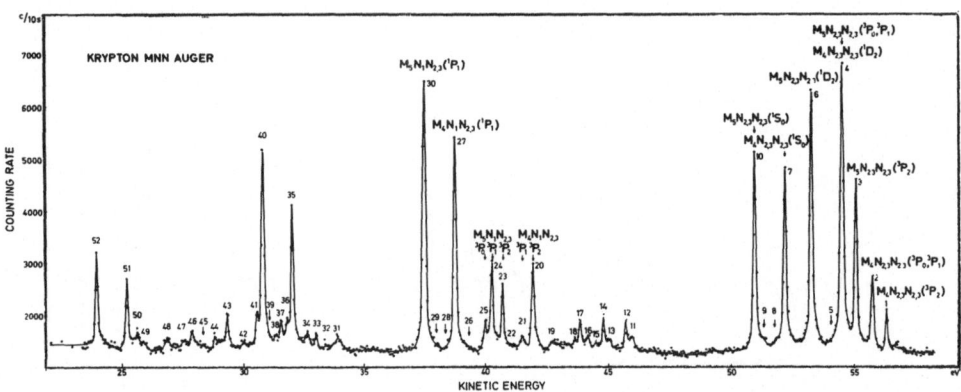

Fig. 9. Krypton $M_{IV,V}NN$ Auger electron spectrum.

conveniently be excited by non-monoenergetic electrons. These lines
are independent of the energy of the exciting radiation. Figures
8-11 give examples of the mentioned types of electron spectra.

Fig. 8 shows an autoionization electron spectrum of Ar excited
by an electron beam. There are characteristic emission- or absorp-
tion-like line profiles and Rydberg series can be identified in this
spectrum with series limits corresponding to a second ionization.
The appearance of the two upper Rydberg series in the figure can be
assigned to spin-orbit splitting of the 3p shell. It is interesting
to note that the energy splitting of 0.18 eV and the intensity ratio
of 2:1 is in close agreement with what is observed in the electron
spectrum when Ar is photoionized by HeI at $h\nu$ = 21 eV (inserted in
Fig. 8).

Fig. 9 shows the krypton $M_{IV,V}NN$ Auger electron spectrum. Most
of the strong lines are identified. In addition to these the spectrum
consists of many satellite lines, some of which have high intensities
The satellites are usually transitions from doubly to triply ionized
states. Such transitions should give rise to a multitude of lines
since both the initial and final states are split into many terms.
The $M_{IV,V}N_{II,III}$ initial vacancies, for example, form 12 different
terms and an Auger transition of the type $M_{IV,V}N_{II,III}-N_{II,III}^3$ could
consist of as many as 60 lines. The satellites at low kinetic energy
occur in pairs with a separation equal to the $M_{IV}-M_V$ spin-orbital
splitting. This suggests that these lines originate from shake-up
in the Auger process itself rather than from shake-off in the primary
ionization.

In order to clarify the latter processes Fig. 10 is given to
demonstrate the main features. Neon is excited by $MgK\alpha$. The main
peak is the Ne1s electron line. The MgK X-ray satellites and $K\beta$

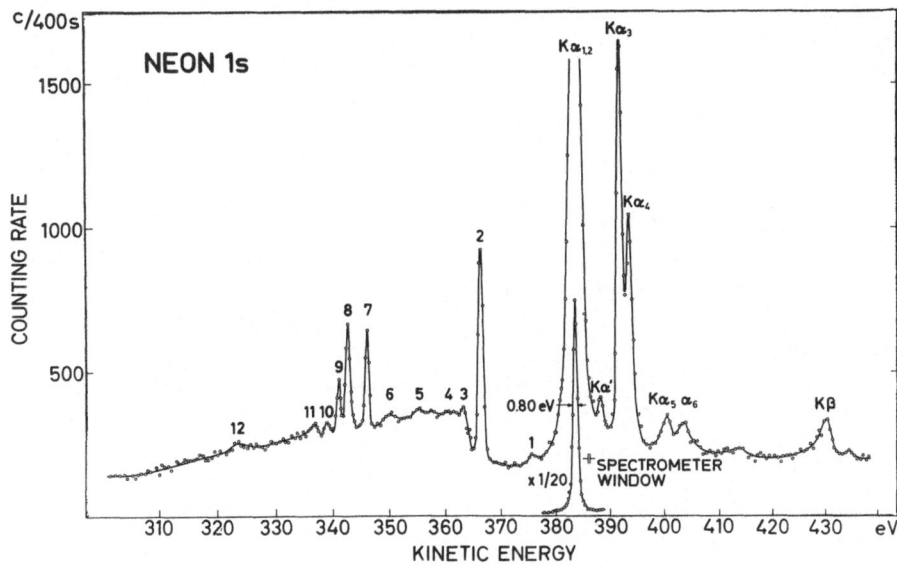

Fig. 10. Neon 1s electron spectrum excited by MgKα. The main peak is the Ne1s line. The MgKα satellites and Kβ give rise to the peaks with higher kinetic energy. The peaks with the lower energies (numbers 1-12) are due to shake-up, shake-off and inelastic scattering.

radiation give peaks with higher kinetic energy. The peaks with the lower energies (numbers 1-12) are due to shake-up, shake-off and inelastic scattering, the latter being discrete energy loss lines in the gas of the primary Ne1s electron line (lines 1-5). The electron lines 5-12 are independent of pressure and are due to ionization processes at which a valence electron is simultaneously excited (shake-up) or ejected (shake-off). Using the "sudden approximation" a theoretical study of the possible shake-up states indicates that only states of the types $1s2s^22p^5np\,^2S$ and $1s2s2p^6ns\,^2S$ have to be considered and that the excitations are due to monopole transitions ($\Delta J = \Delta L = \Delta S = \Delta M_J = \Delta M_L = \Delta M_S = 0$). A Hartree-Fock calculation shows that the lines 7-11 are due to shake-up processes of the first type and line 12 of the second type. The complete shake-off of a 2p electron, i.e. the ionization limit of the first term series, occurs close to line 11 at an excitation energy of 47 eV. The shake-up states have three unpaired electrons. Using a multiconfiguration SCF procedure the term splitting could be calculated for n = 3, 4 and 5. Thus, lines 7 and 8 could be identified as a lower and upper doublet state, respectively, in the term $Ne^+\ 1s2s^22p^53p$, lines 9 and 11 as a lower and upper state in $Ne^+\ 1s2s^22p^54p$ and line 10 as the lower state in $Ne^+\ 1s2s^22p^55p$.

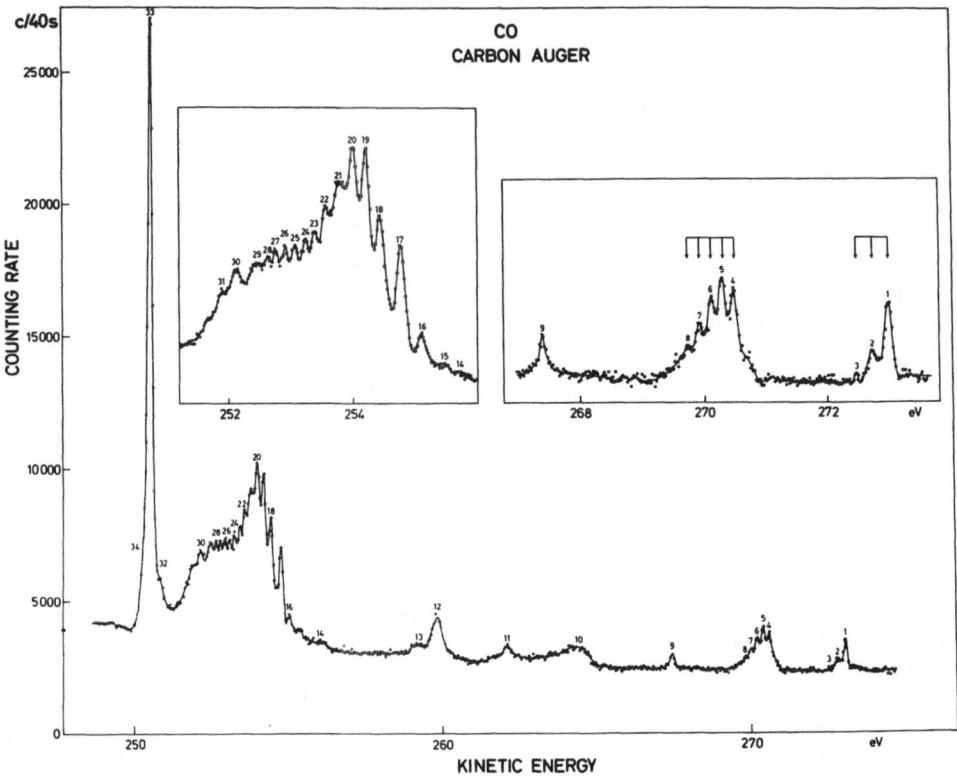

Fig. 11. Molecular carbon Auger and autoionization electron spectrum
from CO excited by electron impact. Inserted are vibrational autoionization lines and Auger vibrational lines.

       For molecules the shake-up transitions which occur may occasion-
ally be as high in intensity as 15-20 % of that of the main line.
They depend on the atom from which the core electron is removed. This
dependence may, in turn, yield information about the charge densities
of the molecular orbitals involved in the shake-up transitions. A
good example is the linear molecule carbon suboxide
$$O=C=C=C=O.$$
The electron spectrum shows that there is no C1s shake-up transition
associated with the central carbon, corresponding to the first O1s
shake-up line, but that there is such a transition associated with
the other carbons. This implies that the valence orbital involved
has a much smaller charge density at the central carbon than at the
other two carbons, i.e. that the molecular orbital has a node at the
center of the molecule and that the population of this orbital at
the central carbon is negligible.

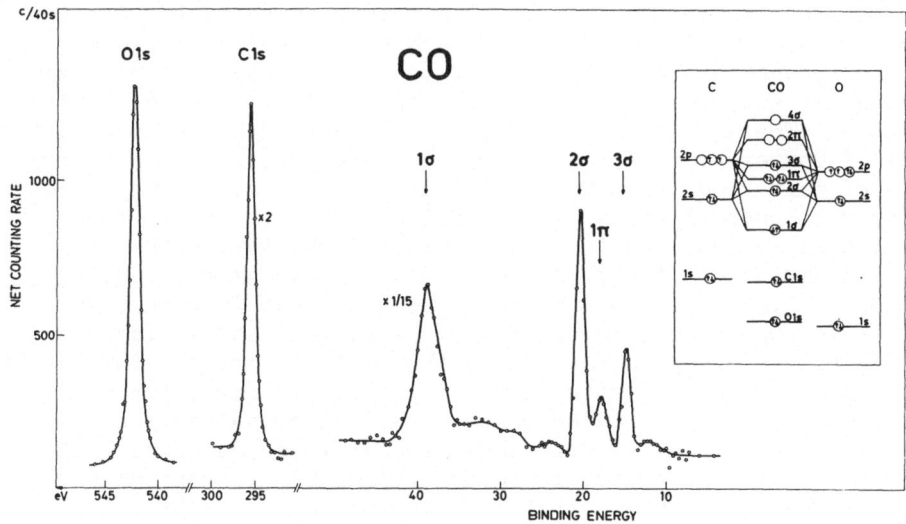

Fig. 12. Core and valence electron spectrum in CO excited by MgKα.

Fig. 11 is an example of a molecular Auger electron spectrum plus autoionization lines. It concerns CO and contains some features of more general interest. The left part of the figure is the C Auger electron spectrum containing the C1s level plus two valence orbitals involved. The width of the strong single electron line at the left is 0.18 eV. This would indicate that, apart from other possible causes of electron line broadenings at the photoexcitation process of CO and at sufficiently high resolution, an improvement in electron linewidth would be possible. One then has to reduce the linewidth of the exciting X-radiation. In the past the best attained resolution (mostly because of the width of the exciting X-radiation) has been ≈ 0.8 eV. At the end of this review the present efforts will be discussed to improve the resolution in ESCA by monochromatization of the AlKα line. The CO spectrum furthermore shows a vibrational structure in the Auger part (lines 15-30) (inserted in the figure). This is also the case for the right part of the spectrum (inserted in the figure) which covers the autoionization electron spectrum of carbon in CO with vibrational quanta around 0.2 eV. Such resolved structures are likely to occur only for simple molecules.

We will continue with the CO molecule. The electron configuration is $(1\sigma)^2(2\sigma)^2(1\pi)^4(3\sigma)^2$. The electron spectrum excited with MgKα shows both the core and valence orbitals, see Fig. 12. Excitation by X-rays emphasizes the s character. A CNDO calculation gives

85 % 2s character for the 1σ orbital and 46 % 2s character for the
2σ orbital, whereas the 3σ ought to have much less 2s character
according to the intensities in the valence electron spectrum. The
1π orbital is suppressed in this electron spectrum because of its
p-character. The 1σ line at 40 eV is much wider than the others, partl
because of Coster-Kronig broadening, a general feature in electron
spectra of this kind. When UV excitation is used one cannot produce
the deeper valence electron lines. Contrary to X-ray excitation
which strongly emphasizes the s-character of the levels the UV-
excitation emphasizes the p-character. In Ne, for example, using
MgKα, the 2s line is three times as intense as the 2p peak even
though there are three times as many 2p electrons as 2s electrons.
The photoionization cross section is thus 9 times greater for a Ne2s
than for a Ne2p electron using MgKα. It appears, from a large body
of experimental data, that the photoionization cross section for
valence electrons in a molecule shows a similar dependence. This
dependence is best described in terms of the LCAO approximation.
Thus for a given molecular orbital we can define a fractional atomic
parentage, $P_i(A\lambda)$, which gives the relative strength of the λ type
atomic orbital, centred on atom A, in the LCAO expansion of the ith
molecular orbital. Using CO as an example, $P_{1\sigma}(C2s)$ is the relative
carbon 2s atomic character of the 1σ molecular orbital. The photoionization cross section for the ith molecular orbital, σ_i, is then
related to the atomic parentage of that orbital by an expression of
the form

$$I_i = \sum_{A\lambda} P_i(A\lambda)\sigma_{A\lambda}^i$$

where $\sigma_{A\lambda}^i$ is the relative photoionization cross section for an $A\lambda$
electron in the ith orbital. If the photoionization cross sections
are known one can predict quantitatively peak intensities using the
atomic parentage obtained from LCAO-SCF calculations. It appears
that for the second-row elements C, N, O, and F, 2s character is
seen approximately ten times as strongly as 2p character. This intensity model can be further elaborated and is useful in discussing
the symmetry properties of measured molecular orbitals in electron
spectroscopy when X-ray excitation is used.

The synchrotron radiation will soon be taken into use for the
excitation of electron spectra. A wide span of photon energies which
furthermore can be continuously varied will form a bridge between
presently used X-ray and UV lines. The synchrotron radiation is
linearly polarized which may be utilized for particular experiments.
In this connection it should also be mentioned that the intensity
distribution of the emitted electrons which is a function of the
angle between the exciting photon and the emitted electron, can be
measured and be used to give information about the symmetry properties of the levels involved.

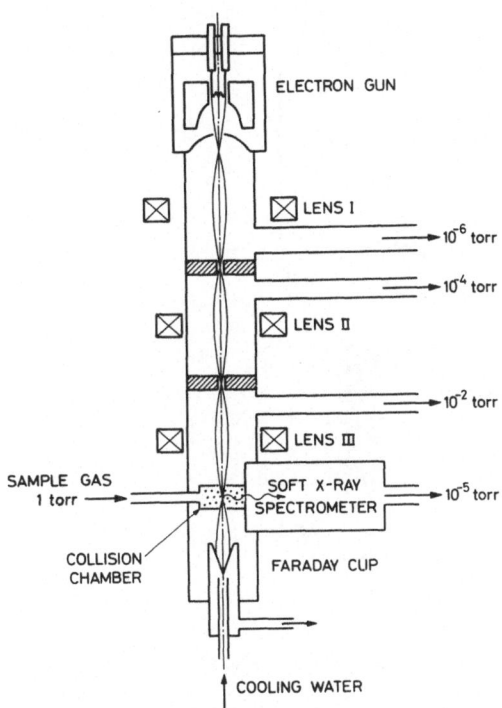

Fig. 13. Arrangement for excitation of X-rays from gases, using a high intensity electron gun, magnetic lenses and differential pumping. A grazing incidence grating spectrometer is used to achieve high resolution in this ultra soft X-ray region.

ULTRA SOFT X-RAY EMISSION FROM THE MOLECULES AS A COMPLEMENT TO ELECTRON SPECTROSCOPY

With CO as an example we will next discuss how the valence electron orbitals of a molecule can be studied by a different method which is presently being developed in our laboratory. If a C1s electron vacancy is created, for example by directing an electron beam into the gas, then, with a small probability compared to Auger effect, the excitation energy will be emitted in the form of X-radiation. Electrons from the various valence orbitals will make quantum jumps to fill the empty 1s level. If sufficiently high resolution - at sufficiently high intensity - can be achieved, fine structure in the CKα X-ray emission line ($\lambda \approx 45$ Å) from free molecules would appear. The width of each component is at least as broad as the 1s level itself, which can be regarded as the window used to inspect the various valence levels. The transition probabilities are governed by selection rules - and only the atomic p-character of a molecular orbital ought to contribute. If extremely high resolution could be

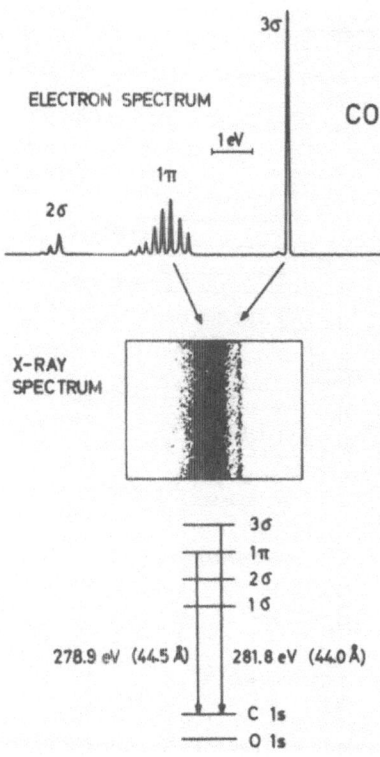

Fig. 14. The carbon X-ray emission spectrum from gaseous CO obtained with the grating spectrometer. For comparison is also shown the
electron spectrum excited by HeI. Compare also Fig. 12.

obtained it might even in principle be possible to observe vibrational structure inside the various line components in the fine
structure. Assuming a magnitude of the vibrational quanta of 0.2 eV
we can estimate the required resolving power ($E_{CK\alpha} \approx 285$ eV) to be
R = 285:0.2 = 1400. Little work has so far been done to try to
realize the adequate experimental conditions for this purpose.
Recently we have designed an apparatus which can still be improved
but which in its present form has been able to resolve completely
valence orbitals from each other in the case of CO. The resolving
power is in principle such that vibrational structure is within
reach. The experimental arrangement is shown in Fig. 13. A high intensity electron gun is used for the excitation. The focussed electron beam is directed by means of magnetic lenses through three
stages of differential pumping into the gas chamber which it passes
through two holes, which are windowless. The gas is continuously
renewed by the pumping. The narrow electron beam excites the gas and
parallel and close to it the X-ray spectrometer slit ($\approx 5\,\mu m$) is situated. A grazing incidence grating spectrometer is used in order to
achieve the high resolution at this wave length. Electrical detection

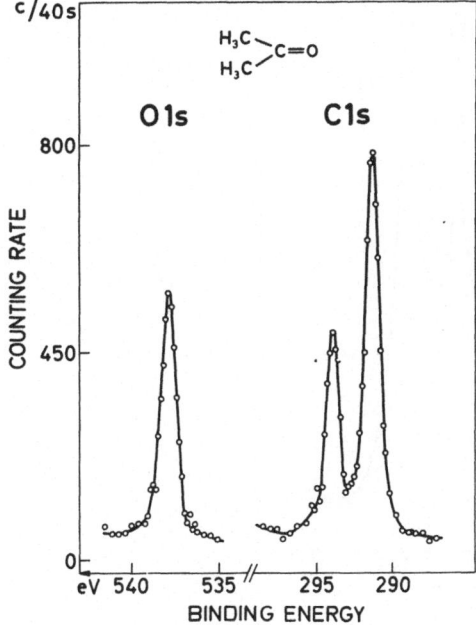

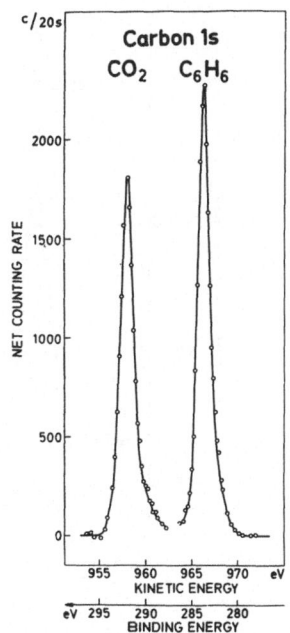

Fig. 15. O1s and C1s electron
spectrum of aceton excited by MgKα,
showing C1s chemical shift.

Fig. 16. Electron spectrum
of a gas mixture of benzene
and carbon dioxide excited by
MgKα, showing a large C1s
chemical shift.

can be used but for scanning purposes it is more convenient with
photographic recording. Fig. 14 shows the result. For comparison a
high resolution electron spectroscopic recording of the valence
orbitals is shown using HeI for excitation. As can be seen the 1π and
the 3σ valence orbitals are completely resolved in the X-ray emission
spectrum. The 1π orbital has considerable vibrational broadening. It
is also interesting to note the quite different intensity ratios in
the three cases when X-ray and UV excitation of electron spectra is
used and when X-ray emission is recorded, as given in Figs. 13 and
14. Using the same apparatus the $L_{II,III}$ X-ray emission spectrum of
argon ($\approx$ 56 Å) was found to contain resolved fine structure due to
shake-up processes accompanying the X-ray emission.

CHEMICAL SHIFTS OF CORE LEVELS

The chemical shifts of core levels mentioned in the introduction
can be used as a local atomistic probe to measure the molecular
charge or the potential distribution resulting from the valence
electron orbital density. Before discussing the various theoretical

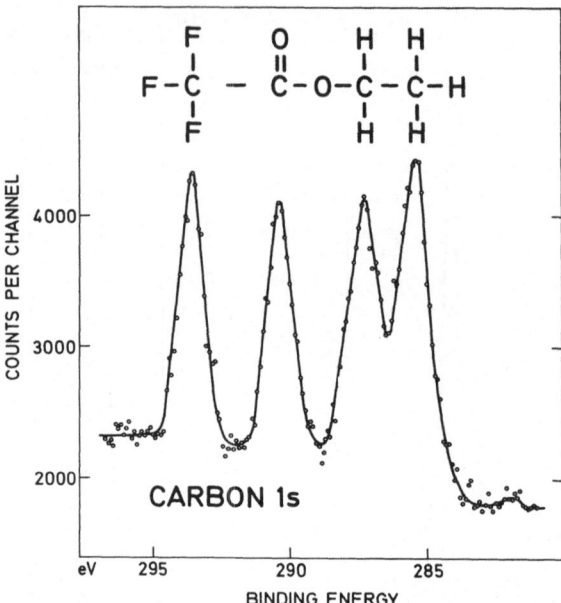

Fig. 17. C1s electron spectrum of ethyltrifluoroacetate excited by MgKα. Due to the different chemical shifts the four carbon atoms are distinguished as four separate lines in the same sequence as given in the structure formula.

treatments a few examples are given of chemical shifts. Fig. 15 shows the 1s electron spectra of oxygen and carbon in gaseous aceton excited by MgKα (one can also solidify the vapour by freezing). The C1s peak is double with a chemical shift of 2.6 eV and with the intensity ratio of 2:1. This indicates that the methylcarbon peak is the one to the right and the keto carbon is the one to the left. The latter carbon atom is chemically bound to oxygen which is strongly electronegative. This carbon will consequently lose part of its valence charge. The inner 1s core electrons will feel a stronger nuclear attraction showing itself in a higher binding energy. The methyl carbon will not be similarly affected since it is bound to 3 hydrogens and one carbon.

Fig. 16 shows a C1s electron spectrum of a gas mixture of carbon dioxide and benzene (a convenient pressure range is ≈ 0.1 Torr). In the former molecule the carbon is bound to two oxygens whereas there is no strongly electronegative atoms in benzene which could attract charge. The chemical shift is therefore in this case quite high, 7.3 eV.

The molecule in Fig. 17 is so designed that the chemical shifts of the four carbon atoms affect the C1s spectrum in such a way that

TABLE I. Binding energies of some levels in metals and graphite, which are suitable for calibration of electron spectra from solid samples.

Level	Binding energy* (eV)
Cu $2p_{3/2}$	932.8(2)
Ag $3p_{3/2}$	573.0(3)
Ag $3d_{5/2}$	368.2(2)
Pd $3d_{5/2}$	335.2(2)
C 1s (Graphite)	284.3(3)
Cu 3s	122.9(2)
Au $4f_{7/2}$	83.8(2)
Pt $4f_{7/2}$	71.0(2)
Pd 4d (Leading edge)	0.0(1)

*Relative to the Fermi level.

TABLE II. Binding energies of levels in gases suitable for calibration of electron spectra from gaseous samples.

Level	Binding energy (eV)
Ne 1s	870.37(9)
F 1s (CF_4)	695.52(14)
O 1s (CO_2)	541.28(12)
N 1s (N_2)	409.93(10)
C 1s (CO_2)	297.69(14)
Ar $2p_{3/2}$	248.62(8)
Kr $3p_{3/2}$	214.55(15)
Kr $3d_{5/2}$	93.80(10)
Ne 2s	48.47
Ne 2p	21.59*
Ar 3p	15.81*

*Weighted mean value.

the four C lines follow in the same sequence as in the structure formula, inserted in the figure. The first carbon is surrounded by 3 fluorines which are known to be very electronegative atoms. This peak is therefore shifted very much towards higher binding energies. Next comes the carbon which is doubly bound to one oxygen and has a single bond to another oxygen. Then one carbon with a single bond to oxygen follows and finally a methyl carbon. The total energy span of the C1s electron spectrum for ethyltrifluoroacetate (in solid form) is around 8 eV.

The chemical shifts are usually much smaller than those chosen here for illustration, frequently of the same order as or even less than the present widths of the core electron lines, $\approx$ 1 eV. A factor of two in improvement of electron line widths is therefore quite important from this point of view, a development which is now soon to be realized.

A large body of experimental material concerning chemical shifts for various elements in different chemical surroundings is now available. The calibration procedures in various cases are subject to particular attention, in particular for solid insulating materials. New methods and more experience of those in use are needed in order

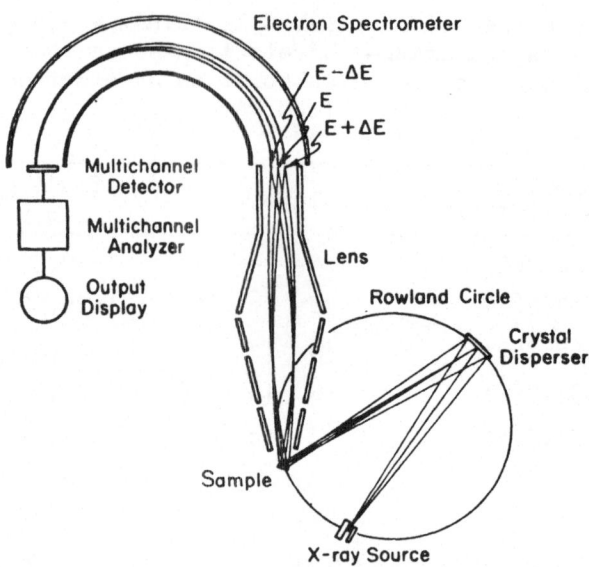

Fig. 18. Principle of an ESCA instrument (HP5950A) using "dispersion compensation" to achieve X-ray monochromatization.

to reach higher accuracy. For gases or vapours the conditions are simpler and higher precision is easier to achieve. Chemical shifts can then presently be given with an accuracy of 0.1 eV. Tables I and II give the binding energies of some levels in metals and graphite (I) and in gases (II) which are suitable for calibration in ESCA.

NEW ESCA INSTRUMENTS WITH MONOCHROMATIZATION

There are several different designs of ESCA instruments, some of which are now available on the market. Fig. 18 shows schematically the principle of "dispersion compensation" which has been utilized in one of the latest ESCA instruments (HP5950A). This is primarily designed for solid samples and is therefore provided with ultrahigh-vacuum and a special surface cleaning arrangement (sputtering) etc. By means of a spherically bent quartz crystal the AlKα line is dispersed along the Rowland circle on the surface of the sample. Then, to each point on the source in the plane of the figure corresponds a certain wavelength and consequently the energy of the expelled photoelectron will vary smoothly along the width of the sample. An electrostatic lens system focusses these electrons at the entrance slit of the electron analyzer, consisting of two concentric hemispheres. The electrons which form the first electron optical picture

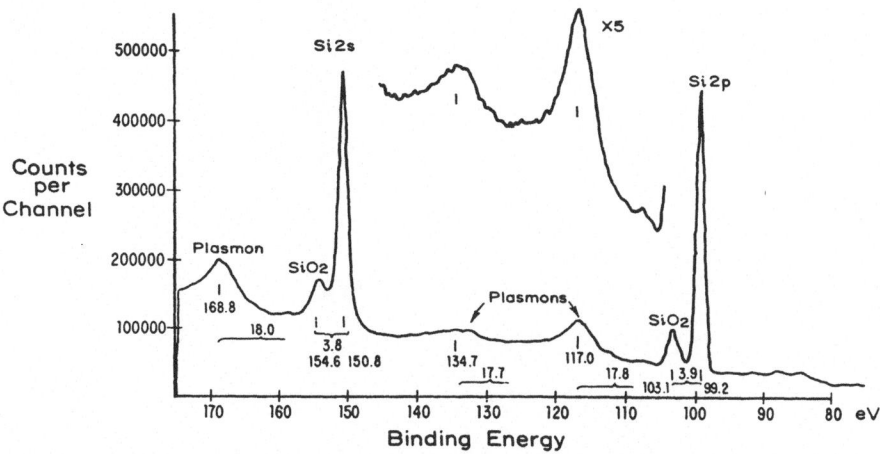

Fig. 19. Silicon, briefly exposed to air studied with the ESCA instrument shown in Fig. 18. One observes Si lines from the 2p and 2s levels and also lines due to a few Å thick surface layer of SiO_2. Some plasmons can furthermore be noticed. Excitation by means of monochromatized $AlK\alpha$.

of the sample at this entrance slit have consequently slightly different energies. One can make the magnitude of this energy dispersion exactly equal to the dispersion of the analyzer and with opposite sign. The final image at the detector is then, disregarding aberration errors, independent of the primary energy spread of the X-ray line and also of the width of the sample. The detector is situated in the focal plane of the spectrometer and consists of a multichannel plate detector with a high amplification factor ($\approx 10^8$). The amplified electron pulses are impinging on a fluorescent screen which in turn is continuously scanned by a TV camera. The pulses from an extended energy interval are in this way registered simultaneously and sorted in a multichannel analyzer. This ESCA instrument has presently a best resolution for solid samples of ≈ 0.5 eV and a good speed of information at the recording of electron spectra. Fig. 19 shows a spectrum taken with this instrument. The sample is a single crystal of Si. One observes the Si2p and 2s electron lines and also plasmon lines. There are also two small intensity lines at the low energy sides of each main line which are characteristic of SiO_2 and correspond to a few Å thick layer after exposure to air. According to Fig. 20 the Si2p line furthermore, because of the monochromatizing system, can be resolved in its two components $2p_{3/2}$ and $2p_{1/2}$ in the intensity ratio of 2:1. The resolution is 0.5 eV according to the figure.

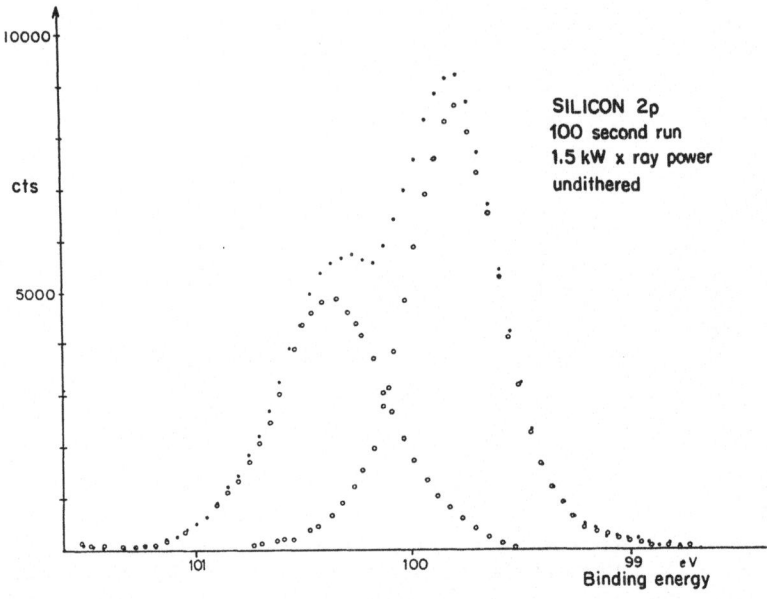

Fig. 20. The Si2p$_{1/2,3/2}$ doublet resolved by means of the ESCA instrument shown in Fig. 18. The observed linewidth after dispersion compensation is according to this study 0.5 eV, including the inherent level linewidth.

    Recently, an ESCA instrument for high resolution studies of gaseous samples has been designed in our laboratory, which has a different monochromatizing system, suitable to get the required intensity to study gases at low pressure (pressure range $\approx 1\text{-}10^{-5}$ Torr). See Fig. 21. A high power electron gun (8 kW) produces a small focus ($\emptyset \approx 2$ mm) at the periphery of a watercooled, swiftly rotating anode (≈ 5000 rpm). The spherically bent quartz crystal sees this focal spot at an angle of 5° and diffracts the AlKα radiation in a vertical direction. Only part of the Kα line will be "illuminated" on the Rowland circle and the photon flux is concentrated within one fifth of the line, corresponding to ≈ 0.2 eV. The X-ray beam is focussed on a horizontal slit on the top of the source chamber and allowed to pass through the gas along the entrance slit of the spectrometer. There is a two-stage, differential pumping so that a gas pressure of 1 Torr can be maintained in the sample chamber at a pressure of $10^{-6}\text{-}10^{-5}$ Torr in the spectrometer. In order to improve the resolution the electrons are retarded before entering the spectrometer. This consists of two concentric spherical condensor plates with a mean radius of 36 cm and with a sector angle of 157°. The detector system is the same as in the previously described ESCA instrument. Also solid or liquid samples can be studied in the gas phase through evaporation by heating the samples and the sample house to a controlled temperature of maximum 200 C° in this design.

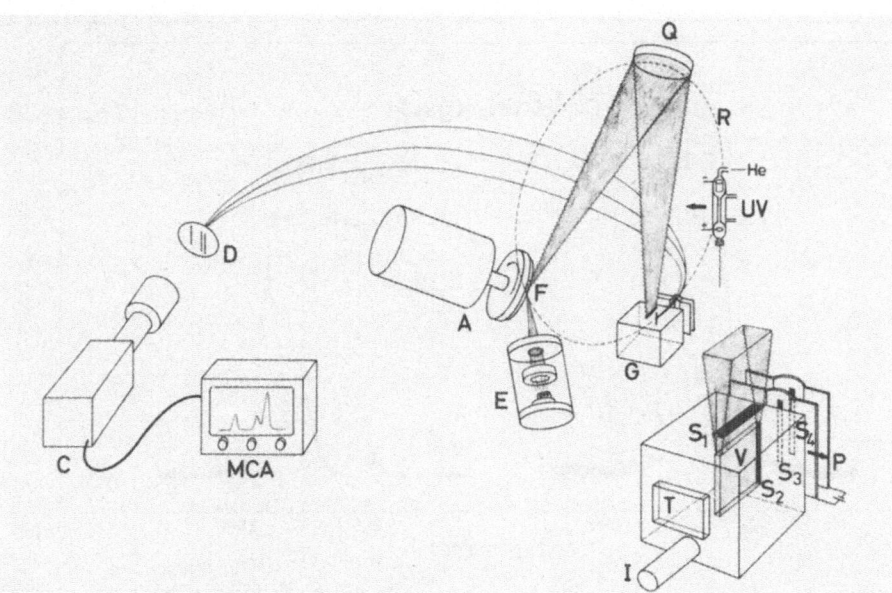

Fig. 21. The new ESCA instrument for studying gases. E = electron gun, A = rotating anode, F = focal spot, Q = spherically bent quartz crystal, R = Rowland circle, UV = helium lamp, G = gas compartment, S_1-S_4 = slits, V = effective irradiated gas volume, T = temperature raising device, I = gas inlet system, P = two stage differential pumping system with an electron retardation step, D = multichannel plate detector, C = television camera, and MCA = multichannel analyzer.

Fig. 22 shows the first spectrum which was recently recorded with this instrument, obtained by directly photographing the analyzer fluorescent screen when the spectrum had grown sufficiently. It shows the core spectrum of O1s and C1s from ethylpropionate. The resolution is markedly improved and there are strong indications that it can be further improved when the electron retardation system is put into operation. Figs. 23-26 show some recent recordings with this new instrument.

THEORETICAL TREATMENT OF SOME OF THE INFORMATION OBTAINED IN ELECTRON SPECTROSCOPY

There are two main types of results concerning molecules which are obtained in electron spectroscopy and which are particularly interesting to compare with theoretical calculations, namely binding energies of molecular orbitals and chemical shifts of core levels. In addition to this, intensity relations in valence spectra are of interest as discussed above in connection with the "intensity model"

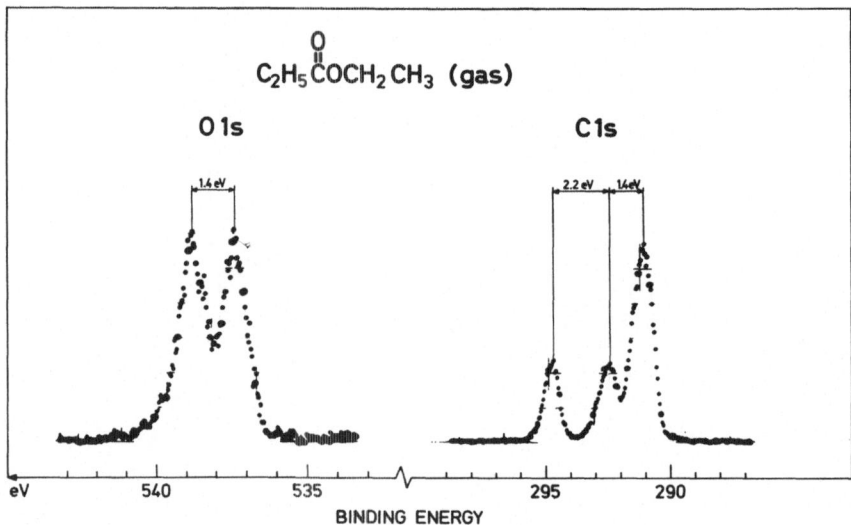

Fig. 22. First electron spectrum recorded with the new ESCA instrument for gaseous samples using monochromatized AlKα (0.2 eV) radiation for excitation. The spectrum shows the core levels of ethyl-propionate as photographed directly from the fluorescence screen. No electron retardation is used in this case. The kinetic energies are ≈ 950 eV and ≈ 1200 eV for the oxygen and carbon electron groups, respectively.

and also other features such as Auger electron transition rates, spin exchange splittings etc. Basically one would require a very precise knowledge of the actual molecular wave function to meet with the precision of present day electron spectroscopy. Remarkable progress in this respect is taking place and it does not seem unreasonable to forecast a most interesting development to occur shortly within the field of computational quantum chemistry associated with big computors. Programs with extended basis sets in H-F-calculations, also including configuration interaction to account for electron correlation effects are presently in the test stage and promise to greatly reduce the previously uncomfortably long computer times. One simplification which has been frequently used so far is to identify the measured ionization energies (the binding energies) of the various orbitals with the energy eigenvalues of the H-F-operator. This is less satisfactory when more precise comparisons between theory and experiment are wanted, since one is then neglecting the orbital relaxation which occurs at the electron emission. This approximation is known under the name of Koopmans' theorem. A more accurate comparison can be made - unfortunately at the expense of more computer time - by computing separately the total energy of the molecule and the molecular ion.

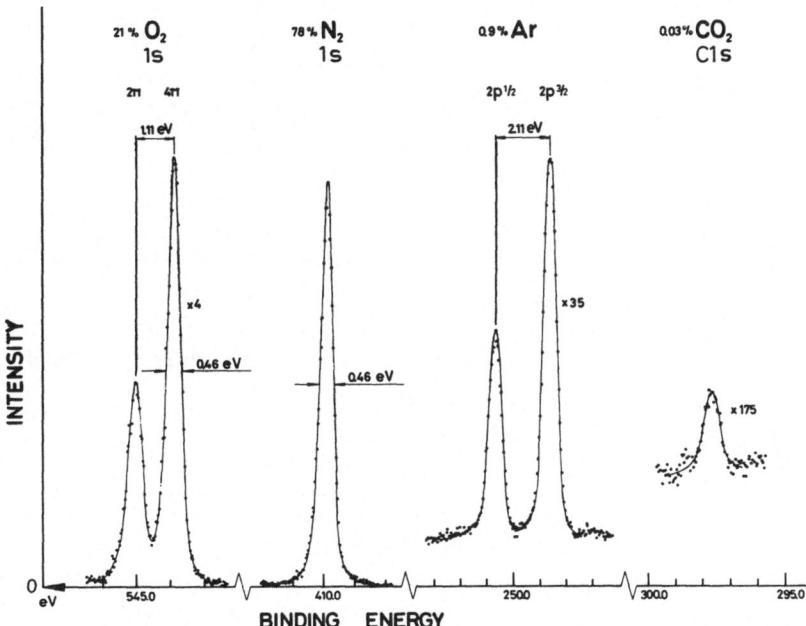

Fig. 23. Electron spectrum of air, as analyzed by the new ESCA-instrument with monochromatized AlKα radiation according to Fig. 21. The 1s line of O_2 is split in the intensity ratio of 2:1. This "spin splitting" is due to the exchange interaction between the remaining 1s electron and the two unpaired electrons in the π_g–2p orbital, responsible for the paramagnetism of this gas. The resulting spin can be either 1/2 or 3/2. The corresponding electrostatic exchange energies can be calculated and correspond well with the measured splitting of 1.11 eV. Apart from oxygen and nitrogen also argon and CO_2 are recorded in spite of the low abundances of the latter gases in air.

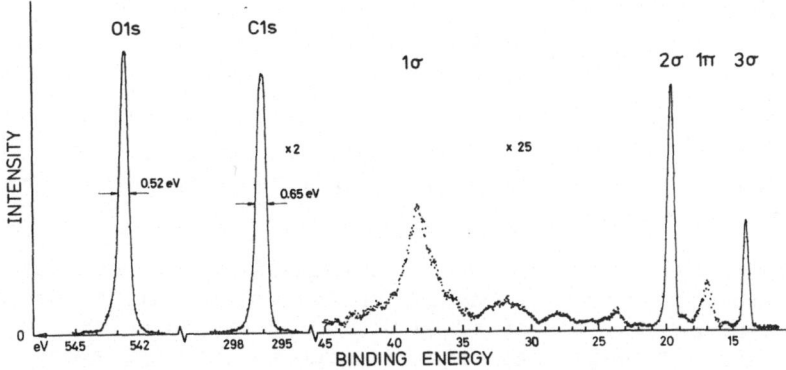

Fig. 24. Valence and core electron spectra of CO recorded with the new instrument (Fig. 21). Compare with Fig. 12.

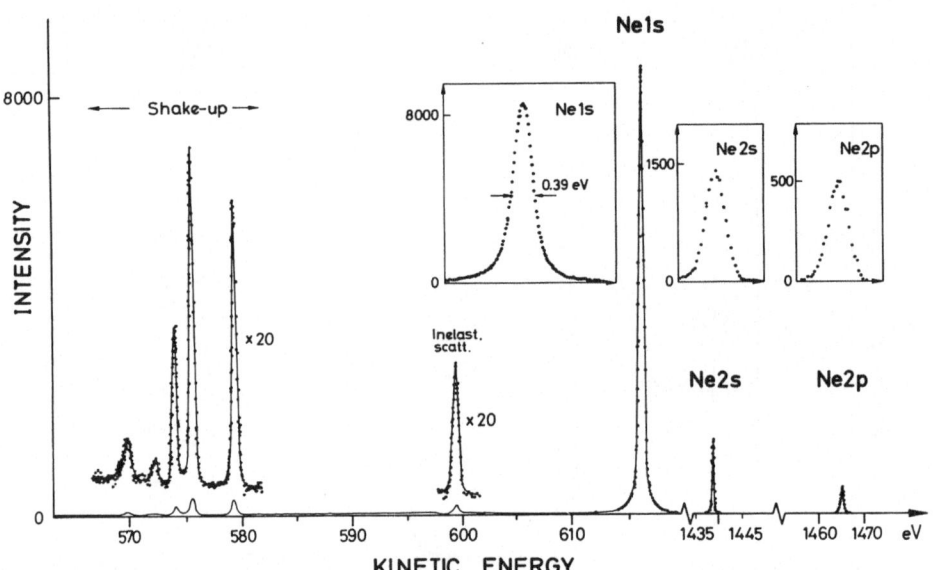

Fig. 25. The Ne spectrum with shake-up lines recorded with the new
instrument (Fig. 21). Compare resolution and background with Fig. 10.
The detailed study of the 1s line shows a linewidth of 0.39 eV which
is the lowest so far recorded. The Lorentzian form indicates that
this halfwidth corresponds rather nearly to the inherent 1s level
width. The widths of the 2s and 2p are instrumental. No electron retardation was used.

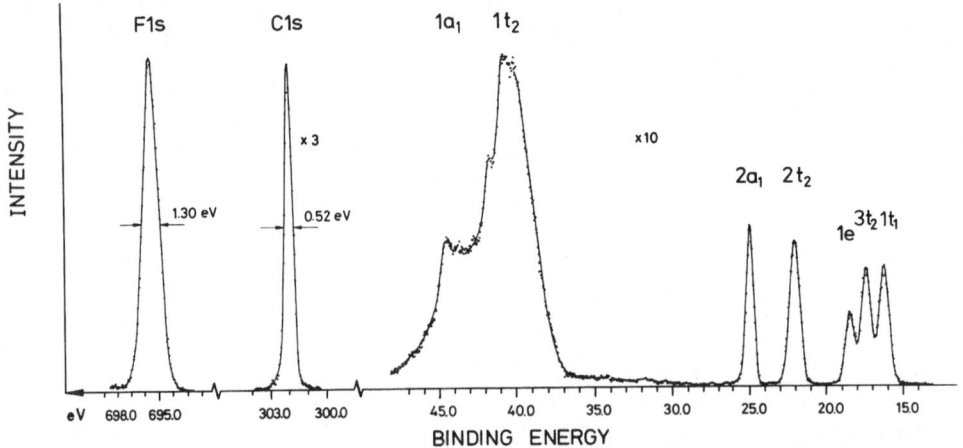

Fig. 26. Valence and core electron spectrum of CF_4 as recorded with
the new instrument (Fig. 21). One can notice a great difference in
widths between the 1s lines of F and C. The 32 valence electrons are
distributed among 16 molecular orbitals. Due to the high symmetry
several of these orbitals are degenerate resulting in 7 different
valence orbitals which are all resolved and identified in the recorded electron spectrum.

The relaxation or reorganization energies of core levels are generally much larger than the chemical shifts. Nevertheless, the experience so far obtained shows that the chemical shifts for an atom usually correlate quite well with the molecular species in their ground states. Thus, the corresponding orbital relaxation energies are fortunately very similar from molecule to molecule. This can be exemplified by plotting the measured chemical shifts for an atom in various molecules as a function of a calculated quantity assuming the molecules to be in their ground states. Instead of ab initio calculations the quantity may more readily be obtained from semiempirical quantum chemical calculations like CNDO or similar schemes. One can even use quite simple electronegativity concepts, for example due to Pauling, when the partial ionic character I of a bond between A and B is obtained from the electronegativities χ_A and χ_B through

$$I = 1 - \exp[-0.25(\chi_A - \chi_B)^2] \; .$$

The "charge" of an atom in a molecule can then be approximately estimated by taking the sum of the partial ionic character of the bonds to the next neighbours with their appropriate signs. Using this approach the observed chemical shifts have been correlated with the calculated charge for elements like nitrogen, carbon, sulphur etc. in large series of compounds with widely varying structures.

A more realistic approach is to use the potential model, briefly mentioned in the introduction. The change of the local potential determining the chemical shift can then be considered as obtained from a superposition of two potentials. The first, which generally is the dominating, originates from the change in electronic distribution around the particular nucleus and is given by q/r, where r is the atomic radius. The second potential, which we may call the molecular potential, is set up by the charge distribution from the rest of the molecule. These transferred charges are considered to be distributed in the molecule on the sites of the various nuclei. These charges can be calculated by ab initio methods within the H-F-scheme or by approximative methods, for example by CNDO. They give rise to a potential $V = \Sigma q_i/r_i$. The shift in the considered core

$$\Delta E = q/r + \Sigma^q i/r_i + l \quad \text{or}$$

$$\Delta E = kq + V + l,$$

where q_i is the charge on atom i,
      r_i is the distance to the nucleus i,
      l is a constant determined by the choice of reference level,
      k is a constant characteristic of the element, which is approximately equal to the electrostatic interaction between a core and a valence electron in a free atom.

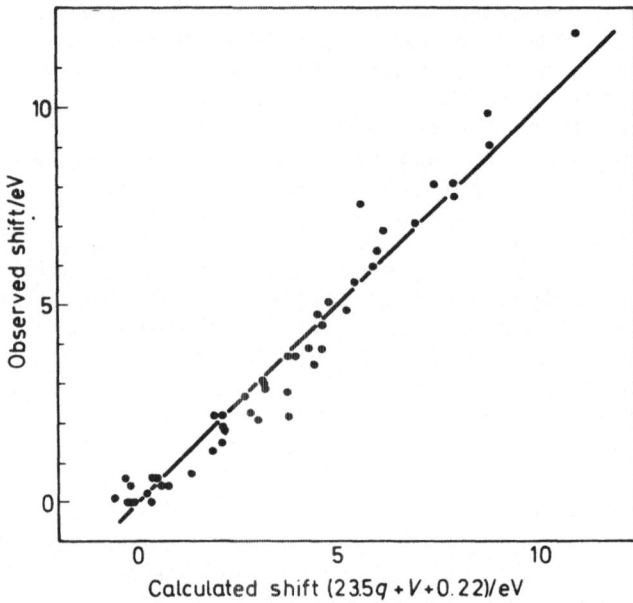

Fig. 27. Correlation between measured chemical shifts of C1s for various organic compounds and the shifts calculated with the potential model using charges obtained from CNDO/2 calculations.

Fig. 27 shows the correlation obtained for the C1s chemical shifts of a large series of compounds using the potential model. k was treated as a parameter and was determined by a least squares fit.

As a final example of what is of current interest on the theoretical side of this field we briefly mention the so called SCF-Xα scattered-wave method (Slater-Johnson-Connolly). One of the features of this method is the introduction of so called "transition" states, representing a one-electron state "half way" between the initial and the final states. This state takes into consideration the orbital relaxation and therefore one can claim that its orbital energy can be identified with the ionization energy. In this scheme one uses a so called "muffin-tin" potential as the one-electron potential. Each atom is surrounded by a sphere, each sphere in the molecule touching its neighbour spheres. Inside each sphere the exact potential is replaced by a spherical average. The potential in the region between the spheres is regarded as constant, namely the average of the exact potential over this interatomic region. The molecule is then surrounded by an outer sphere and the exact potential outside this sphere is correspondingly replaced by a spherical average. This model is interesting and deserves a closer study. It seems possible to develop it further and modify it in various ways. The agreement with ESCA

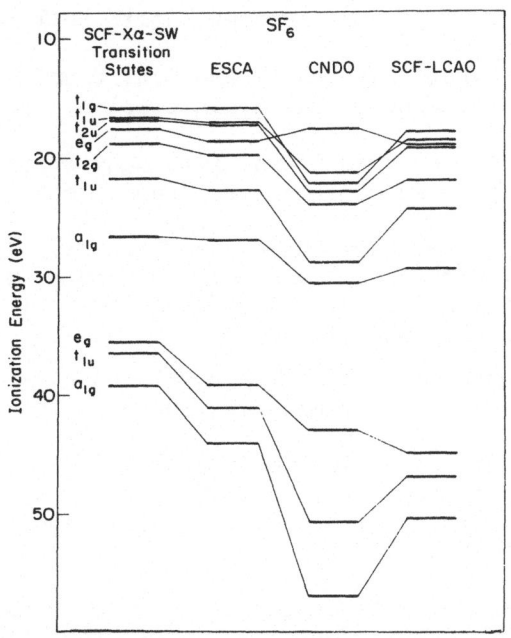

Fig. 28. The ionization energies of the orbitals in SF_6 as calculated by various methods compared to the experimental results of ESCA. The SCF-Xα scattered-wave method (left in the figure) is evidently successful in this case. The various schemes given in the figure are all subject to further developments.

concerning the ionization energies of some recently calculated molecules are impressive as can be seen from Fig. 28 concerning SF_6, where some other schemes are included for comparison. The computer time is also greatly reduced. There still remain some difficulties in the treatment of the chemical shifts in this scheme.

REFERENCES

The new field of electron spectroscopy is presently in a stage of rapid development. The following books on the subject have been printed:

1. K. Siegbahn, C. Nordling, A. Fahlman, R. Nordberg, K. Hamrin, J. Hedman, G. Johansson, T. Bergmark, S.-E. Karlsson, I. Lindgren and B. Lindberg, ESCA: Atomic, Molecular and Solid State Structure Studied by Means of Electron Spectroscopy, Nova Acta Regiae Societatis Scientiarum Upsaliensis, Ser. IV, Vol. 20, 1967.
2. K. Siegbahn, C. Nordling, G. Johansson, J. Hedman, P.F. Hedén, K. Hamrin, U. Gelius, T. Bergmark, L.O. Werme, R. Manne and Y.

Baer, ESCA Applied to Free Molecules, North-Holland Publ. Co., 1969.
3. D.W. Turner, C. Baker, A.D. Baker and C.R. Brundle, Molecular Photo-Electron Spectroscopy, Wiley-Interscience Publ., 1970.

In addition to these books there exist many shorter survey articles of the field, printed in current periodicals.

During the last four years there have been five international conferences devoted to this field, namely in London 1969, Uppsala 1970, Oxford 1970, Asilomar 1971 and Brighton 1972. The following conference reports have been printed:

1. The London Conference in Phil. Trans. of the Royal Society of London (Editors: W.C. Price and D.W. Turner), Vol. 268, Nr. 1184, Nov. 1970.
2. The Asilomar Conference: Electron Spectroscopy (Editor: D.A. Shirley), North-Holland Publ. Co., 1972.
3. The Brighton Conference will be printed in The Chemical Society Faraday Division 1972 (Editor: J.N. Murrell).

An international journal devoted to the field has been started with its first issue appearing Oct. 1972 under the title: Journal of Electron Spectroscopy and Related Phenomena (Editors: C.R. Brundle and T.A. Carlson).

POLARIZED ELECTRON SOURCES

J. Kessler

Physikalisches Institut der Universität Münster

Germany

Abstract

A survey of working methods for producing
polarized electron beams is presented. The utility of
these sources is compared.

Introduction

In the past 10 or 15 years many processes have
been studied or even discovered where free electrons are
produced whose spin directions do not have a statistical
distribution, but a preferential orientation. That's
what we call "polarized electrons". These sometimes
rather complicated processes where polarized electrons
are produced have been frequently studied just because
they are interesting in themselves.

Then it turned out, however, that colleagues from
very different fields of physics were not so much
interested in the basic physical processes happening
there, but they asked: "Would it be possible to use
one or the other of these processes to build an efficient source of polarized electrons for other experiments ?" That's the question I want to answer in this
paper. I also want to give you some figures of merit of
a source of spin polarized electrons so that we can
compare the utility of different sources.

Roughly speaking, we can outline the requirements
as follows: One wants a well collimated beam of high
intensity and high polarization. In certain cases one
has further special wishes; one may want, for example,
a low energy spread of the electrons. The most interesting quality of the beam is its degree of polarization P
which is defined by

$$P = \frac{N_\uparrow - N_\downarrow}{N_\uparrow + N_\downarrow} \, ,$$

where $N_\uparrow$ and $N_\downarrow$ are the numbers of electrons with spin
parallel or antiparallel to the preferential direction.
P can have any value between -1 and +1. P = 1 means,
for example, that all the spins are parallel to the
preferential direction.

The reasons which cause the polarization of the
free electrons can be divided into 2 big groups:

1) We can get polarized electrons if the interaction in
   the process by which the free electrons are produced
   is spin-dependent. An example for this group is
   electron scattering by unpolarized targets where spin-
   orbit interaction causes an orientation of the spins
   of the scattered electrons.

2) We can get polarized electron beams, if at least one
   of the systems which are involved in the production
   process of the electrons is pre-polarized. An
   example for this: We take a ferromagnet and try to
   extract its polarized electrons. Needless to say,
   it can happen that both reasons mentioned work
   together. An example for this is the Fano-effect
   which I will explain later. There the interaction
   is spin-dependent and one of the systems involved
   (the photon-system) is polarized.

Of the many possibilities proposed so far[1] I can
pick out here only those which have already been working
and given good results. I won't discuss all those cases
where somebody suggests that somebody else should try
this or that idea.

Let me now discuss the most important sources and
compare their utility.

Polarization by Scattering from Unpolarized Targets

If an unpolarized electron beam impinges on an atom of fairly high atomic number, then the electrons scattered by rather large angles θ are in general polarized. The direction of the polarization is perpendicular to the scattering plane (Fig. 1).

The degree of polarization is a rather complicated function of electron energy E, scattering angle θ and the atomic number of the scattering atom. Fig. 2 shows

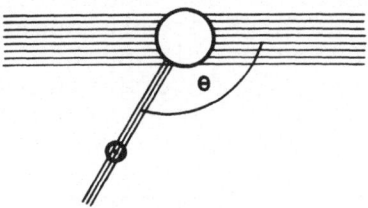

Fig. 1 Polarization by scattering. ⊙ Spin arrow showing out of the scattering plane

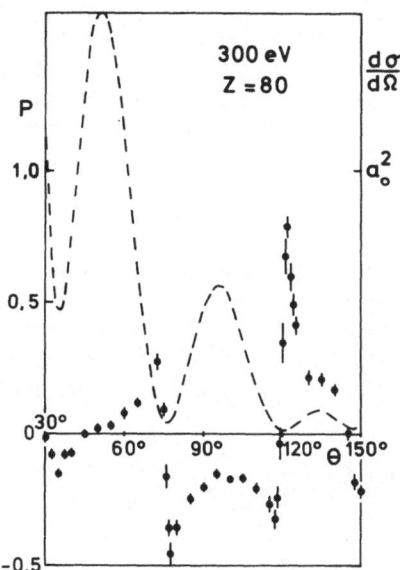

Fig. 2 Degree of polarization[2] ⧫ and differential cross section[3] – – – vs. scattering angle for 300-eV electrons scattered elastically by Hg atoms

an example. You see that near θ = 120° we observed very
high polarization. The satisfaction at this is spoiled,
however, if one looks at the corresponding cross
sections. Wherever we have polarization peaks we have
cross section minima (that means low intensities) and
vice versa. I chose here a combination of parameters
which gives rather favorable values. Let's have a quick
look at other values of the parameters. If the electron
energy is much higher, say in the 100 keV region, we
also get polarization by scattering as has been known
for a long time. However, the polarization is not very
high: At most 50%, but generally smaller. We can also
change the atomic number of the scattering atom. We are
restricted, however, to high atomic numbers, since P
decreases with decreasing atomic number. This is due
to the fact that spin-orbit interaction which is
responsible for the polarization, decreases with the
atomic number.

The fact that spin-orbit interaction causes a
polarization of the scattered electrons can be easily
made plausible: We can regard the incident unpolarized
beam as a mixture of equal numbers of spin-up and spin-
down electrons (Fig. 3). Spin dependent interaction
means that the effective scattering potential depends
on the spin direction. The two kinds of electrons are
therefore scattered with different probabilities into
a certain direction. That means: We observe different
numbers of spin-up and spin-down electrons in this
direction. In other words: The scattered beam is
polarized.

Now let's assume we want to build a source of
polarized electrons on the basis of the data of Fig. 2.
Would it be more favorable to choose a high polarization
P which means inevitably a small scattering intensity,
or would it be better to work at an angle where the
intensity is high but the polarization rather small ?

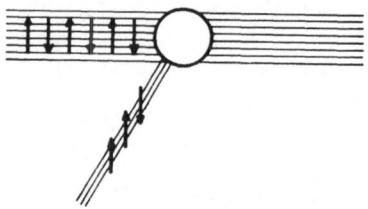

Fig. 3   Schematic diagram of spin-dependent scattering

Figures of Merit for Sources of Polarized Electrons

To answer this question we need some criteria for the utility of a source of polarized electrons. What would you say is a better bargain: A completely polarized beam of a certain intensity I or a beam, having 10 times as much intensity but only 1/10 of the polarization ? I think you have a feeling that the first beam is more valuable, so that it is not simply the product polarization x intensity which determines the utility.

This feeling is correct, indeed, as a simple mathematical calculation shows. For nearly all the experiments one wants to do with polarized electrons can be reduced to the measurement of the spin-dependency of some process. If we have a completely polarized beam (P=1) of intensity I_c, we can make the experiment in a certain time with a certain statistical error. If we have a beam, which is only partially polarized (P < 1), then the question comes up: Which intensity I would we need in order to make with this beam the same experiment in the same time. The simple calculation shows that we would need the intensity $I = \dfrac{I_c}{P^2}$.

If we have only 1/10 of the polarization we need an intensity which is 100 times larger in order to be competitive. It is therefore the product $P^2 I$ which is a reasonable figure of merit for most experiments. One should not use, however, the expression $P^2 I$ uncritically to describe the quality of a source: If the polarization is too small, so that the spin-sensitive information one wants to obtain is masked by the spin-independent phenomena, then it is of no use any more.

Another quality one would like to have is that the source yields a well collimated beam of high current density. A suitable figure of merit for this quality is the brightness. The brightness is the current density emitted into the unit solid angle. The more parallel our beam (that means the smaller its angular divergence) and the higher its current density, the higher is its brightness. One must bear in mind, however, that due to Lagrange's law of electron optics, the angular spread of an electron beam decreases when the electrons are accelerated. If we, therefore, compare the brightness

of various sources with different output energies, we
have to do this at the same energy. Of two sources with
different output energies but equal brightness the
source with the lower energy is preferable, since its
brightness will improve if we bring the electrons to
the higher energy.

Needless to say, there are other figures of merit
for a source of polarized electrons. For example it is
for many experiments important to have a low energy
spread of the electrons. But for the comparison of the
various sources, I will restrict myself to the criteria
mentioned and not apply those criteria which are relevant
for certain experiments and irrelevant for others.

Let's go back now to the method of scattering. What
can we achieve when we use it as a polarized electron
source ? This has been done twice so far: By Wilmers[4]
et al. in Mainz and by Zeman[5] et al. in Stanford. In
both cases an electron beam of several milliamperes has
been scattered by a mercury atomic beam. Table I shows
the results obtained with the various methods. The data
given for the method of scattering are those of Zeman
et al. Wilmers et al. obtained very similar results.
For the brightness I can only give a qualitative rating,
since most of the papers don't give enough data for a
quantitative determination.

Needless to say, sources of this kind can be
varied in many respects. Instead of the mercury beam,
for example, one can take a crystal surface of gold or
tungsten as target and use the polarized electrons one
obtains at large diffraction angles. This has been
proposed, but not yet attempted.

I'm now going to discuss the other methods for
producing polarized electrons. But let me first say
one other word. If you ever hear of a new polarized
electron source, please ask for <u>all</u> the figures of
merit given in Table I. Just one of these criteria is
not enough for judging whether the source is any good
or not. Or to say it more pointed: A single electron
which you may produce is always 100% polarized.

Method	Group	P	Best values I Ampere	P²I Ampere	Brightness
Scattering from Unpolarized Targets	Mainz[4] Karlsruhe[2,3] Stanford[5]	0.20	$3.5 \cdot 10^{-8}$	$1.4 \cdot 10^{-9}$	high
Photoionization of Polarized Atoms	Yale[6,7] Bonn[8] Orsay[9]	0.78	$3 \cdot 10^{-10}$ (average)	$1.8 \cdot 10^{-10}$	medium
Exchange Scattering from Polarized Atoms	Edinburgh[10]	0.5	$8 \cdot 10^{-14}$ (average)	$2 \cdot 10^{-14}$	low
Optically Pumped He Discharge	Rice[11]	0.2	10^{-7}	$4 \cdot 10^{-9}$	high
Collisional Ionisation of Quenched Metastables	Yale[12]	0.33	10^{-11}	10^{-12}	low
Field Emission from Magnetic Materials	Munich[13-15]	0.79	10^{-6}	$6 \cdot 10^{-7}$	high
Photoemission from Magnetic Materials	Zurich[16]	0.54	10^{-9}	$3 \cdot 10^{-10}$	medium
Fano Effect	Karlsruhe-Münster[18,19] Yale[7]	0.65	$1.6 \cdot 10^{-9}$	$4 \cdot 10^{-10}$	medium

Tab. I   Comparison of Various Sources of Polarized Electrons

Photoionization of Polarized Alkali Atoms

I'll now discuss the methods where at least one of
the systems involved is pre-polarized. Let us consider
an alkali atom beam which passes through a Stern-
Gerlach-magnet. This beam will split into two beams
which differ by the spin direction of their valence
electrons. So we get two atomic beams polarized in
opposite directions. If we extract the electrons of one
of these beams, for example by photoionization, we
obtain polarized electrons. This sounds nearly as
simple as the scattering experiment discussed before.
But it is much more complicated in the experimental
details, although Fig. 4 which is only a schematic
diagram does not really show this.

First an atomic beam is produced. It passes
through a six-pole magnet which selects a beam with a
preferential orientation of the spins of the valence
electrons. One then reflects ultra-violet light into
this polarized atomic beam, so that the atoms are
photoionized. So one obtains polarized photoelectrons
which need only to be extracted. It is important to
have the photoionization take place in a magnetic field
which is strong enough to decouple electronic and
nuclear spins of the atoms. Otherwise the electron
polarization would be considerably reduced due to the
interaction with the nuclear spins.

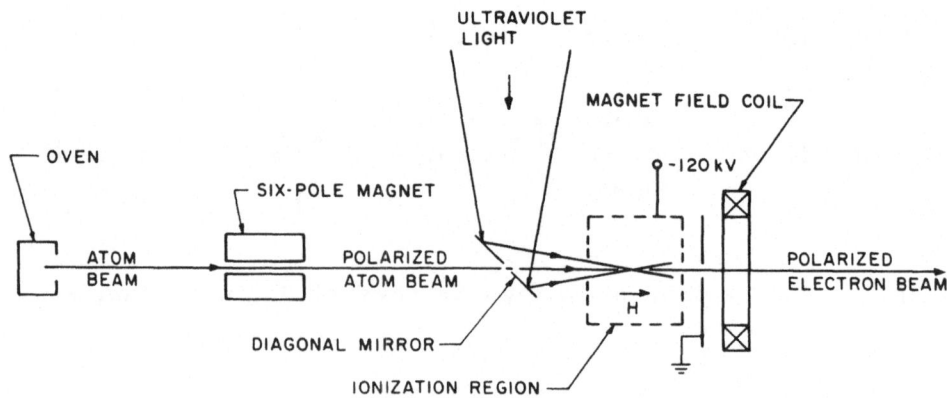

Fig. 4   Polarized electrons from photoionization of
         polarized alkali atoms[8]

At this method people have been working very hard
for more than 10 years. Of all the existing sources
this one has received by far the most intensive
development. Table I shows the results obtained by
Raith[6] and coworkers with lithium atoms. For photoionization a pulsed light source was used, since this
is the way to produce the highest light intensity. One
therefore gets a pulsed beam of polarized electrons. The
maximum intensity was $2 \cdot 10^8$ electrons per pulse. Table I
gives the corresponding average current. Similar results
were achieved with K atoms by Baum and Koch[8] in Bonn.
Coiffet[9] in France obtained lower intensities.

Spin Exchange

Before explaining why the "Fano effect" is highly
competitive with the method just described I want to
discuss three other methods also starting from polarized
atoms. Instead of using photoionization one can remove
the polarized bound electrons by exchange scattering.
For this purpose one has slow electrons pass through
the polarized atomic beam. Since for slow electrons the
cross sections for exchange scattering are very large,
there is a chance that the polarized bound electrons
are replaced by the injected electrons and thus come
out. Figure 5 shows the experimental arrangement.

The polarized atomic beam is produced like before.
An electron pulse is drawn from a conventional gun and
trapped in the region of the atomic beam. The electron
trap is a combination of an electrostatic potential well

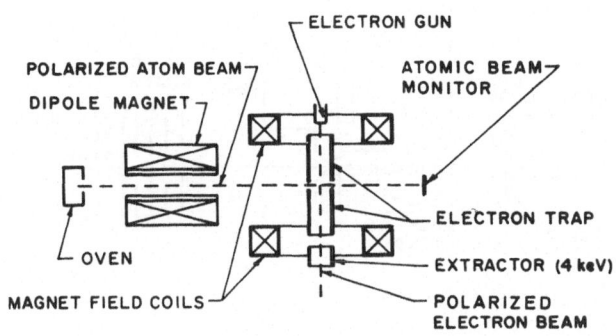

Fig. 5 Polarized electrons by spin exchange[10]

and a uniform magnetic field. During the trapping time
the electrons have much opportunity to make exchange
processes with the atomic electrons. After about 20 ms
the trap is opened. The outcoming electrons can have a
polarization up to 50%. This method also yields a
pulsed electron beam. According to the latest paper of
Farago[18] and co-workers in Edinburgh one obtains 10^4
electrons per pulse. Table I shows the comparison with
the other methods.

Electrons from an Optically Pumped Gas Discharge

In the last two methods the polarized atomic beams
were produced in the same way. The methods differ only
in the way to liberate the bound electrons. But there
are also other ways to produce the polarized atoms.
One of them has been successfully used at
Rice University by Walters[11] and his group which I had
the great pleasure of working with for 6 months.

A weak helium discharge (Fig. 6) is maintained in
a small magnetic field. In such a discharge a certain
fraction of the He-atoms is in metastable states. By
irradiation with circularly polarized light of suitable
wavelength the metastables are excited into higher states.
Since it is circularly polarized light which is used for
the optical pumping one obtains an orientation of the
electron spins in the metastables. The reason is that
due to the selection rules there is a preferential
depopulation of certain magnetic sublevels by the
circularly polarized light.

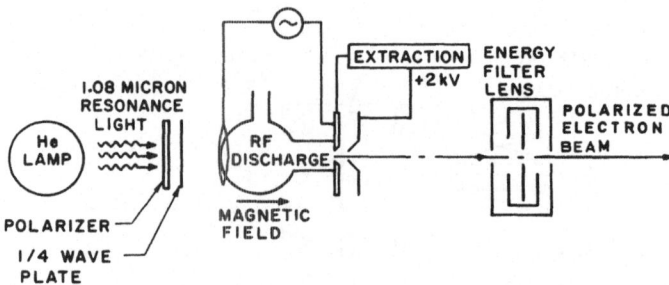

Fig. 6 Extraction of polarized electrons from
        optically pumped helium discharge[11]

So we get polarized metastable He-atoms. The
electrons in these metastables are only loosely bound,
their binding energy being much lower than in the helium
ground state. These loosely bound polarized electrons
are therefore preferentially liberated by the collision
processes in the discharge. Since their spin direction
is in general conserved in the collision processes we
find them as free polarized electrons in the discharge.
All we have to do is to extract them. Needless to say,
there are many other free electrons of different origin
in the discharge, so that the polarization is not very
high. However, since the current is very remarkable a
high value of P^2I is obtained. One gets the best values
by using a pulsed discharge and extracting the electrons
from the afterglow. Table I shows the results. When the
polarization increases, the current decreases rapidly
resulting in smaller P^2I.

Collisional Ionization of Quenched Metastables

Let me briefly mention another technique which also
uses collisional ionization of polarized metastable
atoms. In this method the polarized atoms are produced
by level crossing and quenching.

Figure 7 shows the apparatus of Donnally[12] et al.
A beam of deuterons is passed through Cs-vapor. This

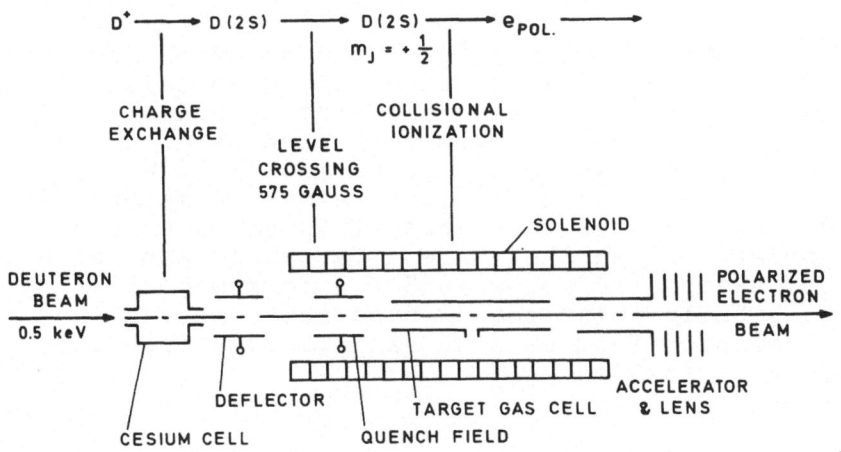

Fig. 7 Collisional ionization of quenched atoms[12]

occasions the deuterons to pick up electrons from
cesium preferentially forming deuterium atoms in the
metastable 2S-state. After passing through an electric
field which removes all charged particles the deuterium
beam containing about 25% metastables in 2S-states
enters an axial magnetic field of 575 G. This is the
field where the spin-down Zeeman level of the metastable
deuterium crosses a P-level. A small electric field
mixes these states and causes the spin-down metastables
to decay to the ground state. The spin-up metastables
together with a great number of ground state atoms
enter a gas cell filled with hydrogen. By the collision
in this cell the loosely bound electrons of the polarized
metastables are preferentially removed. To some extent
also electrons from the ground state atoms are removed,
so that the maximum polarization was 33%. Table I shows
the results. I put them in brackets because no attempt
has yet been made to optimize this method for use as a
source of polarized electrons.

Polarized Electrons from Magnetic Materials

        Instead of utilizing free polarized atoms as
starting point for the polarized electrons one can also
use polarized atoms in solids. One of the most obvious
methods for producing a polarized electron beam is the
extraction of polarized electrons from ferromagnets.
However, for a long time such attempts failed and only
in the past few years successful experiments have been
made.

        There are 2 ways to remove the polarized electrons
from the magnetic materials: Field emission and photoemission. Let us first consider field emission. Tips of
iron and nickel[13] and also of ferromagnetic gadolinium[14]
below the Curie temperature have been used. The
polarization obtained with these materials was in the
10% range. The search for more favorable materials
showed that the chalcogenides of Gd or Eu should be very
appropriate, since they have a favorable band structure.
EuS for example is below 16.5 K a ferromagnetic insulator
with a totally polarized 4f band above the valence band.
Field emission from this 4f band yielded free electrons
of 79% polarization.

This result has been obtained by Müller[15] in Munich whose apparatus is shown in Fig. 8. A field emission tip of tungsten is connected with a liquid helium bath. The tungsten tip is covered with EuS by evaporation from an oven located somewhat off the axis. The temperature of the sample was about 14 K. The magnetic field necessary for saturation depends on the geometry of the particular tip and was a few kilogauss. The electrons extracted by the electric field showed reproducible values of the polarization only if an ultra high vacuum of $3 \cdot 10^{-10}$ Torr was maintained.

Table I gives the results. Comparison with the other sources shows that these results are very impressive. On the other hand, the use of ultra high vacuum and liquid helium and the know-how which is necessary for producing useful EuS tips make the source difficult to handle.

If photoemission is used to remove the polarized electrons from the magnetic material, the apparatus looks like in Fig. 9. It has been utilized by Siegmann[16] et al. to study electron polarization in photoemission from iron, cobalt, nickel, gadolinium and various chalcogenides of Gd and Eu like EuO, EuSe, EuS, EuTe and from GdP. The samples are again produced by evaporation. The oven can be removed for the measurement. The samples are cooled below their Curie point by liquid helium and are magnetized by a superconducting coil generating up to 50 kG. The sample is exposed to UV-light of proper wavelength. Acceleration to a few keV forms a beam out of the photoelectrons. This polarized electron beam is deflected out of the light beam by a cylindrical condenser.

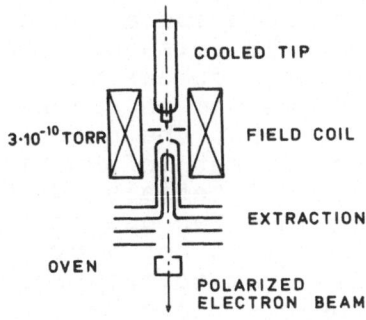

Fig. 8   Field emission of polarized electrons[15]

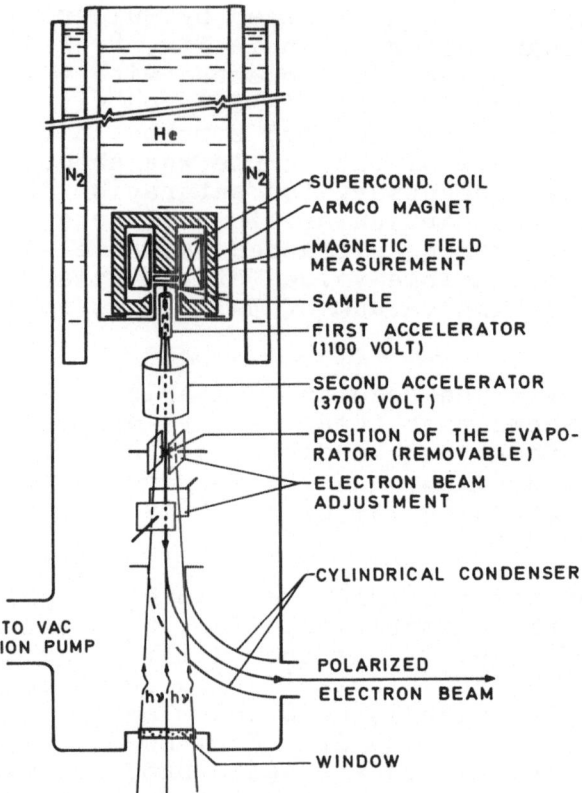

Fig. 9   Photoemission of polarized electrons[16]

In order to get reproducible results the surface
of the photocathode has to be extremely clean. During
the measurement the ultra high vacuum must be 10^{-10} Torr,
during evaporation it must not fall below 10^{-8} Torr. So
the only point where this experiment is easier to handle
than field emission is the preparation of the samples,
the production of the field emission tips being even more
delicate than that of the photocathodes. Table I gives
the best result of this method in brackets. The reason
to put it in brackets is that the photoemission studies
have been made in order to get information on the
structure of the magnetic materials. No attempt has been
made to optimize the method for use as a source of
polarized electrons. So it is not very fair to compare
these data with results from sources which have been
optimized. But nevertheless these data are very good.

Fano Effect

Most of the methods discussed so far use polarized atoms as starting point for the polarized electrons. Only recently it has been found out that one can get by with less effort: Under certain circumstances one can start from unpolarized atoms if circularly polarized light is used for photoionization.

This is by no means trivial. You must not simply think: "Small wonder that I get polarized photoelectrons when photoionizing with polarized photons". For - as a rule - the preferential orientation of the photon spins in circularly polarized light does not cause a preferential orientation of the spins of the photoelectrons produced. But the spin of the absorbed photon normally reappears as orbital angular momentum of the photoelectron just as described by the selection rules for l and for the magnetic quantum numbers m_l.

In particularly favorable cases, however, it may happen that the coupling between spin- and orbital angular momentum of the electrons does produce an orientation of the electron spins in the experiment discussed.

This is illustrated by Fig. 10. Without spin-orbit interaction it would not make a difference whether we photoionize with ordinary light or with circularly polarized light. We would obtain a curve with a deep minimum as it is typical for most of the alkali atoms. But if we take spin-orbit interaction into account, we obtain a difference between the cross section for producing electrons with spin parallel to the photon spins in the circularly polarized light and the cross section for producing electrons with spin antiparallel to that direction. If we therefore photoionize an unpolarized atomic beam by circularly polarized light, we obtain different numbers of spin-up and spin-down electrons, that means, a polarized electron beam.

The polarization can be quite significant, if we irradiate wavelengths near a minimum of one of the curves of Fig. 10a. We then obtain practically only electrons of one spin direction, that means a large polarization as shown in Fig. 10b.

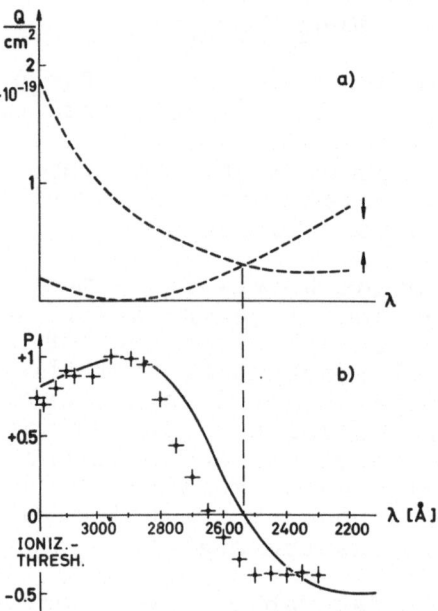

Fig. 10   a) Cross section vs. wavelength for photoeffect
             at the cesium atom. Photoelectron spin in
             and opposite to direction of photon spin,
             respectively.

         b) Spin polarization P of the photoelectrons.
            Theory[17] and experimental points[18]

        The experimental points were obtained with the
apparatus of Fig. 11, consisting of light source, monochromator, circular polarizer, oven for producing a
vapor beam (which is of course unpolarized), extraction
system, polarization detector. This apparatus was not
built to have a source of polarized electrons, but
rather to study the Fano effect in order to get information on the influence of spin-orbit interaction on
photoionization. Nevertheless the apparatus yielded 81%
polarization at $3 \cdot 10^{-11}$A when the spectrum of a mercury
high-pressure arc lamp was irradiated. This showed that
the Fano effect would be a simple and effective basis
for a source of polarized electrons. The groups at Bonn
and Yale started to develop such a source. Their
experimental arrangement has not yet been published.
Table I shows the results of the Yale group.

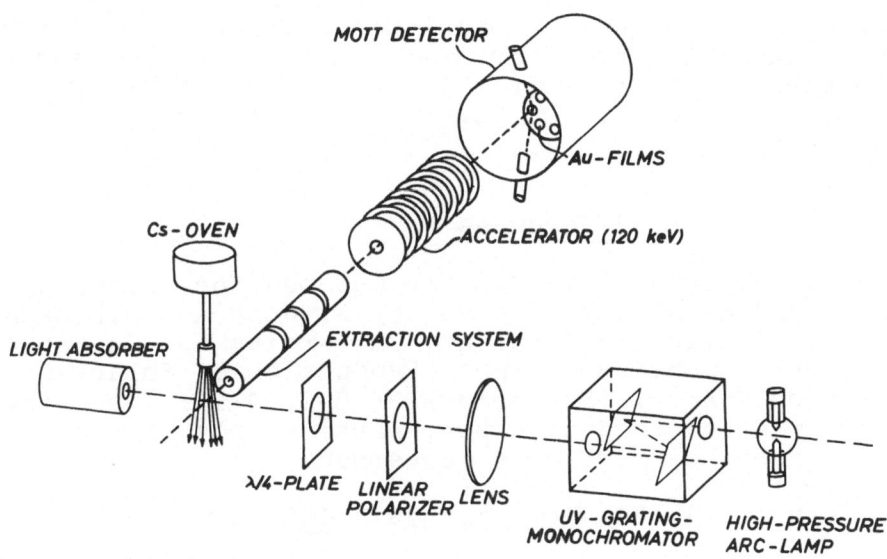

Fig. 11   Apparatus for measuring the polarization of
           photoelectrons coming from an unpolarized
           atomic beam[18]

     Let me finally mention that the Fano effect also
works with solid materials. When we replaced our Cs-beam
by a solid Cs-target the current increased by a factor of
1000 and the polarization decreased to 4%. This is, of
course, too small a polarization to be of practical
interest. What is interesting, however, is that also
solid materials yield polarized electrons when irradiated
with circularly polarized light. A search for materials
yielding higher polarization than cesium would be very
interesting. Before this can be successfully tackled the
Fano effect at solid cesium should be understood quantitatively. That has not yet been achieved.

     Let me conclude with a last look at our table. A lot
of effort has been invested and many good results have
been obtained. However, experiments with polarized
electrons are still troublesome and at the margin of what
can be done today. What is still missing is the simple
gun of polarized electrons which is easy to handle for
everybody, not only for the expert, and which yields a
highly polarized beam of 10 µA or so. A few of the methods

discussed here are quite promising for further development.
Looking back at the rapid progress in the past 10 years
I would guess that in another 10 years experiments with
polarized electrons are no longer a matter for a few
artists only.

References

1 A more complete review on the methods for producing
   polarized electrons is given by K. Jost and H.D.Zeman,
   now Physikalisches Institut, University of Münster,
   in their unpublished paper "Production of Polarized
   Electron Beams". (Stanford HEPL 590,1971). I am very
   much indebted to this paper, particularly as far as
   the figures of merit are concerned.

2 K. Jost and J. Kessler, Z.Physik 195, 1 (1966)

3 W. Eitel, K. Jost, and J. Kessler, Z. Physik 209, 348
   (1968). Review article: J. Kessler, Rev.Mod.Phys. 41,
   3 (1969)

4 M. Wilmers, R. Haug, and H. Deichsel, Z. angew. Physik
   27, 204 (1969)

5 H.D. Zeman, K. Jost, and S. Gilad, Abstracts of the
   VIIth International Conference on the Physics of
   Electronic and Atomic Collisions, Amsterdam 1971,
   North Holland Publishing Company, p. 1005

6 W. Raith, Atomic Physics (Eds. B. Bederson et al.),
   Plenum Press, New York and London, 1969, S. 389

7 V.W. Hughes, R.L. Long, Jr., M.S. Lubell, M. Posner,
   and W. Raith. Phys. Rev. A5, 195 (1972)

8 G. Baum und U. Koch, Nucl. Instr. 71, 189 (1969)

9 P. Coiffet, Journal de Physique, 32, 113 (1971)

10 D.M. Campbell, H.M. Brash, and P.S. Farago, Physics
   Letters 36A, 449 (1971)

11 M.V. McCusker, L.L. Hatfield, and G.K. Walters, Phys.
   Rev. A5, 177 (1972)

12 B. Donnally, W. Raith, and R. Becker, Phys. Rev.
   Letters 20, 575 (1968)

13 W. Gleich, G. Regenfus, and R. Sizmann, Phys. Rev. Letters 27, 1066 (1971)

14 G. Chrobok, M. Hofmann, and G. Regenfus, Phys. Letters 26A, 551 (1968)

15 N. Müller and W. Eckstein, Verhandl. Deutsche Physikal. Gesellschaft 7, 464 (1972)

16 G. Busch, M. Campagna, and H.C. Siegmann, J. Appl. Phys. 41, 1044 (1970) and Phys. Rev. 4B, 746 (1971)

17 U. Fano, Phys. Rev. 178, 131 (1969)

18 U. Heinzmann, J. Kessler, and J. Lorenz, Z. Physik 240, 42 (1970)

19 U. Heinzmann, K. Jost, J. Kessler, and B. Ohnemus, Z. Physik 251, 354 (1972)

PHOTOELECTRIC MEASUREMENTS OF DOUBLET INTENSITY RATIOS IN CESIUM[*]

Gabor F. Fulop and H. Henry Stroke

Department of Physics, New York University

New York, N. Y.   10003

INTRODUCTION

The effects of the spin-orbit perturbation on the alkali fine-structure doublets have a number of manifestations that have been reviewed recently by zu Putlitz.[1] These include the non-zero minimum, as a function of energy, of the photoionization cross sections, polarization of the photoelectrons obtained near this cross-section minimum, and the departure, in the discrete spectrum, from the value 2 of the doublet intensity ratio. The value 2 is expected simply on the basis of the ratio of the statistical weights of the excited $^2P_{3/2}$ and $^2P_{1/2}$ states. The interest in these phenomena was renewed after the theoretical prediction by Fano,[2] based on the perturbation effects by the spin-orbit interaction, that circularly-polarized light could orient the spins of electrons ejected in the photoionization process in cesium. The potential application of the phenomenon to make polarized electron sources was to be exploited soon thereafter.[3,4] The two experimental efforts that were thus made to produce polarized electrons, and to measure the degree of polarization, relied on related, but different, techniques. The photoionization results, extrapolated into the discrete spectrum, made different predictions for the behavior of the resonance-line doublet intensity ratios: in one case[3] the ratio would pass through a maximum, as the principal quantum number, n, of the excited P states increases toward infinity; in the other[4] the ratio would continue to increase as the ionization potential is approached. It is for the purpose of clarifying this discrepancy that we undertook the present direct measurements of the intensity ratios of the cesium doublets for which we present our initial results. Though a number of experimental studies, begun nearly fifty years ago, have been made, the results show much disagreement, or else their precision or n-value range are insufficient to draw conclusions. These measurements have been summarized by Baum, Lubell, and Raith[5] along with their photoionization study results.

THEORY

Intensity Ratios

Consider some of the experimental values for cesium of the intensity ratios ρ_n, defined by

$$\rho_n \equiv \frac{S(n\ P_{3/2})}{S(n\ P_{1/2})} \quad , \tag{1}$$

given in Table I. S is the line strength of the transition $np\ ^2P$ to the $6s\ ^2S_{1/2}$ ground state. We also give the corresponding oscillator strengths, $f(np\ ^2P_{3/2} - 6s\ ^2S_{1/2})$. In the first place it is seen that the departure of the intensity ratio from the value 2 is by a substantial amount. Second, the experimental difficulty of measuring the intensity ratios as a function of n is obvious from an examination of the f values: the intensities decrease rapidly as n increases. In particular, the decrease by a factor ≈ 50 between n=6 and n=7 was suggested already in 1929 by Fermi[6] as a cause of the intensity anomalies for n > 6. He attributed the effect to <u>different</u> admixtures of the various $P_{1/2}$ and $P_{3/2}$ states through the spin-orbit interaction. Thus, a small admixture of the state n=6 with its large f value can have important effects on all the other P-level intensity ratios.

TABLE I. Experimental intensity ratios ρ_n of the resonance lines and oscillator strengths f_n of the $(np\ ^2P_{3/2} - 6s\ ^2S_{1/2})$ transitions for the first five n levels in cesium.

n	ρ_n			f_n	Ref.
	(a)	(b)	(c)		
6	----	-----	2.027	0.796	c
7	5	4.15	4.285	1.33×10^{-2}	b
8	10	7.63	8.0	0.294×10^{-2}	b
9	15.5	11.0	8.1	0.91×10^{-3}	b
10	25	21.5	8	0.417×10^{-3}	b

(a) Sambursky, Ref. 7
(b) Agnew, Ref. 8
(c) Kvater and T. Meister, Leningrad Univ., Vestn. 9, 137 (1952).

We outline the elementary theory below.

For the unperturbed wavefunctions, the transition probability is proportional to the square of the matrix element of the electric dipole operator er_q^1. The electron charge is e, and q denotes the spherical component of the vector (tensor of rank 1) operator $\vec{r}$. Summing over the initial and final magnetic substates m and m' we evaluate the strength

$$S = \sum_{mm'} |\langle n\ell = 1\ s = \tfrac{1}{2}\ j'm'|er_q^1|\ 6\ 0\ \tfrac{1}{2}\ \tfrac{1}{2}\ m\rangle|^2 \qquad (2)$$

between the excited np and 6s $^2S_{1/2}$ ground states. In terms of the radial functions R, we obtain readily with the use of Racah algebra

$$S_n(P_{1/2} \to S_{1/2}) = \frac{4}{9} \left| \int_0^\infty R_{np}\ er\ R_{6s}\ dr \right|^2 \equiv \frac{4}{9} R_o \qquad (3a)$$

and

$$S_n(P_{3/2} \to S_{1/2}) = \frac{8}{9} \left| \int_0^\infty R_{np}\ er\ R_{6s}\ dr \right|^2 \equiv \frac{8}{9} R_o\ . \qquad (3b)$$

These unperturbed wavefunctions (3a) and (3b) are seen to give $\rho_n = 2$. Fermi considered perturbed wavefunctions of a level n

$$\psi'_n = \psi_n + \sum_k{}' \alpha_{nk}\ \psi_k\ , \qquad (4)$$

where the admixture coefficient is

$$\alpha_{nk} = \frac{H'_{kn}}{E_n - E_k}\ , \qquad (5)$$

and the perturbing Hamiltonian that results from the spin-orbit interaction is

$$H' = \frac{1}{2} \left(\frac{\hbar}{mc}\right)^2 \frac{1}{r} \frac{dV}{dr}\ \vec{L} \cdot \vec{S}\ . \qquad (6)$$

E_k denotes the energy of the perturbing level k. We obtain for the mixing coefficient

$$H'_{kn} = \frac{1}{4} \left(\frac{\hbar}{mc} \right)^2 \int_o^\infty \frac{1}{r} \frac{dV}{dr} R_k R_n \, dr \, \delta_{j_k j_n} \, \delta_{m_k m_n} \, \delta_{\ell_k \ell_n}$$

$$\times \left\{ \begin{array}{l} \ell_n \text{ for } j_n = \ell_n + \frac{1}{2} \\ \\ -(\ell_n + 1) \text{ for } j_n = \ell_n - \frac{1}{2} \end{array} \right. . \tag{7}$$

The important result is that the admixture coefficients for the $^2P_{3/2}$ and $^2P_{1/2}$ states have a common factor multiplied, respectively, by 1 and -2, and that they contain the above δ - functions. One can therefore factor out the common angular part of the perturbed and unperturbed contributions to the line strength to obtain[2,3]

$$S'_n \, (P_{1/2} \to S_{1/2}) = \frac{4}{9} \left[R_o - \frac{2}{3} \Delta_n R \right] \equiv \frac{4}{9} R_1$$

$$S'_n \, (P_{3/2} \to S_{1/2}) = \frac{8}{9} \left[R_o + \frac{1}{3} \Delta_n R \right] \equiv \frac{8}{9} R_3 \tag{8}$$

where the perturbation effect on the state n, $\Delta_n R$, involves the radial integrals that appear in (7). It is clear that one may have to consider perturbing states k not only in the discrete spectrum but also in the continuum. Also, relativistic effects, more generally, can lead to differences in the perturbations of the $P_{1/2}$ and $P_{3/2}$ radial functions, and hence to anomalous values of ρ. Fano[2] introduced a parameter x to describe the polarization effects in the photoionization process,

$$x \equiv \frac{2R_3 + R_1}{R_3 - R_1} = \frac{3R_o}{\Delta R} , \tag{9}$$

in terms of which we obtain

$$\rho = 2 \left(\frac{x + 1}{x - 2} \right)^2 . \tag{10}$$

For unperturbed functions, $\Delta R = 0$, so that in this limit x becomes infinite and $\rho = 2$. The question arises: is there a level n in the discrete spectrum for which, or in the vicinity of which there is a pole, i.e. for which x = 2?

Electron Polarization in Photoionization

Following Fano, the photoionization process can be shown to involve the evaluation of the same operators as in the preceding calculation, but now the initial state is the ground state, and the final state is in the continuum. The use of circularly polarized light for photoionization leads to spin orientation of the ejected electrons. To calculate the polarization obtained, we have to evaluate the matrix element, say for σ_+ light, of $x + iy = -\sqrt{2}\, r_1^1$, i.e.

$$- \sqrt{2}\ (1\ m_\ell'\ m_s'\,|r_1^1|\ 0\ 0\ m_s) \qquad . \tag{11}$$

By expanding the required ℓs wave functions in terms of $j\ m_j$ wavefunctions, and labeling further the radial wavefunctions in the continuum by energy ε, photoionization matrices corresponding to various experimental conditions can be calculated in terms of $R(\varepsilon, j = 1/2) = R_1$ and $R(\varepsilon, j = 3/2) = R_3$. In one experiment[4] circularly polarized light was used with unpolarized atoms, and the numbers of electrons with spin up (+) and spin down (−) were measured. The polarization parameter, which can be expressed in terms of x, that was determined is

$$P = \frac{N_+ - N_-}{N_+ + N_-} = \frac{1 + 2x}{2 + x^2} \qquad . \tag{12}$$

In the other experiment[3] polarized atoms were irradiated successively with σ_+ and σ_- light, and the corresponding ion currents I_+ and I_- were measured. The polarization measure here was in terms of

$$Q = \frac{I_+ - I_-}{I_+ + I_-} = \frac{2x - 1}{x^2 + 2} \qquad . \tag{13}$$

The problem is to determine the scale x vs. ε, and in particular near the photoionization threshold. Assuming that $x(\varepsilon)$ continues smoothly from the continuum into the discrete spectrum, the results of the measurements of P indicate extrapolated values in the discrete spectrum x > 2. The measurements of Q, on the other hand, show x < 2 as one crosses the photoelectric threshold, and that x = 2 is reached in the region for principal quantum numbers $10 \le n \le 15$. In the latter case one would therefore have a pole in ρ. Consequently, a maximum in the doublet intensity ratio would be expected to be observed.

EXPERIMENTS

A 1928 experiment of Sambursky[7] was actually the only one to show a peak in the intensity ratio for n = 10. The most recent work, an absorption experiment by Agnew,[8] did not show a peak, though an increase in the error brackets to two standard deviations could also be consistent with the peak observed by Sambursky. An experiment by Beutell[9] also shows the rapid rise in ρ but did not extend to levels with sufficiently high n to then show a decrease in the intensity ratio.

One may think of several possible reasons for inconsistent results. First, as noted, the intensities of the resonance lines from the higher n levels diminish rapidly and one is soon faced with signal-to-noise problems for these weak signals. Second, most of the work was done by photography, and it is difficult to do accurate intensity measurements because of the non-linear detection. Third, it is possible to have self-reversed light sources, particularly if one is trying to operate the lamps at a high exciration level to bring out the weaker lines. Self reversal would lead to erroneous intensity-ratio determinations.

We made measurements with our 10-m focal length Czerny-Turner monochromator, in which a 25-cm blazed diffraction grating is used near autocollimation at about 60 deg. A schematic of the apparatus is shown in Fig. 1. With narrow slits, the limit of resolution is approximately 0.02 cm^{-1}. This is to be compared to the ground-state hyperfine structure in Cs^{133}, $\Delta\nu \approx 0.3$ cm^{-1}, which can thus be resolved easily. The possibility of resolving this hfs allowed

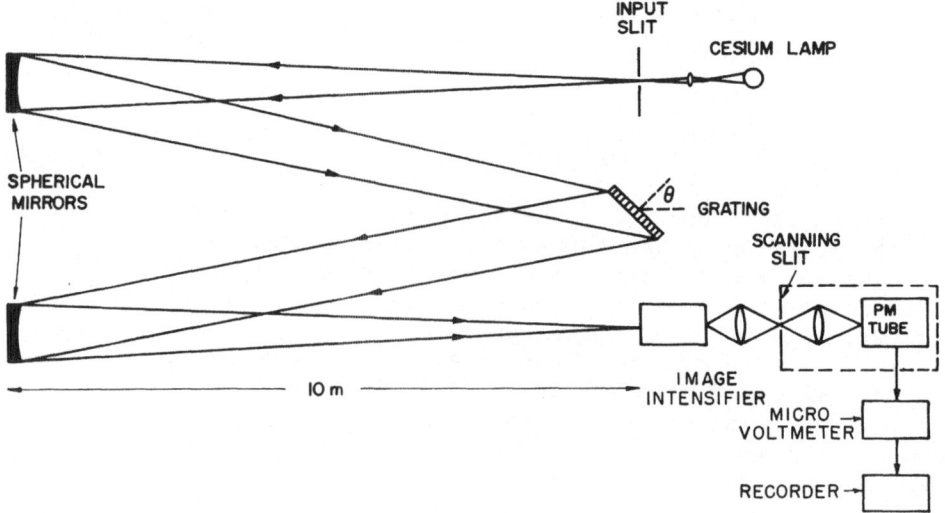

Fig. 1.   Schematic Diagram of Apparatus

us to measure also the ratio of the intensities of the hfs components. If we include the nuclear spin, $I = 7/2$, and characterize the wavefunctions in terms of the total angular momentum quantum number F, Eq. (2) becomes

$$S = \sum_{m_F m'_F} \left| \langle n\ 1\ \tfrac{1}{2}\ j\ I\ F\ m_F | er^1_q | 6\ 0\ \tfrac{1}{2}\ \tfrac{1}{2}\ I\ F'\ m'_F \rangle \right|^2 . \tag{14}$$

We evaluated Eq. (14) and the corresponding spin-orbit perturbation effects, and found that (7) now includes further a factor $\delta_{F_k F_n}$, but that the square brackets in (8) remain unchanged and, significantly, independent of F. We could therefore use the relative intensities of the hfs components as a test for the correct operation of our light source, i.e. verify that there were no self-reversal effects. It was assumed that given correct hfs component intensity ratios, the measured fine structure doublet intensity ratio would also be correct.

Our light source was an electrodeless lamp, excited at approximately 25 MHz.[10] The temperature of the bulb was kept below $\approx 110°C$. At this temperature we were able to ascertain, by monitoring hfs intensity ratios, the absence of self-reversal for n=7 to n=9. With increasing lamp temperature self-reversal sets in and the hfs intensity ratio decreases for a given fine structure level because of the greater absorption of the hfs component connected to the ground F=4 level compared to the one with F=3. The resolution of the ground-state hfs required the use of narrow slits. Because of the diminishing intensities for the higher series members, for n>9 we had to increase the slit widths so that we could no longer resolve the hfs. However, as already described, the temperature of the lamp was kept below its value for which even for n<9 self-reversal was absent. The fine structure intensity ratio measurements were made with wide input and output slits.

Photoelectric detection was used. Because, as seen in Table I, the higher members of the principal series are quite weak, we used an image intensifier (RCA Type 33034) in the image plane of the spectrograph. The spectrum appearing on the intensifier phosphor was scanned with the photomultiplier. The lines were positioned to fall on the same portion of the intensifier photocathode so as to avoid effects caused by non-uniformities in the photosensitive surface. Possible order overlap in the monochromator was eliminated with the use of 100-Å bandwidth pass interference filters.

RESULTS

Preliminary results of our measurements are shown in Fig. 2 as a plot of intensity ratio of the doublets vs. wavenumber. The

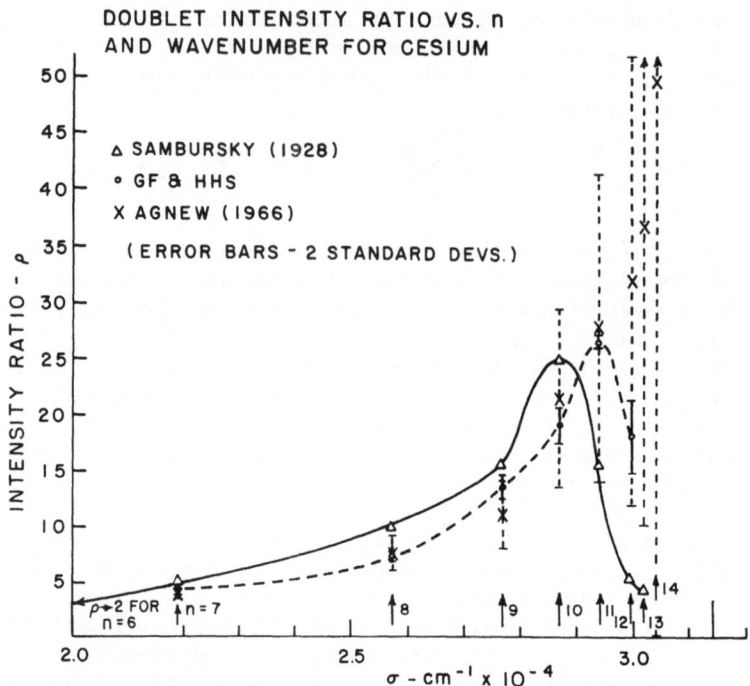

Fig. 2. Doublet intensity ratio,
$\rho = S(np\ ^2P_{3/2} - 6s\ ^2S_{1/2})\ /\ S(np\ ^2P_{1/2} - 6s\ ^2S_{1/2})$, as a function of the wavenumber above the ground state and the principal quantum number n.

corresponding principal quantum numbers are also indicated. There appears to be a maximum in ρ in the vicinity of n=11. We are seeking to confirm further this maximum by a measurement of ρ_{13} which would be expected to have a smaller value than ρ_{12}. Because our present signals for n=13 were masked by shot noise in the photodetectors, we are now instrumenting a more sophisticated detection system to allow this measurement. In Fig. 2 we also show, for comparison, the results of the measurements of Sambursky[7] that appear to be compatible with our own finding. As mentioned earlier, the results of Agnew[8] do not preclude being considered as reasonably consistent with our measurements.

DISCUSSION

In addition to an extension of measurements to higher principal quantum numbers, these first experiments are also being repeated with different light sources. There may be possible differences, for example, in atomic collisional effects on the population of the excited $^2P_{1/2}$ and $^2P_{3/2}$ states,[11] though these differences are expected to produce an effect on ρ relatively unimportant compared to the relativistic effect considered here. The differences between the two photoelectron orientation experiments have been discussed recently by Norcross.[12]

We have discussed the calculation of ρ and of the polarization parameters in the photoionization processes only from the simplest point of view of perturbation by the spin-orbit interaction, based on the work of Fermi and Fano. Several relativistic calculations of the oscillator strengths and of photoabsorption have been made recently. These include the work of Koenig-Luc,[13] in which a parametric potential is used, and the relativistic Hartree-Fock calculation of Chang and Kelly.[14] In the former, transition-probability calculations do not extend to the principal quantum numbers of importance here, and in the latter, the work is being refined with the expectation that the inclusion of electron correlation effects will improve agreement with experiment. Core polarization effects, however, are included, through the use of an effective dipole operator, $\vec{Q}$, in the work of Weisheit and Dalgarno.[15] An effective core radius cutoff, r_c, determines the radial dependence of $\vec{Q}$. The value of r_c that gives agreement with both sets of photoionization experiments[4,5] near $x(\varepsilon) = 0$, and in the case of the results of Ref. 5, for nearly the entire region of ε, from threshold to the maximum measured, leads to the prediction of a pole in ρ for $n \approx 16$. A similar calculation by Norcross,[16] including a relativistic correction, has predicted the absence of a pole for ρ in the discrete spectrum. Without the correction, a pole would be expected. A summary of the experimental and theoretical results for x as a function of photon wavelength λ is shown in Fig. 3.

We are grateful to Professor W. Raith for bringing this problem to our attention, and to Drs. S. Feneuille and S. Gerstenkorn for stimulating discussions. We also wish to thank Dr. L. Agnew for communicating to us unpublished details of his experimental results.

* This work was supported by the National Science Foundation and in part by the James Arthur Endowment Fund at New York University. One of us (H.H.S.) wishes to thank the Faculté des Sciences Université de Paris Sud and the Laboratoire Aimé Cotton, C.N.R.S. II, for their hospitality during Summer 1972.

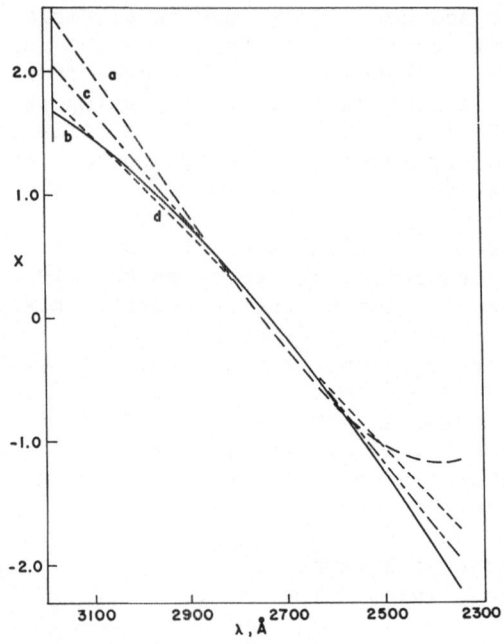

Fig. 3. (Experimental and theoretical determinations of Fano's spin-orbit perturbation parameter x (from Ref. 16) a) Heinzmann, Kessler, and Lorentz (Ref. 4); b) Baum, Lubell and Raith (Ref. 3,5); c) Norcross (Ref. 16); d) Weisheit (Ref. 15).

References

1. G. Zu Putlitz, Comments At. Mol. Phys. 1, 51 (1969).
2. U. Fano, Phys. Rev. 178, 131 (1969) and 184, 250 (1969).
3. G. Baum, M.S. Lubell and W. Raith, Phys. Rev. Lett. 23, 211 (1969) and 25, 267 (1970).
4. U. Heinzmann, J. Kessler and J. Lorentz, Z. Physik 240, 42 (1970).
5. G. Baum, M.S. Lubell and W. Raith, Phys. Rev. 5A, 1073 (1972).
6. E. Fermi, Z. Physik 59, 680 (1930).
7. S. Sambursky, Z. Physik 49, 731 (1928).
8. L. Agnew, Bull. Am. Phys. Soc. 11, 327 (1966).
9. M. Beutell, Ann. Physik 36, 533 (1939).
10. R. Brewer, Rev. Sci. Instr. 32, 1356 (1961).
11. M. Pimbert, J. Physique 33, 331 (1972).
12. D.W. Norcross, private communication, February 1972.
13. E. Koenig-Luc, Physica (to be published).
14. J.J. Chang and H.P. Kelly, Phys. Rev. A5, 1713 (1972).
15. J.C. Weisheit, Phys. Rev. A5, 1621 (1972).
16. D.W. Norcross, private communication, June 1972 (submitted for publication to Phys. Rev.).

STUDY OF e-O_2 AND e-H_2 SCATTERING

BY ELECTRON TIME-OF-FLIGHT SPECTROSCOPY*

James E. Land and Wilhelm Raith[†]

J. W. Gibbs Laboratory, Yale University

New Haven, Connecticut 06520

Electron time-of-flight measurements have been employed to study such fundamental problems as the identification of cathode rays[1], verification of the relativistic mass formula[2], and free-fall studies on single electrons[3]. For studies of electron-atom and electron-molecule scattering, however, time-of-flight spectroscopy has been tried only recently[4,5].

Thus far, all high-resolution electron-scattering spectroscopy has been done with spectrometers based on energy dispersion in static electric or magnetic fields. At energies below about 100 meV, cross section data have been obtained only by swarm experiments[6,7]. While swarm experiments can provide good absolute cross sections, they cannot reveal pronounced structure in the cross section, such as resonances. To complement the swarm data, we are developing electron time-of-flight spectroscopy as a technique for detecting resonances in cross sections. Here we report some first results which indicate that the time-of-flight method is promising, although much more work remains before the advantages and limitations of the method are understood in detail.

The experimental arrangement is shown in Fig. 1. The main component is the 25-cm drift tube, which serves both as the energy-dispersing element and as the gas-cell target. An electron burst of 10-nsec width is produced by sweeping a 200-eV d.c. beam across a narrow aperture. The electrons are abruptly decelerated, and traverse the drift tube with energies in the range 20 to 300 meV. The generation of an electron burst, and the detection of an electron in the electron multiplier, provide two time marks which determine the flight time for a single electron. A time-of-flight spectrum is obtained by measuring the flight times of 10^6 or more elec-

553

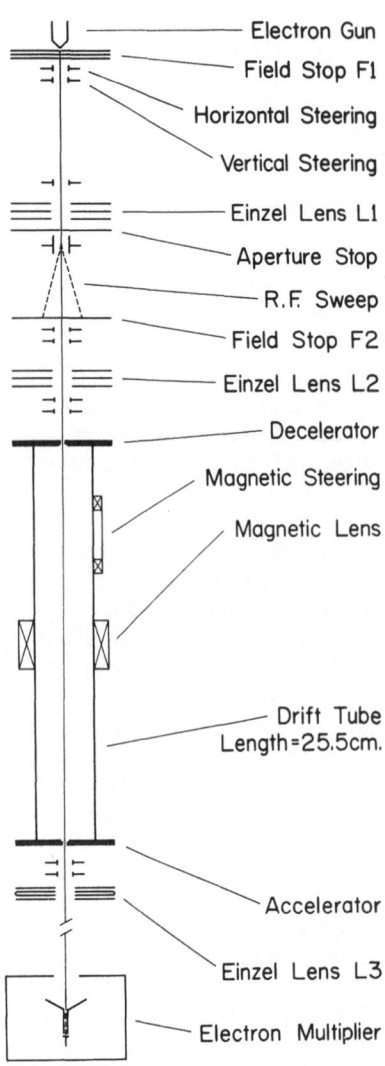

Electron Gun

Field Stop F1

Horizontal Steering

Vertical Steering

Einzel Lens L1

Aperture Stop

R.F. Sweep

Field Stop F2

Einzel Lens L2

Decelerator

Magnetic Steering

Magnetic Lens

Drift Tube
Length=25.5cm.

Accelerator

Einzel Lens L3

Electron Multiplier

Fig. 1

Scale drawing of the time-of-flight spectrometer, radial dimensions enlarged by a factor of two.

trons, using a time-to-amplitude converter and a multi-channel analyzer. Spectra are taken (a) without gas in the target, and (b) with a gas pressure which produces a high attenuation of the transmitted electron intensity.

A principal limitation of the time-of-flight method is that the actual energy resolution can be determined only by resolving narrow structures in electron-scattering cross sections. Estimates of some effects which are known to limit the resolution are shown in Fig. 2. At the lowest energies studied, the principal limitation is probably the patch effect, which produces a variation in potential inside the drift tube. At the highest energies studied, the resolution is limited by the accuracy with which the electrons can be timed, $\delta t = 10$ nsec. Other effects include the difference in trajectory lengths in the drift tube for a beam angle $\alpha = 0.1$ rad, and Doppler broadening due to thermal motion of the molecules in the drift tube at room temperature.

The energy dependence of the total scattering cross section follows from a comparison of the spectra taken with and without gas in the drift tube. The spectrum taken without gas has a distribution $I_0 = f(E)$, determined by the electron-optical parameters of the system. If the spectrum taken with gas in the target, I, is related to I_0 by

$$I = I_0 e^{-NL\sigma},$$

then the negative logarithm of their ratio yields

$$-\log_e I/I_0 = NL\sigma.$$

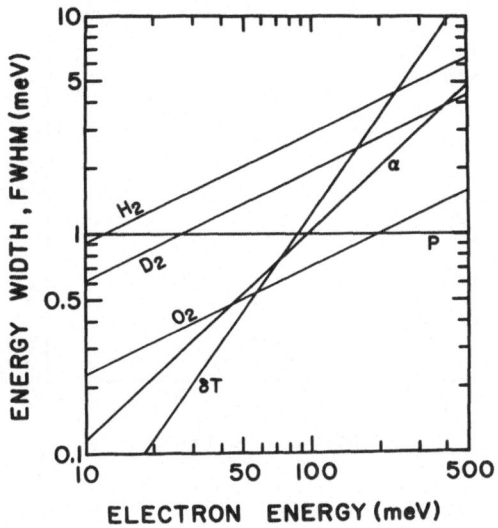

Fig. 2

Estimates of resolution-limiting effects: Finite width of electron
burst (δt); different lengths of trajectories focused by the magnetic lens (α); patch effect (P); Doppler broadening at room temperature for H_2, D_2, and O_2.

This is the quantity plotted in the figures which follow.

     The crucial question is whether the only difference between I
and I_o is due to the scattering described by the cross section σ,
or whether other effects could cause differences which would be mis-
takenly interpreted as resonances. Changes in the contact potential
of the drift tube due to the gas inlet were simulated by varying the
drift-tube potential; no false resonances could be produced, but
small changes in the slope of the cross section cannot yet be ruled
out. In order to minimize pressure-dependent effects outside the
drift tube, the gas flow into the vacuum system was held constant,
with the flow fed either into the target or bypassed directly to the
vacuum chamber. Despite this precaution, small energy shifts were
observed in the low-energy onset of the transmitted-electron spectrum. We therefore rejected data from portions of the spectrum with
large slopes. At the target pressures used, ~ 3 - 10 mTorr, the transmitted-electron intensity was attenuated by a factor of 100 or more
in the drift tube. In order to avoid long data-accumulation times,
it was necessary to increase the cathode temperature when taking
spectra with gas in the target, but tests showed that this produced
no false structures in the spectrum.

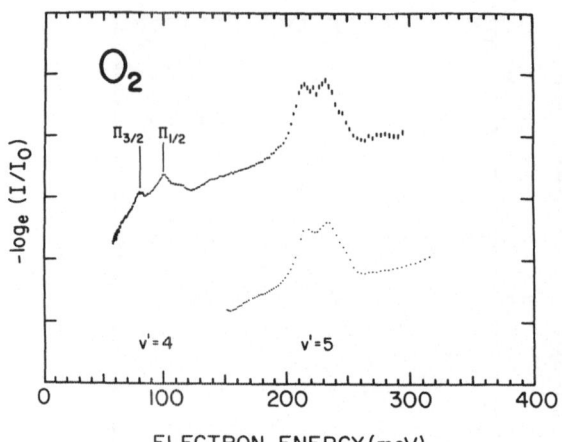

Fig. 3

Transmitted-electron scattering data for O_2. The vertical error
bars give the statistical error from the number of counts in each
channel. I = count rate of electrons transmitted through the drift
tube with a pressure of $\sim 10^{-2}$ Torr; I_0 = count rate with the same
gas flow bypassing the target.

OXYGEN

        Molecular oxygen was chosen for the first measurements with the
time-of-flight spectrometer because it provides, in the energy range
of interest, resonances which are well understood[8-13]. The resonances correspond to vibrationally-excited autoionizing states of O_2^-,
which have been identified as a $^2\Pi_g$ configuration with vibrational
quantum numbers v' = 4,5,...[10,11]. The resonances shown in Fig. 3
are the first two members of the sequence. The doublet structure,
which is due to spin-orbit coupling, has never before been observed.
We measure a fine-structure splitting of ΔE = 20±2 meV, corresponding
to a coupling constant of A = -161 ± 16 cm^{-1}. This can be compared
with a theoretical estimate of A = -150 ± 30 cm^{-1} [14], and with the
measured coupling constant in the analogous $^2\Pi_g$ state of O_2^+, A =
+195 cm^{-1} [15].

        From several measurements similar to Fig. 3, we locate the cen-
ter of the v' = 4 doublet at 91±4 meV. Heretofore the energy scale
was uncertain by ±20 meV [12]. By using Linder and Schmidt's data
for the interval E(v' = 8) – E(v' = 4) and our value of E(v' = 4),
we find that the v = 3 vibrational level of O_2 at 570 meV lies between the v' = 8 doublet of O_2^-. This is in good agreement with
the results of Spence and Schulz[10], who found the levels coincident
within $\sim$ 10 meV. Our energy scale combined with the resonance spac-
ings measured by Linder and Schmidt, extrapolated to the $\Pi_{3/2}$

v' = 0, J' = 3/2 level, gives excellent agreement with the electron-affinity measurement of Celotta et al[13].

HYDROGEN

In 1968 Frommhold [16,17] offered an explanation for anomalies observed in electron swarm experiments at very high pressures and very low electron energies. He postulated the existence of resonances in e–H₂ scattering associated with rotationally-excited autoionizing states of H_2^- in the energy range 10 to 100 meV. Although this hypothesis was able to explain the anomalous swarm data, the existence of a state of H_2^- at these energies is in contradiction to theoretical considerations: (a) The $^2\Sigma_u^+$ state of H_2^- at 2-3 eV is believed to be the ground state of H_2^-; there are no molecular orbitals available for a state at lower energies[18,19]. (b) The close-coupling calculation of Henry and Lane[20] gave no indication of rotational resonances.

In recent swarm experiments[21], the anomalous results persisted at densities too low to be explained by a continuum theory[22,23]. The swarm data of Crompton and Robertson are consistent with the existence of resonances in H₂ at 37 and 60 meV; although unable to demonstrate conclusively the existence of the resonances, Crompton and Robertson concluded that their data support Frommhold's hypothesis.

At present, we have available only the data of a few preliminary measurements on H₂ and D₂ between 20 and 100 meV. These data show that structure in the cross section is indeed present. The most pronounced feature is a peak in the H₂ cross section at 24 meV[24] with an observed width of about 4 meV. No other sharp features were found in H₂, although there is an indication that the cross section has a broad maximum in the energy range 60 to 90 meV. The results for D₂ are qualitatively different, with structure appearing between 26 and 60 meV.

In addition to further measurements on H₂ and D₂, measurements with HD and para-H₂, which we are preparing, should permit a more complete understanding and a more conclusive identification of the resonances.

The authors thank Professor V. W. Hughes for his continued encouragement and support, and Professors W. L. Lichten and A. Herzenberg for several stimulating conversations.

*This work was initiated by support of the Advanced Research Projects Agency, as administered by the Air Force Office of Scientific Research and the Office of Naval Research, and continued with joint funding by the Atmospheric Sciences and the Physics Sections of the National Science Foundation.

†Present Address: Universität Bielefeld, Fakultät für Physik, D-48 Bielefeld, Viktoriastr. 44, Germany.

1. E. Wiechert, Annalen der Physik und Chemie 69, 739 (1899).

2. F. Kirchner, Physik. Zeitschr. 30, 773 (1929).

3. F. C. Witteborn and W. M. Fairbank, Nature 220, 436 (1968).

4. G. C. Baldwin and S. I. Friedman, Rev. Sci. Instr. 38, 519 (1967).

5. M. Y. Nakai, D. A. LaBar, J. A. Harter, and R. D. Birkhoff, Rev. Sci. Instr. 38, 820 (1967).

6. A. V. Phelps, Rev. Mod. Phys. 40, 399 (1968).

7. R. W. Crompton, Advances in Electronics and Electron Physics 27, 1 (1969).

8. R. D. Hake and A. V. Phelps, Phys. Rev. 158, 70 (1967).

9. A. Herzenberg, J. Chem. Phys. 51, 4942 (1969).

10. D. Spence and G. J. Schulz, Phys. Rev. A 2, 1802 (1970).

11. M. J. W. Boness and G. J. Schulz, Phys. Rev. A 2, 2182 (1970).

12. F. Linder and H. Schmidt, Z. Naturforsch. 26a, 1617 (1971).

13. R. J. Celotta, R. A. Bennett, J. L. Hall, M. W. Siegel, and J. Levine, Phys. Rev. A 6, 631 (1972).

14. M. Krauss, National Bureau of Standards, Washington, D. C. (unpublished).

15. D. S. Stevens, Phys. Rev. 38, 1292 (1931).

16. L. Frommhold, Phys. Rev. 172, 118 (1968).

17. D. J. Kouri, W. N. Sams, and L. Frommhold, Phys. Rev. 184, 252 (1969).

18. H. S. Taylor and F. E. Harris, J. Chem. Phys. <u>39</u>, 1012 (1963).

19. T. E. Sharp, Atomic Data <u>2</u>, 119 (1971).

20. R. J. W. Henry and N. F. Lane, Phys. Rev. <u>183</u>, 221 (1969).

21. R. W. Crompton and A. G. Robertson, Austral. J. Phys. <u>24</u>, 543 (1971).

22. W. Legler, Phys. Lett. <u>31A</u>, 129 (1970).

23. A. Bartels, Phys. Rev. Lett. <u>28</u>, 213 (1972).

24. R. W. Crompton has informed us (private communication) that a delta-function resonance at 24 meV gives a worse computer fit to his swarm data than one at 37 meV.

PROGRESS WITH POSITRON BEAMS FOR ATOMIC COLLISIONS

W.C. Keever, B. Jaduszliwer and D.A.L. Paul

Physics Department, University of Toronto

Toronto 181

INTRODUCTION

Positron collision studies at low energies using beams of slow positrons below 1 keV appear to have been limited to the following: an unpublished thesis by Cherry[1] on secondary electron emission from surfaces under positron bombardment; and two published papers on the e^+ - He scattering cross sections by Costello et al.[2] and ourselves[3]. In a second publication of Costello et al.[4] they discuss experimental evidence in favour of postulating a negative work function for slow positrons escaping from a metal surface. In the period 1965 to the present there have been several other attempts by various groups to produce slow positron beams, most of them unsuccessful and unpublished. The more recent work by McGowan and his coworkers and by ourselves has all been directed at obtaining a satis- factory beam of slow positrons for atomic collisions or at studying the characteristics of the beam, partly with a view to obtaining a better beam, and partly so as to understand the slow positron emission process about which we shall make some state- ments in what follows.

Although these studies date from Cherry's work (1958) there does not seem to have been any statement in the literature about the maximum intensity which theoretically should be obtainable for positron beams, using, say, a radioactive source of primary positrons.

THEORETICAL ESTIMATES

The rationale behind present positron beam production for

low-energy collision studies is as follows. There are two
basically differing possibilities for producing a beam of energy
analysed positrons of a few electron volts. In the first, an
energy slice of the spectrum of radioactive or shower positrons
is selected at an energy where the spectrum $\frac{dN}{dE}$ is intense, and
these particles are electromagnetically slowed down. Only a
small fraction of randomly directed fast particles can be slowed
down by conservative fields, the fraction being somewhat less
than $\frac{E_s}{E_i} \frac{\delta\Omega}{4\pi} f$, where E_i is the initial energy and E_s the final
energy of the positrons, $\delta\Omega$ is the solid angle within which the
final beam is to be confined, and f is the fraction of the
original source within the chosen energy slice. The value of
this overall fraction, for a final beam having a 1 eV spread at
1 eV is about 10^{-10}. This method was successfully developed by
Lohnert and Schneider[5] for positron beams having a wide energy
spread at a few keV.

There is strong reason, therefore, to adopt another method
of production, as has been done in all low energy experiments
which we know of to date. In the second method the radioactive
or shower source is partially absorbed in a thin solid moderator,
and a fraction of the positrons thus trapped are found to escape
from the absorber surface after having been slowed down to low
energies in the solid. Such a slow positron source is character-
ised by: 1) type of positron activity (e.g. 20 MeV shower; radioactive sodium-22 source), 2) area of primary source, 3) moderating
material, 4) thickness of moderating layer, 5) backing material
(if any), and 6) condition of surface.

Supposing that the moderator has an absorption coefficient μ
in broad geometry for the positrons of the primary spectrum, then
the number of positrons trapped in the absorber is

$$\tfrac{1}{2} N_o \, (1 - e^{-\mu a})$$

for a source of strength N_o and an absorber thickness 'a', and
neglecting backscattering corrections. The number of positrons
escaping from the outer surface of such a layer we estimate from
diffusion theory, since the problem is essentially the same as
that of neutron diffusion in and escape from a moderating medium.
There are two extreme cases, one in which the positrons reaching
the outer surface at x=a are transmitted into the vacuum with
100% probability, and the other in which the transmission probability P at the boundary is very small. We obtain in the first
case an approximate upper limit to the number of positrons which
can escape into the vacuum of

$$\tfrac{1}{2} \, \mu L e^{-\mu a} \qquad\qquad (1)$$

where L is the diffusion length in the medium[†]. Expression (1) assumes that the absorber is considerably thicker than one diffusion length. In the second case we obtain the fraction

$$\tfrac{1}{4} \, \mu \ell_a P e^{-\mu a} \qquad\qquad (2)$$

where ℓ_a is the absorption length of positrons in the medium. If, as is supposed by Costello et al.[4] and Tong[7], the positrons are more or less thermalized in the medium prior to emission, then ℓ_a in a perfect crystal would be the annihilation mean free path which is $\sim 10^{-3}$ cm. If on the other hand the positrons must be epithermal before they can be emitted, then ℓ_a would be the mean free path for the combined processes of annihilation and slowing down out of the epithermal energy group. We remark at this point that it is well known in real metals that positrons are trapped at vacancy sites[8]. It follows that the trapping rate should be added to the annihilation rate for the purpose of calculating a realistic absorption length ℓ_a. The trapping rate would be largest for the slowest positrons and might be zero for epithermal positrons.

A recent experiment of Pendyala et al.[9] suggests that L is of the order of 500 Å for gold, which sets the upper limit for positron beam production, using (1), at $1.5 \times 10^{-3} \, N_0$ for a sodium-22 source of strength N_0 and a gold moderator.

In practice we have not achieved a beam of more than a few $\times 10^{-7} \, N_0$ which suggests that the reflection coefficient at the emitting surface is indeed always high.

EXPERIMENTAL

The slow positron emitters which we used to produce our first low energy cross sections[3] were smooth evaporated gold layers about 1000 Å thick on mica substrates. The equipment consisted of source, focussing accelerator, 90° spherical spectrometer and a straight scattering chamber, and is illustrated in fig. 1. Typical count rates at that time were 5 coincidences/min using a 14 mCi Na^{22} positron source mounted on a

---

[†]   In deducing this formula (1) we assumed that an extrapolation distance of $^2/_3 \, \lambda_t$, as is used in diffusion theory extrapolations[6], would provide an adequate boundary condition for this case: the parameter λ_t is the transport mean free path.

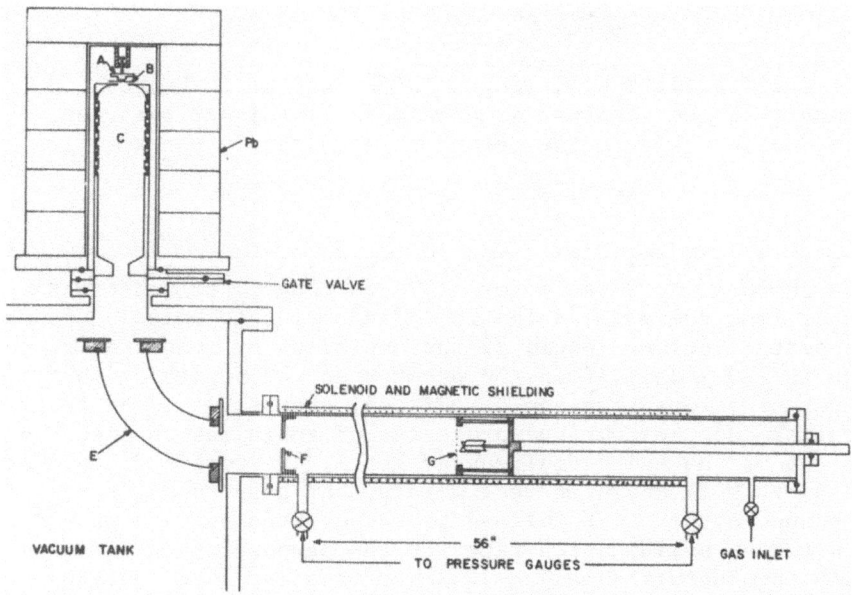

Fig. 1:   Schematic representation of the positron beam apparatus.
A: source mount; B: moderator; C: electrostatic accelerating
and focussing optics; E: electrostatic spherical spectrometer;
F: baffle and entrance hole; G: slow positron annihilation
target. Shadowed areas indicate insulators.

gold button and having a thick protective layer. The 5% coincidence detection efficiency for the scintillation counter
system corresponds to a beam intensity of $6 \times 10^{-9} N_0$ where N_0
is the total positron emission rate of the source. However
this intensity should not be taken as the beam intensity emitted
by the source since the transmission of our spectrometer at a
test energy of 30 eV is at most 30%. The intensity numbers
reported here are all referred to experiments at 30 eV where the
spectrometer transmission would be the same.

    There is now evidence that the beam intensity can be
increased by a large factor by tampering suitably with the physical
characteristics of the metal-vacuum interface.

    The first step in that direction was the observation that
large numbers of positrons were coming out of a 200 cell per inch
gold-plated nickel "Micromesh" grid supplied by E.M.I., even
though the grid transparency was 80% and it therefore inter-
cepted only 20% of the primary beam. An electron micrograph of
that grid shows that it has a very rough surface.

    Figure 2 shows the energy spectrum of the positrons coming
out of the Micromesh grid. During the first 24 hours inside the
vacuum system, the intensity decayed to less than 1/3 of
initial value. This immediately suggests that the presence of
adsorbed and/or absorbed gases is somehow important to enhance
the probability of positron escape. To check that hypothesis,
as well as to identify the agents of the increased emission,
a series of experiments was performed in which the grid sat in
the vacuum until the positron intensity had reached a low
value after which further decay was very slow, and then it was
reactivated by exposing it to some particular gas.

    In fig. 3 (a,b,c) we can see the results of the first set
of such experiments; the intensity in the peak of the positron
energy spectrum was monitored for periods of up to 60 hours
after exposures of 15 hours to air, nitrogen and oxygen. In
each run we observed the presence of two successive exponential
decays with very different lifetimes, corresponding to different
modes of gas attachment to the surface. For N_2, those lifetimes
were 6.8±1.4 hours and 96±20 hours.

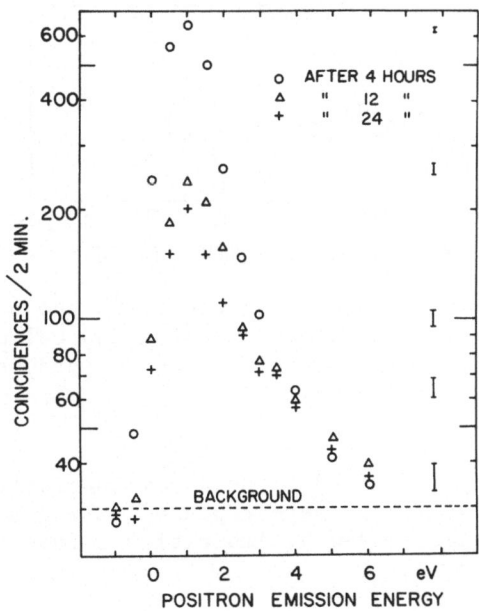

Fig. 2:   Energy spectra of positrons coming out of a 200
cells per inch gold-plated nickel "Micromesh" 4 hours, 12 hours
and 24 hours after introduction to the vacuum system. The
instrumental resolution is about 1 eV full width at half
maximum (F.W.H.M.).

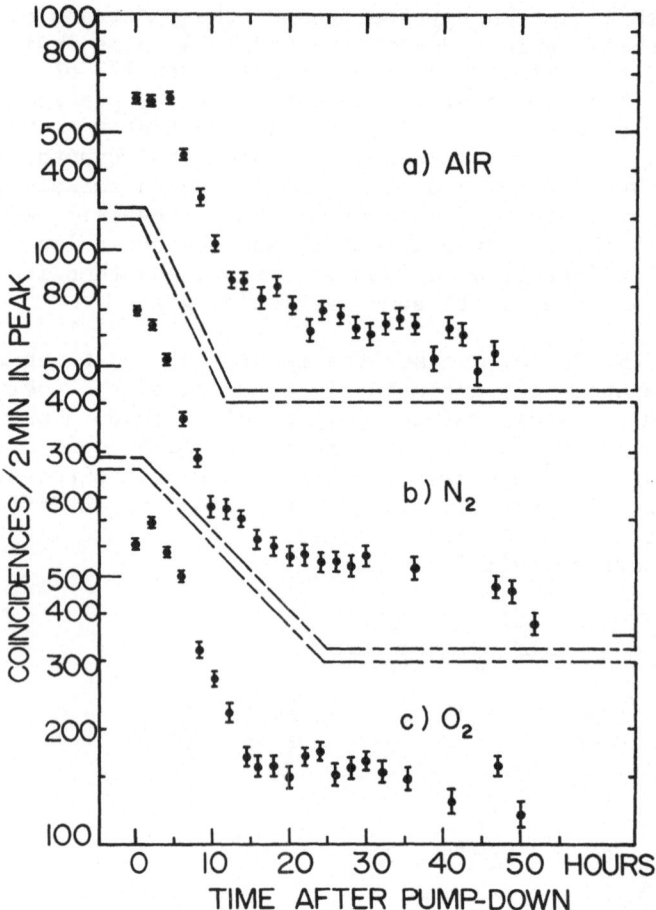

Fig. 3:   Time decay of the number of counts in the peak of the
energy spectrum of positrons coming out of a 200 cells per inch
gold-plated nickel "Micromesh" after 15 hours exposure to
a) air,   b) nitrogen,   c) oxygen.

As these experiments failed to show any conclusive
difference between gases, they were repeated taking some precautions against results caused by impurities common to all of
them.

The results after one hour exposure to N_2, O_2 and He are
shown in fig. 4. Now differences between gases are clearly
visible, both in the maximum available intensity and in the
decays. It was also observed in a separate run that the
presence of water vapor on the grid quenched the slow positron
emission.

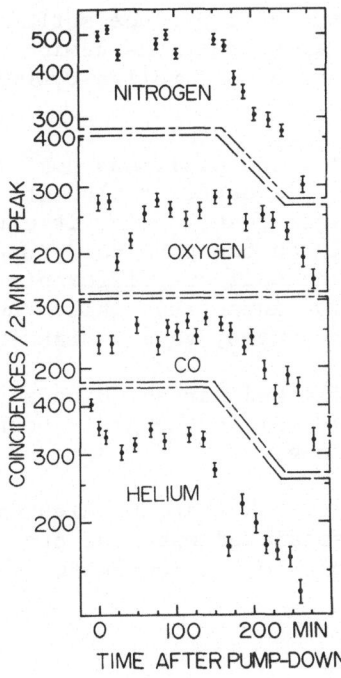

Fig. 4:   Time decay of the number of counts in the peak of the
energy spectrum of positrons coming out of a 200 cells per inch
gold-plated nickel "Micromesh" after one hour exposures to N_2,
O_2, CO and He. The time scale is relative; the existence of
a negative time point for He only means an earlier start.

From the purely pragmatic point of view of obtaining an
intense beam of low energy positrons, even if an increase of the
counting rate by a factor of 50 had been accomplished, this
still was not too satisfactory, because of the rapid decrease in
intensity after a few hours. We hoped to improve that by
finding a gas that would remain longer periods on the metal
surface, but as fig. 4 shows, we had not succeeded.

A series of experiments with other Micromesh grids again
pointed out the importance of the surface conditions for the
positron intensity. A 2000 cell per inch gold-plated nickel
grid, having a transmission of only 20% was expected to yield 4
times as many positrons as the former grid having 80% transmission but gave instead only half as many, while identical
copper and nickel grids, but now unplated, gave a factor of two
increase with respect to the original grid.

Several further attempts to produce surfaces that would reproduce the high numbers of slow positrons coming out of Micromesh
grids all failed. Finally, a cold rolled 0.00015" nickel foil gave
very satisfactory results.

The energy spectrum of the positrons coming out of that foil
can be seen in fig. 5. It is narrower than the spectrum of positrons coming out of the grids, and again, it can be seen that the
intensity depends strongly on the physical properties of the surface. The two faces of the foil are different; one of them,
highly polished, gave a low intensity, while the other, dull looking due to small-grain roughness, gave a high yield.

After etching the polished surface with a nitric acid - acetic
acid solution that attacks mainly the grain boundaries, its emission
increased by a factor of seven.

Further experiments proved that the high emission of the
rough surface was also associated with the presence of adsorbed
and/or absorbed gases, but the time scale for change was much

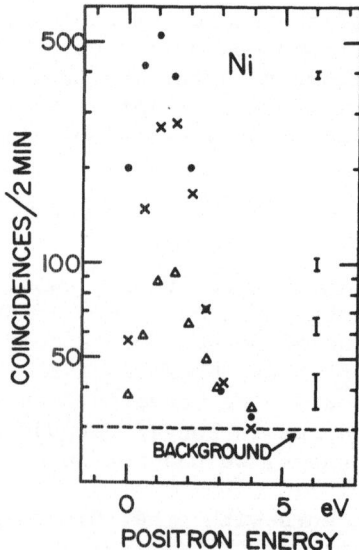

Fig. 5: Energy spectra of positrons coming out of a 0.00015 inch
cold-rolled nickel foil. Triangles correspond to emission from
the polished surface; black circles to emission from the rough
surface. Crosses are for the emission from the polished surface
after etching. The instrumental resolution is about 1 eV F.W.H.M.

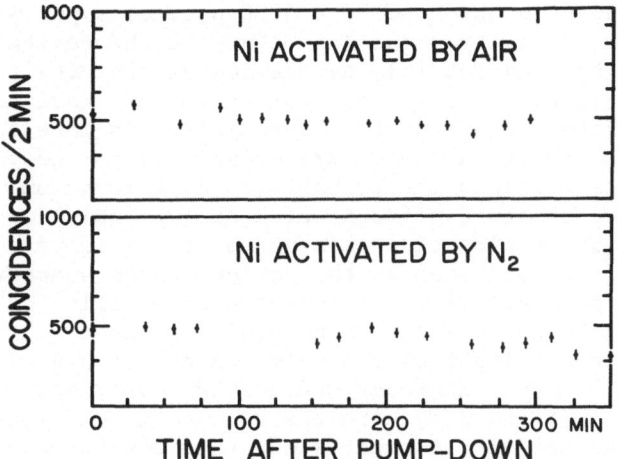

Fig. 6: Time decays of the number of counts in the peak of the
energy spectrum of positrons coming out of the rough surface of a
0.00015 inch cold-rolled nickel foil, after exposure to air and N_2.

larger; in six hours only 15% of the initial intensity was lost
(fig. 6), and while a one hour exposure of N_2 was enough to
reactivate the Micromesh grid material, ten or twelve hours have
been necessary to reactivate the nickel foil.

     As H_2 is soluble in solid nickel, the nickel foil was exposed
to a hydrogen atmosphere which was found to suppress the emission
almost completely. After two weeks alternately spent under vacuum
and in a N_2 atmosphere the initial intensity has not been fully
recovered.

     As compared with the original smooth gold emitter the beam
intensity obtained from the cold-rolled nickel foil is $3 \times 10^{-7} N_o$,
a factor of 50 improvement. Although this is still only 40%
of the highest specific intensity which was obtained using the
gold-plated Micromesh of 200 cells/inch, the long period of
retention of the emission from nickel foil makes this material more
useful for atomic collisions experiments.

                              DISCUSSION

     The results which we have presented are both preliminary and
qualitative. Our equipment was not designed for surface studies
and indeed we feel that we have almost exhausted its possibilities
in the current investigations, at least for the type of work
reported here. A broad, energy-analyzed positron beam can now

be routinely produced in which we have 5000 particles per minute. Nevertheless, this intensity comes far below the theoretical optimum suggested by equation (1), even allowing that the μL product in nickel might be very much smaller than we have assumed for gold. We therefore conclude that formula (2) is more appropriate and that the positrons which are escaping into the vacuum have a rather small transmission probability at the vacuum interface. We have observed effects which are probably due to adsorption and to absorption of common gases on and in metals. The enhancement of the emission in the various cases supports the above notion of a generally low transmission probability at the vacuum interface. We emphasize that none of the experiments reported here sheds any light on the question of the negative work function[4,7,10]. We cannot say whether the positrons escaping in the energy peaks (figs. 2,5) were thermalized in the metal. However these peaks are all somewhat wider than the experimental resolution and in particular the curves of fig. 2 contain a significant fraction of the total intensity at higher energies, up to 6 eV. We note that in the beam decay studies it was generally the 1 eV peak which decayed (fig. 2) leaving the epithermal spectrum rather little changed. If the peak were due to positrons having very low energies in the metal (< 1 eV) one would certainly expect any barrier penetration factor to be altered by adsorbed layers. On the other hand, the epithermal spectrum several eV higher in energy should hardly be affected.

An independent mechanism for the suppression of the emission of free positrons from a metal surface is positronium formation at the surface. On the "jellium" model of a metal there is a "tail" of electron plasma extending into the vacuum at the surface. There is a probability of positronium formation in this layer but we do not know of any calculations predicting the magnitude of this effect. However, weak positronium emission from aluminum has recently been observed in our laboratories in a preliminary experiment. This result has, however, to be confirmed.

REFERENCES

[1]. W.H. Cherry, Ph.D. Thesis, Princeton University (1958).

[2]. D.G. Costello, D.E. Groce, D.F. Herring and J.W. McGowan, Can.J.Phys. 50, 23 (1972).

[3]. B. Jaduszliwer, W.C. Keever and D.A.L. Paul, Can.J.Phys. 50, 1414 (1972).

[4]. D.G. Costello, D.E. Groce, D.F. Herring and J.W. McGowan, Phys.Rev. B, 5, 1433 (1972).

5. G.H. Lohnert and R.T. Schneider, Nuclear Technology, 10, 315 (1971).
6. S. Glasstone and M.C. Edlund, "The Elements of Nuclear Reactor Theory" (Van Nostrand, 1952), chapter 5.
7. B.Y. Tong, Phys.Rev. B, 5, 1436 (1972).
8. B.T.A. McKee, A.G.D. Jost and I.K. McKenzie, Can.J.Phys. 50, 417 (1972).
9. S. Pendyala, J.W. McGowan and P.W. Zitzewitz, Phys. in Canada 28, No. 4 (Congress Issue), abstract EB12 (1972).
10. C.H. Hodges and M.J. Stott, Phys. in Canada 28, No. 4 (Congress Issue), abstract BA3 (1972).

SPIN POLARIZATION IN PROTON-XENON CHARGE EXCHANGE COLLISIONS

Joseph Macek and Robin Shakeshaft

Behlen Laboratory of Physics, The University of

Nebraska, Lincoln, Nebraska 68508

The production of spin polarized particles by collisions between initially unpolarized particles has been of continuing interest since the early theoretical work of Mott[1] on elastic electron scattering. In the subsequent forty years various experiments have been devised to polarize elementary particles, particularly electrons and nucleons, by this method. Compound systems including light nuclei can also be produced by scattering from unpolarized targets. Conspicuously absent from the list of particles which have been spin polarized by scattering are atomic atoms and ions, particularly the hydrogen atom. Their absence is due to the availability of other techniques for polarizing atoms, such as atomic beam and optical pumping techniques, and to the lack of a suitable reaction. In this paper we propose a suitable reaction,[2] and report detailed calculations of the spin polarization.

In order to polarize a particle by a parity conserving interaction, its spin must couple to another axial vector whose direction is determined by the experimental arrangement. In a scattering experiment the classical orbital angular momentum of the projectile provides the necessary axial vector. This vector is parallel or antiparallel to the axial vector $\vec{n}=\vec{p}_i \times \vec{p}_f$, where $\vec{p}_i$ and $\vec{p}_f$ are the momentum vectors of the incoming and outgoing particle. Since the spin of the particle couples to the orbital angular momentum through the spin orbit interaction, particles with spin oriented parallel to $\vec{n}$ scatter with a different angular distribution than those oriented antiparallel to $\vec{n}$. When the spin orbit interaction is weak compared to the electrostatic interaction the resulting spin polarization is small, unless its effect can be enhanced by some "leverage" mechanism,[3] e.g., resonant scattering or scattering near a minima in the angular distribution. Our

proposed reaction achieves a high spin polarization by utilizing
a "leverage" mechanism implied by the Massey adiabatic principle.[4]
This principle states that a reaction involving a conversion of
translational kinetic energy to internal energy by an amount ΔE
peaks when the relative velocity v of the collision partners
equals $b\Delta E/h$, where b is a length of the order of the atomic
dimensions involved.

In the charge exchange reaction

$$H^+ + Xe \rightarrow H + Xe^+(^2P_{1/2}) \tag{1}$$

0.17 eV of translational kinetic energy is converted to internal
energy, while in the competing reaction

$$H^+ + Xe \rightarrow H + Xe^+(^2P_{3/2}) \tag{2}$$

1.3 eV of internal energy is converted into translational kinetic
energy. According to the Massey criteria, with b set equal to
5 a.u., the cross section for reaction (1) peaks at ~25 eV incident proton energy, while the cross section for reaction (2)
peaks at ~2 keV. Thus the small spin orbit splitting of 1.3 eV
between the $Xe^+(^2P_{1/2})$ and $Xe^+(^2P_{3/2})$ states significantly affects
the charge transfer reaction over a 2 keV energy range. This
magnification of the energy scale measures the "leverage" provided by the adiabatic principle in reactions of the type (1) or
(2), and we see that it is very large. Further, the leverage
maximizes approximately when the cross section maximizes in contrast with the mechanism operative in the Fano effect[5] and in
low energy scattering of electrons by heavy atoms.[6]

Consider collisions of slow (less than 100 eV) protons with
xenon. Charge exchange occurs predominantly via reaction (1).
The quantum number corresponding to the reflection operator

$$Y = Pe^{i\pi\hat{n}\cdot\vec{J}}, \tag{3}$$

where P is the internal parity operator and $\vec{J}$ is the internal
electronic angular momentum operator, is conserved. Consequently
the final state of $H + Xe^+(^2P_{1/2})$ consists of H atoms and Xe^+ ions
with the electronic spin of the H atom oriented parallel to the
electronic angular momentum of the Xe^+ ion and both oriented
parallel or anti-parallel to the vector $\vec{n}$. The spin polarization
fraction is then

$$P_s = \frac{\sqrt{8}\,|a_o a_1|\sin\Delta}{|a_o|^2 + 2|a_1|^2}, \tag{4}$$

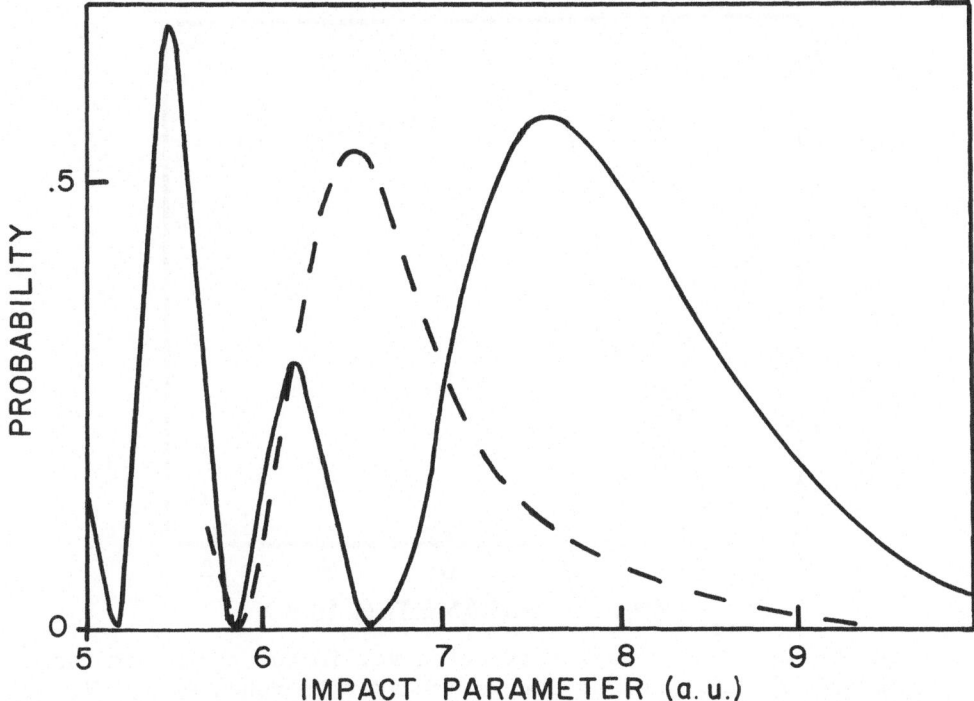

Fig. 1. Charge transfer probabilities vs. impact parameter for
15 eV protons in xenon. The full curve gives the probability for
$|M| = 1$ final states, and the dotted curve gives the probability
for the $M = 0$ final state.

where a_M is the amplitude for finding the HXe$^+$ system in the M'th
magnetic substate, Δ is the phase difference between a_0 and a_1 and
where we have used $a_M=(-)^M a_{-M}$. Since a_0, a_1 and Δ are in general
non-zero, P_s is non-zero and the reaction (1) produces spin polarized H atoms.

Detailed calculations based on the impact parameter approximation[7] and a Herman-Skillman[8] Hartree-Fock model for the outer
5p electrons of Xe and Xe$^+$ incorporating empirical $P_{1/2,3/2}$ splitting of the Xe$^+$ levels predict substantial polarization at 15 eV
incident proton energy. While the calculations are tedious, the
results are quite easy to interpret. Figure 1 shows the charge
transfer probabilities $|a_0|^2$ and $2|a_1|^2$ vs. impact parameter. Note
that $|a_0|^2$ peaks at a smaller impact parameter than does $|a_1|^2$,
but that both are non-zero over a substantial range of impact parameters. We can understand the positions of the peaks if we
assume that the electron is transferred only when the temporarily
formed HXe$^+$ molecule is in a Σ state. At large impact parameters
the electron is transferred only at the distance of closest

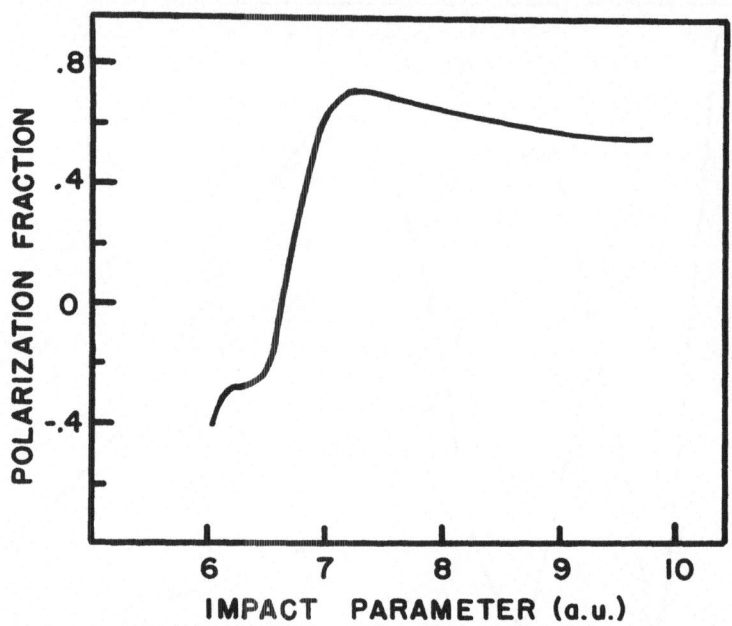

Fig. 2.   Spin polarization fraction vs. impact parameter for
hydrogen atoms formed by charge transfer in xenon at 15 eV.

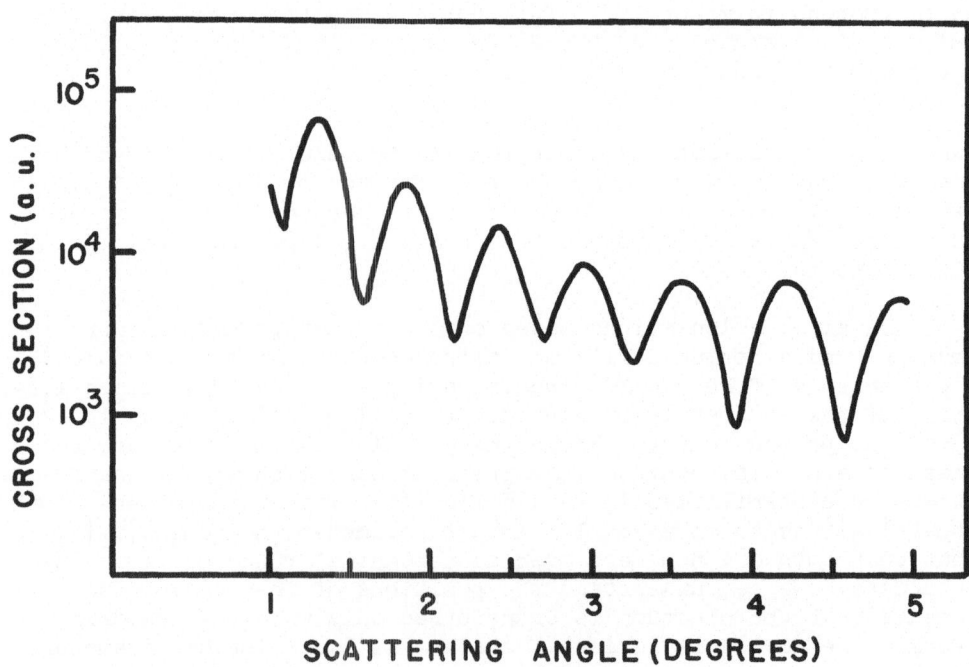

Fig. 3.   Differential charge transfer cross section.

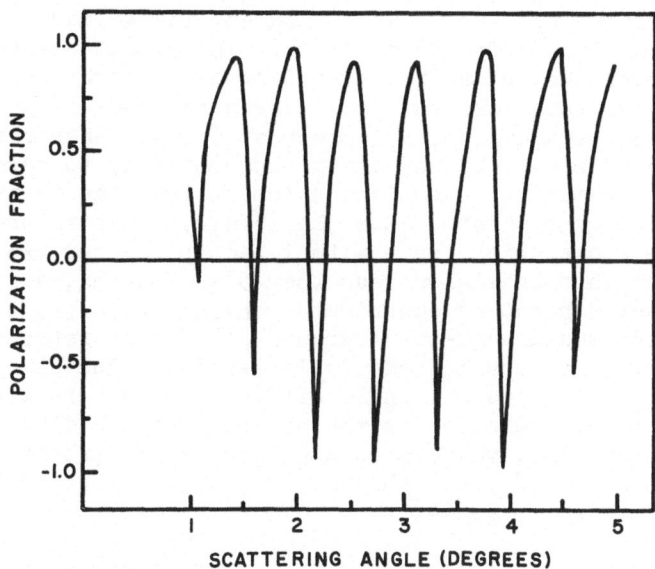

Fig. 4. Spin polarization fraction vs. scattering angle for
hydrogen atoms formed by charge transfer in xenon at 15 eV.

approach when the internuclear axis is perpendicular to the beam
axis. As the H atom and Xe^+ ion separate, the projection of electronic angular momentum maintains the orientation it had at the
instant of transfer, but the internuclear axis rotates through 90°
and the Σ state becomes a π type final state corresponding to a
large value for a_1. At intermediate impact parameters the transfer
takes place when the internuclear axis is oriented along the beam
axis corresponding to a large value of a_0. The spin polarization
vs. impact parameter is shown in Fig. 2 where we note that a spin
polarization of 70% occurs when the charge transfer probability
equals 0.65.

      While the results presented in Figs. 1 and 2 support our con-
tention that reaction (1) produces spin polarized H atoms, they do
not relate to experimentally measurable quantities since the charge
transfer probabilities and polarization are given in terms of the
unobservable impact parameter. Spin polarization and differential
charge transfer cross sections for a given scattering angle may be
obtained using the eikenol integrals of Wilets and Wallace;[9]

$$f_M(\theta) = (-i)^{M+1} \cos^2(\theta/2) \int_o^\infty \rho \, d\rho a_M(\rho) J_M(\hbar^{-1} p_f \rho \sin\theta). \qquad (5)$$

Figure 3 shows the differential cross section and Fig. 4 shows the
spin polarization calculated using eqs. (4) and (5). Note the
occurrence of sharp minima in both curves, with the spin
polarization rising to 100% between the minima.

By referring to Figs. 1 and 2 and eq. (5) we can interpret the sharp minima as diffraction minima. The amplitudes $a_M(\rho)$ in eq. (5) describe the transmission of H atoms through the Xe target. The transmission functions in Fig. 1 are seen to oscillate rapidly up to an impact parameter of the order of 6 a.u., then rise to an abrupt maxima after which they decay exponentially to zero at large impact parameter. Atoms which depart from the scattering region with impact parameter less than 6 a.u. scatter through large angles, while atoms which leave with impact parameter greater than 6 a.u. scatter through angles less than 5°. However, a substantial range of impact parameters contribute coherently to the scattering into one of the small angles. A ground glass disc represents the optical analogue of such a transmission system. The abrupt edge of the disc produces a small angle diffraction pattern with minima separated by $\Delta\theta \approx \lambda/d$, where d is the diameter of the disc. For 15 eV H atoms $\lambda = 2\pi\hbar/mv$, where m is the mass of the proton, $v = .0245$ a.u., $d = 12$ a.u., thus $\Delta\theta \approx .6°$, in good agreement with the results in Figs. 3 and 4.

In summary, spin polarized hydrogen atoms can be produced by charge transfer for H^+ in Xe. The occurrence of diffraction minima and maxima in the angular distribution of the scattered hydrogen atoms tends to increase the maximum value of the spin polarization.

1. N. F. Mott, Proc. Roy. Soc. (London) A124, 425 (1929); A135, 429 (1932).

2. J. Macek and R. Shakeshaft, Phys. Rev. Letters, 27, 1487 (1971).

3. U. Fano, Comm. At. Mol. Phys. 2, 36 (1970).

4. H. S. W. Massey and E. H. S. Burhop, Electronic and Ionic Impact Phenomena (Oxford U.P., New York, 1952), p. 450.

5. U. Fano, Phys. Rev. 178, 131 (1969).

6. J. Kessler, Rev. Mod. Phys. 41, 3 (1969).

7. R. Shakeshaft, J. Phys. B5, 559 (1972).

8. F. Herman and S. Skillman, Atomic Structure Calculations (Prentice Hall, Englewood Cliffs, N.J. (1963).

9. L. Wilets and S. J. Wallace, Phys. Rev. 169, 84 (1968).

SPECTROSCOPY WITH TUNABLE LASERS[*]

Theo W. Hänsch

Department of Physics, Stanford University

Stanford, California 94305

CONTENTS

1. INTRODUCTION

The vigorous development of various types of broadly tunable lasers in the past few years has led to remarkable progress. Organic dye lasers, semiconductor diode lasers, spin-flip Raman lasers and parametric oscillators together with nonlinear mixing devices now cover the entire wavelength range from the near

---

[*]Partly sponsored by the U.S. Office of Naval Research, Washington, D.C.

ultraviolet (2600Å) to the middle infrared (34μm). The tuning range
will certainly be extended in the near future. Single mode
operation and very narrow linewidths have been achieved with each of
these devices in the laboratory. Intensive efforts are underway and
already partly successful to make the wavelength of such lasers very
stable and reliably tunable over any given small spectral region of
interest. Tunable lasers will undoubtedly have a large impact on
the future of atomic and molecular spectroscopy.

      Lasers can be many orders of magnitude superior to conventional
light sources in their intensity per spectral interval and in their
spatial coherence, which determines for instance, how well the
radiation can be focused or collimated. Intense light pulses in the
nano- and picosecond range can be generated. The now available
tunable lasers, moreover, can operate at wavelengths where strong
conventional sources such as hollow cathode lamps are difficult or
impossible to build. And their monochromaticity and tunability can
eliminate the need for costly and bulky spectrometers and
monochromators and can provide impressively increased instrumental
resolution. It is therefore easy to predict that tunable lasers
will greatly increase the potential and convenience of such classic
spectroscopic techniques as absorption spectroscopy, fluorescence
spectroscopy, level crossing, optical-RF double resonance spectroscopy
or optical pumping. The latter techniques will for instance become
easily applicable to rare isotopes or to excited states of atoms,
ions and molecules.

      In addition, however, the unique properties of laser light open
the way to powerful new nonlinear spectroscopic techniques. Various
such techniques have been developed in the past using fixed frequency
lasers of very limited tuning range. Among the most useful are the
various methods of high resolution saturation spectroscopy or Lamb-
Dip spectroscopy, which eliminate Doppler-broadening of resonance
lines in gaseous media by spectral hole burning, and permit
spectroscopic studies of unprecedented resolution in the optical
and infrared region. In the past, these techniques were essentially
restricted to the study of laser transitions or to molecular
transitions in accidental coincidence. The advent of narrowband
widely tunable lasers is now removing these restrictions and will
certainly open the doors to much exciting progress in spectroscopy.

      In the first part of the present paper we will try to summarize
the underlying principles and basic concepts of the various kinds of
presently known tunable lasers and to outline the more recent
developments and the present state of the art in their technology.
In the second part we will demonstrate the wide potential of tunable
lasers as research tools in spectroscopic investigations by reviewing
a number of recent experiments. A comprehensive review is certainly
beyond the scope of this paper. The main emphasis of this summary
will be placed on tunable dye lasers and on spectroscopy near the

visible region, in accordance with the background and the special
interests of the author.

2.   TUNABLE LASERS

2.1 Organic Dye Lasers

Organic dye lasers offer unrivaled versatility and simplicity
in the generation of broadly tunable coherent light from the near
ultraviolet (3400Å) to the near infrared (1.2μm). Stimulated
emission from fluorescent organic dyes in liquid solution was first
discovered in 1966 by Sorokin and Lankard (1) and independently by
Schaefer, Schmidt, and Volze (2). Since then, dye lasers have seen
an almost explosive development and numerous review articles have
been written, which give fairly extensive bibliographies up to 1969
or 1970 (3-8).

The electronically excited energy levels of a laser-dye
molecule appear as continuous rather broad bands due to a complicated
substructure of vibrational and rotational states, as schematically
depicted in Fig. 1. The molecule is optically excited from the
singlet ground state to higher singlet states, using laser or flashlamp pumping. The molecular energy then decays nonradiatively within
10^{-11} to 10^{-12} sec to the lower edge of the first excited singlet

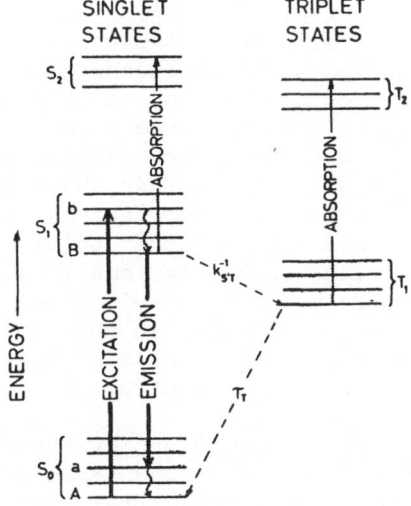

Fig. 1 Energy levels of dye molecule, schematic (from Ref. 4).

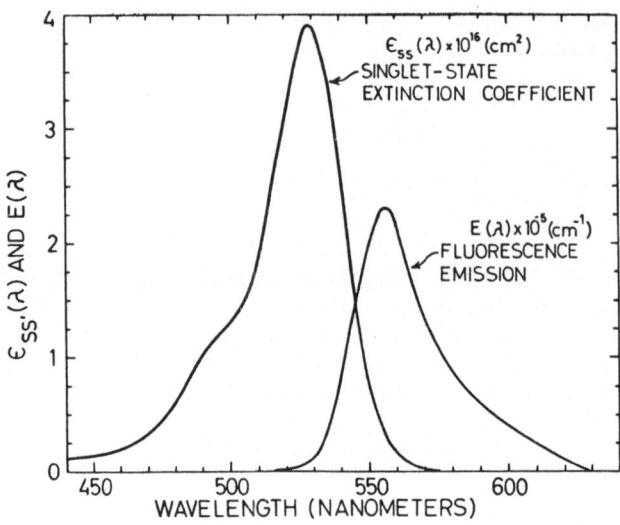

Fig. 2 Absorption and fluorescence spectrum of the laser
dye rhodamine 6G in ethanol solution (from Ref. 4).

state. From here, the molecule can undergo transitions to the
various vibrational-rotational levels of the ground state under
emission of broadband Stokes-shifted fluorescent light, as shown in
Fig. 2. When population inversion is achieved by sufficiently strong
optical pumping, light amplification by stimulated emission of
radiation can be obtained over almost the entire fluorescence band,
except where it is effectively overlapped by the absorption band,
i.e. typically over several hundred Angstroms. The rapid thermalization of vibrational and rotational energy leads to an essentially
homogeneously broadened gain band, and it is possible to channel
almost the entire available energy into a very narrow spectral region
by using a laser cavity with wavelength selective feedback. The
upper laser level has typically a rather short lifetime of only a few
nsec, which prevents a substantial storage of pump-energy and the
generation of giant pulses in Q-switched operation. Nonradiative
singlet-triplet transitions result in complications due to absorption
of excited triplet states at laser wavelengths. Due to the short
phase memory of the active molecules in the liquid solution, it is
possible to understand many details of dye laser operation
quantitatively within the framework of simple rate equations.

To date, laser action has been achieved with hundreds of organic
dyes. Major compilations of laser- and flashlamp-pumped dyes have
been published (5,8,9). Many useful laser dyes belong to the
families of oxazoles, coumarins, xanthenes and polymethines, for

NO.	DYE	FAMILY	STRUCTURE
1	3,3' DIETHYLTHIA-TRICARBOCYANINE IODIDE	POLYMETHINE	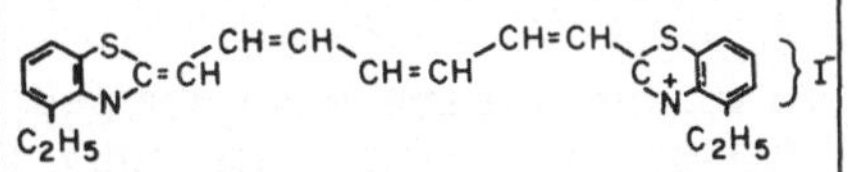
2	RHODAMINE 6G	XANTHENE	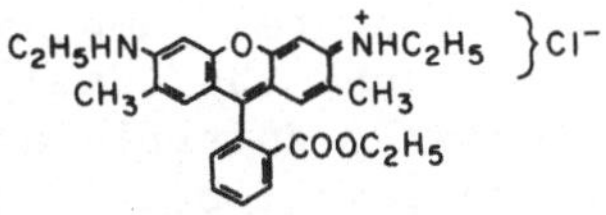
3	4-METHYL-UMBELLIFERONE	COUMARIN	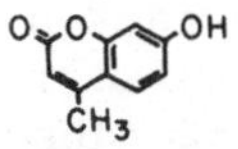
4	2,5 DIPHENYL-OXAZOLE	OXAZOLE	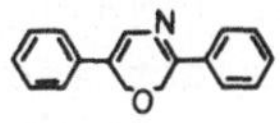

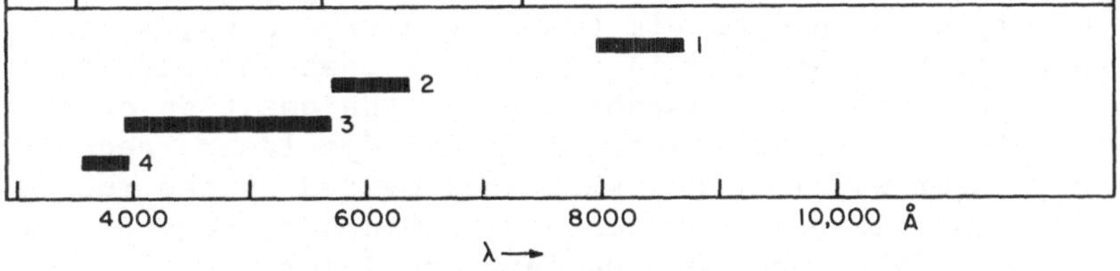

Fig. 3 Examples of laser dyes with their respective wavelength tuning range.

which typical examples with their respective tuning range are shown in Fig. 3. Several workers have investigated the effect of triplet quenching agents, such as cyclooctatetraene, which can markedly improve the performance of long-pulse and continuous dye lasers (10-12). The tuning ranges of certain laser dyes can be extended by adjusting the pH value of the solvent (13), and by using a delicate balance between 4-methyl-umbelliferone and its acidic excited complex form, Shank, et al. (14) have been able to cover the entire range from the near UV to the yellow (390nm - 544nm) with a single dye solution. Dye mixtures have also been used to extend the tuning range or to transfer energy from an available pumpsource to a dye with improper absorption bands (4,15). As our understanding of the mechanisms in organic laser dyes grows, it will become increasingly possible to tailor or "engineer" dyes with specific properties (16).

Solid hosts, in particular polymethylmethacrylate (4) and gelatin (17) have been used successfully instead of liquid solvents.

Attempts to obtain dye laser action in the vapor phase are presently
underway in several laboratories.

The first dye lasers were pumped with Q-switched solid state
lasers (1-3,5-8), in particular ruby (694.3nm) and frequency doubled
Nd+ lasers (530.0nm). Shorter pump wavelengths can be generated
with frequency doubled ruby (347.2nm) and frequency quadrupled Nd+
(265.0nm). A particularly suitable and reliable pump laser source
is the pulsed nitrogen laser, operating at (337.1nm), whose high
pulse repetition rate provides a convenience similar to cw operation
(18-20). The short pump-pulse lengths of 5 - 100 nsec eliminate
the problem of triplet state absorption, and efficiencies up to 50%
have been observed with certain laser pumped dyes. Laser-pumping
certainly provides the easiest access to tunable dye laser radiation
once the pump laser exists. It is worth mentioning that it is
possible to pump organic dyes via two-photon processes, so that the
emitted laser light is shorter in wavelength than the pumplight
(21). Due to photochemical instability and the position of their
absorption bands, most of the known infrared dye lasers require
pumplight of longer wavelengths, such as provided by the ruby laser.
The choice of possible pump sources can be widened, however, by
pumping with the light from intermediate dye lasers.

Different geometries have been utilized for laser-pumping of
dye lasers (Fig. 4). End-pumping requires a dichroic mirror,
transparent for the pumplight, and complicates the insertion of
wavelengh-selectors into the cavity. This problem can be overcome
by using a near-endpumped geometry as first suggested by Bradley
(22). The rectangular beam geometry of the commercially available
superradiant nitrogen laser is particularly suitable for side-pumping
(18-20).

Strong directional 'superradiant' emission of amplified
spontaneous radiation is often observed from laser-pumped dyes in the
absence of an optical resonator (23). Such lightsources provide a
smooth continuous spectrum, which can be useful as background for
absorption measurements.

Many dyes, operating from the UV throughout the visible, can be
pumped successfully with flashlamps (4,9). The excitation is easiest
with flashlamps of short risetimes of the order of 100 nsec or less.
The pump geometry can be coaxial, or similar to the geometries
developed for solid state lasers (Fig. 5). Thermal schlieren effects
can be overcome by rapid circulation of the liquid dye solution.
Flashlamp-pumped dye lasers can be very simple and cheap, so that
they have been recommended even to the hobbyist at home (24).
Efficiencies on the order of 0.05-4% approach or surpass those of
solid state lasers. Output energies of several J/pulse have been
obtained (25), and repetitively pulsed flashlamp pumped dye lasers
have been reported (26,27). Schmidt has obtained 3.5 W average

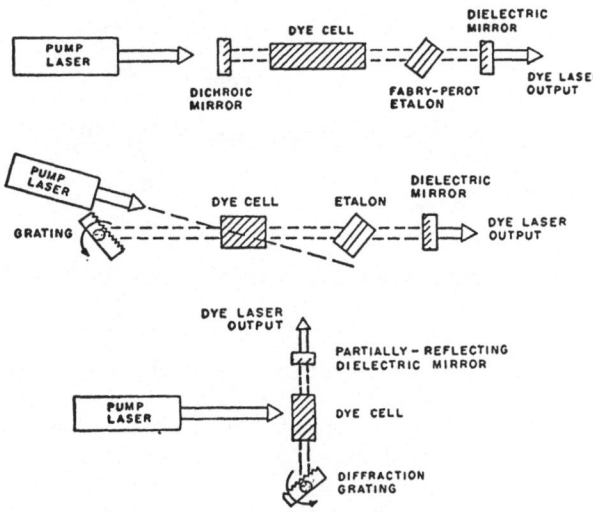

Fig. 4   Various geometries of laser-pumped dye lasers (from Ref. 8).

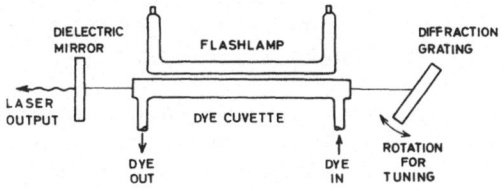

Fig. 5   Flashlamp pumped dye laser (from Ref. 4).

output with a flashlamp pumped laser, operating at 100 pps in the
yellow, using rhodamine 6G, one of the most effective laser dyes.

Continuous operation of a dye laser was first demonstrated by
Peterson, Tuccio and Snavely in 1970 (28) by end-pumping a
hemispherical cavity of only 4.5mm length with the tightly focused
blue-green light of a 1W Argon ion laser. The mirrors were directly
immersed into the rapidly flowing dye medium, a solution of rhodamine
6G in water with added detergent to prevent dimerization. CW dye
lasers with separate dye cell and longer cavities, which facilitate
the insertion of wavelength selectors, modulators or saturable
absorbers, have since been described (29-37). All of the so far
realized cw dye lasers are pumped by Argon ion lasers. The threshold

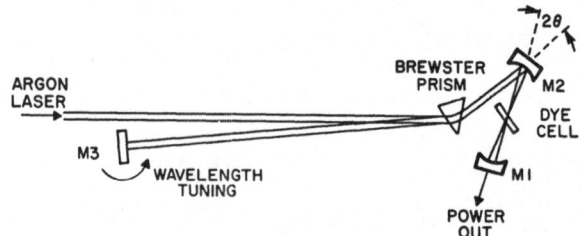

Fig. 6   Continuous wave dye laser (from Ref. 36).

can be as low as 50 mW.   An astigmatically compensated folded three
mirror cavity with short Brewster-angle dye cell, as shown in Fig. 6,
is noteworthy (36).   Such a configuration was also used by
researchers at Bell Telephone Laboratories as a common cavity for
both the Argon pump laser and the dye laser (30).   To date, cw laser
action has been obtained with several dye solutions, spanning the
range from 5220 to 6900 Angstroms (34,11,37).   A continuous output of
about 3W has been achieved with a 10W pump laser (38).

While it is possible to tune organic dye lasers within certain
limits by changing the always present spectral losses, e.g. by
adjusting the dye concentration or dye cell length, by changing the
solvent or by changing the temperature of the medium, monochromatic
radiation from dye lasers has mostly been generated with the help of
dispersive intracavity wavelength selectors.

In cw dye lasers, a narrowing down to a fraction of one Angstrom
can already be achieved with single wavelength selectors of low
dispersion, such as prisms or chromatic lenses (29,31,32), thanks to
the multiple passes of the laser light inside the cavity.   Low losses
of the tuning elements are important, because the single pass gain
available in cw dye lasers seldom exceeds a few percent.   Single mode
operation with a bandwidth between 10-20 MHz, primarily determined by
random perturbations, has been achieved with additional inserted
Fabry-Perot interferometers (33,39).   The technical problem of
continuous smooth wavelength tuning, important for applications in
high resolution spectroscopy, will very likely be solved in the near
future.

Larger losses can be tolerated in flashlamp-pumped lasers.
Gratings, prisms, Fabry-Perot interferometers, or combinations of
these elements have been used successfully as tuning elements and
bandwidths between several Å and less than 0.01Å have been obtained
in this way (3-8,40-42).   Bradley and his group (41A) and Gibson
(42) have been very successful using multiple low loss narrow gap
interferometers as the only tuning elements.   Improved power output
and line narrowing of a laser with grating in Littrow mount has been
obtained by Bjorkholm and coworkers (43) with an additional mirror

of intermediate reflectivity between active medium and grating. A
very stable bandwidth of less than 0.01Å, has been reported by
Walther and Hall (44), who use a voltage tunable Lyot-filter inside
the cavity.

The insertion of tuning elements is easiest in laser pumped dye
lasers, which can tolerate very high insertion losses due to their
large gain, and wavelength tuning of a dye laser with a diffraction
grating was first achieved by Soffer and McFarland in this way (45).
Single mode operation with less than 0.01Å bandwidth has been reported
by Bradley (22), using a near end-pumped infrared dye laser with an
echelle grating in Littrow-mount and an inserted Fabry-Perot
interferometer. Single mode operation without any dispersive elements
has been obtained in end-pumped dye lasers of very short cavity length
(several μm), so that the spacing of cavity resonances exceeds the
useable gain range (46). Electronic tuning over a wide range at
bandwidths of several Å was obtained with an acoustooptical filter
developed by Harris and collaborators (47). One should also mention
the technologically interesting thin film dye lasers with solid hosts,
using periodic changes of the refractive index in a grating-like
structure to obtain wavelength-selective distributed feedback (48-51).
Tunable dye lasers with distributed feedback have been realized,
using pumplight with variable interference fringes (52).

Bandwidths below 10Å seemed to be difficult to achieve with
side-pumped dye lasers, in particular when excited by the superradiant
molecular nitrogen laser (53,54). The problems were partly due to the
limited number of light passes in the cavity during the short
excitation of 5-10 nsec, and partly to the rather small active cross
section, resulting in a substantial angular spread of the emerging
light due to diffraction, which limits the resolution obtainable with
angle-dependent wavelength selectors such as gratings or prisms.
The bandwidth could be reduced to 0.4Å with a holographic grating of
high dispersion near grazing incidence (55). An intracavity prism
used at a high angle of incidence near 90° was successfully used to
narrow the laser bandwidth to 0.9Å. Single mode operation with a
bandwidth of less than 0.01Å has been obtained with a diffraction
grating and additional Fabry-Perot interferometer (46,56) or with a
holographic grating and a Lyot-filter in a short cavity (46). The
wavelength-reproducibility from shot to shot remained unsatisfactory,
however.

At Stanford we have developed a repetitively pulsed dye laser,
pumped by a commercial nitrogen laser (AVCO C950), which offers a
stable and reliably tunable output of only 0.004Å line width
throughout the visible (57). Peak powers in the kilowatt range can
be generated up to repetition rates of 100 pps. Because this laser
appears particularly useful for applications in high resolution
spectroscopy, we will give some details of its construction. The
basic elements are displayed in Fig. 7. A short glass cell (l=10mm)

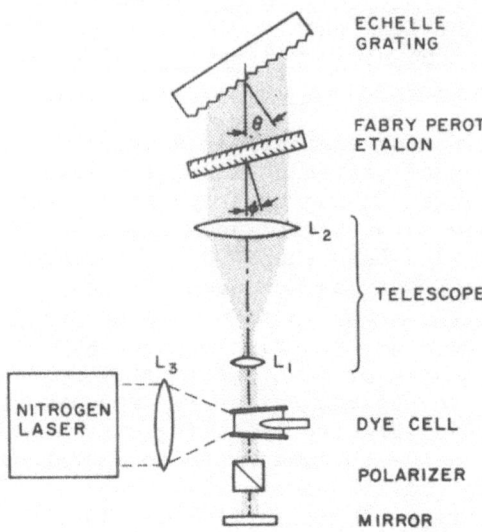

Fig. 7 Repetitively pulsed narrow-band dye laser (from Ref. 57).

with tilted, antireflection coated windows and with transversely
circulating dye solution is side-pumped by the nitrogen laser. The
ultraviolet pumplight is focused with a spherical quartz lens into
a thin line near the inner cell wall, so that the active volume is
a small filament of only about 0.2 mm diameter. The single pass
gain under these conditions can be as large as 1000 per mm. The
cavity is about 40 cm long. A plane dielectrically coated mirror is
used at one end, and a diffraction grating in Littrow mount (blaze
angle 60°) as wavelength selective reflector at the other. A beam
expanding telescope inside the cavity increases the resolution
dramatically, owing to the larger number of illuminated grooves on
the grating. With the telescope and grating alone, the bandwidth
can be reduced to 0.03-0.05Å with negligible loss of power. The
bandwidth can be further reduced to 0.004Å by inserting a tilted
Fabry-Perot interferometer of 0.5 cm^{-1} free spectral range and with
a finesse of 20 into the collimated beam. Walk-off problems are
negligible up to substantial tilt angle due to the large collimated
beam diameter (typically 2 cm), and the laser wavelength can be
tuned continuously over several Angstroms by simultaneously changing
the tilt angles of grating and etalon. The wavelength stability is
primarily limited by temperature changes of the etalon, the laser
has been stable within 0.01Å over many hours without special
precautions however. The output beam is nearly diffraction limited,
and the efficiency can be as high as 20%. Thermal schlieren effects
play no perturbing role at the short pulse widths of 5-10 nsec,

because they are primarily related to a thermal volume expansion of the liquid, which is limited to the speed of sound.

The linewidth can be even further reduced, theoretically without limit, by using an additional confocal Fabry-Perot interferometer as an ultra narrowband passband filter outside the laser cavity. Using a laser spectrum analyzer of 2GHz free spectral range with piezoelectric spacer as filter, we have narrowed the linewidth to 7 MHz or 4.10^{-4}Å. The peakpower is of course reduced in this way to a few watts, and the pulse is stretched from 5-10 nsec to about 30 nsec FWHM. But this power is still sufficient for high resolution saturation spectroscopy of atomic resonance lines, as we shall demonstrate later on.

The power of a narrowband dye laser oscillator can be boosted by one or more external dye laser amplifier stages (56-58). If the amplifier is operating in a highly saturated mode, it will at the same time reduce random amplitude fluctuations of the oscillator. At Stanford we have been able to amplify entire optical images with near diffraction limited resolution and with a high gain up to 30 dB/mm, using nitrogen laser-pumped dye amplifiers (59). Another very effective way to increase the intensity without loss of monochromaticity is the injection of weak monochromatic radiation into a second dye laser oscillator with broadband feedback (60-63). An effective power gain of 4000 and an energy output of 600 mJ/pulse has in this way been achieved by Magyar, et al. with flashlamp pumping (63).

Several workers have attempted to lock the wavelength of an organic dye laser to an atomic resonance line. Sorokin (63A) succeeded in locking a flashlamp-pumped dye laser to the wings of the sodium D lines, utilizing the Faraday rotation near the resonance lines in an intracavity sodium vapor cell with axial magnetic field. Another successful locking experiment was reported recently, using the anomalous reflection at a glass-atomic vapor interface (sodium and rubidium) (63B). The separation between dye laser output and atomic resonance frequency was less than 0.02 cm^{-1} in this latter experiment.

Intense short optical pulses can be generated with dye lasers by mode locking, i.e. by proper phase-synchronization of many simultaneously oscillating cavity modes (64). The wide bandwidth available with dye lasers would theoretically permit extremely short sub-picosecond pulses. Dispersion and other problems have so far set lower limits in the range of a few picoseconds. The earliest mode-locking experiments with dye lasers were performed using mode-locked solid-state lasers as pump sources of a dye laser oscillator with the same intracavity transit time (see e.g. Ref. 65). Flashlamp-pumped dye lasers have been modelocked by intracavity acoustooptic modulators (66), or by saturable absorbers inside the

cavity, i.e. by a dye filter, which is bleached at high light
intensities and exhibits a sufficiently rapid recovery time (4).
Near transform-limited pulses of 3-5 psec width, tunable between 580
and 700 nm, have in this way been generated recently by Bradley,
et al. (67-68). The laser emitted typically a train of about 600
pulses within 2 μsec with a total energy of 20-80 mJ. The pulse
intensities of many megawatts are particularly interesting for
studies of nonlinear optical phenomena and for measurements of fast
relaxation times. Mode locking of a nitrogen laser-pumped dye laser
was achieved with a saturable absorber by von Gutfeld (69).
Continuous operation of a mode locked red Cresyl violet laser was
reported by Runge (70), using a strong, mode-locked He-Ne gas laser
as pump source. Mode locking of cw dye lasers was achieved with
active light modulators inside the cavity (35,35A). Even more
successful were experiments with passive saturable absorbers, and a
continuous stream of 1.5 psec long pulses with a bandwidth of 5Å, 10
nsec pulse separation, and with 100W peak power has been generated
in this way by Shank, Ippen, and Dienes (71).

Various laser- and flashlamp-pumped dye lasers are already
commercially available (9,72). Considering the present wide and
intensive efforts, it will certainly take only a rather short time
until many of the more recent advances are transformed into
commercial instruments.

2.2 Semiconductor Lasers

Laser action in semiconductors, using stimulated emission of
recombination radiation across the gap between valence and conduction
band was obtained as early as 1962. Tunable coherent radiation over
the entire range from 0.62μm to 34μm has been generated to date with
semiconductor diode lasers. To the spectroscopist, these lasers are
so far mainly of interest as tunable infrared sources of relatively
low power. Rather extensive and general review articles with many
references have been published recently by Kressel (73) and Harmann
(74,75). Some very recent advances are summarized in an article by
Hinkley (76).

Population inversion in semiconductor lasers is most conveniently
achieved by a reverse current through a p-n juction. A typical
geometry of such a diode laser is shown in Fig. 8. The optical
cavity can be formed simply by the cleaved surfaces of the small
semiconductor crystal. The advantages of very small size and
simplicity of excitation are often outweighted by the need for
cryogenic cooling, however.

Laser action in the visible and near ultraviolet has been
obtained by electron beam pumping of semiconductors. Optical
pumping is also possible (77,78) and might be of practical interest

in the future for the generation of infrared laser radiation of
high output power and good beam quality.

The commercially available GaAs laser diodes are perhaps the
most widely known semiconductor lasers. GaAs is one of the few
materials in which laser action has been obtained at room temperature.
The wavelength can be tuned between 800 and 900μm by changing the
temperature from 4 - 300°K. The efficiency can be as high as 80%,
peakpowers of several hundred watts or continuous powers of several
hundred mW can be generated with commerical units. Much progress in
the development of tunable semiconductor lasers is due to the use of
ternary alloys, whose bandgap can be changed by adjusting the chemical
composition. A list of the materials which have been used in junction
lasers to date is given in Fig. 9. To avoid thermal excitation of
carriers across the bandgap, cryogenic cooling is necessary for the
lasers operating at longer wavelengths. CW operation with the rather
important material $Pb_xSn_{1-x}Te$, for instance, is possible only near
the temperature of liquid helium. Powers of 0.1-mW have been
obtained in this way. Pulsed operation has been obtained near liquid
nitrogen temperature.

The gain band of an individual laser diode, which may be as
narrow as 0.1 cm^{-1} can be tuned by applying hydrostatic pressure or
an external magnetic field. A tuning range of up to 50 cm^{-1} near
10μm can be obtained simply by changing the discharge current and
hence the junction temperature. Narrowband single mode operation is
often obtained without any external wavelength selectors, on account
of the small cavity dimensions. The cavity resonance frequency can
also be temperature-tuned, since it depends critically on the linear
dimensions and the refractive index. The unavoidable mismatch in
tuning speed between gain band and cavity modes gives rise to mode
jumps and discontinuities, however.

Hinkley and Freed (79) have recently measured a linewidth as
narrow as 54 kHz for a $Pb_xSn_{1-x}Te$ laser, operating near 10μm, by
heterodyning with a single frequency CO_2 laser (79). This resolution
of 2 parts in 100 million is vastly superior to any conventional
spectrometer in this region.

2.3 Sum- and Difference-Frequency Generation,
Parametric Oscillators

Tunable ultraviolet radiation down to 2500Å has been obtained
with tunable dye lasers by second harmonic generation and sum-
frequency generation in nonlinear optical materials. Experiments to
generate tunable radiation in the infrared as the difference
frequency of two laser oscillators have also been successful.

The fundamentals of the vast and rapidly growing field of

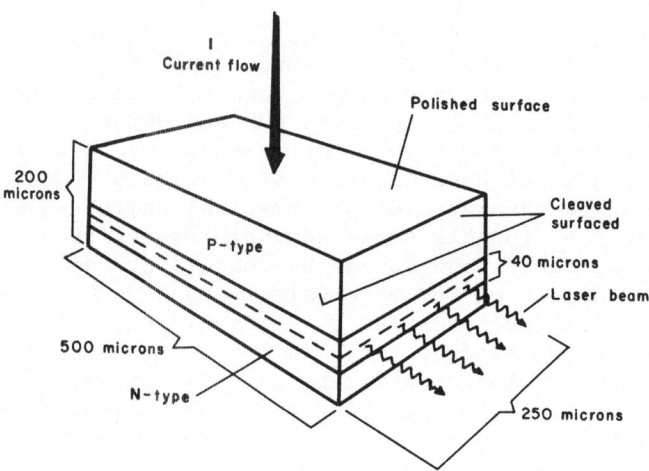

Fig. 8   Geometry of p-n junction diode laser (from Ref. 124).

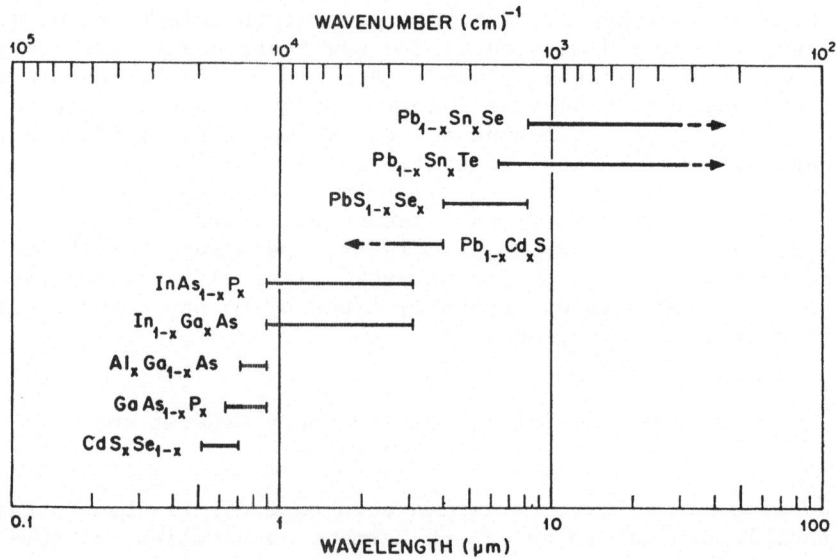

Fig. 9   Tuning ranges for semiconductor diode lasers made from
         different semiconductor alloys (from Ref. 125).

nonlinear optics are described in several good reviews and monographs (80-82). A survey of the presently used materials is compiled in Ref. (91).

Sum- and difference-frequency generation with intense laser light is possible in crystals of the noncentrosymmetric group, where the dielectric polarization exhibits a term quadratic in the inducing field. In order to permit the build-up over a large "coherence length," the momentum vectors of the interacting light fields must satisfy the same additive relationship as the frequencies. This phase matching condition is rather restrictive in practice. It can be fulfilled despite the dispersion of the refractive index with certain birefringent materials by properly choosing angular direction and polarization of the incident light waves or by accurately controlling the crystal temperature.

The generation of tunable ultraviolet radiation from visible dye lasers is closely related to the well established harmonic generation with fixed-frequency lasers. The nonlinear crystal can be placed outside the laser cavity, if peak powers in the kW - MW range are available. Broadband pulsed dye lasers have been used to generate narrowband tunable ultraviolet by symmetric sum frequency generation in KDP (83). Higher efficiencies up to 10 - 20% have been obtained, however, with narrowband tunable dye lasers. Megawatt second harmonic radiation tunable between 2800 and 2900Å has been generated by Bradley, et al. (84) in an angle-tuned ADP crystal, employing an interferometrically narrowed rhodamine 6G dye laser, side-pumped by a frequency-doubled Nd:glass laser. Similar pulses with a bandwidth of 0.05Å, tunable between 3400 and 4000Å have been obtained by doubling the frequency of a ruby-pumped dye laser in a LiIO$_3$ crystal (85). Sum-frequency generation from the output of a ruby laser and a tunable dye laser has also been reported (86). Peak powers in the order of 10W in 0.6 μsec long pulses, tunable from 2500 to 2900Å, have been obtained by Jennings and Varga (87) in 90° phase-matched second harmonic generation with a flashlamp pumped dye laser and a temperature-controlled ADP crystal. An angle-tuned ADP crystal inside the dye laser cavity is used in a commercial tunable UV laser, pumped by a repetitively pulsed frequency doubled Nd:YAG laser (88). The output (up to 200W peak powers, 5 mW average power, 0.2 - 0.4 μsec pulselength, 2 cm^{-1} bandwidth) can be tuned between 2610 and 3150Å. An intracavity doubling crystal of lithium formate has been used in a cw dye laser by Hercher and coworkers (89) for the generation of continuous tunable ultraviolet in the region 2600 - 3250Å.

Although KDP and ADP are transparent down to 2000Å, it is not possible to generate wavelengths below 2500Å by frequency doubling in an efficient 90° phase-matched configuration. Ultraviolet radiation at shorter wavelengths can be generated, however, as the sum frequency of two laser waves of different frequency, as demonstrated by the

generation of coherent radiation at 2120Å as the sum frequency of
the infrared light (1.06μm) of a Nd$^+$-laser and its fourth harmonic
in the ultraviolet (90).

Very promising for the generation of tunable radiation in the
vacuum ultraviolet and possibly the soft x-ray region is the third
harmonic generation or threefold sum frequency generation in metal-
vapors, such as Rb or Cd, in which phase matching can be achieved
by compensating the anomaleous dispersion by an appropriate buffer
gas such as He. This scheme has been proposed by Harris and
coworkers and wavelengths down to 1100Å have already been generated,
using a high-power picosecond Nd-laser with optical solid state
frequency doubler and quadrupler as the primary source (90,91).

Infrared radiation tunable between 3 aand 4μm has been generated
by Dewey and Hocker as the difference frequency of a ruby laser and
a dye laser with an angle-tuned $LiNbO_3$ crystal (92). Peak powers in
the kilowatt range have been obtained at 3-5 cm^{-1} bandwidth. Narrower
spectral widths and an extension of the tuning range towards shorter
wavelengths down to 1.7μm should be possible without difficulties.
Tunable radiation in the middle infrared from 10.1 to 12.7μm has
been generated in a similar fashion in proustite by R. C. Smith and
his group (93) and continuous tuning from 2.5 to 12.5μm with less
than 0.1 cm^{-1} bandwidth and with peak powers in the kilowatt range
seems possible in this way, if one extrapolates from the reported
preliminary results. Tunable far-infrared radiation in the region
from 20 to 38 cm^{-1} has been produced earlier in $LiNbO_3$ as the
difference frequency of two ruby lasers (94).

The process of generating the sum frequency of two incident
laser waves in a nonlinear crystal is reversed in the optical
parametric oscillator, a reactive device, which is closely related
to parametric amplifiers and oscillators in the microwave region.
Optical parametric oscillators are of considerable practical interest
for the efficient generation of tunable coherent radiation in the
visible and, most importantly, in the near and middle infrared where
dye lasers are not available. A thorough review of parametric
oscillators with many references has been written by Harris (95) and
the most recent developments are reviewed by Byer (96).

A typical parametric converter consists of a nonlinear optical
crystal in an optical resonantor (Fig. 10). The pumplight which is
normally provided by a fixed frequency laser enters through one
dichroic end mirror. If the parametric gain exceeds the losses,
oscillation builds up on two different frequencies, the signal and
the idler whose frequencies add up to the pump frequency and which
are mutually phase matched for difference frequency generation with
the pump. The wavelengths of signal and idler can be tuned by
altering the phase matching conditions, i.e. by changing the
orientation or temperature of the crystal or by applying an electric

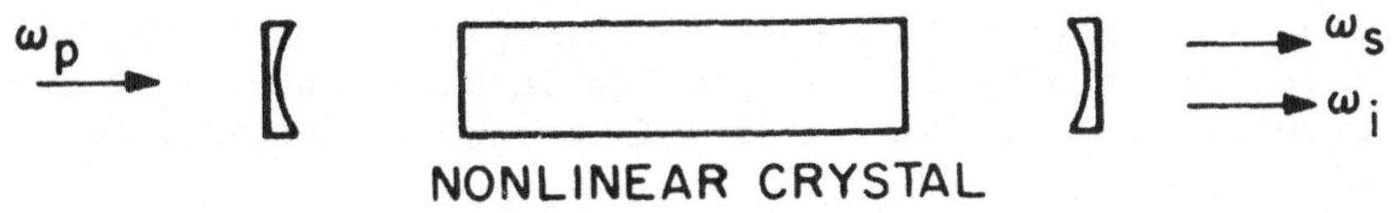

NONLINEAR CRYSTAL

Fig. 10 Scheme of optical parametric oscillator (from Ref. 95).

field. The lowest thresholds are obtained when the cavity is
resonant for both the signal and the idler wave, and cw-operation
has been achieved in this way (97). Large instabilities have so far
precluded the practical use of doubly resonant parametric oscillators,
however, and most oscillators have been used in a pulsed mode with a
singly resonant cavity. The energy conversion can be very efficient,
as is indicated by reported pump depletions as high as 67%.

A commercial parametric oscillator is available which uses a
45 mm long temperature-tuned $LiNbO_3$ crystal in the parametric
converter, and a Nd:YAG laser, operating at several wavelenghts near
1μm with internal $LiIO_3$ frequency doubler as pumpsource (98). This
device can be operated anywhere between 0.60 and 3.7μm with changes
of mirrors on both the oscillator and pump laser. The pulses are
70-250 nsec long, the peak powers in a TEM_{00} mode are between 50 and
700W, the average power at 75 pps repetition rate is between 1 and
10 mW, depending on the wavelength. Higher powers can be generated
with stronger pump sources. The linewidth is normally in the order
of 1 cm^{-1}. A linewidth of less than 0.001 cm^{-1} near 2.5μm has
recently been obtained, however, with an interferometric mode
selector (99). A scheme for locking a parametric oscillator to a
gas absorption line has also been demonstrated by Harris (100).

Rapid tuning of a parametric oscillator in the region from 0.72
to 2.6μm has been accomplished by Wallace (101), using a tunable
rhodamine 6G dye laser as the pump source.

An ADP crystal without cavity, pumped by the fourth harmonic of
a Nd:YAG laser, has been used by Yarborough, et al. as a parametric
converter tunable over the entire range from 4200Å to 7300Å (102).
At 750 MW/cm^2 pump power, the single pass gain in the 5 cm long
crystal was as high as 5.10^{12}. The output pulses of 2 nsec length
has a bandwidth of 5Å and reached 100 kW peak power. The average
power was 5 mW at 30 pps repetition rate.

Future extensions of the wavelength range of parametric
oscillators depend critically on the development of new materials.
An important step in this direction is the achievement of parametric
oscillation in CdSe, pumped by a Nd:YAG laser operated at 1.8μm,

which was recently reported by Byer and his group (103). The oscillator was tunable from 9..8 - 10.4µm in preliminary experiments, and an extension of this range to 8 - 14µm appears feasible in the near future.

2.4 Raman Lasers

Stimulated Raman scattering of laser light in solids, liquids or gases provides a mechanism which can greatly increase the number of available fixed frequency laser wavelengths or extend the wavelength range of tunable lasers (8,104-106). This nonlinear optical process can be described by a polarization cubic in the exciting field. The Stokes shift in liquids such as benzene or carbon disulfide is in the range from 450 to 4200 cm^{-1}, corresponding to the characteristic vibration frequencies. The large number of available scattering media (106), together with the possibility of anti-Stokes shifting or multiple Stokes shifting, offer a wide choice of wavelength shifts. With a 5W Ar^+ laser (5145Å), pumping a long, wave-guiding liquid core of CS_2 in a thin glass capillary, an efficiency of more than 40% was recently obtained for a quasicontinuous Raman laser (107). Dye lasers seem to offer more convenience in the visible region, however.

Intense, very narrowband infrared radiation, tunable over several µm in the vicinity of 10µm and 5µm has been generated efficiently with spin-flip Raman lasers, which have been developed by Patel and coworkers at Bell Telephone Laboratories (108) and by Mooradian and coworkers at Lincoln Laboratories (109). The pump-light has been provided by a fixed frequency CO_2 laser (10.6µm) or CO laser (5 - 6µm). These lasers use stimulated Raman scattering from the conduction electrons of an InSb crystal in an external magnetic field. The Stokes-shift corresponds to the spin-flip energy gbH, where g is the g-factor, b the Bohr magneton and H the magnetic field. Tuning is achieved by varying the magnetic field, which can be as high as 100 kG when provided by a superconducting magnet.

Figure 11 shows the geometry of a spin-flip Raman laser. The InSb crystal has typical linear dimensions of only a few millimeters. The infrared pumplight is focused to a diameter of 50 - 300µm. The polished crystal end-surfaces form the optical cavity. Whereas the pump-threshold for operation near 10µm is of the order of hundreds of watts, a threshold as low as 50 mW, and cw-operation with more than 1W output at 50% efficiency have been obtained near 5µm, due to the much enhanced scattering cross section near resonance (110). The crystal was cooled to 30°K in the latter experiment. The operation of a continuous spin-flip Raman laser in magnetic fields as low as 400 Gauss, and laser action on the anti-Stokes line and the second Stokes line was recently reported by Patel (111).

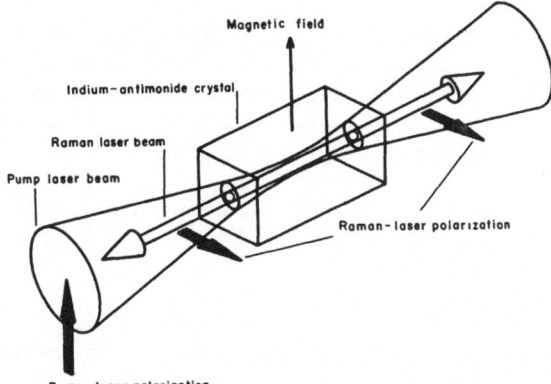

Fig. 11   Geometry of spin-flip Raman laser (from Ref. 124).

Magnetic fine tuning of the cavity resonance frequency is possible, using the effect of mode-pulling by the tuned gainband. Discontinuities arise from mode jumping, however, and an external cavity with wavelength selectors may be more desirable. A bandwidth as long as 1 kHz has been observed by Patel from a spin-flip Raman laser (112). Owing to their substantial output power, spin-flip Raman lasers hold great promise for applications in nonlinear high resolution spectroscopy.

Another type of tunable Raman laser has been described earlier by Pantell, et al. (113), using stimulated polariton scattering in a LiNbO$_3$ crystal, i.e. stimulated Raman scattering from a transverse optical lattice mode, whose energy is partially mechanical and partially electromagnetic. The pump light is produced by a 2 mW q-switched ruby laser, and the Stokes frequency is tuned over some 38 cm^{-1} near 248 cm^{-1} by changing the angle between pumplight and optical resonator axis. Such a system is of interest as a potential source of tunable radiation at the polariton frequency in the far infrared.

2.5   Other Tunable Lasers

High pressure molecular gas lasers, in which adjacent vibration-rotation lines are sufficiently broadened to overlap, hold great promise as powerful sources of tunable radiation, in particular in the infrared. Electron beam excitation can be used for such lasers. An attractive alternative may be optical pumping with laser light, which has recently been successfully demonstrated for the excitation of a CO$_2$ laser (114).

Madey (115) has theoretically analyzed the possibility to generate tunable radiation from the far infrared through the visible into the ultraviolet by stimulated emission of Bremsstrahlung, using a relativistic electron beam, passing through a periodically

alternating transverse magnetic field. Madey has treated this
process theoretically as forward scattering of virtual photons, seen
by the electrons. The frequency of the scattered radiation can be
tuned by changing the electron energy. Finite gain appears possible
under favorable conditions.

Pantell and coworkers (116) are studying the possibility to
generate tunable ultraviolet radiation below 3000Å by stimulated
emission of Cerenkov radiation.

New, so far unexplored concepts may be successful in the future.
Interesting progress can certainly be expected for radiation in the
vacuum ultraviolet and in the x-ray region.

3. APPLICATIONS OF TUNABLE LASERS IN SPECTROSCOPY

In this last chapter, we will look at a number of recent
experiments, which illustrate some of the exciting new possibilities
in optical spectroscopy, opened by the development of wavelength
tunable lasers. Due to the limited space, we will have to omit a
wealth of interesting spectroscopic investigations performed with
lasers of fixed frequency or very limited tuning range. Many such
experiments have been reviewed by Demtröder (117).

3.1 Absorption Spectroscopy

For absorption spectroscopy tunable lasers have a number of
obvious advantages over conventional equipment, in particular their
high spectral brightness, much improved spectral resolution and
their spatial coherence, which permits the use of very long optical
path lengths or of very small samples.

The progress is particularly striking in the infrared region,
where severe instrumental limitations made high resolution
absorption spectroscopy almost impossible in the past. Research
groups at Lincoln Laboratories and at Bell Telephone Laboratories
have impressively demonstrated the potential of tunable diode lasers
and of spin-flip Raman lasers by measuring high resolution absorption
spectra of a number of molecular gases, including NH_3, SF_6, and CH_4
near 10µm (71,118,119), SO_2 at 8.6µm (120), NO near 5µm (121,122)
and CO at 4.7µm (123). The resolution in these studies is only
limited by Doppler broadening. Linewidths as narrow as 100 MHz have
been observed and it has, for instance, been possible to resolve the
the nuclear hyperfine splitting in the vibration-rotation absorption
spectrum of the NO molecule (121).

Much interest has been evoked recently by the possibility to

use such high resolution absorption spectroscopy for the selective
detection of air pollutants (71,118,124,125). The sensitivity for
narrowband trace absorptions can be enhanced by frequency modulation
of the laser source, for instance by superimposing an ac current to
the driving current of an injection laser, and by phase sensitive
detection of a synchronous light intensity modulation after
absorption. Another very sensitive technique uses a microphone inside
a gas absorption cell which detects the sound waves generated by the
absorption of an incident intensity-modulated laser beam. Kreuzer
and Patel (122) have in this way been able to detect NO concentration
as low as 0.01 ppm in a 1 cm^3 sample, using a 50 mW spin-flip Raman
laser.

Very small optical extinctions can be detected in a novel and
unconventional way by using a broadband laser itself as nonlinear
sensor. Several workers have placed discrete absorbers inside the
cavity of flashlamp pumped broadband dye lasers and have observed a
dramatic decrease or complete quenching of the laser action at the
absorbed wavelengths. Na and I_2 vapor (126), liquid $Eu(NO_3)_3$
solution (127), and Sr or Ba^+ in an acetylene air flame (128) were
used as absorbers, and an increase in sensitivity of 100 to 1000,
compared to a single pass measurement, has been reported. In an
experiment performed at Stanford (129) we have observed a sensitivity
increase of 100,000 with I_2 vapor as absorber inside the cavity of
a cw dye laser. To measure the selective quenching of the dye laser
at the narrow (0.03 cm^{-1}) absorption lines of molecular iodine, the
bright fluorescence of a second iodine vapor cell outside the laser
cavity was monitored. The tremendous sensitivity can be explained
by the multiple passes of the light through the absorbing medium and
in addition by the strong competition of the simultaneously
oscillating modes for the available energy. We have obtained a
reasonable quantitative estimate with a rate equation model, taking
the spectral homogeneity but spatial inhomogeneity of the laser
saturation into account.

3.2 Light Scattering

In studies of light scattering, wavelength tunable lasers are
naturally of particular interest for the investigation of resonant
phenomena, such as resonance fluorescence or resonance Raman
scattering. The techniques of nonresonant light scattering, in
particular Raman spectroscopy, have already been forcefully
revitalized if not revolutionized by the development of powerful
fixed frequency lasers (117,130).

Pulsed dye lasers have been used by a number of workers to
excite fluorescence on atomic and molecular resonance lines. Ba atoms
in a flame have been detected in this way (131) and Na atoms have
been detected in concentrations as low as 0.003 ng/cm^3 (132). If the

resonance fluorescence is excited by a cw laser, it should in
principle be possible to obtain almost 10^8 scattered photons per
second from one single Na atom, i.e. it should literally be possible
to 'see' single atoms.

A pulsed dye laser has also been applied to excite resonance
fluorescence from metastable helium atoms in a cold plasma (133),
and it was possible to study different relaxation processes by
monitoring scattering signals on many other lines, excited by
collisional excitation transfer and radiative decay.

The lifetimes of excited atomic and molecular states can be
measured rather directly by resonant excitation with a short laser
pulse and observation of the subsequent exponential decay. The
pulselength of 5-20 nsec available with laser-pumped dye lasers is
well suited for the study of longer living levels, and lifetime
measurements have been performed in this way for electronically
excited states of the molecules NO_2 (134,135), BaO (136), Br_2 (137)
and I_2 (138).

The high intensity of short pulse dye lasers opens the
possibility for the remote detection of resonant light scattering
in a radar-like fashion. This technique has been used by several
groups to study the atomic sodium layer in the earth's atmosphere
at an altitude of about 90 km, employing flashlamp-pumped dye
lasers, tuned to the sodium D-lines (139,140). The same technique
appears suitable to determine other atmospheric constituents.
Laboratory experiments have for instance already confirmed the
possibility to detect OH in concentrations as low as one part in
10^{11} to 10^{13} by excitation with a frequency doubled dye laser,
operating at 2822Å (141).

3.3 Selective Excitation

The intense radiation of tunable lasers can be applied
effectively to excite selected discrete energy states of atoms,
ions, molecules or solids. This possibility is not only of great
interest in spectroscopy, but also in such fields as photochemistry,
gas dynamics or plasma research.

Numerous applications of tunable lasers for selective excitation
have already been demonstrated. McIlrath (143) has used a laser-
pumped pulsed dye laser emitting 30 mJ at 6573Å in 40 nsec to pump
an intercombination transition in atomic Ca vapor and to effectively
equalize the population of the ground state and the excited
metastable 3P(4s4p) level. Absorption spectra from the excited
level were subsequently recorded by use of a flashlamp continuum.
Bradley and coworkers (144) have populated the $5P_{3/2}$ state of atomic

Rb in a similar way and recorded the absorption from the upper state with a second broadband dye laser continuum.

It has also been possible to study photoionization cross sections from atomic states, selectively excited by tunable lasers. Such experiments have been reported for Cs, optically pumped by a GaAs laser (145) and for Mg atoms, excited by a frequency doubled dye laser (146). The selective two-step ionization of atoms opens exciting prospects for applications such as optical isotope separation (147).

Other interesting possibilities are opened in the study of the wavelength dependence of the photodetachment of electrons from negative ions. Smyth and coworkers (148), for instance, have determined the detachment energies for PH_2^- and NH_2^-, using a continuously tunable parametric oscillator and an ion cyclotron resonance spectrometer. Lineberger (149) is studying photodetachment processes for various negative metal ions, using a flashlamp-pumped dye laser.

Strong selective excitation with tunable lasers can be a rather ideal pumping mechanism to obtain population inversion and laser action on atomic or molecular transitions. Such optically pumped lasers provide not only interesting objects for nonlinear spectroscopic investigation, but they can also serve as valuable ultra precise wavelength markers or secondary wavelength standards and they may be useful as coherent light sources at new wavelengths.

A GaAs laser diode has recently been used to pump Cs^{133} vapor (149A), creating a population inversion in the hyperfine levels of the ground state. Sorokin and Lankard (150) have obtained infrared laser action from triplet transitions in atomic Sr vapor in He buffer gas after pumping the singlet resonance line at 4607Å with a pulsed dye laser. In this case, the upper laser states are populated by inelastic collision processes, following the optical excitation. Laser action in atomic vapors has also been obtained by selective resonant two-photon excitation. Korolev, et al. (151) have observed directed stimulated emission in Rb vapor at the transitions $5^2S_{1/2} - 6^2P_{3/2}$ (4202Å) and $5^2S_{1/2} - 6^2P_{1/2}$ (4215Å), when the vapor was pumped by a ruby-pumped dye laser, operating in the range 7730 - 7900Å. Similar results have been reported by other workers for K-vapor (152). Sorokin, et al. have reported infrared laser action in K, Rb and Sr vapor after optical excitation with various strong pulsed lasers (153). The mechanism of excitation in this case was two-photon absorption by diatomic molecules, followed by the photodissociation into excited atomic states.

Numerous visible and near infrared laser transitions have been observed in optically pumped molecular iodine vapor by Byer, Levenson, et al. (154). Selected vibrational-rotational levels of the $(B^3\Pi_{0u}^+ - X^1\Sigma_g)$ electronic transition were excited by the second harmonic output of a Nd:YAG laser. At Stanford, we have subsequently

obtained laser action on many more transitions, using a nitrogen
laser pumped dye laser. The total number of laser transitions in
iodine, which can be excited by various pump frequencies, is in the
order of 10^6.

Varsanyi (155) has reported room temperature visible laser
action in solid $PrCl_3$ and $PrBr_3$, pumping with a tunable pulsed dye
laser directly into the upper laser level. Superradiant laser
action could be obtained with tiny crystals of a few microns in
dimensions, when the pumplight was focused into the cleaved crystal
surface.

In an interesting study in solid state spectroscopy, Holzrichter,
et al. (155A) have used a flashlamp pumped dye laser, tuned to 5420Å,
to excite both excitons and magnons in a crystal of antiferromagnetic
manganese fluoride near liquid helium temperature. The magnetization
corresponding to the spin change of the manganese ions is observed
with a pick-up coil arond the crystal. By stressing the crystal,
the two exciton lines, corresponding to the two antiparallel
sublattices, can be split, and the laser can be tuned to excite
either the spin in one sublattice or in the other. With some
detuning it was also possible to selectively excite spinwave side-
bands and to measure relaxation times for both the excitons and
magnons.

3.4   Saturation Spectroscopy

Doppler broadening of gas resonance lines, which limits the
resolution in ordinary emission or absorption spectroscopy, can be
virtually eliminated with the new nonlinear spectroscopic techniques
of laser saturated absorption, which will be reviewed in the
subsequent summary paper by Hall. A strong monochromatic laser
field, interacting selectively with atoms in a narrow interval of
the Maxwellian velocity distribution, causes energy level population
changes, which can be detected by a second probe field. Until
recently, such studies have been restricted to either gas laser
transitions or to molecular absorption lines in accidental
coincidence with gas laser lines. Tunable lasers generally did not
provide the required spectral resolution and stability.

During the last year the present author, together with I. S.
Shahin and A. L. Schawlow, have performed two experiments (157,158)
applying saturation spectroscopy for the first time to atomic
resonance lines. We have studied the yellow Na D resonance lines
and the red hydrogen Balmer line in saturated absorption, using a
pulsed narrowband dye laser with narrowband passband filter, pumped
by a nitrogen laser, as described in Section 2.1. These experiments
demonstrate that it is now possible, using wavelength tunable dye
lasers, to apply the powerful techniques of saturation spectroscopy

to virtually any atomic or molecular absorption line throughout the visible spectrum.

   We used a particularly sensitive and convenient technique of saturation spectroscopy based on earlier experiments by Hänsch and Toschek (158A) which is well suited for the use of a pulsed laser source and which has been successfully applied in several saturation experiments with gas lasers. A simplified scheme of the saturation spectrometer is shown in Fig. 12. The tunable dye laser with a pulse repetition rate of typically 80 pps provides peak powers of several watts in pulses of 8-30 nsec length at a bandwidth of 7 MHz. The laser output is divided by a beamsplitter into a weak probe beam and a stronger saturating beam, typically of a few mm in diameter, which are sent in nearly opposite directions through a cell containing the absorbing gas sample. When the laser is tuned to the center of the Doppler profile of a given resonance line, both lightwaves can interact simultaneously with the same atoms, i.e. those with essentially zero axial velocity. The saturating beam then bleaches a path for the probe, reducing its absorption. In order to detect small bleaching effects, the saturating beam is periodically blocked by a rotating chopper, synchronized to the pulsed laser, and the resulting amplitude modulation of the probe is detected with a phase sensitive amplifier. The noise due to

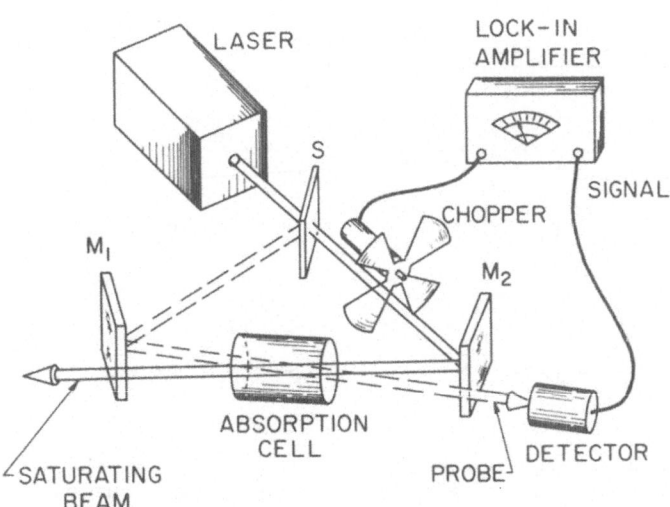

Fig. 12 Scheme of laser saturation spectrometer.

random laser amplitude fluctuations can be further reduced by using
a second dummy probe, which also passes the absorbing gas, but does
not cross the bleached region and by comparing the two probe
intensities in a differential detection scheme.

Some results, obtained in the first experiment with Na vapor
(157) are shown in Fig. 13. The saturation spectra of the two D
lines are compared with the theoretical hyperfine structure of these
lines, given at the bottom. The hyperfine splitting of the ground
state is clearly resolved in both lines, and even the splitting of
the excited $^2P_{1/2}$ state is resolved in the D1 line. The narrowest
observed linewidth is less than 50 MHz. The Doppler width is 1.35
GHz for comparison. The resolution obtained is markedly superior
to the best previous resolution, obtained with atomic beam techniques.
Halfway in between resonance lines sharing a common upper or lower
level, there appear additional cross-over signals, which are artifacts
of the spectroscopic method, but which are theoretically well understood and which can reveal interesting additional information.

In a series of measurements, we delayed the probe pulse up to
700 nsec with respect to the saturating pulse by sending it through
a folded optical delay line. The narrowband saturation resonances
remained observable, despite the delay of many radiative lifetimes.
This phenomenon can be explained as remanent hole burning in the
velocity distributions of the stable groundstate levels due to a
velocity selective optical pumping cycle which also gives rise to
the inverted signal (enhanced probe absorption) at the center of
the saturation spectra.

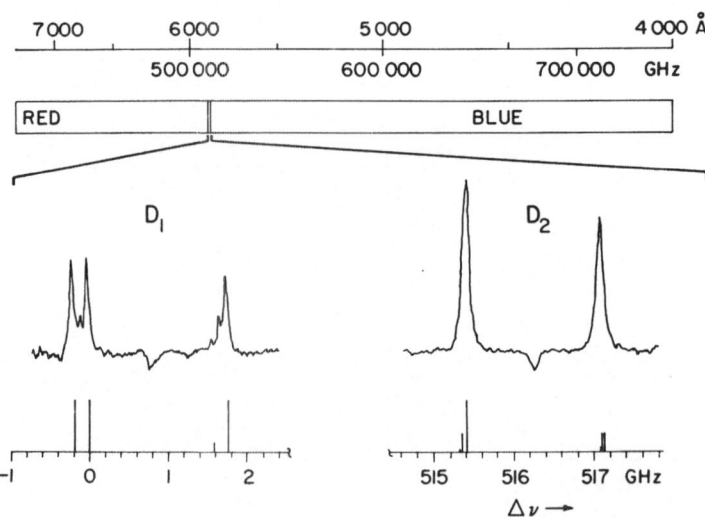

Fig. 13 Saturation spectrum of Na D-lines.

Using a time delayed probe we have also studied the
thermalization of the perturbed atomic velocity distribution by
elastic collisions with Ar and He buffer gas atoms. The lines
broaden relatively slowly with time, indicating a diffusion process
in velocity space rather than sudden random velocity changes due to
hard collisions, in agreement with the known differential scattering
cross section for a Van der Waals potential, which strongly favors
small angle scattering.

Saturation spectra obtained at higher bleaching intensity
exhibit an interesting phenomenon. The linewidth decreases very
noticeably when the probe is delayed by more than the laser pulse
length. The larger width in the presence of the saturating
lightfield can be interpreted as an experimental confirmation of the
theoretically predicted dynamic Stark splitting (159).

The subsequent study of the red Balmer line of atomic hydrogen
by the same method is of special interest, because Doppler
broadening is particularly large for these light atoms, almost 6000
MHz at room temperature, and masks all the finer details of its
visible lines. Our present accurate knowledge of the hydrogen fine
structure is almost entirely based on radiofrequency spectroscopy
and level-crossing techniques. Nonetheless there have been
countless efforts to study the visible Balmer lines with ever
increasing resolution, because an accurate determination of the
wavelength of one of the line components is the basis for the
determination of the Rydberg constant, one of the cornerstones in the
evaluation of the fundamental constants. The Doppler width has been
reduced in previous experiments by using the heavier isotopes D and
T and by cooling a gas discharge tube to the temperature of liquid
nitrogen or helium. Nonetheless it has not been possible to isolate
single fine structure components, and deconvolution processes remain
necessary in the wavelength determination.

In our experiment, the hydrogen atoms were excited in the state
$n = 2$ in a simple Wood gas discharge tube, operating with a continuous
flow of electrolytically generated H_2 gas at low pressure
(0.1 - 0.5 torr). Stark broadening and level mixing due to electric
fields in the plasma could be minimized by observing the saturated
absorption in the afterglow, about one microsecond after stalling the
dc discharge with an electronic switch. This is another example of
the versatility offered by pulsed laser sources. A saturation
spectrum of H_α (6563Å), as obtained by our method, is shown at the
bottom of Fig. 14. The theoretical fine structure components and
their Doppler-envelope at room temperature are for comparison shown
above. The four strongest components and one cross-over signal are
clearly resolved, and the Lamb shift can for the first time be
observed directly in the optical absorption spectrum. The narrowest
linewidth in the spectrum shown is about 250 MHz and is primarily
determined by the natural linewidth and by unresolved hyperfine

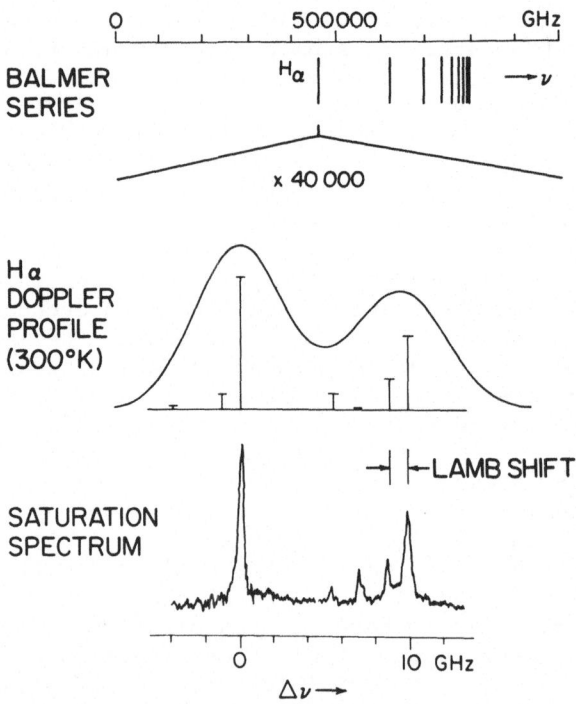

Fig. 14 Hydrogen line H$_\alpha$ with theoretical fine structure
and Doppler-profile at room temperature (center) and
saturation spectrum of H$_\alpha$ (bottom).

splitting. More recent measurements with reduced laser bandwidth
and a smaller crossing angle between the two beams reveal even the
hyperfine splitting of the metastable 2s state (170 MHz) in the
second component from the right.

Time resolved measurements indicate that the lifetime of the 2p
states (1.6 nsec) is effectively lengthened by a factor 1000 owing
to resonance trapping of the emitted ultraviolet Lyman-alpha
radiation. Studies with delayed probe reveal that the velocity
distribution of the p-states is rapidly thermalized by this process,
in contrast to the metastable 2s state.

The known fine structure intervals could be reproduced with the
present measurements within the standard deviations of 2-5 MHz
despite the unresolved hyperfine structure. The results even seem
to confirm for the first time experimentally that the hydrogen state
3P$_{3/2}$ is some 5 MHz above the 3D$_{3/2}$ state, as expected from radiative
corrections. These data indicate that it should be possible, even
with hydrogen atoms in the present somewhat uncontrolled bottle-
environment, to measure the wavelength of one of the line components
in saturated absorption with an accuracy of 10 MHz or 2 parts in 10^8,

i.e. with the accuracy of the present standard of length and we are presently working at such an experiment which will hopefully improve the accuracy of the Rydberg constant by an order of magnitude.

3.5 Other Spectroscopic Applications

A number of other interesting applications of tunable lasers in spectroscopy have already been demonstrated, and many more new applications are likely to be developed in the future.

An interesting technique which will gain wide applicability with the development of narrow band tunable lasers is optical heterodyne spectroscopy (118,160). Light from an incoherent source under investigation is mixed at a photodetector with the radiation from a monochromatic laser, serving as local oscillator. The radiofrequency beat spectrum at the detector output, which can be analyzed by standard electronic equipment, contains all the information about the source spectrum in the vicinity of the laser oscillator.

This technique holds great promise for high resolution emission spectroscopy, in particular from small sources. Applications in astronomy have been discussed by Nieuwenhuijzen (160), and Hinkley and Kelley (118) have suggested its use for the analysis of the thermal infrared radiation emitted from pollution molecules in smoke stacks.

Short intense light pulses, interacting with resonant transitions, can give rise to a number of interesting coherent transient effects, which sometimes have analogies in phenomena observed in the microwave region in studies of nuclear magnetic resonance. Many workers have studied such effects in the past with fixed frequency lasers. With the help of tunable lasers, coherent transient effects can become a useful tool of spectroscopic investigations, revealing dynamic system parameters such as relaxation rates due to phase changing atomic collisions. Bölger and Diels (161) have already reported the observation of photon echoes, excited in Cs vapor by a GaAs laser with nanosecond pulses at 8521Å. Bradley, et al. (162) have observed self-induced transparency and dispersion delays by sending intense short light pulses from a tunable dye laser through potassium vapor.

A new type of nonlinear spectroscopy, related to Raman spectroscopy, has recently been demonstrated by Levenson, et al. (163) and by Wynn (164), using tunable pulsed dye lasers. If two laser beams of two different frequencies ω_1 and ω_2 are sent through a transparent medium a signal is created at the new frequency $2\omega_1 - \omega_2$. By changing the frequency difference, it is possible to map the third order susceptibility for three wave mixing. Levenson has in this way observed a striking resonance in diamond, if the frequency difference is close to a transverse phonon frequency.

We conclude with a rather delightful application of tunable laser radiation, which has been proposed by Ashkin (165). It is possible to selectively manipulate atoms or molecules, using the radiation pressure in the light of a tunable laser. Substantial forces can be exerted near a resonance frequency, where the scattering cross section is very large. With an intense laser field crossing an atomic beam it is in this way possible to deflect certain atomic species or atoms in certain quantum states. Such a scheme might even find practical applications for the separation of isotopes. Walther and coworkers (166) have recently succeeded in demonstrating this effect experimentally.

Acknowledgements

The author would like to thank D. J. Bradley, R. L. Byer, A. Dienes, K. H. Drexhage, S. E. Harris, M. Hercher, P. L. Kelley, D. J. Kuizenga, M. D. Levenson, C. Lineberger, A. Mooradian, O. G. Peterson, A. L. Schawlow, F. Schäfer, C. V. Shank, B. B. Snavely, R. W. Wallace, H. Walther, and H. Welling for many inspiring and helpful discussions.

References

1. P. P. Sorokin and J. P. Lankard, IBM J. Res. Dev. 10, 162 (1966).
2. F. P. Schäfer, W. Schmidt, and J. Volze, Appl. Phys. Letters 9, 306 (1966).
3. P. Sorokin, Sci. Am. 220, 30 (1969).
4. B. B. Snavely, Proc. IEEE 57, 1374 (1969).
5. M. Bass, T. F. Deutsch, and M. J. Weber, in *Lasers*, 3, A. K. Levine and A. J. DeMaria, eds. (Dekker, New York, 1971) p. 269.
6. F. P. Schäfer, Z. Ang. Chemie 82, 25 (1970).
7. W. Schmidt, Laser 2, 47 (1970).
8. M. F. Dewey, Jr., in *Modern Optical Methods in Gas Dynamics Research*, S. S. Dosanjh, ed. (Plenum Press, New York, 1971), p. 221.
9. J. T. Warden and L. Gough, Appl. Phys. Letters 19, 345 (1971).
10. R. Pappalardo, H. Samelson, and A. Lempicki, Appl. Phys. Letters 16, 267 (1970).
11. F. C. Strome and S. A. Tuccio, Opt. Comm. 4, 58 (1971).
12. F. B. Marling, L. L. Wood, and D. W. Gregg, IEEE J. Quant. Electr. QE-7, 498 (1971).
13. R. J. Von Gutfeld, B. Welber, and E. E. Tynan, IEEE J. Quant. Electr. QE-6, 532 (1970).

14. C. V. Shank, A. Dienes, A. M. Trozzolo, and J. A. Myer, Appl.
     Phys. Letters 16, 405 (1970).
15. C. E. Moeller, C. M. Verber, and A. H. Adelmann, Appl. Phys.
     Letters 18, 278 (1971).
16. K. H. Drexhage, in *VII International Quantum Electronics
     Conference, Montreal, Canada, 1972*, Technical Digest p. 8.
17. T. W. Hänsch, M. Pernier, and A. L. Schawlow, IEEE J. Quant.
     Electr. QE-7, 45 (1971).
18. J. R. Lankard and R. J. Von Gutfeld, IEEE J. Quant. Electr. QE-5,
     625 (1969).
19. J. A. Myer, C. L. Johnson, E. Kierstead, R. D. Sharma, and
     I. Itzkan, Appl. Phys. Letters 16, 3 (1970).
20. H. P. Broida and S. C. Haydon, Appl. Phys. Letters 16, 142 (1970).
21. J. G. Kepros and E. M. Eyring, Appl. Phys. Letters 20, 160 (1972).
22. D. J. Bradley, A. J. F. Durrant, G. M. Gale, M. Moore, and
     P. D. Smith, IEEE, J. Quant. Electr. QE-4, 707 (1968).
23. M. A. Mack, Appl. Phys. Letters 15, 166 (1969).
24. J. R. Lankard, Sci. Am. 221, 116 (1970).
25. D. J. Bradley, Physics Bull. 21, 116 (1970).
26. M. E. Mack, Appl. Phys. Letters 19, 108 (1971).
27. W. Schmidt and N. Wittekindt, Appl. Phys. Letters 20, 71 (1972).
28. O. G. Peterson, S. A. Tuccio, and B. B. Snavely, Appl. Phys.
     Letters 17, 245 (1970).
29. M. Hercher and H. A. Pike, Opt. Comm. 3, 65 (1971).
30. R. L. Kohn, C. V. Shank, E. P. Ippen, and A. Dienes, Opt. Comm.
     3, 177 (1971).
31. M. Hercher and Q. Pike, IEEE J. Quant. Electr. QE-7, 13 (1971).
32. S. A. Tuccio, IEEE J. Quant. Electr. QE-7, 12 (1971).
33. M. Hercher and H. A. Pike, Opt. Comm. 3, 346 (1971).
34. M. Hercher and H. A. Pike, IEEE J. Quant. Electr. QE-7, 473 (1971).
35. D. J. Kuizenga, Appl. Phys. Letters 19, 260 (1971).
35A. A. Dienes, E. P. Ippen, and C. V. Shank, Appl. Phys. Letters
     19, 258 (1971).
36. H. Kogelnik, A. Dienes, E. P. Ippen, and C. V. Shank, IEEE J.
     Quant. Electr. QE-8, 373 (1972).
37. A. L. Bloom, Opt. Engrng. 16, 1 (1972).
38. B. B. Snavely, private communication.
39. H. Welling, private communication.
40. F. P. Schäfer and H. Müller, Opt. Comm. 2, 407 (1971).
41. F. C. Strome and J. P. Webb, Appl. Opt. 10, 1348 (1971).
41A. D. J. Bradley, W. G. I. Caughey, J. I. Vukusic, Opt. Comm. 4,
     150 (1971).
42. A. J. Gibson, J. Sci. Instr. 2, 802 (1969).
43. J. E. Bjorkholm, T. C. Damen, and J. Shaw, Opt. Comm. 4,
     283 (1971).
44. H. Walther and J. L. Hall, Appl. Phys. Letters 17, 239 (1970).
45. B. H. Soffer and B. B. McFarland, Appl. Phys. Letters 10,
     266 (1967).
46. T. W. Hänsch and A. L. Schawlow, Bull. Am. Phys. Soc. 15,
     1638 (1970).

47. D. J. Taylor, S. E. Harris, S. T. K. Nieh, and T. W. Hänsch,
     Appl. Phys. Letters 19, 269 (1971).
48. H. Kogelnik and C. V. Shank, Appl. Phys. Letters 18, 152 (1971).
49. I. P. Kaminow, H. P. Weber, and E. A. Chandross, Appl. Phys.
     Letters 18, 497 (1971).
50. J. E. Bjorkholm and C. V. Shank, Appl. Phys. Letters 20,
     306 (1972).
51. R. L. Fork, K. R. German, and E. A. Chandross, Appl. Phys.
     20, 139 (1972).
52. C. V. Shank, J. E. Bjorkholm, and H. Kogelnik, Appl. Phys.
     18, 395 (1971).
53. G. Capelle and D. Phillips, Appl. Opt. 9, 2742 (1970).
54. H. Kogelnik, C. V. Shank, T. P. Sosnowski, and A. Dienes,
     Appl. Phys. Letters 16, 499 (1970).
55. S. A. Myers, Opt. Comm. 4, 187 (1971).
56. I. Itzkan and F. W. Cunningham, IEEE J. Quant. Electr. QE-7,
     14 (1971).
57. T. W. Hänsch, Appl. Opt. 11, 895 (1972).
58. P. Flamant and Y. H. Meyer, Appl. Phys. Letters 19, 491 (1971).
59. T. W. Hänsch, F. Varsanyi, and A. L. Schawlow, Appl. Phys.
     Letters 18, 108 (1971).
60. L. E. Erickson and A. Szabo, Appl. Phys. Letters 18, 433 (1971).
61. B. I. Stepanov and V. A. Batyrev, Zhurnal Prikladnoy
     Spektroskopii 14, 619 (1971).
62. Q. H. F. Vrehen, in VII International Quantum Electronics
     Conference, Montreal, Canada, 1972, Technical Digest p. 9.
63. G. Magyar and H. J. Schneider-Muntau, Appl. Phys. Letters 20,
     406 (1972).
63A. P. P. Sorokin, J. R. Lankard, and V. L. Moruzzi, Appl. Phys.
     Letters 15, 179 (1969).
64B. B. Bölger and C. H. Weysenfeld, in VII International Quantum
     Electronics Conference, Montreal, Canada, 1972, Technical
     Digest p. 11.
64C. A. J. DeMaria, P. A. Stetser, and W. H. Glenn, Jr., Science,
     156, 1557 (1967).
65. B. H. Soffer and J. W. Linn, J. Appl. Phys. 39, 5859 (1968).
66. C. M. Ferrar, IEEE J. Quant. Electr. QE-5, 550 (1969).
67. E. G. Arthurs, D. J. Bradley, and A. G. Roddie, Appl. Phys.
     Letters 19, 480 (1971).
68. E. G. Arthurs, D. J. Bradley, and A. G. Roddie, Appl. Phys.
     Letters 20, 125 (1972).
69. R. J. Von Gutfeld, Appl. Phys. Letters 18, 481 (1971).
70. P. K. Runge, Opt. Comm. 4, 195 (1971).
71. C. V. Shank, E. P. Ippen, and A. Dienes, in VII International
     Quantum Electronics Conference, Montreal, Canada, 1972,
     Technical Digest p. 7.
72. Laser Focus, 1972 Buyer's Guide, p. 129.
73. H. Kressel, in Lasers, 3, A. K. Levine and A. J. DeMaria, eds.
     (Dekker, New York, 1971), p. 2-202.
74. T. C. Harman, in Proceedings of the Conference in Dallas, Texas,

1970, D. L. Carter and R. T. Bates, eds. (Pergamon Press, New York).

75. T. C. Harman, J. Phys. Chem. Solids, Suppl. $\underline{32}$, 363 (1971).

76. E. D. Hinkley, Optoelectronics, to be published.

77. J. A. Rossi, S. R. Chim, and A. Mooradian, Appl. Phys. Letters $\underline{20}$, 84 (1972).

78. D. R. Scifres, N. Holonyak, Jr., H. M. Macksey, and R. D. Dupuis, Appl. Phys. Letters $\underline{20}$, 184 (1972).

79. E. D. Hinkley and C. Freed, Phys. Rev. Letters $\underline{23}$, 277 (1969).

80. N. Bloembergen, in *Nonlinear Optics*, (W. A. Benjamin, Inc.,1965).

81. R. W. Terhune and P. D. Maker, in *Lasers*, $\underline{2}$, A. K. Levine, ed. (Dekker, New York, 1968), p. 295.

82. S. Singh, in *CRC Handbook of Lasers*, R. J. Pressley, ed. (The Chemical Rubber Co., Cleveland, 1971), p. 489.

83. F. M. Johnson and M. W. Swagel, Appl. Opt. $\underline{10}$, 1624 (1971).

84. D. J. Bradley, J. V. Nicholas, and J. R. D. Shaw, Appl. Phys. Letters $\underline{19}$, 172 (1971).

85. S. M. Hamadani and G. Magyar, Opt. Comm. $\underline{4}$, 310 (1971).

86. E. S. Young and C. B. Moore, J. Am. Chem. Soc. $\underline{93}$, 2059 (1971).

87. P. A. Jennings and A. J. Varga, J. Appl. Phys. $\underline{42}$, 5171 (1971).

88. R. W. Wallace, Opt. Comm. $\underline{4}$, 316 (1971).

89. G. Gabel and M. Hercher, in *VII International Quantum Electronics Conference, Montreal, Canada, 1972*, Technical Digest p. 6.

90. A. G. Akhmanov, S. A. Akhmanov, B. V. Zhdanov, A. I. Kovrigin, N. K. Podsotskaya, and R. V. Khoklov, JETP Letters $\underline{10}$, 154 (1969).

91. J. F. Young, G. C. Bjorklund, A. H. Kung, R. B. Miles, and S. E. Harris, Phys. Rev. Letters $\underline{27}$, 155 (1971).

92. C. F. Dewey, Jr. and L. O. Hocker, Appl. Phys. Letters $\underline{18}$, 58 (1971).

93. D. C. Hanna, R. C. Smith, and C. R. Stanley, Opt. Comm. $\underline{4}$, 300 (1971).

94. D. W. Faries, P. L. Richards, Y. R. Shen, and K. H. Yang, Phys. Rev. $\underline{A3}$, 2148 (1971).

95. S. E. Harris, Proc. IEEE $\underline{57}$, 2096 (1969).

96. R. L. Byer, in *Treatise in Quantum Electronics*, H. Raben and C. L. Tang, eds. (Academic Press), to be published.

97. R. L. Byer, M. K. Oshman, J. F. Young, and S. E. Harris, Appl. Phys. Letters $\underline{13}$, 109 (1968).

98. R. W. Wallace, Appl. Phys. Letters $\underline{17}$, 497 (1970).

99. J. Pinard and J. F. Young, Opt. Comm. $\underline{4}$, 425 (1972).

100. S. E. Harris, Appl. Phys. Letters $\underline{14}$, 335 (1969).

101. R. W. Wallace, private communication.

102. J. M. Yarborough and G. A. Massey, Appl. Phys. Letters $\underline{18}$, 438 (1971).

103. R. L. Herbst and R. L. Byer, Appl. Phys. Letters, to be published.

104. C. C. Wang, Phys. Rev. Letters $\underline{16}$, 344 (1966).

105. J. F. Holzrichter and J. McMahon, in *VII International Quantum Electronics Conference, Montreal, Canada, 1972*, Technical Digest p. 55.

106. F. M. Johnson, in *CRC Handbook of Lasers*, R. J. Pressley, ed.

(The Chemical Rubber Co., Cleveland, 1971), p. 526.

107. E. P. Ippen, Appl. Phys. Letters 16, 303 (1970).

108. C. K. N. Patel and E. D. Shaw, Phys. Rev. Letters 24, 451 (1970).

109. A. Mooradian, S. R. J. Brueck, and F. A. Blum, Appl. Phys. Letters 17, 481 (1970).

110. S. R. J. Brueck and A. Mooradian, Appl. Phys. Letters 18, 229 (1971).

111. C. K. N. Patel, Appl. Phys. Letters 19, 7400 (1971).

112. C. K. N. Patel, Phys. Rev. Letters 28, 7649 (1972).

113. J. Gelbwachs, R. H. Pantell, H. E. Puthoff, and J. M. Yarborough, Appl. Phys. Letters 14, 258 (1969).

114. T. Y. Chang and O. R. Wood, in VII International Quantum Electronics Conference, Montreal, Canada, 1972, Technical Digest p. 80.

115. J. M. Madey, J. Appl. Phys. 42, 1906 (1971).

116. R. H. Pantell, Stanford University Microwave Report No. 1954 (1971).

117. W. Demtröder, in Topics in Current Chemistry, 17, (Springer Verlag, Berlin-Heidelberg-New York, 1971).

118. E. D. Hinkley and P. L. Kelley, Science 171, 635 (1971).

119. C. K. N. Patel, E. D. Shaw, and R. J. Kerl, Phys. Rev. Letters 25, 8 (1970).

120. P. L. Kelley, E. D. Hinkley, and A. R. Calawa, in VII International Quantum Electronics Conference, Montreal, Canada, 1972, Technical Digest p. 87.

121. K. W. Nill, F. A. Blum, A. R. Calawa, and T. C. Harman, in VII International Quantum Electronics Conference, Montreal, Canada, 1972, Technical Digest p. 88.

122. L. B. Kreuzer and C. K. N. Patel, Science 173, 45 (1971).

123. K. W. Nill, F. A. Blum, A. R. Calawa, and T. C. Harman, Appl. Phys. Letters 19, 79 (1971).

124. H. R. Schlossberg and P. L. Kelley, Phys. Today (July 1972), p.36

125. I. Melngailis, IEEE Trans. on Geosci. Electro. GE-10, 7 (1972).

126. N. C. Peterson, M. J. Kurylo, W. Braun, A. M. Bass, and R. A. Keller, J. Opt. Soc. Am. 61, 746 (1971).

127. R. A. Keller, E. F. Zalewski, and N. C. Peterson, J. Opt. Soc. Am. 62, 319 (1972).

128. R. J. Thrash, H. von Weyssenhoff, and J. S. Shirk, J. Chem. Phys. 55, 4659 (1971).

129. T. W. Hänsch, A. L. Schawlow, and P. Toschek, IEEE J. Quant. Electr. (October 1972), to be published.

130. A. Mooradian, Science 169, 20 (1970).

131. M. B. Denton and H. V. Malmstadt, Appl. Phys. Letters 18, 485 (1971).

132. J. Kuhl and G. Marovsky, Opt. Comm. 4, 125 (1971).

133. C. F. Burrell and H. J. Kunze, Phys. Rev. Letters 28, 1 (1972).

134. P. B. Sachett and J. T. Yardley, Chem. Phys. Letters 6, 323 (197(

135. K. Sakurai and G. Capelle, J. Chem. Phys. 53, 3764 (1970).

136. S. E. Johnson and H. P. Broida, Bull. Am. Phys. Soc. 15, 1630 (1970).

137. G. Capelle, K. Sakurai, and H. P. Broida, J. Chem. Phys. 54, 1728 (1971).
138. K. Sakurai, G. Capelle, and H. P. Broida, J. Chem. Phys. 54 1220 (1971).
139. C. J. Schuler, C. T. Pike, and H. A. Miranda, Appl. Opt. 10, 1689 (1971).
140. M. C. W. Sandford and A. J. Gibson, J. of Atmospheric and Terrestrial Phys. 32, 1423 (1970).
141. E. L. Baardsen and R. W. Terhune, in *VII International Quantum Electronics Conference, Montreal, Canada, 1972*, Technical Digest p. 10.
143. T. J. McIlrath, Appl. Phys. Letters 15, 41 (1969).
144. D. J. Bradley, G. M. Gale, and P. D. Smith, J. Phys. B. 3, 11 (1970).
145. K. J. Nygaard and R. E. Hebner, U.S. Government R/D Report AD-703 628 (March 1970), p. 36.
146. D. J. Bradley, P. Ewart, J. V. Nicholas, and J. R. D. Shaw, in *VII International Quantum Electronics Conference, Montreal, Canada, 1972*, Technical Digest p. 58.
147. V. S. Letokhov and R. V. Ambartzymian, in *1971 IEEE, Optical Society of America, Conference on Laser Eningeering and Applications, Washington, D.C.*, Technical Digest p. 47.
148. K. C. Smyth, P. T. McIver, Jr., J. I. Baumann, and R. W. Wallace, J. Chem. Phys. 54, 2758 (1971).
149. C. Lineberger, private communication.
149A. G. Singh, P. DiLavove, and C. O. Alley, IEEE J. Quant. Electr. QE-7, 196 (1971).
150. P. P. Sorokin and J. R. Lankard, Phys. Rev. 186, 342 (1969).
151. F. A. Korolev, S. A. Bakhramov, and V. I. Odintsov, JETP Letters 12, 90 (1970).
152. P. Agostini, P. Bensoussan, and J. C. Boulassier, in *VII International Quantum Electronics Conference, Montreal, Canada, 1972*, Technical Digest p. 26.
153. P. P. Sorokin and J. R. Lankard, J. Chem. Phys. 54, 2184 (1971).
154. R. L. Byer, R. L. Herbst, H. Kildal, and M. D. Levenson, Appl. Phys. Letters 20, 463 (1971).
155. F. Varsanyi, Appl. Phys. Letters 19, 169 (1971).
155A. J. F. Holzrichter, R. M. Macfarlane, and A. L. Schawlow, Phys. Rev. Letters 26, 652 (1971).
157. T. W. Hänsch, I. S. Shahin, and A. L. Schawlow, Phys. Rev. Letters 27, 707 (1971).
158. T. W. Hänsch, I. S. Shahin, and A. L. Schawlow, Nature 235, 56 (1972).
158A. T. W. Hänsch and P. Toschek, IEEE J. Quant. Electr. QE-4, 467 (1968).
159. E. V. Baklanov and V. P. Chebotayev, Sov. Phys. JETP 23, 300 (1971).
160. H. Nieuwenhuijzen, Optics Technology 2, 13 (1970); 2, 68 (1970).
161. B. Bölger and J. C. Diels, Phys. Letters 28A, 401 (1968).

162. D. J. Bradley, G. M. Gale, and P. D. Smith, Nature 225, 719 (1970).

163. M. D. Levenson, C. Flytzanis, and N. Bloembergen, in *VII International Quantum Electronics Conference, Montreal, Canada, 1972*, post-deadline paper.

164. J. J. Wynne, in *VII International Quantum Electronics Conference, Montreal, Canada, 1972*, Technical Digest p. 89.

165. A. Ashkin, Phys. Rev. Letters 25, 1321 (1970); A. Ashkin and J. M. Dziedzic, Appl. Phys. Letters 19, 283 (1971).

166. H. Walther, Opt. Comm., to be published.

SATURATED ABSORPTION SPECTROSCOPY WITH APPLICATIONS TO THE 3.39 μm

METHANE TRANSITION

J. L. Hall[*]

Joint Institute for Laboratory Astrophysics

University of Colorado, Boulder, Colorado   80302

In basic research we are grateful to be able to pursue research in profitable directions, e.g., those directions where the boundaries of experimental or theoretical possibilities seem most susceptible to growth. Of course we know of some areas that could be investigated and many numerical data recorded, but we are uneasy in not knowing how to make use of the data. As physicists we especially like to make progress synthesizing specialized concepts into more general forms. We savor the similarities -- and differences -- when ideas developed in one area prove useful in another set of circumstances. Just now it is optical resonance physics that seems to be ripe for explosive growth using new laser techniques and "classical" resonance ideas. This paper represents a direct effort to sketch, in the opinion of a certain class of partisans, "where the action is." We begin with a brief discussion of experiments in which a laser is useful but not necessary, and a consideration of the basic optical facts of life. The bulk of the paper explores the exciting land beyond the Doppler limit.

With all the recent enthusiasm over lasers, sometimes it escapes our attention that very high resolution spectroscopy can be done -- and has been done -- with conventional incoherent sources. Characteristically the non-laser experiments that surpass the Doppler-broadened resolution limit are ones which depend on quantum-mechanical interference between two-state amplitudes in a single atom. We can isolate the interference term by sweeping these energy states through each other using Zeeman-effect tuning or by

[*]Staff Member, Laboratory Astrophysics Division, National Bureau of Standards.

tuning the frequency of an applied oscillating field. Using broad-
line (Doppler-broadened) light sources, each sample atom can find
a part of the exciting spectrum that matches its own particular
Doppler-shifted absorption frequency. Thus, these classical spectroscopic methods such as Hanle effect, level crossing,[1] and optical double resonance[2] can study splittings within a given state
with a resolution limited, basically, by the natural lifetimes involved. But both source and absorber have important Doppler broadening of optical transitions between widely separated states.

The next step in increasing spectral resolution is possible
only because suitable lasers can produce so much intensity per unit
bandwidth that generically non-linear new high-resolution phenomena
are accessible. In this paper we are concerned with using lasers
to produce velocity-resolved changes in the population distributions of atoms and molecules.[3] Our high resolution will come about
because the narrow laser spectral width will allow us to investigate particles with rather sharply-defined Doppler shifts, the
velocity (frequency) resolution being limited variously by atomic
state lifetimes or by the time of interaction with the laser.

We begin with the remark that in general <u>two</u> radiative interactions will be required to attain sub-Doppler line widths, and
that at least one of the interactions must be sufficiently strong
to generate a non-linear response. We picture the Doppler profile
of a gaseous absorber as being composed of many sharp sub-profiles,
or "packets" whose narrow width is due to homogeneous broadening
processes such as lifetime broadening and collisional broadening
mechanisms. The packet width measures the bandwidth of its inter--
action with an applied monochromatic electromagnetic field. The
absorption coefficient of such an absorber, defined by a trans-

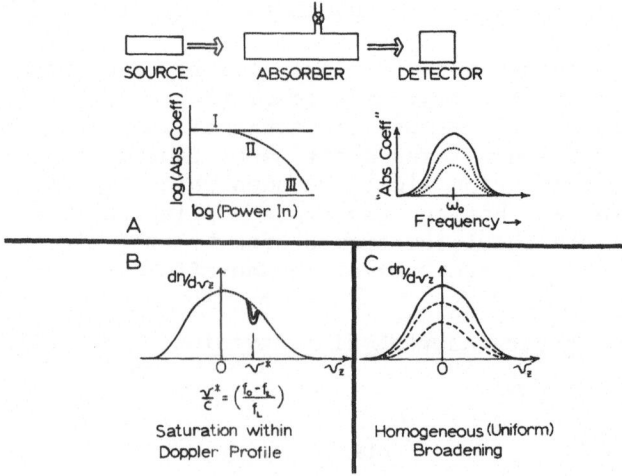

Fig. 1. Measurement of the absorption coefficient.

mission experiment, will depend upon the input intensity in the
general way illustrated in Fig. 1A. In regime I, at low intensity,
the absorption coefficient is unaffected by the measurement. At
higher powers two more-or-less distinct regimes may be noticed.
In regime II, Fig. 1B, we are radiatively putting power into the
absorber's resonant packets so rapidly that we begin to deplete
the population of ground state absorbers having the right velocity
component to be Doppler-shifted into resonance with the applied
field. As the power is increased we can appreciably radiation-
broaden[3] this resonance packet so that it begins to cover
a wider velocity range: the absorption coefficient decreases[4] as
(power)$^{-1/2}$. The saturated absorption experiments to be described
in this paper work in this power regime. Ultimately at sufficiently high power, regime III corresponds to such rapid radiative
processes that the detuning due to the entire Doppler effect is no
longer very significant. The field begins to affect the absorption coefficient over the whole Doppler profile, giving a (power)$^{-1}$
dependence. Of course this behavior is also obtained when the
transition is strongly homogeneously broadened by some other mechanism, such as collisions or the strong radiative processes associated with the vacuum ultraviolet part of the spectrum. This
homogeneously-broadened case is illustrated in Fig. 1C, but we will
not discuss it further.

     In solid state magnetic resonance physics, one typically has
an inhomogeneous broadening situation analogous to Doppler broadening of gaseous absorbers. In the crystal the sharp quantum re-
sonators (nuclei, trapped electrons and holes) are spatially localized, with the inhomogeneous broadening arising from static
strain fields or differing local magnetic environments. Portis[5]
showed that an estimate of the ratio of packet-to-envelope line
widths could be obtained from analysis of the shape of the curve
of absorption coefficient vs power. It is probably only the microwave technology problem that prevented people from using a second
probe frequency to sample the frequency extent of the saturation,
and thus directly measuring the packet width. (Microwave resonant cavities were necessary and a tunable bimodal system is not
perhaps the first choice for simplicity of design.) Only now is
this method (ELDOR)[6] coming into wide use, made practical mainly
by improved broad-band slow-wave structures.

     Now in the infrared and visible regions, the envelope func-
tion due to Doppler broadening may be very wide compared with the
packet line widths. Ratios of 10^4 to 10^5 may exist for suitable
transitions. However our transmission experiment of Fig. 1 gives
no direct evidence of the sharpness of these saturation features:
they are fundamentally unobservable in a static single frequency
experiment. We must arrange to probe the resonance structure with
a second interaction.

Much progress has been made in nonlinear optical spectroscopy
and there are now several well-known methods of making these sharp
resonances visible.[7] For example, we may scan through the narrow
frequency interval near resonance with an auxiliary probe beam of
variable frequency. Such a beam may have a propagation direction
parallel to the saturating beam. In this case, we can study the
relaxation processes that broaden the resonance function by either
phase perturbations or by direct quenching. The second beam may
also be at another wavelength, causing transitions to a third level
from either of the two original strongly perturbed levels.[8] Now
we are relatively more sensitive to population effects. For paral-
lel-running beams only the differential Doppler effect for the
two wavelengths causes broadening.[9] Although these experiments
show very sharp resonance behavior in the differential Doppler
velocity, there is no interesting dependence on the frequency of
the saturating beam within the Doppler profile. To be sensitive
to the absolute frequency we must arrange to nullify the Doppler
shift. An attractive simple proposal uses counter-running beams
of the same frequency. The Doppler shifts have the opposite signs
for the two beams, and thus the two frequencies are viewed as
equivalent only by absorbers free of longitudinal velocity.[10]
This is the usual setup in a laser. The dip in the number of possible emitters, and consequently in the output power, exactly at
line center is termed the Lamb dip, after the important calculation of Willis Lamb in 1963.[11] An example of the Lamb dip (saturated emission dip) is shown in Fig. 2. These data are taken for
the 1.15 µm laser transition of pure neon.[12] The line width in
this case is clearly much less than the Doppler-broadened line
width, but is still appreciably broadened beyond the natural line
width by processes involving collisions modified by ultraviolet
radiation trapping.[13]

Neon[20]   p = 120 mTorr   1.15 µm

C/2L = 465 MHz   P = 1 µW/div

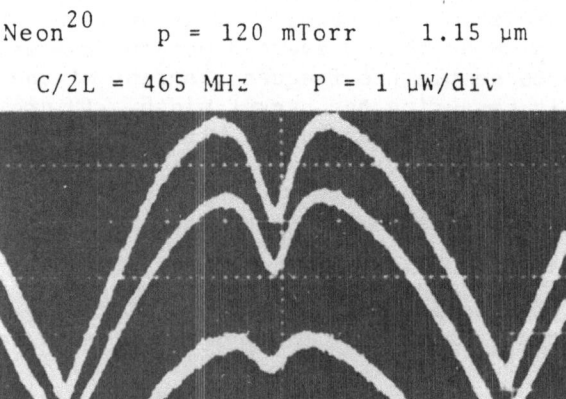

Fig. 2. Lamb dip in single mode laser in pure neon.

The next important step was taken by Paul Lee and M. L. Skolnick when they observed that the amplifier and absorption medium need not be identical.[14] Their configuration employed both a He-Ne gain cell and a pure neon absorption cell within one laser cavity. The saturated absorption peak obtained at the absorber's Doppler center is shown in Fig. 3. Of particular interest is the relative narrowness (~20MHz) of this peak and the differential frequency shift between neon atoms in the two environments, due mainly to collisions with the few Torr of He atoms in the gain cell. Collision effects were studied in detail by Lisitsyn and Chebotayev in Novosibirsk.[15]

The next major step was imagined almost simultaneously at four laboratories.[16] We chose to use molecular rather than atomic absorbers. Suitable molecules can exhibit absorption from the ground state; thus the transition is not broadened by auxiliary transitions from the usual optically-active high-lying levels of an atomic system. Of course we must find an accidental absorber/laser wavelength overlap. In Fig. 4 we show the saturated absorption peak in methane.[17] It is saturated using the 3.39 μm helium-neon laser. The peak is quite narrow -- line widths typically are less than ½ MHz. The apparent accurate centering of this methane absorption feature with the helium-neon gain curve is contrived by adjusting the helium pressure in the laser. This pressure shifts the helium-neon laser into wavelength coincidence with the low pressure methane absorption. Alternatively isotopic neon 22 could be used to achieve the bulk of the shift.[18] The peak size is in the range 3 to 5% and has a line width here of ~300 kHz,

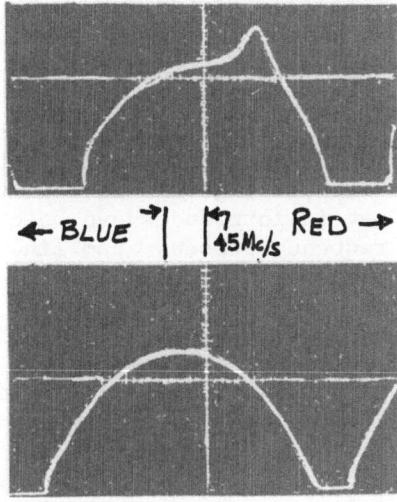

Fig. 3. Saturated absorption in pure neon, from Ref. 14.

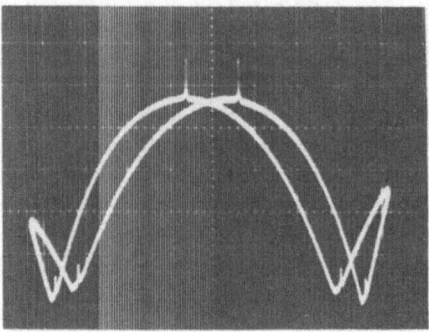

Fig. 4.   Intracavity saturated absorption in CH_4, using He-Ne
laser at 3.39 μm.

basically independent of the natural lifetime. Rather, the line
width is dictated by pressure- and intensity-broadening. When
these broadening sources are minimized to more nearly approach the
very narrow line width (<1kHz) suggested by the long molecular
lifetime, one observes the uncertainty-principle broadening associated with molecular transit through the light spot.[19] Refrigeration
to 77°K gives the expected decrease of effective thermal velocity
and hence line width.

     Another interesting experimental arrangement utilizes counter-
running waves of different frequencies. The saturation effect is
observed when the two frequencies are symmetrically disposed around
the absorber's rest frame center frequency. For this configuration
the saturation peak is formed by particles with definite, non-zero
velocity components along the laser axis, given by $v_z = (\Delta f/f_o) \cdot c$.
Here Δf is the frequency offset of the source from absorber line
center, f_o. This new capability to study absorbers of a variable
velocity (component) may lead to some interesting collision-physics
results[20]; for example, some collisional energy transfer processes
may have a strong velocity dependence. Clearly the second-order
Doppler effect plays an interesting role in this configuration
since the frequency offset it introduces can be controlled to some
extent: the required frequency offsets from line center will not be
quite equal for the two beams.

     It is fashionable at present to speak enthusiastically about
spectroscopy with tunable lasers, for example dye lasers with a few
hundred angstroms tuning range, and the exciting work reported in
this volume by Dr. Hänsch completely justifies this enthusiasm. However there are still some interesting possibilities even if one is
restricted to a Doppler tuning interval, given a suitable laser/molec-
ular absorber overlap.   By now a large number of such wavelength
overlaps are known[21] and, in Table I, we list a few of the more
extensively studied cases.

Table I. Selected Laser Molecular Absorber Wavelength Coincidences

Laser	λ (μm)	Molecule	Transition		Ref.
Ar^+	.4658	Na_2 §	(X,4) → (B,17)	P(38)	
Ar^+	.4727	Na_2 §†	(X,1) → (B,9)	R(37)	
Ar^+	.4765	Na_2 §†	(X,3) → (B,10)	P(13)	
Ar^+	.4880	Na_2 §†	(X,3) → (B,6)	Q(43)	22
Ar^+	.4965	Na_2 §	(X,6) → (B,7)	P(44)	
Ar^+	.5017	Na_2 §	(X,6) → (B,5)	P(38)	
Ar^+	.5145	Na_2 §†	(X,14) → (B,11)	Q(49)	
HeNe	.633	Na_2 §†	(X,2) → (A,14)	P(45)	23
HeNe	.640	Na_2 §†	(X,0) → (A,11)	P(73)	
HeNe	.633	K_2 §	X → B		24
Ar^+	6 lines	Rb_2 §, Cs_2 §	X →		25
Ar^+	.5145	I_2 §	(X,0) → (B,43)	P(12)	26
Kr^+	.568	$(^{127}I)_2$	(X,1) → (B,21)	P(117)	27
HeNe	.633	$(^{127}I)_2$	(X,5) → (B,11)	R(127)	28
			(X,3) → (B,6)	P(33)	
HeNe	.633	$(^{129}I)_2$	X → B		29
HeNe	.633	$(^{79}Br)_2$, $(^{81}Br)_2$			30
Ar^+, Kr^+	11 visible lines	NO_2 §†	not identified		31
HeXe	3.36	H_2CO §	$3_{2,2}$ → $2_{1,1}$ ($\nu_2+\nu_5$)		32
HeNe	3.39	$^{12}CH_4$	$F_1^{(2)}$ of ν_3	P(7)	17,33
HeNe	3.39	$^{12}CH_4$	E of ν_3	P(7)	33,34
HeNe	3.39	$^{13}CH_4$	$F_2^{(2)}$ of ν_3	P(6)	35
HeNe	3.39	CH_3Br	6 lines		
HeNe	3.39	$CH_3C\ell$	2 lines		
HeNe	3.39	CH_3I	2 lines		33,36
HeNe	3.39	C_2H_6	19 lines		
HeNe	3.39	CH_3OH	2 lines		
HeNe	3.39	CH_3F †	J,K = 10,6 → 9,6 (ν_1)		33,36 37
HeNe	3.39	H_2CO §†	$6_{3,3}$ → $5_{2,4}$ ($\nu_2+\nu_5$)		32

Table I (continued)

Laser	λ (μm)	Molecule	Transition	Ref.
HeXe	3.5	H_2CO §	11 lines, 2 assigned (ν_4)	38
HeXe	3.5	HDCO §	10 lines, 7 assigned (ν_1)	38
CO	5.307	NO_2 §	($\nu = 9 \to 8$) P(13)	39
HeNe	7.7015	CH_4 §	ν_4 R(5)	40
HeNe	7.765	CH_4 §	ν_4 P(3)	
CO_2	P(14) 10.54 μm	NH_2D	$(0_a, 4_{04}) \to (1_s, 5_{14})$ ν_2	41
CO_2	R(12) 10.30 μm	NH_2D	$(0_s, 4_{14}) \to (1_a, 5_{24})$ ν_2	42
CO_2	P(20) 10.59 μm	NH_2D	$(0_a, 4_{04}) \to (1_a, 5_{05})$	41,42
CO_2	P(20) 9.55 μm	CH_3F	^QQ(J=12, K=2) ν_3	43
N_2O	P(13) 10.78 μm	$^{14}NH_3$	aQ - (8,7) ν_2	44
CO_2	P(20) 10.56 μm	SF_6	ν_3	45
CO_2	all lines	CO_2	same line	46
CO_2	P(30) 9.64 μm P(32) 9.66 μm	SiF_4	ν_3	47
CO_2	P(26) to P(52) of 10.4 μm branch	CF_2Cl_2	several peaks each line	48
N_2O	P(10) to P(20)	CF_2Cl_2	several peaks each line	
CO_2	P(4) to P(32) of 10.4 μm branch	PF_5	ν_3, ν_5	
N_2O	R(8) to R(23)	PF_5	ν_3	
N_2O	P(20) 10.85 μm	C_2H_4		

§ only linear spectroscopy;   † more than one transition

* For diatomic molecules our convention is (lower electronic state, lower vib. state) →
   (upper electronic state, upper vib. state)P,Q,R(J_{lower}). All of the infrared
   polyatomic molecular transitions are within the ground electronic state.

The application of suitable saturated absorption techniques
can lead to unprecedented spectral resolution. For example, using
the methane 3.39 μm line we have observed resonances as narrow as
25 kHz HWHM,[49] corresponding to a spectral resolution of 1.75×10^9.
To meaningfully employ this kind of resolution, it is necessary to
interest oneself in the technology of stable lasers. The tuning
capability of even gas lasers is an embarrassing wealth -- perhaps
10^4 resonance line widths! In refined work of this type, it is

found that a "two-stage" concept proves attractive[12]: One laser
is servo-stabilized to the saturation peak. We don't ask about
its absolute frequency, only changes of its frequency. The stability
over a few minutes' interval is optimized by a linear combination
of analysis, design, sweat, and inspiration. This frequency stability will then be transferred to the powerful laser of interest
using Frequency-Offset Locking techniques.[12,17] First we briefly
discuss the reference laser question.

    Our typical frequency reference lasers are of the form shown
in Fig. 5. The important feature is the design for low frequency
noise consistent with modular construction for convenience. A
frequency control servomechanism continuously adjusts the average
frequency of the laser to match the apparent maximum of the saturated absorption feature. (In practice, we use the central zero
of the <u>third</u> frequency derivative of the resonance to better suppress baseline problems.[50]) The calculated frequency instability
resulting from the finite signal-to-noise ratio improves with
(averaging time)$^{\frac{1}{2}}$ and has a calculated value of 2 Hz rms for 1 sec
averaging. As will be seen, for various technical reasons we have
not yet achieved frequency stability controlled by this noise
mechanism.

    In Fig. 6 we show the observed frequency stability for three
averaging times: 1 sec, 2 sec, and 45 sec per measurement. It will
be seen that the frequency stability is improving with averaging
time near 1 sec. However, at 45 sec per channel the random noise

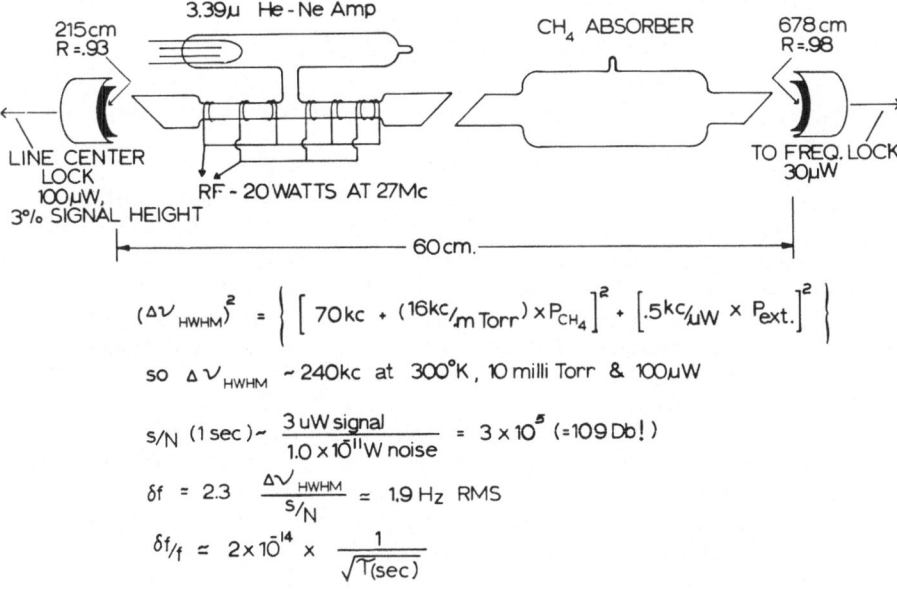

$$(\Delta\nu_{HWHM})^2 = \left\{ \left[70kc + (16kc/_{m\,Torr}) \times P_{CH_4} \right]^2 + \left[.5kc/_{\mu W} \times P_{ext.} \right]^2 \right\}$$

so $\Delta\nu_{HWHM} \sim 240kc$ at $300°K$, 10 milli Torr & $100\mu W$

$$S/N\ (1\,sec) \sim \frac{3\,uW\,signal}{1.0 \times 10^{11}W\,noise} = 3 \times 10^5\ (=109\,Db!\,)$$

$$\delta f = 2.3\ \frac{\Delta\nu_{HWHM}}{S/N} = 1.9\,Hz\ RMS$$

$$\delta f/_f = 2 \times 10^{-14} \times \frac{1}{\sqrt{T(sec)}}$$

Fig. 5. Laser-saturated methane frequency reference.

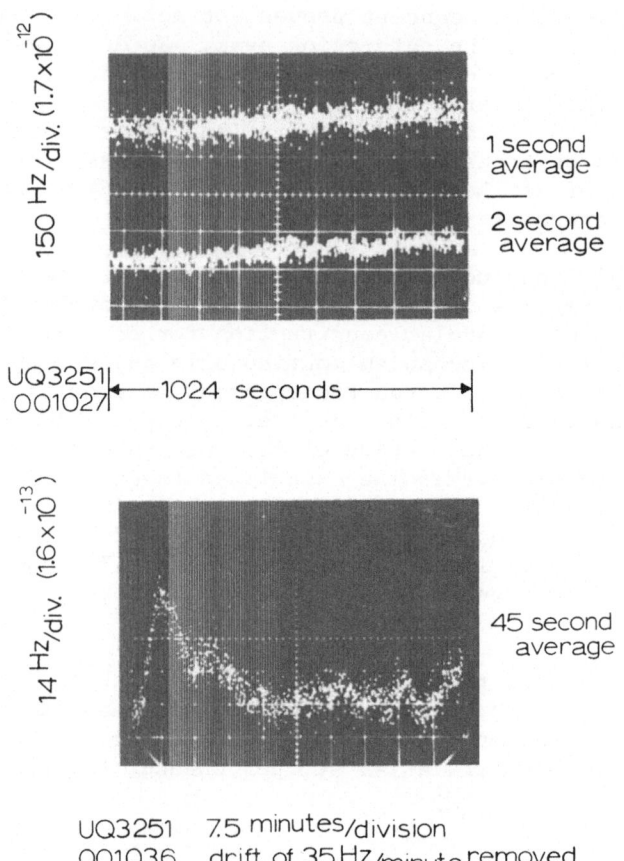

Fig. 6. Optical heterodyne beat frequency between two independent methane-stabilized lasers.

has decreased sufficiently that we become aware of small but systematic effects. These drifts (and changes of systematic offsets from line center) increase with increasing time. They tend to compensate the decreasing random noise at longer integration times, thus giving rise to a region of roughly time-independent frequency stability. A large amount of such data is summarized in Fig. 7, where we show the fractional frequency variation vs the averaging time on the log-log plot described by Allan.[51] The plotted data represent the beat frequency between two independent methane-stabilized lasers. In fact, each stabilized laser was heterodyned with a third, common local oscillator laser to remove any possibility of coupling between the stabilized lasers. The two resulting rf beat notes were heterodyned again after a small offset was introduced in one of them to remove the sign ambiguity near zero frequency. The second (audio) beat is thus free of instability of

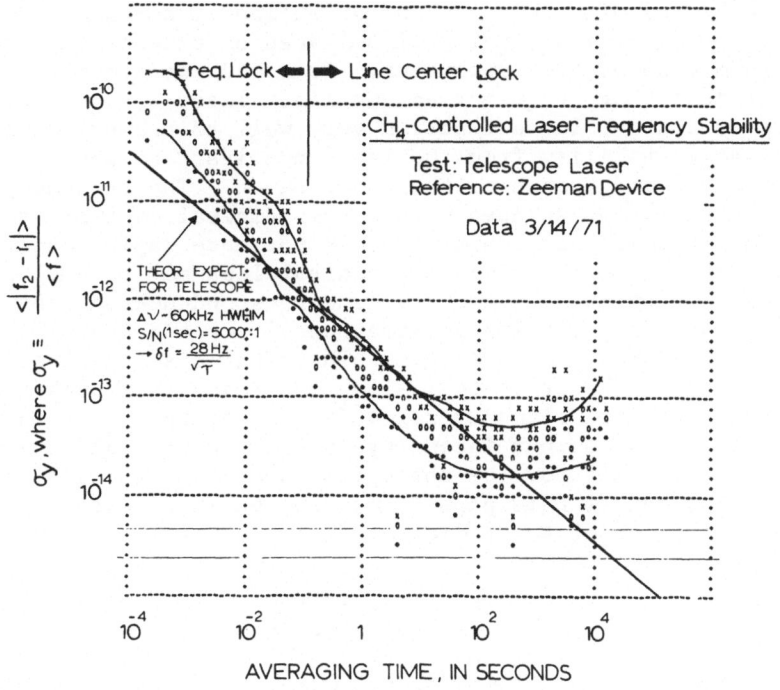

Fig. 7. CH_4 - controlled laser frequency stability.

the optical local oscillator reference. The lasers being compared here were dissimilar. One contained an internal five-power telescope to magnify the beam diameter and thus reduce the transit time line width by a factor of 5. The other laser was optimized for high signal-to-noise ratio. Thus at short times, the random noise level of the telescope laser controls the obtained performance. At long times, drifts of the offsets of the high signal-to-noise reference laser dominate the frequency instability. See Fig. 7. There is evidently some improvement possible, but in view of this rather remarkable frequency stability of a few Hz for averaging times near 100 sec, a variety of new experiments are already feasible. Perhaps the most obvious applications would be optical heterodyne spectroscopy,[52] long-path interferometry,[53] precision line-profile analysis,[54] and laserized versions of the classical optical relativity experiments[55] (Michelson-Morley, Kennedy-Thorndike, etc.).

We turn now to the study of the saturated absorption line shape, first its measurements and later its characterization. As previously mentioned, the high stability displayed in Fig. 7 can form the basis of a powerful spectroscopic concept which we call Frequency Offset Locked Laser Spectroscopy.[56] The high frequency

stability achieved with this methane-stabilized laser is used to
stabilize a powerful laser which illuminates an external absorption
cell. Figure 8 shows this apparatus. To avoid the troublesome
region near zero beat frequency an auxiliary laser is introduced.
The frequency control loops function very well to transfer the
total available stability from the reference laser to the local
oscillator laser and thence to the power laser. The frequency of
the power laser may be scanned with either digital control of the
frequency synthesizer as illustrated, or for more rapid scans,
analog programming in the frequency control loop may be employed.
One thus has a very precise and stable correspondence between
frequency offset from the stable reference laser and channel numbers
of the multichannel signal averager. The total absolute frequency
jitter of the power laser is typically less than 3 kHz peak to peak,
and the frequency variations averaged for 1 sec are less than 30 Hz.
Servo control of the output intensity is used to provide very
satisfactory baselines for the absorption spectra. As will become
apparent at the end of this paper, one must be interested in the
spatial distribution of the radiation field. Consequently we

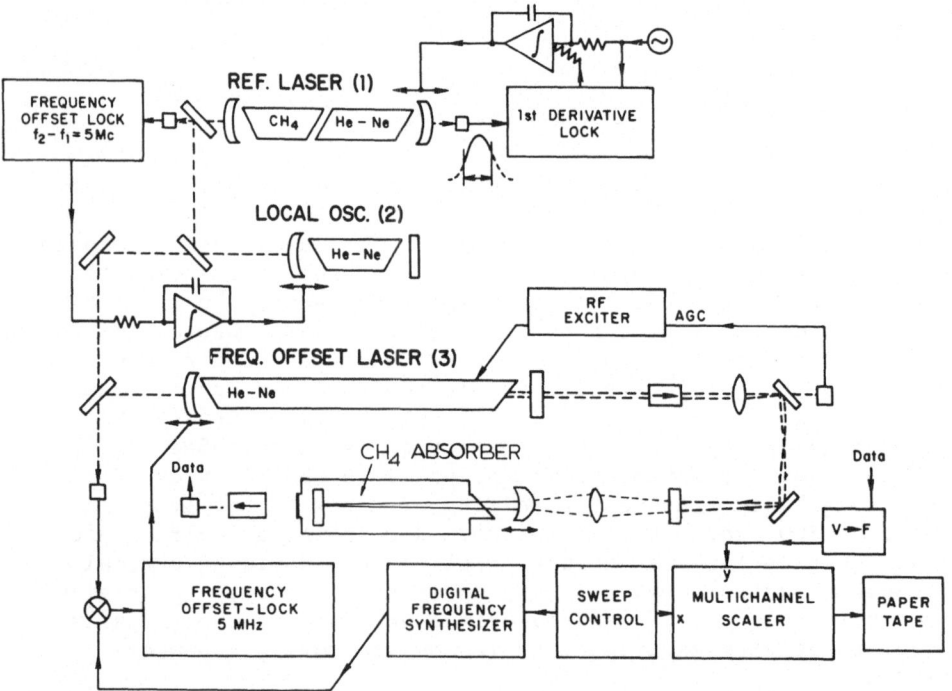

Fig. 8. Frequency-Offset-Locked Laser Spectrometer. Frequency
stability from reference laser (1) is transferred through laser (2)
to the 10 mW power laser (3) with a frequency offset corresponding
to channel number of the signal averager. The present apparatus
uses magnetic tape output, and a third-derivative laser reference
algorithm.

presently employ a curved-mirror interferometer (slaved to the
laser wavelength) to spatially purify the laser output beam. The
absorption cell is inside the optical cavity. Thus the absorbers
see a laser field having a spatial distribution of known diameter
and radius of curvature, with well-prescribed frequency and amplitude. Several interesting problems can be investigated with this
type of apparatus. In Fig. 9 we show a typical experimental line
shape and the least-squares fitted curve based on an assumed
Lorentzian line profile.[58] It may be seen that the Lorentz function
accounts for the essence of the line shape.

From the optical frequency standards viewpoint, it has been
of interest to investigate the possibility of an asymmetry in the
saturated absorption line. Such an asymmetry could result from
some interesting radiation physics (such as absorber recoil from
the photon momentum)[59] or from incipient, but unresolved, structure.
In Fig. 10 we show data of the same type which have been numerically
differentiated. The advantage of this procedure in looking for
asymmetry is evident -- we know precisely what operation is being
performed -- but it results in strong degradation of the signal-
to-noise ratio since the data are heavily oversampled. On a number
of experiments such as this, we have found no asymmetry to the limit
of the experimental uncertainty of a least-squares-fitted asymmetric

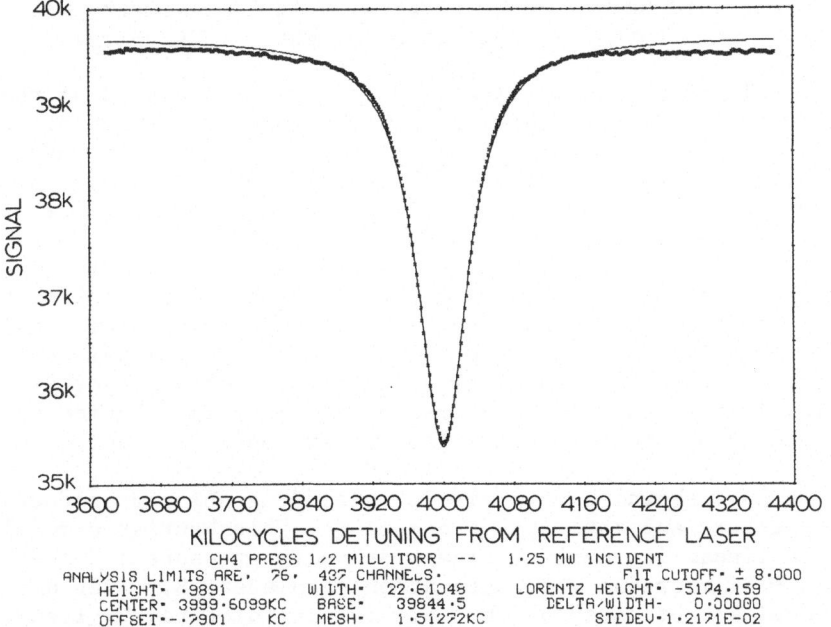

CH4 PRESS 1/2 MILLITORR --   1.25 MW INCIDENT
ANALYSIS LIMITS ARE. 76, 437 CHANNELS.   FIT CUTOFF. ± 8.000
    HEIGHT. .9891   WIDTH.   22.61048   LORENTZ HEIGHT. -5174.159
    CENTER. 3999.6099KC   BASE.   39844.5   DELTA/WIDTH.   0.00000
    OFFSET.-.7901 KC   MESH.   1.51272KC   STDDEV. 1.2171E-02

Fig. 9. Methane saturated absorption resonance in external cell.
Spatial filtering (rather than external resonator) was employed.
The least squares computer program allows a small asymmetry as the
fifth parameter.

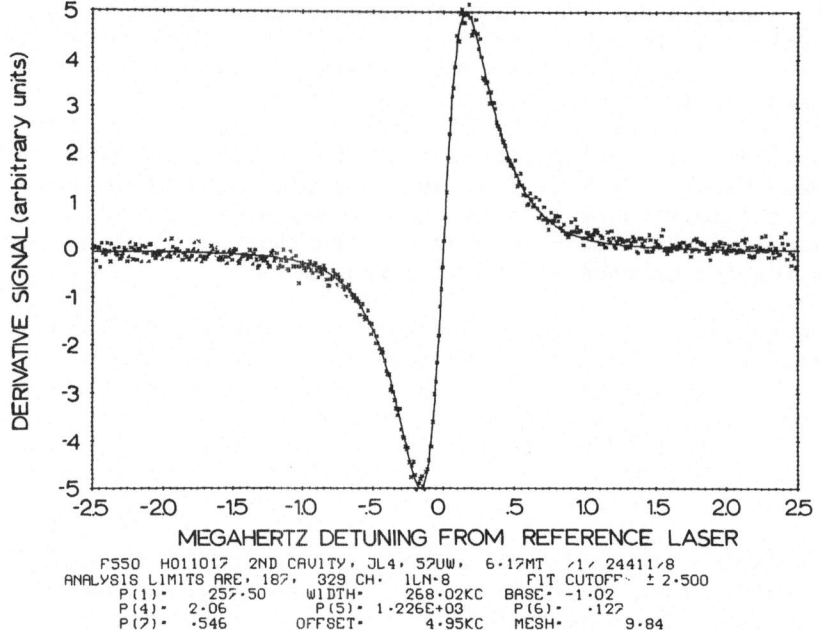

F550   H011017   2ND CAVITY,   JL4, 57UW,   6·17MT   /1/ 24411/8
ANALYSIS LIMITS ARE, 187,   329 CH·   1LN·8   FIT CUTOFF· ± 2·500
         P(1)·   257·50   WIDTH·   268·02KC   BASE· -1·02
         P(4)·   2·06   P(5)· 1·226E+03   P(6)·   ·122
         P(7)·   ·546   OFFSET·   4·95KC   MESH·   9·84

Fig. 10. Saturated absorption resonance power, numerically filtered
and differentiated before curve fitting. Fitted asymmetry was
2 ± 3 units, with Lorentz (derivative) height = 1226 units.

component. Typically this limit is one part per thousand of the
basic Lorentz function. These experiments have been tried at
several pressures and several intensities. Unfortunately very
high resolution data are not yet available under circumstances
where the beam geometry is very precisely known. Better experiments are underway.

     It is of interest to try to model the pressure and intensity
dependence of these resonant line widths. On the one hand, the
parameters of this model may provide a pseudo-experimental testing
ground for the results of a detailed a priori theory. Another
motivation is the possibility of assessing -- by calculation -- the
performance of a potential stabilized-laser configuration.

     As we have shown, a very simple model[54] is able to account
for the pressure and intensity broadening. This heuristic model
supposes a linear increase of line width with pressure and the
usual $\sqrt{1 + S}$ increase with saturation parameter, S. Such a
simple model is motivated by the hole burning picture.[60] Its
success in accounting for the data may be appreciated from Fig. 11
where we show a family of line widths vs pressure parameterized by
different intensities. Experimental data are plotted as "+" and
include the effects of both broadening mechanisms. The reduced
data, corrected for the finite intensity, are plotted as "o."

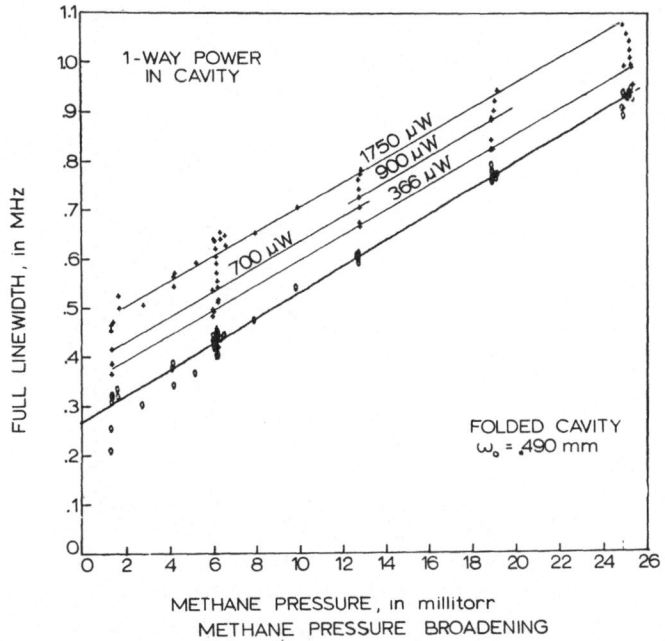

Fig. 11. Observed saturated absorption line widths vs methane pressure. The parameter is (1 way) cavity power. Lower line and symbol "o" are calculated for zero intensity.

It may be seen that the assumed model accounts reasonably well for the power dependence of the experimental data. Similarly in Fig. 12, we show a perpendicular cut of this three-dimensional surface of line width vs intensity and pressure. Here the abscissa is intensity and the parameter is pressure. The experimental line width data are again represented by "+" and the corrected by "o." Again the utility of the model is apparent. The three line shape parameters that result from this experiment, along with those from four other experiments at different apertures, are summarized in Table II. The limiting line width at low pressure and low intensity is represented by the quantity P1. Since this residual broadening is due to transit through the radiation spot, it is useful also in Table II to have a column P1 × spot size, ω. In the lower three experiments it will be seen that this product is near the value 70 kHz · mm, which represents an effective transverse thermal velocity. For comparison, two degrees of freedom lead to an rms kinetic velocity of 5.5×10^4 cm/sec. The pressure broadening coefficient is P_2, roughly constant at 12 kHz/mtorr.[61] Since the interaction time is mainly limited by the spot size, the saturation scaling turns out to be a power, (P_1^2/P_3), rather than an intensity. The first line labelled "EUZ" is included only for reference

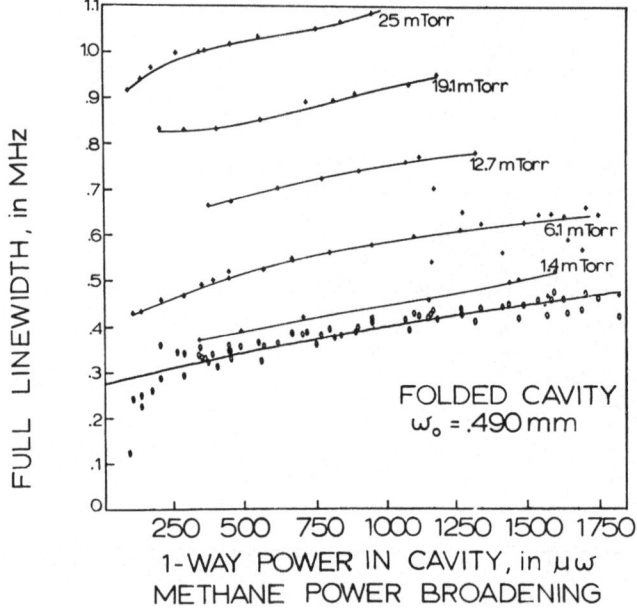

Fig. 12. Intensity broadening of saturated absorption resonances.
Measured points denoted "+"; lower line and symbol "o" calculated
for zero pressure.

Table II. Summary of Line Shape Parameters[*]

Run ID	Beam[e] Radius=ω	P_1 (kHz)	P_2 (kHz/mTorr)	P_3/P_1 (kHz/mW)	P_1^2/P_3 (mW)	$\omega \times P_1$ (kHz×mm)
EUZ[a]	5mm	29.3 ± 2.3	12.66 ± .17	8.9 ± 2.0	3.3 ±1.0	145
First FBTK[b]	5mm	23.28 ± .48	9.92 ± .21	9.10 ± .85	2.56 ± .29	116
Small Spot[c]	85µ	846. ±45.	10.27 ± .91	635 ±162	1.33 ± .41	72
First Cavity[d] Total Data	.706mm	104.0 ± 5.6	15.7 ± .4	59.4 ± 6.0	1.75 ± .27	73
Second Cavity[d]	.490mm	141.3 ± 2.3	13.26 ± .09	148. ± 5.5	0.95 ± .05	70

[*]All entries are ordinary frequency, not radian frequency

    a   passive beam expansion
    b   same as a but better alignment
    c   focussed spot
    d   length-stabilized resonant cavity, $F \sim 40$
    e   for $1/e^2$ intensity

purposes because the wave-fronts used in this experiment were completely inadequate in their flatness and freedom from perturbations. At the end of this paper we return to the question of the phase variation experienced by molecules in their transit through the radiation field.

Now we turn to the topic of spectroscopy; first, Zeeman and Stark spectroscopy of the methane transition itself, and second, some preliminary results on several methyl compounds. In Fig. 13 we show the Zeeman shift of the methane resonance using circular polarization. Since the electronic ground state of methane is a $^1\Sigma$, we are not surprised that the shift in frequency per unit magnetic field corresponds to such a weak magnetic interaction, with a g factor of only 0.31 nuclear magnetons. From knowledge of the sign of the frequency shift, direction of magnetic field, and sense of circular polarization, one finds that the g factor is positive.[57] In Fig. 14 we show the corresponding π transitions. Both $\Delta m = +1$ and -1 transitions are visible as expected. In the center we see a new feature, a three-level resonance corresponding to absorption of σ^+ from one running wave and σ^- from the other wave. Resonances of this type were studied theoretically by Javan and Schlossberg.[7] They showed that such a three-level, or "cross-over," (Raman-like) resonances are independent of the lifetime of the common level and

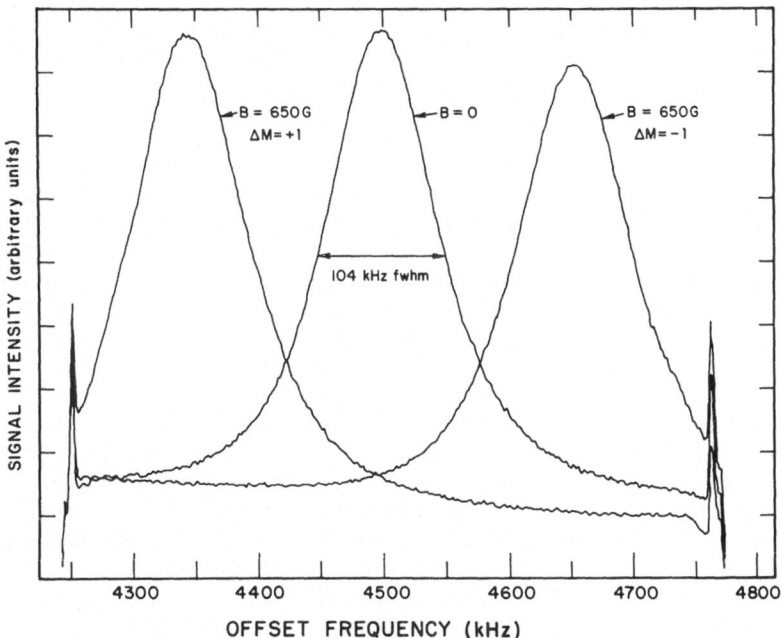

Fig. 13. Zeeman effect in CH_4 using circular polarized light and axial magnetic field. Sharp spikes at end are electrical transients from baseline-suppression circuit.

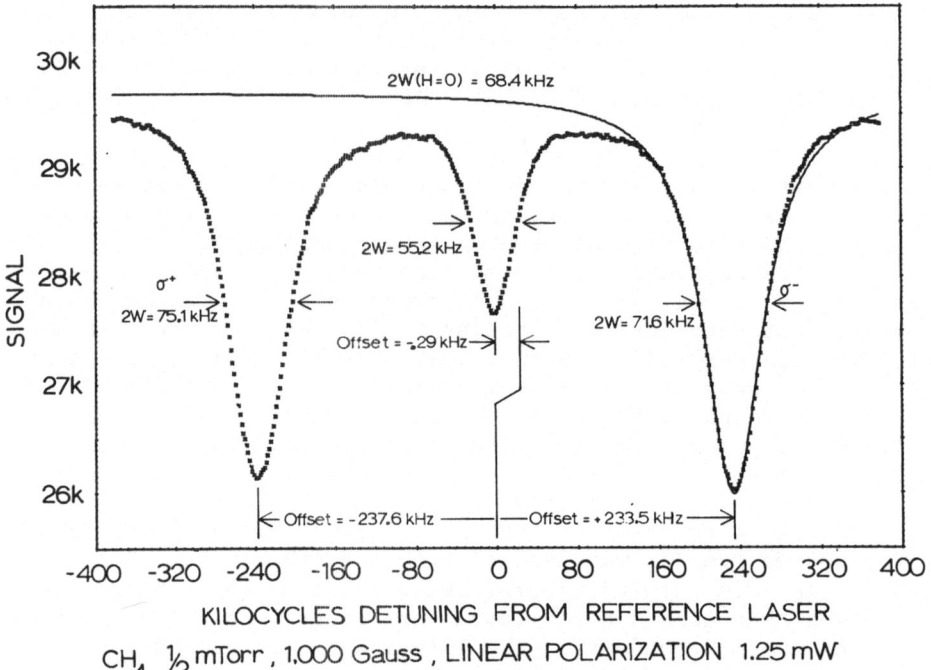

Fig. 14. Zeeman effect in CH_4 with linear polarization and axial field. The central feature is a three-level resonance. The resonance width at zero magnetic field is 68.4 kHz, from Fig. 9.

consequently can be more narrow than their parent resonance, as observed (Fig. 14). However, near equality of lower and upper state g factors tends to inhibit quantitative interpretation. The intensity of three-level resonances is reduced by a factor 2 for reasons having to do with spatial averaging. The computer line-profile fits have allowed study of the quadratic, as well as linear, Zeeman effect.

Stark effect studies using saturation spectroscopy have been very fruitful. For example Brewer and his colleagues have studied Stark spectra in NH_2D,[41] in CH_3F,[43] even in (symmetric-top) methane![34] The high resolution and sensitivity of saturation methods enabled their precision measurement of a dipole moment of only .02 Debye in the excited state of the E Coriolis component of the methane P(7) line.[34] For more usual dipole moments, ~1 Debye, the Stark effect leads to large shifts, thus making a number of useful level crossing and optical-optical double-resonance phenomena available.[7] The latter experiments are particularly interesting since they imply a natural scale of absolute frequencies, which will facilitate line assignment using known ground state dipole moments and will allow measurement of electric dipole moments in optically-excited states.[43] Other workers using single-mode lasers and various saturation methods have studied interesting hyperfine and electric quadrupole splittings in $(^{127}I)_2$,[27,28] $(^{129}I)_2$,[29] and $(Br)_2$.[30]

We now turn to the work in JILA by John Magyar concerned with saturation spectroscopy of methyl halide molecules.[36] The neon of a 3.39 μm He-Ne laser is Zeeman-tuned over the molecular saturated absorption peaks. This work is thus a direct extension of the early linear spectroscopic work of Shimoda, and Gerritsen and Heller.[33] Figure 15 shows some of these results. The tuning rate of the neon transition is about 1.6 MHz per gauss. At the time this illustration was prepared the signs of these Zeeman-induced frequency shifts were not known, since the laser operated on both sigma components. This information has been obtained in later experiments with a scanning interferometer. With the recent success[62] of direct frequency measurements from the cesium frequency standard up to the methane 88 THz (3.39 μm) line by Ken Evenson and his associates at NBS, it is possible to put these frequency measurements on an absolute scale. Thus, in Table III we show the absolute transition frequency of a number of these methyl compounds. Efforts are underway to identify these transitions and to study the interesting (quadrupolar ?) structure visible on some lines, such as the −576 MHz line of CH_3Br.

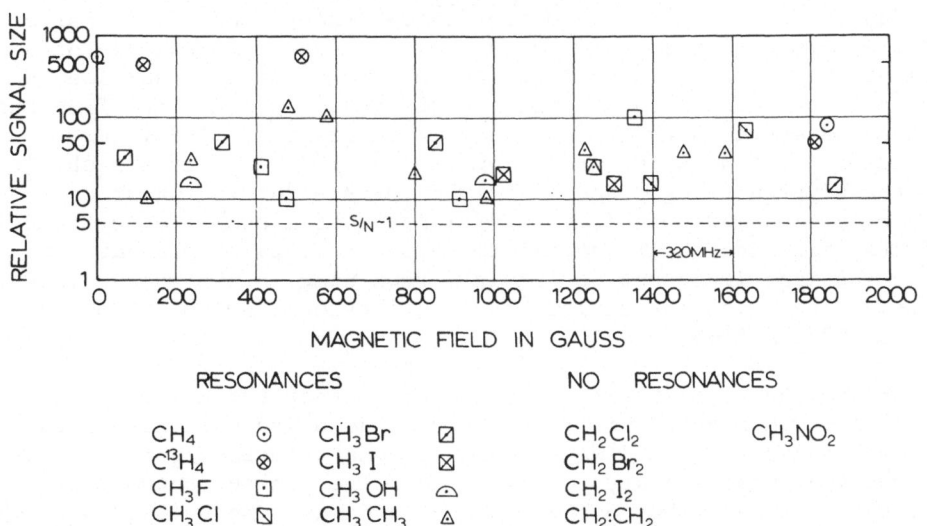

Fig. 15. Saturated absorption resonances observed using Zeeman tuning He-Ne 3.39 μm laser.

Table III. Saturated Absorption Resonances Measured by Optical-Heterodyne Techniques as of May 1972

Compound	Frequency Offset from Reference in MHz	Optical Frequency in THz
$C^{12}H_4$ P7 $F_1(2)$	0.00 Reference	88.37618161 ± .00000025[†]
$C^{13}H_4$ P6 $F_2(2)$*	-936.52 ± .05	88.37524509 ± .00000030
CH_3 F	-696.76 ± .05	88.37548485 ± .00000030
CH_3 Br	-575.96 ± .05	88.37560565 ± .00000030
CH_3 OH	+336.85 ± .10	88.37651846 ± .00000035
$CH_3:CH_3$	-826.22 ± .05	88.37535539 ± .00000030

[†]Private Communication, K. Evenson et al., May 1972

*Private Communication, L. Henry et al., November 1971

Meanwhile, the Infrared Frequency Synthesis experiments have been further refined[63,64] by Evenson, Wells, Peterson, Danielson and Day to a present (in-) accuracy of 50 kHz, or 5.6 parts in 10^{10} and it may be of interest to record here their value of the frequency for the 3.39 μm methane stabilized laser. For the $F_1^{(2)}$ component of the P(7) line of the ν_3 vibration of $C^{12}H_4$, they find $\nu = 88.376\ 181\ 627(50)$. In a coordinated experiment, R. L. Barger and the author have measured[64,65] the wavelength of this transition against the primary standard of length, the krypton transition at 6057 Å. Again the idea of transfer of frequency stability from the methane laser was used, this time to a servo-controlled interferometer.[66] A pointing precision of 2×10^{-4} orders was obtained for the krypton fringes and 2×10^{-5} orders for the laser case. This precision made possible careful study of a number of systematic effects, allowing their removal to an uncertainty of ± 2 parts in 10^9. The most important uncertainty remaining in the measurement is associated with corrections for the intrinsic asymmetry of the krypton 6057 Å line.

Since the asymmetry[67] was discovered after the krypton meter definition was adopted in 1960, there is no universal convention for applying the defined wavelength to some reference point in the (slightly asymmetric) krypton line. For the present purpose we make the arbitrary choice that the defined wavelength (6057.802 105 Å) is applied to the center of gravity of the krypton line. With this choice we find[64,65] $\lambda = 3.392\ 231\ 376$ μm, with an uncertainty of 3.5 parts in 10^9.

Combination of these two results thus yields a definitive value for the speed of light.[64] We find c = 299,792,456.2 (1.1) m/sec. This new value for the speed of light is in agreement with the previous value and is about 100 times less uncertain.

It is interesting that the uncertainty in this present speed of light, 3.5 parts in 10^9, comes directly from the operational realization of the international meter using the krypton discharge lamp. It may be expected that a suitable stabilized laser will someday supplant the present definition of the meter.[68] Such a stabilized laser could be directly adopted as a new basic standard of length, leading to still more precise values for the speed of light as Infrared Frequency Synthesis techniques improve over the next few years. Alternatively we could imagine altogether dispensing with the concept of an arbitrary standard of length, choosing instead to define the meter by adopting a conventional, nominal value for the speed of light.[69] Then the stabilized lasers would be secondary standards of both length and frequency. We note that already such a nominal value for the speed of light is in wide use to convert the dimensions of time delay to those of distance, for example in geophysical distance measurements which use modulated electromagnetic radiation and in astronomical measurements such as planetary radar and laser lunar ranging. No matter how such standards decisions turn out, it is clear that ultraprecise physical measurements made in the interim can be preserved through wavelength or frequency comparison with a suitable stabilized laser such as the 3.39 μm methane device.

Up to now we have been concerned with steady-state experiments, basically using that terrifically sharp saturation hole in the Doppler velocity distribution as made visible by a counter-running probe wave. Another very interesting kind of experiment involves transient saturated absorption phenomena. Although transient effects, such as optical echoes[70-72] have previously been observed with high-power pulsed lasers, detailed interpretation of the experiments has been difficult due to factors such as multimode operation of the lasers and/or incomplete knowledge of the relevant absorber transition assignment. Recently Brewer and Shoemaker[73-75] have studied a variety of transient processes with a single mode cw laser, using voltage pulses to Stark-tune a certain absorber velocity class into resonance with the laser frequency. The CO_2 laser used is almost ideal for this kind of study, its high power output giving rise to a large available ratio of induced transition probability rate to collisional dephasing rate. Thus in the first experiments they observed optical nutation,[73] in addition to coherently-phased spontaneous emission (superradiance).[74] Multiple pulse analogs of NMR techniques were employed, leading to definitive strong optical echoes[73] of the expected type. Later a two-photon superradiant transition was observed.[75]

In all these transient experiments, the Stark effect was used
to pulse-tune the molecules into resonance with the laser.   In our
3.39 μm frequency stabilization experiments with methane, it
became clear that higher modulation rates were desirable and we
began experiments using an intracavity electro-optic phase modulator
crystal (LiNbO$_3$).   Applying sharply-rising voltage steps rapidly
tunes the laser frequency, providing the intended four sampling
points[76] near the saturation resonance (to allow realtime suppression
of frequency offsets[50] from the "free-molecule" transition frequency
due to baseline effects).   Of course, by abruptly switching the
laser frequency, it is possible to observe interference between
the new laser frequency and coherent polarization previously created
in the absorbing gas, equivalent to the experiment of Shoemaker and
Brewer.[74]   The beat of this polarization with the laser's new
frequency may be seen in Fig. 16.   These signals are analogous to
the free induction decay effect in nuclear magnetic resonance.[77]
Comparison of the decay times at two pressures shows the dephasing
effect of collisional perturbations (see below).   The longest-last-
ing part of the "old" polarization belongs to the velocity group
with the appropriate Doppler-shift to be resonant with the old
laser frequency.   It is interesting to observe that the old polarization also carries the spatial dependence of the old laser frequency.
When a standing laser wave is present, there is an important interaction only between co-propagating waves of molecular polarization
(old frequency) and laser field (new frequency):   the energy exchange term for the reverse-wave interaction is spatially periodic
at twice the laser wave number and thus averages to zero in a cell
of practical length.   For co-propagating waves, a related spatial

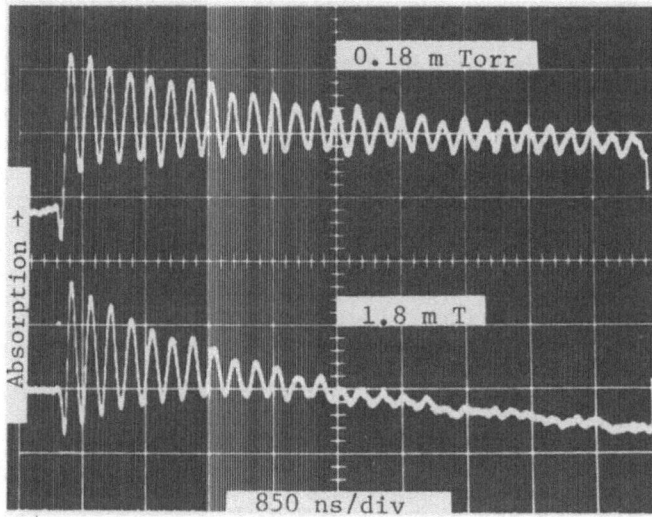

Fig. 16.   Transient saturated absorption beat between decaying
polarization and new laser frequency.   Note collisional contribution
to decay of phase coherence.   See text.

interference effect is also operative for large frequency steps and/or long cells.

Another transient absorption phenomenon is observable in these frequency-switching experiments. After the laser frequency has been switched to a new value, resonant radiative interaction begins with the corresponding Doppler velocity packet. At first there is no dipole moment, only population difference. Later dipole moment has been developed and strong interaction can begin. This is an interesting case of non-exponential decay.[78] The decaying absorption coefficient of this new velocity group may be seen in Fig. 17, along with the phase-beat signal already described.

By now there are two theories of these effects. Hopf, et al.[79] have given a theory for the case of abrupt switching of molecular energy levels (keeping a fixed laser frequency), corresponding to the experiments of Shoemaker and Brewer. The other theory[80] was worked out by G. Kramer while he was visiting JILA from the PTB.

Kramer's theoretical work synthesizes two independently attractive theoretical frameworks: 1) the Feynman-Vernon-Hellwarth geometrical model from magnetic resonance is used to describe the complete two-level absorber problem,[81] and 2) Age Theory from neutron transport calculations serves to describe the absorber collisional histories.[82] As previously mentioned, a very clear distinction can be made between the two types of signals. For example in the top of Fig. 18 we show the calculated absorption decay of the new velocity group. To avoid unnecessary complications and to better model our experimental conditions, we have assumed

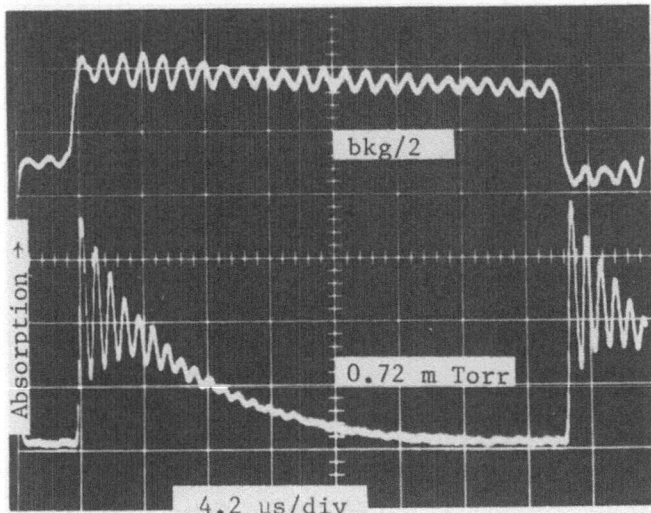

Fig. 17. Transient saturated absorption. Similar to Fig. 16 but longer time scale to display bleaching of opacity of new velocity group.

the transition rate (Rabi frequency) to be much less than the
damping rate, thus excluding transient nutation effects.[73,83,84]
In the bottom of Fig. 18 we have plotted the calculated power
exchanged with the laser field after the switching. The sum of
these two opacity contributions may be compared with the experimental
composite signal of Fig. 17. The beat feature is prominent, damping
down at <u>double</u> the collision broadening rate. Physically this
factor of 2 comes from the preparation/measurement format we must
use to surpass the Doppler resolution limit: phase-spread is prepared into the packet's polarization and additional phase-quenching
events occur during the measurement process. In general, the
observable decay rate would be the sum of preparation and measurement phase decay constants. We note that the size of the observed
beat signal depends on the product of polarization magnitude and
laser field strength. Thus there may be an experimental strategy
which minimizes the packet power-broadening by operating at low

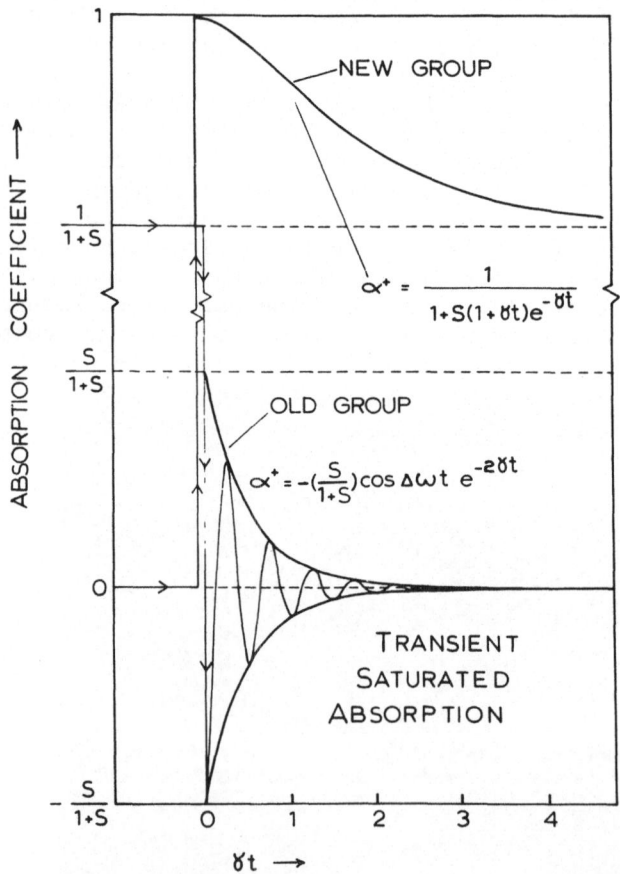

Fig. 18. Transient saturated absorption. Calculated absorption of
the two resonant velocity groups. Note that immediately after
frequency switching the old group is radiating power.

power while the polarization is being prepared, but preserves the signal-to-noise ratio by using a higher "readout" power level.

The study of the dephasing rate as a function of laser aperture may well serve (finally) to elucidate the aperture dependence of the (small-angle) collisional dephasing rate.[61] For example in transient experiments with our telescope laser, we find a pressure slope of the line width, $\Delta\nu_{HWHM} = \frac{1}{2} (1/2\pi T_{obs})$, to be (17.2 ± 0.6) kHz/mTorr (HWHM basis). For comparison, this same laser has a line width-broadening slope of (12.0 ± 0.3) kHz/mTorr in the cw experiments. Study of these effects using cw, frequency-dependent line shape information is complicated by the selectivity in favor of collisionless slow absorbers inherent in non-linear saturation spectroscopy -- we exclude the stronger decay channels from our observations. It will be amusing to try separating the several relaxation processes using the powerful methods of pulse echoes.[85]

The conceptual distinction between these two kinds of signals may be demonstrated experimentally by the technique of "jump and chirp." The inset of Fig. 19 shows this frequency program. For a long time one velocity group is saturated, corresponding to a constant laser frequency. We then abruptly switch the frequency and observe the beat between the old polarization and the new frequency. This new frequency, however, is chirped and does not remain within one packet line width long enough to generate much polarization. Thus, after the second jump we do not expect to see a coherent ringing, but only the bleaching of the opacity associated with the new velocity group. As may be seen from Fig. 19, the chirp rate was not completely sufficient and some saturation was occurring. Refined experiments of this type should prove very valuable in

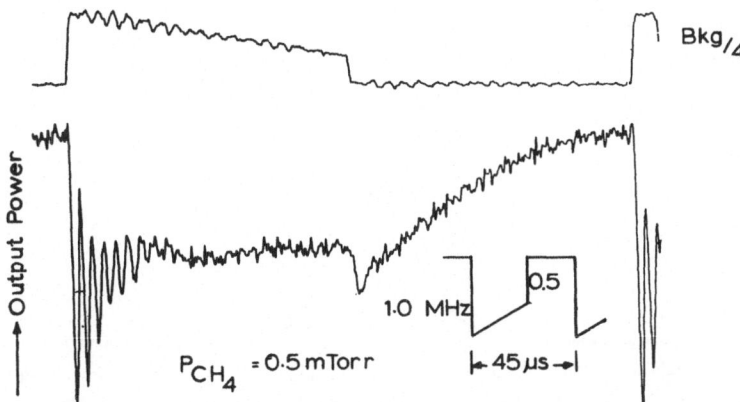

Fig. 19. Separation of two transient saturated absorption signals. Inset shows the frequency "jump and chirp" program. The oscillating signal is from the old polarization; the second half shows bleaching of the new velocity group.

studying collision processes in these gases, as both phase- and
energy-relaxation times are separably involved. Also, the selection process favoring particles of low transverse velocity appear
to enter differently in the two types of experiments.

Returning now from these transient phenomena to the frequency
domain, there are several line shape anomalies which are of
interest. For example, in the pursuit of narrow lines, one may
operate with low pressure and with low intensity. Figure 20 shows
such a curve. Plotted also on this curve is the "best-fitting"
Lorentzian function. It will be noted that the experimental signal
cuts off more quickly in the wings than does the Lorentz function.
This is a manifestation of the free transit of the molecule through
the radiation spot. The Gaussian laser profile in space is transformed to a Gaussian time impulse by the molecular velocity and
consequently into an essentially Gaussian resonance function. (Slow,
strongly saturated molecules will make an anomalous sharp contribution near the resonance center, but without collisions and/or
power broadening there is no source of Lorentz line wings.) At
very low pressures and intensities a strong bias in favor of slow
particles develops. We have seen this "artificial refrigeration"
of the effective transverse velocity by a factor in excess of 2.
The details of this line shape will be interesting to pursue.

The whole subject of the saturated absorption resonance line
shape is, of course, closely related to the question of the phase

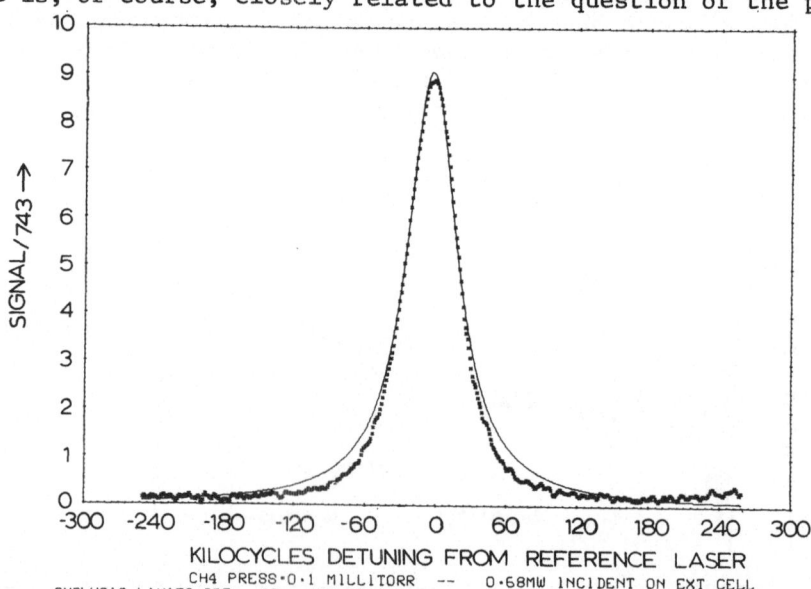

Fig. 20. Low pressure-low intensity saturated absorption resonance.
Smooth curve is "best fitting" Lorentzian, 51.5 kHz FWHM.

history of the particles. For example, there has been an experiment in which the external absorption cell of the Frequency-Offset Locked Laser Spectrometer was illuminated in the first off-axis mode,[86] the so-called "dumbbell" mode. The instantaneous phase of this field distribution has a sign change across the nodal line. Thus a particle that symmetrically enters one region and transits through the other region will experience no net transition probability at the frequency of molecular line center. If the frequency is somewhat detuned, a phase slip of π can develop between the two interaction regions for an absorber entering at a small angle. Thus a maximum in the transition probability will occur for a certain detuning, independent of its sign. In a more general context, with two coherent interactions one expects to see interference fringes -- here, the two slit interference fringes which in resonance physics bear the name of Professor Ramsey.[87] The fringe intensity will be somewhat modest because of the small solid angle factor. In Fig. 21 we show these Ramsey fringes in the residuals after the main ("single loop") resonance has been projected out by a least squares procedure. In the language of the high power laser pulse experiments,[84] this geometry corresponds to the generation of "zero π" pulses. It may well be that the most interesting optical frequency standards of the future will be based on the use of these fringes. Some exciting molecular beam experiments based on this idea are in preparation by H. Hellwig and his associates in the frequency standards group at NBS, Boulder. The potential

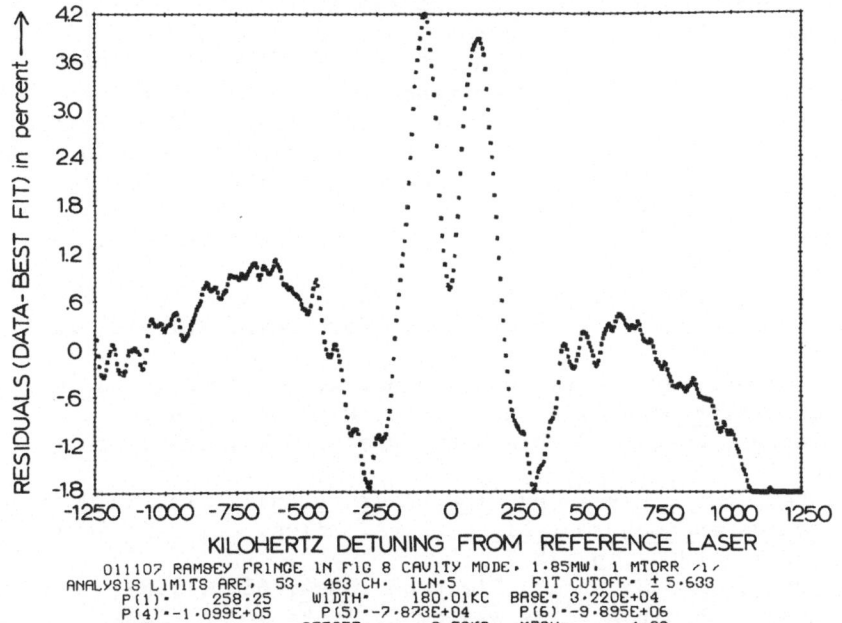

Fig. 21. Ramsey fringes in Fig. 8 cavity mode. Plotted signal is residuals after subtraction of best fitting Lorentzian from data.

is great since the estimated lifetime of this methane transition corresponds to a line width of 100 Hz or a line Q of 10^{12}. Since we have demonstrated optical frequency stabilities of 3 Hz(3×10^{-14}) for a laser with 300 kHz line width, one can only express enthusiasm for the prospects using a 100 Hz line width!

The author is indebted to his longtime colleague, R. L. Barger for expert, useful and pleasant collaboration on many of these experiments. Extended discussions with G. Kramer during his visit at JILA have generated expanded insight into the twin domains of high intensity saturation theory and precision electronics. As usual, P. L. Bender has been helpful in sharing enthusiasm and occasionally tempering it with friendly critical insight.

References

1. F. D. Colegrove, P. A. Franken, R. R. Lewis and R. H. Sands, Phys. Rev. Lett. 3, 420 (1959).
2. J. Brossel, P. Sagalyn and F. Bitter, Phys. Rev. 79, 225A (1950).
3. W. R. Bennett, Jr., Phys. Rev. 126, 580 (1962).
4. E. I. Gordon, A. D. White and J. D. Rigden, Optical Masers (Brooklyn Polytechnic Press, 1963), pp. 309-18.
5. A. M. Portis, Phys. Rev. 104, 584 (1956).
6. Acronymn for "Electron-Electron Double Resonance."
7. H. R. Schlossberg and A. Javan, Phys. Rev. 150, 267 (1966); R. G. Brewer, "Nonlinear Spectroscopy," lecture at Ettore Majorana International School of Quantum Electronics, Erice, Sicily, April 16-29, 1972 (to appear in Science, Oct. 20, 1972). I am indebted to Dr. Brewer for making a preprint of this article available.
8. T. Hänsch and P. Toschek, IEEE J. Quant. Electron. QE4, 467 (1968).
9. H. K. Holt, Phys. Rev. Lett. 19, 1275 (1967); G. E. Notkin, S. G. Rautian and A. A. Feoktistov, Sov. Phys.-JETP 25, 1112 (1967).
10. W. R. Bennett, Jr., Appl. Opt. Suppl. 1, 24 (1962), and in Quantum Electronics III, Paris 1963, P. Grivet and N. Bloembergen, eds. (Columbia Univ. Press, N.Y., 1964), pp. 442-58.
11. W. E. Lamb, Jr., Phys. Rev. 134, A1429 (1964), also in Quantum Electronics and Coherent Light, P. A. Miles, ed. (Academic Press, N.Y., 1964), p. 78.
12. J. L. Hall, IEEE J. Quant. Electron. QE-4, 638 (1968). One of the first Lamb dip studies was by A. Szöke and A. Javan, Phys. Rev. Lett. 10, 921 (1963).

13. Yu. A. Matyugin, A. C. Provorov and V. P. Chebotayev, "Influence of Radiation Trapping on Spectral Profiles in Neon," Academy of Sciencies (USSR), Siberian Branch, Institute of Physics of Semiconductors, Preprint #20, Novosibirsk 1972.

14. P. H. Lee and M. L. Skolnick, Appl. Phys. Lett. 10, 303 (1967). The same concept was pursued in the USSR by N. G. Basov and V. S. Letokhov at the Lebedev Institute and by V. P. Chebotayev and V. N. Lisitsyn in Novosibirsk.

15. V. N. Lisitsyn and V. P. Chebotayev, Sov. Phys.-JETP 27, 227 (1968).

16. See Ref. 4 of Ref. 17 and S. N. Bagaev, Yu. D. Kolomnikov, V. N. Lisitsyn and V. P. Chebotayev, IEEE J. Quant. Electron. QE4, 868 (1968).

17. R. L. Barger and J. L. Hall, Phys. Rev. Letters 22, 4 (1969).

18. K. Sakurai, Y. Ueda, M. Takami and K. Shimoda, J. Phys. Soc. Japan 21, 2090 (1966).

19. J. L. Hall, in Lectures in Theoretical Physics (1969) K. T. Mahanthappa and W. E. Brittin, eds. (Gordon and Breach, N.Y., in press).

20. Experiments of this type, performed in 1970 in collaboration with R. L. Barger, showed anomalously large pressure-broadening effects which still are not well understood.

21. For other good saturation candidates, see the tables by C. B. Moore, in Fluorescence: Theory, Instrumentation and Practice, G. G. Guilbault, ed. (M. Dekker, N.Y., 1967); pp. 133-53; also L. B. Kreuzer, N. D. Keryon and C. K. N. Patel, Science 177, 347 (1972).

22. W. Demtröder, M. McClintock and R. N. Zare, J. Chem. Phys. 51, 5495 (1969).

23. S. E. Johnson, K. Sakurai and H. P. Broida, J. Chem. Phys. 52, 6441 (1970).

24. W. J. Tango, J. K. Link and R. N. Zare, J. Chem. Phys. 49, 4264 (1968).

25. R. M. McClintock and L. C. Balling, Bull. Am. Phys. Soc. 13, 55 (1968).

26. S. Ezekiel and R. Weiss, Phys. Rev. Lett. 20, 91 (1968); R. B. Kurzel and J. I. Steinfeld, J. Chem. Phys. 53, 3293 (1970); nonlinear spectroscopy in both I_2 isotopes, see M. D. Levenson and A. L. Shawlow, Phys. Rev. A 6, 10 (1972).

27. T. W. Hänsch, M. D. Levenson and A. L. Shawlow, Phys. Rev. Lett. 26, 946 (1971).

28. G. R. Hanes and C. E. Dahlstrom, Appl. Phys. Lett. 14, 362 (1969); and Ref. 29.

29. J. D. Knox and Y. M. Pao, Appl. Phys. Lett. 18, 360 (1971).

30. R. S. Eng and J. T. LaTourrette, Bull. Am. Phys. Soc. 16, 43 (1971); R. S. Eng, Thesis, Polytechnic Institute of Brooklyn, 1971.

31. K. Sakurai and H. P. Broida, J. Chem. Phys. 50, 2404 (1969).

32. K. Uehara, T. Shimizu and K. Shimoda, IEEE J. Quant. Electron.

QE4, 728 (1968). This reference reports many Stark effect
data.

33. Zeeman-tuned (linear) laser spectroscopy: H. J. Gerritsen and
M. E. Heller, Appl. Opt. Suppl. #2, 73 (1965); H. J. Gerritsen,
in Physics of Quantum Electronics, P. L. Kelly, B. Lax and
P. E. Tannenwald, eds. (McGraw-Hill, N.Y., 1966). See also
K. Sakurai and K. Shimoda, Japan J. Appl. Phys. 5, 744 (1966).

34. Nonlinear spectroscopy: A. C. Luntz, R. G. Brewer, K. L.
Foster and J. D. Swalen, Phys. Rev. Lett. 23, 951 (1969);
A. C. Luntz and R. G. Brewer, J. Chem. Phys. 54, 3641 (1971).

35. Identification by G. Poussige, N. Husson, M. Dang Nhu and L.
Henry (private communcation, Nov. 1971). Saturation spectroscopy ·reported in Ref. 36.

36. J. A. Magyar and J. L. Hall, Bull. Am. Phys. Soc. 17, 67 (1972).
The number of discrete saturation resonances within $\pm$ 3 GHz
is shown -- many of the lines show additional structure
< 0.5 MHz.

37. A. C. Luntz, J. D. Swalen and R. G. Brewer, J. Chem. Phys.
Lett. 14, 512 (1972). Identification is for the line at
+ 2.25 GHz relative to methane. Three other lines observed,
one provisionally identified.

38. K. Sakurai, K. Uehara, M. Takami and K. Shimoda, J. Phys. Soc.
Japan 23, 103 (1967).

39. A. W. Mantz, E. R. Nichols, B. D. Alpert and K. N. Rao, J.
Mol. Spectry. 35, 325 (1970); A. Kaldor, W. B. Olson and
A. G. Maki, Science 176, 508 (1972).

40. H. Bunet, IEEE J. Quant. Electron. QE2, 382 (1966).

41. R. G. Brewer, M. J. Kelly and A. Javan, Phys. Rev. Lett. 23,
559 (1969); R. G. Brewer and J. D. Swalen, J. Chem. Phys.
52, 2774 (1970).

42. M. J. Kelly, R. E. Francke and M. S. Feld, J. Chem. Phys. 53,
2979 (1970).

43. R. G. Brewer, Phys. Rev. Lett. 25, 1639 (1970).

44. F. Shimizu, J. Chem. Phys. 52, 3572 (1970), about 100 other
Stark-tunable coincidences with CO_2 and NO_2 lasers are given
along with transition identification; F. Shimizu, J. Chem.
Phys. 53, 1149 (1970) gives about 95 such identified coincidences for $^{15}NH_3$. M. Ouhayoun and C. Bordé, Compt. Rend., Acad.
Sci. Paris 274, 411 (1972), report Stark effect on the NH_3
aQ(8,7) using saturation methods.

45. P. Rabinowitz, R. Keller and J. T. LaTourrette, Appl. Phys.
Lett. 14, 376 (1969). See also Ref. 48 for 7 CO_2 laser transitions and 8 N_2O laser transitions that overlap SF_6.

46. C. Freed and A. Javan, Appl. Phys. Lett. 17, 53 (1970).

47. F. R. Peterson and B. L. Danielson, Bull. Am. Phys. Soc. 15,
1324 (1970).

48. C. Bordé, Compt. Rend., Acad. Sci. Paris 271, 371 (1970).

49. J. L. Hall, in Proceedings of the Esfahan Symposium on Basic
and Applied Laser Physics, M. S. Feld, N. A. Kurnitt and A.
Javan, eds. (Wiley, N.Y., in press).

50. J. L. Hall, "Accuracy Capability of Saturated Absorption:
     The Prognosis for Optical Frequency Standards," to be published.
51. D. Allan, Proc. IEEE 54, 221 (1966).
52. A. Javan, E. A. Ballik and W. D. Bond, J. Opt. Soc. Am. 52,
     96 (1962); A. Szöke and A. Javan, Phys. Rev. Lett. 10, 521
     (1963); J. L. Hall and W. W. Morey, Appl. Phys. Lett. 10,
     152 (1967); J. A. Magyar and J. L. Hall, Bull. Am. Phys. Soc.
     17, 67 (1972).
53. J. Levine and J. L. Hall, J. Geophys. Res. 77, 2595 (1972).
54. J. L. Hall and R. L. Barger, Bull. Am. Phys. Soc. 17, 67
     (1972), and in Ref. 49.
55. For example, see T. S. Jaseja, A. Javan, J. Murray and C. H.
     Townes, Phys. Rev. 133, A1221 (1964).
56. See, for example, Refs. 12, 49, and 57.
57. E. E. Uzgiris, J. L. Hall and R. L. Barger, Phys. Rev. Lett.
     26, 289 (1971).
58. The computing skill and insight of J. Levine have been essen-
     tial to this work; we thank him vigorously.
59. No totally satisfactory theory of this effect is yet available.
     See, for example, A. P. Kol'chenko, S. G. Rautian and R. I.
     Sokolovskii, Soc. Phys.-JETP 28, 968 (1969).
60. W. R. Bennett, Jr., Ref. 10 and Comm. Atom. Mol. Phys. 2, 10
     (1970).
61. The aperture-dependent selective effects operating in these
     collisions have been discussed in Ref. 19, and by J. L. Hall,
     in Sixth International Conference on Electronic and Atomic
     Collisions: Abstracts of Papers (MIT Press, Cambridge, Mass.,
     1969), pp. 994-6.
62. K. M. Evenson, G. W. Day, J. S. Wells and L. O. Mullen, Appl.
     Phys. Lett. 20, 133 (1972). The pioneering experiments of this
     type were done by A. Javan and his colleagues at MIT; see for
     example L. O. Hocker, A. Javan, D. R. Rao, L. Frenkel and T.
     Sullivan, Appl. Phys. Lett. 10, 5 (1967).
63. K. M. Evenson, J. S. Wells, F. R. Peterson, B. L. Danielson
     and G. W. Day, Appl. Phys. Lett. (submitted).
64. K. M. Evenson, J. S. Wells, F. R. Peterson, B. L. Danielson,
     G. W. Day, R. L. Barger and J. L. Hall, Phys. Rev. Lett.
     (to appear Nov. 6, 1972).
65. R. L. Barger and J. L. Hall, Appl. Phys. Lett. (submitted);
     and paper in preparation.
66. R. L. Barger and J. L. Hall, in Precision Measurements and
     Fundamental Constants, Proc. of the International Conference
     held at Gaithersberg, Md., August, 1970. Nat. Bur. Std. Spec.
     Publ. 343. Note that the preliminary wavelength value reported
     here was not yet corrected for several systematic offsets.
67. W. R. C. Rowley and J. Hamon, Rev. d'Opt. 42, 519 (1963).
68. J. L. Hall, IEEE J. Quant. Electron. QE4, 638 (1968).
69. D. A. Halford, H. Hellwig and J. S. Wells, Proc. IEEE 60,
     623 (1972).

70. I. D. Abella, N. A. Kurnitt and S. R. Hartmann, Phys. Rev. 141, 391 (1966).
71. J. P. Gordon, C. H. Wang, C. K. N. Patel, R. E. Slusher and W. J. Tomlinson, Phys. Rev. 179, 294 (1969).
72. B. Bölger and J. C. Diels, Phys. Lett. 28A, 401 (1968).
73. R. G. Brewer and R. L. Shoemaker, Phys. Rev. Lett. 27, 631 (1971).
74. R. L. Shoemaker and R. G. Brewer, Bull. Am. Phys. Soc. 17, 67 (1972).
75. R. L. Shoemaker and R. G. Brewer, Phys. Rev. Lett. 28, 1430 (1972).
76. J. L. Hall and G. Kramer, to be published.
77. A. Abragam, The Principles of Nuclear Magnetism (Oxford Univ. Press, 1961), p. 22ff.
78. Unfortunately, relative to the "Neoclassical Electromagnetic Theory" vs "Quantum Electrodynamics" controversy, we also have the rest of the Doppler profile to contribute opacity with an unresolvably fast response speed $\sim 1/\pi \Delta \nu_D \sim 1$ nsec.
79. R. F. Shea, F. A. Hopf and M. O. Scully, quoted in Ref. 75.
80. G. Kramer, to be published.
81. R. P. Feynman, F. L. Vernon, Jr. and R. W. Hellwarth, J. Appl. Phys. 28, 49 (1957).
82. See, for example, Ref. 11, or Elements of Nuclear Reactor Design, S. Glasstone and M. C. Edlund (Van Nostrand, N.Y., 1952), p. 90ff.
83. H. C. Torrey, Phys. Rev. 76, 1059 (1949).
84. G. L. Lamb, Rev. Mod. Phys. 43, 99 (1971).
85. Ref. 77, p. 58ff.
86. H. Kogelnik and T. Li, Appl. Opt. 5, 1550 (1966).
87. N. F. Ramsey, Phys. Rev. 76, 996 (1948), and in Molecular Beams (Oxford Univ. Press, 1956), p. 124ff.

RECENT ADVANCES IN THE

SPECTROSCOPY OF SMALL MOLECULES

W. Demtröder

Fachbereich Physik

Universität Kaiserslautern (Germany)

I. INTRODUCTION

Three main reasons may be considered for the exciting progress achieved during recent years in molecular spectroscopy:

1. The use of lasers as spectroscopic light sources with all their advantages in high resolution spectroscopy.

2. The extension of modern spectroscopic techniques so far only used in atomic spectroscopy to the investigation of molecules (e.g. level crossing, double resonance, optical pumping, spin alignment).

3. The increase of theoretical interest in molecular physics and more rigorous computational studies of small molecules (e.g. computation of more accurate molecular wave functions, calculations of diatomic potential curves and dissociation energies and of triatomic potential surfaces).

The outcome of these combined experimental and theoretical efforts lies in the investigation of finer details of molecular spectra by high resolution spectroscopy and their interpretation, which yielded very accurate molecular constants, Zeeman- and Stark-splittings, dipole-moments and internuclear separations. Even small effects resulting from the coupling of nuclear and electronic motion, or hyperfine structure splittings could be studied with high precision.

These experimental results have made it possible to test theoretical molecular models, to study deviations from the Born Oppen-

647

heimer approximation[1a] and to elucidate the often very complicated coupling schemes of angular momenta and magnetic moments (Zeeman effect!) in molecules.[1b]

One important branch of molecular spectroscopy deals with the investigation of elastic or inelastic collision processes and chemical reactions in order to determine interaction potentials between excited state and ground state atoms or molecules.[1c]

The spectroscopy of small molecules has received increasing interest in astrophysics since absorption and emission spectra of interstellar gases have shown that the interstellar space contains a non-negligible fraction of molecules.

In the following I will give some examples which demonstrate the advances achieved in this field.[1d]

II. MOLECULAR SPECTROSCOPY WITH LASERS

a) Absorption Spectroscopy

The spectacular progress in resolution achieved by saturation spectroscopy with lasers has been discussed already in the preceeding papers by J. L. Hall and T. W. Hänsch. Only one further example shall illustrate the high resolution attainable in molecular absorption spectra:

Using Lamb dip spectroscopy with a He-Ne laser tunable over a free spectral range of 650 MHz, Knox and Pao[2] measured 38 separate absorption lines in the I_2 spectrum. With a resolution of 1 MHz the hyperfine structure of I_2 could be resolved and the absolute reproducibility of the line positions was at least 1 part in 10^9.

Hanes et al.[2b] increased the tunability range by Zeeman-splitting of the laser line. The relative positions of the hfs-components of I_2 could be measured with a precision of 0.36 MHz. This enabled the authors to determine the difference in the nuclear electric quadrupole coupling constants and in the spin-rotation-constants for the two states involved in the optical transitions. The decrease in absorption when the laser line is tuned across the Lamb dip of the Doppler broadened transition can be also detected by observing the corresponding decrease in fluorescence. This method, which is particularly useful at low pressures where the total absorption is small, has been used by Sorem et al.[2c] to measure the hyperfine structure of I_2 inside the tuning range of the argon-laser line $\lambda = 5145\text{Å}$.

Refinement of this saturation spectroscopy in combination with heterodyne methods enable the absolute determination of mole-

cular absorption lines within ±100 kHz[3] with line widths also
down to 100 kHz.

One should point out that with Doppler limited absorption
spectroscopy the resolution generally is 100 times worse.

These high resolution spectra exhibit many narrow spaced structures hidden inside the Doppler width in conventional spectroscopy.
Coriolis interaction, Zeeman splitting of $^1\Sigma$ - ground state molecules[4] and even small hyperfine structure effects[5] are now accessible to experimentalists.

b)   Induced Emission Spectroscopy

Since the wavelength determination of laser lines can be performed with much higher accuracy than in the case of spontaneous
lines (because of higher intensity and smaller line width), measurements of molecular laser lines yield molecular constants with an
accordingly high precision.[6]   Long-path interferometry[7] as well
as heterodyne techniques have been used. One example shall illustrate the latter technique:[8]

The output from two CO_2 lasers, each tuned by a grating on a
definite rotation-vibration transition of CO_2 and stabilized to
the center of a chosen line, was mixed in a Ga-As mixer. The beat
frequencies in the 50-80 GHz region, which yield the energy differences between the rotational levels, were measured to better than
1 MHz. The deduced rotational constants for the relevant vibrational levels are up to 200 times more accurate than the best previous results.

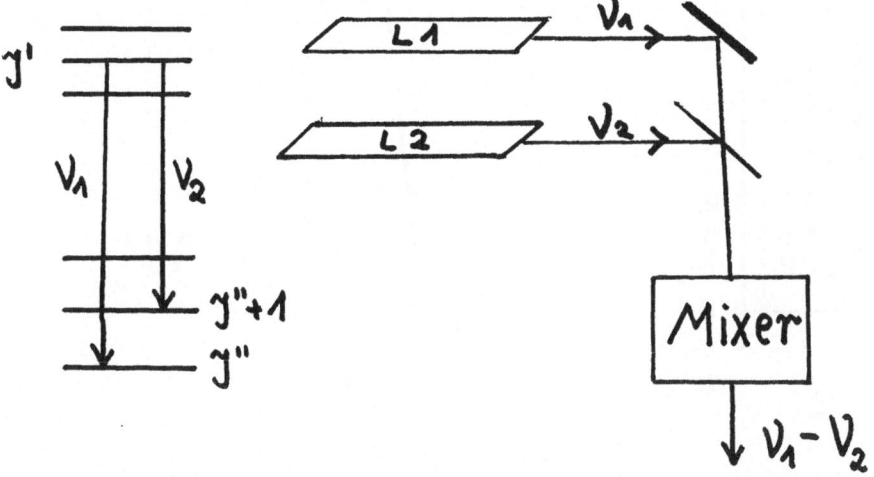

Figure 1: Heterodyne technique for accurate determination of molecular constants.

c)   Laser Induced Fluorescence

By absorption of laser lines single rotation-vibration levels
in electronic ground states or excited states can be selectively
populated. Because of the large laser intensity available, a high
population density is achieved in these excited levels, comparable
to that of the absorbing levels in the ground state.

The excited molecules release their excitation energy either
by spontaneous emission or, at sufficiently high pressures, by inelastic collisions.

The fluorescence spectrum originating from the excited level
is determined by certain selection rules.[9] It is by far simpler
than the absorption spectrum of the molecule.

Precise wavelength measurements enable one to obtain the molecular ground state constants[10] from which the potential curve can
be constructed.[11] Fluorescence progressions terminating on high
vibrational levels v" allow the determination of the dissociation
energy.[12] Analyzing the fluorescence spectrum of sodium-lithium

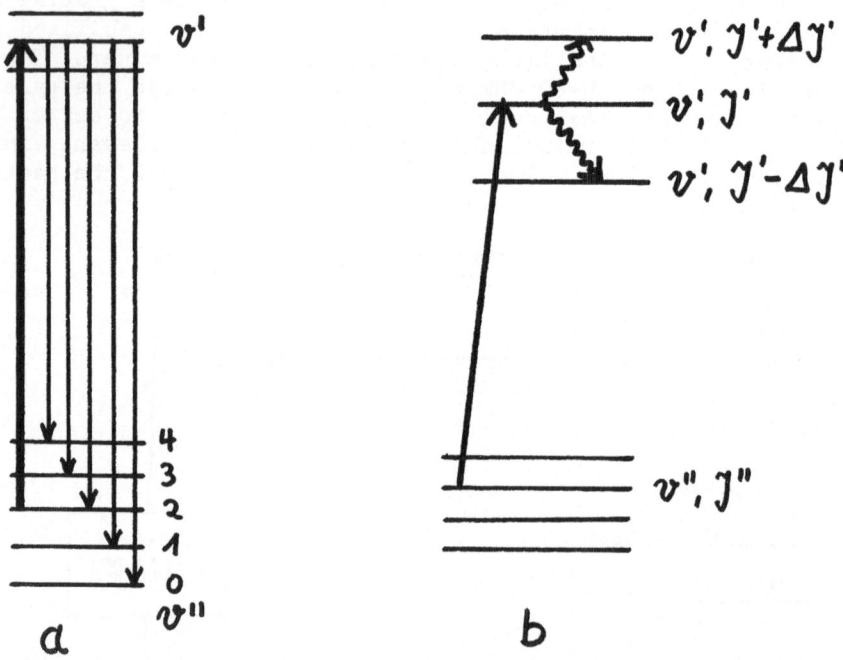

Figure 2: Schematic illustration of laser excited spontaneous
fluorescence series from a single excited level (a) and collision
induced fluorescence (b).

vapor excited by argon laser lines, Hessel could determine the spectroscopic constants of the NaLi molecule[13] which had never been
observed before.

Measurements of the lifetimes of excited levels and their dependence on the vibrational quantum numbers have yielded information about radiative transition probabilities at different internuclear separations.[14] This is a more sensitive test on the radial variation of wavefunctions than one would obtain from wavelength
determinations.

Many papers have been published about the dependence of lifetimes of excited molecular states on the pressure of added gases.[15]
These measurements yield total deactivation cross sections for all
inelastic collision processes which depopulate the excited level.

More detailed information about the collision process can be
obtained, however, by the method of collision induced fluorescence:

If the excited molecule suffers an inelastic collision during
its lifetime, radiationless transitions to neighbouring rotational
or vibrational levels may occur which can be detected by observing
the fluorescence from these collisional populated levels. From the
intensity of the collision induced fluorescence lines absolute cross
sections for the corresponding inelastic collisions can be obtained,
provided the upper state lifetime is known.[16] With this technique
collision-induced rotational and vibrational transitions have been
investigated. The cross sections for rotational energy transfer
turns out to be rather large ($\sigma = 70$ Å^2 for $|\Delta J| = 1$ for Na$_2$-He
collisions at thermal energies.)[17] It should be emphasized that
this technique allows the determination of all relevant parameters
of the collision: Initial and final states are known as well as
the amount of energy transfer and even information about the polarization of both states may be included if excitation by polarized
light is used.

Experiments showed that in the case of excited alkali molecules
the cross section for rotational transitions with $\Delta J = + 1$ differs
remarkably from $\sigma(\Delta J = - 1)$. This asymmetry in the cross section
depends very strongly on the collision partner and can be explained
by the different symmetry of the electronic wave functions in adjacent rotational levels of the excited alkali molecule[18]. Measurements of the asymmetry are therefore well suited for probing the
interaction potential between different collision partners.

Laser-induced fluorescence has also been used to determine the
rotational-vibrational distribution of reaction products in chemical reactions:

A thermal beam of barium atoms intersects another beam of O_2 molecules and reacts to form $BaO(X^1\Sigma) + O(^3P)$. Simultaneously light from a pulsed tunable dye laser passes through the reaction zone. Since the absorption of the laser line by the BaO molecules is porportional to the ground state density in the transition $(v_1''J'' \to v_1'J')$ the dependence of the laser induced fluorescence from the laser wavelength yields information about the rotational-vibrational distribution of the BaO molecules.

III. EXTENSION OF MODERN TECHNIQUES FROM ATOMIC TO MOLECULAR SPECTROSCOPY

a) Radiofrequency Spectroscopy of Free Molecules

An interesting technique for observing magnetic resonance transitions among several Zeeman sublevels of the hyperfine states of H_2^+ in weak magnetic fields has been developed by Dehmelt and his group.[19] The method uses an optical pumping technique:

Free H_2^+ ions are kept for several seconds in an ultra-high vacuum ion trap and are irradiated by a beam of linearly polarized light. Molecules with their dipole moment parallel to the electric vector of the light wave are preferentially excited into the repulsive upper state from where they dissociate. The remaining H_2^+ ions become therefore aligned. These aligned ions are subjected to amplitude and frequency modulated rf-magnetic fields which alter the alignment when the resonance condition is fulfilled. The change in alignment is reflected in the number of protons created through photodissociation. Hyperfine transitions with line widths less than 5 Hz (!) have been observed.[20]

The measurements provide a very exacting test of the ground state molecular wavefunction. Since the H_2^+ ion can be rigorously treated in the framework of the Born-Oppenheimer approximation, the breakdown of this approximation can be studied in detail and interesting small effects such as electron-proton radiative corrections should be revealed.

b) Level-Crossing Spectroscopy of Diatomic Molecules

Level-crossing spectroscopy provides a method for obtaining spectral resolution of the order of natural line widths and outwits, in analogy to saturation spectroscopy, the much larger Doppler width. Zare[21] was the first to point out that this technique, hitherto only used in atomic spectroscopy, could be employed as well for measuring line widths and narrow level splittings in the more complex molecular spectra.

Such couplings as nuclear spin - electron spin and nuclear

rotation – electron spin can often be removed by application of small magnetic fields and level-crossing spectroscopy is well suited to determine the coupling parameters.[22]

Level crossing and optical double resonance in the state of CS has been observed by Klemperer and his group.[23]

From the Hanle effect the excited state lifetime was obtained. The Stark level crossing and the Stark-Zeeman recrossing gave estimates of the dipole moment and the lambda doubling. Optical double resonance provided more accurate values for both molecular parameters.

Very detailed investigations of OH and OD molecules have been performed by Zare and coworkers using high field magnetic level crossing and optical double resonance.[23a] The shift of the crossing points by a dc-electric field parallel to the magnetic field allowed an accurate determination of the electric dipole moment in the excited state.

The combination of Stark tuned level crossing and saturation spectroscopy enabled Luntz et al.[24] to measure the line widths of an excited vibrational level in CH_4, which is linearly dependent on pressure and which can be as narrow as 200 kHz.

IV. SPECTROSCOPY OF VAN DER WAALS MOLECULES

The interesting investigations of very loosely bound diatomic molecules, as may be formed for instance during a collision of two atoms, can be mentioned here only very briefly.[25] The potential well depths of these dimers is of the order of 200 cm^{-1} and vibrational transitions between neighbouring levels lie in the far infrared region. With the development of far infrared technologies (e.g. detectors, Fourier-spectrometers, lasers) the absorption-, fluorescence-, and Raman-spectroscopy of dimers has advanced considerably.

During a collision of two different atoms a dipole moment is developed by the distorted overlapping charge densities.[26] This transient dipole moment of the colliding pair may emit or absorb radiation. The spectral distribution is given, essentially, by the Fourier analysis of the dipole moment pulse.

The investigations of such pressure-induced absorption or emission spectra can give much information about the long-range potential and the distortion of the electronic charge distribution during the collision.

For more detailed information see the review articles by Levine[25] and Foley.[26]

References:

1.a) P. R. Bunker, J. Mol. Spectry. $\underline{5}$, 478 (1972).

   b) W. H. Flygare and R. C. Benson, Mol. Phys. $\underline{20}$, 225 (1971).

   c) H. F. Schaefer, D. Wallach and Ch. F. Bender, J. Chem. Phys.
      $\underline{56}$, 1219 (1972).

   d) For more extensive information see the author's review on
      "High Resolution Spectroscopy with Lasers," Physics Reports
      (in press, 1972).

2.a) J. D. Knox, Yoh Han Pao, Appl. Phys. Letters $\underline{18}$, 360 (1971).

   b) G. R. Hanes, J. Lapierre, P. R. Bunker and K. C. Shotten, J.
      Mol. Spectry. $\underline{39}$, 506 (1971).

   c) M. S. Sorem and A. L. Schawlow, Opt. Commun. $\underline{5}$, 148 (1972).

3.   M. W. Goldberg and R. Yusek, Appl. Phys. Letters $\underline{17}$, 349
     (1970).

4.   E. E. Uzgiris, J. L. Hall and R. L. Barger, Phys. Rev. Letters
     $\underline{26}$, 289 (1971).

5.   G. R. Hanes and C. E. Dahlström, Appl. Phys. Letters $\underline{14}$, 362
     (1969).

6.   J. H. Parks, D. R. Rao and A. Javan, Appl. Phys. Letters $\underline{13}$,
     142 (1968).

7.   W. J. Schade Jr., Opt. Commun. $\underline{4}$, 399 (1972).

8.   T. J. Bridges and T. Y. Chang, Phys. Rev. Letters $\underline{22}$, 811
     (1969).

9.   G. Herzberg, Molecular Spectra II, (Van Nostrand, 1967), p.
     240 ff.

10. W. Demtröder, M. McClintock and R. N. Zare, J. Chem. Phys.
     $\underline{51}$, 5495 (1969); K. Sakurai, S. E. Johnson and H. P. Broida,
     J. Chem. Phys. $\underline{52}$, 1625 (1970).

11. W. C. Stwalley, J. Chem. Phys. $\underline{56}$, 2485 (1972).

12. R. Velasco, Ch. Ottinger and R. N. Zare, J. Chem. Phys. $\underline{51}$
     5522 (1969).

13. M. M. **Hessel**, Phys. Rev. Letters <u>26</u>, 215 (1971).

14. A. W. Johnson and R. G. Fowler, J. Chem. Phys. <u>53</u>, 65 (1970);
    G. Baumgartner, W. Demtröder, M. Stock, Z. Phys. <u>232</u>, 462 (1970).

15. S. E. Johnson, J. Chem. Phys. <u>56</u>, 149 (1972); C. B. Moore,
    Ann. Rev. **Phys**. Chem. <u>22</u>, 387 (1971).

16. Ch. Ottinger, R. Velasco and R. N. Zare, J. Chem. Phys. <u>52</u>,
    1636 (1970); J. I. Steinfield and A. N. Schweid, J. Chem.
    Phys. <u>53</u>, 3304 (1970).

17. K. Bergmann and W. Demtröder, J. Phys. B <u>5</u>, 1386 (1972).

18. K. Bergmann, H. Klar and W. Schlecht, Chem. Phys. Letters <u>12</u>,
    522 (1972).

18a. A. Schultz, H. W. Cruse and R. N. Zare, J. Chem. Phys. (in
    press, 1972).

19. K. B. Jefferts, Phys. Rev. Letters <u>23</u>, 1476 (1969).

20. H. G. Dehmelt and S. C. Menasian, Abstract submitted to **this**
    conference.

21. R. N. Zare, J. Chem. Phys. <u>45</u>, 4510 (1966).

22. D. W. Robinson, J. Mol. Spectry. <u>35</u>, 1 (1970).

23. St. Silvers, Th. Bergmann and W. Klemperer, J. Chem. Phys. <u>52</u>,
    4385 (1970).

23a. E. M. Weinstock and R. N. Zare, J. Chem. Phys. (in press,
    1972); K. R. German, T. H. Bergeman, E. M. Weinstock and R.
    N. Zare, J. Chem. Phys. (in press, 1972).

24. A. C. Luntz <u>et al</u>., Phys. Rev. Letters <u>23</u>, 951 (1969).

25. H. B. Levine, J. Chem. Phys. <u>56</u>, 2455 (1972).

26. H. Foley, Comments Atom. Mol. Phys. <u>1</u>, 189 (1970).

ATOMIC g-FACTORS OF HYDROGEN-LIKE F^{19} *

R. Brenn, H. Calvin, H. Metcalf, G. Sprouse, and L. Young

State University of New York, Stony Brook, N. Y. 11790

We have used a unique method to directly measure both the atomic g-factor of the ground state of hydrogen-like F^{19} ions and part of the energy dependence of the charge state equilibrium. In this experiment, we precess the ion in an applied magnetic field for a fixed period of time and then use the anisotropic angular distribution of nuclear gamma rays to detect the precession.

A beam of F^{19} ions accelerated by the Stony Brook FN tandem to energies between 30 and 60 MeV is incident upon a gold foil 2.5μ thick (5 mg/cm^2). Some of the nuclei are Coulomb excited to the I=5/2, 197 keV level ($t_{1/2}$ = 89nsec, g_I=1.436) and scattered into a solid angle of about 0.13 sr defined by our collimators and centered about a line 40° from the incident direction (see Fig. 1). These ions with excited nuclei travel about 12 cm in vacuum before they are implanted into an annealed copper catcher. The detector is shielded from all decay gammas except those emanating from the catcher. Because the scattering process detected in this experiment does not have symmetry around the beam direction, the excited nuclear population is polarized. If this polarization is preserved in flight to the catcher foil, an anisotropic gamma ray angular distribution can be observed.

The incident energy is chosen so that a large fraction of the nuclei leave the target with only one electron left (hydrogen-like). All electronically excited states of these ions except the $2^2S_{1/2}$ metastable state decay to the ground state in less than the first mm of this flight, so that all the ions are in ^{2}S states with electronic g-factors g_J = 2. (The expected value of the principal quantum number, n, is about 1 at the energies used in this exper-

iment.)   The electronic and nuclear ($I = 5/2$)   angular momenta can
couple together to yield F=3 with g_F = .3340 or F=2 with g_F =-.3324.
The important point is that, since I is much larger than J, the
initial orientation of the nuclei is only slightly reduced by the
hyperfine interaction in flight (a factor of 5/6).

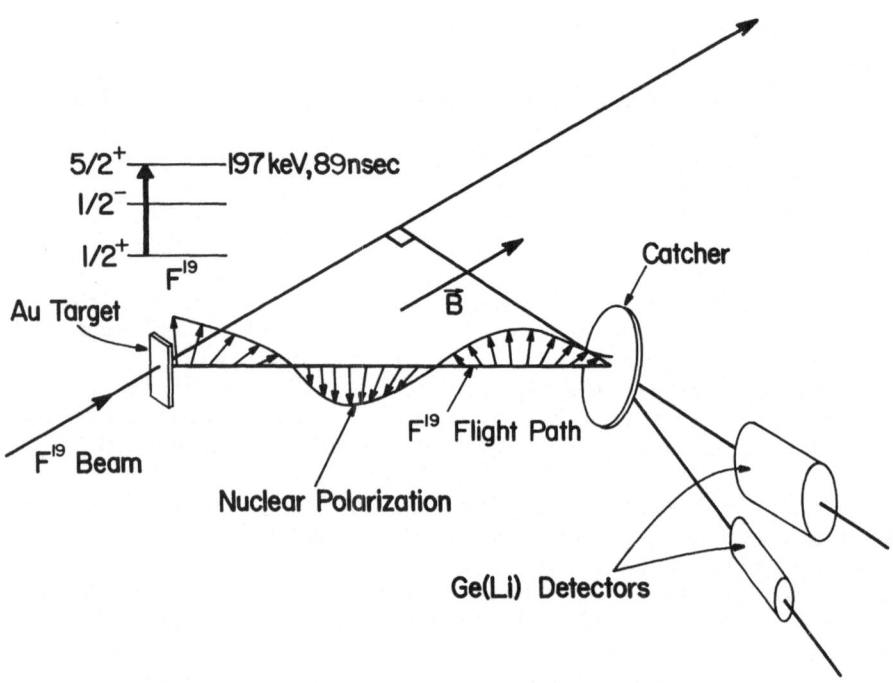

Figure 1 shows a schematic sketch of the experimental arrangement.

A weak magnetic field is applied along the initial beam direction (see Fig. 1).   The angular momenta of the oriented ions
precess around this field in opposite directions at their slightly
different Larmor frequencies.   Since the field is weak compared
to the hyperfine field, the Larmor precession of the electron spin
is slow compared to the precession of the nuclear and electron spins
about F, and the nucleus follows the ionic precession.

     When the ion enters the catcher foil, it quickly slows down and
regains its electron shell, turning off the hyperfine interaction,
and the external field interacts with the small nuclear moment
until the nuclear state decays.   Previous experiments have shown[1]
that only 10-15% of the nuclear polarization is lost in the slowing
down process and in the subsequent interactions in the solid.   Thus
the nuclear orientation is essentially preserved upon implantation

and precessed until it decays emitting a 197 keV gamma. This is
analogous to zero field level crossing in the excited nuclear state
and is called peturbed angular correlation. The rotated axis of
the anisotropic emission pattern of these gammas is determined by
the total angle of precession in flight and the smaller rotation in
the catcher. A Ge(Li) detector in the scattering plane observes
part of the change of the spatial distribution of the gammas as the
precession angle is changed by sweeping the magnetic field from
zero to about 800 G. We observe an oscillatory change of the gamma
count rate with applied field as shown in Fig. 2. Least squares

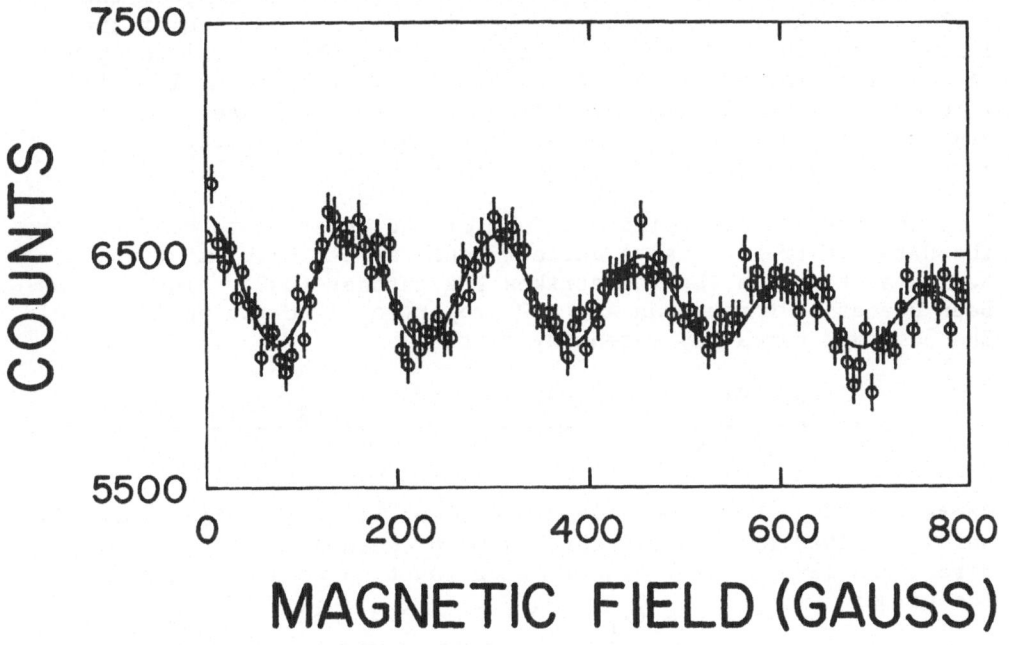

Figure 2 shows the data from a typical run.

fitting allows extraction of the background level, amplitude, and
frequency of these oscillations; the frequency determines the
g-factor.

    Extraction of the g-factor is straightforward. In order to
find the time of flight, we use the energy determined from the NMR
frequency of the 90° analyzing magnet of the accelerator. Relativistic corrections are less than 1/4 %. We determine the thickness
of the gold target using alpha transmission (from Am²⁴¹) to a
precision of about 4%. F¹⁹ nuclei lose about 2.9 MeV per mg/cm²
of gold traversed at about 45 MeV. The nuclei lose about 16.5 MeV
in the foil and emerge with speeds in the range of 1.7 cm/nsec. The

target is tilted so that its normal bisects the incident and final momenta so that nuclei scattered through a particular angle from any point within the target will traverse the same distance in the material. Therefore all nuclei scattered at a particular angle lose the same amount of energy in the target. Because the solid angle of the catcher is finite, it has been curved and canted slightly to assure that the nuclei scattered at various angles will spend approximately the same time in flight. These two precautions have reduced the spread in flight time for various ions to about 2%. Since we can measure the target to catcher distance to about 1%, we can determine the time of flight to about 1.2%. We measure the field along the flight path with NMR in oil, but because of the large volume required for the flight region, the field inhomogeneities limit the precision of this measurement to about 1%. The frequency of oscillations determined from our data has a statistical precision of about 1.5%. Thus the uncertainty associated with our measured value of g is about 2.5%.

In order to determine the form of the function to be fitted to the data, we treat it as a nuclear Hanle effect experiment which begins at the time the ion strikes the catcher (t=0). The ion has been precessed through an angle θ proportional to the applied field. The observed gamma ray intensity is then

$$I = \frac{\cos(2\theta) + 2\omega_N \tau \sin(2\theta)}{1 + (2\omega_N \tau)^2} = \frac{\cos 2(\theta - \emptyset)}{\sqrt{1 + \tan^2(2\emptyset)}}$$

where $\tan(2\emptyset) = 2\omega_N \tau$ and τ is the nuclear lifetime. Since our signal is derived from nuclei which have precessed in either of two directions (F=2 or F=3), the fitting function we use is a sum of two terms:

$$I = A \frac{1.2 \cos 2(\omega_L t - \emptyset + \delta) + 0.8 \cos 2(-\omega_L t - \emptyset + \delta) + C}{\sqrt{1 + \tan^2(2\emptyset)}} + B$$

where δ is some initial phase, ω_L is the Larmor frequency, and the amplitudes are different because of population and coupling differences between the different F states. We ignore the very slight differences between the magnitudes of the g-factors.

We have taken data at many energies with several target thicknesses. A plot of total precession angle vs flight time (Fig. 3) is shown for incident energies ranging from 32 to 61 MeV. The best value of the g-factor resulting from all our data is given by $g_{exp} = 0.345(8)$ and therefore $1000(g_{exp} - g_{theor}) = 12 \pm 8$. We do not consider the discrepancy to be significant.

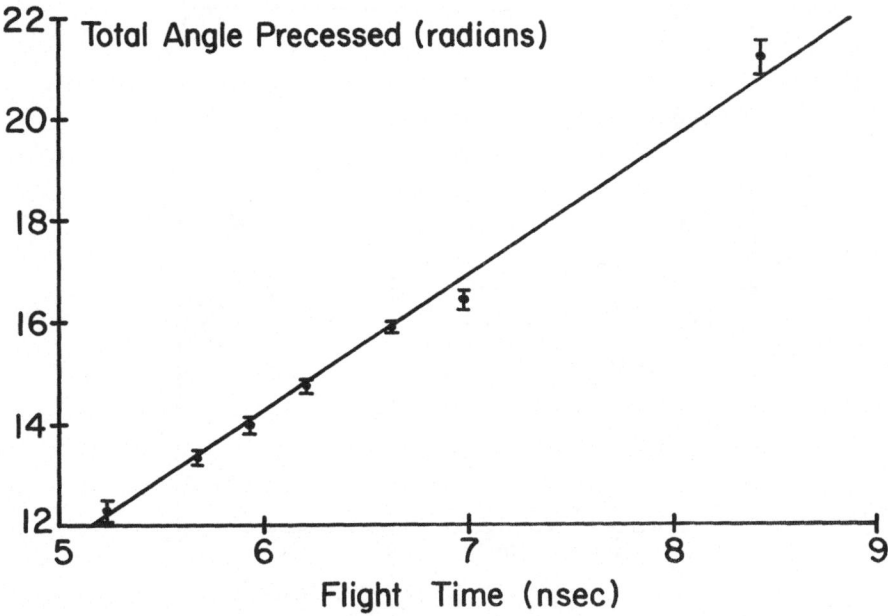

Figure 3. A plot of total precession angle vs flight time. Flight time is varied by changing the energy of incident nuclei.

The fraction of ions traversing the flight path in atomic states with $|g| = 1/3$ can be measured by calculating the expected signal amplitude and comparing with the observed amplitude. The initial nuclear polarization is calculated from the DeBoer-Winther Coulomb excitation program[2] and is corrected for the expected fractional losses of signal from: a) finite solid angle (15%); b) gamma ray background (25%); c) coupling of I to J (16.7%); d) dealignment in the catcher (15%). Comparison with charge state fractions measured with other techniques is shown in Fig. 4.[3] The discrepancy may be attributable to the large uncertainties in the charge state fractions available for fluorine.

At lower energies the charge state equilibrium curves indicate that we should have a sizeable fraction of the beam in the +7 state. These helium-like ions will not show the same oscillatory effect in their ground state because the total electronic angular momentum is zero. The nucleus would precess at the much slower nuclear Larmor frequency (PAC). The metastable 2^1S_0 state also has no hfs interaction, but the 2^3S_1 state ions should produce oscillations. There are three couplings of J=1 with F=5/2; F= 3/2 has $g_F = -4/5$, F = 5/2 has $g_F = 8/35$, and F = 7/2 has $g_F = 4/7$.

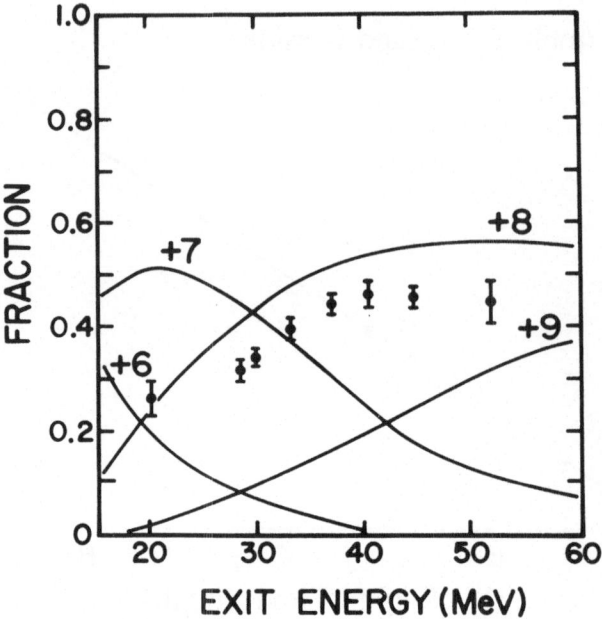

Figure 4. Our measured fractions of $|g_J| = 1/3$ deduced from the amplitude of the oscillatory part of the signal. Data points are measured fractions of ions with $|g_J| = 1/3$. Curves from Ref. 3.

Oscillations of the corresponding frequencies are not easily visible in our data but analysis is not inconsistent with their presence.

*Work supported in part by the National Science Foundation.
1. F. Bosch and O Klepper, Zeit.f.Physik, <u>240</u>, 153 (1970).
2. <u>Coulomb Excitation</u>, Ed. K. Alder, Academic Press, N.Y. (1966).
3. <u>Nuclear Reaction Analysis</u>, J. Marion and F. Young, p 44, North Holland Publishing, Amsterdam (1968).

MODULATION OF MERCURY RESONANCE FLUORESCENCE UNDER PULSED

ELECTRIC FIELDS: MEASUREMENT OF DIFFERENTIAL STARK SHIFTS

W.J. Sandle, M.C. Standage and D.M. Warrington

Physics Department, University of Otago, Dunedin,

New Zealand

The study of the Stark effect dates from 1913 when Stark[1] and Lo Surdo[2] independently observed structure on the Balmer lines from atomic hydrogen subjected to an electric field.   The investigation of Stark shifts in atoms other than hydrogen has required experimental techniques with resolution higher than that first used, and the history of the effect reveals the attention that has been given towards the development of such techniques.

We are here concerned with differential Stark shifts in the 6^3P_1 state of mercury. Figure 1 shows the behaviour of the 6^3P_1 level for an even isotope of mercury in an electric field.   The <u>differential</u> Stark shift is the separation $\Delta E(m = \pm 1) - \Delta E(m = 0)$ of the doubly degenerate level $m = \pm 1$ from the level $m = 0$. Here, the shift $\Delta E(m)$ is equal to $(A + Bm^2)\,\mathcal{E}^2$, where $\mathcal{E}$ is the applied electric field;   the differential Stark shift involves only the coefficient B.

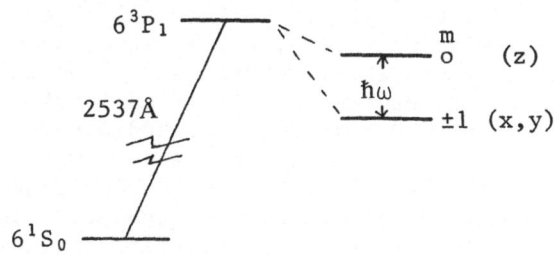

Fig. 1.   Stark shifts in the 6^3P_1 state of ^{198}Hg

The shifts in mercury are relatively small – at the most a few tens of megahertz compared with the Doppler width of the order of one gigahertz – and their measurement requires techniques of the highest resolution: those for which the limit is the fundamental one of the radiative width of the excited state. Indeed, the first measurement of the differential shift in the 6^3P_1 state, by Blamont[3] in 1957, was the occasion of the introduction of such a technique (optical double resonance) to Stark effect studies. This method can of course give only the B coefficient, and to date there appear to have been no experimental determinations of the A coefficient. The other differential Stark effect measurements on this level have been made by Khadjavi, Lurio and Happer[4], utilizing level crossings in even isotopes with parallel electric and magnetic fields, and recently, by Kaul and Latshaw[5] who studied the shift of the high-field level crossing in ^{199}Hg with electric field.

It may seem rather surprising that zero-field crossing in a pure electric field (the electric field analogue of the Hanle effect) has not been exploited. The method has an appeal in principle, in that the electric field is the only perturbation (other than the optical excitation) applied to the atoms. However, as in the magnetic field case, the information obtained from a zero-field level crossing involves the product of the splitting factor with the lifetime. If the splitting factor is to be obtained directly with a pure electric field, either the optical excitation or the electric field must be time dependent.

The technique we wish to present in this paper involves the sudden application of an electric field of magnitude large enough to cause a separation of the m = 0 from the m = ±1 levels which exceeds the radiative width of the levels. The polarization direction of the excitation is chosen so that all three levels are excited; for zero electric field, the excited state is a coherent superposition of the three substates. When the electric field is applied, this coherence makes itself manifest in the beating of the fluorescent radiations associated with decays from the m = 0 and m = ±1 states to the ground state. One thus observes modulation of the fluorescence which, since the coherence lasts for a natural lifetime, is damped at the rate of radiative decay. The frequency of the modulation is just the splitting frequency in the electric field, and herein lies the advantage of the technique over the steady-state level crossing method: the splitting frequency is obtained directly . We note that as in the steady-state method the electric field is (apart from the optical field) the only perturbation present. This pulsed electric field technique relates to the steady-state level crossing method in the same way that the pulsed magnetic-field technique of Dodd, Sandle and Williams[6] relates to the use of the Hanle effect.

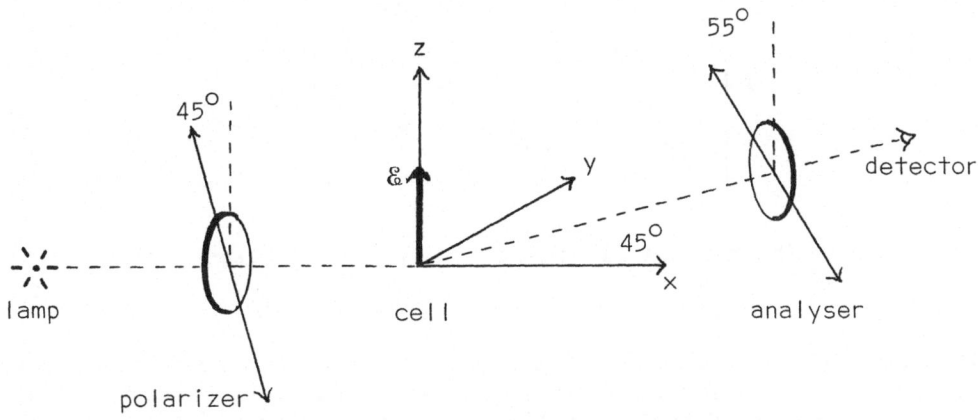

Fig. 2.   Excitation and detection geometry

Let us look at the theory of the effect in a little more detail.   The excitation geometry is shown in Fig. 2.

If we introduce as basis states $|x\rangle$, $|y\rangle$ and $|z\rangle$, where

$$|x\rangle = -(|+1\rangle - |-1\rangle)/\sqrt{2}$$
$$|y\rangle = i(|+1\rangle + |-1\rangle)/\sqrt{2}$$
$$|z\rangle = |0\rangle,$$

to replace the more usual states $|+1\rangle$ $|0\rangle$ $|-1\rangle$ (which correspond to m = 1, 0, and -1), the description of the excitation is simple:

$$|\psi\rangle = (|y\rangle - |z\rangle)/\sqrt{2}$$

for the case of the polarizer at 45° to 0z in the zy plane.

If for the moment we neglect the effect of radiative decay, we find that ψ describes the state of <u>all</u> excited atoms prior to the time (t = 0 say) of application of the electric field.   Now the effect of the electric field is to cause a splitting of the excited state sublevels as shown in Fig. 1.   Thus, after t = 0, the evolution in time associated with $|z\rangle$ will be different from that associated with $|y\rangle$;   we have

$$|\psi(t)\rangle = (|y\rangle - e^{-i\omega t}|z\rangle)/\sqrt{2}$$

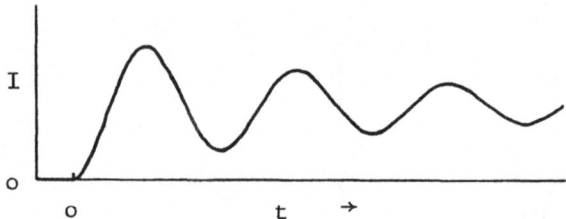

Fig. 3.   Theoretical fluorescent signal ("crossed" analyser)

The state of polarization of the excited state thus goes from linear (along the excitation direction) through circular to linear at right angles (in the yz plane), and then through circular of the opposite sense back to the original linear.

For the case (as in Fig. 2) where observation of the fluorescence is made in the xy plane with an analyser inclined at $\cos^{-1} 1/\sqrt{3}$ to the z axis, the intensity is proportional to $|1 - \exp(-i\omega t)|^2$ i.e. to $(1 - \cos \omega t)$.   The effect of damping is simple provided that ω is reasonably large compared with the decay constant Γ.   Then we find the intensity is proportional to $\{1 - \exp(-\Gamma t)\cos \omega t\}$.   A sketch of this is shown in Fig. 3.

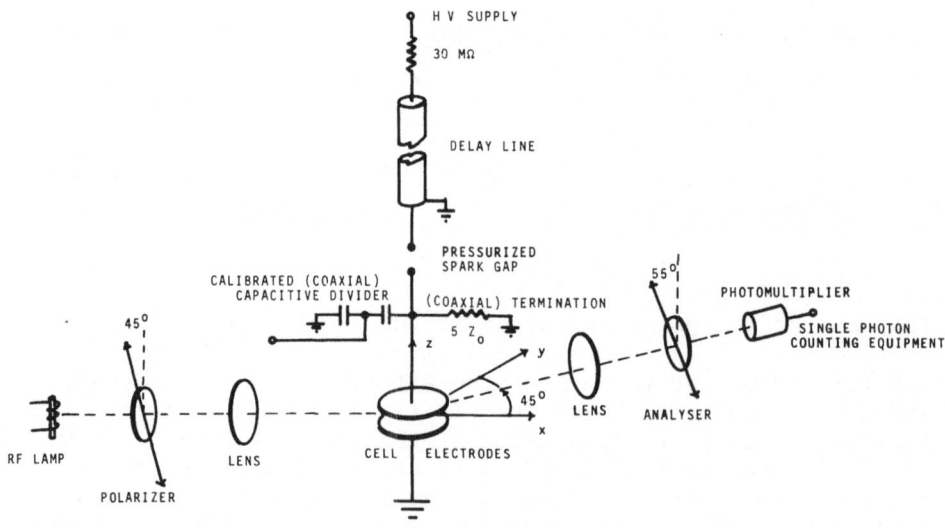

Fig. 4.   The experimental arrangement

The experimental arrangement is shown in Fig. 4.   Features of
the cell design are (i) the electrode assembly is first constructed
as a unit (with care taken to ensure that the electrodes are parallel)
before being incorporated in the cell, and (ii) the spectrosil
windows are reentrant and are placed close to the electrodes in order
to minimize optical absorption and depolarization due to radiation
imprisonment.   The electrode separation (of the order of 2mm) is
known to ½%.   Two cells were used, containing ½ mg of 94% ^{198}Hg and
½ mg of 99.3% ^{199}Hg.   Excitation is by means of an rf lamp containing
the same isotope as in the cell.

The earth's magnetic field is cancelled by Helmholtz coils (not
shown).   The electric field pulse is obtained by the connection of
a charged delay line to the cell electrodes via a pressurized spark
gap fired by "overvolting".   Pulsed voltages in the range of 5 - 20 kV
can be reproducibly obtained by adjustment of the pressure.   An
oscilloscope trace of the voltage wave form is shown in Fig. 5(c).
The risetime is 10 ns, much shorter than the lifetime (117 ns) of the
$^3P_1$ state.   The pulse length, of approximately 100 ns, has been chosen
from three considerations; the need to observe not less than about one
modulation period; the desirability of making as small as practicable
the energy dissipated per pulse; and the importance of suppressing
discharge in the mercury vapour.   Since, even at the high voltages
employed, discharge takes an appreciable time to establish (if it
occurs at all), we can minimize the effects of discharge by keeping
the pulse length short.   Care was taken to operate at electric fields
below that at which discharge occurred (~ 100 kV/cm).   During the
experiments there was no evidence of either significant outgassing
of the electrodes or "clean up" of mercury in the cell even though
the high voltage is applied to the electrodes from a low impedance
source.   In order to obtain a voltage as large as is practicable
at the electrodes for given energy stored in the line, a termination
of several times the characteristic impedance is used.   The effect
of this can be seen in Fig. 5(c) in the appearance of a smaller
reflected pulse.   Single-photon counting equipment is used to
register the distribution in time of photo-electric pulses from the
photo-multiplier.   Briefly, pulses derived from the firing of the
spark gap "start" a time-to-pulse height converter, and pulses
from the photo-multiplier "stop" the converter;   the distribution
of photo-electric pulses in time is thus converted to a distribution
of pulse heights which is recorded by a pulse height analyser.

Typical data for ^{198}Hg are shown in Fig. 5.

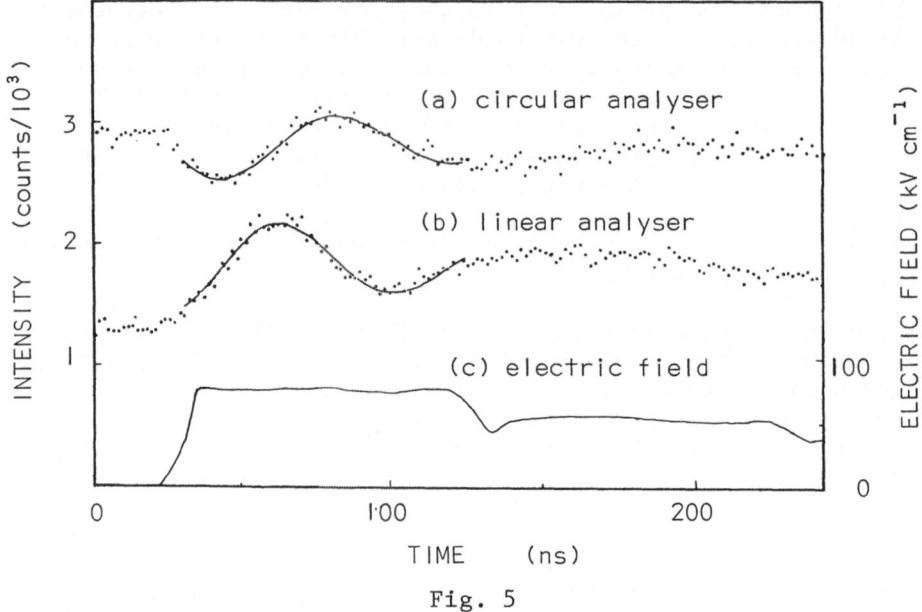

Fig. 5

The data (points) in (b) and (a) represent the time dependence of the fluorescence after linear and circular analysers respectively (with a circular analyser one expects a fluorescent signal of the form exp $(-\Gamma t)$ sin ωt). The lines in (a) and (b) are computer fits of exponentially damped sinusoids to the data. From these the modulation frequency is obtained. At the atom number density used ($\sim 5 \times 10^{11}$ cm^{-3}) the effect of radiation imprisonment on the lifetime is small; the damping factor is assumed to be the natural decay constant. (We find that the fitted period is insensitive to significant changes ($\sim 10\%$) in the assumed damping rate).

Results for modulation frequency versus electric field squared are shown in Fig. 6 for both ^{198}Hg and ^{199}Hg. (In the latter case also, the modulation is characterized by a single frequency – corresponding to the separation of the $\pm \frac{3}{2}$ and $\pm \frac{1}{2}$ levels). Within experimental error, the relationship is linear in both cases. Table I lists our values for the ratio of modulation frequency to electric field squared. The upper part of the table (J=1) also lists for comparison values from refs. 3, 4 and 5; there is good agreement with the recent values. Furthermore, the ratio of our values for ^{199}Hg and ^{198}Hg agrees with the theoretical figure of 2/3 to within experimental error.

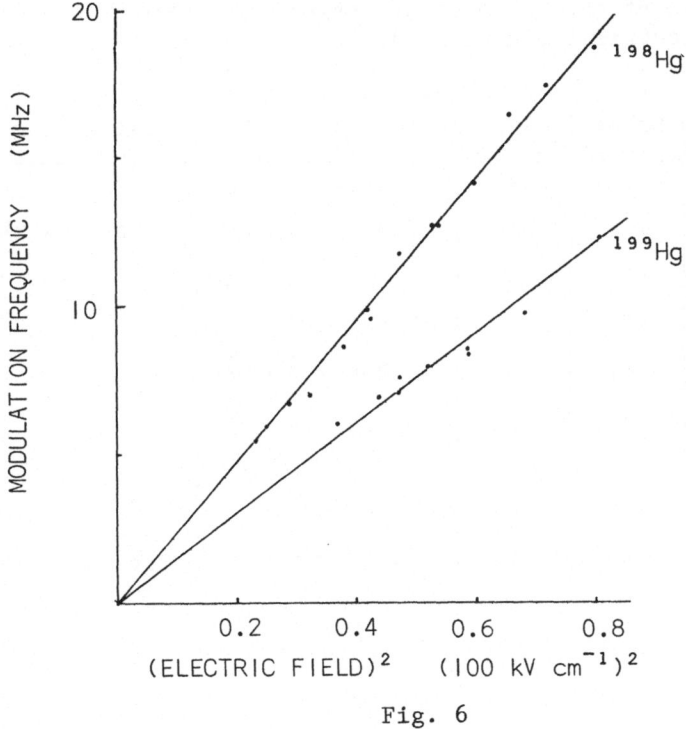

Fig. 6

In conclusion, we have presented a new technique for the measurement of the differential Stark shift for a resonance transition in an atomic vapour.   The technique, which involves the recording of transients in resonance fluorescence following the sudden application of a large electric field pulse, offers advantages both in principle and application: (a) the electric field is the only perturbation (other than the optical excitation) applied to the atoms;   (b) the splitting frequency is measured directly as the modulation frequency of the fluorescence;   and (c) the shortness of the electric field pulse means that the effects of discharge in the vapour are very largely eliminated.

TABLE I.   Values of the ratio of splitting frequency to electric field squared.

Method	Level	Parameter measured		Reference
		$(\div h)$	$\{kHz/(kV\ cm^{-1})^2\}$	
Pulsed electric field in ^{198}Hg	J=1	B	-2.37 ± 0.05	Present work
Double resonance	J=1	B	-2.13 ± 0.05	3
Level crossing	J=1	B	-2.355 ± 0.090	4
High-field level crossing in ^{199}Hg	J=1	B	-2.367 ± 0.024	5
Pulsed electric field in ^{199}Hg	$F=\frac{3}{2}$	$\frac{2}{3}B$	-1.51 ± 0.05	Present work

REFERENCES

1.   J. Stark, Acad. Wiss. Berlin 40, 932 (1913).

2.   A. Lo Surdo, Acad. Linc.Atti. 22, 665 (1913).

3.   J.E. Blamont, Ann. Phys. (Paris) 2, 551, (1957).

4.   A. Khadjavi, A. Lurio and W. Happer, Phys. Rev. 167, 128 (1968).

5.   R.D. Kaul and W.S. Latshaw, J. Opt.Soc.Am. 62, 615, (1972).

6.   J.N. Dodd, W.J. Sandle and O.M. Williams, J. Phys.B.: Atom. molec. Phys. 3, 256 (1970).

SUBJECT INDEX